ICSA Book Series in Statistics

Series Editor

Ding-Geng Chen, College of Health Solutions, Arizona State University, Chapel Hill, NC, USA

The ICSA Book Series in Statistics showcases research from the International Chinese Statistical Association that has an international reach. It publishes books in statistical theory, applications, and statistical education. All books are associated with the ICSA or are authored by invited contributors. Books may be monographs, edited volumes, textbooks and proceedings.

Yichuan Zhao • Ding-Geng Chen
Editors

Statistics in Precision Health

Theory, Methods and Applications

Springer

Editors
Yichuan Zhao
Department of Mathematics and Statistics
Georgia State University
Atlanta, GA, USA

Ding-Geng Chen
College of Health Solutions
Arizona State University
Phoenix, AZ, USA

ISSN 2199-0980 ISSN 2199-0999 (electronic)
ICSA Book Series in Statistics
ISBN 978-3-031-50689-5 ISBN 978-3-031-50690-1 (eBook)
https://doi.org/10.1007/978-3-031-50690-1

This work was supported by National Research Foundation of South Africa (127727), South African National Research Foundation (NRF), South African Medical Research Council (SAMRC), and South African DST-NRF-SAMRC SARChI Research Chair in Biostatistics (114613).

This Springer imprint is published by the registered company Springer Nature Switzerland AG
The registered company address is: Gewerbestrasse 11, 6330 Cham, Switzerland

Preface

This book is to focus on the topics of precision health and aims to provide a comprehensive overview of theory, methods, and applications in this field. The primary goal of this book is to bring together leading experts in the field to discuss the latest developments and challenges in this exciting area of precision medicine and biostatistics research. It intends to stimulate further exploration of precision health-related topics. The book features contributions from outstanding researchers in top-tier universities. The contributors of these book chapters come from diverse backgrounds with their leading authorities, reflecting the interdisciplinary nature of precision health research.

The book covers a wide range of topics related to precision health, including the foundation, theory, statistical procedures, advanced methods, and applications. It also delves into areas such as precision medicine, mobile health research, and big data analysis. Each book chapter underwent a rigorous review process, with two referees' reviewing multiple times in order to ensure its quality. The chapters in the book present advanced statistical methods tailored to personalized medicine and related interdisciplinary fields. The book is timely and valuable, particularly in the big data era. It is expected to have a significant impact on the fields of statistics, biostatistics, precision medicine, and precision health. The book is devoted to a broad audience, ranging from professionals and practitioners to researchers and graduate students. It aims to provide valuable insights and new directions for personalized medicine-related research. Therefore, this book is a comprehensive and authoritative resource for anyone interested in precision health.

The chapters of the book are organized into four parts. Part I serves as an introduction, offering an overview of precision health within the context of the big data era. Part II delves into recent advancements in statistical methods of precision medicine, showcasing their applications through five chapters. Part III focuses on the utilization of precision medicine in clinical trials and its applications to electronic health record (EHR) data, featuring four comprehensive chapters. Part IV addresses the role of precision medicine in survival analysis and genomic studies. Each chapter is self-contained, and references are put at the end of each chapter for easy access. To facilitate readers' use of the new procedures, the book includes computer programs and datasets, making implementation straightforward. Many of these program codes are publicly accessible on the GitHub pages of the respective chapter authors, and the authors are open to readers' requests for R codes, further enhancing the book's practical utility.

The successful publication of this book is impossible without the invaluable support and contributions of many people. We would like to express our heartfelt gratitude to them. First, we extend our sincere appreciation to all the chapter authors listed in the "List of Contributors" for their high-quality contributions to this book. In addition, we would like to acknowledge the diligent efforts of many reviewers listed in the "List of Chapter Reviewers", whose insightful comments and suggestions significantly enhanced the book's presentation and quality. We owe a special debt of gratitude to Dr. Eva Hiripi, Senior Editor for Statistics Books at Springer, and Daniel Jagadisan, Project Coordinator of Books at Springer Nature, for their exceptional guidance and support throughout the entire publication process. Their expertise and dedication were instrumental in bringing this book to fruition.

As we strive for continuous improvement, we welcome feedback from the readers regarding any typos or errors in the book. We are open to receiving suggestions for further enhancements. Please feel free to contact the two editors, Dr. Yichuan Zhao and Dr. Ding-Geng Chen. Your input is invaluable to us, and we look forward to receiving your comments and suggestions on this book.

Atlanta, GA, USA Yichuan Zhao
Phoenix, AZ, USA Ding-Geng Chen

Contents

Editors and Contributors

About the Editors

Yichuan Zhao is a Professor of Statistics at Georgia State University in Atlanta. He has a joint appointment as associate member of the Neuroscience Institute, and he is also an affiliated faculty member of the School of Public Health at Georgia State University. His current research interest focuses on survival analysis, empirical likelihood methods, nonparametric statistics, analysis of ROC curves, bioinformatics, Monte Carlo methods, and statistical modelling of fuzzy systems. He has published more than 100 research articles in statistics and biostatistics, has co-edited 6 books on statistics, biostatistics and data science, and has been invited to deliver more than 200 research talks nationally and internationally. Dr. Zhao has organized the Workshop Series on Biostatistics and Bioinformatics since its initiation in 2012. He also organized the 25th ICSA Applied Statistics Symposium in Atlanta as the chair of the organizing committee to great success. In addition, the 6th ICSA China Conference that he organized as the chair of both the organizing committee and program committee was a huge success. He is currently serving as associate editor, or on the editorial board, for several statistical journals. Dr. Zhao is a Fellow of the American Statistical Association, an elected member of the International Statistical Institute.

Ding-Geng Chen is a fellow of the American Statistical Association and is currently the executive director and professor in biostatistics at the College of Health Solutions, Arizona State University. He is also an extraordinary professor and the SARChI in biostatistics at the University of Pretoria, an honorary professor at the University of KwaZulu-Natal, South Africa. Dr. Chen was a professor in Biostatistics at the University of North Carolina at Chapel Hill, a professor in biostatistics at the University of Rochester Medical School, and the Karl E. Peace Endowed Eminent Scholar Chair in Biostatistics at Georgia Southern University. He is a senior biostatistics consultant for biopharmaceuticals and government agencies with extensive expertise in biostatistics, clinical trials, and public health statistics. Dr. Chen has more than 200 referred professional publications, co-authored 11 books and co-edited 24 books on clinical trial methodology, meta-analysis, data science, causal inference, and public health research. This work is partially supported by the National Research Foundation of South Africa (Grant Number 127727) and the South African National Research Foundation (NRF) and South African Medical Research Council (SAMRC) (South African DST-NRF-SAMRC SARChI Research Chair in Biostatistics, Grant Number 114613).

Contributors

Ding-Geng Chen College of Health Solutions, Arizona State University, Phoenix, AZ, USA
Department of Statistics, University of Pretoria, Pretoria, South Africa

Bin Cheng Department of Biostatistics, Columbia University, New York, NY, USA

Yifan Cui Center for Data Science, Zhejiang University, Hangzhou, China

Chongzhi Di Seattle, WA, USA

Kelly R. Evenson Department of Epidemiology, Gillings School of Global Public Health, University of North Carolina–Chapel Hill, Chapel Hill, NC, USA

Nikki L. B. Freeman Department of Biostatistics, University of North Carolina at Chapel Hill, Chapel Hill, NC, USA

Haoda Fu Eli Lilly and Company, Lilly Corporate Center, Indianapolis, IN, USA

Emily Getzen Department of Biostatistics, Epidemiology and Informatics, University of Pennsylvania, Philadelphia, PA, USA

Kevin Gunn Department of Statistics, North Carolina State University, Raleigh, NC, USA

Beibei Guo Department of Experimental Statistics, Louisiana State University, Baton Rouge, LA, USA

Hyokyoung G. Hong Biostatistics Branch, Division of Cancer Epidemiology and Genetics, NCI/NIH, Bethesda, MD, USA

Feifang Hu Department of Statistics, George Washington University, Washington, DC, USA

Jianchang Hu Institute of Medical Biometry and Statistics, University of Lübeck, Lübeck, Germany

Queen Ikhelowa Department of Statistics, University of California-Riverside, Riverside, CA, USA

Jian Kang Department of Biostatistics, University of Michigan, Ann Arbor, MI, USA

Andrea Z. LaCroix Division of Epidemiology, Herbert Wertheim School of Public Health and Human Longevity Science, University of California, La Jolla, CA, USA

Michael J. LaMonte Department of Epidemiology and Environmental Health, School of Public Health and Health Professions, University at Buffalo–SUNY, Buffalo, NY, USA

Muxuan Liang Department of Biostatistics, University of Florida, Gainesville, FL, USA

Mingming Liu Department of Statistics, University of California-Riverside, Riverside, CA, USA

Yufeng Liu Department of Statistics and Operations Research, University of North Carolina at Chapel Hill, Chapel Hill, NC, USA

Xinyi Li School of Mathematical and Statistical Sciences, Clemson University, Clemson, SC, USA

Yi Li Department of Biostatistics, University of Michigan, Ann Arbor, MI, USA

Yuhan Li Department of Statistics, University of Illinois at Urbana-Champaign, Champaign, IL, USA

Qi Long Department of Biostatistics, Epidemiology and Informatics, University of Pennsylvania, Philadelphia, PA, USA

Wenbin Lu Department of Statistics, North Carolina State University, Raleigh, NC, USA

Shujie Ma Department of Statistics, University of California-Riverside, Riverside, CA, USA

Weibin Mo Purdue University, West Lafayette, IN, USA

Xin Qiu Janssen Pharmaceuticals, Raritan, NJ, USA

Yanping Qiu School of Statistics, Renmin University of China, Beijing, China

Yucheng Ruan University of Pennsylvania, Philadelphia, PA, USA

Tao Shen Department of Statistics and Data Science, National University of Singapore, Singapore, Singapore

Rui Song Department of Statistics, North Carolina State University, Raleigh, NC, USA

Ming T. Tan Department of Biostatistics, Bioinformatics and Biomathematics, Georgetown University, Washington, DC, USA

Robert Tumasian III Department of Biostatistics, Columbia University, New York, NY, USA

Lyle Ungar Department of Computer and Information Science, University of Pennsylvania, Philadelphia, PA, USA

Guangxing Wang US Food and Drug Administration, Silver Spring, MD, USA

Lily Wang Department of Statistics, George Mason University, Fairfax, VA, USA

Shiying Wang Department of Biostatistics, Yale University School of Public Health, New Haven, CT, USA

Wenjie Wang Eli Lilly and Company, Lilly Corporate Center, Indianapolis, IN, USA

Yuanjia Wang Department Biostatistics, 722 W168th Street, Columbia University, New York, NY, USA

Sixuan Wu Expedia Group, Seattle, WA, USA

Leqi Xu Department of Biostatics, Yale University, New Haven, CT, USA

Yanxun Xu Department of Mathematics and Applied Mathematics, Johns Hopkins University, Baltimore, MD, USA

Yuan Yang Newton, MA, USA

Yixuan Ye Program of Computational Biology and Bioinformatics, Yale University, New Haven, CT, USA

Anqi Yin Department of Biostatistics, Bioinformatics and Biomathematics, Georgetown University, Washington, DC, USA

Ao Yuan Department of Biostatistics, Bioinformatics and Biomathematics, Georgetown University, Washington, DC, USA

Ying Yuan Department of Biostatistics, The University of Texas MD Anderson Cancer Center, Houston, TX, USA

Donglin Zeng Department of Biostatistics, University of Michigan, Ann Arbor, MI, USA

Heping Zhang Department of Biostatistics, Yale University, New Haven, CT, USA

Xuan Zhang Department of Biostatistics and Health Data Science, School of Medicine, Indiana University, Indianapolis, IN, USA

Zhiyue Zhang Department of Mathematics and Applied Mathematics, Johns Hopkins University, Baltimore, MD, USA

Hongyu Zhao Program of Computational Biology and Bioinformatics, Yale University, New Haven, CT, USA
Department of Biostatics, Yale University, New Haven, CT, USA

Yichuan Zhao Department of Mathematics and Statistics, Georgia State University, Atlanta, GA, USA

Ying-Qi Zhao Public Health Sciences Division, Fred Hutchinson Cancer Center, Seattle, WA, USA

Wenzhuo Zhou Department of Statistics, University of California Irvine, Irvine, CA, USA

Ruoqing Zhu Department of Statistics, University of Illinois at Urbana-Champaign, Champaign, IL, USA

Chapter Reviewers

Dipankar Bandyopadhyay Department of Biostatistics, School of Medicine at the Virginia Commonwealth University, Richmond, VA, USA

Ding-Geng Chen College of Health Solutions, Arizona State University, Phoenix, AZ, USA
Department of Statistics, University of Pretoria, Pretoria, South Africa

Xiaofei Chen Department of Statistical Science, Southern Methodist University, Dallas, TX, USA

Bin Cheng Department of Biostatistics, Mailman School of Public Health, Columbia University, New York, NY, USA

Yifan Cui Center for Data Science, Zhejiang University, Hangzhou, China

Caiyun Fan School of Statistics and Information, Shanghai University of International Business and Economics, Shanghai, China

Kexin Huang Department of Computer Science, Stanford University, Stanford, CA, USA

Kevin He Department of Biostatistics, University of Michigan, Ann Arbor, MI, USA

Jian Kang Department of Biostatistics, University of Michigan, Ann Arbor, MI, USA

Marta Karas Department of Biostatistics, Harvard T.H. Chan School of Public Health, Harvard University, Boston, MA, USA

Seunggeun Shawn Lee Graduate School of Data Science, Seoul National University, Seoul, Republic of Korea

Yikuan Li Deep Medicine, Oxford Martin School, University of Oxford, Oxford, UK

Yisheng Li Department of Biostatistics, The University of Texas MD Anderson Cancer Center, Houston, TX, USA

Heng Lian Department of Mathematics, City University of Hong Kong, Hong Kong, China

Muxuan Liang Department of Biostatistics, University of Florida, Gainesville, FL, USA

Kristin Linn Department of Biostatistics, Epidemiology, and Informatics, Perelman School of Medicine University of Pennsylvania, Philadelphia, PA, USA

Wenbin Lu Department of Statistics, North Carolina State University, Raleigh, NC, USA

Qiongshi Lu Department of Biostatistics and Medical Informatics at University of Wisconsin-Madison, Madison, WI, USA

Qing Mai Department of Statistics, Florida State University, Tallahassee, FL, USA

Jing Ning Department of Biostatistics, The University of Texas MD Anderson Cancer Center, Houston, TX, USA

Matteo Pedone Department of Statistics, Computer Science, Applications, University of Florence, Florence, Italy

Zhengling Qi Department of Decision Sciences, The George Washington University, Washington, DC, USA

Min Qian Department of Biostatistics, Mailman School of Public Health, Columbia University, New York, NY, USA

Elizabeth Schifano Department of Statistics, University of Connecticut, Storrs, CT, USA

Phillip Schulte Division of Clinical Trials and Biostatistics, Department of Quantitative Health Sciences, Mayo Clinic, Rochester, MN, USA

Juan Shen School of Management, Shanghai, China

Chengchun Shi Department of Statistics, London School of Economics and Political Science, London, UK

Roy Tamura Health Informatics Institute, University of South Florida Tampa, Tampa, FL, USA

Peter Thall Department of Biostatistics, M.D. Anderson Cancer Center, Houston, TX, USA

Lu Wang Department of Biostatistics, University of Michigan, Ann Arbor, MI, USA

Ao Yuan Department of Biostatistics, Bioinformatics and Biomathematics, Georgetown University, Washington, DC, USA

Baqun Zhang School of Statistics and Management, Shanghai University of Finance and Economics, Shanghai, China

Daiwei Zhang Department of Biostatistics, Epidemiology and Informatics, Perelman School of Medicine, University of Pennsylvania, Philadelphia, PA, USA

Min Zhang Department of Biostatistics, University of Michigan, Ann Arbor, MI, USA

Zhiwei Zhang Biostatistics Branch, NCI, Rockville, MD, USA

Lihui Zhao Division of Biostatistics, Department of Preventive Medicine, Feinberg School of Medicine, Northwestern University, Chicago, IL, USA

Ying-Qi Zhao Public Health Sciences Division, Fred Hutchinson Cancer Center, Seattle, WA, USA

Ruoqing Zhu Department of Statistics, University of Illinois at Urbana-Champaign, Champaign, IL, USA

Yichuan Zhao Department of Mathematics and Statistics, Georgia State University, Atlanta, GA, USA

Jin Zhou Department of Biostatistics at UCLA, Los Angeles, CA, USA

Hongjian Zhu Statistical Innovation Group, AbbVie Inc, Virtual Office, Sugar Land, TX, USA

Part I
An Overview of Precision Health in the Big Data Era

Precision Health: Past, Current, and Future

Yichuan Zhao and Ding-Geng Chen

Abstract This chapter is to give an overview of the roadmap from precision medicine to precision health in their past, current, and future. As a common goal for both precision medicine and precision health to improve health for each of us and all of us, this chapter will focus on the introduction of the theory, methods, and applications covered in this book. It includes an overview of precision health in the big data era, new advances in statistical methods of precision medicine, precision medicine in clinic trials, and precision medicine in survival analysis and genomic studies.

Keywords Precision medicine · Precision health · Personalized medicine · Medical treatment · Health interventions

1 The Roadmap from Precision Medicine to Precision Health

The fundamental objective from precision medicine to precision health is the improvement of individual and collective health. Precision medicine, often referred to as personalized medicine, primarily serves the purpose of assisting medical professionals in identifying unique disease risks and tailoring treatments that are most effective for an individual's health.

In contrast to traditional medical approaches, which often employ a one-size-fits-all model to find the "average treatment effect" for the "average patient", precision medicine represents a paradigm shift. It recognizes the inherent diversity among individuals, considering factors such as their genetic makeup, environmental

---

Y. Zhao (✉)
Department of Mathematics and Statistics, Georgia State University, Atlanta, GA, USA
e-mail: yichuan@gsu.edu

D.-G. Chen
College of Health Solutions, Arizona State University, Phoenix, AZ, USA

Department of Statistics, University of Pretoria, Pretoria, South Africa

Y. Zhao, D.-G. Chen (eds.), *Statistics in Precision Health*, ICSA Book Series in Statistics, https://doi.org/10.1007/978-3-031-50690-1_1

influences, and lifestyle choices. This approach acknowledges that what works for one person may not yield the same results for another. Therefore, precision medicine represents an innovative strategy aimed at precisely customizing disease treatments to each patient's unique characteristics and needs, while also considering the optimal timing for intervention.

Significant advancements in the field of precision medicine have already led to groundbreaking discoveries in medical treatments. These advancements are meticulously tailored to the specific attributes of individuals, such as their genetic composition or the distinctive genetic profile of a patient's tumor. A notable example of this progress is the routine use of molecular testing for patients diagnosed with various types of cancer as an integral aspect of their healthcare. This testing equips healthcare professionals with the ability to select treatments with exceptional precision, increasing the chances of survival while simultaneously minimizing the potential for adverse effects.

Expanding on the concept of precision medicine, precision health encompasses approaches and practices that extend beyond the confines of a doctor's office or hospital. In addition to precision medicine's focus on treatment, precision health encompasses disease prevention, health well-being, and health promotion activities. In some contexts, precision health is also referred to as precision public health, representing a significant shift towards proactive and personalized care, empowering individuals to lead healthy lives.

Precision health reimagines medical treatment and practices, centering on the precise prediction, prevention, and cure of disease by leveraging comprehensive health information and data. This includes genetic information, lifestyle behaviors such as exercise and dietary habits, and environmental factors, all integrated into a unified approach. The overarching goal of precision health is to safeguard individual health by measuring these factors, enabling health interventions to be tailored specifically to each person, rather than adopting a uniform approach for everyone.

To achieve the vision of precision health, further research is imperative. This book aims to present the theory, methods, and applications necessary to advance precision health. It encompasses an exploration of precision health in the era of big data, the latest advancements in statistical methodologies applied in precision medicine, the integration of precision medicine into clinical trials, and its role in survival analysis and genomic studies. By doing so, it seeks to contribute to the ongoing evolution of healthcare, where personalized and proactive approaches become the cornerstone of improved health outcomes for individuals and communities alike.

This book is structured into four parts. Each part is dedicated to exploring topics related to precision health within the contexts of the past, present, and future. This organizational framework is designed to provide readers with a comprehensive and cohesive understanding of the evolution and potential of precision health.

2 Part I: An Overview of Precision Health in the Big Data Era (Second Chapter to Seventh Chapter)

Part I includes second chapter to seventh chapter. These six chapters collectively constitute an in-depth exploration of precision health within the context of the big data era. This part serves as a comprehensive overview of how the convergence of vast and complex data sets is reshaping the landscape of healthcare and enabling precise, data-driven approaches to health and well-being. Below, we provide an outline of the key topics covered in these chapters.

In second chapter of this book, titled "A Selective Review of Individualized Decision Making," Mo and Liu deliver a comprehensive overview of the advancements in personalized decision-making, with a specific focus on its applications in precision medicine. The chapter addresses both single-stage and multi-stage decision problems, providing a thorough review of various existing methodologies. Mo and Liu go beyond summarizing these methods; they also critically assess their strengths and weaknesses, offering valuable insights into their practical utility and limitations. Additionally, the authors offer practical recommendations for tackling real-world challenges that frequently arise in the application of these decision-making techniques. This chapter serves as an informative and thought-provoking resource for those seeking a deeper understanding of individualized decision-making within the context of precision medicine, making it a significant contribution to the book's content.

Third chapter, titled "Utilizing Wearable Devices to Enhance Precision in Physical Activity Epidemiology: Sensors, Data, and Analytic Methods," offers a comprehensive review of recent advances in physical activity epidemiology and exercise medicine. The chapter covers topics such as the evolution of sensors, data collection methods, and advanced analytics, all influenced by the development of wearable devices. Additionally, it provides real-world examples of how these technologies are applied in physical activity epidemiological studies, showcasing their practical impact on monitoring and improving physical activity levels. This chapter serves as a valuable resource, highlighting the transformative role of wearable devices in enhancing our understanding of physical activity and its relevance in public health and exercise medicine.

Fourth chapter, titled "Policy Learning for Individualized Treatment Regimes on Infinite Time Horizon," authored by Zhou, Li, and Zhu, offers an insightful overview of reinforcement learning methods within the context of an infinite time horizon setting. The chapter presents a clear and informative exploration of these methods. Additionally, it delves into essential topics such as the modeling framework, generalizability, and interpretability. To enhance understanding, the authors provide practical examples that further illustrate the concepts discussed. This chapter proves to be a valuable resource, offering readers a comprehensive understanding of policy learning for individualized treatment regimes, including its theoretical foundations and practical applications.

Fifth chapter, titled "Q-Learning Based Methods for Dynamic Treatment Regimes," authored by Li, Freeman, and Wang, delivers a comprehensive introduction to Q-learning-based methods for dynamic treatment regimes. The chapter adeptly presents these methods in two essential contexts: dynamic treatment regimes with finite time horizons and those with infinite time horizons. Moreover, the chapter includes a practical illustration of Q-learning implementation using the R programming language. This combination of theoretical insights and hands-on application makes fifth chapter a valuable resource for readers seeking to understand and utilize Q-learning in the context of dynamic treatment regimes.

Sixth chapter, titled "Personalized Medicine with Multiple Treatments," authored by Wang, Zhang, and Fu, offers a comprehensive and impressive review of recent statistical methods. These methods are designed to estimate optimal treatment rules when multiple treatments are considered, taking into account a patient's pretreatment characteristics. This topic holds significant importance in the field of personalized medicine. The chapter serves as a valuable resource for those interested in understanding the evolving statistical approaches used to tailor treatments effectively in scenarios involving multiple treatment options and patient-specific characteristics.

Seventh chapter, titled "Statistical Reinforcement Learning and Dynamic Treatment Regimes," authored by Shen and Cui, provides a clear and insightful introduction to reinforcement learning in the context of modern-day dynamic treatment regimes. The chapter outlines the concept and procedures of reinforcement learning and its application to dynamic treatment regimes. It covers various methods, including policy iteration, temporal difference learning, and policy gradient, for dynamic treatment regimes in scenarios involving both point exposure and time-varying settings. Additionally, the chapter includes a discussion of causal reinforcement learning. To enhance understanding, the authors illustrate these methods with several practical examples, making it a valuable resource for readers interested in the intersection of statistical reinforcement learning and healthcare decision-making.

3 Part II: New Advances in Statistical Methods of Precision Medicine and the Applications (Eighth Chapter to Twelfth Chapter)

Part II of the book encompasses eighth chapter through twelfth chapter, focusing on the exploration of new advances in statistical methods within the realm of precision medicine and their practical applications. This part serves as a critical examination of cutting-edge methodologies and their relevance in the field. Below is a brief overview of the key topics covered in this part.

Eighth chapter, titled "Integrative Learning to Combine Individualized Treatment Rules from Multiple Randomized Trials," authored by Qiu, Zeng, and Wang, presents a novel integrative learning approach aimed at combining insights from multiple clinical trials. The objective is to develop an integrative individualized treatment rule (ITR) that enhances precision in healthcare decision-making. Within this chapter, the authors provide a detailed exploration of their methodology, emphasizing the importance of combining evidence from various sources to create more precise ITRs. To demonstrate the effectiveness of their approach, the authors conduct extensive simulation studies. These studies serve to showcase how the integrative ITRs outperform single-study ITRs in terms of precision, underlining the potential benefits of this integrative approach in advancing healthcare decision-making.

Ninth chapter, titled "Adaptive Semi-supervised Learning for Optimal Treatment Regime Estimation with Application to EMR Data," introduces a novel approach to improve the estimation of optimal treatment regimes using electronic medical record (EMR) data. This method leverages "unlabeled" individuals within the data to enhance the efficiency of treatment regime estimation. Through extensive simulations, the authors compare the performance of this semi-supervised method with a fully supervised approach, demonstrating its potential advantages. Furthermore, the chapter provides a practical illustration of this approach using real EMR data, highlighting its relevance and applicability in optimizing treatment decisions and leveraging electronic medical records for improved healthcare outcomes.

Tenth chapter, titled "Estimation and Inference for Individualized Treatment Rules Using Efficient Augmentation and Relaxation Learning (EARL)," authored by Liang and Zhao, introduces a doubly-robust learning approach called EARL. This method, along with its extensions tailored for high-dimensional data, facilitates the estimation of optimal individualized treatment plans. Within this chapter, the authors not only develop EARL as a valuable tool but also create an inferential framework for it. This framework enables the calculation of p-values and confidence intervals, offering insights into how various predictors can influence the determination of the best treatment decisions. In essence, tenth chapter provides a robust methodology for estimating individualized treatment rules and offers a means to statistically assess the impact of predictor variables on treatment decisions.

Eleventh chapter, titled "Subgroup Analysis Using Doubly Robust Semiparametric Procedures," authored by Yin, Yuan, and Tan, explores an alternative approach to the doubly robust estimator (DRE). The chapter begins with a concise review of the DRE and its enhanced version. It then proceeds to introduce the enhanced doubly robust procedure for subgroup analysis. The authors utilize this method to identify specific patient subgroups that derive significantly greater benefits from a treatment compared to a control group. This identification process is demonstrated through both simulation studies and an analysis of a smoking cessation study. In essence, eleventh chapter focuses on using advanced statistical procedures to uncover meaningful subgroups of patients who experience enhanced treatment benefits, offering valuable insights into precision medicine applications.

Twelfth chapter, titled "A Selective Overview of Fusion Penalized Learning in Latent Subgroup Analysis for Precision Medicine," authored by Liu, Ikhelowa, and Ma, offers a review of a pairwise fusion penalized learning approach used in latent subgroup analysis. The authors introduce the fusion penalized method, which effectively categorizes patients into distinct groups and estimates the relevant parameters. The chapter provides detailed algorithms for implementing this method and showcases its applications, particularly in the context of longitudinal data analysis. Additionally, the authors offer a practical demonstration of the proposed method through its application to biomedical data, highlighting its relevance and effectiveness in advancing precision medicine and subgroup analysis.

4 Part III: Precision Medicine in Clinic Trials and the Applications to EHR Data (Thirteenth Chapter to Sixteenth Chapter)

Part III of the book, comprising thirteenth chapter through sixteenth chapter, delves into the integration of precision medicine into clinical trials and its practical applications within electronic health record (EHR) data. This part explores the critical intersection of precision medicine and clinical trial research to HER data. Below is a brief overview of the key topics covered in this part.

Thirteenth chapter, titled "Mining for Health: A Comparison of Word Embedding Methods for Analysis of EHRs Data," authored by Getzen, Ruan, Ungar, and Long, presents an innovative application of word embedding methods in electronic health records (EHRs) data analysis. The authors extend these methods to embed structured medical codes from a patient's complete medical history and utilize these embeddings to construct predictive disease models, aiming to enhance healthcare outcomes. The chapter offers a valuable comparative analysis of multiple embedding methods using the Medical Information Mart for Intensive Care database, providing insights into their performance and guidance for their use in EHR data analysis. This research demonstrates the potential of word embedding techniques in healthcare analytics, facilitating disease prediction and data-driven decision-making in the medical field.

Fourteenth chapter, titled "Adaptive Designs for Precision Medicine in Clinical Trials: A Review and Some Innovative Designs," authored by Qiu and Hu, offers an insightful exploration of adaptive designs in the context of precision medicine clinical trials. The chapter begins by elucidating the mathematical framework behind adaptive randomization and provides a comprehensive review of prevalent adaptive designs that incorporate covariates. Additionally, the authors introduce a novel class of covariate-adjusted response-adaptive procedures aimed at optimizing treatment allocation. To further enhance understanding, the chapter delves into the theoretical properties of these innovative designs, shedding light on their potential benefits and implications in precision medicine research and clinical trials.

Fifteenth chapter, titled "Maximum Likelihood Estimation and Design and Inference Considerations for Sequential Multiple Assignment Randomized Trials," authored by Tumasian III and Cheng, provides a comprehensive discussion of the sequential multiple assignment randomized trial (SMART) design. The chapter elucidates the advantages of SMARTs in comparison to alternative design options and thoroughly explores the statistical analyses associated with SMARTs. Additionally, the authors delve into inference techniques and address the critical aspect of sample size calculation within the SMART framework. This chapter serves as a valuable resource for readers seeking a deep understanding of SMART design, statistical analysis, and considerations for conducting sequential multiple assignment randomized trials.

Sixteenth chapter, titled "Precision Medicine Designs for Cancer Clinical Trials," offers a comprehensive review of Bayesian adaptive designs within the context of precision oncology clinical trials. The chapter, authored by Guo and Yuan, highlights the application of the Bayesian optimal interval phase I-II (BOIN12) design to illustrate how it can effectively capture the multi-dimensional nature of clinical outcomes. Additionally, the authors introduce a novel Bayesian adaptive design, showcasing how it can be utilized to identify subgroup optimal biological doses (OBD) while accounting for patient heterogeneity—a crucial consideration in precision medicine. This chapter provides valuable insights into the use of Bayesian adaptive designs in cancer clinical trials, emphasizing their potential to enhance the accuracy and effectiveness of precision medicine approaches.

5 Part IV: Precision Medicine in Survival Analysis and Genomic Studies (Seventeenth Chapter to Twenty-First Chapter)

Part IV of the book comprises seventeenth chapter through twenty-first chapter, focusing on the integration of precision medicine principles within the domains of survival analysis and genomic studies. This section explores the critical intersection of precision medicine, genetics, and the analysis of survival data. Below is a brief overview of the key topics covered in this part.

Seventeenth chapter, titled "Variant Selection and Aggregation of Genetic Association Studies in Precision Medicine," authored by Hu, Wang, and Zhang, offers a comprehensive review of set-based association methods with a specific emphasis on the selection and aggregation of genetic variants, particularly those with low frequency and signal variants. Within this chapter, the authors present a unified framework and a detailed classification of methods, providing readers with insights into the various approaches and techniques used to identify and aggregate relevant genetic variants in the pursuit of precision medicine. This chapter serves as a valuable resource for understanding the methodologies involved in harnessing genetic data for personalized healthcare applications.

Eighteenth chapter, titled "Leveraging Functional Annotations Improves Cross-population Genetic Risk Prediction," authored by Ye, Xu, and Zhao, explores the advantages of incorporating functional annotations to enhance the accuracy of polygenic risk score (PRS) methods for predicting genetic risks across different populations. The authors utilize data from unrelated individuals in the UK Biobank to demonstrate the superior performance of more sophisticated PRS models, specifically the P+T PRS, compared to standard PRS models based on a Bayesian framework. This chapter sheds light on how leveraging functional annotations can significantly improve the precision of genetic risk prediction, particularly in cross-population studies, making it a valuable contribution to the field of precision medicine.

Nineteenth chapter, titled "A Soft-Thresholding Operator for Sparse Time-Varying Effects in Survival Models," authored by Yang, Kang, and Li, introduces an innovative soft-thresholding operator designed for modeling sparse, piecewise smooth, and continuous time-varying coefficients in the Cox time-varying effects model. The authors also develop an efficient algorithm for making inferences within the target functional space. Furthermore, the chapter demonstrates the effectiveness of this approach through comprehensive simulation studies and real data analysis involving the Boston Lung Cancer Survivor Cohort. This research showcases the utility of the soft-thresholding operator in enhancing the modeling of time-varying effects in survival models, with practical applications in precision medicine.

Twentieth chapter, titled "Discovery of Gene-specific Time Effects on Survival," introduces an innovative approach focused on detecting gene-specific time effects on survival outcomes. In this chapter, Hong proposes a method that emphasizes the divergence in survival functions across strata defined by candidate covariates. The author further develops a comprehensive class of screening criteria capable of identifying biomarkers with either early or delayed impacts on survival distributions. This procedure offers new insights and perspectives for high-dimensional survival analysis, enabling the discovery of genes or biomarkers that play a crucial role in influencing survival outcomes over time.

Twenty-first chapter, titled "Modeling and Optimizing Dynamic Treatment Regimens in Continuous Time," authored by Xu and Zhang, introduces a model-based decision framework designed to optimize personalized clinical decisions within a continuous time frame. This innovative approach involves scheduling follow-up visits and determining the optimal treatment at each visit. The method combines Bayesian techniques with sequential clinical decision optimization using policy optimization methods. The chapter illustrates the application of this novel approach using electronic medical records from patients who have undergone kidney transplantation. This research contributes to the development of more precise and dynamic treatment regimens, particularly in the context of continuous healthcare management.

6 Further Discussions

In the opening chapter for this book, our primary aim is to provide readers with a comprehensive overview of the evolutionary journey from precision medicine to the broader concept of precision health. We delve into the historical, current, and future perspectives of these fields, highlighting their shared mission: to enhance the health and well-being of both individuals and entire populations. This chapter serves as a foundational introduction, laying the groundwork for the theories, methodologies, and practical applications that will be explored throughout this book based on the 20 contributed chapters.

Specifically, our focus encompasses several critical areas. First and foremost, we navigate the landscape of precision health in the era of big data from the second chapter to seventh chapter in Part I, shedding light on how the proliferation of vast datasets has revolutionized healthcare. We elucidate the pivotal role of advanced analytics, machine learning, and data-driven insights in shaping the future of medicine and health management.

Furthermore, we delve into the latest advancements in statistical methods within the realm of precision medicine from eighth chapter to twelfth chapter in Part II, emphasizing the importance of robust statistical approaches for achieving accurate and meaningful results. This exploration will underscore the indispensable role of statistical methodologies in driving progress in precision healthcare.

Additionally, we come to the practical applications of precision medicine in clinical trials and the application to HER data with thirteenth chapter to sixteenth chapter in Part III. We elucidate how personalized treatments and patient stratification are poised to redefine the outcomes of clinical research, ultimately translating into more effective and tailored healthcare interventions.

Lastly, we explore the integration of precision medicine in survival analysis and genomic studies with seventeenth chapter to twenty-first chapter in Part IV, with a keen focus on its implications for diseases such as cancer. By examining how genomics and genetic information intersect with precision health strategies, we aim to provide readers with a comprehensive understanding of the multifaceted landscape of precision medicine and its profound potential for improving health outcomes for individuals and communities at large.

In summary, this chapter serves as a pivotal starting point for our journey into the intricate and transformative world of precision health. It sets the stage by providing essential summary and insights, offering a solid foundation upon which we can build our understanding of the multifaceted concepts, methodologies, and applications in precision health.

A Selective Review of Individualized Decision Making

Weibin Mo and Yufeng Liu

Abstract With fast development in data-driven decision science, we have seen great advances in individualized decision making, including precision medicine, personalized marketing, and policy making. In particular, for precision medicine, due to the inherent heterogeneity among people, different patients may respond differently to a particular treatment. Compared to the one-size-fits-all strategy, it is more desirable to identify the optimal treatment among available options for every individual. Given data with covariates, treatment assignments and outcomes, a key goal is to find the optimal individualized decision rule that maps the individual characteristics or contextual information to the treatment assignment, such that the overall expected outcome is optimized. In this article, we provide a comprehensive review of the development on individualized decision making, where we particularly focus on the applications in precision medicine. Both single-stage and multi-stage problems are considered, with a selective review on several existing methods in the literature. We also discuss strengths and weaknesses of these methods, and offer some suggestions in handling real-world applications.

Keywords Direct-search method · Multi-stage decision making · Precision medicine · Regression-based method · Single-stage decision making

Data-driven individualized decision making problems are commonly seen in practice and have been studied intensively in the literature. In public policy administration, the decision maker may decide who should be treated based on the individual characteristics to maximize the overall social welfare (Kitagawa & Tetenov, 2018). In a web-based recommender system, users are typically presented

---

W. Mo
Purdue University, West Lafayette, IN, USA
e-mail: harrymok@purdue.edu

Y. Liu (✉)
University of North Carolina at Chapel Hill, Chapel Hill, NC, USA
e-mail: yfliu@email.unc.edu

Y. Zhao, D.-G. Chen (eds.), *Statistics in Precision Health*, ICSA Book Series in Statistics, https://doi.org/10.1007/978-3-031-50690-1_2

with personalized promotion, which are customer, content and context-aware recommendation (Aggarwal, 2016). In diabetic disease management, the doctor may decide whether to maintain or switch an ongoing glycemic therapy based on the patient's demographic information, disease and medical treatment history, as well as the responses of other similar patients to different treatment options (Bertsimas et al., 2017). In these applications, the underlying decision maps the individual characteristics or contextual information to the treatment assignment. One common goal is to optimize the overall expected outcome. Depending on how many times the decision maker can assign treatments, one may work with a single-or multi-stage problem. When the patient is assigned with some treatment only once before his/her final outcome is observed, it is a single-stage decision making problem. When several treatments are assigned sequentially before the final outcome is obtained, this becomes a multi-stage problem. In this article, we discuss several existing approaches in the single-stage and multi-stage settings in Sects. 1 and 2 respectively. Some further discussions are provided in Sect. 3.

1 Single-Stage Decision Problems

For a single-stage decision problem, each data point consists of a covariate vector $\boldsymbol{X} \in \mathcal{X} \subseteq \mathbb{R}^p$ incorporating the individual characteristics or contextual information, a treatment assignment $A \in \mathcal{A}$ and an outcome $Y \in \mathcal{Y} \subseteq \mathbb{R}$. Assume without loss of generality that a larger outcome Y is better. One important goal is to find the optimal Individualized Treatment Rule (ITR) $d : \mathcal{X} \to \mathcal{A}$ that maximizes the expected outcome (Manski, 2004):

$$d^\star \in \operatorname*{argmax}_{d:\mathcal{X}\to\mathcal{A}} \Big\{ \overbrace{\mathcal{V}(d) := \mathbb{E}[Y|A = d(\boldsymbol{X})]}^{\text{single-stage value function}} \Big\}. \tag{1}$$

Here, $\mathbb{E}[Y|A = d(\boldsymbol{X})] = \mathbb{E}\big\{\mathbb{E}[Y|\boldsymbol{X}, A = d(\boldsymbol{X})]\big\}$, where the outer expectation is taken over $\boldsymbol{X}$.

In this section, we first consider methods to find an optimal ITR. We categorize selected methods from the literature into the regression-based and direct-search approaches, which we will discuss in Sects. 1.1 and 1.2 respectively. In Sect. 1.3, we further discuss some other research goals, including the utilities beyond expectation and the generalizability concern, and the related methods.

1.1 Regression-Based Methods

One approach for finding an optimal ITR is to estimate the mean outcome conditional on covariates and treatment. This is known as the regression-based approach. In Sect. 1.1.1, we introduce Q-Learning and A-Learning, two main regression-based methods in the literature (Schulte et al., 2014). In Sect. 1.1.2, we further discuss an extended framework that was initially developed for subgroup identification but can unify many related methods in the literature.

1.1.1 Q-Learning and A-Learning

Define the Q-function $Q(\boldsymbol{x}, a) := \mathbb{E}(Y|\boldsymbol{X} = \boldsymbol{x}, A = a)$ as a function of the covariates $\boldsymbol{x} \in \mathcal{X}$ and treatment $a \in \mathcal{A}$. Then the corresponding optimal ITR is given by $d^\star(\boldsymbol{x}) = \text{argmax}_{a\in\mathcal{A}} Q(\boldsymbol{x}, a)$. This motivates Q-Learning to estimate the Q-function for finding an optimal ITR (Watkins, 1989; Qian & Murphy, 2011). Specifically, we first estimate $Q(\boldsymbol{X}, A)$ by regressing Y over $(\boldsymbol{X}, A)$. Then we obtain the optimal ITR as $\widehat{d}_n(\boldsymbol{X}) \in \text{argmax}_{a\in\mathcal{A}} \widehat{Q}_n(\boldsymbol{X}, a)$. Such an approach is also known as the predict-then-optimize strategy in the Contextual Stochastic Optimization (CSO) literature (Bertsimas & Kallus, 2020; Elmachtoub & Grigas, 2022).

In the binary treatment case $\mathcal{A} = \{0, 1\}$, it is sufficient to estimate the Conditional Average Treatment Effect (CATE) $C(\boldsymbol{X}) := Q(\boldsymbol{X}, 1) - Q(\boldsymbol{X}, 0)$. The corresponding optimal ITR is $\widehat{d}_n(\boldsymbol{X}) = \mathbb{I}[\widehat{C}_n(\boldsymbol{X}) \geqslant 0]$. Such an approach is known as A-Learning (Murphy, 2003). Related research focuses on the covariate-treatment interaction effect $a \times C(\boldsymbol{X}) = Q(\boldsymbol{X}, a) - Q(\boldsymbol{X}, 0)$ (Royston & Sauerbrei, 2008), which is also known as the blip-to-zero function in the following Structural Mean Model (SMM) (Robins, 1994):

$$Y = m(\boldsymbol{X}) + A \times C(\boldsymbol{X}; \boldsymbol{\beta}) + \epsilon; \quad \mathbb{E}(\epsilon|\boldsymbol{X}, A) = 0. \tag{2}$$

Here, $C(\boldsymbol{X}; \boldsymbol{\beta})$ is an assumed parametric model for the CATE with $\boldsymbol{\beta} \in \mathbb{R}^p$ as the p-dimensional parameter vector of interest. When p is large, a sparse linear model $C(\boldsymbol{X}; \boldsymbol{\beta}) = \boldsymbol{X}_\mathcal{S}^\intercal \boldsymbol{\beta}_\mathcal{S}$ associated with relevant variable subset $\mathcal{S} \subseteq \{1, 2, \cdots, p\}$ can be assumed (Imai & Ratkovic, 2013; Lu et al., 2013). The treatment-free effect function $m(\boldsymbol{X})$ is nuisance and assumed nonparametric. To obtain the semiparametric efficient estimate of $\boldsymbol{\beta}$, Lu et al. (2013), Nie and Wager (2021) considered Robinson (1988)'s transformation as follows:

$$Y - \mathbb{E}(Y|\boldsymbol{X}) = [A - \mathbb{E}(A|\boldsymbol{X})]C(\boldsymbol{X}; \boldsymbol{\beta}) + \epsilon.$$

Denote the nuisance functions as $\mu(\boldsymbol{x}) := \mathbb{E}(Y|\boldsymbol{X})$ and $\pi(\boldsymbol{x}) := \mathbb{E}(A|\boldsymbol{X})$. Let $(\widehat{\mu}_n, \widehat{\pi}_n)$ be the corresponding finite-sample estimates. Then the CATE parameter vector $\boldsymbol{\beta}$ can be estimated from the following least-squares problem:

$$\min_{\boldsymbol{\beta}\in\mathbb{R}^p} \frac{1}{2}\mathbb{E}_n\Big\{Y - \widehat{\mu}_n(\boldsymbol{X}) - [A - \widehat{\pi}_n(\boldsymbol{X})]C(\boldsymbol{X}; \boldsymbol{\beta})\Big\}^2. \tag{3}$$

The corresponding estimating function is equivalent to a G-estimating function for the single-stage SMM:

$$\boldsymbol{\phi}(\boldsymbol{\beta}) = \Big\{Y - \widehat{\mu}_n(\boldsymbol{X}) - [A - \widehat{\pi}_n(\boldsymbol{X})]C(\boldsymbol{X}; \boldsymbol{\beta})\Big\}[A - \widehat{\pi}_n(\boldsymbol{X})]\frac{\partial}{\partial \boldsymbol{\beta}}C(\boldsymbol{X}; \boldsymbol{\beta}). \tag{4}$$

A G-estimate $\widehat{\boldsymbol{\beta}}_n$ solves $\mathbb{E}_n[\boldsymbol{\phi}(\widehat{\boldsymbol{\beta}}_n)] = \boldsymbol{0}$. If either $\widehat{\mu}_n \to \mu$ or $\widehat{\pi}_n \to \pi$, then $\widehat{\boldsymbol{\beta}}_n \to \boldsymbol{\beta}$, which is known as the doubly robust property. In other words, if either one of $\widehat{\mu}_n$ and $\widehat{\pi}_n$ is consistent, then we have the consistency of $\widehat{\boldsymbol{\beta}}_n$. This can allow at most one misspecification of $\widehat{\mu}_n$ or $\widehat{\pi}_n$.

In order to establish standard $\sqrt{n}$-asymptotic properties for $\widehat{\boldsymbol{\beta}}_n$, both $\widehat{\mu}_n$ and $\widehat{\pi}_n$ need to be consistent. In particular, it requires that $(\widehat{\mu}_n, \widehat{\pi}_n) = (\mu, \pi) + o_{\mathbb{P}}(n^{-1/2})$, so that the plug-in bias due to $(\widehat{\mu}_n, \widehat{\pi}_n)$ is $\sqrt{n}$-negligible. In this case, the $\sqrt{n}$-asymptotic properties of $\widehat{\boldsymbol{\beta}}_n$ is not affected if $(\widehat{\mu}_n, \widehat{\pi}_n)$ are replaced by (μ, π). To relax such a condition, Chernozhukov et al. (2018), Nie et al. (2021) leveraged the cross-fitting technique to establish the $\sqrt{n}$-asymptotic properties under a weaker condition: $(\widehat{\mu}_n, \widehat{\pi}_n) \to (\mu, \pi)$ and $\left\{\mathbb{E}[\widehat{\mu}_n(\boldsymbol{X}) - \mu(\boldsymbol{X})]^2\right\}^{1/2} \left\{\mathbb{E}[\widehat{\pi}_n(\boldsymbol{X}) - \pi(\boldsymbol{X})]^2\right\}^{1/2} = o(n^{-1/2})$. The last condition only requires $\widehat{\mu}_n(\boldsymbol{X}) = \mu(\boldsymbol{X}) + \mathcal{O}_{\mathbb{P}}(n^{-\zeta_1})$ and $\widehat{\pi}_n(\boldsymbol{X}) = \pi(\boldsymbol{X}) + \mathcal{O}_{\mathbb{P}}(n^{-\zeta_2})$ with $\zeta_1 + \zeta_2 > 1/2$, which is known as the rate-doubly robust property (Rotnitzky et al., 2021). If (i) $(\widehat{\mu}_n, \widehat{\pi}_n) \to (\mu, \pi)$, (ii) the plug-in bias is $\sqrt{n}$-negligible, and (iii) $\mathbb{E}(\epsilon^2|\boldsymbol{X}, A) = \mathbb{E}(\epsilon^2)$, then the least-squares estimating function (4) is semiparametrically efficient (Tsiatis, 2007). That is, the $\sqrt{n}$-asymptotic variance of $\widehat{\boldsymbol{\beta}}_n$ is the smallest in the positive semi-definite sense among all Regular and Asymptotically Linear (RAL) estimates under Model (2). However, when any of these conditions are violated, the $\sqrt{n}$-asymptotic properties can be challenging to study.

Dukes et al. (2021) considered the case when either $\widehat{\mu}_n \to \check{\mu} \neq \mu$ or $\widehat{\pi}_n \to \check{\pi} \neq \pi$. In order to establish the $\sqrt{n}$-convergence of $\widehat{\boldsymbol{\beta}}_n$ under the rate-doubly robust condition, they further estimated the biases $\check{\mu} - \mu$ and $\check{\pi} - \pi$ nonparametrically and refined the plug-in estimates $(\widehat{\mu}_n, \widehat{\pi}_n)$. However, the $\sqrt{n}$-asymptotic variance of $\widehat{\boldsymbol{\beta}}_n$ under such nuisance model misspecification was not studied. To address this issue, Mo and Liu (2021) considered either one or both of $\widehat{\mu}_n \to \check{\mu} \neq \mu$ (treatment-free effect misspecification) and $\mathbb{E}(\epsilon^2|\boldsymbol{X}, A) \neq \mathbb{E}(\epsilon^2)$ (heteroscedasticity). They introduced the Efficient-Learning (E-Learning) estimating function with an additional nuisance variance function. When $\widehat{\mu}_n \to \check{\mu} \neq \mu$ or/and $\mathbb{E}(\epsilon^2|\boldsymbol{X}, A) \neq \mathbb{E}(\epsilon^2)$, the $\sqrt{n}$-asymptotic variance of $\widehat{\boldsymbol{\beta}}_n$ remains to be the smallest among the class of G-estimates. Moreover, E-Learning was developed based on multiple treatment options using the angle-based representation as in Zhang and Liu (2014), which extends the work based on binary treatment.

Estimating a parametric model for the Q-function or CATE may suffer from potential model misspecification. It can be desirable to approximate it using flexible nonparametric regression or machine learning approaches. This problem has been studied intensively in the causal inference literature (Dorie et al., 2019; Guo et al., 2020; Kennedy et al., 2022). Many flexible modeling methods have been proposed. Some main examples include nonparametric regression of the doubly robust transformed outcome (Kennedy, 2020), the index models (Liang & Yu, 2022), the local regression (Bertsimas & Kallus, 2020), the classification and regression tree (CART) (Bertsimas et al., 2019), the random regression forest models (Wager & Athey, 2018), the Bayesian additive regression tree (BART) (Hahn et al., 2020), the Gaussian process (Alaa & van der Schaar, 2017), the neural network (Shalit et al., 2017; Yoon et al., 2018), and the meta learners (Künzel et al., 2019).

1.1.2 Weighted Loss Minimization

Motivated from the least-squares problem (3) of A-Learning, the following weighted loss minimization can unify several related methods (Chen et al., 2017):

$$\min_{f:\mathcal{X}\to[-1,1]} \mathbb{E}\left\{w(\boldsymbol{X}, A)\ell\Big(Y, (A-1/2)\times f(\boldsymbol{X})\Big)\right\}, \tag{5}$$

where $w(\boldsymbol{X}, A) \geqslant 0$ is a weight function, and $\ell(y, \widehat{y}) \geqslant 0$ is a general loss function. With $f^\star$ as the solution to Problem (5), the corresponding optimal ITR is $d^\star(\boldsymbol{x}) = \mathbb{I}[f^\star(\boldsymbol{x}) \geqslant 0]$.

Problem (5) requires certain conditions to result in a sign-consistent ITR estimate. In particular, the weight function $w(\boldsymbol{X}, A)$ needs to satisfy the following balancing condition (Wallace & Moodie, 2015):

$$w(\boldsymbol{x}, 1)\pi(\boldsymbol{x}) = w(\boldsymbol{x}, 0)[1-\pi(\boldsymbol{x})]; \quad \forall \boldsymbol{x} \in \mathcal{X}. \tag{6}$$

Examples include the Inverse Probability Weight (IPW) function $w(\boldsymbol{x}, 1) = \pi(\boldsymbol{x})^{-1}$, $w(\boldsymbol{x}, 0) = [1-\pi(\boldsymbol{x})]^{-1}$ (Tian et al., 2014), and the overlap weight function $w(\boldsymbol{x}, 1) = 1-\pi(\boldsymbol{x})$, $w(\boldsymbol{x}, 0) = \pi(\boldsymbol{x})$ (Crump et al., 2006). Here, if we define $\mathbb{P}_w \ll \mathbb{P}$ such that $\mathrm{d}\mathbb{P}_w/\mathrm{d}\mathbb{P} := w(\boldsymbol{X}, A)$, then $\mathbb{P}_w(A=1|\boldsymbol{X}) = \mathbb{P}_w(A=0|\boldsymbol{X}) = 1/2$, and for any $h: \mathcal{X} \to \mathbb{R}$, we have $\mathbb{E}_w[h(\boldsymbol{X})|A=1] = \mathbb{E}_w[h(\boldsymbol{X})|A=0] = \mathbb{E}_w[h(\boldsymbol{X})]$. That is, the weight $w(\boldsymbol{X}, A)$ is a specific form of Rosenbaum and Rubin (1983)'s balancing score such that $\boldsymbol{X} \perp\!\!\!\perp A$ under $\mathbb{P}_w$. Consequently, the balancing condition (6) can correspond to the general class of Covariate Balancing Propensity Score (CBPS) in Imai and Ratkovic (2014), Li et al. (2018).

Problem (5) with the squared loss $\ell(y, \widehat{y}) = (1/2)(y-\widehat{y})^2$ corresponds to the solution $f^\star(\boldsymbol{x}) = C(\boldsymbol{x})$, the CATE in Model (2). In this case, Wallace and Moodie (2015) considered the following weighted least-squares problem:

$$\min_{\boldsymbol{\beta},\boldsymbol{\eta}} \frac{1}{2}\mathbb{E}_n\left\{\widehat{w}_n(\boldsymbol{X}, A)\left[Y - \boldsymbol{X}^\top\boldsymbol{\eta} - (A-1/2)\boldsymbol{X}^\top\boldsymbol{\beta}\right]^2\right\}. \tag{7}$$

Here, $\boldsymbol{X}^\top\boldsymbol{\eta}$ and $\boldsymbol{X}^\top\boldsymbol{\beta}$ are the parametric models for the treatment-free effect and CATE respectively, $\widehat{w}_n(\boldsymbol{X}, A)$ is a finite-sample estimate of $w(\boldsymbol{X}, A)$ that satisfies (6). The weighted least-squares estimate of $\boldsymbol{\beta}$ based on (7) is consistent if either one of the estimated weight function $\widehat{w}_n(\boldsymbol{X}, A)$ and the working linear model $\boldsymbol{X}^\top\boldsymbol{\eta}$ for the treatment-free effect is correctly specified. This suggests the double robustness of (7).

Chen et al. (2017) considered the general loss function $\ell(y, \widehat{y})$ in (5) that satisfies the following two conditions.

- The score function $\mathcal{S}(y, \widehat{y}) := (\partial/\partial\widehat{y})\ell(y, \widehat{y})$ is strictly increasing in $\widehat{y}$ for every $y \in \mathcal{Y}$.
- The utility function $\mathcal{U}(y) := -(\partial/\partial\widehat{y})\ell(y, 0)$ is strictly monotone in y.

Without loss of generality, assume that the utility function $\mathcal{U}(y)$ is strictly increasing in y. Then the solution $f^\star$ to Problem (5) satisfies that for any $\boldsymbol{x} \in \mathcal{X}$, $f^\star(\boldsymbol{x}) \geqslant 0$ if and only if $\mathbb{E}[\mathcal{U}(Y)|\boldsymbol{X} = \boldsymbol{x}, A = 1] \geqslant \mathbb{E}[\mathcal{U}(Y)|\boldsymbol{X} = \boldsymbol{x}, A = 0]$. As a special case, suppose the distribution of Y belongs to the exponential family with the canonically parametrized negative log-likelihood function $\ell(y, \eta) := -y\eta + \psi(\eta) - \log h(y)$, where η is the canonical parameter and $\psi(\eta) = \log \int h(y)e^{\eta y}\mathrm{d}y$ is the log-partition function. Then the corresponding utility function is $\mathcal{U}(Y) = Y - \psi'(0)$, so that $\mathsf{sign}[f^\star(\boldsymbol{x})] = \mathsf{sign}[C(\boldsymbol{x})]$. Therefore, the solution $f^\star$ to Problem (5) is sign-consistent with the optimal ITR. In fact, Problem (5) with the negative log-likelihood loss function $\ell(y, \widehat{y})$ is equivalent to the Maximal Likelihood Estimate (MLE) of the working model $\mathbb{E}(Y|\boldsymbol{X}, A) = \psi'[(A - 1/2) \times f(\boldsymbol{X})]$. Such a working model needs not be correctly specified, while we can still establish the sign consistency. Specifically, we have $\psi'[(1/2)f^\star(\boldsymbol{x})] - \psi'[-(1/2)f^\star(\boldsymbol{x})] = C(\boldsymbol{x})$, where the function $\eta \mapsto \psi'(\eta/2) - \psi'(-\eta/2)$ is strictly increasing and equals to 0 at $\eta = 0$. Therefore, $f^\star(\boldsymbol{x}) \geqslant 0$ if and only if $C(\boldsymbol{x}) \geqslant 0$.

The general formulation (5) is a supervised learning problem. Under such a framework, Tian et al. (2014), Xu et al. (2015), Chen et al. (2017), Qi and Liu (2018), Qi et al. (2020) can consider the ITR problem for continuous, binary and survival outcomes with a flexible function class for the decision function f.

1.2 Direct-Search Methods

Other than the regression-based methods, another approach for estimating the optimal ITR is to directly maximize an estimate of the value function $\mathcal{V}(d)$, which is known as the direct-search approach. To be specific, for every fixed ITR d, we first consider an unbiased estimate $\widehat{\mathcal{V}}_n(d)$ of $\mathcal{V}(d)$. Let $\mathcal{D}$ be the class of ITRs for training. Then the corresponding direct-search ITR estimate is given by $\widehat{d}_n \in \operatorname{argmax}_{d\in\mathcal{D}} \widehat{\mathcal{V}}_n(d)$. In this line of research, the Inverse Probability Weighted Estimate (IPWE) and Augmented IPWE (AIPWE) are two main estimates motivated from the analysis of missing data (Tsiatis, 2007). We will discuss both of them in Sects. 1.2.1 and 1.2.2 respectively.

1.2.1 Inverse Probability Weighted Estimate (IPWE)

For a given ITR $d \in \mathcal{D}$, the IPWE of its value $\mathcal{V}(d)$ is given by

$$\widehat{\mathcal{V}}_n(d) = \mathbb{E}_n\left\{\frac{\mathbb{I}[d(\boldsymbol{X}) = A]}{\pi(\boldsymbol{X}, A)}Y\right\},$$

where $\pi(\boldsymbol{x}, a) := \mathbb{P}(A = a|\boldsymbol{X} = \boldsymbol{x})$. Based on the IPWE, Laber and Zhao (2015) considered $\mathcal{D}$ as the class of decision trees, and introduced some splitting criteria that maximize the IPWE in the corresponding CART algorithm. Kitagawa and Tetenov (2018) considered $\mathcal{D}$ as a general Vapnik-Chervonenkis (VC) class of ITRs, and established the $\sqrt{n}$-regret bound and the minimax rate optimality for $\widehat{d}_n$. For implementation, they considered $\mathcal{D}$ as a class of linear decision functions, and solved a Mixed Integer Programming (MIP) problem.

When $\widehat{\mathcal{V}}_n(d)$ is the IPWE, the optimization problem for $\max_{d\in\mathcal{D}} \widehat{\mathcal{V}}_n(d)$ can be nonconvex in d, which is challenging to solve. In order to overcome this challenge, Zhao et al. (2012) reformulated the IPWE maximization into the following Outcome Weighted Learning (OWL) problem:

$$\min_{f\in\mathcal{F}} \left\{ \mathbb{E}_n \left(\frac{Y}{\pi(\boldsymbol{X}, A)} \phi[(2A-1)f(\boldsymbol{X})] \right) + \lambda_n J(f) \right\},$$

where ϕ is a margin-based convex surrogate loss function, $\mathcal{F} \subseteq \{f : \mathcal{X} \to \mathbb{R}\}$ is a pre-specified decision function class, and $\lambda_n J(f)$ is a functional penalty. Many related methods have been developed under the OWL framework. In particular, OWL has been extended to the general types of outcomes, such as the binary (Huang & Fong, 2014) and survival (Zhao et al., 2015b; Cui et al., 2017) outcomes. It can be equipped with many supervised learning methods, such as the bagging and neural network (Mi et al., 2019). To reduce the finite sample variance, Zhou et al. (2017), Liu et al. (2018) further proposed the Residual Weighted Learning (RWL) with Y replaced by $Y - g(\boldsymbol{X})$ for some function $g : \mathcal{X} \to \mathbb{R}$. To handle the negative outcomes and gain more robustness in presence of covariate outliers, Huang and Fong (2014), Zhou et al. (2017) considered the nonconvex ramp loss function for ϕ. In particular, the weighted loss functions $\frac{Y}{\pi(\boldsymbol{X},A)}\phi[(2A-1)f(\boldsymbol{X})]$ and $\frac{|Y|}{\pi(\boldsymbol{X},A)}\phi[(2A-1)\mathsf{sign}(Y)f(\boldsymbol{X})]$ are equivalent in this case. Moreover, the ramp loss with a properly tuned bandwidth parameter can converge to the 0–1 loss. When the number of variables p is large, penalties that induce sparsity can be incorporated in the OWL framework (Song et al., 2015; Xu et al., 2015). For multiple and continuous treatment problems, the extensions of OWL were further studied in Chen et al. (2016), Chen et al. (2018), Kallus and Zhou (2018), Fu et al. (2019), Zhang et al. (2020), Meng et al. (2020), Fan et al. (2022).

1.2.2 Augmented Inverse Probability Weighted Estimate (AIPWE)

In observational studies, the propensity score function $\pi(\boldsymbol{X}, A)$ needs to be estimated from data. In order to protect the potential misspecification of the propensity score model, the AIPWE of $\mathcal{V}(d)$ was proposed:

$$\begin{aligned}\widehat{\mathcal{V}}_n(d) &= \mathbb{E}_n \left\{ \frac{\mathbb{I}[d(\boldsymbol{X})=A]}{\widehat{\pi}_n(\boldsymbol{X}, A)} Y - \frac{\mathbb{I}[d(\boldsymbol{X})=A] - \widehat{\pi}_n(\boldsymbol{X}, A)}{\widehat{\pi}_n(\boldsymbol{X}, A)} \widehat{Q}_n(\boldsymbol{X}, d) \right\} \\ &= \mathbb{E}_n \left\{ \widehat{Q}_n(\boldsymbol{X}, d) + \frac{\mathbb{I}[d(\boldsymbol{X})=A]}{\widehat{\pi}_n(\boldsymbol{X}, A)} [Y - \widehat{Q}_n(\boldsymbol{X}, A)] \right\},\end{aligned}$$

where $Q(\boldsymbol{x}, d) := \sum_{a\in\mathcal{A}} Q(\boldsymbol{x}, a)\mathbb{I}[d(\boldsymbol{x}) = a]$. The first definition can be obtained from the efficient influence function under the missing data framework (Tsiatis, 2007) or Targeted Minimum Loss-based Estimation (TMLE) (van der Laan & Rubin, 2006; van der Laan & Rose, 2018). Based on the second equivalent definition, maximizing the AIPWE is equivalent to $\max_{d:\mathcal{X}\to\mathcal{A}} \mathbb{E}_n \left\{\widehat{\mathcal{V}}_n(d) - \widehat{\mathcal{V}}_n(-d) = \Gamma \times [2d(\boldsymbol{X}) - 1]\right\}$, where

$$\Gamma := \widehat{Q}_n(\boldsymbol{X}, 1) - \widehat{Q}_n(\boldsymbol{X}, 0) + \frac{2A - 1}{\widehat{\pi}_n(\boldsymbol{X}, A)}[Y - \widehat{Q}_n(\boldsymbol{X}, A)]$$

is also known as the doubly robust score in Athey and Wager (2021) and the doubly robust pseudo outcome of Kennedy (2020). In particular, Γ is a function of $(\boldsymbol{X}, A, Y)$, while for every fixed $h : \mathcal{X} \to \mathbb{R}$, we have either $\widehat{Q}_n \to Q$ or $\widehat{\pi}_n \to \pi$ implies that $\mathbb{E}_n[\Gamma \times h(\boldsymbol{X})] \to \mathbb{E}[C(\boldsymbol{X})h(\boldsymbol{X})]$. This suggests the doubly robust property of the AIPWE (Dudík et al., 2011; Zhang et al., 2012b).

Based on the AIPWE, Zhang et al. (2012b) considered the direct maximization of $\widehat{\mathcal{V}}_n(d_\eta)$ over a parametric ITR class $\mathcal{D}_\eta = \{d_\eta : \eta \in \Xi\}$ using the genetic algorithm. Alternatively, Zhang et al. (2015) considered the class of decision lists for $\mathcal{D}$ and proposed an approximation algorithm. Zhang et al. (2012a) proposed C-Learning to minimize the weighted misclassification error $\mathbb{E}_n\{|\Gamma| \times \mathbb{I}[d(\boldsymbol{X}) \neq \mathbb{I}(\Gamma \geqslant 0)]\}$. In a slightly different way, Dudík et al. (2011) also considered cost-sensitive classification algorithms but based on $\widehat{\mathcal{V}}_n(d)$ directly.

Similar to the regression-based doubly robust estimates, the cross-fitting technique can also be leveraged to establish a sharp regret bound under weak conditions, where the estimation effects of $(\widehat{Q}_n, \widehat{\pi}_n)$ can be negligible. In this line of research, Athey and Wager (2021) proposed to maximize the cross-fitting AIPWE over the ITR class of decision trees. The $\sqrt{n}$-regret bound and its optimality can be further established. Alternatively, Zhao et al. (2019a) proposed the Efficient Augmentation and Relaxation Learning (EARL) framework with a convex surrogate loss relaxation analogous to OWL. Similarly, based on the surrogate loss relaxation, Bennett and Kallus (2020) proposed the Efficient Surrogate Policy Risk Minimization (ESPRM) that solves a variational method-of-moment problem, in which case the $\sqrt{n}$-regret bound and its optimality can also be obtained.

1.3 Other Considerations

The research problems discussed in Sects. 1.1 and 1.2 may not cover all needs from the practical point of view. In this section, we focus on two additional topics: the utilities beyond expectation, and the generalizability issue.

1.3.1 Risk-Averse Utilities

The objective in Sects. 1.1 and 1.2 is to maximize the value function defined from the expected outcome in (1). Other outcome utilities have also been studied in the literature. Consider the illustrating example in Table 1. Suppose we have three possible contexts x_1, x_2, x_3. For each context, a decision maker can choose to invest ($A = 1$) or not invest ($A = 0$) a project. The payoff Y is 1 or -1 if a project is launched, while the payoff is 0 if no investment is launched. Suppose the success probability of x_1, x_2, x_3 are $1, 0.55, 0.95$ respectively. The expected payoffs for x_1, x_2, x_3 are $1, 0.1, 0.9$ respectively, which are greater than the no-investment payoff 0. Therefore, in terms of the expected outcome, investment ($A = 1$) is always superior to no investment ($A = 0$) across all possible covariate values $X \in \{x_0, x_1, x_2\}$. However, for $X = x_1, x_2$, the invested payoffs are negative with probabilities $0.45, 0.05$ respectively, in which case the no-investment payoffs are superior. Although investment is superior for $X = x_1$ in expectation, a risk-averse decision maker may prefer no investment to protect the downside risk.

Given an ITR d, we define the Cumulative Distribution Function (CDF) $F^d(y) := \mathbb{P}[Y \leqslant y | d(\boldsymbol{X}) = A] = \mathbb{E}\{\mathbb{P}[Y \leqslant y | \boldsymbol{X}, d(\boldsymbol{X}) = A]\}$ for $y \in \mathbb{R}$ when the treatment A is assigned by $d(\boldsymbol{X})$ at the individual level. Notice that F^d represents the marginal distribution of Y. Also define $F(y|\boldsymbol{X}, A) := \mathbb{P}(Y \leqslant y|\boldsymbol{X}, A)$ as the conditional CDF of Y given the covariates and treatment. We will discuss those utilities as some functionals of the marginal outcome distribution $F^d(\cdot)$ or the conditional outcome distribution $F(\cdot|\boldsymbol{X}, d(\boldsymbol{X}))$. For the example in Table 1, we compute the marginal outcome distribution for each possible decision rule in Table 2.

One popular risk-averse utility is the τ-quantile $\mathcal{Q}_\tau(d) := \inf\{y : F^d(y) \geqslant \tau\}$ for a fixed $\tau \in [0, 1]$. In the example of Tables 1 and 2, suppose that $p_1 = p_2 = 0.5$. Then for $\tau \leqslant 0.225$, any decision rule d such that context x_1 is invested ($d(x_1) = 1$) is suboptimal in the τ-quantile, since $\mathcal{Q}_\tau(d) = -1$, while the optimal τ-quantile utility is 0 when x_1 is not invested ($d(x_1) = 0$). In this way, the risk of negative payoff for $X = x_1$ is prevented.

To obtain the quantile-optimal ITR, Linn et al. (2017) took a regression-based approach that extends Q-Learning, while Wang et al. (2018) proposed a

Table 1 An illustrating example

	$X = x_0$	$X = x_1$	$X = x_2$
	Marginal distribution of X		
	$1 - p_1 - p_2$	p_1	p_2
	Conditional distribution of $Y \mid (\boldsymbol{X}, A)$		
$A = 1$	$Y = 1$	$Y = \begin{cases} 1, & \text{with probability } 0.55; \\ -1, & \text{with probability } 0.45 \end{cases}$	$Y = \begin{cases} 1, & \text{with probability } 0.95; \\ -1, & \text{with probability } 0.05 \end{cases}$
$A = 0$	$Y = 0$	$Y = 0$	$Y = 0$

Note: $p_1, p_2 \geqslant 0$, $p_1 + p_2 \leqslant 1$

Table 2 Marginal outcome distribution $F^d(y)$

$(d(x_0), d(x_1), d(x_2))$	$F^d(-1)$	$F^d(0)$	$F^d(1)$
(1, 1, 1)	$0.45p_1 + 0.05p_2$	$0.45p_1 + 0.05p_2$	1
(1, 0, 1)	$0.05p_2$	$p_1 + 0.05p_2$	1
(1, 1, 0)	$0.45p_1$	$0.45p_1 + p_2$	1
(1, 0, 0)	0	$p_1 + p_2$	1
(0, 1, 1)	$0.45p_1 + 0.05p_2$	$1 - 0.55p_1 - 0.95p_2$	1
(0, 0, 1)	$0.05p_2$	$1 - 0.95p_2$	1
(0, 1, 0)	$0.45p_1$	$1 - 0.55p_1$	1
(0, 0, 0)	0	1	1

direct-search method. Directly maximizing the τ-quantile utility is computationally challenging. Qi et al. (2023) proposed to consider a closely related utility, the Conditional Value at Risk (CVaR) (Artzner et al., 1999) $\texttt{CVaR}_\tau(d) := \frac{1}{\tau}\mathbb{E}[Y\mathbb{I}(Y \leqslant \mathcal{Q}_\tau(d))|d(\boldsymbol{X}) = A] = \frac{1}{\tau}\int_0^\tau \mathcal{Q}_t(d)\mathrm{d}t$. One technical advantage of such a utility is due to the representation $\texttt{CVaR}_\tau(d) = \sup_{q\in\mathbb{R}}\left\{q - \frac{1}{\tau}\mathbb{E}[(q-Y)^+|d(\boldsymbol{X}) = A]\right\}$, which is attained at $q = \mathcal{Q}_\tau(d)$. Then we have

$$\max_{d\in\mathcal{D}} \texttt{CVaR}_\tau(d) = \max_{d\in\mathcal{D}, q\in\mathbb{R}} \left\{q - \frac{1}{\tau}\mathbb{E}\left[\frac{\mathbb{I}(d(\boldsymbol{X}) = A)}{\pi(\boldsymbol{X}, A)}(q-Y)^+\right]\right\}. \tag{8}$$

Based on this formulation, the estimation and optimization strategies in Sect. 1.2.1 can be leveraged to handle the corresponding learning problem. Note that the CVaR utility is actually a special case of the Optimized Certainty Equivalent (OCE) utility (Ben-Tal & Teboulle, 1986) $\mathcal{O}_\phi(d) := \sup_{q\in\mathbb{R}}\{q + \mathbb{E}[\phi(Y-q)|d(\boldsymbol{X}) = A]\}$, where $\phi : \mathbb{R} \to \mathbb{R}$ is a concave non-decreasing function satisfying certain regularity conditions. This can suggest similar learning strategies to obtain the CVaR-optimal and OCE-optimal ITRs.

Consider the conditional τ-quantile function $Q_\tau(\boldsymbol{X}, A) := \inf\{y : F(y|\boldsymbol{X}, A) \geqslant \tau\}$. Another risk-averse utility is the Average Conditional Quantile (ACQ) (Liu & Kennedy, 2021) $\bar{\mathcal{Q}}_\tau(d) := \mathbb{E}[Q_\tau(\boldsymbol{X}, d(\boldsymbol{X}))]$, where the conditional τ-quantile function is averaged over all covariates. Notice that the quantile utility $\mathcal{Q}_\tau(d)$ and the ACQ $\bar{\mathcal{Q}}_\tau(d)$ are not directly related. In particular, different from the expected value $\mathbb{E}[Y|d(\boldsymbol{X}) = A] = \mathbb{E}\{\mathbb{E}[Y|\boldsymbol{X}, d(\boldsymbol{X}) = A]\}$, we have $\mathcal{Q}_\tau(d) \neq \mathbb{E}[Q_\tau(\boldsymbol{X}, d(\boldsymbol{X}))] = \bar{\mathcal{Q}}_\tau(d)$. While the quantile $\mathcal{Q}_\tau(d)$ can incorporate the distributional heterogeneity across the covariates $\boldsymbol{X}$, the ACQ $\bar{\mathcal{Q}}_\tau(d)$ targets the conditional distribution $F(\cdot|\boldsymbol{X}, d(\boldsymbol{X}))$. More detailed comparisons can be found in Liu and Kennedy (2021, Section 4.2). For the example in Tables 1 and 2, suppose that $p_1 = 0.1$ and $p_2 = 0.9$. Let $\tau = 0.1$. We compare policies d_1, d_2 such that $d_1(x_0) = d_2(x_0) = d_1(x_2) = d_2(x_2) = 1$, and $d_1(x_1) = 1$, $d_2(x_1) = 0$. That is, both d_1 and d_2 invest for contexts x_0 and x_2. In addition, d_1 invests for context x_1 while d_2 does not invest for x_1. On one hand, $F^{d_1}(0) = 0.09$, $F^{d_2}(0) = 0.145$, $F^{d_2}(-1) = 0.045$. Then $\mathcal{Q}_\tau(d_1) = 1$, $\mathcal{Q}_\tau(d_2) = 0$, and d_1 is superior to d_2 under the τ-quantile utility. On the other hand, $F(-1|x_1, 1) = 0.45$, $F(0|x_1, 0) = 1$, $F(-1|x_1, 0) = 0$. Then $Q_\tau(x_1, 1) = -1$, $Q_\tau(x_1, 0) = 0$, and d_2 is superior to d_1 under the conditional τ-quantile utility. In this particular example, the different preferences of the τ-quantile and conditional τ-quantile is mainly due to the covariate distributional heterogeneity, that is, $p_1 = \mathbb{P}(X = x_1) = 0.1$ while $p_2 = \mathbb{P}(X = x_2) = 0.9$. Such an effect has overridden the risk preference from the conditional outcome distribution in the τ-quantile utility, while the conditional τ-quantile can avoid such a problem.

For binary treatment $A \in \{0, 1\}$, learning an ACQ-optimal ITR is closely related to estimating the Conditional Quantile Treatment Effect (CQTE) $C_\tau(\boldsymbol{X}) := Q_\tau(\boldsymbol{X}, 1) - Q_\tau(\boldsymbol{X}, 0)$ in the causal inference literature (Lee, 2003; Belloni et al., 2019). That is, we can first estimate the CQTE function $C_\tau(\boldsymbol{X})$, and then obtain the ACQ-optimal ITR as $d_\tau^\star(\boldsymbol{X}) = \mathbb{I}[C_\tau(\boldsymbol{X}) \geqslant 0]$. In fact, such an approach is an A-Learning analog for the ACQ utility. Xiao et al. (2019), Cho and Ghosh (2021) took this approach in the ITR problem.

Define the conditional OCE function $O_\phi(\boldsymbol{X}, A) := \sup_{q\in\mathbb{R}}\{q + \mathbb{E}[\phi(Y - q)|\boldsymbol{X}, A]\}$. Qi et al. (2019) considered the optimized Covariate-Dependent Equivalent (CDE) $\bar{O}_\phi(d) := \mathbb{E}[O_\phi(\boldsymbol{X}, d(\boldsymbol{X}))] = \sup_{q:\mathcal{X}\to\mathbb{R}}\{\mathbb{E}[q(\boldsymbol{X}) + \phi(Y - q(\boldsymbol{X}))| d(\boldsymbol{X}) = A]\}$. When $\phi(u) := -\frac{1}{\tau}(-u)^+$, we have $\bar{O}_\phi(d) = \mathbb{E}[\texttt{CVaR}_\tau(\boldsymbol{X}, d(\boldsymbol{X}))]$, where $O_\phi(\boldsymbol{X}, A) = \texttt{CVaR}_\tau(\boldsymbol{X}, A) := \frac{1}{\tau}\mathbb{E}[Y\mathbb{I}(Y \leqslant Q_\tau(\boldsymbol{X}, A))] = \frac{1}{\tau}\int_0^\tau Q_t(\boldsymbol{X}, A)\mathrm{d}t$. In this case, the supremum of $O_\phi(\boldsymbol{X}, A)$ with respect to $q \in \mathbb{R}$ is attained at $q = Q_\tau(\boldsymbol{X}, A)$, and the supremum of $\bar{O}_\phi(d)$ with respect to $q : \mathcal{X} \to \mathbb{R}$ is attained at $q(\boldsymbol{X}) = Q_\tau(\boldsymbol{X}, d(\boldsymbol{X}))$. Based on the CDE extension of (8), Qi et al. (2019) further considered a difference-of-convex surrogate loss relaxation to learn the CDE-optimal ITR.

For fairness and equality concern, Kitagawa and Tetenov (2021) considered the rank-dependent utility $\mathcal{U}_\Lambda(d) := \int_0^{+\infty} \Lambda(F^d(y))\mathrm{d}y$, where $\Lambda : [0, 1] \to [0, 1]$ is a non-increasing convex function with $\Lambda(0) = 1$ and $\Lambda(1) = 0$. It can be equivalently written as $\mathcal{U}_\Lambda(d) = \int_0^1 \omega(\tau)\mathcal{Q}_\tau(d)\mathrm{d}\tau$ where $\omega(\tau) := \frac{\mathrm{d}[1-\Lambda(\tau)]}{\mathrm{d}\tau}$. When $\Lambda(t) = (1 - t/\tau)^+$, $\mathcal{U}_\Lambda(d) = \texttt{CVaR}_\tau(d)$. When $\Lambda(t) = (1 - t)^2$, $\mathcal{U}_\Lambda(d)$ becomes the Gini utility (Blackorby & Donaldson, 1978). Kitagawa and Tetenov (2021) further proposed an IPWE-based method to learn the $\mathcal{U}_\Lambda$-optimal ITR under the empirical welfare maximization framework.

1.3.2 Generalizability

Most literature in the ITR problem assume that the training and testing distributions are identical. The generalizability concern arises when this assumption is violated.

Recall that the value function based on the expected outcome is

$$\mathcal{V}(d) = \mathbb{E}[Y|d(\boldsymbol{X}) = A] = \mathbb{E}\Big\{\mathbb{E}[Y|\boldsymbol{X}, A = d(\boldsymbol{X})]\Big\},$$

where the inner expectation is taken with respect to the conditional outcome distribution of Y given $(\boldsymbol{X}, A)$, while the outer expectation is taken with respect to the covariate distribution of $\boldsymbol{X}$. When either of these distributions are not identical across training and testing, an optimal ITR under the training distribution can be sub-optimal under the testing distribution. From now on, we denote $\mathbb{P}$ and $\mathbb{Q}$ as the training and testing distributions respectively. In particular, $\mathbb{P}_{\boldsymbol{X}}$ and $\mathbb{Q}_{\boldsymbol{X}}$ are the marginal distributions of $\boldsymbol{X}$, while $\mathbb{P}_{Y|(\boldsymbol{X},A)}$ and $\mathbb{Q}_{Y|(\boldsymbol{X},A)}$ are the conditional distributions of $Y|(\boldsymbol{X}, A)$.

In the case of $\mathbb{Q}_{\boldsymbol{X}} = \mathbb{P}_{\boldsymbol{X}}$ but $\mathbb{Q}_{Y|(\boldsymbol{X},A)} \neq \mathbb{P}_{Y|(\boldsymbol{X},A)}$, the CVaR special case of the CDE-optimal ITR (Qi et al., 2019) can guarantee the worst-case testing performance to certain extent. Specifically, by Qi et al. (2023, Theorem 6.2), we have the following Distributionally Robust Optimization (DRO) formulation:

$$\mathtt{CVaR}_\tau^{\mathbb{P}}(\boldsymbol{X}, A) = \inf\left\{\mathbb{E}^{\mathbb{Q}}(Y|\boldsymbol{X}, A) : \mathbb{Q}_{Y|(\boldsymbol{X},A)} \ll \mathbb{P}_{Y|(\boldsymbol{X},A)},\ 0 \leqslant \frac{\mathrm{d}\mathbb{Q}_{Y|(\boldsymbol{X},A)}}{\mathrm{d}\mathbb{P}_{Y|(\boldsymbol{X},A)}} \leqslant \frac{1}{\tau}\right\}. \tag{9}$$

Here, the notations $\mathtt{CVaR}_\tau^{\mathbb{P}}$ and $\mathbb{E}^{\mathbb{Q}}$ emphasize the underlying distributions $\mathbb{P}$ and $\mathbb{Q}$ respectively. Equation (9) suggests that $\mathtt{CVaR}_\tau^{\mathbb{P}}(\boldsymbol{X}, A)$ is the lower bound of $\mathbb{E}^{\mathbb{Q}}(Y|\boldsymbol{X}, A)$ among all possible $\mathbb{Q}_{Y|(\boldsymbol{X},A)}$ such that it is absolutely continuous with respect to $\mathbb{P}_{Y|(\boldsymbol{X},A)}$, and the density ratio $\frac{\mathrm{d}\mathbb{Q}_{Y|(\boldsymbol{X},A)}}{\mathrm{d}\mathbb{P}_{Y|(\boldsymbol{X},A)}}$ is upper bounded by $1/\tau$. Therefore, the problem of the CVaR-specific CDE-optimal ITR can be formulated as the following DRO:

$$\begin{aligned}
&\max_{d\in\mathcal{D}} \mathbb{E}^{\mathbb{P}}[\mathtt{CVaR}_\tau^{\mathbb{P}}(\boldsymbol{X}, d(\boldsymbol{X}))] \\
&\quad = \max_{d\in\mathcal{D}} \inf\left\{\mathbb{E}^{\mathbb{Q}}[Y|d(\boldsymbol{X}) = A] : \mathbb{Q}_{\boldsymbol{X}} = \mathbb{P}_{\boldsymbol{X}},\ \mathbb{Q}_{Y|(\boldsymbol{X},A)} \ll \mathbb{P}_{Y|(\boldsymbol{X},A)},\right. \\
&\quad\quad \left. 0 \leqslant \frac{\mathrm{d}\mathbb{Q}_{Y|(\boldsymbol{X},A)}}{\mathrm{d}\mathbb{P}_{Y|(\boldsymbol{X},A)}} \leqslant \frac{1}{\tau}\right\}.
\end{aligned} \tag{10}$$

That is, the worst-case value function is optimized when the underlying distribution $\mathbb{Q}$ is different from the training distribution $\mathbb{P}$.

When $\mathbb{Q}_{\boldsymbol{X}} \neq \mathbb{P}_{\boldsymbol{X}}$ while $\mathbb{Q}_{Y|(\boldsymbol{X},A)} = \mathbb{P}_{Y|(\boldsymbol{X},A)}$, Mo et al. (2021a) took a DRO formulation analogous to (10) and considered

$$\max_{d\in\mathcal{D}} \inf\left\{\begin{array}{l}\mathbb{E}^{\mathbb{Q}}[Y|d(\boldsymbol{X}) = A] - \mathbb{E}^{\mathbb{Q}}(Y|A = \mathtt{random}) : \\ \mathbb{Q}_{Y|(\boldsymbol{X},A)} = \mathbb{P}_{Y|(\boldsymbol{X},A)},\ \mathbb{Q}_{\boldsymbol{X}} \ll \mathbb{P}_{\boldsymbol{X}},\ \left\|\frac{\mathrm{d}\mathbb{Q}_{\boldsymbol{X}}}{\mathrm{d}\mathbb{P}_{\boldsymbol{X}}}\right\|_{L^k(\mathbb{P})} \leqslant c\end{array}\right\}.$$

Here, "$A = \mathtt{random}$" means that $\mathbb{P}(A = 1|\boldsymbol{X}) = \mathbb{P}(A = 0|\boldsymbol{X}) = 1/2$, $\|Z\|_{L^k(\mathbb{P})} = [\mathbb{E}^{\mathbb{P}}(Z^k)]^{1/k}$ for $k \in (1, +\infty)$ and $\|Z\|_{L^\infty(\mathbb{P})} = \inf\{M : \mathbb{P}(Z \leqslant M) = 1\}$, and $c \geqslant 1$ quantifies the largest distance of $\mathbb{Q}$ from $\mathbb{P}$. When $k = +\infty$ and $c = 1/\tau$, this becomes a CVaR-related formulation. Zhao et al. (2019b) took a different approach to handle the covariate distributional change from training to testing. Specifically, they assumed multivariate normal covariate distributions over the domains $\{\boldsymbol{x} \in \mathcal{X} : d^{\star}(\boldsymbol{x}) = a\}$ for $a \in \{0, 1\}$ respectively. These coveriate distributions can change from training to testing with certain constraints on the first and second moments. Then they minimized the worst-case testing regret among all possible distributional changes over these domains to find the optimal ITR.

When both $\mathbb{Q}_X \neq \mathbb{P}_X$ and $\mathbb{Q}_{Y|(X,A)} \neq \mathbb{Q}_{Y|(X,A)}$, the covariate and conditional outcome distributions need not be distinguished. In this case, Qi et al. (2023, Theorem 2.5) showed that the CVaR-optimal ITR (8) is equivalent to the following DRO formulation, while Si et al. (2020), Kallus et al. (2022) also considered a similar DRO but used the Kullback-Leibler divergence to measure the distance of $\mathbb{Q}$ from $\mathbb{P}$:

$$\max_{d \in \mathcal{D}} \inf \left\{ \mathbb{E}^{\mathbb{Q}}[Y|d(\boldsymbol{X}) = A] : \mathbb{Q} \ll \mathbb{P}, \begin{array}{ll} 0 \leqslant \frac{\mathrm{d}\mathbb{Q}}{\mathrm{d}\mathbb{P}} \leqslant \frac{1}{\tau} & \text{(Qi et al., 2023)} \\ \mathbb{E}^{\mathbb{Q}}\left[\log\left(\frac{\mathrm{d}\mathbb{Q}}{\mathrm{d}\mathbb{P}}\right)\right] \leqslant \delta & \text{(Si et al., 2020;} \\ & \text{Kallus et al., 2022)} \end{array} \right\}.$$

The aforementioned methods in this section assume that we have no access to the testing distribution during training. Therefore, all these methods took a worst-case optimization strategy. When a sample from testing is available and can be combined with the training data, more targeted learning methods can be developed. Details were discussed in Mo et al. (2021b).

2 Multi-Stage Decision Problems

For a T-stage decision problem, each data point consists of a longitudinal trajectory $(\boldsymbol{X}_t, A_t, Y_t : 1 \leqslant t \leqslant T)$, with the time-varying covariates $\boldsymbol{X}_t \in \mathcal{X}_t \subseteq \mathbb{R}^{p_t}$, treatment $A_t \in \mathcal{A}_t$ and outcome $Y_t \in \mathcal{Y}_t \subseteq \mathbb{R}$ for $1 \leqslant t \leqslant T$. A Dynamic Treatment Regime (DTR) is defined as a sequence of stagewise decision rules $\boldsymbol{d}_{1:T} = (d_1, d_2, \cdots, d_T) \in \mathcal{D}_1 \times \mathcal{D}_2 \times \cdots \times \mathcal{D}_T = \boldsymbol{\mathcal{D}}_{1:T}$, where $\mathcal{D}_t = \{d_t : \mathcal{H}_t \to \mathcal{A}_t\}$ consists of all mappings from the stage-t pre-treatment history $\boldsymbol{H}_t := \big((\boldsymbol{X}_s, A_s, Y_s : 1 \leqslant s \leqslant t-1), \boldsymbol{X}_t\big) \in \mathcal{H}_t$ to the stage-t treatment assignment $A_t \in \mathcal{A}_t$. The goal is to find the optimal DTR that maximizes the expected cumulative outcome

$$\boldsymbol{d}^{\star}_{1:T} \in \operatorname*{argmax}_{\boldsymbol{d}_{1:T} \in \boldsymbol{\mathcal{D}}_{1:T}} \Bigg\{ \underbrace{\mathcal{V}(\boldsymbol{d}_{1:T}) := \sum_{t=1}^{T} \mathbb{E}[Y_t|A_t = d_t(\boldsymbol{H}_t)]}_{T\text{-stage value function}} \Bigg\}. \tag{11}$$

One key challenge of the DTR problem is that the stage-t treatment A_t can have time-varying effects on the post-treatment variables $\big(Y_t, (\boldsymbol{X}_u, A_u, Y_u : t+1 \leqslant u \leqslant T)\big)$. However, a standard regression analysis conditioning on the observed trajectory may cut off all indirect effects such as $A_t \to \boldsymbol{X}_{t+1} \to Y_{t+1}$ (Almirall et al., 2010). As a consequence, $\text{argmax}_{a_t \in \mathcal{A}_t} \mathbb{E}\left\{\sum_{u=t}^{T} Y_u \middle| \boldsymbol{H}_t, A_t = a_t\right\}$ is the not same as the stage-t optimal decision rule $d_t^\star(\boldsymbol{H}_t)$. Therefore, the observed data should be adjusted to remove those time-varying treatment effects across stages.

Similar to the single-stage problem, we will further discuss the regression-based and direct-search methods for the multi-stage problem in Sects. 2.1 and 2.2 respectively.

2.1 Regression-Based Methods

The first approach to adjust for the cross-stage treatment effects is to perform stagewise model-based outcome transformation. Specifically, consider the Bellman equations (Bellman, 1966) that recursively define the state-value functions, also known as the V-functions, $\{V_t(\boldsymbol{H}_t)\}_{t=1}^{T}$, and the state-action value functions, also known as the Q-functions, $\{Q_t(\boldsymbol{H}_t, A_t)\}_{t=1}^{T}$:

$$V_t(\boldsymbol{H}_t) := \max_{a_t \in \mathcal{A}_t} \underbrace{\mathbb{E}\Big\{Y_t + V_{t+1}(\boldsymbol{H}_{t+1}) \Big| \boldsymbol{H}_t, A_t = a_t\Big\}}_{:=Q_t(\boldsymbol{H}_t, a_t)}; \quad t = T, \cdots, 1. \tag{12}$$

Here, we denote $V_{T+1}(\boldsymbol{H}_{T+1}) = 0$ for convenience. The stage-t optimal decision rule can be induced by $d_t^\star(\boldsymbol{H}_t) = \text{argmax}_{a_t \in \mathcal{A}_t} Q_t(\boldsymbol{H}_t, a_t)$. It remains to estimate the Q-functions from data.

The stage-T estimation problem with data $(\boldsymbol{H}_T, A_T, Y_T)$ can be handled by any single-stage methods discussed in Sect. 1.1. For the stage-$t(\leqslant T-1)$ problem, assume that $Q_{t+1}(\boldsymbol{H}_{t+1}, A_{t+1})$ is obtained. Define the stage-t q-outcome as $Y_t^{(\text{q})} := Y_t + Q_{t+1}\big(\boldsymbol{H}_{t+1}, d_{t+1}^\star(\boldsymbol{H}_{t+1})\big)$. By (12), we have $\mathbb{E}(Y_t^{(\text{q})} | \boldsymbol{H}_t, A_t) = Q_t(\boldsymbol{H}_t, A_t)$. That is, the stage-$t$ estimation problem can be solved as a single-stage problem with data $(\boldsymbol{H}_t, A_t, Y_t^{(\text{q})})$. Based on this strategy, we can estimate $Q_t(\boldsymbol{H}_t, A_t)$ for $t = T, \cdots, 1$ backward stagewise. Such an approach gives the T-stage Q-Learning method (Watkins, 1989; Zhu et al., 2019). If we further assume that the Q-functions are stationary $\mathbb{E}(Y_t^{(\text{q})} | \boldsymbol{H}_t, A_t) = Q(\boldsymbol{X}_t, A_t)$ across stages $1 \leqslant t \leqslant T$, then Q-Learning can be extended to the infinite-horizon setting with $T = +\infty$ (Sutton & Barto, 2018).

In the binary treatment case $\mathcal{A}_t = \{0, 1\}$, we define the stagewise CATE functions as $C_t(\boldsymbol{H}_t) := Q_t(\boldsymbol{H}_t, 1) - Q_t(\boldsymbol{H}_t, 0)$ for $1 \leqslant t \leqslant T$. Then the optimal DTR becomes $d_t^\star(\boldsymbol{H}_t) = \mathbb{I}[C_t(\boldsymbol{H}_t) \geqslant 0]$ $(1 \leqslant t \leqslant T)$. We further introduce the stagewise g-outcome as $Y_t^{(\text{g})} := \sum_{u=t}^{T} Y_u - \sum_{u=t+1}^{T} \{A_u - \mathbb{I}[C_u(\boldsymbol{H}_u) \geqslant 0]\} C_u(\boldsymbol{H}_u)$. It can be shown that $\mathbb{E}(Y_t^{(g)}|\boldsymbol{H}_t, A_t) = Q_t(\boldsymbol{H}_t, A_t) = Q_t(\boldsymbol{H}_t, 0) + A_t C_t(\boldsymbol{H}_t)$. Then the stage-$t$ problem can be solved based on the single-stage semiparametric model:

$$Y_t^{(g)} = m_t(\boldsymbol{H}_t) + A_t \times C_t(\boldsymbol{H}_t; \boldsymbol{\beta}_t) + e_t^{(g)}; \quad \mathbb{E}(e_t^{(g)}|\boldsymbol{H}_t, A_t) = 0. \tag{13}$$

This modeling approach is an instance of the optimal Structural Nested Mean Model (SNMM) (Robins, 2004).

There are three different estimation strategies for the optimal DTR based on the SNMM. The first strategy, known as the stagewise A-Learning (Blatt et al., 2004; Shi et al., 2018), is implemented analogously to the Q-Learning. Specifically, at stage $t(\leqslant T-1)$, we obtain the estimated g-outcome $\widehat{Y}_t^{(g)} := \sum_{u=t}^T Y_u - \sum_{u=t+1}^T \{A - \mathbb{I}[C_u(\boldsymbol{H}_u; \widehat{\boldsymbol{\beta}}_{u,n}) \geqslant 0]\} C_u(\boldsymbol{H}_u; \widehat{\boldsymbol{\beta}}_{u,n})$ based on the estimated CATE parameters $\{\widehat{\boldsymbol{\beta}}_{u,n}\}_{u=t+1}^T$ from the subsequent stages. Then we consider the single-stage semiparametric regression problem (13) with $\widehat{Y}_t^{(g)}$ as the response. Following this strategy, Shi et al. (2018) considered the linear working model $C_t(\boldsymbol{H}_t; \boldsymbol{\beta}_t) := \boldsymbol{H}_t^\top \boldsymbol{\beta}_t$ and the single-stage estimating function

$$\boldsymbol{\phi}_t(\boldsymbol{\beta}_t) = [\widehat{Y}_t^{(g)} - m_t(\boldsymbol{H}_t; \widehat{\boldsymbol{\eta}}_t) - A_t \boldsymbol{H}_t^\top \boldsymbol{\beta}_t][A_t - \pi_t(\boldsymbol{H}_t; \widehat{\boldsymbol{\alpha}}_t)]\boldsymbol{H}_t; \quad 1 \leqslant t \leqslant T, \tag{14}$$

where $m_t(\boldsymbol{H}_t; \boldsymbol{\eta}_t)$ and $\pi_t(\boldsymbol{H}_t; \boldsymbol{\alpha}_t)$ are the parametric models for $\mathbb{E}[Y_t^{(g)}|\boldsymbol{H}_t, A_t = 0]$ and $\mathbb{P}(A_t = 1|\boldsymbol{H}_t)$ respectively, and $(\widehat{\boldsymbol{\eta}}_t, \widehat{\boldsymbol{\alpha}}_t)$ are the corresponding estimates. On the other hand, Wallace and Moodie (2015) considered the balancing weight function (6) and proposed the dynamic Weighted Ordinary Least Squares (dWOLS) that solves the weighted least-squares problem (7) using $\widehat{Y}_t^{(g)}$ as the response and $(\boldsymbol{H}_t, A_t \boldsymbol{H}_t)$ as the covariates for $t = T, \cdots, 1$ backward stagewise. Both the stagewise A-Learning and the dWOLS can be shown doubly robust.

The second strategy to estimate the optimal SNMM, known as the regret regression (Murphy, 2003; Almirall et al., 2010; Henderson et al., 2010), exploits the following representation that can be shown equivalent to the joint model of (13) across T stages:

$$\sum_{t=1}^T Y_t = V_0 + \sum_{t=1}^T \{A_t - \mathbb{I}[C_t(\boldsymbol{H}_t; \boldsymbol{\beta}_t) \geqslant 0]\} C_t(\boldsymbol{H}_t; \boldsymbol{\beta}_t) + \sum_{t=1}^{T+1} \Delta M_t(\boldsymbol{H}_t),$$

subject to $\mathbb{E}[\Delta M_{t+1}(\boldsymbol{H}_{t+1})|\boldsymbol{H}_t, A_t] = 0; \quad 0 \leqslant t \leqslant T.$

Here, we denote $(\boldsymbol{H}_0, A_0) := \emptyset$ and $\boldsymbol{H}_{T+1} := (\boldsymbol{H}_T, A_T, Y_T)$ for convenience. Then the following fully parametric least-squares problem can be considered:

$$\min_{\left((\boldsymbol{\beta}_t,\boldsymbol{\eta}_t:1\leqslant t\leqslant T),v_0\right)} \mathbb{E}_n \left\{ \sum_{t=1}^{T} Y_t - v_0 - \sum_{t=1}^{T} \left\{A_t - \mathbb{I}[C_t(\boldsymbol{H}_t;\boldsymbol{\beta}_t) \geqslant 0]\right\} C_t(\boldsymbol{H}_t;\boldsymbol{\beta}_t) - \sum_{t=1}^{T} \Delta M_t(\boldsymbol{H}_t;\boldsymbol{\eta}_t) \right\}^2,$$

where $\{\Delta M_t(\boldsymbol{H}_t;\boldsymbol{\eta}_t)\}_{t=1}^T$ are the corresponding parametric models subject to $\mathbb{E}[\Delta M_t(\boldsymbol{H}_t;\boldsymbol{\eta}_t)|\boldsymbol{H}_{t-1}, A_{t-1}] = 0$ for $1 \leqslant t \leqslant T$. If the nuisance models $\{\Delta M_t(\boldsymbol{H}_t;\boldsymbol{\eta}_t)\}_{t=1}^T$ are correct, the regret regression can enjoy better efficiency than the stagewise A-Learning. However, the regret regression can be vulnerable to nuisance model misspecifications.

The third strategy is the G-Estimation (Robins, 2004) for the optimal SNMM. Under Model (13), the stage-t working g-outcomes is $Y_t^{(\mathrm{g})}(\boldsymbol{\beta}_{(t+1):T}) = \sum_{u=t}^{T} Y_u - \sum_{u=t+1}^{T} \left\{A_u - \mathbb{I}[C_u(\boldsymbol{H}_u;\boldsymbol{\beta}_u) \geqslant 0]\right\} C_u(\boldsymbol{H}_u;\boldsymbol{\beta}_u)$, and the characterizing moment condition of (13) is

$$\mathbb{E}\left\{ \overbrace{\left[Y_t^{(\mathrm{g})}(\boldsymbol{\beta}_{(t+1):T}) - A_t C_t(\boldsymbol{H}_t;\boldsymbol{\beta}_t)\right]}^{:=\mathfrak{H}_t(\boldsymbol{\beta}_{t:T})} - \mathbb{E}\left[Y_t^{(\mathrm{g})}(\boldsymbol{\beta}_{(t+1):T}) - A_t C_t(\boldsymbol{H}_t;\boldsymbol{\beta}_t) \middle| \boldsymbol{H}_t\right] \middle| \boldsymbol{H}_t, A_t \right\} = 0.$$

For $1 \leqslant t \leqslant T$, let $\boldsymbol{G}_t : \mathcal{H}_t \times \mathcal{A}_t \to \mathbb{R}^p$ be some user-defined instrument function, where $p := \sum_{t=1}^{T} p_t$ is the dimension of the parameter $\boldsymbol{\beta}_{1:T}$. Then the corresponding G-estimating function is

$$\boldsymbol{\phi}(\boldsymbol{\beta}_{1:T}) := \sum_{t=1}^{T} \left\{\mathfrak{H}_t(\boldsymbol{\beta}_{t:T}) - \mathbb{E}[\mathfrak{H}_t(\boldsymbol{\beta}_{t:T})|\boldsymbol{H}_t]\right\}\left\{\boldsymbol{G}_t(\boldsymbol{H}_t, A_t) - \mathbb{E}[\boldsymbol{G}_t(\boldsymbol{H}_t, A_t)|\boldsymbol{H}_t]\right\}.$$

At stage t, the required nuisance functions include: the treatment-free effect $m_t(\boldsymbol{H}_t) := \mathbb{E}[\mathfrak{H}_t(\boldsymbol{\beta}_{t:T})|\boldsymbol{H}_t]$, and the propensity score $\pi_t(\boldsymbol{h}_t, a_t) := \mathbb{P}(A_t = a_t|\boldsymbol{H}_t = \boldsymbol{h}_t)$ in evaluating $\mathbb{E}[\boldsymbol{G}_t(\boldsymbol{H}_t, A_t)|\boldsymbol{H}_t]$. If either of the models of $m_t(\boldsymbol{H}_t)$ and $\pi_t(A_t|\boldsymbol{H}_t)$ is correct for $1 \leqslant t \leqslant T$, then we have $\mathbb{E}[\boldsymbol{\phi}(\boldsymbol{\beta}_{1:T})] = \mathbf{0}$ at the true parameter $\boldsymbol{\beta}_{1:T}$. This gives the double robustness of G-Estimation. If the nuisance models for $m_t(\boldsymbol{H}_t)$ and $\pi_t(\boldsymbol{H}_t, A_t)$ for $1 \leqslant t \leqslant T$ are all correct, then there exists optimal instrument functions $\{\boldsymbol{G}_{\mathrm{eff},t}(\boldsymbol{H}_t, A_t)\}_{t=1}^T$ such that the corresponding estimating function $\boldsymbol{\phi}(\boldsymbol{\beta}_{1:T})$ is semiparametric efficient (Robins, 1994, 2004). However, the closed forms of such $\{\boldsymbol{G}_t(\boldsymbol{H}_t, A_t)\}_{t=1}^T$ are hardly tractable unless we assume the stronger condition $\mathsf{Var}[\mathfrak{H}_t(\boldsymbol{\beta}_{t:T})|\boldsymbol{H}_t, A_t] = \mathsf{Var}[\mathfrak{H}_t(\boldsymbol{\beta}_{t:T})|\boldsymbol{H}_t]$ for

$1 \leqslant t \leqslant T$. Even in this case, the optimal instrument functions $\{\boldsymbol{G}_t(\boldsymbol{H}_t, A_t)\}_{t=1}^T$ can involve complicated vector-valued nuisance functions and be hard to evaluate correctly (Vansteelandt & Joffe, 2014).

The G-estimating equations need to be solved simultaneously across stages. If we assume that the t-th block of the estimating function $\boldsymbol{\phi}_t(\boldsymbol{\beta}_{1:T})$ corresponding to $\boldsymbol{\beta}_t$ takes the following form:

$$\boldsymbol{\phi}_t(\boldsymbol{\beta}_{1:T}) = \boldsymbol{\phi}_t(\boldsymbol{\beta}_{t:T}) = \sum_{u=t}^{T} \left\{\mathfrak{H}_u(\boldsymbol{\beta}_{u:T}) - \mathbb{E}[\mathfrak{H}_u(\boldsymbol{\beta}_{u:T})|\boldsymbol{H}_u]\right\} \times \left\{\boldsymbol{G}_{tu}(\boldsymbol{H}_u, A_u) - \mathbb{E}[\boldsymbol{G}_{tu}(\boldsymbol{H}_u, A_u)|\boldsymbol{H}_u]\right\},$$

then G-Estimation is equivalent to a backward stagewise procedure in Robins (2004, Section 7.2) and Moodie et al. (2007, Section 3.3.2). Specifically, at stage $t(\leqslant T-1)$, we obtain the estimated g-outcomes $\{Y_u^{(g)}(\widehat{\boldsymbol{\beta}}_{(u+1):T})\}_{u=t}^T$ for the current and subsequent stages based on the estimated parameters $\{\widehat{\boldsymbol{\beta}}_u\}_{u=t+1}^T$. Then $\{\mathfrak{H}_u(\widehat{\boldsymbol{\beta}}_{u:T})\}_{u=t+1}^T$ can be computed, and $\mathfrak{H}_t(\boldsymbol{\beta}_t, \widehat{\boldsymbol{\beta}}_{(t+1):T}) = Y_t^{(g)}(\widehat{\boldsymbol{\beta}}_{(t+1):T}) - A_t C_t(\boldsymbol{H}_t; \boldsymbol{\beta}_t)$ is expressed as a function of $\boldsymbol{\beta}_t$ only. The stage-t estimate $\widehat{\boldsymbol{\beta}}_t$ is obtained by solving the estimating equations $\mathbb{E}_n[\boldsymbol{\phi}_t(\boldsymbol{\beta}_t, \widehat{\boldsymbol{\beta}}_{(t+1):T})] = \mathbf{0}$. If we further assume that $\boldsymbol{G}_{tu}(\boldsymbol{H}_u, A_u) = A_t \times (\partial/\partial\boldsymbol{\beta}_t)C_t(\boldsymbol{H}_t; \boldsymbol{\beta}_t)\mathbb{I}(u = t)$ for $1 \leqslant t, u \leqslant T$, then G-Estimation is equivalent to the stagewise A-Learning (14) in this case (Schulte et al., 2014).

The q-outcome and g-outcome are both nonsmooth functions of the parameters of interest. Specifically, under the semiparametric model (13), for $1 \leqslant t \leqslant T-1$, we have $Y_t^{(q)}(\boldsymbol{\beta}_{t+1}) = Y_t + m_{t+1}(\boldsymbol{H}_{t+1}) + \mathbb{I}[C_{t+1}(\boldsymbol{H}_{t+1}; \boldsymbol{\beta}_{t+1}) \geqslant 0]C_t(\boldsymbol{H}_t; \boldsymbol{\beta}_t)$ and $Y_t^{(g)}(\boldsymbol{\beta}_{(t+1):T}) = \sum_{u=t}^T Y_u - \sum_{u=t+1}^T \left\{A_u - \mathbb{I}[C_u(\boldsymbol{H}_u; \boldsymbol{\beta}_u) \geqslant 0]\right\} C_u(\boldsymbol{H}_u; \boldsymbol{\beta}_u)$. In particular, the indicator function $\mathbb{I}[C_u(\boldsymbol{H}_u; \boldsymbol{\beta}_u) \geqslant 0]$ is nonsmooth in $\boldsymbol{\beta}_u$ no matter how $C_u(\boldsymbol{H}_u; \boldsymbol{\beta}_u)$ depends on $\boldsymbol{\beta}_u$. This can result in an exceptional law if $\mathbb{P}[C_u(\boldsymbol{H}_u; \boldsymbol{\beta}_u) = 0] > 0$ that leads to a biased parameter estimate of G-Estimation (Robins, 2004; Moodie & Richardson, 2010). The same nonregularity was also studied for Q-Learning (Chakraborty et al., 2010; Laber et al., 2014).

2.2 *Direct-Search Methods*

Besides the regression-based approach, another framework, known as the Marginal Structural Mean Model (MSMM) (Robins, 1998), allows the direct estimation of the value function (11). Fix a DTR $\boldsymbol{d}_{1:T} \in \mathcal{D}_{1:T}$. Define the $\boldsymbol{d}_{1:T}$-specific V-functions

$\{V_t^{\boldsymbol{d}}(\boldsymbol{H}_t)\}_{t=1}^T$ and Q-functions $\{Q_t^{\boldsymbol{d}}(\boldsymbol{H}_t, A_t)\}_{t=1}^T$ recursively from the following Bellman equations:

$$\begin{aligned} Q_t^{\boldsymbol{d}}(\boldsymbol{H}_t, A_t) &:= \mathbb{E}[Y_t + V_{t+1}^{\boldsymbol{d}}(\boldsymbol{H}_{t+1})|\boldsymbol{H}_{t+1}, A_t]; \\ V_t^{\boldsymbol{d}}(\boldsymbol{H}_t) &:= Q_t^{\boldsymbol{d}}(\boldsymbol{H}_t, d_t(\boldsymbol{H}_t)); \quad t = T, \cdots, 1. \end{aligned} \tag{15}$$

Here, $V_{T+1}^{\boldsymbol{d}}(\boldsymbol{H}_{T+1}) = 0$. We also define $\Delta M_{t+1}^{\boldsymbol{d}}(\boldsymbol{H}_{t+1}) := Y_t + V_{t+1}^{\boldsymbol{d}}(\boldsymbol{H}_{t+1}) - Q_t^{\boldsymbol{d}}(\boldsymbol{H}_t, A_t)$ as the $\boldsymbol{d}$-specific Bellman error. Then the stage-1 marginal value $V_0^{\boldsymbol{d}} := \mathbb{E}[V_1^{\boldsymbol{d}}(\boldsymbol{H}_1)]$ gives the T-stage value function $\mathcal{V}(\boldsymbol{d}_{1:T})$ in (11).

2.2.1 T-Stage IPWE

Denote $Y_{s:t} := \sum_{u=s}^t Y_u$ and $W_{s:t}^{\boldsymbol{d}} := \prod_{u=s}^t \frac{\mathbb{I}[d_u(\boldsymbol{H}_u)=A_u]}{\pi_u(\boldsymbol{H}_u, A_u)}$ for $1 \leqslant s \leqslant t \leqslant T$. Murphy et al. (2001) studied the estimation of the MSMM $V_0^{\boldsymbol{d}}(\boldsymbol{Z}; \boldsymbol{\beta}) := \boldsymbol{Z}^\top \boldsymbol{\beta}$ using the IPW estimating function $\boldsymbol{\phi}^{\boldsymbol{d}}(\boldsymbol{\beta}) = W_{1:T}^{\boldsymbol{d}}(Y_{1:T} - \boldsymbol{Z}^\top \boldsymbol{\beta})\boldsymbol{Z}$, where $\boldsymbol{Z}$ is the subject-specific baseline covariate vector, and $\boldsymbol{\beta}$ is the parameter of interest. As a special case, if $Z = 1$, then the estimating function of the MSMM $V_0^{\boldsymbol{d}}(\beta) := \beta$ can be simplified as $\phi^{\boldsymbol{d}}(\beta) = W_{1:T}^{\boldsymbol{d}}(Y_{1:T} - \beta)$. The corresponding estimate is $\widehat{\mathcal{V}}_{\text{Hájek-IPWE},n}(\boldsymbol{d}_{1:T}) = [\mathbb{E}_n(W_{1:T}^{\boldsymbol{d}})]^{-1}[\mathbb{E}_n(W_{1:T}^{\boldsymbol{d}} Y_{1:T})]$, which gives the T-stage Hájek IPWE of $\mathcal{V}(\boldsymbol{d}_{1:T})$.

To obtain the optimal DTR from the IPWE, Orellana et al. (2010) proposed to first estimate the $\boldsymbol{\beta}$-parametric MSMM $V_0^{\boldsymbol{d}_\eta}(\boldsymbol{Z}; \boldsymbol{\beta})$ for every given DTR $\boldsymbol{d}_\eta \in \boldsymbol{\mathcal{D}}_\eta = \{\boldsymbol{d}_\eta : \eta \in \Xi\}$, and then solve $\widehat{\eta}(\boldsymbol{Z}) \in \operatorname{argmax}_{\eta\in\Xi} V_0^{\boldsymbol{d}_\eta}(\boldsymbol{Z}; \widehat{\boldsymbol{\beta}}_\eta)$ for the optimal DTR parameter. Despite that the estimation based on MSMM is more straightforward than SNMM, the parametric DTR class $\boldsymbol{\mathcal{D}}_\eta$ is typically restrictive and cannot handle too complicated functional forms.

To learn the optimal DTR of flexible functional forms, Zhao et al. (2015a) proposed the T-stage extension of OWL that directly maximizes $\widehat{\mathcal{V}}_n(\boldsymbol{d}_{1:T}) = \mathbb{E}_n(W_{1:T}^{\boldsymbol{d}} Y_{1:T})$. They considered two strategies to relax the T-stage nonsmooth nonconvex function $\boldsymbol{d}_{1:T} \mapsto W_{1:T}^{\boldsymbol{d}}$. The first approach, known as the Backward OWL (BOWL), solves T single-stage OWL problems in a backward stagewise manner:

$$\begin{aligned} \widehat{f}_{t,n} \in \operatorname*{argmin}_{f_t\in\mathcal{F}_t} \Bigg\{ \mathbb{E}_n \left(\frac{W_{(t+1):T}^{\widehat{\boldsymbol{f}}_n} V_{t:T}}{\pi_t(\boldsymbol{H}_t, A_t)} \phi_t[(2A_t - 1) f_t(\boldsymbol{H}_t)] \right. \\ \left. + \lambda_{t,n} J_t(f_t) \right\}; \quad t = T, \cdots, 1, \end{aligned}$$

where $W^{\widehat{f}_n}_{(t+1):T} := \prod_{u=t+1}^{T} \frac{\mathbb{I}[(2A_u-1)\widehat{f}_{u,n}(\boldsymbol{H}_u)\geqslant 0]}{\pi_u(\boldsymbol{H}_u, A_u)}$ based on $\{\widehat{f}_{u,n}\}_{u=t+1}^{T}$ obtained from the subsequent stages, ϕ_t is the stage-t loss surrogate loss function, $\mathcal{F}_t$ is the stage-t decision function class, and $\lambda_{t,n} J_t(f_t)$ is the corresponding functional penalty. The second approach, known as the Simultaneous OWL (SOWL), utilizes a multivariate surrogate loss $\phi : \mathbb{R}^T \to \mathbb{R}_+$ to approximate the multivariate 0-1 loss function $\boldsymbol{u}_{1:T} \mapsto 1 - \prod_{t=1}^{T} \mathbb{I}(u_t \geqslant 0)$ for $\boldsymbol{u}_{1:T} = (u_1, \cdots, u_T)^\intercal \in \mathbb{R}^T$. The SOWL solves the following multi-dimensional large-margin classification problem:

$$\widehat{\boldsymbol{f}}_{1:T,n} \in \underset{\boldsymbol{f}_{1:T} \in \boldsymbol{\mathcal{F}}_{1:T}}{\operatorname{argmin}} \left\{ \mathbb{E}_n \left[\frac{Y_{1:T}}{\prod_{t=1}^{T} \pi_t(\boldsymbol{H}_t, A_t)} \phi \begin{pmatrix} (2A_1 - 1) f_1(\boldsymbol{H}_1) \\ \vdots \\ (2A_T - 1) f_T(\boldsymbol{H}_T) \end{pmatrix} \right] + \lambda_n J(\boldsymbol{f}_{1:T}) \right\},$$

where $\boldsymbol{f}_{1:T} = (f_1, \cdots, f_T)^\intercal \in \boldsymbol{\mathcal{F}}_{1:T} = \mathcal{F}_1 \times \cdots \times \mathcal{F}_T$, and $\lambda_n J(\boldsymbol{f}_{1:T})$ is a functional penalty for $\boldsymbol{\mathcal{F}}_{1:T}$. When considering the multivariate hinge loss $\phi(\boldsymbol{u}_{1:T}) = 1 + \min_{1 \leqslant t \leqslant T}(u_t - 1)^+$, the dual problem of SOWL can be solved by Quadratic Programming (QP).

2.2.2 T-Stage AIPWE

Analogous to the single-stage problem, the T-stage IPWE can correspond to an efficient estimating function, which gives the T-stage AIPWE as in the following Theorem 1. The same result was also used for Optimal Policy Evaluation (OPE) in the reinforcement learning literature (Jiang & Li, 2016; Thomas & Brunskill, 2016; Kallus & Uehara, 2020).

Theorem 1 (T-Stage AIPWE) *Consider a semiparametric model with the parameter of interest $\beta \in \mathbb{R}$ defined from the IPW estimating function $\phi^{\boldsymbol{d}}(\beta) = W^{\boldsymbol{d}}_{1:T}(Y_{1:T} - \beta)$. Then the efficient estimating function is*

$$\phi^{\boldsymbol{d}}_{\mathit{eff}}(\beta) = V^{\boldsymbol{d}}_1(\boldsymbol{H}_1) + \sum_{t=1}^{T} W^{\boldsymbol{d}}_{1:t} \Delta M^{\boldsymbol{d}}_{t+1}(\boldsymbol{H}_{t+1}) - \beta.$$

The corresponding semiparametric efficient estimate is the AIPWE

$$\widehat{\beta}_{\mathit{eff},n} = \widehat{\mathcal{V}}_n(\boldsymbol{d}_{1:T}) = \mathbb{E}_n \left\{ V^{\boldsymbol{d}}_1(\boldsymbol{H}_1) + \sum_{t=1}^{T} W^{\boldsymbol{d}}_{1:t} \Delta M^{\boldsymbol{d}}_{t+1}(\boldsymbol{H}_{t+1}) \right\}.$$

Based on the AIPWE, Zhang et al. (2013) proposed to directly maximize $\widehat{\mathcal{V}}_n(\boldsymbol{d}_\eta)$ over a parametric DTR class $\boldsymbol{\mathcal{D}}_\eta = \{\boldsymbol{d}_\eta : \eta \in \Xi\}$. However, the nuisance functions $V^{\boldsymbol{d}}_1(\boldsymbol{H}_t)$ and $\{\Delta M^{\boldsymbol{d}}_{t+1}(\boldsymbol{H}_{t+1})\}_{t=1}^{T}$ also depend on $\boldsymbol{d}_{1:T}$, which can result in a challenging computational problem when maximizing $\widehat{\mathcal{V}}_n(\boldsymbol{d}_{1:T})$ over $\boldsymbol{d}_{1:T} \in \boldsymbol{\mathcal{D}}_{1:T}$.

Jiang and Li (2016), Zhang and Zhang (2018) pointed out that $\widehat{\mathcal{V}}_n(\boldsymbol{d}) = \mathbb{E}_n(\widehat{V}_1^{\boldsymbol{d}})$, where $\widehat{V}_1^{\boldsymbol{d}}$ can be computed from $\widehat{V}_{T+1}^{\boldsymbol{d}} := 0$ and $\widehat{V}_T^{\boldsymbol{d}}, \cdots, \widehat{V}_1^{\boldsymbol{d}}$ in the following backward stagewise manner according to (15):

$$\widehat{V}_t^{\boldsymbol{d}} := V_t^{\boldsymbol{d}}(\boldsymbol{H}_t) + W_t^{d_t}\left[Y_t + \widehat{V}_{t+1}^{\boldsymbol{d}} - Q_t^{\boldsymbol{d}}(\boldsymbol{H}_t, A_t)\right]; \quad t = T, \cdots, 1.$$

Here, $\mathbb{E}_n(\widehat{V}_t^{\boldsymbol{d}})$ can be interpreted as an AIPWE of $\mathbb{E}[V_t^{\boldsymbol{d}}(\boldsymbol{H}_t)]$. At stage t, if $\boldsymbol{d} = (\boldsymbol{d}_{1:t}, \boldsymbol{d}_{(t+1):T})$ is replaced by $\boldsymbol{d} = (\boldsymbol{d}_{1:t}, \boldsymbol{d}^\star_{(t+1):T})$, then we further have

$$\begin{aligned}
\widehat{V}_t^{d_t, \boldsymbol{d}^\star_{(t+1):T}} &= Q_t(\boldsymbol{H}_t, d_t(\boldsymbol{H}_t)) + W_t^{d_t}\left[Y_t + \widehat{V}_{t+1}^{\boldsymbol{d}^\star} - Q_t(\boldsymbol{H}_t, A_t)\right]; \\
d_t^\star &\in \operatorname*{argmax}_{d_t \in \mathcal{D}_t} \mathbb{E}\big(\widehat{V}_t^{d_t, \boldsymbol{d}^\star_{(t+1):T}}\big);
\end{aligned} \qquad t = T, \cdots, 1.$$

Here, the Q-function $Q_t(\boldsymbol{H}_t, A_t)$ is defined from (12) instead of (15). This can lead to a method of stagewise AIPWE maximization: $\widehat{V}_{T+1} = 0$,

$$\begin{aligned}
\widehat{d}_{t,n} &\in \operatorname*{argmax}_{d_t \in \mathcal{D}_t} \mathbb{E}_n \left\{Q_t(\boldsymbol{H}_t, d_t) + W_t^{d_t}\left[Y_t + \widehat{V}_{t+1} - Q_t(\boldsymbol{H}_t, A_t)\right]\right\}; \\
\widehat{V}_t &:= Q_t(\boldsymbol{H}_t, \widehat{d}_{t,n}) + W_t^{\widehat{d}_{t,n}}\left(Y_t + \widehat{V}_{t+1} - Q_t(\boldsymbol{H}_t, A_t)\right); \\
& \qquad t = T, \cdots, 1.
\end{aligned} \tag{16}$$

In particular, the nuisance functions $\{Q_t(\boldsymbol{H}_t, A_t)\}_{t=1}^T$ can be estimated from Q-Learning in Sect. 2.1.

Based on the stagewise computation of the AIPWE, Liu et al. (2018) proposed the Augmented OwL (AOL) framework that can be equivalent to (16). Specifically, the stage-t AOL problem is

$$\begin{aligned}
\widehat{f}_{t,n} \in \operatorname*{argmax}_{f_t \in \mathcal{F}_t} &\left\{\mathbb{E}_n\left(\tfrac{|Y_t + \widehat{V}_{t+1} - g_t(\boldsymbol{H}_t)|}{\pi_t(\boldsymbol{H}_t, A_t)}\phi_t\left\{(2A_t - 1)\right.\right.\right. \\
&\left.\left.\left.\mathsf{sign}\left[Y_t + \widehat{V}_{t+1} - g_t(\boldsymbol{H}_t)\right] f_t(\boldsymbol{H}_t)\right\}\right) + \lambda_t J_t(f_t)\right\}; \\
\widehat{V}_t &= Q_t\left(\boldsymbol{H}_t, \mathbb{I}[\widehat{f}_{t,n}(\boldsymbol{H}_t) \geqslant 0]\right) + \tfrac{\mathbb{I}[(2A_t - 1)\widehat{f}_{t,n}(\boldsymbol{H}_t) \geqslant 0]}{\pi_t(\boldsymbol{H}_t, A_t)}\left(Y_t + \widehat{V}_{t+1} - Q_t(\boldsymbol{H}_t, A_t)\right).
\end{aligned}$$

Here, $g_t : \mathcal{H}_t \to \mathbb{R}$ can be an arbitrary function for efficiency augmentation, and $g_t(\boldsymbol{H}_t) = \pi_t(\boldsymbol{H}_t, 0)Q_t(\boldsymbol{H}_t, 1) + \pi_t(\boldsymbol{H}_t, 1)Q_t(\boldsymbol{H}_t, 0)$ corresponds to the case that $\widehat{f}_{t,n}$ maximizes the AIPWE.

When considering the statistical inference of the estimated value at the optimal DTR $\widehat{\mathcal{V}}_n(\widehat{\boldsymbol{d}}_{1:T,n})$, the nonregularity problem can also occur. The construction of valid confidence intervals was further studied in van der Laan and Luedtke (2015), Luedtke and van der Laan (2016).

3 Discussion

In this article, we discuss several existing methods for the single-stage and multi-stage individualized decision making problems. The introduced methods are categorized into the regression-based and direct-search approaches respectively. For the regression-based approach, such as Q-Learning, A-Learning and G-Estimation, it can be straightforward for the researcher to incorporate model assumptions and obtain interpretable results in terms of the underlying data generating process. However, the violation of model assumptions can be a typical concern. For the direct-search approach, such as OWL and AIPWE-maximization, the estimation of an optimal ITR/DTR is cast as a supervised learning problem, and many flexible machine learning techniques can be applied. Nevertheless, the resulting decision rules may be less interpretable in terms of how the data are generated than the model-based methods. When the true data generating process is relatively simple, the direct-search methods may be less efficient than the regression based methods with correct model specification.

The main goal in this article is to optimize the value functions (1) or (11) based on the expected outcome, with the assumption that the training and testing distributions are identical. We also discuss some other research goals in Sect. 1.3. In particular, we review literature on risk-averse utilities based on some functionals of the marginal or conditional outcome distributions. The optimal ITR under such utilities can still be estimated using certain regression-based or direct-search formulations. These methods can be useful for many risk-averse applications. We also survey the generalizability concern for the ITR problem, and discuss several approaches to protect the testing performance against different aspects of distributional changes via worst-case optimization. When the data contain complicated structures and latent heterogeneity, these methods can be particularly useful.

Appendix: Proofs

Proof of Theorem 1

$$\phi^{d}_{\text{eff}}(\beta) = \phi^{d}(\beta) - \sum_{t=1}^{T}\left\{\mathbb{E}[\phi^{d}(\beta)|\boldsymbol{H}_t, A_t] - \mathbb{E}[\phi^{d}(\beta)|\boldsymbol{H}_t]\right\} \tag{17}$$

$$= W^{d}_{1:T}(Y_{1:T} - \beta)$$

$$-\sum_{t=1}^{T}\left\{W^{d}_{1:t}[Y_{1:(t-1)} + Q^{d}_{t}(\boldsymbol{H}_t, A_t) - \beta] - W^{d}_{1:(t-1)}[Y_{1:(t-1)} + V^{d}_{t}(\boldsymbol{H}_t) - \beta]\right\} \tag{18}$$

$$= W_{1:T}^{\boldsymbol{d}}[Y_T - Q_T^{\boldsymbol{d}}(\boldsymbol{H}_T, A_T)] + \sum_{t=1}^{T-1} W_{1:t}^{\boldsymbol{d}}[Y_t - Q_t^{\boldsymbol{d}}(\boldsymbol{H}_t, A_t)$$

$$+ V_{t+1}^{\boldsymbol{d}}(\boldsymbol{H}_{t+1})] + V_1^{\boldsymbol{d}}(\boldsymbol{H}_1) - \beta$$

$$= \sum_{t=1}^{T} W_{1:t}^{\boldsymbol{d}} \Delta M_{t+1}^{\boldsymbol{d}}(\boldsymbol{H}_{t+1}) + V_1^{\boldsymbol{d}}(\boldsymbol{H}_1) - \beta,$$

where (17) follows from Robins (2000, Theorem 4.2), and (18) follows from

$$\begin{aligned}
\mathbb{E}(W_{1:T}^{\boldsymbol{d}} | \boldsymbol{H}_t, A_t) &= W_{1:t}^{\boldsymbol{d}}; \\
\mathbb{E}(W_{1:T}^{\boldsymbol{d}} Y_{1:T} | \boldsymbol{H}_t, A_t) &= W_{1:t}^{\boldsymbol{d}} \times [Y_{1:(t-1)} + Q_t^{\boldsymbol{d}}(\boldsymbol{H}_t, A_t)]; \\
\mathbb{E}(W_{1:T}^{\boldsymbol{d}} | \boldsymbol{H}_t) &= W_{1:(t-1)}^{\boldsymbol{d}}; \\
\mathbb{E}(W_{1:T}^{\boldsymbol{d}} Y_{1:T} | \boldsymbol{H}_t) &= W_{1:(t-1)}^{\boldsymbol{d}} \times [Y_{1:(t-1)} + V_t^{\boldsymbol{d}}(\boldsymbol{H}_t)].
\end{aligned}$$

□

References

Aggarwal, C. C. (2016). *Recommender systems: The textbook*. Springer.

Alaa, A. M., & van der Schaar, M. (2017). Bayesian inference of individualized treatment effects using multi-task Gaussian processes. In I. Guyon, U. V. Luxburg, S. Bengio, H. Wallach, R. Fergus, S. Vishwanathan & R. Garnett (Eds.), *Advances in neural information processing systems* (Vol. 30). Curran Associates.

Almirall, D., Ten Have, T., & Murphy, S. A. (2010). Structural nested mean models for assessing time-varying effect moderation. *Biometrics, 66*(1), 131–139.

Artzner, P., Delbaen, F., Eber, J.-M., & Heath, D. (1999). Coherent measures of risk. *Mathematical Finance, 9*(3), 203–228.

Athey, S., & Wager, S. (2021). Policy learning with observational data. *Econometrica, 89*(1), 133–161.

Bellman, R. (1966). Dynamic programming. *Science, 153*(3731), 34–37.

Belloni, A., Chernozhukov, V., & Kato, K. (2019). Valid post-selection inference in high-dimensional approximately sparse quantile regression models. *Journal of the American Statistical Association, 114*(526), 749–758.

Ben-Tal, A., & Teboulle, M. (1986). Expected utility, penalty functions, and duality in stochastic nonlinear programming. *Management Science, 32*(11), 1445–1466.

Bennett, A., & Kallus, N. (2020). Efficient policy learning from surrogate-loss classification reductions. In H. Daumé III & Singh, A. (Eds.), *Proceedings of the 37th International Conference on Machine Learning*, volume 119 of *Proceedings of Machine Learning Research* (pp. 788–798). PMLR.

Bertsimas, D., Dunn, J., & Mundru, N. (2019). Optimal prescriptive trees. *INFORMS Journal on Optimization, 1*(2), 164–183.

Bertsimas, D., & Kallus, N. (2020). From predictive to prescriptive analytics. *Management Science, 66*(3), 1025–1044.

Bertsimas, D., Kallus, N., Weinstein, A. M., & Zhuo, Y. D. (2017). Personalized diabetes management using electronic medical records. *Diabetes Care, 40*(2), 210–217.

Blackorby, C., & Donaldson, D. (1978). Measures of relative equality and their meaning in terms of social welfare. *Journal of Economic Theory, 18*(1), 59–80.

Blatt, D., Murphy, S. A., & Zhu, J. (2004). A-learning for approximate planning. Technical Report 04-63, The Methodology Center, Pennsylvania State University. http://people.seas.harvard.edu/~samurphy/papers/Alearning2004.pdf

Chakraborty, B., Murphy, S., & Strecher, V. (2010). Inference for non-regular parameters in optimal dynamic treatment regimes. *Statistical Methods in Medical Research, 19*(3), 317–343.

Chen, G., Zeng, D., & Kosorok, M. R. (2016). Personalized dose finding using outcome weighted learning. *Journal of the American Statistical Association, 111*(516), 1509–1521.

Chen, J., Fu, H., He, X., Kosorok, M. R., & Liu, Y. (2018). Estimating individualized treatment rules for ordinal treatments. *Biometrics, 74*(3), 924–933.

Chen, S., Tian, L., Cai, T., & Yu, M. (2017). A general statistical framework for subgroup identification and comparative treatment scoring. *Biometrics, 73*(4), 1199–1209.

Chernozhukov, V., Chetverikov, D., Demirer, M., Duflo, E., Hansen, C., Newey, W., & Robins, J. (2018). Double/debiased machine learning for treatment and structural parameters. *The Econometrics Journal, 21*(1), C1–C68.

Cho, Y., & Ghosh, D. (2021). Quantile-based subgroup identification for randomized clinical trials. *Statistics in Biosciences, 13*(1), 90–128.

Crump, R. K., Hotz, V. J., Imbens, G. W., & Mitnik, O. A. (2006). Moving the goalposts: Addressing limited overlap in the estimation of average treatment effects by changing the estimand. Working Paper 0330, National Bureau of Economic Research. http://www.nber.org/papers/t0330

Cui, Y., Zhu, R., & Kosorok, M. (2017). Tree based weighted learning for estimating individualized treatment rules with censored data. *Electronic Journal of Statistics, 11*(2), 3927–3953.

Dorie, V., Hill, J., Shalit, U., Scott, M., & Cervone, D. (2019). Automated versus do-it-yourself methods for causal inference: Lessons learned from a data analysis competition. *Statistical Science, 34*(1), 43–68.

Dudík, M., Langford, J., & Li, L. (2011). Doubly robust policy evaluation and learning. In L. Getoor & T. Scheffer (Eds.), *Proceedings of the 28th International Conference on Machine Learning*, ICML '11 (pp. 1097–1104). ACM.

Dukes, O., Vansteelandt, S., & Whitney, D. (2021). On doubly robust inference for double machine learning. arXiv preprint arXiv:2107.06124.

Elmachtoub, A. N., & Grigas, P. (2022). Smart "predict, then optimize". *Management Science, 68*(1), 9–26.

Fan, Y., Lu, X., Zhao, J., Fu, H., & Liu, Y. (2022). Estimating individualized treatment rules for treatments with hierarchical structure. *Electronic Journal of Statistics, 16*(1), 737–784.

Fu, S., He, Q., Zhang, S., & Liu, Y. (2019). Robust outcome weighted learning for optimal individualized treatment rules. *Journal of Biopharmaceutical Statistics, 29*(4), 606–624.

Guo, R., Cheng, L., Li, J., Hahn, P. R., & Liu, H. (2020). A survey of learning causality with data: Problems and methods. *ACM Computing Surveys, 53*(4), 75.

Hahn, P. R., Murray, J. S., & Carvalho, C. M. (2020). Bayesian regression tree models for causal inference: Regularization, confounding, and heterogeneous effects (with discussion). *Bayesian Analysis, 15*(3), 965–1056.

Henderson, R., Ansell, P., & Alshibani, D. (2010). Regret-regression for optimal dynamic treatment regimes. *Biometrics, 66*(4), 1192–1201.

Huang, Y., & Fong, Y. (2014). Identifying optimal biomarker combinations for treatment selection via a robust kernel method. *Biometrics, 70*(4), 891–901.

Imai, K., & Ratkovic, M. (2013). Estimating treatment effect heterogeneity in randomized program evaluation. *The Annals of Applied Statistics, 7*(1), 443–470.

Imai, K., & Ratkovic, M. (2014). Covariate balancing propensity score. *Journal of the Royal Statistical Society: Series B (Statistical Methodology), 76*(1), 243–263.

Jiang, N., & Li, L. (2016). Doubly robust off-policy value evaluation for reinforcement learning. In M. F. Balcan & K. Q. Weinberger (Eds.), *Proceedings of The 33rd International Conference on Machine Learning*, volume 48 of *Proceedings of Machine Learning Research* (pp. 652–661). PMLR.

Kallus, N., Mao, X., Wang, K., & Zhou, Z. (2022). Doubly robust distributionally robust off-policy evaluation and learning. arXiv preprint arXiv:2202.09667.

Kallus, N., & Uehara, M. (2020). Double reinforcement learning for efficient off-policy evaluation in Markov decision processes. *Journal of Machine Learning Research, 21*(167), 1–63.

Kallus, N., & Zhou, A. (2018). Policy evaluation and optimization with continuous treatments. In A. Storkey & F. Perez-Cruz (Eds.), *Proceedings of the Twenty-First International Conference on Artificial Intelligence and Statistics*, volume 84 of *Proceedings of Machine Learning Research* (pp. 1243–1251). PMLR.

Kennedy, E. H. (2020). Optimal doubly robust estimation of heterogeneous causal effects. arXiv preprint arXiv:2004.14497.

Kennedy, E. H., Balakrishnan, S., & Wasserman, L. (2022). Minimax rates for heterogeneous causal effect estimation. arXiv preprint arXiv:2203.00837.

Kitagawa, T., & Tetenov, A. (2018). Who should be treated? Empirical welfare maximization methods for treatment choice. *Econometrica, 86*(2), 591–616.

Kitagawa, T., & Tetenov, A. (2021). Equality-minded treatment choice. *Journal of Business & Economic Statistics, 39*(2), 561–574.

Künzel, S. R., Sekhon, J. S., Bickel, P. J., & Yu, B. (2019). Metalearners for estimating heterogeneous treatment effects using machine learning. *Proceedings of the National Academy of Sciences, 116*(10), 4156–4165.

Laber, E. B., Lizotte, D. J., Qian, M., Pelham, W. E., & Murphy, S. A. (2014). Dynamic treatment regimes: Technical challenges and applications. *Electronic Journal of Statistics, 8*(1), 1225.

Laber, E. B., & Zhao, Y.-Q. (2015). Tree-based methods for individualized treatment regimes. *Biometrika, 102*(3), 501–514.

Lee, S. (2003). Efficient semiparametric estimation of a partially linear quantile regression model. *Econometric Theory, 19*(1), 1–31.

Li, F., Morgan, K. L., & Zaslavsky, A. M. (2018). Balancing covariates via propensity score weighting. *Journal of the American Statistical Association, 113*(521), 390–400.

Liang, M., & Yu, M. (2022). A semiparametric approach to model effect modification. *Journal of the American Statistical Association, 117*(538), 752–764.

Linn, K. A., Laber, E. B., & Stefanski, L. A. (2017). Interactive Q-learning for quantiles. *Journal of the American Statistical Association, 112*(518), 638–649.

Liu, L., & Kennedy, E. H. (2021). Median optimal treatment regimes. arXiv preprint arXiv:2103.01802.

Liu, Y., Wang, Y., Kosorok, M. R., Zhao, Y.-Q., & Zeng, D. (2018). Augmented outcome-weighted learning for estimating optimal dynamic treatment regimens. *Statistics in Medicine, 37*(26), 3776–3788.

Lu, W., Zhang, H. H., & Zeng, D. (2013). Variable selection for optimal treatment decision. *Statistical Methods in Medical Research, 22*(5), 493–504.

Luedtke, A. R., & van der Laan, M. J. (2016). Statistical inference for the mean outcome under a possibly non-unique optimal treatment strategy. *The Annals of Statistics, 44*(2), 713.

Manski, C. F. (2004). Statistical treatment rules for heterogeneous populations. *Econometrica, 72*(4), 1221–1246.

Meng, H., Zhao, Y.-Q., Fu, H., & Qiao, X. (2020). Near-optimal individualized treatment recommendations. *Journal of Machine Learning Research, 21*(183), 1–28.

Mi, X., Zou, F., & Zhu, R. (2019). Bagging and deep learning in optimal individualized treatment rules. *Biometrics, 75*(2), 674–684.

Mo, W., & Liu, Y. (2021). Efficient learning of optimal individualized treatment rules for heteroscedastic or misspecified treatment-free effect models. *Journal of the Royal Statistical Society: Series B (Statistical Methodology)*. https://doi.org/10.1111/rssb.12474

Mo, W., Qi, Z., & Liu, Y. (2021a). Learning optimal distributionally robust individualized treatment rules. *Journal of the American Statistical Association, 116*(534), 659–674. https://doi.org/10.1080/01621459.2020.1796359

Mo, W., Qi, Z., & Liu, Y. (2021b). Rejoinder: Learning optimal distributionally robust individualized treatment rules. *Journal of the American Statistical Association, 116*(534), 699–707. https://doi.org/10.1080/01621459.2020.1866581

Moodie, E. E. M., & Richardson, T. S. (2010). Estimating optimal dynamic regimes: Correcting bias under the null. *Scandinavian Journal of Statistics, 37*(1), 126–146.

Moodie, E. E. M., Richardson, T. S., & Stephens, D. A. (2007). Demystifying optimal dynamic treatment regimes. *Biometrics, 63*(2), 447–455.

Murphy, S. A. (2003). Optimal dynamic treatment regimes. *Journal of the Royal Statistical Society: Series B (Statistical Methodology), 65*(2), 331–355.

Murphy, S. A., van der Laan, M. J., Robins, J. M., & Conduct Problems Prevention Research Group (2001). Marginal mean models for dynamic regimes. *Journal of the American Statistical Association, 96*(456), 1410–1423.

Nie, X., Brunskill, E., & Wager, S. (2021). Learning when-to-treat policies. *Journal of the American Statistical Association, 116*(533), 392–409.

Nie, X., & Wager, S. (2021). Quasi-oracle estimation of heterogeneous treatment effects. *Biometrika, 108*(2), 299–319.

Orellana, L., Rotnitzky, A., & Robins, J. M. (2010). Dynamic regime marginal structural mean models for estimation of optimal dynamic treatment regimes, part I: Main content. *The International Journal of Biostatistics, 6*(2), 8.

Qi, Z., Cui, Y., Liu, Y., & Pang, J.-S. (2019). Estimation of individualized decision rules based on an optimized covariate-dependent equivalent of random outcomes. *SIAM Journal on Optimization, 29*(3), 2337–2362.

Qi, Z., Liu, D., Fu, H., & Liu, Y. (2020). Multi-armed angle-based direct learning for estimating optimal individualized treatment rules with various outcomes. *Journal of the American Statistical Association, 115*(530), 678–691.

Qi, Z., & Liu, Y. (2018). D-learning to estimate optimal individual treatment rules. *Electronic Journal of Statistics, 12*(2), 3601–3638.

Qi, Z., Pang, J.-S., & Liu, Y. (2023). On robustness of individualized decision rules. *Journal of the American Statistical Association, 118*(543), 2143–2157.

Qian, M., & Murphy, S. A. (2011). Performance guarantees for individualized treatment rules. *The Annals of Statistics, 39*(2), 1180–1210.

Robins, J. M. (1994). Correcting for non-compliance in randomized trials using structural nested mean models. *Communications in Statistics - Theory and Methods, 23*(8), 2379–2412.

Robins, J. M. (1998). Marginal structural models. In *1997 Proceedings of the Section on Bayesian Statistical Science* (pp. 1–10). American Statistical Association.

Robins, J. M. (2000). Marginal structural models versus structural nested models as tools for causal inference. In M. E. Halloran & D. Berry (Eds.), *Statistical models in epidemiology, the environment, and clinical trials*, volume 116 of *The IMA volumes in mathematics and its applications* (pp. 95–133). Springer.

Robins, J. M. (2004). Optimal structural nested models for optimal sequential decisions. In D. Y. Lin & P. J. Heagerty (Eds.), *Proceedings of the Second Seattle Symposium in Biostatistics*, volume 179 of *Lecture Notes in Statistics* (pp. 189–326). Springer.

Robinson, P. M. (1988). Root-N-consistent semiparametric regression. *Econometrica, 56*(4), 931–954.

Rosenbaum, P. R., & Rubin, D. B. (1983). The central role of the propensity score in observational studies for causal effects. *Biometrika, 70*(1), 41–55.

Rotnitzky, A., Smucler, E., & Robins, J. M. (2021). Characterization of parameters with a mixed bias property. *Biometrika, 108*(1), 231–238.

Royston, P., & Sauerbrei, W. (2008). Interactions between treatment and continuous covariates: A step toward individualizing therapy. *Journal of Clinical Oncology: Official Journal of the American Society of Clinical Oncology, 26*(9), 1397–1399.

Schulte, P. J., Tsiatis, A. A., Laber, E. B., & Davidian, M. (2014). Q-and A-learning methods for estimating optimal dynamic treatment regimes. *Statistical Science, 29*(4), 640–661.

Shalit, U., Johansson, F. D., & Sontag, D. (2017). Estimating individual treatment effect: Generalization bounds and algorithms. In D. Precup & Y. W. Teh (Eds.), *Proceedings of the 34th International Conference on Machine Learning*, volume 70 of *Proceedings of Machine Learning Research* (pp. 3076–3085). PMLR.

Shi, C., Fan, A., Song, R., & Lu, W. (2018). High-dimensional A-learning for optimal dynamic treatment regimes. *The Annals of Statistics, 46*(3), 925–957.

Si, N., Zhang, F., Zhou, Z., & Blanchet, J. (2020). Distributionally robust policy evaluation and learning in offline contextual bandits. In H. Daumé III & A. Singh (Eds.), *Proceedings of the 37th International Conference on Machine Learning*, volume 119 of *Proceedings of Machine Learning Research* (pp. 8884–8894). PMLR.

Song, R., Kosorok, M., Zeng, D., Zhao, Y.-Q., Laber, E., & Yuan, M. (2015). On sparse representation for optimal individualized treatment selection with penalized outcome weighted learning. *Stat, 4*(1), 59–68.

Sutton, R. S., & Barto, A. G. (2018). *Reinforcement learning: An introduction* (2nd ed.). Adaptive Computation and Machine Learning Series. The MIT Press.

Thomas, P., & Brunskill, E. (2016). Data-efficient off-policy policy evaluation for reinforcement learning. In M. F. Balcan & K. Q. Weinberger (Eds.), *Proceedings of The 33rd International Conference on Machine Learning*, volume 48 of *Proceedings of Machine Learning Research* (pp. 2139–2148). PMLR.

Tian, L., Alizadeh, A. A., Gentles, A. J., & Tibshirani, R. (2014). A simple method for estimating interactions between a treatment and a large number of covariates. *Journal of the American Statistical Association, 109*(508), 1517–1532.

Tsiatis, A. (2007). *Semiparametric theory and missing data*. Springer Series in Statistics. Springer.

van der Laan, M. J., & Luedtke, A. R. (2015). Targeted learning of the mean outcome under an optimal dynamic treatment rule. *Journal of Causal Inference, 3*(1), 61–95.

van der Laan, M. J., & Rose, S. (2018). *Targeted learning in data science: Causal inference for complex longitudinal studies*. Springer Series in Statistics. Springer.

van der Laan, M. J., & Rubin, D. (2006). Targeted maximum likelihood learning. *The International Journal of Biostatistics, 2*(1), 11.

Vansteelandt, S., & Joffe, M. (2014). Structural nested models and G-estimation: The partially realized promise. *Statistical Science, 29*(4), 707–731.

Wager, S., & Athey, S. (2018). Estimation and inference of heterogeneous treatment effects using random forests. *Journal of the American Statistical Association, 113*(523), 1228–1242.

Wallace, M. P., & Moodie, E. E. M. (2015). Doubly-robust dynamic treatment regimen estimation via weighted least squares. *Biometrics, 71*(3), 636–644.

Wang, L., Zhou, Y., Song, R., & Sherwood, B. (2018). Quantile-optimal treatment regimes. *Journal of the American Statistical Association, 113*(523), 1243–1254.

Watkins, C. J. C. H. (1989). *Learning from Delayed Rewards*. PhD thesis, King's College, Cambridge, UK.

Xiao, W., Zhang, H. H., & Lu, W. (2019). Robust regression for optimal individualized treatment rules. *Statistics in Medicine, 38*(11), 2059–2073.

Xu, Y., Yu, M., Zhao, Y.-Q., Li, Q., Wang, S., & Shao, J. (2015). Regularized outcome weighted subgroup identification for differential treatment effects. *Biometrics, 71*(3), 645–653.

Yoon, J., Jordon, J., & van der Schaar, M. (2018). GANITE: Estimation of individualized treatment effects using generative adversarial nets. In *International Conference on Learning Representations*.

Zhang, B., Tsiatis, A. A., Davidian, M., Zhang, M., & Laber, E. (2012a). Estimating optimal treatment regimes from a classification perspective. *Stat, 1*(1), 103–114.

Zhang, B., Tsiatis, A. A., Laber, E. B., & Davidian, M. (2012b). A robust method for estimating optimal treatment regimes. *Biometrics, 68*(4), 1010–1018.

Zhang, B., Tsiatis, A. A., Laber, E. B., & Davidian, M. (2013). Robust estimation of optimal dynamic treatment regimes for sequential treatment decisions. *Biometrika, 100*(3), 681–694.

Zhang, B., & Zhang, M. (2018). C-learning: A new classification framework to estimate optimal dynamic treatment regimes. *Biometrics, 74*(3), 891–899.

Zhang, C., Chen, J., Fu, H., He, X., Zhao, Y., & Liu, Y. (2020). Multicategory outcome weighted margin-based learning for estimating individualized treatment rules. *Statistica Sinica, 30*(4), 1857–1879.

Zhang, C., & Liu, Y. (2014). Multicategory angle-based large-margin classification. *Biometrika, 101*(3), 625–640.

Zhang, Y., Laber, E. B., Tsiatis, A., & Davidian, M. (2015). Using decision lists to construct interpretable and parsimonious treatment regimes. *Biometrics, 71*(4), 895–904.

Zhao, Y.-Q., Laber, E. B., Ning, Y., Saha, S., & Sands, B. E. (2019a). Efficient augmentation and relaxation learning for individualized treatment rules using observational data. *Journal of Machine Learning Research, 20*(48), 1–23.

Zhao, Y.-Q., Zeng, D., Laber, E. B., & Kosorok, M. R. (2015a). New statistical learning methods for estimating optimal dynamic treatment regimes. *Journal of the American Statistical Association, 110*(510), 583–598.

Zhao, Y.-Q., Zeng, D., Laber, E. B., Song, R., Yuan, M., & Kosorok, M. R. (2015b). Doubly robust learning for estimating individualized treatment with censored data. *Biometrika, 102*(1), 151–168.

Zhao, Y.-Q., Zeng, D., Rush, A. J., & Kosorok, M. R. (2012). Estimating individualized treatment rules using outcome weighted learning. *Journal of the American Statistical Association, 107*(499), 1106–1118.

Zhao, Y.-Q., Zeng, D., Tangen, C. M., & Leblanc, M. L. (2019b). Robustifying trial-derived optimal treatment rules for a target population. *Electronic Journal of Statistics, 13*(1), 1717–1743.

Zhou, X., Mayer-Hamblett, N., Khan, U., & Kosorok, M. R. (2017). Residual weighted learning for estimating individualized treatment rules. *Journal of the American Statistical Association, 112*(517), 169–187.

Zhu, W., Zeng, D., & Song, R. (2019). Proper inference for value function in high-dimensional Q-learning for dynamic treatment regimes. *Journal of the American Statistical Association, 114*(527), 1404–1417.

Utilizing Wearable Devices to Improve Precision in Physical Activity Epidemiology: Sensors, Data and Analytic Methods

Chongzhi Di, Guangxing Wang, Sixuan Wu, Kelly R. Evenson, Michael J. LaMonte, and Andrea Z. LaCroix

Abstract In physical activity (PA) epidemiology, precisely quantifying benefits of PA in terms of intensity, frequency and duration has been challenging, partly due to the difficulty and complexity in measurement. Traditional self-reported questionnaires are known to be subject to reporting bias and substantial measurement error. Enabled by technological advances in the past decades, wearable devices (including research-grade accelerometers, heart rate monitors and fitness trackers, etc.) have been increasingly adopted in large-scale epidemiological studies, recording massive amounts of high-resolution data. To take advantage of the rich data resources from these studies, advanced analytic tools are needed to rigorously quantify health benefits of patterns of PA and to inform effective intervention strategies. We first provided a review of a few recent developments in this area, covering topics on analytic methods for various aspects of accelerometry data processing and analysis, including standard analysis based on summarized data, alternative acceleration-

C. Di (✉)
Program of Biostatistics, Public Health Sciences Division, Fred Hutchinson Cancer Center, Seattle, WA, USA
e-mail: cdi@fredhutch.org

G. Wang
US Food and Drug Administration, Silver Spring, MD, USA

S. Wu
Expedia Group, Seattle, WA, USA

K. R. Evenson
Department of Epidemiology, Gillings School of Global Public Health, University of North Carolina at Chapel Hill, Chapel Hill, NC, USA

M. J. LaMonte
Department of Epidemiology and Environmental Health, School of Public Health and Health Professions, University at Buffalo-SUNY, Buffalo, NY, USA

A. Z. LaCroix
Department of Family Medicine and Public Health, University of California at San Diego, La Jolla, CA, USA

Y. Zhao, D.-G. Chen (eds.), *Statistics in Precision Health*, ICSA Book Series in Statistics, https://doi.org/10.1007/978-3-031-50690-1_3

based metrics from high resolution raw data and machine learning approaches for PA behavior classification. We then introduced a flexible functional data analysis approach for quantifying PA-health association across varying PA intensity levels. To showcase how the real world application of these methods, we applied them to the Objective Physical Activity And Cardiovascular Health (OPACH) Study to quantify does-response relationships between PA intensity and cardiometabolic risk factors among older women. We also discussed possible directions for future research.

Keywords Wearable devices · Physical activity · Functional data analysis · Machine Learning · Mobile health

1 Introduction

1.1 Physical Activity Epidemiology

It is well known that engaging in regular physical activity (PA) improves overall health and prevents many adverse health outcomes (Dishman et al., 2012), including obesity, diabetes, cardiovascular diseases and cancer (Kohl, 2001; Taylor et al., 2004; Warburton et al., 2006; McTiernan et al., 2003; Monninkhof et al., 2007; Wolin et al., 2009; Physical Activity Guidelines Advisory Committee et al., 2018; Torres et al., 2018). However, there is insufficient scientific evidence regarding what intensity, duration and frequency of PA is optimal for substantial health benefits and chronic disease prevention. One of the main challenges for precise quantification of health benefits of PA dose and patterns is the difficulty and complexity in measurement.

PA is a complex multi-dimensional construct, which involves activities of varying intensity, frequency, duration and types, and it is subject to daily, weekly and seasonal variations (Dishman et al., 2012). In PA epidemiology, a common practice is to divide activity into four distinct categories based on intensity level in METs (metabolic equivalents), where 1 MET is equivalent to energy expenditure of sitting quietly (Dishman et al., 2012). These categories are sedentary behavior (SB; 1–1.5 METs), light PA (LPA; 1.5–3 METs), moderate PA (MPA; 3–6 METs) and vigorous PA (VPA; 6+ METs). Moderate and vigorous physical activities (MVPA) are often combined as a single category. Cumulative time spent in each intensity-specific categories are commonly used summary measures of PA.

Traditionally, PA is measured by self-reported questionnaires, which collect information on frequency, duration and types of various activities based on recalls and assign intensity levels to each activity based on a compendium (Ainsworth et al., 2011). However, it has been recognized that self-reported PA are subject to reporting bias and substantial measurement error (Albanes et al., 1990; Ferrari et al., 2007; Prince et al., 2008; Neuhouser et al., 2013). In fact, it is shown that self-reported PA explains only 10–15% of the variance in objectively measured PA (Troiano et al., 2008; Neuhouser et al., 2013). In the past decade, efforts have been made to utilize objective PA measurements based on accelerometers or other wearable devices, in order to improve measurement accuracy and to understand measurement error of self-reports (Ainsworth et al., 2012; Janz, 2006; Matthews et al., 2012). These studies offer great potential to improve our understanding of PA and to devise optimal strategies for public health interventions.

1.2 Wearable Devices

The field of PA epidemiology is an early adopter of wearable devices. An increasing number of studies have utilized wearables to objectively measure PA, with accelerometers being the most popular devices adopted in large epidemiological studies (Troiano et al., 2008; Freedson et al., 1998, 2011). There are research-grade devices with high resolution, including ActiGraph and ActivPAL accelerometers, as well as consumer-grade devices, such as Fitbits, smartphones and smartwatches. They are small, lightweight and unobtrusive, and thus have recently become popular tools for objectively measuring PA in free living settings in large-scale studies. As an example, we have been involved in several PA epodemiological studies within the Women's Health Initiative (WHI), which collected 7-day PA measurement using high-resolution ActiGraph GT3X+ accelerometers among thousands of women. These tri-axial accelerometers have sensors that record acceleration of movements into electrical signals proportional to the force producing motion. These recorded signals are then processed to summarize magnitude of movements. Participants typically wear these tri-axial accelerometers to record PA continuously for a certain period of time (e.g., seven days), which offer opportunities to fully characterize participants' detailed habitual PA patterns.

In recent years, with technology rapidly evolving, other wearable devices become available to continuously record heart rates, including wearable ECG patch monitors (e.g., Zio patch or Cardiac Insight Solo; McConnell et al., 2018). In addition, multi-sensor wearable monitors are now available to simultaneously record information on movement, heart rates, GPS, etc. These new wearable monitors will further improve PA measurement and monitoring.

1.3 Motivating Studies

We describe a few PA-related ancillary studies that we have been actively involved in within the Women's Health Initiative (WHI). The WHI consists of an observational study involving 93,676 post-menopausal women and three clinical trials involving 68,132 post-menopausal women (Anderson et al., 2003; Prentice, 2010; Neuhouser et al., 2013). The WHI main studies completed in 2005, but its extensions and ancillary studies are still ongoing.

The *Objective PA and Cardiovascular Health Study* (OPACH; PI: Dr. Andrea LaCroix) is an ancillary study to the WHI on 6489 women aged 63–99 years old in 2012–2014 (LaCroix et al., 2017). It collected 7-day PA using high-resolution tri-axial ActiGraph GT3X+ accelerometers (30 Hz), as well as self-reported questionnaires. The primary aim of OPACH is to determine the associations of objectively measured PA with risks of incident cardiovascular diseases and mortality. A unique design of the OPACH is that it included a calibration study with 200 women (Evenson et al., 2015), who wore accelerometers at the hip and wrist while performing eight types of PA tasks during two-hour laboratory visits. In addition, they wore Oxycon portable calorimeters to measure oxygen intake for estimating energy expenditure along side the accelerometers. The OPACH calibration study collected data to calibrate accelerometer measured PA with energy expenditure among older adults. We have successfully quantified associations of various PA intensity measures (e.g., MVPA, LPA or SB) with many health outcomes using the OPACH data, including cardiovascular diseases, mortality, physical function, falls and diabetes (LaCroix et al., 2019; Bellettiere et al., 2019; LaMonte et al., 2017; Buchner et al., 2017; LaMonte et al., 2018).

The *Sedentary Time and Aging Program* (STAR; PI: Dr. Andrea LaCroix) is a National Institute on Aging supported program project to study immediate, intermediate and long-term effects of SB on aging outcomes in its three projects, respectively. The first and second projects are randomized trials evaluating effects of interventions that break SB bouts. The third project of STAR aims to understand effects of SB time and patterns measured by accelerometry on trajectories of physical function and mortality over eight to ten years among OPACH women. A key component of STAR is to develop novel machine learning methods on raw accelerometry for PA behavior classification (Rosenberg et al., 2017; Jain et al., 2021). The three projects recorded high-resolution accelerometry in $7,367$ participants in total.

The *Women's Health Accelerometry Collaboration* (WHAC, PI: Dr. Kelly Evenson) aims to investigate refined patterns of PA and SB especially for cancer outcomes, by combining two cohorts, the WHI OPACH and the Women's Health Study (WHS), which collectively had accelerometry on 23,000 women (Evenson et al., 2021).

The *WHI Strong and Healthy Trial* (WHISH; PIs: Drs. Marcia Stefanick, Andrea LaCroix, Charles Kooperberg). The WHISH Trial (Stefanick et al., 2021) is a randomized trial to test a centralized PA intervention involving 52,500 women since 2015. All women have PA data from self-reported questionnaire data. In a substudy of 2300 women, accelerometer data are collected at three different visits during the trial, at 6, 18 and 36 months. These women are selected from OPACH participants, so each has up to four repeated accelerometry within 6–8 years.

In these studies, we utilized ActiGraph GT3X+, which allows recording of high-resolution raw data as well as standard resolution summarized counts. They were worn on the hip, and in the OPACH calibration study (Evenson et al., 2015), an additional accelerometer was worn on their non-dominant wrist simultaneously.

1.4 *Analytic Challenges for Accelerometry*

With the size and complexity of wearable device data (e.g., accelerometry) exploding, there are serious challenges for statistical analysis (Staudenmayer et al., 2012). Accelerometer output can be downloaded as standard resolution activity counts data (AC; in pre-specified epochs, e.g., 1-minute or 15-second epochs) or high-resolution raw data (10–100 Hz), with the latter becoming available on newer generation devices (i.e., ActiGragh's devices later than GT3X+). In the literature, most studies only explore the AC data and reduce the AC trajectories over seven days into a few summary variables (e.g, time in SB, LPA and MVPA), which dramatically oversimplifies and overlooks potentially relevant features of PA.

For standard resolution AC data, feature extraction methods are needed to extract information that might be omitted by standard analysis that reduces the time series into a few simple summary measures. Functional data analysis techniques will be helpful to extract information from such data. Furthermore, there is a need for better methods to process the high-resolution raw data. For example, several raw acceleration-based metrics and machine learning approaches have been proposed to process on high-resolution raw data to extract more information not captured by the AC data (Staudenmayer et al., 2009; Strath et al., 2012). There are also computational challenges in analyzing accelerometer data from large-scale studies, such as the WHI, due to the massive data size. Despite developments in many of these areas in recent years, new analytic methods that are interpretable, robust, and computationally efficient are still needed to take advantage of the rich information in the data, extracting PA features and patterns and quantifying their dose-response relationships with health outcomes.

There has been substantial developments in accelerometry processing and analysis in the past decades, with a few excellent review papers summarizing some of these efforts (Staudenmayer et al., 2012; Karas et al., 2019; Zhang et al., 2019). Rather than providing a comprehensive review of the literature, the scope of this paper is to focus on methods for processing and analyzing accelerometry in large-scale epidemiological studies that primarily aims to quantify associations between PA and health outcomes. We especially focus on processing, summarization and association analysis that have been directly motivated by and applied to our WHI-related studies.

2 Accelerometry Data and Analytic Workflow

In our WHI-related motivating studies, seven-day PA were measured by research grade ActiGraph GT3X+ accelerometers. Two types of data were downloaded, corresponding to different resolution and formats.

The first type is called activity count (AC) data, usually processed by device manufacturer (e.g., ActiGraph) provided algorithms, contains time series reflecting

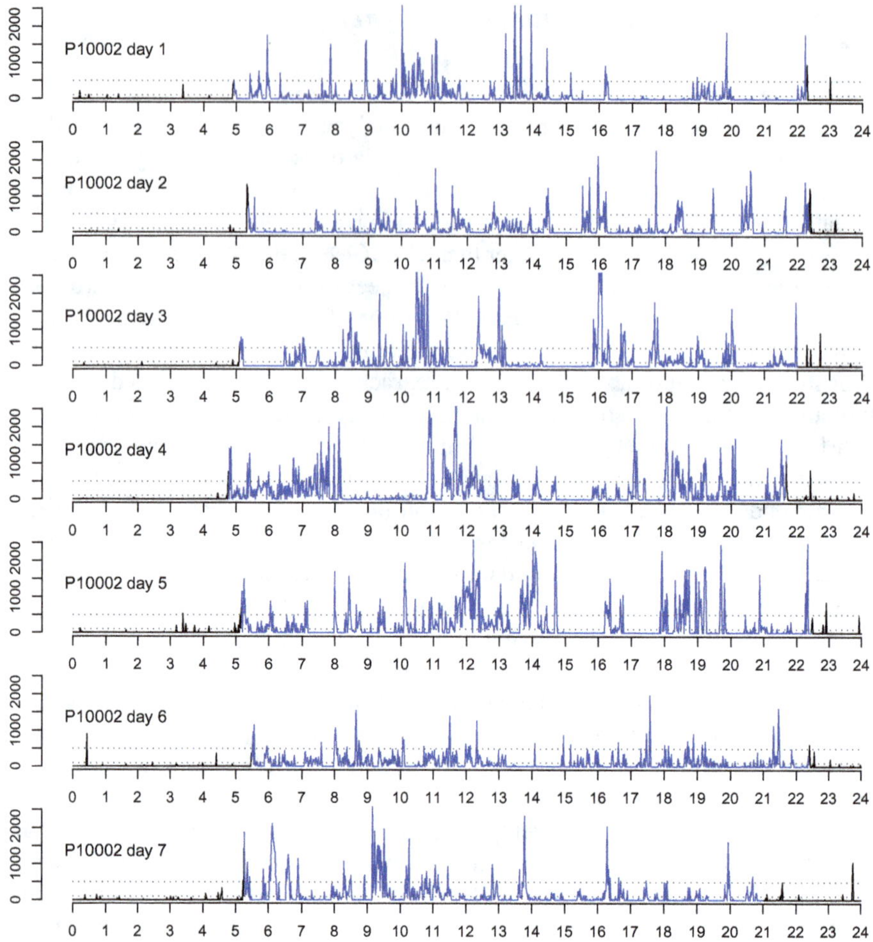

Fig. 1 An example of 7-day accelerometry activity count (AC) data from a subject in the WHI OPACH Study, in 15-second epochs. The blue and black lines correspond to awake and sleep periods, respectively

magnitude of movement in each epoch (e.g., 1 minute or 15 seconds) per subject. For ActiGraph devices, their algorithm for calculating AC were proprietary until 2022, when the current counts algorithm in ActiGraph's ActiLife and CentrePoint software were published as a standalone Python package for research use (Neishabouri et al., 2022). The output data for a subject over 7 days is a three-dimensional time series with 10,080 data points in 1-minute epochs, reflecting movement in three axes (vertical, horizontal and lateral). The vector magnitude AC is often used to summarize movement in all directions, defined by the squart of the sum of three axis AC squared. Figure 1) shows an example of vector magnitude AC series data from an OPACH participant. For older accelerometer models (e.g., ActiGraph GT1M

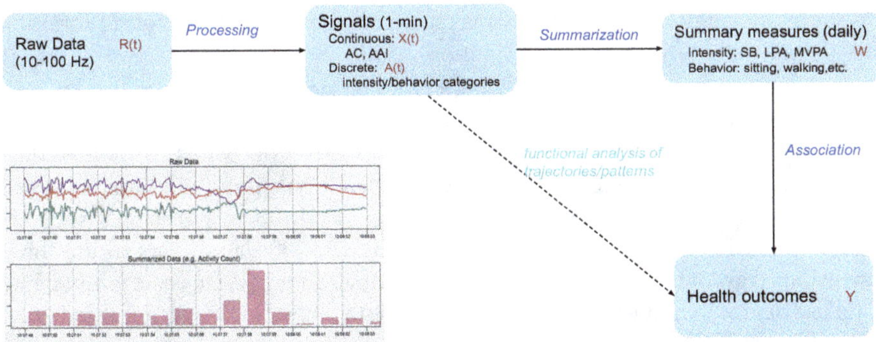

Fig. 2 Illustration of a typical analytic workflow for accelerometry data: processing, summarization and association analysis. The left bottom figure shows an example of processing, from raw data to processed AC data. The dashed arrow reflects functional approaches to be introduced in Sect. 3.4, which do not need an intermediate summarization step

or GT3X models, prior to 2011), AC data is the only data format available, and some older models only measure the vertical axis (e.g., ActiGraph GT1M). Thus, PA studies that utilized older ActiGraph models only analyzed data in this format.

The second data type is high-resolution raw data (10–100 Hz), which is massive in size and available in newer generation research grade devices (e.g., ActiGraph GT3X+ or newer models). The raw accelerometer data can contain up to 18,000 data points per minute, compared to one data point per minute for standard resolution data (e.g., vector magnitude AC in 1-minute epochs), generating 6–60 million data points per subject in a 7-day period. Raw data is useful and important for two reasons. First, they are recorded at high resolution and provide detailed information that can be utilized for PA behavior recognition (e.g., walking, sitting) and detailed quantification of these activities (e.g., walking cadence, sit-to-stand transitions). Second, AC data are often provided by device manufacturers based on applying proprietary algorithms to raw data, making it difficult to compare studies that use different devices. It may be possible to aggregate raw data in different forms that are transparent, complementary and informative. In fact, the PA research community has strongly urged manufacturers to provide raw data for these reasons (Staudenmayer et al., 2012; Wijndaele et al., 2015).

For the goal of quantifying associations between accelerometry-measured PA and health outcomes, an analytic workflow (Fig. 2) consists of three steps: (1) data processing or signal extraction, which transforms high-resolution raw data to meaningful signal series in standard resolution (e.g., AC in 1-minute epochs using device manufacturer provided software); (2) summarization, which generates low-dimensional PA summary measures from standard resolution signal series; and (3) association analysis, which investigates relationships between PA summary measures and health outcomes via regression models.

We now introduce notations for various data types. Let $\tilde{R}(t)$ denote the three-dimensional raw accelerometry time series over the study period $t \in [0, T]$, where

T is the study length. In practice, $\tilde{R}(\cdot)$ is recorded on a set of grid points (18 million time points for 7 days in 30 Hz resolution), $\mathcal{T}_{raw} = \{t_j : j = 0, 1, \cdots, N_{raw}\}$ where $0 = t_0 < t_1 < \cdots < t_{N_{raw}} = T$. The processing step will take $\tilde{R}(\cdot)$ as input data and generate standard resolution signal series, $X(t)$ on a subset of $\mathcal{T}_{raw}$, denoted as $\mathcal{T}$, which is a more sparse set of N grid points. If the raw data is in 30 Hz and the epoch length for standard resolution signals is 1-minute, it can be shown that $N_{raw}/N = 30\ sec^{-1} \times 60\ sec = 1800$, so the signal for a 1-minute epoch is calculated based on $1800 \times 3 = 5400$ data points in the three dimensional raw data points from the same window. Taking AC as an example, one could use device manufacturer provided software to obtain AC in pre-specified (e.g., 1-minute) epochs, which yields a continuous time series with length N (see Fig. 1 as an example). In recent years, several open-source raw acceleration based metrics have been proposed, which serves as alternative processing methods to generate continuous signal series. On the other hand, some preprocessing approaches use either $\tilde{R}(\cdot)$ or $X(\cdot)$ as input to generate discrete signal series, $A(\cdot)$, e.g., PA intensity categories (e.g., SB, LPA and MVPA) or behavior categories (e.g., sitting, standing, daily life movement and walking) in each epoch. Note that the resolution for $X(\cdot)$ or $A(\cdot)$ are often chosen to be scientifically meaningful, with popular choices including 1-minute, 15-second or 10-second.

The summarization step takes either $X(\cdot)$ or $A(\cdot)$ series as input and generates a few scientific meaningful summary measures, denoted as $\tilde{W}$. One example is the total PA volume, which is the sum of AC over the study period. Another example is cumulative time spent in each of the PA intensity categories, including SB, LPA and MVPA. In association analysis, one investigates the relationship between summary measures $\tilde{W}$ and a health outcome, Y, via regression models.

3 Accelerometry Processing, Summarization and Association Analysis

In this section, we first provide a review of standard analytic approaches for accelerometry, followed by further methods developments in data processing, summarization and analysis, especially those motivated by our motivating studies.

3.1 Standard Analytic Approaches

The most commonly adopted accelerometry analysis plan in the PA literature uses device software processed AC data $X(\cdot)$ as input, summarizes AC trajectories into PA intensity measures (e.g., daily SB/LPA/MVPA minutes) via cut points (Schmidt et al., 2003; Troiano et al., 2008), and applies regression analysis for association modeling.

In the summarization step, standard analysis reduces data into a few scientifically meaningful summary indices, such as total AC per day, time spent in SB, LPA and MVPA (Troiano et al., 2008). In fact, any pre-defined feature can be represented as a function of $X(\cdot)$. For example, a total activity volume is defined by $W^{total}\{X(\cdot)\} = \int_0^T X(t)\,dt$. Most studies choose a set of cut points, (c_1, c_2), to categorize PA into SB (less than c_1), LPA (between c_1 and c_2) and MVPA (greater than c_2). These cut points are often established by lab-based calibration studies that are assumed to be applicable to the general population (Troiano et al., 2008). Using this definition, time spent in SB, LPA and MVPA can be represented by $W^{SB}\{X(\cdot)\} = \int_0^T I\{X(t) < c_1\}\,dt$, $W^{LPA}\{X(\cdot)\} = \int_0^T I\{c_1 \le X(t) < c_2\}\,dt$ and $W^{MVPA}\{X(\cdot)\} = \int_0^T I\{X(t) \ge c_2\}\,dt$, respectively, where $I(\cdot)$ is the indicator function. These scientifically meaningful features are then used as univariate or multivariate predictor variables in subsequent regression analysis with health outcomes. Depending on data types of the outcomes, linear models, generalized linear models, Cox proportional models and longitudinal models can be used for continuous, discrete, time-to-event and longitudinal outcomes, respectively.

The choice of AC cut points to define PA intensity categories, (c_1, c_2), is a common issue for adopting this approach. While there are several sets of cut points available in the literature (e.g., Troiano cut points for general adults Troiano et al., 2008), they might not work well for specific populations, e.g., older adults. In the WHI OPACH study, we were able to derive more precise cut points for older women using data collected in the internal calibration study. Specifically, utilizing measurements including accelerometry and energy expenditure via indirect calorimetry in 200 women, we conducted receiver operating charateristics (ROC) analysis to identify optimal cutpoints specifically for older women (Evenson et al., 2015), which were then applied to derive PA intensity summary variables and conduct subsequent association analysis (LaCroix et al., 2019; Bellettiere et al., 2019; LaMonte et al., 2017; Buchner et al., 2017; LaMonte et al., 2018). This provides improvement in the *summarization* step of the analytic workflow (Fig. 2), compared to applying general cut points from the literature.

Missing data due to nonwear is another common issue for accelerometry. Data often contain long stretches of zero counts due to nonwear. Several algorithms have been proposed and validated to identify nonwear periods, including the Choi algorithm (Choi et al., 2011, 2012). Two approaches are commonly used to handle missingness due to nonwear. Most studies perform analysis on complete wear time data only, but it may produce biased estimates. To account for potential biases, we developed a hot-deck multiple imputation approach to account for missing data (Butera et al., 2019).

In the remaining subsections, we will review a few analytic approaches for accelerometry that improve upon standard processing, summarization and association analysis steps. These methods were directly motivated by and applied to objective PA measurements from the WHI studies.

3.2 Raw Accelerometry-Based Metrics (AC Alternatives)

Since the raw accelerometry data become available, a few open-source methods have been proposed to provide summary metrics in standard resolution epochs (e.g., 1-minute), as alternatives to the device manufacturer provided AC. Examples include an accelerometry activity index (AAI) (Bai et al., 2013; Bai et al., 2016), Euclidean distance minus one (Van Hees et al., 2013), mean amplitude deviation (Aittasalo et al., 2015) and monitor-independent movement summary index (John et al., 2019). These methods generally consider non-overlapping time windows of length H and reduce the $3H$ raw acceleration points in the window to a single scalar. We focus on the AAI here, as it is the only metric that has been calibrated for PA intensity from a calibration study to our best knowledge.

The AAI was introduced to provide an alternative, transparent metric to the AC (Bai et al., 2016). Based on observations that the variability of raw acceleration signals in short epochs (e.g., 1 second) removes gravity and provides a summary measure of movement intensity, the AAI was defined based on a combination of three within-axis standard deviations of the raw accelerometry signal. Its formula is given by

$$AI_i(t, H) = \sqrt{max\left(\frac{1}{3}\left\{\sum_{m=1}^{3} \frac{\sigma_{im}^2(t; H) - \bar{\sigma}_i^2}{\bar{\sigma}_i^2}\right\}, 0\right)},$$

where $\mu_m(t; H) = \frac{1}{H}\sum_{h=0}^{H-1} R_m(t + h)$, $\sigma_{im}^2(t; H) = \frac{1}{H}\sum_{h=0}^{H-1}\{R_m(t + h) - \mu_m(t; H)\}^2$ are mean and variance for the ith participant's acceleration signals along axis $m(m = 1, 2, 3)$ in the window of length H starting at t, and $\bar{\sigma}_i^2$ denotes the systematic noise variance when the device does not move. The AAI was shown to enjoy three desirable properties: ease of implementation, additivity, and rotational invariance.

A natural question is whether raw acceleration based metrics like AAI have additional advantages over AC, other than being transparent and thus allowing cross device comparisons. To answer this question, we validated the AAI and compared AAI to AC in terms of predicting PA intensity in the OPACH calibration study (Bai et al., 2016; Wang et al., 2022). Receiver operating characteristic (ROC) analyses indicated that AAI was more sensitive to sedentary and light intensity activities than AC. AAI had higher predictive capacity for PA intensity. These results suggest that AAI provides a better measurement of PA intensity than AC in older women in the OPACH Study.

3.3 *Machine Learned PA Behaviors Based on Raw Accelerometry (Beyond AC)*

Another way of utilizing the rich information from high-resolution raw data is to predict PA behaviors using machine learning approaches (Staudenmayer et al., 2009, 2012; Bai et al., 2012), which is not possible using standard resolution standard AC data. In the context of predicting PA behavior among older women, we have developed a classification algorithm (Kerr et al., 2016; Rosenberg et al., 2017) for predicting sitting, standing, daily life movement, walking and running in 1-minute epochs. The Two Level Behavior Classification algorithm (TLBC; Fig. 3) consists of three major steps: feature extraction from raw data, classification based on random forests, and hidden Markov model smoothing. The TLBC was trained on a data set where participants wore both accelerometers and SenseCam (ground truth annotation) devices, with images annotated and time matched to accelerometer data. In two validations on older women, the algorithms had 82.2% balanced accuracy in one of the studies, and achieved 87.9% accuracy specifically for predicting walking in another study. It demonstrated that free-living algorithms for walking and sedentary time yielded high levels of accuracy and concurrent validity.

A more recent example of the application of machine learning is a deep learning Convolutional Neural Network Hip Accelerometer Posture (CHAP) algorithm to predict sitting vs non-sitting behaviors from tri-axial ActiGraph accelerometers, trained against concurrent thigh-worn ActivPal posture data as the ground truth (Greenwood-Hickman et al., 2021; Bellettiere et al., 2022). Features in the raw ActiGraph data (30 Hz) are learned using a Convolutional Neural Network model, and then smoothed temporally via a bi-directional long short-term memory network. The algorithms use a split-sample paradigm, namely training and performing model selection on a random training set, and then assessing model performance on an independent test set. It was demonstrated that CHAP had better predictive performance than existing methods, e.g., AC-based cut points or previous machine learning (e.g., TLBC) approaches. For sitting vs. non-sitting classification, CHAP performed better (93% agreement with criterion) than alternative methods (74–83% agreement). For detecting sit-to-stand transitions, a key metric for assessing interruptions in sedentary patterns, CHAP also exhibited high sensitivity (cut points 73%; CHAP 83%) and positive predictive (cut points 30%, CHAP 83%). The CHAP algorithm is being applied to more accurately classify sitting epochs in accelerometer data collected in the OPACH, STAR and WHAC studies, and we plan to explore accumulation patterns of sitting as predicted by CHAP and study its relationships with health outcomes.

3.4 *Summarization and Association Analysis: Flexible Functional Approaches*

Next, we consider functional data analysis approaches (Ramsay & Silverman, 2002) that link the accelerometer signal series $X(\cdot)$ directly with health outcomes, without needing an intermediate summarization step. Let $X_i(t)$ denote accelerometer trajectories for the ith subject, with discretized data $\tilde{X}_i = \{X_i(t_j) : j = 1, 2, \cdots, N\}$. The signal can be AC or other raw acceleration based metrics (e.g., AAI). We assume that the range of $X_i(t)$ is $[0, M]$, with M being the maximum possible value. In recent years, functional data analysis have been adopted directly on trajectories $X_i(\cdot)$ to study temporal patterns over time (e.g., a day or a week) and identify

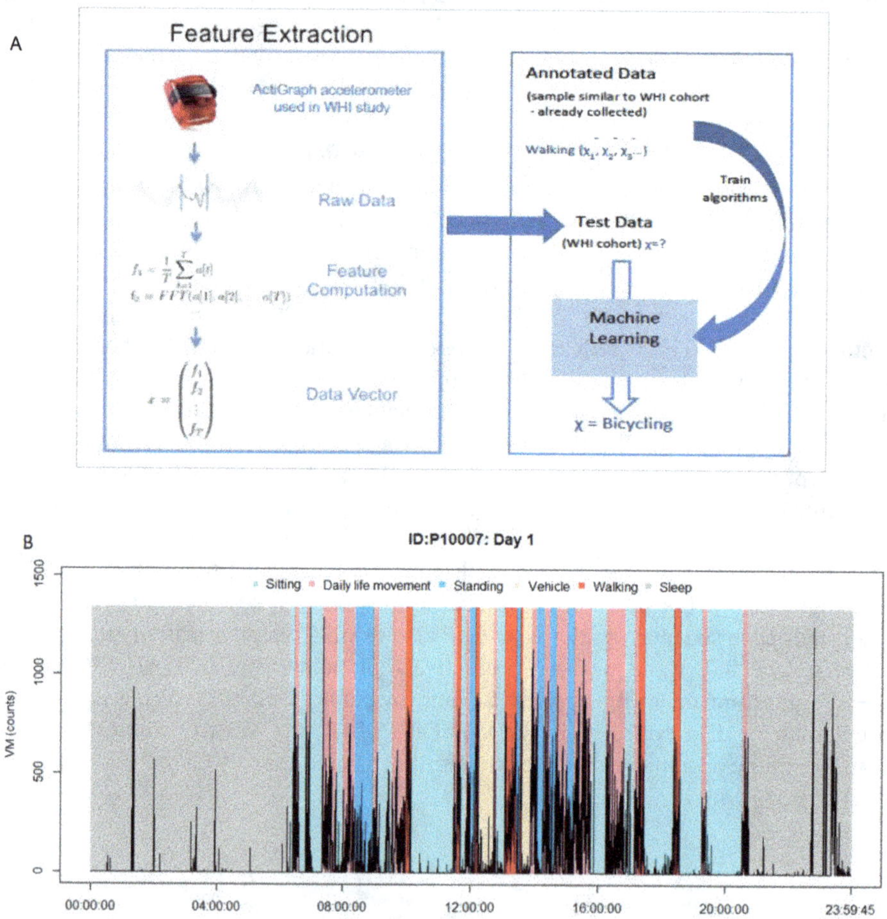

Fig. 3 Diagram of machine learning approaches for predicting PA behaviors from raw accelerometry (**a**) and an illustrative example of predicted behavior labels for one participant (**b**)

patterns that predict health outcomes (Zhang et al., 2019; Xu et al., 2019; Goldsmith et al., 2016). However, this type of analysis often focus on information regarding timing of PA, which might play less important roles than other aspects of PA (e.g., intensity, duration, and frequency) in terms of health benefits. Besides, it has other limitations in terms of interpretation and time synchronization or registration.

To address these challenges, we propose to first transform the functional data into the domain of PA intensity rather than timing, which in turn allows us to directly investigate dose-response relationships with intensity, one of the most important aspect of PA in association with health. Specifically, let $S_i(z)$ denote the complementary cumulative distribution function for the ith subject's activity data $\tilde{Z}_i$. Note that $S_i(z) = \Pr(\tilde{Z}_i > z)$ is also called survival function in time-to-event data, or reliability function in engineering. The value of $S_i(z)$ at any z is the proportion of time spent on activities more intense than the threshold z. It is closely related to commonly used summary measures, and yet provides a natural extension to describe the full activity profile. In fact, specific points on the curve $S_i(z)$ are closely related to standard summary statistics. For example, if one chooses a cut point c_2 for LPA versus MVPA, the proportion of time that subject i spent on MVPA is exactly $S_i(c_2)$. Choosing another cut point c_1 for SB versus LPA, the proportion of time spent on SB and LPA activities are $1-S_i(c_1)$ and $S_i(c_2)-S_i(c_1)$, respectively. Thus, standard intensity-specific summary indices correspond to a few specific points on the $S_i(z)$ curve, which provides snapshots of the profile. Instead, we aim to capture the complete picture by proposing to use the entire function $S_i(z)$ that describes the full distribution of PA accumulated at all intensity levels.

To investigate the association between PA profiles and health outcomes, we consider functional regression models using the functional measure $S_i(z)$ as the main predictor. The goal is to investigate the dose-response relationship between PA accumulated at varying intensity levels and health outcomes. Let Y_i denote the outcome (continuous or discrete) and V_i denote additional covariates. The proposed model is a generalized functional linear model,

$$g[\,\mathrm{E}\{Y_i \mid S_i(\cdot), V_i\}\,] = \beta_0 + V_i\beta_1 + \int_0^M S_i(z)\gamma(z)\,dz\ , \qquad (1)$$

where g is a link function and the coefficient function $\gamma(z)$ allows activities of different intensity to have varying effects on health (see Fig. 2). As a special case, if one chooses $\gamma(z)$ to be a constant function $\gamma(z) = \gamma_0$, then $\int_0^M S_i(z)\gamma(z)\,dz = \gamma_0 \int_0^M S_i(z)dz = \gamma_0 W^{total}$, which implies that the model reduces to a standard regression model with total PA volume as a scalar covariate. Or, if one assumes $\gamma(z) = \gamma_1 I(z > c_2)$ for a given MVPA cut point c_2, then $\int_0^M S_i(z)\gamma(z)\,dz = \gamma_1 \int_0^M S_i(z)I(z > c_2)dz = \gamma_1 W^{MVPA}$, which means the proposed model reduces to a generalized linear model with the MVPA volume as a predictor variable. However, using MVPA alone may lead to counter-intuitive implications, e.g., activities of 2.9 METs have no health benefits while activities of 3.1 METs have major health benefits. In contrast, we assume that the coefficient function $\gamma(\cdot)$ is a continuous and smooth function of PA intensity and propose to estimate it flexibly.

Estimation and inference of this model have been studied in the literature, with two popular methods including basis expansions based on pre-specified or data-adaptive functional principal component basis (Ramsay & Silverman, 2002; Crainiceanu et al., 2009; Ruppert et al., 2003; Goldsmith et al., 2011; Swihart et al., 2014). The estimated $\gamma(\cdot)$ quantifies the dose-response relationship between PA intensity and health outcomes. A constant $\gamma(\cdot)$ implies that PA intensity accumulations does not matter, while a monotonically increasing $\gamma(\cdot)$ means that PA accumulated at higher intensity have additional health benefits. Discoveries on these dose-response relationships have important translational impact, as they provide direct scientific evidence for informing future PA guidelines.

4 Application to the OPACH: A Novel Dose-Response Analysis of PA Intensity and Cardiovascular Health

In this section, we showcase an example of novel analysis of accelerometry using the WHI OPACH data, which improves upon each of three steps of the analytic workflow (Fig. 2) over standard approaches. As a consequence, this analysis allows more precise and detailed quantification of the dose-response relationships between PA intensity and cardiometabolic risk factors.

As described earlier, the OPACH Study collected accelerometry on 6489 women aged 63–99 years old in 2012–2014. In this analysis, the outcome variables are cardiometabolic risk factors including body mass index (BMI) and high density lipoprotein (HDL) cholesterol collected at the same time as accelerometry. The main predictor is PA measured by accelerometry, with variables including age, race and ethnicity, education, and BMI as potential confounders. To investigate relationships between objectively measured PA and these cardiometabolic risk factors, we first conducted standard analysis using SB, LPA and MVPA as PA summary measures (LaMonte et al., 2017), and later performed an AAI-based analysis using AAI-calibrated SB, LPA and MVPA as main predictors (Wang et al., 2022). In the following, we first discuss differences between these analyses (published previously), and then present a new analysis from an AAI-based functional approach. Table 1 lists the main differences between standard analysis (LaMonte et al., 2017), AAI-based analysis (Wang et al., 2022) and the proposed AAI-based functional analysis.

Table 1 Comparison among three analytic approaches: standard, AAI-based, and AAI-based functional approaches

	AC-based (standard)	AAI-based	AAI-based functional
Data source	Activity counts (1-minute)	Raw data (30 Hz)	Raw data (30 Hz)
Processing	–	AAI	AAI
Summarization	Intensity categories	Intensity categories	Continuous intensity
Association	Linear regression	Linear regression	Functional regression

4.1 *Processing and Summarization: AAI vs. AC*

In terms of data source, standard analysis utilize AC data in 15-second epochs. The epoch length was chosen to be 15-second instead of more commonly used 1-minute in order to capture more short bout movements among older women (LaCroix et al., 2017), who are generally more sedentary than younger adults. On the other hand, AAI-based approaches processed the high-resolution raw data directly to generate AAI series in 15-second epochs.

We first evaluate the difference in the processing step, AAI versus AC, using the OPACH calibration study data. Briefly, the calibration study enrolled 200 women in total and 193 women had complete raw data available, which were used in our analysis. Participants performed a variety of PA tasks in a laboratory while wearing Actigraph accelerometers as well as a portable Oxycon calorimeter (Oxycon Mobile; CareFusion, Rolle, Switzerland), the latter of which recorded oxygen uptake (VO2) that can be used to estimate PA intensity in METs. The list of activity tasks include: watching DVD while sitting quietly, assembling puzzle while sitting, washing dishes while standing, doing laundry while standing, 400-meter walking, dust mopping while standing, and treadmill walking at various speeds. This study provides valuable resource to investigate relationships between accelerometry measures (AAI or AI) and PA intensity.

Figure 4 showed scatterplots of AAI and AC against measured METs. The AAI and AC were both calibrated for PA intensity and then compared from a few perspectives. The comparisons were reported in our prior papers (Bai et al., 2016; Wang et al., 2022) and we briefly summarized them below. First, a calibration equation was derived to predict PA intensity in METs based on AAI in 15-second epochs. After exploring various transformations of AAI, it was found that the square root transformation performed better than the original scale or logarithm

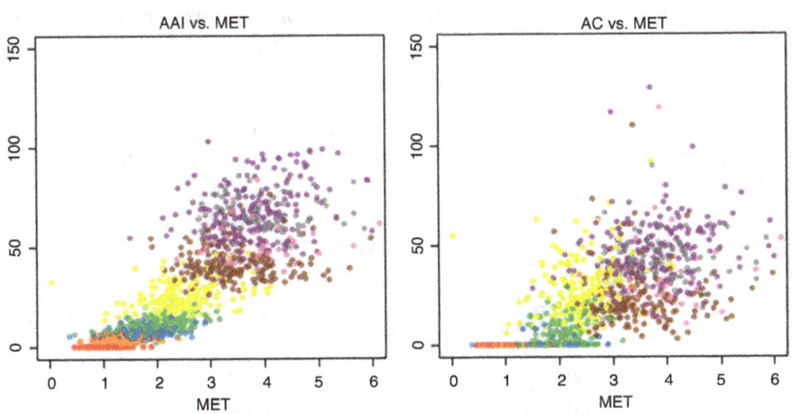

Fig. 4 Scatterplots of AAI vs. MET and AC vs. MET in OPACH calibration study. Different colors represent different types of activity

transformation. We obtained a calibration equation by fitting a linear regression model on the square root of AAI, namely, $MET = 1 + 0.079\sqrt{AAI}$, with $R^2 = 0.74$. Note that we forced the predicted intensity to be 1 MET when $AAI = 0$, as it corresponds to the scenario of no movement. The predictive performance of AAI is substantially higher than that of AC ($R^2 = 0.54$). Second, we conducted ROC analysis to derive AAI cut points for classifying a 15-seconds epoch into SB, LPA or MVPA (Wang et al., 2022). The areas under the ROC curve (AUC) for SB vs. LPA and LPA vs. MVPA are 0.93 and 0.97, respectively. They are also noticeably higher than AUCs for using AC for similar classifications (with AUC 0.84 and 0.90, respectively). These evidence collectively suggest that AAI is a more precise measurement of the PA intensity than AC for older women in the OPACH.

4.2 *Association with PA Intensity Categories: AAI vs. AC*

The next step aimed to investigate the relationships between PA intensity categories derived by AAI and AC data and two outcomes, BMI and HDL. Since different PA intensity measures are compositional in nature (e.g., negatively correlated), we fitted regression models to estimate substitutional effects between two PA categories, for example, reallocation from SB to LPA or from LPA to MVPA (Mekary et al., 2009; Stamatakis et al., 2015). For HDL cholesterol outcomes, the model adjusted for awake wear time, age, race and ethnicity, education, and BMI. In models where BMI is the outcome variable, we adjusted for average awake wear time, age, race and ethnicity, and education.

Table 2 showed estimated effects associated with reallocating 30 minute per day between two activity intensity categories, as previously reported (Wang et al., 2022). For AAI-based and AC-based analyses, reallocating 30 minute per day from SB to LPA was significantly associated with lower BMI and higher HDL levels. On the other hand, the two approaches yielded different conclusions for the substitutional effects between LPA and MVPA. The effects were stronger in AAI-based analysis

Table 2 Estimated effects of 30 minute per day reallocation between PA intensity categories (SB, LPA and MVPA) and two outcomes (BMI and HDL cholesterol) for OPACH Study *Note: reprinted from Table 6 of Wang et al. (2022), Calibration of an accelerometer activity index among older women and its association with cardiometabolic risk factors, Journal for the Measurement of Physical Behavior, 2022 Sep;5(3):145–155*

		AAI		AC	
Outcome		SB → LPA	LPA → MVPA	SB → LPA	LPA → MVPA
BMI	Coef	−0.49	−0.81	−0.62	−0.10
	P-value	0.00	0.00	0.00	0.32
HDL	Coef	0.60	0.96	0.68	0.48
	P-value	0.00	0.00	0.00	0.07

and statistically significant, where those from AC-based analysis were weaker and not statistically significant.

From measurement error perspective, both AAI-based and AC-based intensity measures can be viewed as noisy measurements of PA intensity. For linear regression models, it is well known that regression coefficients are generally attenuated in the presence of measurement error and that the magnitude of attenuation bias increases with the magnitude of measurement error (Carroll et al., 2006). Calibration results in Sect. 4.1 suggest that AAI-based measures are associated with substantially less measurement errors than AC-based measures, and thus it is expected that regression based on using AAI-based variables are closer to effects of true activity intensity measures. We plan to conduct more rigorous measurement error correction for these models in a future project.

4.3 Association with Continuous PA Intensity: An AAI-Based Functional Approach

Standard summarization methods reduce continuous data on activity intensity into several categories using the cut points. However, this approach may lead to loss of statistical information embedded in the continuous data. Next, we aim to analyze PA intensity continuously via the functional approach as outlined in Sect. 3.4. This approach allows for the use of the entire profile of AAI trajectories over the study period, rather than a few summary measures.

Applying the methods described above, we first calculated the estimated CCDF for each participant based on their AAI trajectories. Functional linear regression models were then used to investigate the association between activity intensity with BMI and HDL cholesterol. For HDL cholesterol, we adjusted for average awake wear time, age, race-ethnicity, education and BMI. For BMI, we included above variables except for BMI to adjust the models.

Figure 5 displayed the estimated regression coefficient function $\gamma(\cdot)$, as well as its corresponding 95% confidence regions, with respect to METs from 1.5 to 6 and corresponding AAI values. The overall relationship is positive and monotonically increasing for HDL cholesterol. It suggests that longer time spent in PA across all intensity levels (i.e., a more active lifestyle) is associated with higher average total cholesterol. The monotonically increasing shape indicates that the effect of PA varies by intensity, with PA accumulated at higher intensity levels has a larger impact on HDL while controlling for the total activity volume. The estimated coefficient function $\gamma(\cdot)$ is always negative, indicating an inverse association between PA and BMI across intensity levels. The magnitude of association is monotonically increasing from 1.5 to 3 METs (LPA), and plateaus above 3 METs (MVPA). It implies that for LPA, activity accumulated at higher intensity levels have stronger inverse association with BMI. For MVPA, on the other hand, activity accumulated at different intensity levels tend to have similar effects. It is worth noting that the confidence bands for these coefficients gets wider for high intensity

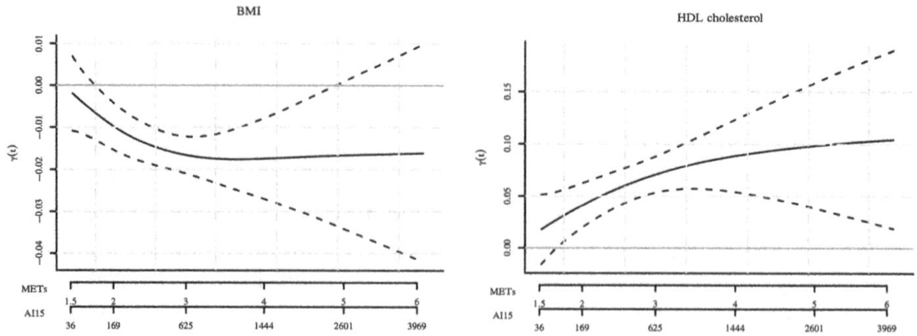

Fig. 5 Estimated functional regression coefficients $\gamma(\cdot)$ for BMI and HDL for varying intensity levels (from 1.5 to 6 MET). The solid lines are point estimates, while dashed lines indicate 95% confidence bands

levels (especially above 5 METs), due to fewer OPACH participants accumulating meaningful minutes at or above these levels.

It is worth noting the AAI-based functional analysis enables us to flexibly estimate detailed dose-response relationships between PA accumulated from varying intensity levels and health outcomes, which is not possible from standard analysis via PA intensity categories. Standard summary indices only provide crude measures of activity intensity, but omit many refined features and patterns of PA. First, by dividing PA into intensity categories using cut points, intensity information within categories is lost. Second, analysis might be sensitive to choices of cut points, which can be subjective and population specific. In contrast, the proposed functional measure provides additional information by characterizing PA accumulation patterns across all intensity levels, and does not depend on choices of cut points. To summarize, standard summary measures only provide a few snapshots of the entire PA profile, while the proposed functional approach capture the complete picture of PA intensity distribution.

5 Discussion

In this paper, we first introduced wearable sensors that are increasingly adopted in PA epidemiological studies and described data types and analytic challenges. Motivated by PA studies within the WHI, we then discussed several analytic methods for various aspects of accelerometry data processing and analysis, including standard analysis based on AC data, alternative acceleration-based metrics from high resolution raw data, machine learning approaches for PA behavior classification and a flexible functional approach for quantifying PA-health association across varying PA intensity levels. We also used the WHI OPACH Study as an illustrative example to showcase how the real world application of these methods led to improved knowledge on the association between PA intensity and cardiometabolic risk factors.

To further advance the field of physical activity epidemiology and exercise medicine, there are several directions that needs future research. The *2018 PA Guidelines for Americans* (PAG2018) states that "regular PA is essential for healthy aging" and makes specific recommendations in terms of PA dose for health benefits (Piercy et al., 2018). While PAG2018 was made based on the best available scientific evidence by 2018 (Torres et al., 2018), it pointed out lack of rigorous evidence for optimal dose of PA in terms of intensity, frequency and duration for most health benefits. The PAG2018 Advisory Committee Scientific Report (Physical Activity Guidelines Advisory Committee et al., 2018) identified several areas as top priorities that need future research for informing the next set of physical activity guidelines, including compositional patterns of 24-hour activity cycle and bout accumulation patterns in relationship to various health outcomes.

To fill in these gaps, new analytic methods are needed in a few areas. First, measurement error correction methods are needed, especially in the context of functional regressions in Sects. 4.2 and 4.3. Another issue is how activity bout accumulation patterns in each activity category are associated with health, e.g., on whether prolonged SB/LPA/MVPA increase or decrease a person's risk for adverse health outcomes compared to the same activity accumulated in short bouts. It will be useful to develop functional data analysis methods to flexibly model distributions of bout lengths, which do not depend on thresholds or parametric distributional assumptions. Third, flexible compositional data analysis models for jointly model PA, SB and sleep in 24-hour activity cycle are needed for drawing robust scientific conclusions on compositional patterns and health. In addition, it is critical to develop methods for studying temporal changes of 24-hour activity compositional patterns.

Finally, a related promising area of research enabled by technology is mHealth-based interventions (Hardeman et al., 2019). Utilizing a combination of wearable sensors and smartphone-based mHealth tools makes it possible to design and implement individualized just-in-time PA interventions that are more effective than traditional intervention strategies in promoting a more active and less sedentary lifestyle. These types of studies present unique challenges for study design and statistical analysis, both of which need statistical methodological research (Murphy, 2005; Boruvka et al., 2018; Seewald et al., 2019).

Acknowledgments This work was partially supported by the National Institute of Health grant to through grants R01HL 130483 to CD (National Heart, Lung, and Blood Institute), P01AG052352 to AZL (National Institute on Aging) and R01CA227122 to KRE (National Cancer Institute, Office of the Director, Office of Disease Prevention, and Office of Behavioral and Social Sciences Research). The WHI program is funded by the National Heart, Lung, and Blood Institute, National Institutes of Health, US Department of Health and Human Services through 75N92021D00001, 75N92021D00002, 75N92021D00003, 75N92021D00004, 75N92021D 00005. All funding agencies were not involved in the design of the study, data collection, analysis, data interpretation, or manuscript writing.

Data Availability The data used in this paper are available upon request. Since the data was collected in an ancillary study of the Women's Health Initiative (WHI), data sharing follows the WHI's established policies and procedures (https://www.whi.org/md/working-with-whi-data). Requests can be made to the WHI Publications and Presentations Committee directly or through contacting the corresponding author.

References

Ainsworth, B. E., Haskell, W. L., Herrmann, S. D., Meckes, N., Bassett, D. R., Tudor-Locke, C., Greer, J. L., Vezina, J., Whitt-Glover, M. C., & Leon, A. S. (2011). 2011 compendium of physical activities: A second update of codes and met values. *Medicine and Science in Sports and Exercise, 43*(8), 1575–1581. PMCID: N/A (not peer-reviewed).

Ainsworth, B. E., Caspersen, C. J., Matthews, C. E., Mâsse, L. C., Baranowski, T., & Zhu, W. (2012). Recommendations to improve the accuracy of estimates of physical activity derived from self report. *Journal of Physical Activity & Health, 9*(Supp 1), S76–84. PMCID: PMC3544158.

Aittasalo, M., Vähä-Ypyä, H., Vasankari, T., Husu, P., Jussila, A. M., & Sievänen, H. (2015). Mean amplitude deviation calculated from raw acceleration data: A novel method for classifying the intensity of adolescents' physical activity irrespective of accelerometer brand. *BMC Sports Science, Medicine and Rehabilitation, 7*(1), 1–7.

Albanes, D., Conway, J. M., Taylor, P. R., Moe, P. W., & Judd, J. (1990). Validation and comparison of eight physical activity questionnaires. *Epidemiology, 1*(1), 65–71. PMCID: N/A (precedes mandate).

Anderson, G., Manson, J., & Wallace, R. (2003). Implementation of the WHI Study design. *Annals of Epidemiology, 13*, S5–S17. PMCID: N/A (precedes mandate).

Bai, J., Di, C., Xiao, L., Evenson, K. R., LaCroix, A. Z., Crainiceanu, C. M., & Buchner, D. M. (2016). An activity index for raw accelerometry data and its comparison with other activity metrics. *PLoS One, 11*(8), e0160644.

Bai, J., Goldsmith, A., Caffo, B., Glass, T., & Crainiceanu, C. (2012). Movelets: A dictionary of movement. *Electronic Journal of Statistics, 6*, 559–578.

Bai, J., He, B., Shou, H., Zipunnikov, V., Glass, T. A., & Crainiceanu, C. M. (2013). Normalization and extraction of interpretable metrics from raw accelerometry data. *Biostatistics, 15*(1), 102–116.

Bellettiere, J., LaMonte, M. J., Evenson, K. R., Rillamas-Sun, E., Kerr, J., Lee, I.-M., Di, C., Rosenberg, D. E., Stefanick, M. L., Buchner, D. M., et al. (2019). Sedentary behavior and cardiovascular disease in older women: The OPACH Study. *Circulation, 139*(8), 1036–1046.

Bellettiere, J., Nakandala, S., Tuz-Zahra, F., Winkler, E. A., Hibbing, P. R., Healy, G. N., Dunstan, D. W., Owen, N., Greenwood-Hickman, M. A., Rosenberg, D. E., et al. (2022). Chap-adult: A reliable and valid algorithm to classify sitting and measure sitting patterns using data from hip-worn accelerometers in adults aged 35+. *Journal for the Measurement of Physical Behaviour, 5*(4), 215–223.

Boruvka, A., Almirall, D., Witkiewitz, K., & Murphy, S. A. (2018). Assessing time-varying causal effect moderation in mobile health. *Journal of the American Statistical Association, 113*, 1112–1121.

Buchner, D. M., Rillamas-Sun, E., Di, C., LaMonte, M. J., Marshall, S. W., Hunt, J., Zhang, Y., Rosenberg, D. E., Lee, I.-M., Evenson, K. R., et al. (2017). Accelerometer-measured moderate to vigorous physical activity and incidence rates of falls in older women. *Journal of the American Geriatrics Society, 65*(11), 2480–2487.

Butera, N. M., Li, S., Evenson, K. R., Di, C., Buchner, D. M., LaMonte, M. J., LaCroix, A. Z., & Herring, A. (2019). Hot deck multiple imputation for handling missing accelerometer data. *Statistics in Biosciences, 11*(2), 422–448.

Carroll, R., Ruppert, D., Stefanski, L., & Crainiceanu, C. (2006). *Measurement error in nonlinear models: A modern perspective*. CRC Press.

Choi, L., Liu, Z., Matthews, C. E., & Buchowski, M. S. (2011). Validation of accelerometer wear and nonwear time classification algorithm. *Medicine and Science in Sports and Exercise, 43*(2), 357.

Choi, L., Ward, S. C., Schnelle, J. F., & Buchowski, M. S. (2012). Assessment of wear/nonwear time classification algorithms for triaxial accelerometer. *Medicine and Science in Sports and Exercise, 44*(10), 2009.

Crainiceanu, C., Staicu, A., & Di, C. (2009). Generalized multilevel functional regression. *Journal of the American Statistical Association, 104*(488), 1550–1561. PMCID: PMC2897156.

Dishman, R. K., Heath, G. W., & Lee, I.-M. (2012). *Physical activity epidemiology*. Human Kinetics.

Evenson, K. R., Wen, F., Herring, A. H., Di, C., LaMonte, M. J., Tinker, L. F., Lee, I.-M., Rillamas-Sun, E., LaCroix, A. Z., & Buchner, D. M. (2015). Calibrating physical activity intensity for hip-worn accelerometry in women age 60 to 91 years: The women's health initiative OPACH calibration study. *Preventive Medicine Reports, 2*, 750–756.

Evenson, K. R., Bellettiere, J., Cuthbertson, C. C., Di, C., Dushkes, R., Howard, A. G., Parada Jr., H., Schumacher, B. T., Shiroma, E. J., Wang, G., et al. (2021). Cohort profile: The women's health accelerometry collaboration. *BMJ Open, 11*(11), e052038.

Ferrari, P., Friedenreich, C., & Matthews, C. E. (2007). The role of measurement error in estimating levels of physical activity. *American Journal of Epidemiology, 166*(7), 832–840. PMCID: N/A (precedes mandate).

Freedson, P. S., Melanson, E., & Sirard, J. (1998). Calibration of the computer science and applications, INC. accelerometer. *Medicine and Science in Sports and Exercise, 30*(5), 777–781.

Freedson, P. S., Lyden, K., Kozey-Keadle, S., & Staudenmayer, J. (2011). Evaluation of artificial neural network algorithms for predicting mets and activity type from accelerometer data: Validation on an independent sample. *Journal of Applied Physiology, 111*(6), 1804–1812.

Goldsmith, J., Bobb, J., Crainiceanu, C. M., Caffo, B., & Reich, D. (2011). Penalized functional regression. *Journal of Computational and Graphical Statistics, 20*(4), 830–851.

Goldsmith, J., Liu, X., Jacobson, J., & Rundle, A. (2016). New insights into activity patterns in children, found using functional data analyses. *Medicine and Science in Sports and Exercise, 48*(9), 1723.

Greenwood-Hickman, M. A., Nakandala, S., Jankowska, M. M., Rosenberg, D. E., Tuz-Zahra, F., Bellettiere, J., Carlson, J. A., Hibbing, P. R., Zou, J., Lacroix, A. Z., et al. (2021). The CNN hip accelerometer posture (chap) method for classifying sitting patterns from hip accelerometers: A validation study. *Medicine and Science in Sports and Exercise, 53*(11), 2445.

Hardeman, W., Houghton, J., Lane, K., Jones, A., & Naughton, F. (2019). A systematic review of just-in-time adaptive interventions (JITAIs) to promote physical activity. *International Journal of Behavioral Nutrition and Physical Activity, 16*, 4.

Jain, P., Bellettiere, J., Glass, N., LaMonte, M. J., Di, C., Wild, R. A., Evenson, K. R., & LaCroix, A. Z. (2021). The relationship of accelerometer-assessed standing time with and without ambulation and mortality: The WHI OPACH study. *The Journals of Gerontology: Series A, 76*(1), 77–84.

Janz, K. (2006). Physical activity in epidemiology: Moving from questionnaire to objective measurement. *British Journal of Sports Medicine, 40*(3), 191–192. PMCID: N/A (precedes mandate).

John, D., Tang, Q., Albinali, F., & Intille, S. (2019). An open-source monitor-independent movement summary for accelerometer data processing. *Journal for the Measurement of Physical Behaviour, 2*(4), 268–281.

Karas, M., Bai, J., Straczkiewicz, M., Harezlak, J., Glynn, N. W., Harris, T., Zipunnikov, V., Crainiceanu, C., & Urbanek, J. K. (2019). Accelerometry data in health research: Challenges and opportunities. *Statistics in Biosciences, 11*(2), 210–237.

Kerr, J., Patterson, R. E., Ellis, K., Godbole, S., Johnson, E., Lanckriet, G., & Staudenmayer, J. (2016). Objective assessment of physical activity: Classifiers for public health. *Medicine and Science in Sports and Exercise, 48*(5), 951.

Kohl, H. W. (2001). Physical activity and cardiovascular disease: Evidence for a dose response. *Medicine and Science in Sports and Exercise, 33*(6), S472–S483. PMCID: N/A (precedes mandate).

LaCroix, A. Z., Rillamas-Sun, E., Buchner, D., Evenson, K. R., Di, C., Lee, I.-M., Marshall, S., LaMonte, M. J., Hunt, J., Tinker, L. F., et al. (2017). The objective physical activity and cardiovascular disease health in older women (OPACH) Study. *BMC Public Health, 17*(1), 192.

LaCroix, A. Z., Bellettiere, J., Rillamas-Sun, E., Di, C., Evenson, K. R., Lewis, C. E., Buchner, D. M., Stefanick, M. L., Lee, I.-M., Rosenberg, D. E., et al. (2019). Association of light physical activity measured by accelerometry and incidence of coronary heart disease and cardiovascular disease in older women. *JAMA Network Open, 2*(3), e190419–e190419.

LaMonte, M. J., Lewis, C. E., Buchner, D. M., Evenson, K. R., Rillamas-Sun, E., Di, C., Lee, I.-M., Bellettiere, J., Stefanick, M. L., Eaton, C.B., et al. (2017). Both light intensity and moderate-to-vigorous physical activity measured by accelerometry are favorably associated with cardiometabolic risk factors in older women: The Objective Physical Activity and Cardiovascular Health (OPACH) Study. *Journal of the American Heart Association, 6*(10), e007064.

LaMonte, M. J., Buchner, D. M., Rillamas-Sun, E., Di, C., Evenson, K. R., Bellettiere, J., Lewis, C. E., Lee, I.-M., Tinker, L. F., Seguin, R., et al. (2018). Accelerometer-measured physical activity and mortality in women aged 63 to 99. *Journal of the American Geriatrics Society, 66*(5), 886–894.

Matthews, C. E., Moore, S. C., George, S. M., Sampson, J., & Bowles, H. R. (2012). Improving self-reports of active and sedentary behaviors in large epidemiologic studies. *Exercise and Sport Sciences Reviews, 40*(3), 118–126. PMCID: PMC3388604.

McConnell, M. V., Turakhia, M. P., Harrington, R. A., King, A. C., & Ashley, E. A. (2018). Mobile health advances in physical activity, fitness, and atrial fibrillation: Moving hearts. *Journal of the American College of Cardiology, 71*(23), 2691–2701.

McTiernan, A., Kooperberg, C., White, E., Wilcox, S., Coates, R., Adams-Campbell, L. L., Woods, N., & Ockene, J. (2003). Recreational physical activity and the risk of breast cancer in postmenopausal women. *JAMA: The Journal of the American Medical Association, 290*(10), 1331–1336. PMCID: N/A (precedes mandate).

Mekary, R. A., Willett, W. C., Hu, F. B., & Ding, E. L. (2009). Isotemporal substitution paradigm for physical activity epidemiology and weight change. *American Journal of Epidemiology, 170*(4), 519–527.

Monninkhof, E. M., Elias, S. G., Vlems, F. A., van der Tweel, I., Schuit, A. J., Voskuil, D. W., van Leeuwen, F. E., et al. (2007). Physical activity and breast cancer: A systematic review. *Epidemiology, 18*(1), 137–157. PMCID: N/A (precedes mandate).

Murphy, S. A. (2005). An experimental design for the development of adaptive treatment strategies. *Statistics in Medicine, 24*, 1455–1481.

Neishabouri, A., Nguyen, J., Samuelsson, J., Guthrie, T., Biggs, M., Wyatt, J., Cross, D., Karas, M., Migueles, J. H., Khan, S., et al. (2022). Quantification of acceleration as activity counts in actigraph wearables. *Scientific Reports, 12*(1), 11958.

Neuhouser, M. L., Di, C., Tinker, L. F., Thomson, C., Sternfeld, B., Mossavar-Rahmani, Y., Stefanick, M. L., Sims, S., Curb, J. D., Lamonte, M., et al. (2013). Physical activity assessment: Biomarkers and self-report of activity-related energy expenditure in the WHI. *American Journal of Epidemiology, 177*(6), 576–585. PMCID: PMC3626043.

Physical Activity Guidelines Advisory Committee et al. (2018). 2018 physical activity guidelines advisory committee scientific report.

Piercy, K. L., Troiano, R. P., Ballard, R. M., Carlson, S. A., Fulton, J. E., Galuska, D. A., George, S. M., & Olson, R. D. (2018). The physical activity guidelines for Americans. *JAMA, 320*(19), 2020–2028.

Prentice, R. (2010). Chronic disease prevention research methods and their reliability, with illustrations from the women's health initiative. *Journal of the American Statistical Association, 105*(492), 1431–1443.

Prince, S., Adamo, K., Hamel, M., Hardt, J., Gorber, S., & Tremblay, M. (2008). A comparison of direct versus self-report measures for assessing physical activity in adults: A systematic review. *International Journal of Behavioral Nutrition and Physical Activity, 5*(1), 56.

Ramsay, J. O., & Silverman, B. W. (2022). *Functional data analysis* (2nd ed.).

Rosenberg, D., Godbole, S., Ellis, K., Di, C., LaCroix, A. Z., Natarajan, L., & Kerr, J. (2017). Classifiers for accelerometer-measured behaviors in older women. *Medicine and Science in Sports and Exercise, 49*(3), 610.

Ruppert, D., Wand, M. P., & Carroll, R. J. (2003). *Semiparametric regression*. Cambridge University Press. PMCID: N/A (precedes mandate).

Schmidt, M. D., Freedson, P. S., & Chasan-Taber, L. (2003). Estimating physical activity using the CSA accelerometer and a physical activity log. *Medicine and Science in Sports and Exercise, 35*(9), 1605–1611. PMCID: N/A (precedes mandate).

Seewald, N. J., Smith, S. N., Lee, A. J., Klasnja, P., & Murphy, S. A. (2019). Practical considerations for data collection and management in mobile health micro-randomized trials. *Statistics in Biosciences, 11*(2), 355–370.

Stamatakis, E., Rogers, K., Ding, D., Berrigan, D., Chau, J., Hamer, M., & Bauman, A. (2015). All-cause mortality effects of replacing sedentary time with physical activity and sleeping using an isotemporal substitution model: A prospective study of 201,129 mid-aged and older adults. *International Journal of Behavioral Nutrition and Physical Activity, 12*(1), 121.

Staudenmayer, J., Pober, D., Crouter, S., Bassett, D., & Freedson, P. (2009). An artificial neural network to estimate physical activity energy expenditure and identify physical activity type from an accelerometer. *Journal of Applied Physiology, 107*(4), 1300–1307. PMCID: PMC2763835.

Staudenmayer, J., Zhu, W., & Catellier, D. J. (2012). Statistical considerations in the analysis of accelerometry-based activity monitor data. *Medicine & Science in Sports & Exercise, 44*(suppl 1), S61–7. PMCID: PMC3388604.

Stefanick, M. L., King, A. C., Mackey, S., Tinker, L. F., Hlatky, M. A., LaMonte, M. J., Bellettiere, J., Larson, J. C., Anderson, G., Kooperberg, C. L., et al. (2021). Women's health initiative strong and healthy pragmatic physical activity intervention trial for cardiovascular disease prevention: Design and baseline characteristics. *The Journals of Gerontology: Series A, 76*(4), 725–734.

Strath, S. J., Kate, R. J., Keenan, K. G., & Swartz, A. M. (2012). Estimating type and energy cost of physical activity using multiple accelerometers and machine learning techniques. In *Medicine and science in sports and exercise* (Vol. 44, pp. 218–218). Lippincott Williams & Wilkins.

Swihart, B. J., Goldsmith, J., & Crainiceanu, C. M. (2014). Restricted likelihood ratio tests for functional effects in the functional linear model. *Technometrics, 56*(4), 483–493.

Taylor, R. S., Brown, A., Ebrahim, S., Jolliffe, J., Noorani, H., Rees, K., Skidmore, B., Stone, J. A., Thompson, D. R., & Oldridge, N. (2004). Exercise-based rehabilitation for patients with coronary heart disease: Systematic review and meta-analysis of randomized controlled trials. *The American Journal of Medicine, 116*(10), 682–692. PMCID: N/A (precedes mandate).

Torres, A., Tennant, B., Ribeiro-Lucas, I., Vaux-Bjerke, A., Piercy, K., & Bloodgood, B. (2018). Umbrella and systematic review methodology to support the 2018 physical activity guidelines advisory committee. *Journal of Physical Activity and Health, 15*(11), 805–810.

Troiano, R. P., Berrigan, D., Dodd, K. W., Mâsse, L. C., Tilert, T., McDowell, M., et al. (2008). Physical activity in the united states measured by accelerometer. *Medicine and Science in Sports and Exercise, 40*(1), 181. PMCID: N/A (no NIH support cited).

Van Hees, V. T., Gorzelniak, L., Dean León, E. C., Eder, M., Pias, M., Taherian, S., Ekelund, U., Renström, F., Franks, P. W., Horsch, A., et al. (2013). Separating movement and gravity components in an acceleration signal and implications for the assessment of human daily physical activity. *PloS One, 8*(4), e61691.

Wang, G., Wu, S., Evenson, K. R., Kang, I., LaMonte, M. J., Bellettiere, J., Lee, I.-M., Howard, A. G., LaCroix, A. Z., & Di, C. (2022). Calibration of an accelerometer activity index among older women and its association with cardiometabolic risk factors. *Journal for the Measurement of Physical Behavior, 5*(3), 145–155.

Warburton, D. E., Nicol, C. W., & Bredin, S. S. (2006). Health benefits of physical activity: The evidence. *Canadian Medical Association Journal, 174*(6), 801–809. PMCID: N/A (precedes mandate).

Wijndaele, K., Westgate, K., Stephens, S. K., Blair, S. N., Bull, F.C., Chastin, S. F., Dunstan, D. W., Ekelund, U., Esliger, D. W., Freedson, P. S., et al. (2015). Utilization and harmonization of adult accelerometry data: Review and expert consensus. *Medicine and Science in Sports and Exercise, 47*(10), 2129.

Wolin, K., Yan, Y., Colditz, G., & Lee, I. (2009). Physical activity and colon cancer prevention: A meta-analysis. *British Journal of Cancer, 100*(4), 611–616. PMCID: PMC2653744 .

Xu, S. Y., Nelson, S., Kerr, J., Godbole, S., Johnson, E., Patterson, R. E., Rock, C. L., Sears, D. D., Abramson, I., & Natarajan, L. (2019). Modeling temporal variation in physical activity using functional principal components analysis. *Statistics in Biosciences, 11*(2), 403–421.

Zhang, Y., Li, H., Keadle, S. K., Matthews, C. E., & Carroll, R. J. (2019). A review of statistical analyses on physical activity data collected from accelerometers. *Statistics in Biosciences, 11*(2), 465–476.

Policy Learning for Individualized Treatment Regimes on Infinite Time Horizon

Wenzhuo Zhou, Yuhan Li, and Ruoqing Zhu

Abstract With the recent advancements of technology in facilitating real-time monitoring and data collection, "just-in-time" interventions can be delivered via mobile devices to achieve both real-time and long-term management and control. Reinforcement learning formalizes such mobile interventions as a sequence of decision rules and assigns treatment arms based on the user's status at each decision point. In practice, real applications concern a large number of decision points beyond the time horizon of the currently collected data. This usually refers to reinforcement learning in the infinite horizon setting, which becomes much more challenging. This article provides a selective overview of some methodologies on this topic. We discuss their modeling framework, generalizability, and interpretability and provide some use case examples. Some future research directions are discussed in the end.

Keywords Individualized treatment rule · Infinite horizon · Reinforcement learning · R package for personalized medicine

1 Background and Challenges

The main purpose of this chapter is to review the statistical methods developed under the infinite horizon setting. Under a finite horizon setting, the optimal treatment regimes can only be estimated within a fixed period with limited decision points

---

Wenzhuo Zhou and Yuhan Li authors contributed equally to this work.

W. Zhou
Department of Statistics, University of California Irvine, Irvine, CA, USA
e-mail: wenzhuz3@uci.edu

Y. Li · R. Zhu (✉)
Department of Statistics, University of Illinois Urbana-Champaign, Champaign, IL, USA
e-mail: yuhanli8@illinois.edu; rqzhu@illinois.edu

Y. Zhao, D.-G. Chen (eds.), *Statistics in Precision Health*, ICSA Book Series in Statistics, https://doi.org/10.1007/978-3-031-50690-1_4

(Chakraborty & Murphy, 2014). However, with the rapid development of sensor technology and the increasing popularity of various wearable devices, a tremendous amount of personal health data are available to healthcare providers and researchers (Steinhubl et al., 2015). For example, such devices may record personal health information over an extremely long period (Silva et al., 2015), making it possible to construct personalized treatment plans for managing chronic diseases and other health issues over an infinite horizon with a large number of decision points. Hence, research on mobile health (mHealth) has emerged (Rehg et al., 2017) in recent years. The focus is to use mobile technologies to improve healthcare quality regardless of location and time. Research and practice in this field face challenges that span many disciplines and fields, including but not limited to engineering, computer science, statistics, bioinformatics, medicine, health policy, and privacy.

To date, mHealth has been used extensively in managing stress, treating depression, and other chronic diseases such as diabetes and cardiovascular diseases. It plays an essential role in assisting healthcare providers to better monitor and treat patients. For instance, the AutoSense suite can detect physiological signals (ECG, respiration, etc.) at a high frequency (approximately 5 million samples per day) and allow machine learning models to convert the raw data into biomarkers of health, behavior, and environment with smartphone software (Sarker et al., 2017). Interventions can be delivered to users promptly if these biomarkers suggest an excessive risk of stress or other health conditions. However, certain limitations exist in the current development of mHealth. Most existing methods consider only the finite horizon setting and are not directly applicable to the infinite horizon setting. Additionally, such methods may require strong assumptions that are difficult to justify or validate in real-life (Kosorok & Laber, 2019). Recent literature mainly borrows tools developed in reinforcement learning (Sutton and Barto, 2018) to address the infinite horizon problem, although certain limitations still exist, such as data coverage, function approximation, Markovian, and many others Fujimoto et al. (2019), Zhan et al. (2022), Cheung et al. (2020).

This book chapter mainly reviews some statistical methods for estimating the optimal treatment regimes under an infinite horizon setting. We will introduce several popular approaches and related concepts, such as Q-learning, temporal difference learning, and the residual gradient algorithm. We also discuss recent developments motivated by the limitations of these classical methods, including the greedy gradient Q-learning, V-learning, and a newly proposed temporal consistent learning framework. Our review is selective in the sense that we focus mostly on methods that can be directly applied to an already collected dataset from an observational study. Hence, they are offline and off-policy. Extensions to online settings are briefly discussed. Some examples are used to demonstrate existing applications. We conclude this chapter with a discussion of open issues on this topic.

2 Preliminaries

2.1 A Motivating Example

Glucose management for patients with diabetes is one of the most widely used applications of mHealth. By continuously monitoring the glucose level, food intake, and physiological information, a series of Just-in-time interventions, such as insulin injection, can be delivered to patients to improve long-term benefits (Muralidharan et al., 2017). An application example is the OhioT1DM study (Marling & Bunescu, 2020), which contains 12 patients with Type 1 Diabetes. This dataset includes a continuous glucose monitoring (CGM) of blood glucose levels every 5 minutes over eight weeks for each patient, along with self-reported times of meal, exercise, sleep, work, and stress. The time points of insulin doses are also recorded. Figure 1 provides a snapshot of the fluctuation of glucose level, insulin injections, meals, exercise, and heart rate of a patient during a 100 hour time interval.

Patients may respond differently to insulin injections due to differences in age, gender, and other personal activity and health measurements. We want to develop a framework to understand the dynamic interactions between insulin injection and glucose level. This also allows us to determine a personalized treatment policy that could best control the glucose level of each individual.

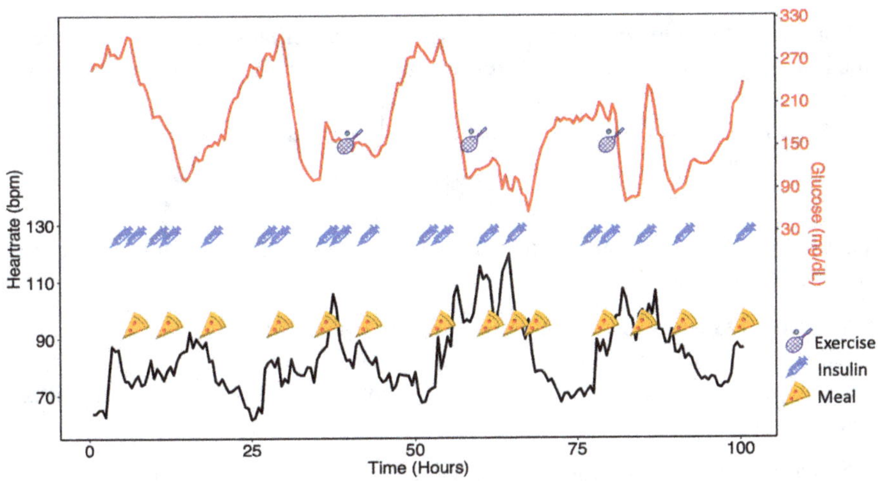

Fig. 1 OhioT1DM Data: A longitudinal observation of a patient

2.2 Notation

To better formulate this problem, we introduce some important notations and concepts. Denote $S^t \in \mathbb{R}^p$ as the patient information collected at time t, which may include all related covariates such as age, gender, and lab results. The values of these covariates are likely to change between time points, reflecting the patient's instantaneous condition. We denote $A^t \in \mathcal{A}$ as the treatment assigned to the patient at time t based on the information S^t, which may include the drug choice and dosing level. Here, $\mathcal{A}$ could be either a discrete or continuous space for possible treatments. Hence, the collected dataset is $\{S_i^0, A_i^0, S_i^1, \ldots, S_i^T, A_i^T, S_i^{T+1}\}_{i=0}^n$, which compromises n i.i.d. trajectories with $T + 1$ time points of treatment decisions. We further assume an observed immediate reward R^t that may depend on S^{t+1}, A^t, S^t, measured after choosing treatment A^t at time t. Specifically, it measures how well the patient reacts to treatment A^t as the covariates change from S^t to S^{t+1}. Typically, a larger value of reward means a better treatment outcome.

In the previous motivating example, S^t can be the patient's glucose level, physical activity, age, gender, dietary intake, and other related information contained in the dataset. A^t is the indicator or dosing of insulin injection, and R^t can be defined as a measure of closeness of the patient's glucose level from the optimal glucose level range. Our goal is to maximize the overall reward, to be defined later, for each individual by learning the optimal strategy of insulin injection. We first introduce several essential concepts.

Definition 1 (Markov Property) Suppose $s^t \in \mathcal{S}$ denotes the observed state at time t, $a^t \in \mathcal{A}$ denotes the action taken at time t, then under the Markov property :

$$\begin{aligned} P(S^{t+1} &= s^{t+1} | S^t = s^t, \ldots, S^1 = s^1, A^t = a^t, \ldots, A^0 = a^0) \\ &= P(S^{t+1} = s^{t+1} | S^t = s^t, A^t = a^t). \end{aligned}$$

The Markov property suggests that the current state and action are sufficient to predict the next state.

Definition 2 (Markov Decision Process, MDP) A Markov decision process (MDP) is an environment in which all states satisfy the Markov property, and a decision maker can interact with the environment using actions. Specifically, it is a tuple $< \mathcal{S}, \mathcal{A}, \mathbf{P}, R >$ where

- $\mathcal{S}$ is a set of states called state space, which may be finite or infinite;
- $\mathcal{A}$ is a set of actions called action space, which may be finite or infinite;
- $\mathbf{P}(s^{t+1}, a^t, s^t) = P(S^{t+1} = s^{t+1} | S^t = s^t, A^t = a^t)$ is the probability that action a^t in state s^t leads to state s^{t+1} in the next time point;
- R^t is the immediate reward received after receiving a^t at s^t. It may also use information from s^{t+1}.

MDP is generally assumed for all methods under infinite time horizon. In an MDP, we often assume that a policy or treatment π is used to make the decision. This is a distribution over actions given states

$$\pi(s) = P(A^t = a | S^t = s).$$

To balance the immediate reward and long-term reward, a discount factor $\gamma \in [0, 1]$ is often used. Our goal is to find a policy π that will maximize some cumulative version of rewards over time t:

$$\mathbb{E}\left[\sum_{k=0}^{\infty} \gamma^k R^{t+k}\right],$$

where we choose $a^t = \pi(s^t)$. Hence, when γ is small, we only focus on the eventual reward when t approaches infinity, while when γ approaches 1, after proper normalization, we are interested in the long-term average reward (Meyn, 1997; Liao et al., 2020). Based on these definitions, we introduce the value function and the Q-function, which are fundamental concepts in a reinforcement learning task.

Definition 3 (Value Function) The state-value function $V_t^\pi(s)$ is the expected cumulative reward starting from state s, and then following policy π

$$V_t^\pi(s) = \mathbb{E}\left[\sum_{k=0}^{\infty} \gamma^k R^{t+k} | S^t = s\right].$$

Definition 4 (Action Value Function) The action-value function $Q_t^\pi(s, a)$ is the expected cumulative reward starting from state s, taking action a and then following policy π

$$Q_t^\pi(s, a) = \mathbb{E}\left[\sum_{k=0}^{\infty} \gamma^k R^{t+k} | S^t = s, A^t = a\right].$$

Based on these definitions, we can treat both V-function and Q-function as estimates of how good a policy is for a patient in any given state. The only difference is whether we know the action at time t. In an MDP, $V_t^\pi(s)$ and $Q_t^\pi(s, a)$ do not depend on t and are finite when $\gamma < 1$ (Hinderer & Waldmann, 2005). By solving a policy that maximizes these quantities, we essentially achieve the goal of personalized medicine. However, this is not trivial considering the dynamics over a long period, which can be difficult to model. Hence, the *Bellman equation* becomes an important tool. The following are two versions of the Bellman equation corresponding to the value and Q-functions. Both of them use the property that they can be written in an integrative fashion in the MDP setting:

$$V^{\pi}(s) = \mathbb{E}[R^t + \gamma V^{\pi}(S^{t+1})|S^t = s],$$

$$Q^{\pi}(s, a) = \mathbb{E}[R^t + \gamma Q^{\pi}(S^{t+1}, A^{t+1})|S^t = s, A^t = a].$$

The best treatment policy can be defined using either the value function, with $V^*(s) = \max_{\pi} V^{\pi}(s)$, or the Q-function with $Q^*(s, a) = \max_{\pi} Q^{\pi}(s, a)$. The two are connected via $V^*(s) = \max_a Q^*(s, a)$. Additionally, $Q^*(s, a)$ is unique and must satisfy the well-known *Bellman optimality equation* (Puterman, 2014):

$$Q^*(s, a) = \mathbb{E}_{S^{t+1}|s,a}\left[R^t + \gamma \max_{a'} Q^*(S^{t+1}, a')|S^t = s, A^t = a\right]. \tag{1}$$

We refer to the policy π^* that attains these maximums as the optimal policy, hence $Q^{\pi^*}(s, a) = Q^*(s, a)$. Equivalently, we can define the Bellman operator $\mathcal{B}$ of the value function as

$$\mathcal{B}V(s) = \max_a \mathbb{E}_{S^{t+1}|s,a}\left[R^t + \gamma V(S^{t+1})|S^t = s, A^t = a\right]. \tag{2}$$

Then $V^{\pi^*}(s)$ is the fixed point of $\mathcal{B}$, and π^* can be solved correspondingly.

Before concluding this section, we want to remark that the decision maker's behavior in MDP during the data generating process is also crucial. Some conditions are needed to ensure that we can adequately estimate the treatment effects and make casual conclusions (Hernán & Robins, 2010). However, due to the scope of this book chapter, we do not discuss causality issues in detail.

- **Positivity:** $P(A^t = a^t|S^t = s^t) > 0$ for all $a^t \in \mathcal{A}$, $s^t \in \mathcal{S}$, and all t.
- **Consistency:** Let $R(a^t)$ denote the potential outcome at time t if receiving action a^t, then $R^t = R(a^t)$ if a^t is actually used.
- **Strong Ignorability:** $\{R(a^t) : a^t \in \mathcal{A}\} \perp\!\!\!\perp A^t \mid S^t$ for all t.

As a brief interpretation, we note that the positivity assumption ensures that all actions in the action space could be chosen at each time t with a probability larger than 0. This allows the comparison among different actions. The consistency assumption guarantees that the estimates are valid, and the strong ignorability assumption suggests that there are no unmeasured confounders in the dataset.

3 Q-learning for Finite Horizon

Q-learning is one of the most important breakthroughs in reinforcement learning (Sutton & Barto, 2018). In the context of personalized medicine, it has been widely used in estimating the optimal treatment regime (Murphy, 2005; Robins, 2004; Zhao et al., 2009; Nahum-Shani et al., 2012; Kosorok & Moodie, 2015) of finite-horizon problems where the number of decision stages is finite. Q-learning is also a popular approach in infinite horizon settings and achieves success in mobile health applications. Dynamic programming is a major technique to solve Q-learning problems without explicitly modeling the entire environment dynamics. In the following, we shall introduce Q-learning and its applications in the finite horizon setting. Then, we review some extensions of Q-learning in the infinite-horizon settings.

3.1 Additional Notation

In finite-horizon settings, data concerning personalized medicine are usually collected from a multi-stage clinical trial or longitudinal observational studies on the disease of interest (Clifton & Laber, 2020). When the number of stages is small, the Markov property is unnecessary. However, standard causal assumptions highlighted previously (positivity, consistency, and strong ignorability) are still assumed (Goldberg & Kosorok, 2012) to make estimation and inference valid. An optimal dynamic treatment regime (DTR) aims to find a sequence of decision rules that assign treatments at each stage based on a patient's baseline characteristics and historical information. Some additional notations should be introduced to avoid confusion and allow higher-order dependencies during the decision process.

Suppose we have pre-specified finite T decision points, indexed by $t = 1, 2, \ldots, T$. Note that in this setting, the time index starts with 1 rather than 0, consistent with the literature. We denote $\bar{A}^t = (A^1, A^2, \ldots, A^t)$ as a possible treatment trajectory administered through decision point t, and denote $\bar{S}^t = (S^1, S^2, \ldots, S^t)$ as the historical information of the patient leading up to t. The realized treatment trajectory and historical information are denoted as $\bar{a}^t = (a^1, a^2, \ldots, a^t)$ and $\bar{s}^t = (s^1, s^2, \ldots, s^t)$, respectively. The immediate observed reward R^t is slightly different here because the Markov property is not assumed. It considers all the previous history leading up to t. Another difference is that $\gamma = 1$ under the finite horizon, as the cumulative reward will surely converge. Therefore, the goal is to find the optimal DTR to maximize the cumulative reward from $t = 1$ to T, i.e., $\sum_{t=1}^{T} R^t$.

A dynamic treatment regime, in this case, is defined as $\boldsymbol{\pi} = (\pi_1, \ldots, \pi_T)$, which forms a sequence of policies to treat a patient over time. Here, the decision rules may depend on t, which is different from Markov settings. They can incorporate all information leading up to t. The optimal DTR $\boldsymbol{\pi}^* = (\pi_1^*, \pi_2^*, \ldots, \pi_T^*)$ is naturally defined as the DTR that satisfies, for all $\boldsymbol{\pi}$,

$$\mathbb{E}\left(\sum_{t=1}^{T} R_{\pi_t}^t\right) \leq \mathbb{E}\left(\sum_{t=1}^{T} R_{\pi_t^*}^t\right). \tag{3}$$

3.2 Backward Induction

To estimate $\boldsymbol{\pi}^*$, Q-learning adopts a backward induction mechanism, moving from the final decision stage back to the first stage. We shall describe this process conceptually instead of giving details of the sample version since the finite-horizon setting is not the focus. Essentially, the algorithm is motivated by two facts. First, the best policy that maximizes the potential reward at the final stage can be learned by knowing the covariate at the $(T-1)$th stage. Secondly, the optimal treatment decision can be induced at any given time point by knowing the best decision at the next time point (Zhao et al., 2015). To better illustrate how the algorithm works, we define the following quantities. They are analogs of similar quantities introduced previously. At the final stage T, we have

$$\begin{aligned} Q_T(\bar{s}^T, \bar{a}^T) &= \mathbb{E}\left(\sum_{t=1}^{T} R^t | \bar{S}^T = \bar{s}^T, \bar{A}^T = \bar{a}^T\right), \\ \pi_T^*(\bar{s}^T, \bar{a}^{T-1}) &= \text{argmax}_{a^T} Q_T(\bar{s}^T, \bar{a}^{T-1}, a^T), \\ \text{and} \quad V_T(\bar{s}^T, \bar{a}^{T-1}) &= \max_{a^T} Q_T(\bar{s}^T, \bar{a}^{T-1}, a^T). \end{aligned}$$

For stages $t = T-1, \ldots, 1$, we perform the inductions:

$$\begin{aligned} Q_t(\bar{s}^t, \bar{a}^t) &= \mathbb{E}\left[V_{t+1}(\bar{s}^t, S^{t+1}, \bar{a}^t) | \bar{S}^t = \bar{s}^t, \bar{A}^t = \bar{a}^t\right] \qquad (4) \\ \pi_t^*(\bar{s}^t, \bar{a}^{t-1}) &= \text{argmax}_{a^t} Q_t(\bar{s}^t, \bar{a}^{t-1}, a^t), \\ \text{and} \quad V_t(\bar{s}^t, \bar{a}^{t-1}) &= \max_{a^t} Q_t(\bar{s}^t, \bar{a}^{t-1}, a^t). \end{aligned}$$

To briefly explain these notations, we start from the final stage T and posit a model for $Q_T(\bar{s}^T, \bar{a}^T)$, such as a linear model, SVM, or random forests (Zhao et al., 2009). We treat the observed cumulative reward $\sum_{t=1}^{T} R^t$ as the response, and $\bar{s}^T, \bar{a}^{T-1}$ as the covariates. After obtaining the model estimation using the observed data, we compare different treatment options to obtain $\hat{\pi}_T^*$ for a patient with history $\bar{S}^T = \bar{s}^T, \bar{A}^{T-1} = \bar{a}^{T-1}$ at decision point T. This allows us to estimate the best value function if only the last time-point requires a decision. Similar procedures can be adopted for the previous time-point $T-1$. Note that by Eq. (4), the Q-function can be estimated with the already estimated value function in the next stage. Hence, we can compare and estimate the best treatment $\hat{\pi}_{T-1}^*$ at a previous time point. Recursively performing these operations, the estimated optimal treatment regime is then $\hat{\boldsymbol{\pi}}^* = (\hat{\pi}_1^*, \ldots, \hat{\pi}_T^*)$.

In the finite horizon setting, Q-learning has gained tremendous popularity due to its simplicity of implementation and overall good performance (Kosorok & Laber, 2019). However, it has limitations. For example, it is sensitive to model misspecifications. If the posited model for Q-functions is misspecified, the performance may suffer significantly. This is because the misspecification errors at each stage will be backpropagated and accumulated back to the first stage. Some alternatives have been proposed, such as Advantage Learning (A-learning, Murphy, 2003), which estimates the optimal regime by modeling the difference in outcomes between two treatment options (Blatt et al., 2004). Thus, it does not require the entire Q-function to be specified, making it more robust to model misspecification (Schulte et al., 2014). Other methods include the robust Q-learning (Ertefaie et al., 2021) that allows estimating nuisance parameters using data-adaptive techniques to address residual confounding and efficiency loss.

The backward induction algorithm for finite-horizon Q-learning can be extended to infinite horizon settings in principle. This again requires accurate estimations of the Q-function. Hence, it is more prevalent in an environment where we can quickly and easily obtain new samples through experiments under a specific policy. Such a strategy may not be realistic for a practical situation in medical science. To deal with this issue, in Sect. 5, we will review other Q-learning algorithms in infinite horizon settings, often referred to as the infinite-horizon Q-learning, first proposed by Watkins and Dayan (1992).

4 Temporal Difference Learning

There are three representative approaches for policy learning: dynamic programming, Monte-Carlo, and Temporal Difference (TD) learning (Sutton, 1988). Among them, TD learning is a classical and powerful approach and often serves as the foundation of modern reinforcement learning. It has achieved great success in theoretical foundations and real applications (Dann et al., 2014). Instead of modeling the entire dynamic for policy learning, TD learning is a model-free approach that can directly perform policy evaluation and optimization given any unknown transition Markov kernel. The advantage of a model-free approach is the high sample efficiency, which is highly desirable in policy learning.

4.1 On-Policy vs. Off-Policy Learning

It would be helpful to clarify the difference between two notions, on-policy and off-policy since a specific TD learning approach depends on the setting. The difference can be roughly understood as if the target of our estimation is based on the observed treatment decision behavior or some other treatment strategy, e.g., the optimal one. For example, the Q-learning algorithms we introduced previously are off-policy learning. Their estimation target is the Q-function under the optimal treatment decision function π, which is rarely realized in an observational study. In contrast, another algorithm called State–Action–Reward–State–Action (SARSA) is interested in learning the Q-function if we follow the current observed behavior of treatment decision-making. Hence, it is evident that SARSA cannot be used to learn the optimal policy unless the patient is already implementing it in practice. Off-policy is likely the dominating case in medical-related studies since patients and doctors usually do not know the optimal policy in advance. However, in online gaming, the on-policy setting is quite frequent since one can manipulate machines to implement any new decision rule. This allows on-policy algorithms such as SARSA to update the policy and the corresponding Q-functions progressively. But in offline settings where data are already collected before any analysis, Q-learning can be more useful than SARSA. We shall discuss the online learning setting in Sect. 8.

One difficulty involved in off-policy learning is that one must account for the change of distribution of states S (distribution shift) when the target policy being estimated is no longer the one being used. This is especially true for offline settings since the observed (or implemented) policy can not be changed once data have been collected. Fundamentally, this is an observational study problem that often appears in statistics. One may expect some form of propensity score adjusting approaches to be implemented to handle the mismatch of the distributions. These ideas will be illustrated in detail in Sect. 4.3. In the following, we shall first focus on the on-policy setting to introduce the intuition behind TD learning and then discuss its off-policy generalization via importance sampling.

4.2 On-Policy TD Learning

In this section, we introduce the on-policy TD learning, where the data is collected in an online fashion. In the on-policy setting, the agent can directly interact with the data-generating environment and continuously collect data from the environment online. This is distinguished from the off-policy setting, where the data is pre-collected, and the agent cannot interact with the environment. Typically, we observe a single sequence of streaming data consisting of the current state, next state, current action, and immediate reward at each time-point t.

The reinforcement learning problem can also be viewed as a traditional regression framework. The goal of TD learning is to minimize the discrepancy between the true value function V^{π} and its function approximation counterpart V_{θ}^{π}. One of the most commonly used loss functions for measuring the discrepancy is the squared error (SE),

$$\mathrm{SE}_t(\theta) = \left[V_{\theta}^{\pi}\left(s^t\right) - V^{\pi}\left(s^t\right)\right]^2. \tag{5}$$

Using a stochastic gradient descent with step size α, the parameter update can be understood as,

$$\begin{aligned} \theta_{t+1} &= \theta_t - \alpha \nabla\, \mathrm{SE}_t(\theta) \\ &= \theta_t - \alpha \left[V_{\theta}^{\pi}\left(s^t\right) - V^{\pi}\left(s^t\right)\right] \nabla_{\theta_t} V_{\theta_t}\left(s^t\right). \end{aligned} \tag{6}$$

Still, Eq. (6) can not be directly computed as the true value function V^{π} is unknown. One natural solution is to collect Monte-Carlo samples to approximate V^{π}. In particular, one can repeat the experiments by implementing the policy π and collecting a total of m sample trajectories given state s^t. This is likely only possible in an online setting, as we could directly interact with the environment and collect the data. If we denote the empirical discounted sum of rewards from state s^t as $\{V_i^{\pi}(s^t)\}_{i=1}^m$, then Eq. (5) is feasible to be optimized and reformed to be

$$\text{SE-MC}_t(\theta) = \frac{1}{m}\sum_{i=1}^{m}\left[V_{\theta}^{\pi}(s^t) - V_i^{\pi}(s^t)\right]^2. \tag{7}$$

However, Sutton (1988) indicates that $V_i^{\pi}(s^t)$ can be difficult to collect. To explicitly perform this, the experiments have to be repeated m times for each state which is only valid in simulated environments but not in practice due to the cost and ethics concerns, especially for medical-related studies. Another main drawback of this Monte-Carlo-based approach is the curse of high variance. The variance of the samples $\{V_i^{\pi}(s^t)\}_{i=0}^m$ tends to be very large for a large number of decision time points. This is particularly severe in infinite-horizon settings.

A stable and reliable alternative is needed. Sutton (1988) propose to view the true value function $V^{\pi}(s^t)$ at the decision step t as the composition of the immediate reward R^t and the discounted value function approximation $\gamma V_{\theta}^{\pi}(s^{t+1})$ at step $t+1$. This composition is also called the value function one-step prediction, i.e.,

$$V^{\pi}\left(s^t\right) \approx R^t + \gamma V_{\theta}^{\pi}\left(s^{t+1}\right). \tag{8}$$

The main idea of this update is motivated by the well-known Bellman equation. As the approximation of the true value function is obtained, one can proceed to minimize equation (5) with the approximation form Eq. (8). The gradient descent can be calculated as follows,

$$\begin{aligned}\theta_{t+1} &= \theta_t - \alpha \nabla \, \text{SE}_t(\theta) \\ &\approx \theta_t - \alpha\left[V_{\theta_t}^{\pi}\left(s^t\right) - R^t - \gamma V_{\theta_t}^{\pi}\left(s^{t+1}\right)\right]\nabla_{\theta_t} V_{\theta_t}^{\pi}\left(s^t\right),\end{aligned} \tag{9}$$

where θ_t is the updated parameter at step t and α is the learning rate. Here the term

$$\delta^t = V_{\theta_t}^{\pi}\left(s^t\right) - \left[R^t + \gamma V_{\theta_t}^{\pi}\left(s^{t+1}\right)\right]$$

is called temporal difference error, and $e_t := \nabla_{\theta_t} V_{\theta_t}\left(s^t\right) = \phi\left(s^t\right)$ is the feature basis. At each updating step, the temporal difference δ^t indirectly quantifies the discrepancy between the true value function and its function approximation counterpart. Consequently, Eq. (9) turns out to be

$$\theta_{t+1} = \theta_t - \alpha \delta^t e_t. \tag{10}$$

One main advantage of this strategy is to circumvent the curse of high variance within the Monte-Carlo sampling methods. From the gradient accumulation point of view, this step-wise gradient updating $\nabla_{\theta_t} V_{\theta_t}$ is actually propagating the gradient from the end decision step t_{end} during the learning process to the staring decision step $t_{\text{start}} = 0$.

4.3 Generalization to Off-Policy Learning

The TD-learning method discussed in the last section is usually referred to as on-policy TD-learning. The notion of the on-policy represents the reinforcement learning tasks that target estimating the value function V^{π} in the scenario that the observed samples are collected when the policy π is executed.

Another strand of the reinforcement learning task is off-policy learning, which is the dominating case in clinical trials and precision medicine. The observed samples collected are not following the policy π but other exploration policies such as the randomization policy. Although the goal of estimating the value function V^{π} remains the same as in on-policy learning, off-policy learning faces additional challenges, especially for distribution mismatching and the curse of high variance in infinite horizon settings (Jiang & Li, 2016).

The most important tool for off-policy learning is importance sampling. To further explain this notion, let us consider the problem of estimating $\mathbb{E}_p[f(S)]$ of a function $f \in \mathcal{F}$ with the input $S \sim p$, where p is any distribution. If we can get access to the Monte-Carlo samples following the distribution p, the expectation $\mathbb{E}_p[f(S)]$ can be approximated by its mean estimator. However, this fails if the samples can not be collected with respect to the distribution p but only for another distribution q. The importance sampling is a key tool for correcting the sample distribution p to the distribution q. Denote the Monte-Carlo samples as $\{s_i\}_{i=1}^{m}$ following the distribution q, then the expectation $\mathbb{E}_p[f(S)]$ can be re-weighting according to the "mismatch" of the density for two distributions, i.e.,

$$\mathbb{E}_p[f(S)] = \mathbb{E}_q\left[\frac{p(S)}{q(S)} f(X)\right].$$

Note that the right-hand side of the expectation is taken with respect to the distribution q. An immediate approximation of the above expectation is

$$\frac{1}{m}\sum_{i=1}^{m}\frac{p(s_i)}{q(s_i)} f(s_i).$$

If the positive assumption holds, that is $q(s)$ is positive for any $s \in \mathcal{S}$, the "mismatch" of the two density $p(s_i)/q(s_i)$ is also known as the inverse probability weighting. We illustrate the inverse probability weighting in Fig. 2.

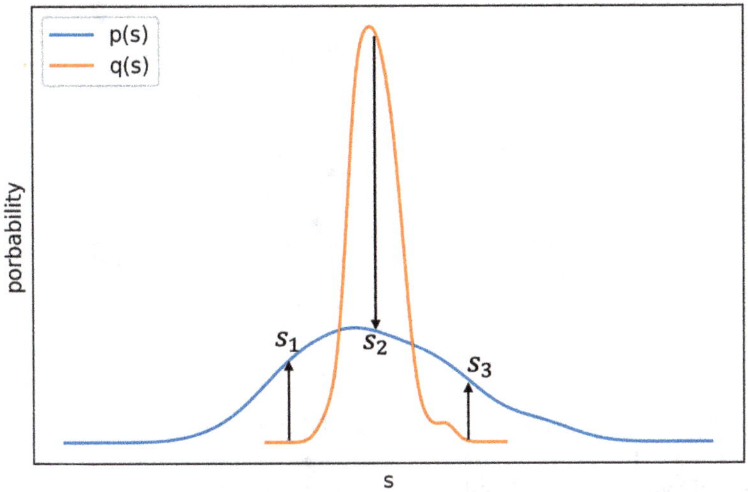

Fig. 2 Importance Sampling: Samples drawn from $q(x)$ are re-weighted by the importance weight $p(s_i)/q(s_i)$ for $i = 1, 2, 3$ to behave like samples from $p(x)$

Challenges of Off-Policy Estimation Off-policy estimation also raises many difficulties. In the on-policy setting, as the behavior policy distribution is equivalent to the target policy distribution, i.e., $\pi_B = \pi$, the state distribution d involved under the target policy d^π is the same as the state distribution involves under the behavior policy d^{π_B}, i.e., $d = d^\pi = d^{\pi_B}$. This implies that d is following a stationary distribution. In contrast, in the off-policy setting that $\pi \neq \pi_B$, the state distributions under the target and behavior policy are different, i.e., $d^\pi \neq d^{\pi_B}$. Hence, the state distribution d is not stationary in general. This makes the estimation quite unstable. Even worse, the estimated value function under off-policy learning could be divergent from the one in an on-policy learning (Dann et al., 2014). A straightforward example is the off-policy mean squared projected Bellman error (MSPBE),

$$\text{MSPBE}(\theta) := \left\| V_\theta - \mathcal{B}^\pi V_\theta \right\|^2_{D_{\pi_B}} . \tag{11}$$

If the state distribution d^{π_B} is different from the state distribution d^π, the estimation results will exhibit significant discrepancies. In the following, we will review an generalization of the TD-learning to the off-policy setting via importance sampling.

Off-Policy TD Learning Recall the parameter updates in an on-policy TD learning (10). At decision step t, one needs to approximate the expectation

$\mathbb{E}_{\pi,\mathcal{P},d^\pi}[\delta_t\phi(s^t)]$. The same updates can still be generalized to off-policy settings if the samples follow π_B. This motivates the following equations

$$
\begin{aligned}
\mathbb{E}_{\pi,\mathcal{P},d^\pi}\left[\delta_t\phi\left(s^t\right)\right] &= \mathbb{E}_{s^{t+1}|s^t,a^t;\, s^t\sim d^\pi}\left[\pi\left(a^t \mid s^t\right) d^\pi\left(s^t\right)\delta_t\phi\left(s^t\right)\right] \\
&= \mathbb{E}_{s^{t+1}|s^t,a^t;\, s^t\sim d^{\pi_B}}\left[\pi\left(a^t \mid s^t\right) d^{\pi_B}\left(s^t\right)\delta_t\phi\left(s^t\right)\right] \\
&= \mathbb{E}_{s^{t+1}|s^t,a^t;\, s^t\sim d^{\pi_B}}\left[\pi_B\left(a^t \mid s^t\right) d^{\pi_B}\left(s^t\right)\frac{\pi\left(a^t \mid s^t\right)}{\pi_B\left(a^t \mid s^t\right)}\delta_t\phi\left(s^t\right)\right] \\
&= \mathbb{E}_{\pi_B,\mathcal{P},d^{\pi_B}}\left[\frac{\pi\left(a^t \mid s^t\right)}{\pi_B\left(a^t \mid s^t\right)}\delta_t\phi\left(s^t\right)\right]. \qquad (12)
\end{aligned}
$$

Through the above procedure, the expectation taken with respect to the target policy π is corrected to the expectation taken with respect to the behavior policy π_B. This is connected by the Radon–Nikodym derivative $\pi\left(a^t \mid s^t\right)/\pi_B\left(a^t \mid s^t\right)$. To execute the off-policy TD learning parameter updates, we have the following, which is the off-policy counterpart of the parameter updates in (10),

$$
\theta_{t+1} = \theta_t - \alpha\frac{\pi\left(a^t \mid s^t\right)}{\pi_B\left(a^t \mid s^t\right)}\delta_t e_t,
$$

Note that we approximate the true value function V^π by V_θ^π, TD learning can diverge or be unstable when adopting a non-linear function approximation, especially in off-policy learning (Tsitsiklis & Van Roy, 1997). Hence, TD learning is also called a semi-gradient method. True gradient methods (Zhang et al., 2020), such as residual gradient methods, have also been proposed and extended to off-policy settings to avoid this drawback. We will review one of the typical true gradient methods in Sect. 6. Before that, we will review a few recent developments in the statistical literature that are closely connected with some of the properties we have introduced already.

5 Temporal Difference and Estimating Equations

The temporal difference error and its associated properties are extensively used. Two recent methods (Ertefaie & Strawderman, 2018; Luckett et al., 2020) utilize this tool and formulate the optimal treatment regime problem in an estimating equation fashion. Note that following the Bellman equation (1), the optimal Q-function satisfies

$$
Q^*(s,a) = \mathbb{E}_{S^{t+1}|s,a}\left[R^t + \gamma\max_{a'} Q^*(S^{t+1},a')|S^t = s, A^t = a\right].
$$

The optimal policy is $\pi^*(s) = arg\,max_a Q^*(s, a)$. Instead of iteratively updating the Q function as we introduced earlier, Ertefaie and Strawderman (2018) proposed a Greedy Gradient Q-learning (GGQ). The key insight is to utilize the temporal difference error defined as

$$\delta^{t+1}(\theta) = R^t + \gamma \max_{a'} Q^*(S^{t+1}, a') - Q^*(s, a),$$

Suppose $Q(s, a) = \theta^T \varphi(s, a)$, where $\varphi(s, a)$ is a set of feature basis over state-action pair (s, a), θ is a p-dimensional parameter vector of parameters. Let θ^* be the true parameter vector that attains the optimal value $Q^*(s, a) = \theta^{*T} \varphi(s, a)$. Then inspired by the property of temporal difference error that $\mathbb{E}[\delta^{t+1}(\theta^*)|S^t = s, A^t = a] = 0$ for any $t \geq 0$ under the true θ^*, the estimating equations of θ can be directly obtained as

$$\mathbb{E}\left\{\sum_{t=0}^{T} \delta^{t+1}(\theta) \nabla_\theta Q_\theta(S^t, A^t)\right\} = 0, \tag{13}$$

where $\nabla_\theta Q_\theta(s, a)$ accounts for the influence of basis. Based on the linear assumption, $\nabla_\theta Q_\theta(s, a) = \varphi(s, a)$. Hence we can solve the sample version of the sum of trajectory loss

$$\sum_{i=1}^{n} \sum_{t=0}^{T} \delta_i^{t+1}(\theta) \varphi(S_i^t, A_i^t)$$

to obtain an unbiased estimator $\hat{\theta}$ for θ^*, and the corresponding estimated optimal policy is $\hat{\pi}^*(s) = arg\,max_a Q_{\hat{\theta}}(s, a)$.

GGQ allows the construction of confidence intervals for the mean outcome difference between the optimal policy and any other policy. Another advantage is that the significance of variables can be tested in the optimal policy by checking whether the confidence interval for the corresponding parameter contains zero. However, the estimation equation (13) contains an non-smooth max operator, which makes the estimation difficult without large amounts of data (Laber et al., 2014). Essentially, the Q-learning framework is a greedy method as it always tends to select the best arm for each decision stage. When the dynamic environment is complex, the greedy action often leads to sub-optimality (Dann et al., 2014).

Luckett et al. (2020) developed another approach called the V-learning. Although it is fundamentally another application of the Bellman equation and the temporal difference error, it maximizes the value function rather than the action-value function. It can also be viewed as an estimating equation approach. Denote the propensity score at stage t as $\mu^t(a^t, s^t) = P(A^t = a^t|S^t = s^t)$. In a micro-randomized trial, $\mu^t(a^t, s^t)$ is known. However, it must be estimated from the data for an observational study. Under assumptions specified in Sect. 2.2, Luckett et al. (2020) define the value function as

$$V^{\pi}(s) = \sum_{k=0}^{\infty} \mathbb{E}\left[\gamma^k R^{t+k}\left\{\prod_{v=0}^{k} \frac{\pi(A^{v+t}, S^{v+t})}{\mu^{v+t}(A^{v+t}, S^{v+t})}\right\}\middle| S^t = s\right]. \tag{14}$$

For any t, the following importance-weighting variant of the Bellman optimality equation holds,

$$\mathbb{E}\left[\frac{\pi(A^t, S^t)}{\mu^t(A^t, S^t)}\{R^t + \gamma V^{\pi}(S^{t+1}) - V^{\pi}(S^t)\}\phi(S^t)\right] = 0, \tag{15}$$

where ϕ denotes the weight function of any state. Similar to GGQ, we can parameterize the state-value function $V^{\pi}(S_i)$ with θ. Then the estimating equation is given by

$$\Lambda_n(\pi, \theta) = \frac{1}{n}\sum_{i=1}^{n}\sum_{t=0}^{T} \frac{\pi(A_i^t, S_i^t)}{\mu^t(A_i^t, S_i^t)}\{R_i^t + \gamma V_{\theta}^{\pi}(S_i^{t+1}) - V_{\theta}^{\pi}(S_i^t)\}\nabla_{\theta} V_{\theta}^{\pi}(S_i^t), \tag{16}$$

where $\nabla_{\theta} V_{\theta}^{\pi}(S_i)$ is the gradient of $V_{\theta}^{\pi}(S_i)$.

V-learning avoids the non-smooth max operator, which makes the optimization more stable. Compared with GGQ, V-learning considers a stochastic policy distribution, which can be more robust to unexpected situations (Zhou et al., 2021a). Although both GGQ and V-learning are pioneer works in infinite horizon DTR estimations, many open problems remain, such as finding the most appropriate features for the parametric model, addressing the computational difficulties for big data, and quantifying uncertainty for policy evaluations.

6 Residual Gradient Algorithm

TD learning has achieved great success in both theoretical developments and real-world applications (Dann et al., 2014). As we can see from Sects. 4–5, the temporal difference error can be used in various ways. It usually requires approximations of the true value function. Tsitsiklis and Van Roy (1997) show that TD learning approaches guarantee the convergence to the optimal policy with discrete state space and linear function approximation (Tsitsiklis & Van Roy, 1997). Unfortunately, TD learning can often be divergent in the case of non-linear function approximation, such as deep neural networks. In fact, there is no explicit objective function minimized for parameter updates in TD-learning.

Baird (1995) proposed a true gradient approach, aiming to minimize an explicit objective function for the true value function estimation under the function approximations. The goal is to estimate the optimal value function V by the function approximator V_θ. Define the Bellman operator as

$$\mathcal{B}V(s) := \max_a \mathbb{E}_{s' \sim P(\cdot|s,a)} \left[R + \gamma V\left(s'\right) \mid s, a \right],$$

where s' is the next state for the current state s. Then Baird (1995) proposed to minimize the mean square Bellman residual,

$$\text{MSBE} = \mathbb{E}_{s \sim \mu} \left[\{\mathcal{B}V(s) - V(s)\}^2 \right],$$

where μ is an observed state distribution. With function approximations and substituting the unknown transition kernel by its one-sample counterpart, the empirical MSBE with m observed samples is

$$\widehat{\text{MSBE}}(\theta) = \frac{1}{m} \sum_i \left[\widehat{\mathcal{B}}V_\theta(s_i) - V_\theta(s_i) \right]^2 \tag{17}$$

$$= \frac{1}{m} \sum_i \left[\{R_i + \gamma V_\theta\left(s_i'\right)\} - V_\theta(s_i) \right]^2, \tag{18}$$

where s' is the next state of the state s. The gradient descent can be formulated as

$$\begin{aligned} \theta_{t+1} &= \theta_t - \alpha \nabla \widehat{\text{MSBE}}(\theta) \\ &= \theta_t - \alpha \left[R_i + \gamma V_{\theta_t}\left(s'\right) - V_{\theta_t}(s) \right] \left[\gamma \nabla_{\theta_t} V_{\theta_t}\left(s'\right) - \nabla_{\theta_t} V_{\theta_t}(s) \right], \end{aligned} \tag{19}$$

where t is the iterative step different from the decision stage step in TD learning. According to the Bellman optimality equation, the MSBE will attain zero if the value function is optimal. Therefore, performing gradient descent on MSBE guarantees that MSBE will eventually converge to a local minimum.

However, in a stochastic environment, the residual gradient algorithm encounters *double sampling issues*. The double sampling issue leads to an empirical objective function biased from its population version.

$$\mathbb{E}_{s \sim d^{\pi_B}} \left[\widehat{\text{MSBE}}(\theta) \right] = \text{MSBE}(\theta) + \mathbb{E}_{s \sim d^{\pi_B}} \left[\text{var}(\widehat{\mathcal{B}}V(s) \mid s) \right] \neq \text{MSBE}(\theta). \tag{20}$$

We will discuss more details of this double sample issue in Sect. 7. As a result, the residual gradient is a biased algorithm but can be guaranteed to converge with non-linear function approximation in a stochastic environment. To correct the bias, Baird (1995) propose to sample the next state twice or more times given any current state. However, this is not practical in the real world. The next section will review a novel approach for resolving the double sampling issue using a functional space embedding framework.

Remark 6.1 For the algorithm implementation, the value function V is usually linearly approximated as a weighted sum of a set of given basis functions $\phi_1, \ldots, \phi_p$: $V_\theta(s) = \sum_{i=1}^{p} \theta_i \phi_i(s)$. The next step is to learn the set of weights $\theta = [\theta_1, \ldots, \theta_p]$ corresponding to an approximation of the optimal value function V, and the value function is linear in θ. The linear function approximations are widely applicable in reinforcement learning tasks as they are suitable for gradient descent and possess a quadratic error surface with a sole minimum. The flexibility is preserved by the choices of the basis function, including linear, polynomial, and Gaussian basis, to represent a very complex functional space. In addition to linear approximations, the value function can also be approximated using deep neural networks and random forests to gain more representation power.

7 Proximal Temporal Consistency Learning

As we have discussed in the last section, one of the biggest challenges for true gradient descent reinforcement learning is the double sampling issue (Baird, 1995). To solve this issue, Zhou et al. (2021a) leveraged the property of functional space embedding to resolve the double sampling issue very efficiently.

In addition to the double sampling issue, some limitations exist for the previously reviewed methods for the applications to solve real-world problems in certain situations. First, Q-learning is a greedy algorithm and thus yields sub-optimal solutions. Meanwhile, policies induced by Q-learning are deterministic, which means they can only provide a single recommendation action at each decision point. This gives little flexibility to select treatments if that option is unavailable due to a temporary medication shortage or not available for other reasons. A possible remedy is to model the policy distribution stochastically. However, existing works typically model the policy distribution by Boltzmann distribution, which can quickly decay to a uniform distribution when the number of treatment options is too large (Min et al., 2014; Zhou et al., 2021a). This is a common issue in all existing approaches.

We will discuss and address these two issues with the newly proposed Proximal Temporal Consistency Learning (pT-Learning) framework proposed by Zhou et al. (2021a).

7.1 Temporal Consistency and Functional Space Embedding

To address the double sampling issue, we need to correct the bias from the one-sample approximation with unknown Markov transitions. Following the Bellman optimality equation, if π^* is the optimal policy, and V^* is the corresponding optimal value function, then (V^*, π^*) is a solution of the following temporal consistency equation for any s and a:

$$\mathbb{E}_{S^{t+1}}\big[R^t + \gamma V(S^{t+1})\big] - V(s) = 0, \tag{21}$$

The intuitive idea is to minimize the square loss of the temporal consistency error, which is the residual gradient algorithm. However, per our discussion, replacing the unknown transition kernel with its sample counterparts will lead to the double sampling issue. Instead, we can embed a Lebesgue measurable function class to the averaged temporal consistency error. Specifically, let's define a critic function $h(s, a)$ and formulate a novel embedding loss $\mathcal{L}_{\text{weight}}$ as follows,

$$\mathcal{L}_{\text{weight}}(V, \pi, h) := \mathbb{E}_{S^t, A^t}\Big[h(S^t, A^t)\big\{R^t + \gamma V(S^{t+1}) - V(S^t)\big\}\Big]. \tag{22}$$

To better understand this embedding, the critic function introduced here is actually to fit the discrepancy of the left-hand side of Eq. (21), and simultaneously promotes those sampled transition pairs with large temporal consistency error. Intuitively, the embedding loss circumvents the double sampling issue since it does not involve the additional variance (bias term) from one-sample bootstrapping. Fundamentally, the embedding loss $\mathcal{L}_{\text{weight}}$ offers compensation for insufficient sampling on $P(s'|s, a)$. The marginal distribution of the state-action pairs $d^\pi(s, a)$ and the transition kernel $P(s'|s, a)$ can be aggregated to joint distribution $p_\pi(s', s, a)$ by the linearity of expectations. Therefore, the embedding loss can be approximated using the sample-path transition pairs drawn from the joint distribution instead of directly approximating the unknown transition kernel. This explains why the embedding loss can resolve the double sampling issue.

Besides these facts, the critic function has nice properties in solving the optimization problem. First, $\mathcal{L}_{\text{weight}}(V, \pi)$ is a valid loss which can identify the optimal value function and optimal policy (V^*, π^*) in that $\mathcal{L}_{\text{weight}}(V^*, \pi^*, h) = 0$. Furthermore, if there exists $(\tilde{V}, \tilde{\pi})$ such that $\mathcal{L}_{\text{weight}}(\tilde{V}, \tilde{\pi}, h) = 0$, then $(\tilde{V}, \tilde{\pi})$ satisfies Eq. (21). Noticing that $\mathcal{L}_{\text{weight}}(V^*, \pi^*, h) = 0$ holds for any h in a bounded L^2 space, such property immediately transforms the original problem to the following minimax optimization problem

$$\min_{V,\pi} \max_{h} \mathcal{L}^2_{\text{weight}}(V, \pi, h). \tag{23}$$

The minimax optimization of (23) gives a clear direction for estimating (V^*, π^*). However, it is still not tractable, as the critic function could be arbitrary, making it infeasible to find a proper representation of $h(\cdot)$. However, if the observed reward R^t is a function continuous over (s^t, a^t), then the critic function $h^*(s, a)$ that maximize $\mathcal{L}^2_{\text{weight}}(V, \pi, h)$ is also continuous. Therefore, if we assume a continuous reward function, we can represent $h^*(s, a)$ in a bounded reproducing kernel Hilbert space

(RKHS), and the minimax problem of (23) can be decoupled into the following single-stage minimization problem

$$\min_{V,\pi} \mathcal{L}_U = \mathbb{E}_{S^t,\tilde{S}^t,A^t,\tilde{A}^t,S^{t+1},\tilde{S}^{t+1}}\Big[\zeta\big\{\tilde{\mathcal{T}}_\pi(S^t, A^t, V) - V(S^t)\big\}$$

$$K(S^t, A^t, \tilde{S}^t, \tilde{A}^t)\big\{\tilde{\mathcal{T}}_\pi(S^t, A^t, V) - V(\tilde{S}^t)\big\}\Big], \tag{24}$$

where $\tilde{\mathcal{T}}_\pi(S^t, A^t, V) = R^t + \gamma V(S^{t+1}) - \psi(\pi(A^t|S^t))$, $(\tilde{S}^t, \tilde{A}^t, \tilde{S}^{t+1})$ is an independent copy of the transition pair (S^t, A^t, S^{t+1}), and $K(\cdot, \cdot)$ is the universal kernel reproducing h^*. This property was explored by Zhou et al. (2021a).

Given the observed data $\mathcal{D}_n$ with a length T trajectory, the above problem can be solved using a trajectory-based order-2 U-statistic estimator. Consequently, the total loss $\mathcal{L}_U$ can be aggregated as the empirical mean of n $i.i.d.$ within-trajectory loss. The proposed sample estimator $\tilde{\mathcal{L}}_U$ is consistent, and its gradient can be approximated by the sampled transitions and optimized using any gradient descent algorithm. Thus, the proposed method achieves much flexibility regarding function approximations of (V^*, π^*), allowing both linear and nonlinear approximations without the risk of divergence from the optimal solution.

7.2 A Proximal Bellman Operator with Sparse Policies

We have shown how the pT-Learning method resolves the double sampling issue. Equipped with this new tool, we turn to solve another practical issue when the number of treatment options is large. This involves constructing a proximal Bellman operator to facilitate smooth optimization and induce a sparse policy distribution. We first revisit the Bellman optimality equation in (2) from a policy explicit view. Suppose the policy π follows a stochastic distribution, then the Bellman optimality equation can be rewritten as

$$\mathcal{B}V(s) := \max_{\pi} \mathbb{E}_{a\sim\pi(.|s),S^{t+1}|s,a}[R^t + \gamma V(S^{t+1})|S^t = s, A^t = a] = V(s). \tag{25}$$

A natural idea is to jointly optimize V and π for minimizing the difference between $\mathcal{B}V(s)$ and $V(s)$ for any s. However, this equation is nonlinear and contains a non-smooth max operator again. This makes the estimation very difficult without a large number of samples. A proximal Bellman operator could circumvent this difficulty. Specifically, the operator adds a strongly convex and continuous component $d(\pi)$ to the dual of (25):

$$\mathcal{B}_\lambda V_\lambda(s) = \max_{\pi} \sum_a [\langle \mathbb{E}_{S^{t+1}\sim p(.|s,a)}[R^t + \gamma V_\lambda(S^{t+1})], \pi(a|s)\rangle$$

$$- \mu(\pi(a|s)) + \lambda d(\pi(a|s))]$$

$$= \max_{\pi} \mathbb{E}_{a \sim \pi(.|s)}[Q_\lambda(s,a) + \lambda\phi(\pi(a|s))], \tag{26}$$

where $\mu(\cdot)$ is a constant function always equal to 0 in our case, $\phi(x) = d(x)/x$ with the proximity function $d(\cdot)$ and $Q_\lambda(s,a) = \mathbb{E}_{S^{t+1}|s,a}[R^t + \gamma V_\lambda(S^{t+1})]$. In particular, Zhou et al. (2021a) consider a special case $d(x) = x\phi(x) = -\frac{x}{2}\log_{\kappa=0}(x)$ for achieving both smoothness and sparsity properties, where $\log_\kappa(x) = \frac{1}{1-\kappa}\left(x^{1-\kappa} - 1\right)$.

Remark 7.1 In general, the proximity function satisfying the following properties is valid for smoothness purposes. 1. Monotonicity: $\phi(x)$ is non-increasing; 2. Non-negativity: $\phi(1) = 0$; 3. Differentiability: $\phi(x)$ is differentiable; 4. Expected Concavity: $x\phi(x) = d(x)$ is strictly concave. Many choices of $d(x)$ satisfy the above conditions. For example, when $d(x) = x\phi(x) = -x\log x$, the well-known Shannon entropy is recovered.

There are several properties that we should highlight. First, the proximal Bellman operator (26) is a valid approximation for the Bellman operator $\mathcal{B}$, where the approximation bias can be bounded by

$$\mathcal{B}_\lambda V(s) - \mathcal{B}V(s) = \frac{\lambda}{2} - \frac{\lambda}{2|\mathcal{K}(s)|}, \tag{27}$$

where $|\cdot|$ denotes the cardinality of a set, and $\mathcal{K}(s)$ represents the action set at state s, that is $\mathcal{K}(s) = \{a_{(i)} \in \mathcal{A} : Q_\lambda(s, a_{(i)}) > \frac{1}{i}\sum_{j=1}^{1} Q_\lambda(s, a_{(j)}) - \frac{\lambda}{i}\}$. Here $a_{(i)}$ is the action with the i-th largest state-action value, and $|\mathcal{K}(s)| \leq |\mathcal{A}|$ always holds for any s.

Additionally, $\mathcal{B}_\lambda$ is a smooth approximation for $\mathcal{B}V(s)$, which is a direct result of applying the Lagrange multiplier. Specifically, $\mathcal{B}_\lambda$ has a closed-form equivalence

$$\mathcal{B}_\lambda V_\lambda(s) = \frac{\lambda}{2}\left\{1 - \sum_{a\in\mathcal{K}(s)}\left[\left(\frac{\sum_{a\in\mathcal{K}(s)} Q_\lambda(s,a)}{\lambda|\mathcal{K}(s)|} - \frac{1}{|\mathcal{K}(s)|}\right)^2 - \left(\frac{Q_\lambda(s,a)}{\lambda}\right)^2\right]\right\} \tag{28}$$

By transforming the Bellman operator into a smooth operator, great convenience could be gained when solving the optimization problem later.

Besides gaining advantages in solving the optimization problem, the unique advantage of introducing $\mathcal{B}_\lambda$ lies in that it induces a sparse optimal policy distribution $\pi_\lambda(a|s)$, whose support set is a sparse subset of action space. Specifically, by applying $\mathcal{B}_\lambda$, we could gain a stochastic policy that only includes optimal and near-optimal policies while removing those options far from optimal. This can be illustrated by presenting $\pi_\lambda(a|s)$ in terms of the state-action value function $Q_\lambda(s,a)$ analytically, that is

$$\pi_\lambda(a|s) = \left(\frac{Q_\lambda(s,a)}{\lambda} - \frac{\sum_{a\in\mathcal{K}(s)} Q_\lambda(s,a)}{\lambda\mathcal{K}(s)} + \frac{1}{|\mathcal{K}(s)|}\right)^+. \tag{29}$$

Notice that (29) gives a well-defined probability mass function in that $\sum_a \pi_\lambda(a|s) = 1$. Also, the policy $\pi_\lambda(a|s)$ tends to assign a large probability to the action according to the rank of the state-action values. Moreover, sparsity can be shown by analyzing the support set of $\pi_\lambda(\cdot|s)$, which satisfies

$$\Big\{a_{(i)} \in \mathcal{A} : Q_\lambda(s, a_{(i)}) > \frac{1}{i}\sum_{j=1}^{i} Q_\lambda(s, a_{(j)}) - \frac{\lambda}{i}\Big\}. \tag{30}$$

The inequality (30) indicates the sparsity parameter λ controls the margins between the smallest action value and the others included in the support set. To better demonstrate how the mechanism works, we consider a toy example where there are three available actions at a decision point. When λ is sufficiently small, such that $\lambda < Q_\lambda(s, a_{(1)}) - Q_\lambda(s, a_{(2)})$, only $a_{(1)}$ will be contained in the support set. If λ increases and falls into the range $[Q_\lambda(s, a_{(1)}) - Q_\lambda(s, a_{(2)}), \sum_{i=1}^{2} Q_\lambda(s, a_{(i)}) - 2Q_\lambda(s, a_{(3)})]$, the support set would contain both $a_{(1)}$ and $a_{(2)}$ but eliminates action $a_{(3)}$.

Therefore, the cardinality of the support set increases as λ increases. This allows us to tune between the deterministic and stochastic policy models by selecting an appropriate λ value. Furthermore, only treatments with rewards close to the optimal treatment will be included in the support set. Besides, the sparse property may also benefit healthcare providers and hospitals. First, when there are many treatment options, it may not be convenient to store all medications and resources simultaneously for the hospital. Only storing the most commonly used optimal and near-optimal treatments would be enough, allowing better medical management. Secondly, two or more medications may have similar effectiveness in some situations. As other methods can hardly show such properties, treatment plans with sparse property could give healthcare providers more flexibility in choosing the most appropriate treatment for the patient.

Plugging in the proximal Bellman Operator $\mathcal{B}_\lambda$ and the induced V_λ to the framework we discussed in Sect. 7.1, we can derive an proximal temporal consistency equation For any s, a, and $\lambda \in (0, \infty)$, if π_λ^* is the optimal policy, and V_λ^* is the corresponding optimal value function, then $(V_\lambda^*, \pi_\lambda^*)$ is a solution of the following proximal temporal consistency equation:

$$\mathbb{E}_{S^{t+1}|s,a}\big[R^t + \gamma V_\lambda(S^{t+1})\big] - \lambda\phi(\pi_\lambda(a|s)) - \Psi(s) + \psi(a|s) - V_\lambda(s) = 0, \tag{31}$$

where $\Psi(s) \in [-\lambda/2, 0]$ and $\psi(a|s) \in [0, \infty)$ are Lagrangian multipliers with $\psi(a|s)\pi_\lambda(a|s) = 0$. We call the discrepancy on the left-hand side pT-error.

Note that Eq. (31) holds for any state-action pair (s, a), the policy π_λ can be estimated directly over the observed transition pairs. This local consistency property makes it easy to incorporate off-policy data. By leveraging this proximal temporal consistency property and combine with the functional embedding framework in

Sect. 7.1, Zhou et al. (2021a) proposed to minimize the following objective function to estimate the optimal policy and value function, that is

$$\begin{aligned}\min_{V_\lambda,\pi_\lambda,\Psi,\psi} \mathcal{L}_U = \mathbb{E}_{S^t,\tilde{S}^t,A^t,\tilde{A}^t,S^{t+1},\tilde{S}^{t+1}}\big[&(\tilde{\mathcal{T}}_{\pi_\lambda}(S^t,A^t,V_\lambda) - \Psi(S^t)\\ &+ \psi(A^t|S^t) - V_\lambda(S^t))\\ &\cdot \zeta K(S^t,A^t,\tilde{S}^t,\tilde{A}^t)(\tilde{\mathcal{T}}_{\pi_\lambda}(S^t,A^t,V_\lambda) - \Psi(\tilde{S}^t)\\ &+ \psi(\tilde{A}^t|\tilde{S}^t) - V_\lambda(\tilde{S}^t))\big],\\ s.t.\quad \pi_\lambda(a|s)\cdot\psi(a|s) = 0 \quad &\text{and} \quad -\lambda/2\\ &\le \Psi(s) \le 0 \text{ for all } s \in \mathcal{S}, a \in \mathcal{A}.\end{aligned} \tag{32}$$

For the implementation, Zhou et al. (2021a) reformulate the minimization problem (24) to be a nonlinear programming problem. The objective function is converted to a quadratic term with a nonlinear equality constraint and a nonlinear inequality constraint. Some computational burden is also alleviated by imposing certain restrictions so that the two Lagrangian functions Ψ, ψ satisfy the constraints automatically. The corresponding algorithm is implemented in the R package `proximalDTR`.

7.3 *OhioT1DM Data Analysis*

7.3.1 A Mobile Health Study: OhioT1DM

We finally apply the pT-Learning method (Zhou et al., 2021a) to the OhioT1DM[1] dataset presented in Sect. 2.1. The dataset contains two groups of individuals with Type-1 diabetes, while each group contains six individuals aged 20–60. All patients were on insulin pump therapy with continuous glucose monitoring (CGM) sensors, and the life-event data were collected via a mobile phone app. Specifically, the dataset includes CGM blood glucose levels measured per 5 minutes, insulin dose level delivered to the patient, meal intakes, and corresponding carbohydrate estimates. Other features, such as self-reported times of work, sleep, and stress, are also available for all patients.

However, the physiological data of the first cohort were collected by *Basis* sensor bands, while the second cohort used *Empatica* sensor band. As a result, the heart rate measured per 5 minutes is available only for patients in the first cohort. The magnitude of acceleration aggregated per minute is only available for patients in the second cohort.

[1] The mHealth dataset OhioT1DM can be requested and downloaded in https://ohio.qualtrics.com/jfe/form/SV_02QtWEVm7ARIKIl.

Based on the preliminary investigation, each patient has distinct blood glucose dynamics. Therefore, we follow the pre-processing strategy used in Zhu et al. (2020b) and treat the data of each patient as a single dataset. Additionally, we assume that we treat each day as an independent trajectory for each patient and estimate the optimal policy based on these trajectories. Therefore, we fully consider the difference between individuals and develop a personalized treatment plan for each patient.

We summarize the measurements over 30-min intervals such that the length of each trajectory is $T = 48$. After removing missing samples and outliers, each dataset contains $n = 15$ trajectories on average. For both cohorts, we study the individualized dose-finding problem by discretizing the continuous dose level into 14 disjoint intervals for intervention options, i.e., $A_i^t \in \{0 = A_{(1)} < \ldots < A_{(14)} = \max(A)\}$, and the grid is determined by the data corresponding to each patient. For the patient in first cohort, patient's states at each stage include the average blood glucose levels $S_{i,1}^t$, the average heart rate $S_{i,2}^t$ and the total carbohydrates intake $S_{i,3}^t$ from time $t-1$ to t. For patients in the second cohort, the states are the same as in the first cohort, except that the average magnitude of acceleration replaces the average heart rate.

The reward for this problem is a measure of the health status of the patient's glucose level, and we define the immediate reward as the average of the index of glycemic control,

$$R_i^t = -\frac{\mathbb{1}(S_{i,1}^t > 140)|S_{i,1}^t - 140|^{1.1} + \mathbb{1}(S_{i,1}^t < 80)(S_{i,1}^t - 80)^2}{30},$$

where R_i^t is non-positive, and a larger value is preferred. For this problem, when we summarize the reward over 30-min intervals, we would use the average reward during that 30-min period. Our goal is to maximize the discounted sum of reward $\mathbb{E}\sum_{t\geq 1}\gamma^{t-1}R^t$.

7.3.2 ProximalDTR Package and Analysis Results

We have specified the states, action space, and reward for this example. We use `proximalDTR`[2] package for implements the pT-Learning method discussed in this section.[3] Applying the algorithm to the dataset, we solve for the optimal treatment policy. The algorithm generates the recommended treatment policy that enjoys the sparse property when given a new state. We show the usage of the package for one patient from the dataset (1st patient in the first cohort) in this section.

[2] The `proximalDTR` package is in https://github.com/vincentskywalkers/ProximalDTR-.

[3] The implementation code for analyses in this chapter is in https://github.com/liyuhan529/Policy-Learning-for-Individulized-Treatment-Recommendation.

After processing the data as described in Sect. 7.3.1, we obtain a matrix X with the state at each time point and two vectors showing the action chosen A and observed reward R at each time point. For this patient, if $A^t = 0$, no insulin injection is taken at time t. If $A^t = i$, then the dose unit for insulin injection is in the interval $[9(i-1)/14, 9i/14]$. The format of the data is shown below.

```
> X[1:7,]
      glucose    heart meal
49  343.00000 65.50000    0
97  122.16667 60.33333    0
145 156.16667 71.66667    0
193 187.66667 71.16667    0
241  84.00000 74.66667    0
385  88.66667 65.16667    0
433  92.66667 67.00000    0
> A[1:7]
[1] 0 0 0 0 0 0 0
> R[1:7]
[1] -11.5114   0.0000  -0.7128  -2.3391  0.0000  -1.4444  0.0000
```

We make several remarks. First, the longitudinal state matrix X should follow a certain format to be used for the package. Specifically, suppose there are n trajectories in the dataset, the first n rows of the matrix should be the first state, i.e., S_i^0 $(i = 1, 2, \ldots, n)$ for the n trajectories, the $(n+1)$th to $2n$th row should be the second state for each trajectory, etc. Secondly, X has one more stage than A and R since at the end of the stage, we only observe X but not A or R. Finally, for this problem, the glucose and acceleration level in matrix X and reward R is the average value during the 30-min when transforming from the original dataset containing a measurement every 5 minutes.

After transforming the data into the required form, we use the proximalDTR function to fit the data:

```
> fit1 = proximalDTR(X = X, A=A, R=R, n_ID = 33, stage=48, gamma
    =0.9,
+         lambda.set = c(2),step.beta = 0.001, step.omega = 0.001,
+         desc.rate = 0.001, max.iter = 3000, max.iter.cv = 1000,
    bw= 1,
+         cv=F, trace =TRUE)
```

Here, n_ID is the number of trajectories, stage is the length of each trajectory, gamma is the discount factor, and lambda.set is the candidate values of λ that we will choose from by cross-validation. We can also specify a certain λ value to obtain an appropriate sparsity level for the induced policy. Other parameters in the function are used to control convergence.

When given a new state for the patient, the algorithm will induce a treatment policy that enjoys sparse property by predict_rl function. An example is shown below: if the patient is in the state where the glucose level is 200 mg/dL, the heart rate is 90 bpm, and the food intake is 0, the recommended best treatment for that patient is the 6th treatment (insulin injection with dose 3.21–3.86 units). The corresponding probability is to choose the 6th treatment with a probability of

0.8110 and the 1st treatment (insulin injection with a dose level less than 0.64 unit) with a probability of 0.1890.

```
> pred.fit2 <- predict_rl(fit1, c(200,90,0))
> pred.fit2$prob
 [1] 0.0000 0.1890 0.0000 0.0000 0.0000 0.0000 0.8110 0.0000
     0.0000 0.0000
[11] 0.0000 0.0000 0.0000 0.0000
> pred.fit2$recommend.trt
[1] 6
```

To visualize the results, We take the 1st patient in the first cohort as an example. Figures 3 and 4 show the induced treatment policy given a new state for that patient. From Fig. 3, we can see that when there is no food intake and the heart rate is normal (90 bpm), the higher the glucose level for the patient, a higher dose of insulin injection would be recommended. Specifically, when the glucose level is 100 mg/dL, which is still in the normal range, the recommended dose level would be relatively small. When the glucose level is slightly higher than the normal range (200 mg/dL), the first and 6th dose levels are recommended, while the 6th level has a higher probability compared with the case when the glucose level is 100. When the glucose level is well above the normal range (300mg/dL), the 8th dose level would also be recommended, while the probability of assigning the 1st dose level is significantly reduced.

Note that in Fig. 3, the highest dose level recommended is the 8th level when there is no food intake, and the heart rate is normal. However, in Fig. 4, a higher dosage may be recommended when the patient is having meals and the heart rate is above average. For instance, comparing the left and middle panels in Fig. 4, it seems that a higher heart rate will increase the recommended dose level when the glucose level and carbohydrate intake (15 representing a snack) are the same. Additionally,

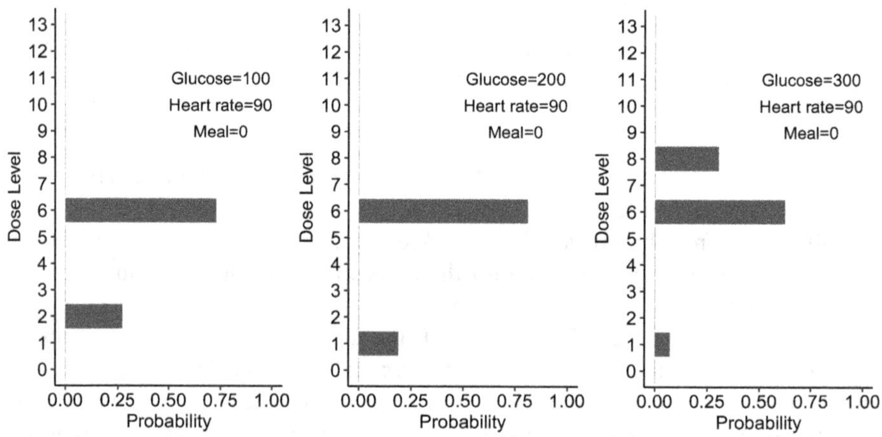

Fig. 3 Induced treatment policy for the 1st patient in the 1st cohort given that there is no food intake and the heart rate is normal

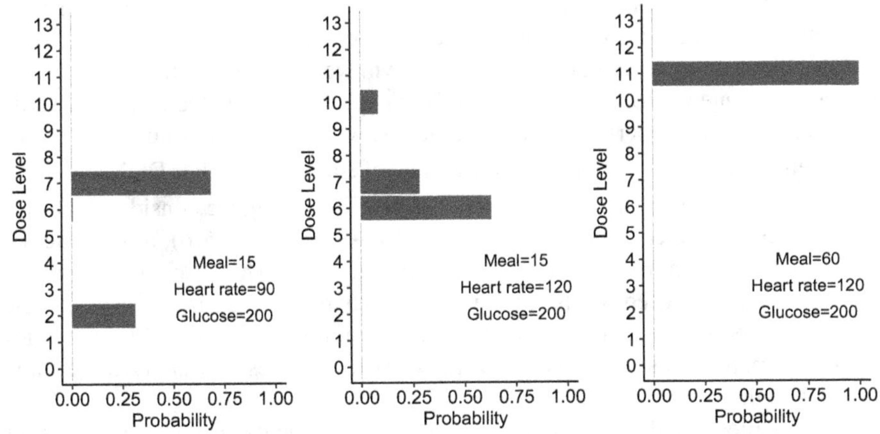

Fig. 4 Induced treatment policy for 1st patient in the 1st cohort given that glucose level is slightly above average, while there are food intake and the heart rate may also be above average

from the right panel of Fig. 4, when the heart rate and carbohydrate intakes (60 representing a regular meal) are both above average, an even higher dose of insulin injection would be recommended.

In summary, the recommended treatment policy we get from the algorithm coincides with what we would expect for a patient with diabetes. A higher insulin injection dose would be recommended when the glucose level is higher. At the same time, a higher heart rate and carbohydrate intake could also lead to a higher dose of insulin injection. Such results show our proposed method is practicable and has a clear benefit when applied to real-world settings. Besides, the sparse property of induced policy gives the healthcare provider more flexibility when selecting the treatment.

We also compare pT-Learning with two other popular methods in reinforcement learning: Deep-Q network (DQN) (Mnih et al., 2016) and Soft actor-critic (SAC) (Christodoulou, 2019). DQN approximates the Q-function with a neural network and adopts the TD learning framework to solve the optimal policy. The induced deterministic optimal policy would select the action with the largest state-action value at each decision point. Meanwhile, SAC uses a separate network to model the optimal policy besides Q-function and further adds the Shannon entropy of the optimal policy as a bonus term to the V-function. Therefore, SAC would induce an optimal stochastic policy and is directly comparable to pT-learning by using a proximity function $d(x) = \frac{1}{2}x(1-x)$ to the V-function.

However, we should mention that both DQN and SAC use a neural network to approximate the Q-function, hence needing a large amount of data for model training, which is hardly available in medical settings. For this case study, we use a shallow two-layer perceptron architecture with two linear layers with 64 hidden units to approximate the Q-function for both DQN and SAC. But these methods may not be suitable for a small dataset and do not always have desired statistical properties such as guaranteed convergence and consistency.

Since the data-generating process is unknown, we follow Luckett et al. (2020) to utilize the Monte Carlo approximation of the expected discounted sum of rewards to evaluate the model performance. Specifically, we compare the value of $\hat{V}(S_i^0)$ of each method, where S_i^0 denotes the initial state of ith trajectory. For DQN and SAC, $\hat{V}(\cdot)$ refers to the estimated value function. For pT-learning, we consider the lower bound of the estimated value function i.e. $\hat{V}_{\hat{\lambda}}(\cdot) - (1-\gamma)^{-1}\hat{\lambda}\phi(0)$, to mitigate the effect of the sparsity parameter λ on the value function. The observed discounted sum of rewards is used as the baseline for each patient. To better evaluate the stability and performance of each method, we randomly select 10 or 15 trajectories from each patient based on available trajectories 50 times and apply these three methods to the selected data.

The boxplots of the improvements on the six patients in the second cohort are presented in Fig. 5. pT-learning generally has the best performance across all patients. As DQN uses the semi-gradient method, it may diverge under off-policy settings, which explains its largest variance and poor performance compared with SAC and pT-learning. SAC can match the performance of pT-learning on patients 7 and 10 but underperforms on the other four patients. Note that the induced policy in SAC follows a Boltzmann distribution, in which non-negligible probability would be assigned to all actions. This may lead to its unstable performance, especially

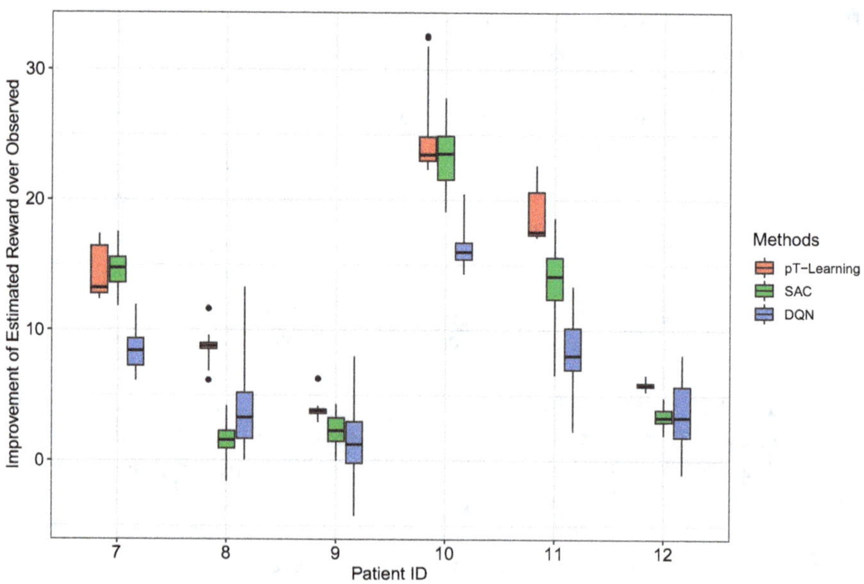

Fig. 5 Boxplots of improvements on Monte Carlo discounted sum of rewards relative to the baseline of observed discounted sum of rewards. The results included the six patients in the second cohort, and are summarized over 50 simulations with $\gamma = 0.9$

when the cardinality of the action space is large. On the contrary, pT-learning enjoys the best and most stable performance compared with other methods. This shows that the flexibility of pT-learning to choose deterministic and stochastic policy adaptively has a clear benefit over other methods, especially when a large number of treatment options are available.

8 Online Experiment and Policy Learning

This section will review some popular adaptive clinical trial designs and online extensions of current policy learning methods. The adaptive clinical trial is distinguished from traditional randomized or observational studies, although they are usually in a finite-horizon setting. It allows the treatment policy to be updated according to historical information during an ongoing clinical trial. Besides, the ethical and sample efficiency concerns also promote the investigation of developing adaptive clinical trials via online policy updating (Hu & Rosenberger, 2006; Thall et al., 2013; Chow, 2014; Hu et al., 2015). For example, Thall et al. (2013) proposed a Bayesian framework to design a trial with multiple combined phases and learn the optimal online policy.

Additionally, their proposed method considers both the sample efficiency and drug toxicity. The limitation is that this method only estimates the average treatment effect and thus fails to account for the patient's heterogeneity information. In contrast, (Kim et al., 2011; Renfro et al., 2016) considered the subgroup treatment effects via enrichments, but their goal is not estimating the optimal policy or online updating of the clinical trial. Designing more advanced adaptive clinical trials would be helpful for practitioners.

On the other hand, one of the most popular online learning frameworks for an infinite-horizon setting is the ε-greedy strategy. This is developed based on incremental information and tends to balance the exploration and exploitation (Lattimore & Szepesvári, 2020). In detail, the ε-greedy strategy assigns the treatment arm to patients following the estimated policy with the probability $1 - \varepsilon$ but lets the ε chance to select a randomized arm with probability ε. This ε-greedy strategy can avoid sub-optimal solutions and yield an optimal online decision rule. This usually requires ε to decay to zero as time progresses. There is a trade-off between the large and small ε. Large ε injects more randomness to select the treatment arm and thus improves the exploration of the policy learning. Small ε gives attention to the estimated policy, increasing exploitations and sample efficiency.

In recent literature, Luckett et al. (2020) follows the ε-greedy to propose an online V-learning framework for updating the estimated policy during the training process. This can also be extended to adaptive clinical trials to improve outcomes. The basic idea of online V-learning is related to offline V-learning, which we have reviewed in Sect. 5. In online settings, Luckett et al. (2020) propose to update the estimated optimal policy $\hat{\pi}_n^t$ using the data collected up to time t. Then adopt the ε-greedy strategy to assign the treatment arm to balance the exploration and exploitation. Most recently, Chen et al. (2021) developed an online and robust ordinary least squares to learn a reward model of different actions given the contextual information and then maximize the long-term reward. Their framework also adapts the ε-greedy paradigm and establishes the asymptotic normality to quantify the uncertainty for the optimal decision rule. We may also be interested in updating the estimated optimal policy with patient-specific data. Briefly, the estimated policy $\widehat{\pi}_n^i$ can be indexed by the patient i, which indicates only trained using the data from the patient i. This strategy allows more flexibility for learning policies for each patient. Randomness can also be adopted. Finally, we note that an alternative to the ε-greedy strategy to encourage exploration is the upper confidence bound (UCB) sampling. We refer reader to Auer (2002) for more details.

9 Discussion

This chapter reviewed the literature on designing and estimating infinite-horizon dynamic treatment regimes. There are still many unsolved problems. We briefly discuss several potential directions for future research.

The Markov property is fundamental to policy optimizations and evaluations in a finite horizon setting. However, in a real-life environment, outcomes may depend on decisions made much earlier than the immediate, previous time point. This requires a higher-order chain of the Markov decision process. One of the direct applications is to identify the lag-effect variables and conditional independent window for the validity of the Markov property (Tsitsiklis & Van Roy, 1996). However, the literature is very sparse on how to validate such properties (Shi et al., 2020a). Overall, it is important to develop new tools that address violations of the current assumptions in the data-generating process. Moreover, when the Markov decision process is only partially observed, additional techniques could be used (Ross et al., 2011; Ghavamzadeh et al., 2016). Many of which were developed under a Bayesian framework.

Most work on developing personalized treatment decisions has focused on a finite number of treatment options. However, learning the individualized decision rule becomes challenging when the treatment space incorporates continuous treatment options. One example is choosing insulin doses from a continuous domain to control blood glucose levels. When learning the optimal dose rules, we may need to account for non-monotone and nonlinear relationships. One of the potential directions is to design an optimal policy that can make continuous action recommendations. Kennedy et al. (2017) considered a method for estimating the average dose-effect with flexible doubly robust covariates adjustment. But the method is not intended for optimal dose-finding problems. Rich et al. (2014) proposed an adaptive strategy, and more recently Schulz and Moodie (2021) developed a doubly robust estimation approach based on a linear model. Chen et al. (2016) proposed to optimize the individualized dosing rule by maximizing a local approximation of the value function, solved under the framework of outcome weighted learning (Zhao et al., 2012). Recently, kernel-assisted learning with linear dimension reduction (Zhou et al., 2021b; Zhu et al., 2020a) for directly learning the optimal dose show good potential. However, all existing methods are limited to single or finite decision stages. Novel methodology frameworks under the infinite horizon setting are much needed.

For diseases such as cancer or HIV infection (Hieke et al., 2015; Simoneau et al., 2020; Xue et al., 2021), survival data often arise. Developing an optimal policy is critical under the survival analysis framework. The main challenge is that treatment and covariate information from patients could be censored in follow-up stages. Also, the true survival time might be unknown for censored patients. Recent developments of the optimal policy estimation under the survival data framework include (Goldberg & Kosorok, 2012; Zhao et al., 2014, 2020; Xue et al., 2021; Hager et al., 2018). However, extending existing approaches to an infinite horizon setting is challenging.

Typically, the goodness of a policy is measured by its value function. Therefore, it is essential to quantify the uncertainty and make statistical inferences about the policy's value function. Specifically, one may construct a confidence interval for the action-value function and its integrated value concerning a given reference distribution. The existing literature in this direction is rare, including the works of Jiang and Li (2016), Kallus and Uehara (2019). Another recent work (Shi et al., 2020b) discussed a recursively updating method to quantify the uncertainty for the off-policy value function. In addition to constructing confidence intervals, validating the policy's value function and measuring how close the policy's value is to the optimal one are also important research questions.

Confounding and causality are other essential issues in infinite horizon policy evaluation and optimization. The observed behavior policy could involve existing estimated policies and ad hoc human interventions to maximize the long-term rewards. Besides classical literature on causal inference, some recent methods can deal with unmeasured confounders and evaluate treatment effects. Tchetgen et al. (2020) proposed a proximal g-computation algorithm in single-stage and two-stage studies. Shi et al. (2020c) proposed to learn the average treatment effect (ATE) with double-negative control adjustment. A few works considered reinforcement learning with confounded datasets in the computer science literature. For example, Wang et al. (2020a) considered learning an optimal policy in an episodic confounded MDP setting. Bennett et al. (2021) introduced an optimal balancing algorithm for off-policy evaluation in a confounded MDP without requiring mediators to exist. It can be promising and practically useful to develop methods that address confounding issues of infinite horizon policy learning.

In precision medicine, data is usually pre-collected as an offline dataset. It implies that no online interaction with the environment is allowed and thus raises an issue on how to exploit the dataset without further exploration. Due to such a lack of exploration, any algorithm for infinite horizon DTR estimations or offline RL suffers from insufficient coverage of the dataset (Fujimoto et al., 2019; Wang et al., 2020b; Agarwal et al., 2020). As a result, an extrapolation error further propagates through each iteration of the algorithm, leading to failure in learning an optimal policy. To tackle this issue, pessimistic methods (Laroche et al., 2019; Jaques et al., 2019; Yu et al., 2020; Nair et al., 2020; Yang et al., 2020; Siegel et al., 2020; Kidambi et al., 2020; Jin et al., 2021) provide various algorithms that enjoy theoretical or empirical success. We may roughly summarize them into two categories: (1) constrain the policy to avoid visiting the states and actions which are less covered by the dataset; (2) penalize the action-value function on such states and actions. In the future, developing more novel methods in policy evaluation and optimization in the principle of pessimism is also crucial.

Besides the limitations and challenges mentioned above, mHealth requires development in many other aspects. On the practical side, some risk factors can be challenging to measure with wearable devices, and available biomarkers may only serve as surrogates. For instance, when monitoring the mood change of patients with depression, a mobile device cannot collect information as accurate as questionnaires (Furukawa, 2010). Missing and censoring are quite standard in the mHealth setting (Goldberg & Kosorok, 2012). There are issues regarding privacy (Bhuyan et al., 2017; Sunyaev et al., 2015). Improving the acceptance of mHealth technology for patients is also essential for facilitating the communication between patients and doctors (Silva et al., 2015; Garavand et al., 2017).

References

Agarwal, R., Schuurmans, D., & Norouzi, M. (2020). An optimistic perspective on offline reinforcement learning. In *International Conference on Machine Learning* (pp. 104–114). PMLR.

Auer, P. (2002). Using confidence bounds for exploitation-exploration trade-offs. *Journal of Machine Learning Research, 3*, 397–422.

Baird, L. (1995). Residual algorithms: Reinforcement learning with function approximation. In *Machine Learning Proceedings 1995* (pp. 30–37). Elsevier.

Bennett, A., Kallus, N., Li, L., & Mousavi, A. (2021). Off-policy evaluation in infinite-horizon reinforcement learning with latent confounders. In *International Conference on Artificial Intelligence and Statistics* (pp. 1999–2007). PMLR.

Bhuyan, S. S., Kim, H., Isehunwa, O. O., Kumar, N., Bhatt, J., Wyant, D. K., Kedia, S., Chang, C. F., & Dasgupta, D. (2017). Privacy and security issues in mobile health: Current research and future directions. *Health Policy and Technology, 6*(2), 188–191.

Blatt, D., Murphy, S. A., & Zhu, J. (2004). A-learning for approximate planning. *Ann Arbor, 1001*, 48109–2122.

Chakraborty, B., & Murphy, S. A. (2014). Dynamic treatment regimes. *Annual Review of Statistics and Its Application, 1*, 447–464.

Chen, G., Zeng, D., & Kosorok, M. R. (2016). Personalized dose finding using outcome weighted learning. *Journal of the American Statistical Association, 111*(516), 1509–1521.

Chen, H., Lu, W., & Song, R. (2021). Statistical inference for online decision making: In a contextual bandit setting. *Journal of the American Statistical Association, 116*(533), 240–255.

Cheung, W. C., Simchi-Levi, D., & Zhu, R. (2020). Reinforcement learning for non-stationary Markov decision processes: The blessing of (more) optimism. In *International Conference on Machine Learning* (pp. 1843–1854). PMLR.

Chow, S.-C. (2014). Adaptive clinical trial design. *Annual Review of Medicine, 65*, 405–415.

Christodoulou, P. (2019). Soft actor-critic for discrete action settings. arXiv preprint arXiv:1910.07207.

Clifton, J., & Laber, E. (2020). Q-learning: Theory and applications. *Annual Review of Statistics and Its Application, 7*, 279–301.

Dann, C., Neumann, G., Peters, J., et al. (2014). Policy evaluation with temporal differences: A survey and comparison. *Journal of Machine Learning Research, 15*, 809–883.

Ertefaie, A., McKay, J. R., Oslin, D., & Strawderman, R. L. (2021). Robust q-learning. *Journal of the American Statistical Association, 116*(533), 368–381.

Ertefaie, A., & Strawderman, R. L. (2018). Constructing dynamic treatment regimes over indefinite time horizons. *Biometrika, 105*(4), 963–977.

Fujimoto, S., Meger, D., & Precup, D. (2019). Off-policy deep reinforcement learning without exploration. In *International Conference on Machine Learning* (pp. 2052–2062). PMLR.

Furukawa, T. A. (2010). Assessment of mood: Guides for clinicians. *Journal of Psychosomatic Research, 68*(6), 581–589.

Garavand, A., Samadbeik, M., Kafashi, M., & Abhari, S. (2017). Acceptance of health information technologies, acceptance of mobile health: A review article. *Journal of Biomedical Physics & Engineering, 7*(4), 403.

Ghavamzadeh, M., Mannor, S., Pineau, J., & Tamar, A. (2016). Bayesian reinforcement learning: A survey. arXiv preprint arXiv:1609.04436.

Goldberg, Y., & Kosorok, M. R. (2012). Q-learning with censored data. *Annals of Statistics, 40*(1), 529.

Hager, R., Tsiatis, A. A., & Davidian, M. (2018). Optimal two-stage dynamic treatment regimes from a classification perspective with censored survival data. *Biometrics, 74*(4), 1180–1192.

Hernán, M. A., & Robins, J. M. (2010). Causal inference.

Hieke, S., Kleber, M., König, C., Engelhardt, M., & Schumacher, M. (2015). Conditional survival: A useful concept to provide information on how prognosis evolves over time. *Clinical Cancer Research, 21*(7), 1530–1536.

Hinderer, K., & Waldmann, K.-H. (2005). Algorithms for countable state Markov decision models with an absorbing set. *SIAM Journal on Control and Optimization, 43*(6), 2109–2131.

Hu, F., & Rosenberger, W. F. (2006). *The theory of response-adaptive randomization in clinical trials* (Vol. 525). John Wiley & Sons.

Hu, J., Zhu, H., & Hu, F. (2015). A unified family of covariate-adjusted response-adaptive designs based on efficiency and ethics. *Journal of the American Statistical Association, 110*(509), 357–367.

Jaques, N., Ghandeharioun, A., Shen, J. H., Ferguson, C., Lapedriza, A., Jones, N., Gu, S., & Picard, R. (2019). Way off-policy batch deep reinforcement learning of implicit human preferences in dialog. arXiv preprint arXiv:1907.00456.

Jiang, N., & Li, L. (2016). Doubly robust off-policy value evaluation for reinforcement learning. In *International Conference on Machine Learning* (pp. 652–661). PMLR.

Jin, Y., Yang, Z., & Wang, Z. (2021). Is pessimism provably efficient for offline RL? In *International Conference on Machine Learning* (pp. 5084–5096). PMLR.

Kallus, N., & Uehara, M. (2019). Efficiently breaking the curse of horizon in off-policy evaluation with double reinforcement learning. arXiv preprint arXiv:1909.05850.

Kennedy, E. H., Ma, Z., McHugh, M. D., & Small, D. S. (2017). Non-parametric methods for doubly robust estimation of continuous treatment effects. *Journal of the Royal Statistical Society: Series B (Statistical Methodology), 79*(4), 1229–1245.

Kidambi, R., Rajeswaran, A., Netrapalli, P., & Joachims, T. (2020). Morel: Model-based offline reinforcement learning. *Advances in Neural Information Processing Systems, 33*, 21810–21823.

Kim, E. S., Herbst, R. S., Wistuba, I. I., Lee, J. J., Blumenschein, G. R., Tsao, A., Stewart, D. J., Hicks, M. E., Erasmus, J., Gupta, S., et al. (2011). The battle trial: Personalizing therapy for lung cancer. *Cancer Discovery, 1*(1), 44–53.

Kosorok, M. R., & Laber, E. B. (2019). Precision medicine. *Annual Review of Statistics and Its Application, 6*, 263–286.

Kosorok, M. R., & Moodie, E. E. (2015). *Adaptive treatment strategies in practice: Planning trials and analyzing data for personalized medicine*. SIAM.

Laber, E. B., Lizotte, D. J., Qian, M., Pelham, W. E., & Murphy, S. A. (2014). Dynamic treatment regimes: Technical challenges and applications. *Electronic Journal of Statistics, 8*(1), 1225.

Laroche, R., Trichelair, P., & Des Combes, R. T. (2019). Safe policy improvement with baseline bootstrapping. In *International Conference on Machine Learning* (pp. 3652–3661). PMLR.

Lattimore, T., & Szepesvári, C. (2020). *Bandit algorithms*. Cambridge University Press.

Liao, P., Qi, Z., & Murphy, S. (2020). Batch policy learning in average reward Markov decision processes. arXiv preprint arXiv:2007.11771.

Luckett, D. J., Laber, E. B., Kahkoska, A. R., Maahs, D. M., Mayer-Davis, E., and Kosorok, M. R. (2020). Estimating dynamic treatment regimes in mobile health using v-learning. *Journal of the American Statistical Association, 115*(530), 692–706.

Marling, C., & Bunescu, R. (2020). The ohiot1dm dataset for blood glucose level prediction: Update 2020. *KHD@ IJCAI*.

Meyn, S. P. (1997). The policy iteration algorithm for average reward Markov decision processes with general state space. *IEEE Transactions on Automatic Control, 42*(12), 1663–1680.

Min, M. R., Ning, X., Cheng, C., & Gerstein, M. (2014). Interpretable sparse high-order Boltzmann machines. In *Artificial Intelligence and Statistics* (pp. 614–622). PMLR.

Mnih, V., Badia, A. P., Mirza, M., Graves, A., Lillicrap, T., Harley, T., Silver, D., & Kavukcuoglu, K. (2016). Asynchronous methods for deep reinforcement learning. In *International Conference on Machine Learning* (pp. 1928–1937).

Muralidharan, S., Ranjani, H., Anjana, R. M., Allender, S., & Mohan, V. (2017). Mobile health technology in the prevention and management of type 2 diabetes. *Indian Journal of Endocrinology and Metabolism, 21*(2), 334.

Murphy, S. (2005). A generalization error for q-learning. *Journal of Machine Learning Research: JMLR, 6*, 1073–1097.

Murphy, S. A. (2003). Optimal dynamic treatment regimes. *Journal of the Royal Statistical Society: Series B (Statistical Methodology), 65*(2), 331–355.

Nahum-Shani, I., Qian, M., Almirall, D., Pelham, W. E., Gnagy, B., Fabiano, G. A., Waxmonsky, J. G., Yu, J., & Murphy, S. A. (2012). Q-learning: A data analysis method for constructing adaptive interventions. *Psychological Methods, 17*(4), 478.

Nair, A., Gupta, A., Dalal, M., & Levine, S. (2020). AWAC: Accelerating online reinforcement learning with offline datasets. arXiv preprint arXiv:2006.09359.

Puterman, M. L. (2014). *Markov decision processes: Discrete stochastic dynamic programming*. John Wiley & Sons.

Rehg, J. M., Murphy, S. A., & Kumar, S. (2017). *Mobile health*. Springer.

Renfro, L. A., Mallick, H., An, M.-W., Sargent, D. J., & Mandrekar, S. J. (2016). Clinical trial designs incorporating predictive biomarkers. *Cancer Treatment Reviews, 43*, 74–82.

Rich, B., Moodie, E. E., & Stephens, D. A. (2014). Simulating sequential multiple assignment randomized trials to generate optimal personalized warfarin dosing strategies. *Clinical Trials, 11*(4), 435–444.

Robins, J. M. (2004). Optimal structural nested models for optimal sequential decisions. In *Proceedings of the Second Seattle Symposium in Biostatistics* (pp. 189–326). Springer.

Ross, S., Pineau, J., Chaib-draa, B., & Kreitmann, P. (2011). A Bayesian approach for learning and planning in partially observable Markov decision processes. *Journal of Machine Learning Research, 12*(48), 1729–1770.

Sarker, H., Hovsepian, K., Chatterjee, S., Nahum-Shani, I., Murphy, S. A., Spring, B., Ertin, E., Al'Absi, M., Nakajima, M., & Kumar, S. (2017). From markers to interventions: The case of just-in-time stress intervention. In *Mobile health* (pp. 411–433). Springer.

Schulte, P. J., Tsiatis, A. A., Laber, E. B., & Davidian, M. (2014). Q-and a-learning methods for estimating optimal dynamic treatment regimes. *Statistical Science: A Review Journal of the Institute of Mathematical Statistics, 29*(4), 640.

Schulz, J., & Moodie, E. E. (2021). Doubly robust estimation of optimal dosing strategies. *Journal of the American Statistical Association, 116*(533), 256–268.

Shi, C., Wan, R., Song, R., Lu, W., & Leng, L. (2020a). Does the Markov decision process fit the data: Testing for the Markov property in sequential decision making. In *International Conference on Machine Learning* (pp. 8807–8817). PMLR.

Shi, C., Zhang, S., Lu, W., & Song, R. (2020b). Statistical inference of the value function for reinforcement learning in infinite horizon settings. arXiv preprint arXiv:2001.04515.

Shi, X., Miao, W., Nelson, J. C., & Tchetgen Tchetgen, E. J. (2020c). Multiply robust causal inference with double-negative control adjustment for categorical unmeasured confounding. *Journal of the Royal Statistical Society: Series B (Statistical Methodology), 82*(2), 521–540.

Siegel, N. Y., Springenberg, J. T., Berkenkamp, F., Abdolmaleki, A., Neunert, M., Lampe, T., Hafner, R., Heess, N., & Riedmiller, M. (2020). Keep doing what worked: Behavioral modelling priors for offline reinforcement learning. arXiv preprint arXiv:2002.08396.

Silva, B. M., Rodrigues, J. J., de la Torre Díez, I., López-Coronado, M., & Saleem, K. (2015). Mobile-health: A review of current state in 2015. *Journal of Biomedical Informatics, 56*, 265–272.

Simoneau, G., Moodie, E. E., Nijjar, J. S., Platt, R. W., Scottish Early Rheumatoid Arthritis Inception Cohort Investigators et al. (2020). Estimating optimal dynamic treatment regimes with survival outcomes. *Journal of the American Statistical Association, 115*(531), 1531–1539.

Steinhubl, S. R., Muse, E. D., & Topol, E. J. (2015). The emerging field of mobile health. *Science Translational Medicine, 7*(283), 283rv3–283rv3.

Sunyaev, A., Dehling, T., Taylor, P. L., & Mandl, K. D. (2015). Availability and quality of mobile health app privacy policies. *Journal of the American Medical Informatics Association, 22*(e1), e28–e33.

Sutton, R. S. (1988). Learning to predict by the methods of temporal differences. *Machine Learning, 3*(1), 9–44.

Sutton, R. S., & Barto, A. G. (2018). *Reinforcement learning: An introduction*. MIT Press.

Tchetgen, E. J. T., Ying, A., Cui, Y., Shi, X., & Miao, W. (2020). An introduction to proximal causal learning. arXiv preprint arXiv:2009.10982.

Thall, P. F., Nguyen, H. Q., Braun, T. M., & Qazilbash, M. H. (2013). Using joint utilities of the times to response and toxicity to adaptively optimize schedule–dose regimes. *Biometrics, 69*(3), 673–682.

Tsitsiklis, J. N., & Van Roy, B. (1996). Feature-based methods for large scale dynamic programming. *Machine Learning, 22*(1), 59–94.

Tsitsiklis, J. N., & Van Roy, B. (1997). An analysis of temporal-difference learning with function approximation. *IEEE Transactions on Automatic Control, 42*(5), 674–690.

Wang, L., Yang, Z., & Wang, Z. (2020a). Provably efficient causal reinforcement learning with confounded observational data. arXiv preprint arXiv:2006.12311.

Wang, R., Foster, D. P., & Kakade, S. M. (2020b). What are the statistical limits of offline RL with linear function approximation? arXiv preprint arXiv:2010.11895.

Watkins, C. J., & Dayan, P. (1992). Q-learning. *Machine Learning, 8*(3–4), 279–292.

Xue, F., Zhang, Y., Zhou, W., Fu, H., & Qu, A. (2021). Multicategory angle-based learning for estimating optimal dynamic treatment regimes with censored data. *Journal of the American Statistical Association, 117*(539), 1438–1451.

Yang, M., Nachum, O., Dai, B., Li, L., & Schuurmans, D. (2020). Off-policy evaluation via the regularized Lagrangian. *Advances in Neural Information Processing Systems, 33*, 6551–6561.

Yu, T., Thomas, G., Yu, L., Ermon, S., Zou, J. Y., Levine, S., Finn, C., & Ma, T. (2020). Mopo: Model-based offline policy optimization. *Advances in Neural Information Processing Systems, 33*, 14129–14142.

Zhan, W., Huang, B., Huang, A., Jiang, N., & Lee, J. (2022). Offline reinforcement learning with realizability and single-policy concentrability. In *Conference on Learning Theory* (pp. 2730–2775). PMLR.

Zhang, S., Boehmer, W., & Whiteson, S. (2020). Deep residual reinforcement learning. In *Proceedings of the 19th International Conference on Autonomous Agents and MultiAgent Systems* (pp. 1611–1619).

Zhao, Y., Kosorok, M. R., & Zeng, D. (2009). Reinforcement learning design for cancer clinical trials. *Statistics in Medicine, 28*(26), 3294–3315.

Zhao, Y., Zeng, D., Rush, A. J., & Kosorok, M. R. (2012). Estimating individualized treatment rules using outcome weighted learning. *Journal of the American Statistical Association, 107*(499), 1106–1118.

Zhao, Y.-Q., Zeng, D., Laber, E. B., & Kosorok, M. R. (2015). New statistical learning methods for estimating optimal dynamic treatment regimes. *Journal of the American Statistical Association, 110*(510), 583–598.

Zhao, Y.-Q., Zeng, D., Laber, E. B., Song, R., Yuan, M., & Kosorok, M. R. (2014). Doubly robust learning for estimating individualized treatment with censored data. *Biometrika, 102*(1), 151–168.

Zhao, Y.-Q., Zhu, R., Chen, G., & Zheng, Y. (2020). Constructing dynamic treatment regimes with shared parameters for censored data. *Statistics in Medicine, 39*(9), 1250–1263.

Zhou, W., Zhu, R., & Qu, A. (2021a). Estimating optimal infinite horizon dynamic treatment regimes via pt-learning. arXiv preprint arXiv:2110.10719.

Zhou, W., Zhu, R., & Zeng, D. (2021b). A parsimonious personalized dose-finding model via dimension reduction. *Biometrika, 108*(3), 643–659.

Zhu, L., Lu, W., Kosorok, M. R., & Song, R. (2020a). Kernel assisted learning for personalized dose finding. In *Proceedings of the 26th ACM SIGKDD International Conference on Knowledge Discovery & Data Mining* (pp. 56–65).

Zhu, L., Lu, W., & Song, R. (2020b). Causal effect estimation and optimal dose suggestions in mobile health. In *International Conference on Machine Learning* (pp. 11588–11598). PMLR.

Q-Learning Based Methods for Dynamic Treatment Regimes

Xinyi Li, Nikki L. B. Freeman, and Lily Wang

Abstract Precision medicine seeks to find the optimal treatments tailored to individual patient characteristics. Dynamic treatment regimes are sequences that formalize the process of decision-making by mapping from patients' observable information to a recommended treatment. Q-learning is a popular approach for estimating an optimal treatment regime. It is closely related to the regression-based analysis in statistics and belongs to the family of reinforcement learning methods. In this chapter, we provide an introduction of Q-learning based methods for the estimation of dynamic treatment regimes. We start with the formal precision medicine framework, followed by the introduction of reinforcement learning. We then delve into Q-learning based methods for dynamic treatment regime in the finite time horizon, including both single-decision setting and multistage decision setting, and infinite time horizon. To concretize the concepts discussed, we present a simple example of Q-learning implementation for the two-stage setting using the R statistical programming language.

Keywords Precision medicine · Reinforcement learning · Causal inference · Markov decision process · Data-driven decision support

---

X. Li
Clemson University, School of Mathematical and Statistical Sciences, Clemson, SC, USA
e-mail: lixinyi@clemson.edu

N. L. B. Freeman
University of North Carolina at Chapel Hill, Department of Biostatistics, Chapel Hill, NC, USA
e-mail: nlbf@live.unc.edu

L. Wang (✉)
George Mason University, Department of Statistics, Fairfax, VA, USA
e-mail: lwang41@gmu.edu

Y. Zhao, D.-G. Chen (eds.), *Statistics in Precision Health*, ICSA Book Series in Statistics, https://doi.org/10.1007/978-3-031-50690-1_5

1 Introduction

The goal of precision medicine is to "match the right treatment to the right patient at the right time" (Ashley, 2015; Collins & Varmus, 2015). Unlike more traditional approaches used to generate evidence for health and healthcare decision-making, the precision medicine paradigm focuses on learning from data the best way to tailor treatment or a sequence of treatments to individuals based on their specific characteristics. This is often referred to as "leveraging heterogeneity." The idea is that by learning how to optimally tailor treatments to individuals, individual outcomes and thereby population outcomes can be improved over those from one-size-fits-all or non-data-driven treatment strategies.

In precision medicine, the formal structure for matching patients to treatments is the dynamic treatment regime (DTR), which is also known as adaptive treatment strategies (Lavori et al., 2000; Murphy, 2005b), adaptive interventions (Collins et al., 2004; Nahum-Shani et al., 2012), treatment policies (Lunceford et al., 2002; Wahed & Tsiatis, 2004), or individualized treatment rules (Petersen et al., 2007). DTRs are sets of treatment rules, one for each key decision point, that map from patient features to an available treatment. The features may include disease and treatment history, clinical, prognostic, and/or genomic characteristics, behaviors, and response to past treatments. In the language of DTRs, the goal of precision medicine then is to learn an optimal DTR, a decision rule that, if followed, would favorably optimize the expected value of a target outcome over the patient population of interest.

In practice, DTRs can be widely applied to the study of chronic diseases or disorders, and are particularly well-poised to support decision-making in disease settings that may require a sequence of medical interventions; examples include cancer (Thall et al., 2007; Zhao et al., 2011), attention deficit hyperactivity disorder (ADHD; Pelham and Fabiano (2008)), human immunodeficiency virus (HIV) infection/acquired immune deficiency syndrome (AIDS) (Cain et al., 2010), and diabetes (Luckett et al., 2020). The schematic in Fig. 1, which is used to describe the possible medical interventions for the treatment of breast cancer in Kidwell (2016), can help us understand DTRs more concretely. As shown in Fig. 1, a patient diagnosed with breast cancer may initially be treated with surgery followed by chemotherapy and radiation. Depending on the toxicity of chemotherapy for the individual patient, the provider will modify the treatment strategy: a break from or reduced dose of chemotherapy if chemotherapy is too toxic for the patient, augmented chemotherapy with a targeted agent if the chemotherapy is not too toxic, and the cancer is not yet eradicated, or an additional aromatase inhibitor if the patient's cancer is eradicated and they are hormone receptor-positive and post-menopausal. Patients whose cancer is eradicated after chemotherapy and are not hormone receptor-positive, or are hormone receptor-positive and not post-menopausal, are surveilled for recurrence but not given any further active therapies.

To illustrate how a DTR might support clinical decision-making, let us reconsider the case of a 60-year-old, hormone receptor-positive and post-menopausal woman

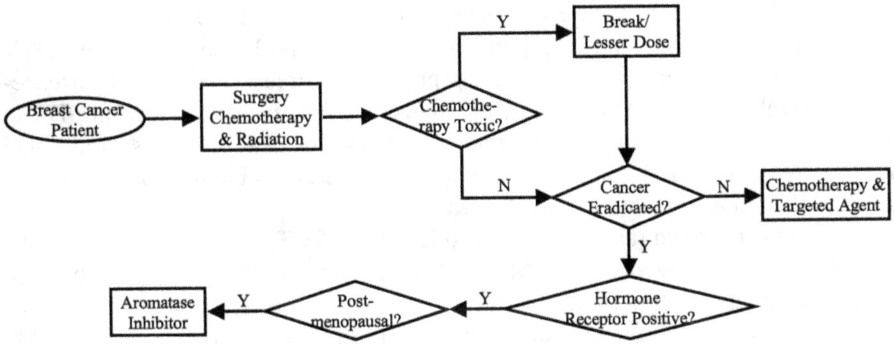

Fig. 1 Schematic depiction of DTR for breast cancer patients reproduced from Kidwell (2016). A patient diagnosed with breast cancer may get the surgery, followed by chemotherapy and radiation at the first stage. The physician will modify this to take a break from chemotherapy and radiation or give a lesser dose if these treatment actions are too toxic to the patient. If the cancer is eradicated by the surgery or these treatment actions, and if the patient is hormone receptor-positive and post-menopausal, an aromatase inhibitor will be prescribed. If the cancer has not been eradicated yet or there is evidence of cancer, the physician may suggest additional chemotherapy and a targeted agent

with stage II breast cancer using the DTR depicted in Fig. 1. For this woman, the treatment recommendations are as follows (Kidwell, 2016):

> Following surgery, treat with chemotherapy for six cycles. If there is no evidence of cancer and the lymph nodes are negative, treat with an aromatase inhibitor for five years. If there is evidence of cancer following chemotherapy, continue chemotherapy for another six cycles. If the patient experiences a grade III or higher toxicity on the prescribed chemotherapy, switch to another chemotherapy.

We note that the DTR provides guidance on how to tailor treatments based on the woman's health and disease status (post-menopausal, lymph node involvement, cancer persistence or eradication, hormone receptor-positive) and response to previous treatment (degree of chemotherapy toxicity)—in other words, how to tailor treatment to this particular patient.

The example DTR outlined in Fig. 1 can also give us an idea of the data and statistical elements needed to learn such a DTR. We observe that in the example, there are multiple decision points at which treatment decisions need to be made and possible treatments for each decision point. These include the decision whether to continue and/or augment chemotherapy and the decision whether to prescribe an aromatase inhibitor. The recommendations rendered by the DTR at each of these decision points, in turn, depend on individual patient features and responses to previous therapy. For example, the decision to continue and/or augment chemotherapy depends on whether the chemotherapy is too toxic for the patient and whether or not the patient's cancer has responded to the previous treatment of surgery in combination with chemotherapy and radiation. Similarly, the decision to prescribe an aromatase inhibitor for patients whose cancer has been eradicated depends on the patient's cancer characteristics, specifically, whether the cancer is hormone receptor-positive. While not explicitly illustrated in Fig. 1, the goal of the DTR is to optimize patient outcomes such as progression-free survival or recurrence-free survival. Thus, at a minimum, the statistical framework for learning DTRs will include decision points, patient features, treatments, and outcomes. Additional features may be included in model-specific settings or problem attributes, but in general, the main elements remain the same.

Data collected from randomized controlled trials (RCTs) and sequentially multiple adaptive randomized trials (SMARTs), as well as observational data, may be used to estimate DTRs. Generally, by design, data from RCTs are suitable for learning DTRs with a single decision point, and data from SMARTs are suitable for learning DTRs with multiple decision points. For each, the decision points correspond to the randomization point(s), and the possible treatments correspond to the treatments being randomized at the randomization points. Point exposure or multiple exposure observational data, too, may be used, although additional assumptions are needed to identify key statistical estimands related to DTRs. For additional discussion on trial and study designs that generate data appropriate for DTR estimation, see Kosorok and Moodie (2016).

Various statistical analysis methods have been developed in support of data-driven decision-making (Kosorok & Laber, 2019; Kosorok & Moodie, 2016; Tsiatis et al., 2019), including outcome weighted learning (OWL) (Zhao et al., 2015, 2012) and related methods, inverse probability weighted (IPW) (Horvitz & Thompson, 1952; Hernán et al., 2004), and doubly robust augmented inverse probability weighted (AIPWE) estimators (Zhang et al., 2012, 2013), among others (see Kosorok & Laber (2019); Kosorok & Moodie (2016); Tsiatis et al. (2019)). Among the methods for learning optimal DTRs, Q-learning is one of the widely used approaches and is our focus in this chapter. Q-learning, where "Q" stands for "quality," is a strategy for learning optimal DTRs by positing models for and estimating Q-functions. As we will see, Q-functions are conditional expectations, which in turn are closely related to the widely adopted regression-based methods in statistics.

We proceed as follows: In Sect. 2, we lay out the precision medicine framework, providing the notation and language of precision medicine used throughout this chapter. This framework sets the statistical goal of learning optimal DTRs. Section 3 formalizes Q-learning in precision medicine and contrasts it with reinforcement learning. Sections 4 and 5 introduce Q-Learning for optimal DTRs in the finite and infinite time horizon settings, respectively. In Sect. 6, we provide a simple example with code to illustrate how to implement Q-learning for the finite horizon setting in R.

2 Introduction to the Precision Medicine Framework

The idea of "matching the right treatment to the right person at the right time" can be formalized concisely as the goal of learning an optimal DTR. Put otherwise, the goal is to learn a sequence of functions, one for each key decision point, that maps from observed patient features to a recommended treatment. An optimal DTR is a DTR that, if followed, would optimize outcomes on average in the target population. In this section, we develop this idea formally.

2.1 Notation and Potential Outcomes

We begin by discussing causal inference and the potential outcomes, or counterfactual, framework. The concept of potential outcomes was first proposed by Neyman (1923) for causal effects in the context of completely randomized experiments, and was then extended to a general framework for causation by Rubin (1974) in both observational and randomized studies. In the standard causal inference setting, or as one might describe as the single timepoint decision setting, a potential outcome is defined as the outcome a subject would have had under a particular treatment. For each subject, there is a potential outcome associated with each possible treatment. At most, only one potential outcome is realized for each individual, and the not-realized potential outcomes are thus counter to fact. Formally, we let $\mathscr{A}$ denote the set of treatment options and $a \in \mathscr{A}$ denote a particular treatment option. Then, we denote $Y^*(a)$ as the potential outcome that would be observed if a subject were assigned treatment $a \in \mathscr{A}$, and $\{Y^*(a)\}_{a \in \mathscr{A}}$ is the set of all potential outcomes. A more detailed overview of the potential outcomes framework and causal inference can be found in Hernán and Robins (2020) and Pearl (2009).

In the precision medicine setting, however, we are often interested in clinical decision-making over time rather than a single timepoint and thus need a more general notion of potential outcomes that can accommodate potential outcomes under sequences of treatments. We let $t = 1, \ldots, T$ denote the sequence of timepoints at which a treatment decision is made and $T < \infty$ the total number of decision points. Later we will consider the case with infinitely many or an indefinite number of decision points (Sect. 5). Potential outcomes in the multiple decision point (multiple stage) setting are slightly more complicated than those in the single decision point (single stage) setting. Most obviously, instead of considering a single treatment assigned at a single decision point, we will need to consider entire sequences of treatments. To denote a sequence of treatments, we will use an overline, for example, $\overline{\boldsymbol{a}}_t = (a_1, \ldots, a_t)$. Additionally, we define $\overline{\boldsymbol{a}} \equiv \overline{\boldsymbol{a}}_T = (a_1, \ldots, a_T)$. Because the treatments that are available at each decision point may differ, the set of possible treatments available at time t is denoted by $\mathscr{A}_t$ and we denote the set of possible treatments that could be realized by the sequence $\overline{\boldsymbol{a}}_t$ as $\mathscr{A}_1 \times \cdots \times \mathscr{A}_t$. Letting Y denote the outcome of interest measured after the T-th treatment decision, we let $Y^*(\overline{\boldsymbol{a}})$ denote the potential outcome we would observe if an individual received a particular sequence of treatments $\overline{\boldsymbol{a}} \in \mathscr{A}_1 \times \cdots \times \mathscr{A}_T$.

In addition to treatments and final outcomes, precision medicine often involves incorporating patient features that evolve over time as well as proximal outcomes that are collected in close proximity to when a treatment is administered. An example of proximal outcomes can be found in mobile health (mHealth) studies (Liao et al., 2021). In mHealth interventions, mobile devices can deliver treatments/interventions to individuals during daily lives, and those outcomes collected in the near time the treatment is taken are considered proximal outcomes, such as stress or physical activity. Because the treatments applied at time t affect both proximal outcomes and interim measurements, when extending the potential outcomes framework to multiple treatments over time, we will need to consider both potential proximal outcomes and potential interim measurements. We let $A_t \in \mathscr{A}_t$ denote the assigned treatment, Y_t denote a proximal outcome measured after the treatment at stage t, and let $Y_t^*(\overline{\boldsymbol{a}}_t)$ be the potential proximal outcome we would observe if $\overline{\boldsymbol{a}}_t \in \mathscr{A}_1 \times \cdots \times \mathscr{A}_t$ had been assigned for $t = 1, \ldots, T-1$. Further, we let $\mathbf{X}_1 \in \mathscr{X}_1$ denote baseline patient information at time $t = 1$, and for $t = 2, \ldots, T$, $\mathbf{X}_t \in \mathscr{X}_t$ stand for interim patient information collected after the $(t-1)$-th decision and before the t-th decision is to be made. Baseline patient information, also referred to as baseline covariates or baseline features, might include co-morbid conditions, biomarkers, disease status, and other patient characteristics observed at the time of the first decision. Interim patient information might include patient features, or covariates, that evolve over time and possibly in response to prior treatments. We let $\mathbf{X}_t^*(\overline{\boldsymbol{a}}_{t-1})$ denote the potential interim measurements of patient information at time t we would observe if treatment sequence $\overline{\boldsymbol{a}}_{t-1} \in \mathscr{A}_1 \times \cdots \times \mathscr{A}_{t-1}$ had been assigned.

Thus far, we have considered potential outcomes for sequences of treatments rather than treatment rules which are the central target of precision medicine. To extend from sequences of treatments to treatment rules, the history notation is convenient. Define $\mathbf{H}_t$ as the history data at time t, so $\mathbf{H}_1 = \mathbf{X}_1$ and $\mathbf{H}_t = (\mathbf{H}_{t-1}, A_{t-1}, Y_{t-1}, \mathbf{X}_t)$; that is, from t to $t+1$, we update the history with the treatment and response collected at t, and the patient covariates collected between t and $t+1$. An important technical note is that not all treatments in $\mathscr{A}_t$, $t = 1, \ldots, T$, may be feasible for all patients. This may be because of drug interactions, past history of drug intolerability, a previous therapy that precludes a future therapy (e.g., amputation of a limb precludes future amputation of the same limb), or other contraindications to a particular treatment or course of therapy. Using the history notation we formalize feasible treatments; letting $\mathscr{H}_t$ be the space of all possible histories, we denote the set of feasible treatments for a patient presenting with $\mathbf{H}_t = \boldsymbol{h}_t$ at time t as $\psi_t(\boldsymbol{h}_t) \in \mathscr{A}_t$.

2.2 DTRs and Optimal DTRs

With some notation established, we can now formally define a dynamic treatment regime (DTR). A DTR is a sequence of mappings $\boldsymbol{d} = (d_1, \ldots, d_T)$ such that for $t = 1, \ldots, T$, $d_t : \mathscr{H}_t \to \mathscr{A}_t$, and $d_t(\boldsymbol{h}_t) \in \psi_t(\boldsymbol{h}_t)$ for all $\boldsymbol{h}_t$. That is, at each decision point, a DTR maps from known patient features and history up to that point to a treatment. Our interest is in finding the optimal DTR, the DTR that, if followed, would lead to the greatest expected value of the outcome (assuming that larger outcomes are better). DTRs may take many forms, ranging from simple decision lists to complicated non-linear functions of features and treatments. We will denote the class of regimes under consideration as $\mathscr{D}$. The optimal DTR is how the precision medicine idea of "leveraging heterogeneity" is operationalized. Patient outcomes may be improved by assigning treatments tailored to patients' unique features and disease histories, and the optimal DTR is the rule that tells us how to make optimal tailored treatment assignments.

We formalize the concepts of treatment regime and interim information using the example of a two-stage setting. In this setting, treatments are assigned according to a treatment regime $\boldsymbol{d} = (d_1, d_2)$, and we denote the corresponding interim information and final outcome as follows.

$$\mathbf{X}_2^*(d_1) = \sum_{a_1 \in \mathscr{A}_1} \mathbf{X}_2^*(a_1) I\{d_1(\mathbf{X}_1) = a_1\},$$
$$Y^*(\boldsymbol{d}) = \sum_{(a_1, a_2) \in \mathscr{A}_1 \times \mathscr{A}_2} I\{d_1(\mathbf{X}_1) = a_1, d_2(\mathbf{X}_1, a_1, \mathbf{X}_2^*(a_1)) = a_2\} Y^*(a_1, a_2), \tag{1}$$

where $I(\cdot)$ is an indicator function. For an arbitrary T, the potential interim information and potential outcome can be defined analogously.

Define the (marginal) value function of a regime $\boldsymbol{d}$ as

$$V(\boldsymbol{d}) = \mathbb{E}\{Y^*(\boldsymbol{d})\}, \tag{2}$$

the expected value of the outcome if all individuals received treatment assignment according to $\boldsymbol{d}$. Finally, an optimal treatment regime $\boldsymbol{d}^{\text{opt}}$ maximizes the expectation of a prespecified cumulative outcome measure Y and satisfies:

(a) for $t = 1, \ldots, T$, $\forall \boldsymbol{h}_t \in \mathscr{H}_t$, $d_t^{\text{opt}}(\boldsymbol{h}_t) \in \psi_t(\boldsymbol{h}_t)$;
(b) for $t = 1, \ldots, T$, $\forall \boldsymbol{h}_t \in \mathscr{H}_t$, for $\boldsymbol{d}$ satisfying $d_t(\boldsymbol{h}_t) \in \psi_t(\boldsymbol{h}_t)$, $\mathbb{E}\{Y^*(\boldsymbol{d}^{\text{opt}})\} \geq \mathbb{E}\{Y^*(\boldsymbol{d})\}$.

That is, an optimal dynamic treatment regime is a regime that, if followed, would yield the greatest expected outcome when compared to other regimes in $\mathscr{D}$.

2.3 *Identification*

In precision medicine, our goal is to use the observed data to learn or estimate $\boldsymbol{d}^{\text{opt}}$. Thus far, however, we have described the decision-making framework and the value of DTRs in terms of potential outcomes. Because patients can receive only one treatment at each decision point, only the potential outcomes corresponding to realized treatment histories are observed, and the rest are unobserved. Without further assumptions, $V(\boldsymbol{d})$ cannot be identified from the observed data. Fortunately, the application of the g-computation algorithm (Robins, 1986) from the causal inference literature shows how $V(\boldsymbol{d})$ can be identified from the data under certain assumptions.

Suppose the observed data consists of n i.i.d. trajectories of the longitudinal data $\{(\mathbf{X}_t, A_t, Y_t)\})_{t=1}^{T}$, presented in the form $\{(\mathbf{X}_{1,i}, A_{1,i}, Y_{1,i}, \ldots, \mathbf{X}_{T,i}, A_{T,i}, Y_{T,i})\}_{i=1}^{n}$. To identify the value of a regime using observed data, we will assume that the following identifying assumptions hold:

1. Stable unit treatment value assumption (SUTVA): also referred to as the consistency assumption (Rubin, 1978), which can be formalized as

$$\mathbf{X}_t = \mathbf{X}_t^* \left(\overline{\mathbf{A}}_{t-1}\right) = \sum_{\overline{\boldsymbol{a}}_{t-1} \in \mathscr{A}_1 \times \cdots \times \mathscr{A}_{t-1}} \mathbf{X}_t^* \left(\overline{\boldsymbol{a}}_{t-1}\right) I\left(\overline{\mathbf{A}}_{t-1} = \overline{\boldsymbol{a}}_{t-1}\right), \quad t = 2, \ldots, T,$$

$$Y_t = Y_t^* \left(\overline{\mathbf{A}}_t\right) = \sum_{\overline{\boldsymbol{a}}_t \in \mathscr{A}_1 \times \cdots \times \mathscr{A}_t} Y_t^* \left(\overline{\boldsymbol{a}}_t\right) I\left(\overline{\mathbf{A}}_t = \overline{\boldsymbol{a}}_t\right), \quad t = 1, \ldots, T.$$

   In other words, the potential outcome is consistent with the observed outcome under the treatment option. SUTVA also implies that there is no interference between subjects and any treatment variation is irrelevant.
2. Sequential randomization assumption (SRA): also known as the no unmeasured confounders (Robins, 1986, 1997), sequential ignorability, or exchangeability assumption in the multiple decision setting, which can be formalized as for any regime $\overline{\boldsymbol{a}}_t$, $t = 1, \ldots, T$,

$$\{\mathbf{X}_{t+1}^*(\overline{\boldsymbol{a}}_t), \ldots, \mathbf{X}_T^*(\overline{\boldsymbol{a}}_{T-1}), Y_t^*(\overline{\boldsymbol{a}}_t), \cdots, Y^*(\overline{\boldsymbol{a}}_T)\} \perp A_t | \mathbf{H}_t.$$

   In other words, for any possible regime $\overline{\boldsymbol{a}}_t$, the current treatment A_t one receives is independent of all future potential covariates or outcomes given the observed histories $\mathbf{H}_t$, $t = 1, \ldots, T$, respectively. In randomized trials like SMARTs, the treatments are assigned at random; thus, SRA is automatically satisfied. However, in observational studies, the SRA is untestable because it is impossible to verify from the observed data, and to tell from the data at hand if there are additional associated variables not recorded. SRA can be formalized in different ways, ranging from weak to strong versions. For further discussion, see Tsiatis et al. (2019).
3. Positivity: the positivity assumption guarantees the positive probability of each treatment for each possible realized history. Without positivity, the effect of the treatment regime might not be estimable. The positivity assumption can be formalized as: for all treatments a_t that are feasible for the history $\boldsymbol{h}_t$,

$$P\left(A_t = a_t | \mathbf{H}_t = \boldsymbol{h}_t\right) > 0, \quad t = 1, \ldots, T.$$

With these identifiability assumptions, we can introduce the conditional expectation for the outcome given covariates and treatments, that is, Q-function, which we focus on and give a detailed introduction through the following of the chapter.

3 Q-Learning for Precision Medicine

In this section, we formalize Q-learning in precision medicine and contrast with reinforcement learning. We provide a conceptual overview of reinforcement learning, and illustrate how it can be linked with Q-learning as well as the distinctions between the two.

3.1 An Incomplete Conceptual Overview of Reinforcement Learning to Q-Learning

In computer science, reinforcement learning (RL) is a type of machine learning characterized by a learning agent and an environment that the agent wants to learn about, as illustrated in Fig. 2. At each decision point, the agent observes the state of their current environment and chooses an action in the action set. Then, the environment responds to the action by transitioning to a new state, and the agent observes a reward that corresponds to the immediate desirability of the action the agent chose. RL problems can be characterized in many different ways. One key descriptor is the time horizon or the number of stages (decision points) and thereby actions an agent takes. When the number of stages is finite, it is called a finite horizon problem, and when the number of stages is infinite, it is called an infinite horizon problem. Much of the RL literature has traditionally focused on online learning, where the agent interacts with the environment and uses experiences from those interactions to improve their decision-making. However, in many medical contexts, this is not feasible, and optimal decisions must be learned from a sample of previously collected data. When optimal policies are learned solely from previously collected data, the RL problem is described as offline.

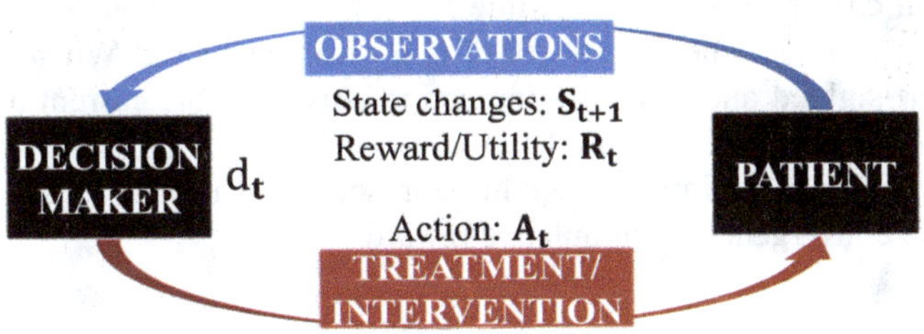

Fig. 2 Reinforcement learning flowchart

Table 1 Corresponding terminologies between the dynamic treatment regimes and reinforcement learning literature (the original table was provided in Clifton and Laber (2020))

Generic	Reinforcement learning	Dynamic treatment regimes	Symbol
Observation unit index	Trajectory	Patient	i
Time index	Time (or time point)	Stage	t
Utility	Reward (R_t)	Outcome	Y, y
Context	State ($\mathbf{S}_t$)	Covariate	X, x
Decision	Action	Treatment	A, a
Decision strategy	Policy	(Dynamic) treatment regime	d

We will mostly use the notation previously established in Sect. 2 throughout our formal development of RL. As the language of RL differs from that of statistics and precision medicine, before continuing, we will comment on some of these differences. In the parlance of RL, the environment in which the agent is making decisions is referred to as the set of states, usually denoted as $\mathbf{S}_t$; in precision medicine, these states correspond to patient features $\mathbf{X}_t$. Similarly, the action the agent takes at time t in the RL setting corresponds to treatment decisions A_t in the precision medicine setting, and rewards (usually denoted as R_t) correspond to proximal outcomes Y_t along with the terminal outcome $Y \equiv Y_T$. The functions mapping from the history of previous states, previous actions, and the current state to the next action are called policies in RL and correspond to DTRs d. However, it is worth noting that the term DTR is commonly used to refer to finite time horizon policies exclusively. In Table 1, we summarize the corresponding terminology between RL and statistical precision medicine, which was originally introduced in Clifton and Laber (2020).

In RL, rewards are conceptualized as known functions y of the history, current action, and the next state; that is,

$$Y_t = y_t(\mathbf{H}_t, A_t, \mathbf{X}_{t+1}) = y_t(\overline{\mathbf{X}}_t, \overline{\mathbf{A}}_t, \mathbf{X}_{t+1}), \tag{3}$$

and as before, we will assume that larger outcomes are better. The goal of RL is to learn the map, i.e., policy, from state space $\mathscr{H}_t$ to action space $\mathscr{A}_t$ to maximize the total (discounted) reward. For ease of exposition, we let $T = \infty$ for the remainder of this subsection as we develop the main ideas from RL, and we will return to the finite horizon case in Sect. 3.2. For time t, $1 \leq t \leq \infty$, the goal can be written as $\max \sum_{k \geq 0} \gamma^k Y_{t+k}$, where $\gamma \in [0, 1]$ is a discount factor. When $\gamma = 0$, the agent is short-sighted and evaluates the action only with the current award; when γ approaches 1, the agent learns the action in each iteration more based on the long-term reward. In the infinite time horizon setting, it is required that $\gamma < 1$ to guarantee the convergence of cumulative reward.

An important function in RL is the state-value function, the total expected future rewards of an agent in a particular state if the agent were to follow the given policy thereafter. Letting d be a policy from the class of all policies $\mathscr{D}$, we can write the value function at time t for state $\boldsymbol{h}_t$ as

$$V_t^d(\boldsymbol{h}_t) = \mathbb{E}_d\left(\sum_{k\geq 0}\gamma^k Y_{t+k}\middle|\mathbf{H}_t = \boldsymbol{h}_t\right), \tag{4}$$

where $\mathbb{E}_d$ denotes that the expectation is taken with respect to the trajectories that would be generated under policy d. Immediately, it can be seen that one could find the optimal policy, the policy that yields the highest (discounted) reward, starting at time t in state $\boldsymbol{h}_t$ by taking the policy that yields the highest value of (4). Except for the simplest of settings, using (4) in this way is impractical. Instead, in RL, one typically uses a recursive formulation of the state-value function to learn optimal policies:

$$\begin{aligned}
V_t^d(\boldsymbol{h}_t) &= \mathbb{E}_d\left(\sum_{k\geq 0}\gamma^k Y_{t+k}\middle|\mathbf{H}_t = \boldsymbol{h}_t\right)\\
&= \mathbb{E}_d\left(Y_t\middle|\mathbf{H}_t = \boldsymbol{h}_t\right) + \mathbb{E}_d\left(\sum_{k\geq 1}\gamma^k Y_{t+k}\middle|\mathbf{H}_t = \boldsymbol{h}_t\right)\\
&= \mathbb{E}_d\left(Y_t\middle|\mathbf{H}_t = \boldsymbol{h}_t\right) + \mathbb{E}_d\left\{\mathbb{E}\left(\gamma\sum_{k\geq 0}\gamma^k Y_{(t+1)+k}\middle|\mathbf{H}_{t+1}\right)\middle|\mathbf{H}_t = \boldsymbol{h}_t\right\}\\
&= \mathbb{E}_d\left(Y_t\middle|\mathbf{H}_t = \boldsymbol{h}_t\right) + \gamma\mathbb{E}_d\left\{V_{t+1}^d\left(\mathbf{H}_{t+1}\middle|\mathbf{H}_t = \boldsymbol{h}_t\right)\right\}\\
&= \mathbb{E}_d\left\{Y_t + \gamma V_{t+1}^d\left(\mathbf{H}_{t+1}\right)\middle|\mathbf{H}_t = \boldsymbol{h}_t\right\}.
\end{aligned} \tag{5}$$

As we will see, this recursive formulation of the state-value function, and later the action-value function, will be essential for learning optimal policies in the infinite time horizon setting and gives us a general strategy for learning optimal policies.

In terms of the time t state-value function, we can express our goal of learning an optimal policy as

$$V_t^{\text{opt}}(\boldsymbol{h}_t) = \max_{d\in\mathscr{D}} V_t^d(\boldsymbol{h}_t). \tag{6}$$

Put otherwise, an optimal policy is the policy in $\mathscr{D}$ that if we start in state h_t maximizes the total (discounted) future reward, and we let V^{opt} denote the optimal state-value. We say "an" optimal policy as opposed to "the" optimal policy since more than one policy in $\mathscr{D}$ may optimize the state-value function. Optimal polices are related to the state-value function via the Bellman optimality equation:

$$
\begin{aligned}
V_t^{\text{opt}}(h_t) &= \max_{a_t \in \mathscr{A}_t} \mathbb{E}_{d^{\text{opt}}} \left(\sum_{k \geq 0} \gamma^k Y_t \middle| \mathbf{H}_t = \boldsymbol{h}_t, A_t = a_t \right) \\
&= \max_{a_t \in \mathscr{A}_t} \mathbb{E} \left\{ Y_t + \gamma \sum_{k \geq 1} \gamma^k Y_{(t+1)+k} \middle| \mathbf{H}_t = \boldsymbol{h}_t, A_t = a_t \right\} \\
&= \max_{a_t \in \mathscr{A}_t} \mathbb{E}_{d^{\text{opt}}} \left\{ Y_t + \gamma V_{t+1}^{\text{opt}}(\mathbf{H}_{t+1}) \middle| \mathbf{H}_t = \boldsymbol{h}_t, A_t = a_t \right\}.
\end{aligned} \tag{7}
$$

One can see that the optimal policy satisfies

$$
d_t^{\text{opt}}(h_t) \in \arg\max_{a_t \in \mathscr{A}_t} \mathbb{E} \left\{ Y_t + \gamma V_{t+1}^{\text{opt}}(\mathbf{H}_{t+1}) \middle| \mathbf{H}_t = \boldsymbol{h}_t, A_t = a_t \right\}. \tag{8}
$$

In precision medicine, we will often be interested in the marginal value function

$$
V^d = \mathbb{E}_{\mathbf{H}_1} \left\{ V^d(\boldsymbol{h}_1) \right\}, \tag{9}
$$

where $\mathbb{E}_{\mathbf{H}_1}$ denotes expectation with respect to the baseline distribution of patient features. Averaging over the baseline features can be thought of as taking the average value of a policy over the population to which it is applied. The marginal value function corresponds to the value function from the precision medicine framework defined in Sect. 2.2.

Before going further, it is worth mentioning that the first class of methods developed to solve multistage decision problems is called dynamic programming and was introduced by Bellman (1957). Dynamic programming is not the most practical strategy because it depends on fully knowing the system dynamics of the learning environment, although work by Bertsekas and Tsitsiklis (1996), Ernst et al. (2005), and Powell (2007) have adapted dynamic programming to environments with unknown transition dynamics. A breakthrough in RL came in 1989 when Watkins developed Q-learning, a way to learn optimal policies based on sample data trajectories. This approach is known as off-policy learning as it allows for learning policies outside of the policy that generated the observed trajectories. Broadly speaking, in Q-learning, one seeks to find the action that yields the high expected reward given the current state rather than determining the expected value of a given state. Q-learning methods focus on the Q-function, which returns the quality of an action. The Q-function, also referred to as the action-value function, at time t, is defined as

$$
Q_t^d(\boldsymbol{h}_t, a_t) = \mathbb{E}_d \left(\sum_{k \geq 0} \gamma^k Y_{t+k} \middle| \mathbf{H}_t = \boldsymbol{h}_t, A_t = a_t \right). \tag{10}
$$

The Q-function at time t yields the total (discounted) expected return if one were in state $\boldsymbol{h}_t$, took action a_t, and then followed policy d thereafter. Like the value function, the Q-function can be defined recursively

$$Q_t^d(\boldsymbol{h}_t, a_t) = \mathbb{E}_d \left\{ Y_t + \gamma Q_{t+1}^d(\mathbf{H}_{t+1}, A_{t+1}) \middle| \mathbf{H}_t = \boldsymbol{h}_t, A_t = a_t \right\}. \tag{11}$$

Similar to the development of the optimal state-value function, the optimal Q-function relates the recursive formulation of the Q-function to an optimal policy

$$Q^{\text{opt}}(\boldsymbol{h}_t, a_t) = \mathbb{E}_{d^{\text{opt}}} \left\{ Y_t + \gamma \max_{a_{t+1} \in \mathscr{A}_{t+1}} Q^{\text{opt}}(\mathbf{H}_{t+1}, a_{t+1}) \middle| \mathbf{H}_t = \boldsymbol{h}_t, A_t = a_t \right\}, \tag{12}$$

and the optimal decision at time t can be derived directly from the Q-function as

$$d_t^{\text{opt}} = \arg\max_{a_t \in \mathscr{A}_t} Q_t^{\text{opt}}(\boldsymbol{h}_t, a_t). \tag{13}$$

Before connecting RL to the precision medicine context, we mention that the value function is closely related to the Q-function, i.e., $V_t^d(\mathbf{h}_t) = \max_{a_t \in \mathscr{A}_t} Q_t(\mathbf{h}_t, a_t)$.

3.2 Q-Learning in the Context of Precision Medicine

So far, we have described RL generally. While the framework echoes medical decision-making, some aspects differ from the precision medicine framework described in Sect. 2. For example, we described medical decision-making in terms of potential outcomes which are generally not part of the RL setting. Moreover, we have purposefully, to this point, avoided making any explicit assumptions about the transition probabilities between states, and some readers familiar with the RL literature will have certainly noticed that we have not (yet) invoked the Markov Decision Process (MDP) structure (although we will shortly). In fact, we implicitly allowed for non-Markovian transition probabilities as we wrote treatment regimes as functions of the entire accrued history of patients. Susan Murphy's seminal works (Murphy, 2003, 2005a) united the RL literature with the potential outcomes framework, and regression modeling. Her work provided an algorithm for the finite horizon precision medicine setting to learn the optimal dynamic treatment by recursively estimating Q-functions, one for each decision point. This method avoids the need to know or estimate the transition dynamics, making it possible to use Q-learning for optimal policy guidance without assuming the overall MDP structure or knowing the specific transition probability from one stage to another.

For the remainder of this section, we will describe the optimal Q-functions for learning optimal policies in the precision medicine context, first in the finite horizon setting and then in the infinite horizon setting. Learning the optimal Q-functions is essential for learning optimal policies using Q-learning, and we will discuss some of the nuances of applying Q-learning in the precision medicine framework. Later, Sects. 4 and 5 will focus on the implementation of Q-learning in the finite horizon and infinite horizon settings.

For precision medicine problems in the finite horizon setting, the optimal Q-functions are defined as follows:

$$Q_T^{\text{opt}}(\boldsymbol{h}_T, a_T) = \mathbb{E}\left(Y_T | \mathbf{H}_t = \boldsymbol{h}_t, A_T = a_T\right), \text{ and}$$
$$Q_t^{\text{opt}}(\boldsymbol{h}_t, a_t) = \mathbb{E}\left\{Y_t + \gamma \max_{a \in \psi(\mathbf{H}_{t+1})} Q_{t+1}(\mathbf{H}_{t+1}, a) \Big| \mathbf{H}_t = \boldsymbol{h}_t, A_t = a_t\right\} \tag{14}$$

for $t = 1, \ldots, T-1$. We observe that once Q_T^{opt} is defined, the $t = 1, \ldots, T-1$ optimal Q-functions are defined recursively as in (12). The maximization in (14) is taken over the feasible treatments $\psi(\mathbf{H}_{t+1}) \in \mathscr{A}_{t+1}$ rather than $\mathscr{A}_{t+1}$ as in (12). This is to reflect the realities of medical decision-making where not all treatments may be feasible for all patients, e.g., due to an allergy or a known counterindication for a particular treatment, a situation that is generally not encountered in general reinforcement learning problems. Often, in the finite horizon precision medicine setting, γ is taken to be 1 so that the reward is the undiscounted sum of the proximal and terminal outcomes. Futhermore, it is often the case that there is only one terminal outcome of interest Y_T, and we take $Y_t = 0$ for $t = 1, \ldots, T$. As we describe in Sect. 4, the algorithm for learning the optimal DTR using (14) proceeds iteratively via backwards induction.

To estimate optimal DTRs in the infinite time horizon setting in precision medicine, additional modeling assumptions are required. A common strategy is to model decision-making as an MDP as in the traditional RL literature. An MDP assumes constant states that describe an individual's health status and constant actions available at each time point $t = 1, \ldots, T$, denoted as $\mathscr{X}$ and $\mathscr{A}$, respectively. It is also assumed that the reward function $y : \mathscr{X} \times \mathscr{A} \times \mathscr{X} \to \mathbb{R}$ is constant, for $t = 1, \ldots, T$. That means, for an individual with health status $\boldsymbol{x}$, receiving treatment a, and transition to health status $\boldsymbol{x}'$, the proximal reward, $y(\boldsymbol{x}, a, \boldsymbol{x}')$, is constant. Moreover, an MDP assumes that the transitions between states are time-homogeneous and Markovian for all t so that formally $P(\cdot|\mathbf{H}_t, A_t) = P(\cdot|\mathbf{X}_t, A_t)$ (Sutton & Barto, 2018). In this setting, the optimal infinite-horizon Q-function is defined as

$$Q^{\text{opt}}(\boldsymbol{x}, a) = \mathbb{E}_{d^{\text{opt}}}\left(\sum_{t=1}^{\infty} \gamma^{t-1} Y_t \Big| \mathbf{X}_1 = \boldsymbol{x}, A_1 = a\right), \tag{15}$$

where as before $\gamma \in [0, 1)$ is a discount factor. Because the optimal infinite-horizon Q-function is the Q-function that corresponds to the optimal policy d^{opt}, a general strategy of Q-learning in this setting is to estimate Q^{opt} and thereby learn the optimal DTR. Rather than using the optimal infinite-horizon Q-function directly, a recursive formulation is used to estimate Q^{opt}. Observe that

$$
\begin{aligned}
Q^{\text{opt}}(\boldsymbol{x}, a) &= \mathbb{E}_{d^{\text{opt}}}\left(\sum_{t=1}^{\infty} \gamma^{t-1} Y_t \middle| \mathbf{X}_1 = \boldsymbol{x}, A_1 = a\right) \\
&= \mathbb{E}\left\{Y_1 + \gamma \max_{a'} \mathbb{E}_{d^{\text{opt}}}\left(\sum_{t=1}^{\infty} \gamma^{t-1} Y_{t+1} | \mathbf{X}_{t+1}, A_1 = a'\right) \middle| \mathbf{X}_1 = \boldsymbol{x}, A_1 = a\right\} \\
&= \mathbb{E}_{d^{\text{opt}}}\left\{Y_1 + \gamma \max_{a'} Q^{\text{opt}}(\mathbf{X}_2, a') \middle| \mathbf{X}_1 = \boldsymbol{x}, A_1 = a\right\},
\end{aligned}
$$

so that in general

$$
Q^{\text{opt}}(\boldsymbol{x}, a) = \mathbb{E}\left\{Y_t + \gamma \max_{a'} Q^{\text{opt}}(\mathbf{X}_{t+1}, a) \middle| \mathbf{X}_t = \boldsymbol{x}, A_t = a\right\},
$$

which is precisely the MDP analogue of (12). We let B denote the Bellman optimality operator defined as

$$
(Bf)(\boldsymbol{x}, a) = \mathbb{E}\left\{Y_t + \gamma \max_{a' \in \mathscr{A}} f(\mathbf{X}_{t+1}, a') | \mathbf{X}_t = \boldsymbol{x}_t, A_t = a\right\} \tag{16}
$$

for a function f, then the optimal infinite-horizon Q-function can be written as $Q^{\text{opt}}(\boldsymbol{x}, a) = (BQ^{\text{opt}})(\boldsymbol{x}, a)$. From this representation, we observe that Q^{opt} is a fixed point and motivates what is called the value-iteration algorithm: $Q_{t+1} \rightarrow BQ_t$ for $T = 1, 2, \ldots$, until convergence. We will see in Sect. 5 how this recursive expression and value iteration are used to estimate the optimal infinite-horizon Q-function.

4 Q-Learning for Optimal DTRs in the Finite Time Horizon Setting

To describe how Q-learning is used in the finite-horizon settings, we start with the single-decision setting ($T = 1$), which is highly instructive before generalizing to the multistage decision setting ($2 \leq T < \infty$). Although we only have one decision point, we still refer to the learned regime as "dynamic" because it still tailors treatments to patient features and patient history, unlike a "static" regime in which the treatment(s) is fixed in advance with a priori.

4.1 Q-Learning for Optimal Single-Stage DTRs

For the single-decision setting, a DTR is a function $d : \mathscr{X} \to \mathscr{A}$ that satisfies $d(\boldsymbol{x}) \in \psi(\boldsymbol{x})$ for all $\boldsymbol{x} \in \mathscr{X}$. As in (1), we denote the potential outcome that would be observed if treatments were assigned according to $d \in \mathscr{D}$ as

$$Y^*(d) = \sum_{a \in \mathscr{A}} Y^*(a) I\{d(\mathbf{X}) = a\}.$$

The value of a DTR and the definition of the optimal DTR are given analogously in Sect. 2.2.

Under the requisite identifying assumptions, the optimal Q-function in the single-decision setting is defined as in (14). Since the first stage is the terminal stage, we simply have

$$Q(\boldsymbol{x}, a) = \mathbb{E}(Y|\mathbf{X} = \boldsymbol{x}, A = a),$$

which suggests $d^{\text{opt}} = \arg\max_{a \in \psi(\boldsymbol{x})} Q(\boldsymbol{x}, a)$ and that the value of regime d^{opt} is

$$V(d^{\text{opt}}) = \mathbb{E}\left\{\max_{a \in \mathscr{A}} Q(\mathbf{X}, a)\right\}.$$

Because the Q-function in this case is a conditional mean model, it is reasonable to propose to estimate the Q-function using regression modeling (Tsiatis et al., 2019). That is, if we posit a regression model for $Q(\boldsymbol{x}, a)$ and suppose the model is correctly specified, we can construct a regression estimator $\widehat{Q}_n(\boldsymbol{x}, a)$ of $Q(\boldsymbol{x}, a)$, and plug it in to get $\widehat{d}^{\text{opt}}(\boldsymbol{x}) = \arg\max_{a \in \psi(\boldsymbol{x})} \widehat{Q}_n(\boldsymbol{x}, a)$.

As a simple illustration, suppose $\mathscr{X} \subset \mathbb{R}$, $\mathscr{A} = \{0, 1\}$, $\psi(x) = \mathscr{A}$ for all $x \in \mathscr{X}$, and Y is continuous. We posit the following parametric regression model for the Q-function:

$$Q(x, a; \boldsymbol{\beta}) = \beta_0 + x\beta_1 + a\beta_2 + ax\beta_3, \tag{17}$$

where $\boldsymbol{\beta} = (\beta_0, \beta_1, \beta_2, \beta_3)^\top$ are regression parameters. We let $\widehat{\boldsymbol{\beta}} = (\widehat{\beta}_0, \widehat{\beta}_1, \widehat{\beta}_2, \widehat{\beta}_3)^\top$ denote estimates of the regression parameters that can be obtained via least-squares regression or other M-estimation methods. Then $\widehat{Q}_n(x, a; \widehat{\boldsymbol{\beta}})$ is our estimator for the Q-function, and the optimal DTR is

$$\begin{aligned}\widehat{d}^{\text{opt}}(x) &= \arg\max_{a \in \psi(x)} \widehat{Q}_n(x, a; \widehat{\boldsymbol{\beta}}) \\ &= \arg\max_{a \in \{0,1\}} \widehat{\beta}_0 + x\widehat{\beta}_1 + a\widehat{\beta}_2 + ax\widehat{\beta}_3\end{aligned}$$

$$= \underset{a \in \{0,1\}}{\arg\max}\, \widehat{\beta}_0 + x\widehat{\beta}_1 + a\left(\widehat{\beta}_2 + x\widehat{\beta}_3\right)$$
$$= I\left(\widehat{\beta}_2 + x\widehat{\beta}_3 > 0\right).$$

Furthermore, $V(d^{\text{opt}})$ can be estimated by

$$\widehat{V}(d^{\text{opt}}) = \mathbb{P}_n \max_{a \in \psi(x)} Q(x_i, a; \widehat{\boldsymbol{\beta}})$$
$$= \mathbb{P}_n \left\{\widehat{\beta}_0 + x_i\widehat{\beta}_1 + I\left(\widehat{\beta}_2 + x_i\widehat{\beta}_3 > 0\right)\left(\widehat{\beta}_2 + x_i\widehat{\beta}_3\right)\right\}.$$

Of note, in Q-learning, the class of treatment regimes under consideration is determined by the Q-function. This is illustrated in (17); the class of regimes under consideration are those of the form $I(\beta_2 + x\beta_3 > 0)$ and indexed by the parameters β_2 and β_3.

4.2 Additional Considerations for the Single-Stage Setting

Before proceeding, it is worth mentioning the relationship between the DTRs and optimal DTRs in the single-decision setting and the conditional average treatment effect (CATE) often encountered in the causal inference and other related literature. To illustrate this relationship, suppose momentarily that $\mathscr{A} = \{0, 1\}$. In this scenario, the CATE is defined as

$$\Delta(\boldsymbol{x}) = \mathbb{E}\{Y^*(1) - Y^*(0)|\mathbf{X} = \boldsymbol{x}\}.$$

The CATE represents the difference in the average outcome that would be achieved if individuals with baseline features $\boldsymbol{x}$ were treated with $A = 1$ and the average outcome that would be achieved if individuals with baseline features $\boldsymbol{x}$ were treated with $A = 0$. In the parlance of DTRs, treating everyone with $A = 1$ is equivalent to the particular DTR $d(\mathbf{X}) \equiv 1$, and the value of this DTR is $\mathbb{E}\{Y^*(1)\}$; similarly, treating everyone in the population with $A = 0$ is equivalent to the particular DTR $d(\mathbf{X}) \equiv 0$ and the value of this particular DTR is $\mathbb{E}\{Y^*(0)\}$. Thus the expected value of the CATE taken over the covariates is the difference of the value of the regime $d(\mathbf{X}) \equiv 1$ and $d(\mathbf{X}) \equiv 0$, i.e., $\mathbb{E}\{\Delta(\mathbf{X})\} = V\{d(\mathbf{X}) \equiv 1\} - V\{d(\mathbf{X}) \equiv 0\}$. Generally, if the optimal treatment regime is an indicator function of the linear combination of the covariates, CATE will give a good estimate for d^{opt}; see more discussions and examples in Kosorok and Laber (2019).

Statistical inference and large sample properties of DTRs and related estimators are of great statistical interest. Consider the single-stage decision setting. In this case, estimating the optimal DTR using Q-learning is relatively straightforward since regression-based analysis may be employed. Asymptotic consistency and asymptotic normality of the regression coefficients are achievable under standard regularity conditions. However, the conventional asymptotic theory cannot be applied directly to the estimators for the maximum value of the regime, even in a simple case. Let's revisit the simple example in (17). It is evident that the asymptotic distributions of $\widehat{\beta}_0$, $\widehat{\beta}_1$, $\widehat{\beta}_2$ and $\widehat{\beta}_3$ are jointly normally distributed; however, $\widehat{V}(d^{\mathrm{opt}})$ is not asymptotically normally distributed due to the non-differentiability of the indicator function. In general, because of the involvement of the non-smooth maximum operator in definition, it is difficult to provide inference for $V(d^{\mathrm{opt}})$, and the difficulty extends to the multistage decision setting; see Tsiatis et al. (2019) for more discussion.

4.3 *Q-Learning for Optimal Finite-Stage DTRs*

In the multistage decision setting ($2 \leq T < \infty$), where there are multiple times at which treatment decisions need to be made, it is crucial to account for both the immediate and long-term impacts of treatment (Kosorok & Laber, 2019). As we will see in this section, the backward induction algorithm can help us do so when there are multiple decisions and thus sequences of treatments at hand.

The characterization of the optimal regime in the multistage decision setting is more complicated than that in the single-decision setting, and the backward induction algorithm is usually employed to define the optimal regime. The backward inductive argument starts at the end and moves backwards through the stages. To demonstrate the reasoning, we first consider a simple two-stage case, that is, $T = 2$. In this case, the Q-functions are defined as

$$\begin{aligned} Q_2(\boldsymbol{h}_2, a_2) &= \mathbb{E}\left(Y|\mathbf{H}_2 = \boldsymbol{h}_2, A_2 = a_2\right), \\ Q_1(\boldsymbol{h}_1, a_1) &= \mathbb{E}\left\{\max_{a_2} Q_2(\boldsymbol{h}_2, a_2)|\mathbf{H}_1 = \boldsymbol{h}_1, A_1 = a_1\right\}, \end{aligned} \tag{18}$$

or equivalently,

$$\begin{aligned} Q_2(\boldsymbol{x}_1, a_1, \boldsymbol{x}_2, a_2) &= \mathbb{E}\left(Y|\mathbf{X}_1 = \boldsymbol{x}_1, A_1 = a_1, \mathbf{X}_2 = \boldsymbol{x}_2, A_2 = a_2\right), \\ Q_1(\boldsymbol{x}_1, a_1) &= \mathbb{E}\left\{\max_{a_2} Q_2(\mathbf{X}_1, A_1, \boldsymbol{x}_2, a_2)|\mathbf{X}_1 = \boldsymbol{x}_1, A_1 = a_1\right\}. \end{aligned}$$

It follows from dynamic programming (Bellman, 1957) that

$$d_2^{\mathrm{opt}}(\boldsymbol{h}_2) = \underset{a_2 \in \psi_2(\boldsymbol{h}_2)}{\arg\max}\ Q_2(\boldsymbol{h}_2, a_2), \quad d_1^{\mathrm{opt}}(\boldsymbol{h}_1) = \underset{a_1 \in \psi_1(\boldsymbol{h}_1)}{\arg\max}\ Q_1(\boldsymbol{h}_1, a_1).$$

Similar to the illustration in the single-decision setting, we consider the simple example that $\mathscr{X} \subset \mathbb{R}$, $\mathscr{A} = \{0, 1\}$, Y is continuous, and the Q-functions are regression models with

$$Q_2(\boldsymbol{h}_2, a_2; \boldsymbol{\beta}_{21}, \boldsymbol{\beta}_{22}) = \boldsymbol{h}_2^\top \boldsymbol{\beta}_{21} + a_2 \left(\boldsymbol{h}_2^\top \boldsymbol{\beta}_{22}\right),$$

$$Q_1(h_1, a_1; \beta_{11}, \beta_{12}) = h_1\beta_{11} + a_1 (h_1\beta_{12}).$$

By solving M-estimating equations, the following steps are involved to obtain the coefficient estimators:

Step I. Stage-2 regression coefficients

$$(\widehat{\boldsymbol{\beta}}_{21}, \widehat{\boldsymbol{\beta}}_{22}) = \underset{\boldsymbol{\beta}_{21}, \boldsymbol{\beta}_{22}}{\arg\min}\, \mathbb{P}_n \left\{Y - \boldsymbol{h}_2^\top \boldsymbol{\beta}_{21} - a_2 \left(\boldsymbol{h}_2^\top \boldsymbol{\beta}_{22}\right)\right\}^2.$$

Step II. Stage-1 regression coefficients

$$(\widehat{\beta}_{11}, \widehat{\beta}_{12}) = \underset{\beta_{11}, \beta_{12}}{\arg\min}\, \mathbb{P}_n \left[\max_{a_2} \left\{\boldsymbol{h}_2^\top \widehat{\boldsymbol{\beta}}_{21} + a_2 \left(\boldsymbol{h}_2^\top \widehat{\boldsymbol{\beta}}_{22}\right)\right\} - h_1\beta_{11} - a_1 (h_1\beta_{12})\right]^2.$$

Once the Q-functions have been estimated, the optimal rule can be estimated by

$$\begin{aligned}
\widehat{d}_2^{\text{opt}}(\boldsymbol{h}_2) &= \underset{a_2}{\arg\max}\, Q_2 \left(\boldsymbol{h}_2, a_2; \widehat{\boldsymbol{\beta}}_{21}, \widehat{\boldsymbol{\beta}}_{22}\right) \\
&= I\left\{Q_2(\boldsymbol{h}_2, 1; \widehat{\boldsymbol{\beta}}_{21}, \widehat{\boldsymbol{\beta}}_{22}) > Q_2(\boldsymbol{h}_2, 0; \widehat{\boldsymbol{\beta}}_{21}, \widehat{\boldsymbol{\beta}}_{22})\right\} = I\left(\boldsymbol{h}_2^\top \widehat{\boldsymbol{\beta}}_{22} > 0\right), \\
\widehat{d}_1^{\text{opt}}(h_1) &= \underset{a_1}{\arg\max}\, Q_1 \left(h_1, a_1; \widehat{\beta}_{11}, \widehat{\beta}_{12}\right) \\
&= I\left\{Q_1(h_1, 1; \widehat{\beta}_{11}, \widehat{\beta}_{12}) > Q_1(h_1, 0; \widehat{\beta}_{11}, \widehat{\beta}_{12})\right\} = I\left(h_1\widehat{\beta}_{12} > 0\right).
\end{aligned}$$

The procedure for determining the optimal regime can be extended to the case where there are more than two decisions, $T > 2$, and (18) can be generalized as

$$\begin{aligned}
Q_T(\boldsymbol{h}_T, a_T) &= \mathbb{E}\left(Y|\mathbf{H}_T = \boldsymbol{h}_T, A_T = a_T\right), \\
Q_t(\boldsymbol{h}_t, a_t) &= \mathbb{E}\left\{\max_{a_{t+1}} Q_{t+1}(\boldsymbol{h}_{t+1}, a_{t+1})|\mathbf{H}_t = \boldsymbol{h}_t, A_t = a_t\right\}, \quad t = 1, \ldots, T-1.
\end{aligned}$$

The estimated optimal DTR is

$$\widehat{d}_t^{\text{opt}} = \underset{a_t}{\arg\max}\, \widehat{Q}_t(\boldsymbol{h}_t, a_t), \quad t = 1, \ldots, T.$$

4.4 Additional Considerations for the General Finite-Stage Setting

Obviously, the success of this approach relies on having a high-quality estimator of the Q-function. An alternative choice to Q-learning is A-learning or advantage learning (Murphy, 2003), which also can be viewed as a special case of g-estimation (Robins, 2004; Stephens, 2016). Both Q-learning and A-learning rely on the Bellman equation to learn the optimal policy as opposed to directly optimizing an objective function, and are sometimes referred to as indirect methods. A-learning does not require full knowledge of the Q-function, i.e., the conditional mean of the outcome; instead, A-learning models contrast functions to construct the estimate. Under the model misspecification for Q-function, A-learning is more robust; and when the model of Q-function is correctly specified, Q-learning is more efficient. One can refer to Schulte et al. (2014) for more details on comparisons, discussions and implementations for these two methods. Finally, we note that as in the single-decision setting, inference for the value of a multistage decision rule is complicated and in general non-regular.

5 Q-Learning for Optimal DTRs in the Infinite Time Horizon Setting

For some diseases, decision support for disease management and prevention requires many similar decisions to be made over time, as illustrated in Fig. 2. For example, individuals living with diabetes make decisions about exercise, food, and insulin dosage multiple times every day to manage their blood glucose. In these cases, an infinite time horizon formulation of the decision problem is used. As previously described in Sect. 3, infinite-horizon problems are modeled as MDPs. Letting $\psi(\boldsymbol{x}_t) \in \mathscr{A}$ denote the feasible treatments for an individual with health status $\boldsymbol{x}_t$ as in the finite time horizon setting, a dynamic treatment regime is a map $d : \mathscr{X} \to \mathscr{A}$ such that $d(\boldsymbol{x}) \in \psi(\boldsymbol{x})$ for all $\boldsymbol{x} \in \mathscr{X}$. Note that by construction, d is stationary, and thus can be applied for all t including $t > T$. As in the finite time horizon setting, the potential outcomes framework is used to characterize the decision-making problem. Then we can write the potential health status if regime d had been followed as

$$\mathbf{X}_t^*(d) = \sum_{\overline{\boldsymbol{a}}_{t-1} \in \mathscr{A}_1 \times \cdots \times \mathscr{A}_{t-1}} \mathbf{X}_t^*(\overline{\boldsymbol{a}}_{t-1}) \prod_{k=1}^{t-1} I\left[d\{\mathbf{X}_k^*(\overline{\boldsymbol{a}}_{k-1})\} = a_k\right],$$

and the potential momentary outcome for that would be observed under d as $Y_t^*(d) = y(\mathbf{X}_t^*(d), d(\mathbf{X}_t^*(d)), \mathbf{X}_{t+1}^*(d))$, where $y(\cdot, \cdot, \cdot)$ is the reward function defined in Sect. 3.2.

We assume that the observed data used to learn d^{opt} are n i.i.d. trajectories $\{(\mathbf{X}_{1,i}, A_{1,i}, \mathbf{X}_{2,i}, \ldots, \mathbf{X}_{T-1,i}, A_{T-1,i}, \mathbf{X}_{T,i})\}_{i=1}^{n}$. Although data are collected over a finite time horizon T, the Markovian structure of the MDP enables us to learn an optimal DTR that can be applied repeatedly over time. Under the same identifying assumptions as in the finite horizon setting, the optimal infinite-horizon Q-function is identified, and we repeat its Bellman equation representation here:

$$Q^{d^{\mathrm{opt}}}(\boldsymbol{x}, a) = \mathbb{E}_{d^{\mathrm{opt}}}\left(\sum_{t=1}^{\infty} \gamma^{t-1} Y_t \middle| \mathbf{X}_1 = \boldsymbol{x}, A_1 = a\right) = (BQ^{\mathrm{opt}})(\boldsymbol{x}, a).$$

With the increasing prevalence of mHealth and sensors requiring ongoing management, there is growing interest in the infinite-horizon precision medicine problem, and it is an open area of research. Following Clifton and Laber (2020), we describe two such methods that have been proposed using Q-functions. The first is an estimating equation approach proposed by Ertefaie and Strawderman (2018). They model the optimal infinite-horizon Q-function with a differentiable parametric model $Q(\cdot; \boldsymbol{\theta})$. Substituting this into the Bellman equation yields

$$Q(\boldsymbol{x}, a; \boldsymbol{\theta}) = \mathbb{E}\left\{Y_t + \gamma \max_{a'} Q(\mathbf{X}_{t+1}, a'; \boldsymbol{\theta}) | \mathbf{X}_t = \boldsymbol{x}, A_t = a\right\}.$$

Then it follows that

$$\begin{aligned}
&\mathbb{E}\left\{Y_t + \gamma \max_{a'} Q(\mathbf{X}_{t+1}, a'; \boldsymbol{\theta}) - Q(\boldsymbol{x}, a; \boldsymbol{\theta}) | \mathbf{X}_t = \boldsymbol{x}, A_t = a\right\} = 0 \\
&\Rightarrow \mathbb{E}\Big[\left\{Y_t + \gamma \max_{a'} Q(\mathbf{X}_{t+1}, a'; \boldsymbol{\theta}) - Q(\boldsymbol{x}, a; \boldsymbol{\theta})\right\} \\
&\nabla_{\boldsymbol{\theta}} Q(\boldsymbol{x}, a; \boldsymbol{\theta}) | \mathbf{X}_t = \boldsymbol{x}, A_t = a\Big] = \mathbf{0} \\
&\Rightarrow \mathbb{E}\left[\sum_{t=1}^{T-1}\left\{Y_t + \gamma \max_{a'} Q(\mathbf{X}_{t+1}, a'; \boldsymbol{\theta}) - Q(\boldsymbol{x}, a; \boldsymbol{\theta})\right\}\right. \\
&\left.\nabla_{\boldsymbol{\theta}} Q(\boldsymbol{x}, a; \boldsymbol{\theta}) | \mathbf{X}_t = \boldsymbol{x}, A_t = a\right] = \mathbf{0},
\end{aligned}$$

yielding an estimating equation for $\boldsymbol{\theta}$. The empirical estimating equation is given by

$$\mathbb{P}_n\left(\sum_{t=1}^{T}\left[\left\{Y_t + \gamma \max_{a'} Q(\mathbf{X}_{t+1}, a'; \boldsymbol{\theta}) - Q(\boldsymbol{x}, a; \boldsymbol{\theta})\right\} \nabla_{\boldsymbol{\theta}} Q(\boldsymbol{x}, a; \boldsymbol{\theta})\right]\right) = \mathbf{0}.$$

For a linear model for Q^{opt}, Ertefaie and Strawderman (2018) showed that the solution to the estimating equation is consistent and asymptotically normal under some regularity conditions.

In Ernst et al. (2005), an alternative approach to learning the optimal dynamic treatment regime in the infinite-horizon setting using Q-learning was proposed, called fitted Q-iteration (FQI). Similar in spirit to the Q-learning in the finite-horizon setting, FQI fits a sequence of candidate Q-functions. Let $\mathscr{Q}_0, \mathscr{Q}_1, \ldots$ be a sequence of candidate Q-functions, then let $\widehat{Q}_0 \leftarrow \arg\min_{Q_0 \in \mathscr{Q}_0} \mathbb{P}_n[\sum_{t=1}^{T}\{Y_t - Q_0(\mathbf{X}_t, A_t)\}^2]$. The algorithm proceeds as follows:

for $k = 0, 1, \ldots,$ until convergence,

$$Q_{k+1} \leftarrow \underset{Q_{k+1} \in \mathscr{Q}_{k+1}}{\arg\min} \; \mathbb{P}_n \left[\sum_{t=1}^{T} \left\{ Y_t + \gamma \max_a \widehat{Q}_k(\mathbf{X}_t, a) - Q_{k+1}(\mathbf{X}_t, A_t) \right\}^2 \right].$$

FQI allows for highly flexible models to be used for the Q-functions; however, it suffers from bias and nonregularity. To address these limitations, Chakraborty et al. (2008) offered a variant of FQI with linear function approximations, which helps to reduce bias.

6 An Illustrative Example

To concretize finite-horizon Q-learning and the backward induction approach, we present a simple example of Q-learning implementation for the two-stage setting in the R statistical programming language. We will use the `bmiData` dataset from the `DynTxRegime` package (Holloway et al., 2020), a package that features a number of functions and methods for implementing Q-learning and other precision medicine methods. Although the `DynTxRegime` package includes a function for Q-learning, `qLearn`, we will use primitive R functions for tractability and ease of exposition in the following example.

The `bmiData` dataset mimics a two-stage randomized clinical trial that studied obesity in adolescents. At the first randomization point, adolescents in the study are randomized to meal replacement shakes "MR" or their conventional diet "CD". The trial collected data to study the effect of meal replacement shakes on adolescent obesity, with the primary endpoint being body mass index (BMI) defined as body weight (kg) divided by squared height (m^2). Throughout, we assume that there is no missing data, there were no participants lost to follow-up, and all study participants were compliant with their assigned treatment. We begin by loading the `DynTxRegime` package. Once loaded, we load the `bmiData` dataset and use the function `head` to take a look at the first five observations in the dataset.

```
# Load the DynTxRegime package
library(DynTxRegime)

# Load the bmiData dataset
data("bmiData")

# Take a peek at the bmiData
head(bmiData, 5)
```

```
  gender race parentBMI baselineBMI month4BMI month12BMI A1 A2
1      0    1  31.59683    35.84005  34.22717   34.27263 CD MR
2      1    0  30.17564    37.30396  36.38014   36.38401 CD MR
3      1    0  30.27918    36.83889  34.42168   34.41447 MR CD
4      1    0  27.49256    36.70679  32.52011   32.52397 CD CD
5      1    1  26.42350    34.84207  33.72922   33.73546 CD CD
```

We see that this dataset contains the baseline features gender, race, parentBMI, and baselineBMI. Variables A1 and A2 record the treatments assigned at the first and second randomization point, respectively, month4BMI is a BMI measurement after the first treatment assignment and before the second treatment assignment, and month12BMI is the final BMI measurement at the end of the study period. Our analysis will focus on the percent change in BMI between the baseline BMI measurement and the measurement taken after 12 months from the initial randomization. We add the variable pctBMIchange to bmiData in the following. Additionally, we negate the percent change in BMI change so that larger percent changes are better.

```
dplyr::mutate(bmiData,
  pctBMIchange = -100 * (month12BMI - baselineBMI)/baselineBMI)
```

To learn an optimal DTR using Q-learning, we proceed using the backward induction. This is a two-stage problem, so we begin by positing a model for the Stage 2 Q-function $Q_2(\boldsymbol{h}_2, a_2)$ given in (18):

$$
\begin{aligned}
Q_2(\boldsymbol{h}_2, a_2; \boldsymbol{\beta}_2) &= \mathbb{E}\left(pctBMIchange|\boldsymbol{h}_2, a_2; \boldsymbol{\beta}_2\right) \\
&= \beta_{2,0} + \beta_{2,1} gender + \beta_{2,2} race + \beta_{2,3} parentBMI + \beta_{2,4} baselineBMI \\
&\quad + \beta_{2,5} A1 + \beta_{2,6} month4BMI + \beta_{2,7} A2 + \beta_{2,8} A2 \times baselineBMI \\
&\quad + \beta_{2,9} A2 \times month4BMI + \beta_{2,10} A2 \times parentBMI + \beta_{2,11} A2 \times A1.
\end{aligned} \tag{19}
$$

The model specified in (19) is a linear model. As specified, gender, race, parentBMI, baselineBMI, first-stage treatment (A1), and month4BMI are specified as treatment effect modifiers; that is, they affect the conditional mean of the percent change in BMI. Additionally, baselineBMI, month4BMI, parentBMI, and A1 are specified as prescriptive variables meaning that the derived decision rule will recommend treatment based on those four variables. In particular, the form of the decision rules under consideration is

$$I(\beta_{2,7} + \beta_{2,8} baselineBMI + \beta_{2,9} month4BMI + \beta_{2,10} parentBMI \\ + \beta_{2,11} A1 > 0),$$

where $\beta_{2,7}, \ldots, \beta_{2,11}$ are real-valued parameters, and if the treatment rule is valued as 1, then meal replacement therapy is recommended at Stage 2. In R, we use the `lm` function to estimate the parameters in (19).

```
# Fit the second stage regression model (Q2)
stage2mod <- lm(pctBMIchange ~ gender + race + parentBMI +
                baselineBMI + A1 + month4BMI + A2 +
                A2:baselineBMI + A2:month4BMI + A2:parentBMI +
                A2:A1, data = bmiData)
```

Using the Stage 2 model, we can use the predicted values to estimate the Stage 2 optimal treatment rule $\widehat{d}_2^{\text{opt}}(\boldsymbol{h}_2)$. In particular, we set `A2` to "MR" for all study participants and compute the predicted values for this modified dataset, and then we repeat the procedure by setting `A2` to "CD". Next, we compare the predicted values. For those whose predicted response is greater when the Stage 2 treatment is set to "MR" than to "CD", the optimal Stage 2 treatment recommendation is "MR". Similarly for those whose predicted response is greater when the Stage 2 treatment is set to "CD", the optimal Stage 2 treatment is "CD".

```
# Fitted values when A2 == "MR"
stage2MRfitted <- predict(stage2mod,
                      data = dplyr::mutate(bmiData, A2 = "MR"))

# Fitted values when A2 == "CD"
stage2CDfitted <- predict(stage2mod,
                      data = dplyr::mutate(bmiData, A2 = "CD"))

# Optimal Stage 2 treatment recommendations
d2opt <- dplyr::if_else(stage2MRfitted > stage2CDfitted,
                      "MR", "CD")

# Allocation of Stage 2 treatments under the optimal rule
table(d2opt)
```

```
  CD  MR
 109 101
```

Thus, at the second stage decision, the optimal rule recommends meal replacement for 101 study participants and conventional diet for 109 study participants.

Using the estimated coefficients for the Stage 2 Q-model, we can write down the estimated stage-decision rule as

$$\widehat{d}_2(\boldsymbol{h}_2; \widehat{\boldsymbol{\beta}}_2) = I(-0.68 - 0.13 \times baselineBMI + 0.13 \times month4BMI \\ +0.04 \times parentBMI - 0.30 \times I(A1 = \text{“}MR\text{”}) > 0),$$

where $\widehat{d}_2(\boldsymbol{h}_2; \widehat{\boldsymbol{\beta}}_2) = 1$ recommends treatment with meal replacement at stage 2. With the Stage 2 decision rule now learned, we can now work backwards to learn the optimal decision rule for the Stage 1 decision. As with second-stage analysis, we posit a model for the Stage 2 Q-function $Q_1(\boldsymbol{h}_1, a_1)$ given in (18). One notable difference between the Stage 2 and Stage 1 Q-functions is that, for the Stage 2 Q-function, we modeled the conditional mean of the response variable, whereas, for the Stage 1 Q-function, we will model the conditional mean of the predicted value estimated from the model posited for Q_2 assuming that optimal treatment was assigned in Stage 2. Denoting this fitted value as $\widehat{Y}_1$, we posit the following model:

$$Q_1(\boldsymbol{h}_1, a_1; \boldsymbol{\beta}_1) = \mathbb{E}[\widehat{Y}_1 | \boldsymbol{h}_1, a_1; \boldsymbol{\beta}_1] \\ = \beta_{1,0} gender + \beta_{1,1} race + \beta_{1,2} parentBMI + \beta_{1,3} baselineBMI \\ + \beta_{1,4} A1 + \beta_{1,5} A1 \times baselineBMI + \beta_{1,6} A1 \times parentBMI. \tag{20}$$

For simplicity, we have again chosen a linear Q model though more diverse models could be considered. Gender, race, parental BMI, and baseline BMI are specified as modifiers of the first-stage treatment. By interacting baseline BMI and parental BMI with the first-stage treatment, they are the variables by which the first-stage treatment rule will tailor treatments. The form of the decision rules under consideration is $I(\beta_{1,4} + \beta_{1,5} baselineBMI + \beta_{1,6} parentBMI)$, where $\beta_{1,4}$, $\beta_{1,5}$, and $\beta_{1,6}$ are real-valued parameters, and if the value of the function is 1, then the recommended treatment is meal replacement. As before, we use the `lm` function in R to estimate the parameters in (20).

```
# Compute yhat
yhat <- dplyr::if_else(stage2MRfitted > stage2CDfitted,
                       stage2MRfitted, stage2CDfitted)

# Append yhat to bmiData
bmiData <- cbind(bmiData, yhat)

# Fit the first stage regression model (Q1)
stage1mod <- lm(yhat ~ gender + race + parentBMI + baselineBMI +
                A1 + A1:baselineBMI + A1:parentBMI,
                data = bmiData)
```

As we did with the Stage 2 analysis, we can use the predicted values from the Stage 1 estimated Q-function to estimate the Stage 2 optimal treatment rule

$\widehat{d}_1^{\text{opt}}(\boldsymbol{h}_1)$. As before, we compare the predicted values generated by the estimated Stage 1 model with Stage 1 treatment set to "MR" to Stage 1 treatment set to "CD".

```
# Fitted values when A1 == "MR"
stage1MRfitted <- predict(stage1mod,
                          data = dplyr::mutate(bmiData, A1 = "MR"))

# Fitted values when A1 == "CD"
stage1CDfitted <- predict(stage1mod,
                          data = dplyr::mutate(bmiData, A1 = "CD"))

# Optimal Stage 1 treatment recommendations
d1opt <- dplyr::if_else(stage1MRfitted > stage1CDfitted,
                         "MR", "CD")

# Allocation of Stage 1 treatments under the optimal rule
table(d1opt)
```

```
  CD  MR
 109 101
```

Coincidentally, the same number of participants (but not necessarily the same participants) are recommended to receive meal replacement at the first stage as the second stage under the optimal dynamic treatment regime. Using the regression coefficient estimates, we can write down the estimated optimal Stage 1 decision rule

$$\widehat{d}_1^{\text{opt}}(\boldsymbol{h}_1; \boldsymbol{\beta}_1) = I(11.11 + 0.74 \times baselineBMI - 1.23 \times parentBMI),$$

where $\widehat{d}_1^{\text{opt}}(\boldsymbol{h}_1; \boldsymbol{\beta}_1) = 1$ recommends meal replacement and $\widehat{d}_1^{\text{opt}}(\boldsymbol{h}_1; \boldsymbol{\beta}_1) = 0$ recommends conventional diet. Together, $\widehat{d}_1^{\text{opt}}$ and $\widehat{d}_2^{\text{opt}}$ form the estimated optimal dynamic treatment regime.

We have illustrated the two-stage treatment regime using basic R functions (and a few functions from `dplyr` just to keep things tidy), but the `DynTxRegime` package offers functions that automate the steps we have outlined, as well as automatically calculate the value of the estimated optimal DTRs on the training data and test data, as shown below. This is important as it allows for the assessment of the generalizability of the estimated DTRs, and whether it can effectively predict outcomes for new patient populations.

```
# Q-learning to learn the optimal dynamic treatment regime
# - using DynTxRegime package
y12 <- -100*(bmiData[,6L] - bmiData[,4L])/bmiData[,4L]
moMain <- buildModelObj(model = ~parentBMI+month4BMI +gender +
                race + baselineBMI +A1, solver.method = 'lm')
moCont <- buildModelObj(model = ~baselineBMI + parentBMI +
                month4BMI + A1, solver.method = 'lm')
fitSS <- qLearn(moMain = moMain, moCont = moCont,
                data = bmiData, response = y12,  txName = 'A2')
```

```
First step of the Q-Learning Algorithm.

Outcome regression.
Combined outcome regression model: ~ parentBMI + month4BMI +
    gender + race + baselineBMI + A1 + A2 +
    A2:(baselineBMI + parentBMI + month4BMI + A1).
Regression analysis for Combined:

Call:
lm(formula = YinternalY ~ parentBMI + month4BMI +
    gender + race + baselineBMI + A1 + A2 +
    baselineBMI:A2 + parentBMI:A2 + month4BMI:A2 +
    A1:A2, data = data)

Coefficients:
     (Intercept)          parentBMI          month4BMI           gender
         7.94094           -0.04681           -2.66904         -0.10585
            race        baselineBMI               A1MR             A2MR
         0.19576            2.49094            0.16042         -0.67686
baselineBMI:A2MR    parentBMI:A2MR    month4BMI:A2MR        A1MR:A2MR
        -0.13449            0.04189            0.12747         -0.30170


Recommended Treatments:
 CD  MR
109 101

Estimated value: 6.682276
```

```
moMain <- buildModelObj(model = ~parentBMI + baselineBMI +
                race + gender, solver.method = 'lm')
moCont <- buildModelObj(model = ~parentBMI+baselineBMI,
                solver.method = 'lm')
fitFS <- qLearn(moMain = moMain, moCont = moCont,
                data = bmiData, response = fitSS,  txName = 'A1')
```

```
Step 2 of the Q-Learning Algorithm.

Outcome regression.
Combined outcome regression model: ~ parentBMI + baselineBMI +
    race + gender + A1 + A1:(parentBMI + baselineBMI).
Regression analysis for Combined:

Call:
lm(formula = YinternalY ~ parentBMI + baselineBMI + race +
    gender + A1 + parentBMI:A1 + baselineBMI:A1, data = data)

Coefficients:
  (Intercept)      parentBMI        baselineBMI              race
     -2.66451       -0.06937            0.34328          -0.46095
       gender           A1MR     parentBMI:A1MR  baselineBMI:A1MR
      0.26356       11.10540           -1.22588           0.73594
```

```
Recommended Treatments:
 CD  MR
109 101

Estimated value: 9.416704
```

The R code above is available for access in the specified GitHub repository: https://github.com/FIRST-Data-Lab/Q-learning.

7 Summary/Conclusions

In this chapter, we have described how Q-learning is used in precision medicine to estimate optimal dynamic treatment regimes in both finite-horizon and infinite-horizon settings. Given the breadth of the precision medicine literature, there are a number of Q-learning extensions that were not included. For example, we have mainly focused on the case where $Y \in \mathbb{R}$, but statistical Q-learning strategies have been developed for more complicated outcomes such as right-censored data (Goldberg & Kosorok, 2012).

Q-learning is an attractive method for learning optimal DTRs since in many cases Q-functions can be modeled using off-the-shelf software routines. Nonetheless, the method depends on having high-quality estimators for the Q-function. Despite these challenges, there is still much exciting work to be done with Q-learning in the precision medicine setting. Inference for DTRs and the value function continues to be an active area of research, and methods for infinite-horizon medical decision-making are being developed for mHealth applications. Overall, Q-learning is an effective strategy for learning optimal treatment recommendations and is likely to be an important strategy in the future.

Acknowledgments We express our sincere gratitude to the reviewers and the editors for their insightful comments and suggestions that greatly improved the quality of this chapter. Research reported in this publication was partially supported by the National Institute Of General Medical Sciences of the National Institutes of Health under Award Number P20GM139769 (Xinyi Li), National Science Foundation awards DMS-2210658 (Xinyi Li) and DMS-2203207 (Li Wang). The content is solely the responsibility of the authors and does not necessarily represent the official views of the National Institutes of Health.

References

Ashley, E. (2015). The precision medicine initiative: A new national effort. *Journal of the American Medical Association, 313*, 2119–2120.
Bellman, R. (1957). *Dynamic programming*. Princeton University Press.
Bertsekas, D. P., & Tsitsiklis, J. (1996). *Neuro-dynamic programming*. Athena Sci.

Cain, L. E., Robins, J. M., Lanoy, E., Logan, R., Costagliola, D., & Hernán, M. A. (2010). When to start treatment? A systematic approach to the comparison of dynamic regimes using observational data. *The International Journal of Biostatistics, 6*, 18.

Chakraborty, B., Strecher, V., & Murphy, S. (2008). Bias correction and confidence intervals for fitted Q-iteration. In *Workshop on model uncertainty and risk in reinforcement learning (NIPS 2008)*. https://cs.uwaterloo.ca/~ppoupart/nips08-workshop/accepted-papers/nips08paper01-final.pdf

Clifton, J., & Laber, E. B. (2020). Q-learning: Theory and applications. *Annual Review of Statistics and Its Application, 7*, 279–301.

Collins, F., & Varmus, H. (2015). A new initiative on precision medicine. *The New England Journal of Medicine, 372*, 793–795.

Collins, L. M., Murphy, S. A., & Bierman, K. (2004). A conceptual framework for adaptive preventive interventions. *Prevention Science Journal, 5*, 185–196.

Ernst, D., Geurts, P., & Wehenkel, L. (2005). Tree-based batch mode reinforcement learning. *Journal of Machine Learning Research, 6*, 503–556.

Ertefaie, A., & Strawderman, R. L. (2018). Constructing dynamic treatment regimes over indefinite time horizons. *Biometrika, 105*, 963–977.

Goldberg, Y., & Kosorok, M. R. (2012). Q-learning with censored data. *Annals of Statistics, 40*, 529–560.

Hernán, M. A., & Robins, J. M. (2020). *Causal inference: What if.* Chapman & Hall/CRC.

Hernán, M. A., Hernández-Díaz, S., & Robins, J. M. (2004). A structural approach to selection bias. *Epidemiology*, 615–625.

Holloway, S. T., Laber, E. B., Linn, K. A, Zhang, B., Davidian, M., & Tsiatis, A. A. (2020). DynTxRegime: Methods for estimating optimal dynamic treatment regimes. https://CRAN.R-project.org/package=DynTxRegime

Horvitz, D. G., & Thompson, D. J. (1952). A generalization of sampling without replacement from a finite universe. *Journal of the American statistical Association, 47*(260), 663–685.

Kidwell, K. M. (2016). DTRs and SMARTs: Definitions, designs, and applications. *Kosorok & Moodie, 2016*, 7–24.

Kosorok, M. R., & Laber, E. B. (2019). Precision medicine. *Annual Review of Statistics and Its Application, 6*, 263–286.

Kosorok, M. R., & Moodie, E. E. M. (2016). *Adaptive treatment strategies in practice: Planning trials and analyzing data for personalized medicine*. SIAM.

Lavori, P. W., Dawson, R., & Rush, A. J. (2000). Flexible treatment strategies in chronic disease: Clinical and research implications. *Biological Psychiatry, 48*, 605–614.

Liao, P., Klasnja, P., & Murphy, S. (2021). Off-policy estimation of long-term average outcomes with applications to mobile health. *Journal of the American Statistical Association, 116*, 382–391.

Luckett, D. J., Laber, E. B., Kahkoska, A. R., Maahs, D. M., Mayer-Davis, E., & Kosorok, M. R. (2020). Estimating dynamic treatment regimes in mobile health using V-learning. *Journal of the American Statistical Association, 115*, 692–706.

Lunceford, J. K., Davidian, M., & Tsiatis, A. A. (2002). Estimation of survival distributions of treatment policies in two-stage randomization designs in clinical trials. *Biometrics, 58*, 48–57.

Murphy, S. A. (2003). Optimal dynamic treatment regimes (with Discussion). *Journal of the Royal Statistical Society Series B, 66*, 331–366.

Murphy, S. A. (2005a). A generalization error for Q-learning. *Journal of Machine Learning Research, 6*, 1073–1097.

Murphy, S. A. (2005b). An experimental design for the development of adaptive treatment strategies. *Statistics in Medicine, 24*, 1455–1481.

Nahum-Shani, I., Qian, M., Almiral, D., Pelham, W., Gnagy, B., Fabiano, G., Waxmonsky, J., Yu, J., & Murphy, S. A. (2012). Experimental design and primary data analysis methods for comparing adaptive interventions. *Psychological Methods, 17*, 457–477.

Neyman, J. (1923). On the application of probability theory to agricultural experiments. Essay in principles. Section 9 (translation published in 1990). *Statistical Science, 5*, 472–480.

Pearl, J. (2009). Causal inference in statistics: an overview. *Statistics Surveys, 3*, 96–146.

Pelham, W. E., & Fabiano, G. A. (2008). Evidence-based psychosocial treatments for attention-deficit/hyperactivity disorder. *Journal of Clinical Child & Adolescent Psychology, 37*, 184–214.

Petersen, M. L., Deeks, S. G., & van der Laan, M. J. (2007). Individualized treatment rules: Generating candidate clinical trials. *Statistics in Medicine, 26*, 4578–4601.

Powell, W. B. (2007). *Approximate dynamic programming: Solving the curses of dimensionality*. Wiley.

Robins, J. M. (1986). A new approach to causal inference in mortality studies with a sustained exposure period to control of the healthy worker survivor effect. *Mathematical Model, 7*, 1393–1512.

Robins, J. M. (1997). Causal inference from complex longitudinal data. In M. Berkane (Ed.), *Latent variable modeling and applications to causality: Lecture notes in statistics* (pp. 69–117). Springer.

Robins, J. M. (2004). Optimal structural nested models for optimal sequential decisions. In D. Y. Lin & P. J. Heagerty (Eds.), *Proceedings of the Second Seattle Symposium in Biostatistics* (pp. 189–326). Springer.

Rubin, D. B. (1974). Estimating causal effects of treatments in randomized and nonrandomized studies. *Journal of Educational Psychology, 66*, 688–701.

Rubin, D. B. (1978). Bayesian inference for causal effects: The role of randomization. *Annals of Statistics, 6*, 34–58.

Schulte, P. J., Tsiatis, A. A., Laber, E. B., & Davidian, M. (2014). Q- and A-learning methods for estimating optimal dynamic treatment regimes. *Statistical Science, 29*, 640–661.

Stephens, D. A. (2016). G-estimation for dynamic treatment regimes in the longitudinal setting. *Kosorok & Moodie, 2016*, 89–117.

Sutton, R. S., & Barto, A. G. (2018). *Reinforcement learning: an introduction*. MIT Press.

Thall, P. F., Wooten, L. H., Logothetis, C. J., Millikan, R. E., & Tannir, N. M. (2007). Bayesian and frequentist two-stage treatment strategies based on sequential failure times subject to interval censoring. *Statistics in Medicine, 26*, 4687–4702.

Tsiatis, A. A., Davidian, M., Holloway, S. T., & Laber, E. B. (2019). *Dynamic treatment regimes: Statistical methods for precision medicine*. CRC Press.

Wahed, A. S., & Tsiatis, A. A. (2004). Optimal estimator for the survival distribution and related quantities for treatment policies in two-stage randomized designs in clinical trials. *Biometrics, 60*, 124–133.

Zhang, B., Tsiatis, A. A., Laber, E. B., & Davidian, M. (2012). A robust method for estimating optimal treatment regimes. *Biometrics, 68*, 1010–1018.

Zhang, B., Tsiatis, A. A., Laber, E. B., & Davidian, M. (2013). Robust estimation of optimal dynamic treatment regimes for sequential treatment decisions. *Biometrika, 100*, 681–694.

Zhao, Y. Q., Zeng, D., Laber, E. B., & Kosorok, M. R. (2015). New statistical learning methods for estimating optimal dynamic treatment regimes. *Journal of the American Statistical Association, 110*, 583–598.

Zhao, Y. Q., Zeng, D., Rush, A. J., & Kosorok, M. R. (2012). Estimating individualized treatment rules using outcome weighted learning. *Journal of the American Statistical Association, 107*, 1106–1118.

Zhao, Y. Q., Zeng, D., Socinski, M. A., & Kosorok, M. R. (2011). Reinforcement learning strategies for clinical trials in non-small cell lung cancer. *Biometrics, 67*, 1422–1433.

Personalized Medicine with Multiple Treatments

Wenjie Wang, Xuan Zhang, and Haoda Fu

Abstract Personalized medicine is an intriguing topic in academia and the pharmaceutical industry. A major branch of research focuses on estimating individualized treatment rules that map patient pretreatment covariates to the space of available treatments to maximize the expected clinical outcome. Most existing research focused on scenarios where only two treatment options are applicable. Nevertheless, it is not uncommon to have more than two treatment options available to patients in practice. Learning the optimal treatment rules among multiple candidate treatments given the patient's pretreatment characteristics is thus of great interest and impact in practice, which can be a much more challenging problem than the problem with only binary treatments. In this chapter, we provide a selective review of the methods proposed in the recent literature for estimating the optimal treatment rules with multiple treatments.

Keywords Individualized treatment rule · Multi-arm clinical trial · Q-learning · Policy search · Large-margin classification

1 Introduction

Personalized medicine is an important topic in the pharmaceutical industry and the scientific community. In particular, a significant amount of research aims to learn individualized treatment rules (ITR) to maximize the expected patients' health outcomes.

Various statistical methods have been proposed on this topic with applications to various domains in the literature. One class of methods (often called *Q-learning*)

W. Wang (✉) · H. Fu
Eli Lilly and Company, Indianapolis, IN, USA
e-mail: wang@wwenjie.org; fu_haoda@lilly.com

X. Zhang
Department of Biostatistics, Indiana University, Indianapolis, IN, USA
e-mail: xz76@iu.edu

Y. Zhao, D.-G. Chen (eds.), *Statistics in Precision Health*, ICSA Book Series in Statistics, https://doi.org/10.1007/978-3-031-50690-1_6

aims to estimate the expected health outcome given the assigned treatment, the patient's pretreatment covariates, and their interactions by using regression models. From a fitted regression model, one can estimate the expected health outcomes if any particular treatment rule was implemented conditional on the patient's covariates, and the optimal treatment rule maximizing the expected health outcomes depends on the interaction terms between treatments and covariates. For instance, Qian and Murphy (2011) proposed a two-step procedure named l_1-PLS to model the clinical outcomes given treatments and covariates by using l_1 penalized regression and obtain the optimal treatment options by comparing the estimated outcomes under different treatment options given the covariates. The Q-learning methods emphasize the prediction of the conditional clinical rewards and the performance of the resulting ITR estimates heavily relies on the prediction accuracy. Qian and Murphy (2011) related the reduction in value function to excess prediction error of the quality function, which suggests that the estimated ITR will be sensible and have a large value if we have a sensible estimation of the Q-function. However, it can be hard to justify the posited parametric or semi-parametric model structure in practice. An inappropriate model specification can result in substantial bias/overfitting and thus produce suboptimal or misleading decision rules. Another closely related method is called the *A-learning* (Murphy, 2003; Blatt et al., 2004), which makes parametric specification only for the quantities measuring the contrast between treatment outcomes. See Robins (2004) and Moodie et al. (2009) for more examples among others.

More recently, statistical learning approaches have been proposed in the literature to target the treatment rules directly without the need to specify the underlying outcome model. One seminal work was the *outcome-weighted learning* (OWL) framework (Zhao et al., 2012, also called the *O-learning*), which converted the problem of estimating ITR to a weighted classification problem and introduced surrogate loss functions to replace the indicator function in the Q-function given in Qian and Murphy (2011). Another branch of the works that aims to learn the decision function by direct search includes Tian et al. (2014), Qi and Liu (2018), and Qi et al. (2020), all of which utilize regression methods to estimate the optimal decision rules without the conversion to a weighted classification problem as the OWL does.

The majority of the existing works considered the problem with binary treatment options only. Very few methods are directly applicable to the scenario under which more than two treatment options are available. Nevertheless, it is not uncommon for multiple treatment options to be available to patients in practice. One obvious example is multi-arm clinical trials. It is thus of great importance and interest to develop methods for estimating ITR with multiple treatments. In this chapter, we focus on the scenarios where there are multiple treatment options available to patients and review a variety of statistical methods that have been proposed in the literature recently. Under such scenarios, an ITR maps the individual's pretreatment characteristics to a particular treatment option (or combination) among those candidate treatments. We refer to "pretreatment covariates" as simply "covariates" when there is no ambiguity in this chapter.

1.1 Outcome-Weighted Learning for Binary Treatments

We briefly review the outcome-weight learning for binary treatments in this section. One substantial contribution of Zhao et al. (2012) is that they showed the equivalency between an optimal treatment rule for binary treatment and a binary classification problem with weights depending on the treatment assignment scheme and the health outcomes.

Consider a two-arm randomized trial, where the treatment assignment random variable A follows a known prior probability distribution satisfying $A \in \mathcal{A} = \{-1, 1\}$ and is independent of the subject's prognostic covariates. Let $\mathbf{X} \in \mathbb{R}^p$ denote the (pretreatment) covariates. Suppose that the clinic outcome or reward denoted by R takes nonnegative real values and a larger value is more desirable. Our goal is to find an optimal ITR that maximizes the expected reward. Suppose the collected data are $\{(\boldsymbol{x}_i, a_i, r_i)\}$ from an unknown distribution $P(\mathbf{X}, A, R)$. The likelihood function of $(\mathbf{X}, A, R)$ under P is then $f_0(\boldsymbol{x})\Pr(a \mid \boldsymbol{x})f_1(r \mid \boldsymbol{x}, a)$, where $f_0(\cdot)$ is the unknown probability density function of $\mathbf{X}$, $\Pr(a \mid \boldsymbol{x})$ is the probability of the treatment a is assigned given the covariates $\boldsymbol{x}$, and $f_1(\cdot)$ is the unknown probability density function of the reward given the covariates and the treatment assignment. We let P^D denote the joint distribution of $\{\mathbf{X}, A, R\}$ where the treatment A is determined by a particular treatment rule $D(\cdot)$, i.e., $A = D(\mathbf{X})$ holds almost surely. Then the likelihood function under P^D is $f_0(\boldsymbol{x})\mathbf{1}\{a = D(\boldsymbol{x})\}f_1(r \mid \boldsymbol{x}, a)$, where $\mathbf{1}\{\cdot\}$ is the indicator function. Assume that $\pi_a(\boldsymbol{x}) := \Pr(A = a \mid \boldsymbol{x}) > 0$ for $a \in \{-1, 1\}$. We may verify that the measure P^D is absolutely continuous with respect to the measure P, and the Radon-Nikodym derivative $\mathrm{d}P^D/\mathrm{d}P$ satisfies that $\mathrm{d}P^D/\mathrm{d}P = \mathbf{1}\{a = D(\boldsymbol{x})\}/\pi_a(\boldsymbol{x})$. Thus, the expectation of the reward can be written as follows:

$$
\begin{aligned}
\mathbb{E}^D(R) &= \int R\,\mathrm{d}P^D = \int R\frac{\mathrm{d}P^D}{\mathrm{d}P}\,\mathrm{d}P = \int R\frac{\mathbf{1}\{A = D(\mathbf{X})\}}{\pi_A(\mathbf{X})}\,\mathrm{d}P \\
&= \mathbb{E}\left[\frac{\mathbf{1}\{A = D(\mathbf{X})\}}{\pi_A(\mathbf{X})}R\right],
\end{aligned} \tag{1}
$$

where the last expectation is taken with respect to the measure P. In the literature, the quantity $\mathbb{E}^D(R) = \mathbb{E}[R \mid A = D(\mathbf{X})]$ is called the *value* of D and often denoted by $V(D)$. An optimal ITR denoted by $D^*(\mathbf{X})$ satisfies that

$$
D^*(\mathbf{X}) \in \arg\max_D \mathbb{E}\left[\frac{\mathbf{1}\{A = D(\mathbf{X})\}}{\pi_A(\mathbf{X})}R\right]. \tag{2}
$$

Equivalent to (2), $D^*(\mathbf{X})$ minimizes the following regret if it is not implemented:

$$
D^*(\mathbf{X}) \in \arg\min_D \mathbb{E}\left[\frac{\mathbf{1}\{A \neq D(\mathbf{X})\}}{\pi_A(\mathbf{X})}R\right],
$$

which resembles the 0–1 misclassification loss for classification tasks. The term $R/\pi_A(\mathbf{X})$ can be regarded as weights. With observations, we approximate the loss function by its sample mean,

$$\widehat{D}^* \in \arg\min_D \frac{1}{n}\sum_{i=1}^{n} \frac{\mathbf{1}\{a_i \neq D(\boldsymbol{x}_i)\}}{\pi_a(\boldsymbol{x}_i)} r_i,$$

where $r_i/\pi_a(\boldsymbol{x}_i)$ are the observational weights.

For binary treatments, it is equivalent to write $D(\cdot) = \text{sign}\{f(x)\}$ for some decision function $f(\cdot)$. To avoid ambiguity when the decision function equals zero, we define $\text{sign}(0) = 1$ in this chapter. The optimal decision function f^* that results in the optimal ITR $D^*(\mathbf{X})$ is thus the solution to the following minimization problem,

$$\widehat{f}^* \in \arg\min_f \frac{1}{n}\sum_{i=1}^{n} \frac{\mathbf{1}\left[a_i \neq \text{sign}\{f(\boldsymbol{x}_i)\}\right]}{\pi_a(\boldsymbol{x}_i)} r_i. \tag{3}$$

Directly solving (3) can be an NP-hard problem because of the discontinuity and non-convexity of the indicator function. A commonly used approach is to introduce a surrogate loss function. In particular, Zhao et al. (2012) adopted the hinge loss function used in support vector machine (SVM) and proposed the treatment rule solving the following minimization problem,

$$\frac{1}{n}\sum_{i=1}^{n} \frac{r_i}{\pi_a(\boldsymbol{x}_i)}\{1 - a_i f(\boldsymbol{x}_i)\}_+ + \lambda\|f\|^2, \tag{4}$$

where $(x)_+ = \max(x, 0)$ is the positive part of x, and $\|f\|$ represents some norm of the f. Therefore, we can apply the resulting decision function $\widehat{f}$ to $\boldsymbol{x}$ and estimate the optimal treatment plan by $\text{sign}\{\widehat{f}(\boldsymbol{x})\}$. Casting the problem of finding an optimal ITR into a weighted classification problem makes it possible to apply well-studied classification methods. In addition, this approach is considered nonparametric as it does not impose any explicit model specifications for modeling the reward as a function of the treatment and patients' covariates, which makes it more attractive in practice.

Following the existing SVM techniques, Zhao et al. (2012) introduced how to model the decision function via linear learning and kernel learning, respectively. They showed that the hinge loss provides Fisher consistency for binary cases, and the proposed outcome-weighted learning approaches outperformed the competing Q-learning methods in several simulation examples when the sample size is small.

1.2 Augmented Outcome-Weighted Learning

The clinical outcome R was assumed to be nonnegative in Zhao et al. (2012), which, however, may not hold in practice. Notice that the optimal rule in (2) is invariant of the location shift of R in the sense that the optimal rule would remain the same if R is substituted by $R + c$ for any finite constant c. Therefore, for negative rewards, Zhao et al. (2012) recommended shifting all the rewards by some large constant so that all of them become nonnegative after the shifting, and using the shifted rewards in the proposed O-learning framework. Nevertheless, Chen et al. (2018) showed that the performance of the OWL given in (4) depends on the choice of the shifting constant and thus the estimation procedure can be numerically unstable. Several follow-up works have been proposed to relax the nonnegative assumption of the rewards. For example, Liu et al. (2018) proposed the *augmented outcome-weighted learning* (AOL) to extend the OWL to allow negative rewards for better numerical stability. In particular, the optimal decision rule learned from AOL is expected to be a consistent estimator of the optimal rule. We provide a brief review of the AOL in this section.

The main idea of AOL is to replace the reward with some surrogate variable. Note that for any $x \in \mathbb{R}$, $x = |x|\text{sign}(x)$, and the negative part of x satisfies that $(x)_- := -\min(0, x) = \mathbf{1}(x < 0)|x|$. For the decision rule $D(\boldsymbol{x}) = \text{sign}\{f(\boldsymbol{x})\}$, we have

$$
\begin{aligned}
\mathbb{E}^D(R) &= \mathbb{E}\left[\frac{\mathbf{1}\{Af(\mathbf{X}) > 0\}}{\pi_A(\mathbf{X})}R\right] \\
&= \mathbb{E}\left[\frac{\mathbf{1}\{Af(\mathbf{X}) > 0\}}{\pi_A(\mathbf{X})}\{R - s(\mathbf{X})\}\right] + \mathbb{E}\{s(\mathbf{X})\} \\
&= \mathbb{E}\left(\frac{\mathbf{1}\left[Af(\mathbf{X})\text{sign}\{R - s(\mathbf{X})\} > 0\right]}{\pi_A(\mathbf{X})}|R - s(\mathbf{X})|\right) - \mathbb{E}\left[\frac{\{R - s(\mathbf{X})\}_-}{\pi_A(\mathbf{X})}\right] \\
&\quad + \mathbb{E}\{s(\mathbf{X})\},
\end{aligned}
$$

where $s(\cdot)$ is any given integrable function. One may verify that the last equation holds by letting $X_1 = R - s(\mathbf{X})$, $X_2 = Af(\mathbf{X})$, and observing that $\text{sign}(X_1)\mathbf{1}(X_2 > 0) = \mathbf{1}\{X_2\text{sign}(X_1) > 0\} - \mathbf{1}(X_1 < 0)$ almost surely for $X_1, X_2 \in \mathbb{R}$. Equivalent to directly maximizing $\mathbb{E}^D(R)$, AOL maximizes the following term,

$$
\mathbb{E}\left[\frac{\mathbf{1}\{Af(\mathbf{X})\text{sign}(\widetilde{R}) > 0\}}{\pi_A(\mathbf{X})}|\widetilde{R}|\right],
$$

where $\widetilde{R} = R - s(\mathbf{X})$ is the surrogate reward variable and the estimation of the decision function $f(\mathbf{X})$ can follow the OWL approach. The resulting weight $|\widetilde{R}_i|/\pi_i$ is clearly nonnegative. Liu et al. (2018) proposed estimating the reward R by $\widehat{s}(x)$ using an ordinary least square regression or a penalized least square regression for high-dimensional data. After that, we compute the surrogate rewards by $\widetilde{R} = R - \widehat{s}(\mathbf{X})$. Again, the removal of the main effect $\widehat{s}(\mathbf{X})$ does not make any difference upon the estimation of the optimal treatment rule. Intuitively, for subjects having large values of $|\widetilde{R}|$, the optimal treatment is likely to remain the same as the observed treatment assignment if $\widetilde{R} > 0$, or to be the alternative treatment otherwise.

1.3 Direct Learning for Binary Treatments

Qi and Liu (2018) proposed a straightforward one-step procedure named *direct learning* (D-learning) for binary treatments. The proposed method directly estimates a new effect measure quantifying the treatment effect. More specifically, it requires neither a model specification for estimating the conditional clinical outcomes nor a weight specification as in the outcome-weighted learning, which thus mitigates the risk of misspecification.

Qian and Murphy (2011) showed that an optimal ITR satisfies $D^*(\mathbf{X}) \in \arg\max_{a\in\mathcal{A}} Q(\mathbf{X}, a)$ almost surely, where $Q(\mathbf{X}, a) = \mathbb{E}(R \mid \mathbf{X}, A = a)$ measures the "quality" of treatment a for covariate vector $\mathbf{X}$. Thus, we rewrite the optimal ITR for binary treatments in (2) as follows:

$$D^*(\mathbf{X}) = \text{sign}\{Q(\mathbf{X}, 1) - Q(\mathbf{X}, -1)\},$$

which implies the following optimal decision function

$$\begin{aligned} f^*(\mathbf{X}) &= Q(\mathbf{X}, 1) - Q(\mathbf{X}, -1) \\ &= \mathbb{E}(R \mid \mathbf{X}, A = 1) - \mathbb{E}(R \mid \mathbf{X}, A = -1) \\ &= \mathbb{E}\left\{\frac{RA}{\pi_A(\mathbf{X})} \,\Big|\, \mathbf{X}, A = 1\right\}\pi_1(\mathbf{X}) + \mathbb{E}\left\{\frac{RA}{\pi_A(\mathbf{X})} \,\Big|\, \mathbf{X}, A = -1\right\}\pi_{-1}(\mathbf{X}) \\ &= \mathbb{E}\left\{\frac{RA}{\pi_A(\mathbf{X})} \,\Big|\, \mathbf{X}\right\}. \end{aligned}$$

Clearly, the optimal ITR satisfies that $D^*(\mathbf{X}) = \text{sign}\{f^*(\mathbf{X})\}$. Motivated by such a relationship, the essential idea of the D-learning is to estimate the optimal decision function $f^*(\mathbf{X})$ directly through its equivalency to the conditional expectation $\mathbb{E}\{RA/\pi_A(\mathbf{X}) \mid \mathbf{X}\}$. Then we can estimate the optimal ITR as $\text{sign}\{\widehat{f}^*(\mathbf{X})\}$.

One may verify that the optimal decision function is invariant to a constant shift $c(\mathbf{X})$ to the outcome R, where $c(\mathbf{X})$ only depends on the covariates $\mathbf{X}$.

$$\mathbb{E}\left[\frac{\{R + c(\mathbf{X})\}A}{\pi_A(\mathbf{X})} \,\Big|\, \mathbf{X}\right] = \mathbb{E}\left\{\frac{RA}{\pi_A(\mathbf{X})} \,\Big|\, \mathbf{X}\right\} + \mathbb{E}\left\{\frac{c(\mathbf{X})A}{\pi_A(\mathbf{X})} \,\Big|\, \mathbf{X}\right\}$$

$$= \mathbb{E}\left\{\frac{RA}{\pi_A(\mathbf{X})} \,\Big|\, \mathbf{X}\right\} + \mathbb{E}\left\{c(\mathbf{X}) - c(\mathbf{X})\right\}$$

$$= \mathbb{E}\left\{\frac{RA}{\pi_A(\mathbf{X})} \,\Big|\, \mathbf{X}\right\} = f^*(\mathbf{X}).$$

Hence, the D-learning method allows negative rewards. In addition, Qi and Liu (2018) provided an insightful illustration of the D-learning method through an example. Assume that the underlying model of the reward variable R is

$$R = m(\mathbf{X}) + \delta(\mathbf{X})A + \epsilon, \tag{5}$$

where $m(\mathbf{X})$ is the main effect due to the covariates $\mathbf{X}$ regardless of the treatment option, $\delta(\mathbf{X})$ measures the relative treatment effect, and ϵ represents the random error with expectation zero. We have

$$\begin{aligned}
f^*(\mathbf{X}) &= \mathbb{E}\left\{\frac{RA}{\pi_A(\mathbf{X})} \,\Big|\, \mathbf{X}\right\} \\
&= \mathbb{E}\left\{\frac{m(\mathbf{X}) + \delta(\mathbf{X})A + \epsilon}{\pi_A(\mathbf{X})} \,\Big|\, \mathbf{X}, A = 1\right\}\pi_1(\mathbf{X}) \\
&\quad + \mathbb{E}\left\{-\frac{m(\mathbf{X}) + \delta(\mathbf{X})A + \epsilon}{\pi_A(\mathbf{X})} \,\Big|\, \mathbf{X}, A = -1\right\}\pi_{-1}(\mathbf{X}) \\
&= m(\mathbf{X}) + \delta(\mathbf{X}) + \mathbb{E}(\epsilon) - \{m(\mathbf{X}) - \delta(\mathbf{X}) + \mathbb{E}(\epsilon)\} \\
&= 2\delta(\mathbf{X}),
\end{aligned}$$

which suggests that the optimal decision function from the D-learning is indeed capable to quantify the heterogeneity between treatment options without involving the main effect $m(\mathbf{X})$.

Under the assumption of interchange between differentiation and expectation, Qi and Liu (2018) showed that

$$f^*(\mathbf{X}) \in \arg\min_{f} \mathbb{E}\left\{\frac{1}{\pi_A(\mathbf{X})}(2RA - f(\mathbf{X}))^2\right\}, \tag{6}$$

which enables us to utilize existing regression methods to estimate the optimal decision function $f^*(\mathbf{X})$ suggested by the D-learning. To see this, we may let the partial derivative of the objective function with respect to f (that is treating f as a "variable") be zero as follows:

$$-2\mathbb{E}\left[\frac{1}{\pi_A(\mathbf{X})}\{2RA - f(\mathbf{X})\}\right] = -2\mathbb{E}_{\mathbf{X}}\left[\mathbb{E}\left\{\frac{2RA}{\pi_A(\mathbf{X})} - \frac{f(\mathbf{X})}{\pi_A(\mathbf{X})} \,\Big|\, \mathbf{X}\right\}\right]$$

$$= -2\mathbb{E}_{\mathbf{X}}\left[\mathbb{E}\left\{\frac{2RA}{\pi_A(\mathbf{X})}\,\Big|\,\mathbf{X}\right\} - 2f(\mathbf{X})\right]$$

$$= -4\mathbb{E}_{\mathbf{X}}\left[\mathbb{E}\left\{\frac{RA}{\pi_A(\mathbf{X})}\,\Big|\,\mathbf{X}\right\} - f(\mathbf{X})\right] = 0.$$

Suppose that we restrict our attention to the linear decision rules $\mathcal{F} := \{f(\mathbf{X}) = \boldsymbol{\beta}^\top \mathbf{X}, \boldsymbol{\beta} \in \mathbb{R}^p\}$. From (6), we estimate $\boldsymbol{\beta}$ by minimizing the weighted squared loss as follows:

$$\widehat{\boldsymbol{\beta}} = \arg\min_{\boldsymbol{\beta}} \frac{1}{n} \sum_{i=1}^{n} \frac{1}{\pi_{a_i}(\boldsymbol{x}_i)} (2r_i a_i - \boldsymbol{\beta}^\top \boldsymbol{x}_i)^2,$$

which corresponds to an ordinary least squared estimator for a regression problem where the responses are $(2r_i a_i)$ with observational weights $\{1/\pi_{a_i}(\boldsymbol{x}_i)\}$. For high dimensional problems where the number of the covariates is large, Qi and Liu (2018) proposed to impose regularization to prevent overfitting and enhance the model interpretability by variable selection. Different from the l_1-PLS and O-learning approach, the D-learning method directly estimates the decision boundary by the least square loss. Thus, the selection of the tuning parameter may follow the commonly used procedure such as grid search with cross-validation for regularized regression to minimize the mean square error (MSE) of the validation set,

$$\text{MSE}(\widehat{f}) = \mathbb{E}_n\left\{(2RA - \widehat{f})^2\right\},$$

where $\mathbb{E}_n$ represents the empirical average.

In addition, one may consider non-linear decision rules and estimate $f^*(\mathbf{X})$ by non-linear learners such as *kernel ridge regression* (Cristianini & Shawe-Taylor, 2000) and *component selection and smoothing operator* (Lin & Zhang, 2006) as introduced in Qi and Liu (2018). A non-linear decision function corresponds to a non-linear treatment contrast function $\delta(x)$ in (5). We continue our introduction to the methods based on the D-learning for multiple treatments in Sect. 2.3.

1.4 Multi-Category Classification

Following the introduction in Sect. 1.1, one may naturally project an ITR problem to a multi-category classification problem under the outcome-weighted learning framework. Thus, we briefly review those commonly used approaches for multi-category classification in this section.

By definition, multi-category (or multi-class) classification refers to the problem of classifying instances into one of three or more classes, which has broad applications in many areas including biology and medicine. Some examples are cell type annotation by using single-cell RNA-sequencing data, small round blue cell tumor classification, and handwritten digit recognition.

As multi-category classification is an extension to binary classification, a few strategies have been proposed to transform the multi-category classification problem into a sequence of binary classification problems. In particular, the *one-versus-rest* and *one-versus-one* approaches are widely used in practice and implemented in statistical software packages. The one-versus-rest method is to train a series of binary classifiers for each class by taking the samples from that class as positive and the rest as negative. It requires some kind of decision scores such as probability estimates from the binary classifier. The final classifications are the classes with the largest decision scores from those binary classifiers. The one-versus-one approach is to train binary classifiers for every combination of two distinct classes and draw the final classifications by a voting scheme. Despite their simplicity and popularity in practice, these two sequential approaches are inefficient and can provide suboptimal classification results. One drawback of the one-versus-rest approach is that the binary classifications are almost always performed using unbalanced data, which is the case even when the original multi-category data are completely balanced. While for the one-versus-one approach, it is possible to have the same number of votes for the final classifications and it is unclear how to wisely break the ties.

More sophisticated methods are desired to generalize binary classifiers to multi-category data. Let us consider a dataset that consists of n independent and identically distributed samples $\{(\boldsymbol{x}_i^\top, y_i)\}$ from an unknown distribution $\mathcal{P}(\boldsymbol{x}, y)$, where $\boldsymbol{x}_i \in \mathcal{S} \subset \mathbb{R}^p$ is the predictor vector for the ith sample, $y_i \in \{1, \ldots, K\}$ is the ith response, $i \in \{1, \ldots, n\}$, $K \geq 2$ is a positive integer representing the number of categories, and p is the dimensionality of $\mathcal{S}$.

One popular choice is the support vector machine. Several multi-category SVM methods have been proposed in the literature. For example, the standard multi-category SVM (Lee et al., 2004) solves the following minimization problem,

$$\underset{\boldsymbol{\beta}}{\text{minimize}}\ \frac{1}{n}\sum_{i=1}^{n}\sum_{k=1}^{K}\mathbf{1}(y_i \neq k)\left(\beta_{k,0} + \boldsymbol{\beta}_k^\top \boldsymbol{x}_i + 1\right)_+ + \lambda \sum_{k=1}^{K}\sum_{j=1}^{p}\beta_{k,j}^2,$$

subject to

$$\sum_{k=1}^{K}\beta_{k,0} = 0,\ \sum_{k=1}^{K}\boldsymbol{\beta}_k = \mathbf{0},$$

where λ is the tuning parameter.

Another widely used method is the multinomial logistic regression, which assumes that the log odds follow a linear model:

$$\log \frac{p_k(\boldsymbol{x}_i)}{p_K(\boldsymbol{x}_i)} = \beta_{k,0} + \boldsymbol{\beta}_k^\top \boldsymbol{x}_i.$$

Thus, we have

$$p_k(\boldsymbol{x}_i) = \frac{\exp\left(\beta_{k,0} + \boldsymbol{\beta}_k^\top \boldsymbol{x}_i\right)}{1 + \sum_{l=1}^{K-1} \exp\left(\beta_{l,0} + \boldsymbol{\beta}_l^\top \boldsymbol{x}_i\right)},$$

$$p_K(\boldsymbol{x}_i) = \frac{1}{1 + \sum_{l=1}^{K-1} \exp\left(\beta_{l,0} + \boldsymbol{\beta}_l^\top \boldsymbol{x}_i\right)},$$

where the K-th category is treated as the *reference class* or the *side constraint.* Alternatively, the symmetric side constraint can be imposed as follows:

$$p_k(\boldsymbol{x}_i) = \frac{\exp\left(\beta_{k,0} + \boldsymbol{\beta}_k^\top \boldsymbol{x}_i\right)}{\sum_{l=1}^{K} \exp\left(\beta_{l,0} + \boldsymbol{\beta}_l^\top \boldsymbol{x}_i\right)},$$

subject to

$$\sum_{k=1}^{K} \beta_{k,0} = 0, \; \sum_{k=1}^{K} \boldsymbol{\beta}_k = \boldsymbol{0},$$

which can be more appropriate when regularization is considered (Zahid & Tutz, 2013).

2 Personalized Medicine with Multiple Treatments

We introduce various ITR methods that have been recently proposed in the literature for the scenario under which multiple treatment options are available. For simplicity, we assume that the mechanism of treatment assignment is known given the covariates. We mainly focus on nominal multi-category problems where one of the treatment options can be assigned to patients. In Sect. 2.4, we extend our attention to the multi-label problems where the treatment combinations that consist of multiple non-exclusive treatments can be assigned to patients.

2.1 *Sequential Outcome-Weighted Learning*

To extend the outcome-weighted learning for multiple treatments, Zhou et al. (2018) proposed the *sequential outcome-weighted multi-category* (SOM) learning. They established the asymptotic results regarding the SOM learning, and showed the optimal ITR estimated by the SOM learning is Fisher consistent suggesting that the estimated ITR is the same as the underlying optimal ITR as the sample size goes to infinity. The proposed SOM learning procedure utilizes the O-learning for binary treatments repeatedly by one sort of one-versus-rest strategy. As a result, it is straightforward to implement the SOM procedure by utilizing the existing computational routines intended for binary classification problems.

The main idea of SOM learning is based on the one-versus-rest comparison. However, it involves a more sophisticated data arrangement than the naive one-versus-rest strategy that we introduced in Sect. 1.4. Suppose we have K different treatment options. Without loss of generality, we assume that the treatment options are arranged in decreasing order by their prevalence in the sample data; the first treatment option is the most popular, while the Kth treatment is the least popular option. The SOM learning algorithm can be decomposed into two nested loops. Along the outer loop, the target treatment iterates from the Kth option to the first option. The objective of a complete outer iteration is to estimate an optimal ITR for the corresponding target treatment. Inside the inner loop, we perform a sequence of learning tasks intended for binary treatments. The target treatment is kept last in the sequential procedure, while those more popular treatment options (arranged before the target treatment) are regarded as the alternatives. The inner loop goes through all possible permutations of those alternative treatments for the sequential procedure. We walk through the major steps of SOM learning algorithm more specifically in the following paragraphs.

In the very first iteration, the Kth option is the target treatment. Among all the possible permutations of the alternative treatments other than the target treatment, we start from one particular permutation $\{j_1, \ldots, j_{K-1}\}$ of $\{1, \ldots, K-1\}$, and let $j_K = K$ representing the target treatment. The first task of the sequential procedure is to learn a rule to decide if a subject should be treated by j_1 for a better health outcome. Mathematically, the objective is to find an optimal decision function $f^*_{j_1}(\boldsymbol{x})$ to maximize the following expected reward:

$$\mathbb{E}\left[\frac{\mathbf{1}\{Z_{j_1} f_{j_1}(\boldsymbol{x}) > 0\}}{\pi_A(\boldsymbol{x})} R\right],$$

where $Z_{j_1} = 2\mathbf{1}(A = j_1) - 1$ takes the value of 1 or -1, respectively, if the treatment j_1 is assigned or not. According to the AOL (Sect. 1.2), it is equivalent to maximize the following to allow negative rewards:

$$\mathbb{E}\left[\frac{\mathbf{1}\{Z_{j_1} \operatorname{sign}(R) f_{j_1}(\boldsymbol{x}) > 0\}}{\pi_A(\boldsymbol{x})} |R|\right].$$

As we discussed before, it can be hard to solve the above maximization problem directly. Thus, we apply the O-learning for binary treatments instead. In addition, Zhou et al. (2018) suggested weighting the observations with treatment j_1 by $(K-1)/K$ and the others by $1/K$ to mitigate the imbalanced comparison between the treatment j_1 and the rest $K-1$ treatments. To be more specific, we estimate the decision function $f_{j_1}(\boldsymbol{x})$ for the optimal treatment rule by minimizing the following empirical loss,

$$
\begin{aligned}
L_{j_1}(f) =& \frac{1}{n}\sum_{i=1}^{n}\Bigg[\frac{|r_i|}{\pi_{j_1}(\boldsymbol{x}_i)}\mathbf{1}\{Z_{ij_1}\operatorname{sign}(r_i)=1\}\{1-f(\boldsymbol{x}_i)\}_+\\
&\left\{\frac{K-1}{K}\mathbf{1}(r_i>0)+\frac{1}{K}\mathbf{1}(r_i\le 0)\right\}\\
&+\frac{|r_i|}{\pi^*_{j_1}(\boldsymbol{x}_i)}\mathbf{1}\left\{Z_{ij_1}\operatorname{sign}(r_i)=-1\right\}\{1+f(\boldsymbol{x}_i)\}_+\\
&\left\{\frac{1}{K}\mathbf{1}(r_i>0)+\frac{K-1}{K}\mathbf{1}(r_i\le 0)\right\}\Bigg]+\lambda_{j_1}\|f\|^2,
\end{aligned}
$$

where $z_{ij_1}=2\mathbf{1}(a_i=j_1)-1, \pi^*_{j_1}(\boldsymbol{x}_i)=\sum_{l=2}^{K}\mathbf{1}(a_i=j_l)\pi_{j_l}(\boldsymbol{x}_i)$, and λ_{j_1} is a tuning parameter for the penalty term. Suppose that the learned treatment rule suggests some of the subjects receiving the treatment rule j_1. For the remaining subjects who are not advised to receive the treatment j_1, we further learn if receiving the treatment j_2 would result in larger rewards by minimizing the following loss function,

$$
\begin{aligned}
L_{j_2}(f) =& \frac{1}{n}\sum_{i\in\mathbb{J}_2}\Bigg[\frac{|r_i|}{\pi_{j_2}(\boldsymbol{x}_i)}\mathbf{1}\{Z_{ij_2}\operatorname{sign}(r_i)=1\}\{1-f(\boldsymbol{x}_i)\}_+\\
&\left\{\frac{K-2}{K-1}\mathbf{1}(r_i>0)+\frac{1}{K-1}\mathbf{1}(r_i\le 0)\right\}\\
&+\frac{|r_i|}{\pi^*_{j_2}(\boldsymbol{x}_i)}\mathbf{1}\left\{Z_{ij_2}\operatorname{sign}(r_i)=-1\right\}\{1+f(\boldsymbol{x}_i)\}_+\\
&\left\{\frac{1}{K-1}\mathbf{1}(r_i>0)+\frac{K-2}{K-1}\mathbf{1}(r_i\le 0)\right\}\Bigg]+\lambda_{j_2}\|f\|^2,
\end{aligned}
$$

where $\mathbb{J}_2=\{i \mid a_i\ne j_1, \widehat{f}_{j_1}(\boldsymbol{x}_i)<0\}$, $z_{ij_2}=2\mathbf{1}(a_i=j_2)-1$, $\pi^*_{j_2}(\boldsymbol{x}_i)=\sum_{l=3}^{K}\mathbf{1}(a_i=j_l)\pi_{j_l}(\boldsymbol{x}_i)$, and λ_{j_2} is a tuning parameter. We repeat such a one-versus-rest procedure to learn if receiving the treatment j_h for $h=3,\ldots,K-1$ would result in better outcomes sequentially for those subjects who are not suggested taking a unique alternative treatment. The corresponding objective function is summarized as follows:

$$
\begin{aligned}
L_{j_h}(f) =& \frac{1}{n}\sum_{i\in\mathbb{J}_h}\Bigg[\frac{|r_i|}{\pi_{j_h}(\boldsymbol{x}_i)}\mathbf{1}\{Z_{ij_h}\operatorname{sign}(r_i)=1\}\{1-f(\boldsymbol{x}_i)\}_+\\
&\left\{\frac{K-h}{K-h+1}\mathbf{1}(r_i>0)+\frac{1}{K-h+1}\mathbf{1}(r_i\le 0)\right\}
\end{aligned}
$$

Algorithm 1 Sequential outcome-weighted multi-category learning

```
for a target treatment s ∈ {K, ..., 2} do                                   ▷ backward outer loop
    for a permutation {j_1, ..., j_{s-1}} of {1, ..., s − 1} do                    ▷ inner loop
        initialize an empty index set S
        for l ∈ {j_1, ..., j_{s-1}} do                                       ▷ sequential procedure
            S ← S ∪ l
            Binary outcome-weighted learning between the target treatment l and the
            alternative treatments {j_1, ..., j_{s-1}} \ S for those subjects who have not yet
            been suggested unique treatment options.
        end for
    end for
end for
```

$$+ \frac{|r_i|}{\pi^*_{j_h}(\boldsymbol{x}_i)} \mathbf{1}\left\{Z_{ij_h}\text{sign}(r_i) = -1\right\} \{1 + f(\boldsymbol{x}_i)\}_+$$

$$\left\{\frac{1}{K-h+1}\mathbf{1}(r_i > 0) + \frac{K-h}{K-h+1}\mathbf{1}(r_i \le 0)\right\}\Bigg] + \lambda_{j_h}\|f\|^2,$$

where $\mathbb{J}_h = \{i \mid a_i \neq j_l, \widehat{f}_{j_l}(\boldsymbol{x}_i) < 0, l = 1, \ldots, h-1\}$, $z_{ij_h} = 2\mathbf{1}(a_i = j_h) - 1$, $\pi^*_{j_h}(\boldsymbol{x}_i) = \sum_{l=h+1}^{K} \mathbf{1}(a_i = j_l)\pi_{j_l}(\boldsymbol{x}_i)$, and λ_{j_h} is a tuning parameter. The sequential procedure would continue until we obtain unique treatment options suggested for every individual. As a result, we would recommend the target treatment K if and only if none of the alternative treatments is suggested along the sequential procedure.

Given the target treatment, the inner loop iterates over all the possible permutations of the alternative treatments. Let $\widehat{D}^k_{(j_1,\ldots,j_{k-1})}(\boldsymbol{x}) = 1$ indicate that the target treatment k is recommended for the permutation $(j_1, \ldots, j_{k-1})$, and let Π_s denote all the permutations of $(1, \ldots, s)$. Define

$$\widehat{D}^k(\boldsymbol{x}) = \max_{(j_1,\ldots,j_{k-1})\in\Pi_{k-1}} \widehat{D}^k_{(j_1,\ldots,j_{k-1})}(\boldsymbol{x}).$$

We would recommend the target treatment to an individual if $\widehat{D}^k(\boldsymbol{x}) = 1$, which means the target treatment is suggested as long as it is suggested for any one of the permutations. This step concludes the first complete iteration of the outer loop. The target treatment becomes the treatment $K-1$ in the second iteration of the outer loop. We exclude those patients receiving treatment K or advised to receive the treatment K from the subsequent iterations. Then one may apply the same sequential procedure to the remaining data until the second treatment. At last, we suggest the remaining patients receive the first treatment.

We summarize the major steps of SOM learning in Algorithm 1. The whole procedure involves training a sequence of weighted binary SVM classifiers. The total number of classifiers that need training is up to $\sum_{l=2}^{K}(l-1)!(l-1) = K! - 1$ suggesting that the computational cost explodes exponentially as the number of treatment categories. However, as the outer loop goes, the computational cost decreases as the sample size for the sequential procedure becomes smaller.

2.2 Multi-Category Outcome-Weighted Margin-Based Learning

As we introduced in Sect. 1.4, several methods for simultaneous multi-category classification have been proposed in the literature. Recall that both the multi-class SVM (Lee et al., 2004) and the multinomial logistic model (with the symmetric side constraint) require sum-to-zero constraints on the coefficients. In contrast to those two methods, Zhang and Liu (2014) proposed the *multi-category angle-based large-margin classification* method and showed it can be more computationally efficient without the explicit sum-to-zero constraints. In addition, they showed that the proposed *angle-based classifiers* (ABC) with differentiable loss functions from the *large-margin unified machine* family (Liu et al., 2011) are Fisher consistent. Based on the angle-based classifiers, Zhang et al. (2020) proposed the *outcome-weighted margin-based learning* (MOML) for multi-category treatment scenarios. They provided the asymptotic properties of the proposed MOML and showed it outperformed the naive OWL based on the one-versus-one and one-versus-rest strategies. In the remainder of this section, we provide a review of the angle-based classifiers, large-margin unified machine, and the corresponding multi-category outcome-weighted learning algorithm. We adopt the notations given in Sect. 1.4 for the following introduction.

The essential idea of the ABC is to project the K categories to K vertices of a simplex in $\mathbb{R}^{K-1}$ as studied in Lange and Tong Wu (2008) and draw the predictions based on the angles between the estimated decision function and those vertices. The center of the simplex is the origin and the inner products between any two vertices are the same. We define the angle between two vectors in $[0, \pi]$. Then the predicted category corresponds to the projected vertex having the least angle with the estimated decision function. We provide illustrations for the cases where $K = 2$, $K = 3$, and $K = 4$, respectively, in Fig. 1. We denote the origin by O and the estimated decision function by $\widehat{f}$. For the binary case, we project two categories to the point W_1 and W_2, respectively, in $\mathbb{R}$. The example decision function shown in Fig. 1a lies in the number line $\mathbb{R}$ and predicts the first category as it has a smaller angle between W_1 than W_2. For $K = 3$, we project those three categories to three points in $\mathbb{R}^2$. The estimated decision function given in Fig. 1b has the smallest angle with W_1 and thus predicts the first category. Similarly, for $K = 4$, we represent those four categories by four points in $\mathbb{R}^3$ as illustrated in Fig. 1c. However, it is not easy to visually determine the smallest angle. Notice that the angle between a vertex and the decision function being the smallest is equivalent for them to have the largest inner product.

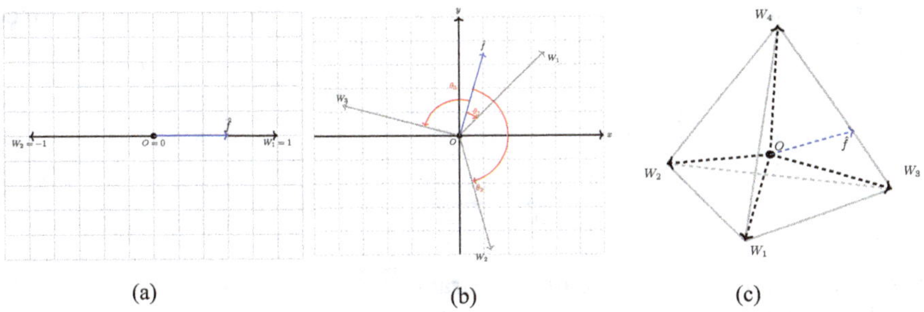

Fig. 1 Geometry interpretation of the angle-based classification. (**a**) $K = 2$. (**b**) $K = 3$. (**c**) $K = 4$

Mathematically, the multi-category angle-based classifier solves the following minimization problem:

$$\underset{f \in F}{\text{minimize}} \frac{1}{n} \sum_{i=1}^{n} \ell\{\langle f(\boldsymbol{x}_i), W_{y_i} \rangle\} + \lambda J(f), \tag{7}$$

where $\ell(\cdot)$ is the loss function, $\langle \cdot, \cdot \rangle$ represents the inner product, $f(\boldsymbol{x}_i) \in \mathbb{R}^{K-1}$ is the classification function of the i-th sample, $J(f)$ is the regularization term, λ is the tuning parameter, and $\{\mathbf{W}_k\}$ consists of K vectors in $\mathbb{R}^{K-1}$ satisfying

$$\mathbf{W}_k = \begin{cases} (K-1)^{-1/2}\zeta & \text{if } k = 1 \\ -(1+K^{1/2})/\{(K-1)^{3/2}\}\zeta + \{K/(K-1)\}^{1/2} e_{k-1} & \text{if } 2 \le k \le K, \end{cases}$$

where $\zeta = \mathbf{1}_{K-1}$ with all elements being one, and $e_k \in \mathbb{R}^{K-1}$, every element of which is zero except the k-th element is one. Notice that $\sum_{k=1}^{K} \mathbf{W}_k = \mathbf{0}$, which means that the set $\{\mathbf{W}_k\}$ forms a simplex in $\mathbb{R}^{K-1}$ whose center is at the origin. One may verify that each $\mathbf{W}_k$ has unit Euclidean norm and the angles between two of $\mathbf{W}_k$ and $\mathbf{W}_{k'}$ are equal for any $k \neq k', k, k' \in 1, \ldots, K$.

The classification function f can be modeled by linear models or with the kernel trick. For instance, Zhang et al. (2016) proposed the *angle-based support vector machines* with linear learning and kernel learning. For linear learning, we have $f_i := f(\boldsymbol{x}_i) = (\beta_{0,1} + \boldsymbol{x}_i^\top \boldsymbol{\beta}_1, \ldots, \boldsymbol{\beta}_{0,K-1} + \boldsymbol{x}_i^\top \boldsymbol{\beta}_{K-1})^\top$, where $\beta_{0,k}$ and $\boldsymbol{\beta}_k \in \mathbb{R}^p$ are the kth intercept and coefficient vector, respectively.

Zhao et al. (2012) considered SVM with the hinge loss function $\ell(u) = (1-u)_+$ as the surrogate loss function. Other choices of the surrogate loss function include the deviance loss $\ell(u) = \log\{1 + \exp(-u)\}$ for the logistic regression and the exponential loss $\ell(u) = \log$ approximating boosting (Hastie et al., 2009). In particular, the large-margin unified machine (LUM) proposed by Liu et al. (2011) contains a broad class of loss functions as special cases and serves as a bridge between hard and soft classification. More specifically, the LUM family is defined as follows:

$$\ell(u) = \begin{cases} 1 - u & \text{if } u < \dfrac{c}{1+c} \\[2ex] \dfrac{1}{1+c}\left\{\dfrac{a}{(1+c)u - c + a}\right\}^a & \text{if } u \geq \dfrac{c}{1+c}, \end{cases}$$

where $a > 0$ and $c \geq 0$ are the tuning parameters. When $a > 0$ and $c \to \infty$, the LUM loss function reduces to the hinge loss. While $a \to \infty$ and any fixed $c \geq 0$, it becomes an interesting hybrid loss function of the hinge loss and exponential loss:

$$\ell(u) = \begin{cases} 1 - u & \text{if } u < \dfrac{c}{1+c} \\[2ex] \dfrac{1}{1+c}\exp\left[-\{(1+c)u - c\}\right] & \text{if } u \geq \dfrac{c}{1+c}, \end{cases}$$

We visualize the hybrid loss function with parameter $c = 0$ in both panels of Fig. 2. The left panel reveals the close relationship of the hybrid loss with the hinge loss and exponential loss. The hybrid loss is equivalent to the exponential loss for $u \geq 0$ and coincides with the hinge loss for $u < 0$. As a result, the hybrid loss is smooth in the sense that its first derivative is continuous and increasingly approaches zero from the negative one. The right panel presents the comparison between the hybrid loss and the logistic deviance loss, which suggests that the former is bounded by the latter from below, and their difference diminishes from at most one to zero as u

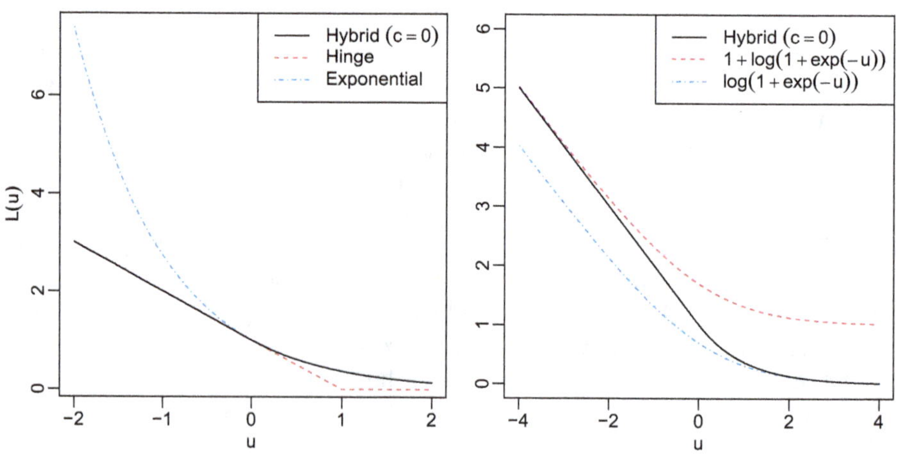

Fig. 2 Hybrid hinge-boost loss function from the LUM family with comparison to the hinge loss, the exponential loss, and the logistic deviance loss

decreases. It is straightforward to verify our observation mathematically. We refer readers to Liu et al. (2011) for discussion on choices of the surrogate loss functions.

There are various choices for the regularization term $J(f)$. For example, Zhang and Liu (2014) considered the elastic-net penalty (Zou & Hastie, 2005) for the linear learning, which includes the lasso penalty (Tibshirani, 1996) and the ridge penalty as special cases. Zhang et al. (2020) applied the following lasso penalty to perform variable selection and showed variable selection consistency under certain assumptions,

$$\lambda J(f) = \lambda \sum_{i>0}^{p} \sum_{j=1}^{K-1} |\beta_{i,j}|,$$

where $\beta_{i,j}$ denote the entry in the ith row and the jth column of the coefficient matrix $\boldsymbol{\beta}$.

To allow negative rewards using the OWL method for binary treatments, Zhang et al. (2020) proposed inverting the loss function for negative rewards in the following optimization problem.

$$\underset{f \in F}{\text{minimize}} \frac{1}{n} \sum_{i=1}^{n} \frac{|r_i|}{\pi_{a_i}(\boldsymbol{x}_i)} \ell_{r_i}\{a_i f(\boldsymbol{x}_i)\} + \lambda J(f), \tag{8}$$

where $\ell_{r_i}(u) = \ell(u)$ if $r_i \geq 0$ and $\ell_{r_i}(u) = \ell(-u)$ if $r_i < 0$. Notice that $\ell_{r_i}(u) = \ell\{\text{sign}(r_i)u\}$, which suggests that inverting the loss functions is equivalent to the augmented OWL (Liu et al., 2018) with $s(x) = 0$ (see Sect. 1.1).

For multi-category treatments, Zhang et al. (2020) adopted the essentials of the ABC and proposed the following minimization problem:

$$\underset{f \in F}{\text{minimize}} \frac{1}{n} \sum_{i=1}^{n} \frac{|r_i|}{\pi_{a_i}(\boldsymbol{x}_i)} \ell_{r_i}\{\langle \mathbf{W}_{a_i}, \boldsymbol{f}(\boldsymbol{x}_i)\rangle\} + \lambda J(\boldsymbol{f}),$$

which is equivalent to

$$\underset{f \in F}{\text{minimize}} \frac{1}{n} \sum_{i=1}^{n} \frac{|r_i|}{\pi_{a_i}(\boldsymbol{x}_i)} \ell\{\langle \mathbf{W}_{a_i}, \text{sign}(r_i)\boldsymbol{f}(\boldsymbol{x}_i)\rangle\} + \lambda J(\boldsymbol{f}). \tag{9}$$

It is not hard to find that (9) reduces to (8) for $K = 2$. Similar to the OWL for binary treatments, the term $|r_i|/\pi_{a_i}(\boldsymbol{x}_i)$ can be regarded as the observational weights. Therefore, the solution to (9) is equivalent to a weighted angle-based classifier with the inverting loss for negative rewards.

2.3 Multi-Category Direct Learning

We reviewed the direct learning method intended for binary treatments in Sect. 1.3. In this section, we introduce two approaches that extend the binary D-learning method for scenarios with multiple treatments.

The first method was introduced in Qi and Liu (2018) and named the *pairwise D-learning* by Qi et al. (2020). Recall the value of the treatment regime D given in (1) for binary treatments. We may extend it to multiple treatments as follows:

$$
\begin{aligned}
V(D) &= \mathbb{E}\left[\frac{\mathbf{1}\{A = D(\mathbf{X})\}}{\pi_A(\mathbf{X})} R\right] \\
&= \mathbb{E}_{\mathbf{X}}\left(\mathbb{E}_{A|\mathbf{X}}\left[\frac{\mathbf{1}\{A = D(\mathbf{X})\}}{\pi_A(\mathbf{X})} \mathbb{E}(R \mid \mathbf{X}, A)\right]\right) \\
&= \mathbb{E}_{\mathbf{X}}\left[\sum_{k=1}^{K} \mathbf{1}\{D(\mathbf{X}) = k\} \mathbb{E}(R \mid \mathbf{X}, A = k)\right].
\end{aligned}
$$

The corresponding optimal ITR is thus

$$
D^*(\mathbf{X}) = \underset{k \in \{1,\ldots,K\}}{\arg\max}\ \mathbb{E}(R \mid \mathbf{X}, A = k), \tag{10}
$$

which can be further rewritten as

$$
\begin{aligned}
D^*(\mathbf{X}) &= \underset{k \in \{1,\ldots,K\}}{\arg\max}\ K\mathbb{E}(R \mid \mathbf{X}, A = k) - \sum_{k'=1}^{K} \mathbb{E}(R \mid \mathbf{X}, A = k') \\
&= \underset{k \in \{1,\ldots,K\}}{\arg\max} \sum_{k' \neq k}^{K} \left[\mathbb{E}(R \mid \mathbf{X}, A = k) - \mathbb{E}(R \mid \mathbf{X}, A = k')\right] \\
&= \underset{k \in \{1,\ldots,K\}}{\arg\max} \sum_{k' \neq k}^{K} \mathbb{E}\left[\frac{R A_{kk'}}{\pi_{A_{kk'}}(\mathbf{X})} \,\middle|\, \mathbf{X}, A \in \{k, k'\}\right] \\
&:= \underset{k \in \{1,\ldots,K\}}{\arg\max} \sum_{k' \neq k}^{K} f_{kk'}(\mathbf{X}) \\
&:= \underset{k \in \{1,\ldots,K\}}{\arg\max}\ f_k(\mathbf{X}),
\end{aligned}
$$

where $A_{kk'} \in \{1, -1\}$ is a binary random variable satisfying

$$
\Pr(A_{kk'} = 1 \mid \mathbf{X}) = \Pr(A = k \mid \mathbf{X}, A \in \{k, k'\}) = \frac{\pi_k(\mathbf{X})}{\pi_k(\mathbf{X}) + \pi_{k'}(\mathbf{X})}.
$$

The corresponding decision function $f_k(\mathbf{X})$ named the *effect measure* is the sum of the pairwise decision functions $\{f_{kk'}(\mathbf{X})\}$ between the kth and k'th treatments for all $k' \neq k$. Therefore, for each $k \in \{1, \ldots, K\}$, one can repeatedly apply the binary D-learning introduced in Sect. 1.3 to estimate the pairwise decision functions between the kth treatment and every alternative treatment. Notice that $A_{k'k} = -A_{kk'}$ given $\mathbf{X}$ and thus $f_{k'k}(\mathbf{X}) = -f_{kk'}(\mathbf{X})$ by definition. Then it would be sufficient to estimate $K(K-1)/2$ pairwise decision functions among the total K treatments. In addition, Qi and Liu (2018) revealed an insightful relationship between the pairwise D-learning approach and the one-versus-one strategy for multi-category classification (see Sect. 1.4). The major difference is that the pairwise D-learning utilizes the effect measure rather than the majority vote to determine the best treatment option. Another interesting observation is that $\sum_{k=1}^{K} f_k(\mathbf{X}) = 0$ by definition implying one of the effect measures is redundant, which is similar to the sum-to-zero constraints for those simultaneous multi-category classification methods such as multi-category SVM and multinomial logistic model introduced in Sect. 1.4.

Although the effect measure from the pairwise D-learning can quantify the relative advantage of each treatment over the alternatives for given covariates, it can be suboptimal and inefficient without taking into account all the available treatment options simultaneously. To resolve this issue, Qi et al. (2020) proposed the *angle-based direct learning* (AD-learning) for multi-armed treatments by using multiple response regression, which is the second D-learning approach that we are going to cover in this section. Similar to the MOML method introduced in Sect. 2.2, we encode those K treatments in the AD-learning by the K vertices forming a simplex in a $(K-1)$-dimensional Euclidean space. For continuous rewards, we rewrite the optimal ITR given in (10) as

$$\begin{aligned}
D^*(\mathbf{X}) &= \underset{k\in\{1,\ldots,K\}}{\arg\max}\ \mathbb{E}\{R \mid \mathbf{X}, A=k\} \\
&= \underset{k\in\{1,\ldots,K\}}{\arg\max}\ \{1-c(K)\}\mathbb{E}\{R \mid \mathbf{X}, A=k\} \\
&= \underset{k\in\{1,\ldots,K\}}{\arg\max}\left[\{1-c(K)\}\mathbb{E}\left(R \mid \mathbf{X}, A=k\right) + c(K)\sum_{k'=1}^{K}\mathbb{E}\left(R \mid \mathbf{X}, A=k'\right)\right] \\
&= \underset{k\in\{1,\ldots,K\}}{\arg\max}\left\{\mathbb{E}\left(R \mid \mathbf{X}, A=k\right) + c(K)\sum_{k'\neq k}^{K}\mathbb{E}\left(R \mid \mathbf{X}, A=k'\right)\right\} \\
&= \underset{k\in\{1,\ldots,K\}}{\arg\max}\left\{\mathbf{W}_k^\top\mathbb{E}\left(R\mathbf{W} \mid \mathbf{X}, A=k\right) + \mathbf{W}_k^\top\sum_{k'\neq k}^{K}\mathbb{E}\left(R\mathbf{W} \mid \mathbf{X}, A=k'\right)\right\} \\
&= \underset{k\in\{1,\ldots,K\}}{\arg\max}\ \mathbf{W}_k^\top\mathbb{E}\left\{\frac{R\mathbf{W}}{\pi_A(\mathbf{X})} \,\middle|\, \mathbf{X}\right\} := \underset{k\in\{1,\ldots,K\}}{\arg\max}\ \mathbf{W}_k^\top \boldsymbol{f}(\mathbf{X}),
\end{aligned} \tag{11}$$

where $c(K) = \mathbf{W}_k^\top \mathbf{W}_{k'} < 1$ is a constant that only depends on K and remains the same for any $k \neq k'$, $\mathbf{W}$ is a random vector taking values from $\{\mathbf{W}_1, \ldots, \mathbf{W}_K\}$ satisfying $\mathbf{W} = \mathbf{W}_k \mid (\mathbf{X}, A = k)$ almost surely, and $\boldsymbol{f}(\mathbf{X})$ represents a function mapping from $\mathbb{R}^p$ to $\mathbb{R}^{K-1}$. This expression motivates the AD-learning to estimate the decision function vector $\boldsymbol{f}(\mathbf{X})$ by using all the data simultaneously. Similar to the multi-category angle-based classifiers, the AD-learning approach has an attractive geometric interpretation. The optimal treatment option corresponds to the vertex having the largest inner product or equivalently the least angle with the decision function. As pointed out by Qi et al. (2020), the AD-learning is quite different from the MOML (Zhang et al., 2020) introduced in Sect. 2.2, though they utilize the same simplex encoding.

Similar to the binary D-learning we reviewed in Sect. 1.3, we may illustrate the AD-learning by deriving the optimal ITR for the following reward model,

$$R = m(\mathbf{X}) + \sum_{k'=1}^{K} \delta_{k'}(\mathbf{X})\mathbf{1}(A = k') + \epsilon,$$

where $m(\mathbf{X})$ is the main effect depending on $\mathbf{X}$, $\delta_k(\mathbf{X})$ represents the interaction effect between the covariates and the kth treatment, and ϵ is the random error term that is independent of $\mathbf{X}$ and A with the mean zero. Notice that

$$\mathbb{E}\left\{\frac{\boldsymbol{W}}{\pi_A(\mathbf{X})} \,\Big|\, \mathbf{X}\right\} = \sum_{k'=1}^{K} \frac{\boldsymbol{W}_{k'}}{\pi_{k'}(\mathbf{X})}\pi_{k'}(\mathbf{X}) = \mathbf{0},$$

$$\mathbb{E}\left\{\frac{\boldsymbol{W}\mathbf{1}(A = k')}{\pi_A(\mathbf{X})} \,\Big|\, \mathbf{X}\right\} = \boldsymbol{W}_k',$$

and thus we have

$$\begin{aligned}
\mathbb{E}\left\{\frac{R\boldsymbol{W}}{\pi_A(\mathbf{X})} \,\Big|\, \mathbf{X}\right\} &= m(\mathbf{X})\mathbb{E}\left\{\frac{\boldsymbol{W}}{\pi_A(\mathbf{X})} \,\Big|\, \mathbf{X}\right\} + \sum_{k'}^{K} \delta_{k'}(\mathbf{X})\mathbb{E}\left\{\frac{\boldsymbol{W}\mathbf{1}(A = k')}{\pi_A(\mathbf{X})} \,\Big|\, \mathbf{X}\right\} \\
&\quad + \mathbb{E}\left\{\frac{\epsilon \boldsymbol{W}}{\pi_A(\mathbf{X})} \,\Big|\, \mathbf{X}\right\} \\
&= \sum_{k'}^{K} \delta_{k'}(\mathbf{X})\boldsymbol{W}_{k'} + \mathbb{E}(\epsilon \mid \mathbf{X})\mathbb{E}\left\{\frac{\boldsymbol{W}}{\pi_A(\mathbf{X})} \,\Big|\, \mathbf{X}\right\} \\
&= \sum_{k'}^{K} \delta_{k'}(\mathbf{X})\boldsymbol{W}_{k'}.
\end{aligned}$$

Plug it into (11) and we get the following optimal ITR,

$$\begin{aligned} D^*(\mathbf{X}) &= \underset{k\in\{1,\ldots,K\}}{\arg\max}\ \boldsymbol{W}_k^\top \mathbb{E}\left\{\frac{RW}{\pi_A(\mathbf{X})}\,\middle|\,\mathbf{X}\right\} \\ &= \underset{k\in\{1,\ldots,K\}}{\arg\max}\ \boldsymbol{W}_k^\top \sum_{k'=1}^{K} \delta_{k'}(\mathbf{X})\boldsymbol{W}_{k'} \\ &= \underset{k\in\{1,\ldots,K\}}{\arg\max}\ c(K)\sum_{k'=1}^{K} \delta_{k'}(\mathbf{X})\boldsymbol{W}_{k'} + \{1-c(K)\}\,\delta_k(\mathbf{X}) \\ &= \underset{k\in\{1,\ldots,K\}}{\arg\max}\ \delta_k(\mathbf{X}), \end{aligned}$$

which does suggest the comparison among those treatment interaction effects with the covariates as one would expect.

To estimate the decision function vector given in (11), Qi et al. (2020) showed that $\boldsymbol{f}(\mathbf{X})$ is an optimal solution to the following minimization problem under the condition that differential is interchangeable with expectation.

$$\boldsymbol{f} \in \underset{\boldsymbol{f}\in\mathbb{R}^{K-1}}{\arg\min}\,\mathbb{E}\left[\frac{1}{\pi_A(\mathbf{X})}\{KRW - \boldsymbol{f}(\mathbf{X})\}^\top \boldsymbol{\Sigma}\{KRW - \boldsymbol{f}(\mathbf{X})\}\right], \tag{12}$$

where $\boldsymbol{\Sigma}$ is a positive definite matrix characterizing the dependency among the response variables. Qi et al. (2020) suggested letting $\boldsymbol{\Sigma}$ be the identity matrix $\boldsymbol{I}_{K-1}$ when no prior knowledge regarding the dependency is available. With the sample data, we minimize the corresponding empirical loss,

$$\underset{f\in\mathcal{F}}{\arg\min}\ \frac{1}{n(K-1)}\sum_{i=1}^{n}\frac{1}{\pi_A(\boldsymbol{x}_i)}\{Kr_i\boldsymbol{w}_i - \boldsymbol{f}(\boldsymbol{x}_i)\}^\top\{Kr_i\boldsymbol{w}_i - \boldsymbol{f}(\boldsymbol{x}_i)\},$$

where $\mathcal{F}$ represents the decision function class and $\boldsymbol{w}_i$ is the realization of the random variable W for the ith sample. For instance, one may focus on the linear decision functions from $\mathcal{F} = \{\boldsymbol{f}(\boldsymbol{x}) = \boldsymbol{B}^\top\boldsymbol{x}, \boldsymbol{B} \in \mathbb{R}^{p\times(K-1)}\}$. Then minimizing the empirical loss is equivalent to solving a multivariate linear regression problem, where the response vector takes values from $(Kr_i\boldsymbol{w}_i)$. The resulting decision rule from the AD-learning is $D^*(\boldsymbol{x}) = \arg\max_{k\in\{1,\ldots,K\}} \boldsymbol{W}_k^\top\boldsymbol{B}^\top\boldsymbol{x}$. Similar to the binary D-learning, we may incorporate the regularization to the objective function to prevent over-fitting and obtain better interpretability as follows:

$$\underset{f\in\mathcal{F}}{\arg\min}\ \frac{1}{n(K-1)}\sum_{i=1}^{n}\frac{1}{\pi_A(\boldsymbol{x}_i)}\{Kr_i\boldsymbol{w}_i - \boldsymbol{f}(\boldsymbol{x}_i)\}^\top\{Kr_i\boldsymbol{w}_i - \boldsymbol{f}(\boldsymbol{x}_i)\} + \lambda J(\boldsymbol{B}),$$

where $J(\boldsymbol{B})$ is the penalty term and $\lambda > 0$ is the tuning parameter. For linear decision rules, Qi et al. (2020) proposed to apply the group lasso penalty (Yuan & Lin, 2006) or the nuclear norm penalty (sum of all the singular values of the coefficient matrix $\boldsymbol{B}$). The former performs variable selection by treating each row of the coefficient matrix associated with individual covariates as one group, which is mostly intended for cases where some important covariates play roles in interacting with all the treatments. The latter gives a low-rank representation of the coefficient matrix and implies latent factors in the covariates. Both penalties remain invariant to any orthogonal transformation to the coefficient matrix, which is a necessary property for the decision function to be invariant to any orthogonal rotation to the simplex encoding.

The methods we have introduced so far are limited to continuous outcomes. In practice, it is not uncommon to encounter various outcomes such as binary or survival outcomes. Thus, it is necessary to have more general methods capable of estimating ITR for multi-armed treatment settings as well as various health outcomes. For instance, Qi et al. (2020) borrowed the idea of Tian et al. (2014), and extended the proposed AD-learning to various types of outcomes including binary and survival responses. We discuss the main idea behind the extension in the remainder of this section.

We assume that the differential and expectation operators are interchangeable henceforth. Similar to (12), $\boldsymbol{f}(\mathbf{X})$ is also an optimal solution to the following minimization problem,

$$\underset{f \in \mathcal{F}}{\arg\min}\, \mathbb{E}\left[\frac{1}{\pi_A(\mathbf{X})}\left\{\frac{K}{K-1}R - \boldsymbol{W}^\top \boldsymbol{f}(\mathbf{X})\right\}^2\right].$$

The part involving the clinical outcome R can be regarded as the square error loss with the working model,

$$\frac{K}{K-1}R = \boldsymbol{W}^\top \boldsymbol{f}(\mathbf{X}) + \epsilon.$$

where ϵ represents the random error term. This observation motivates the main idea behind extending the AD-learning to various types of outcomes, which is to treat $\boldsymbol{W}^\top \boldsymbol{f}(\mathbf{X})$ as the predictor for the outcome R and plug it into the corresponding objective function suitable for the type of R. For linear decision functions, we have

$$\boldsymbol{W}^\top \boldsymbol{f}(\mathbf{X}) = \boldsymbol{W}^\top \boldsymbol{B}^\top \mathbf{X} = \mathrm{tr}\{\boldsymbol{B}^\top(\mathbf{X}\boldsymbol{W}^\top)\} = \mathrm{tr}\left(\boldsymbol{B}^\top \mathbf{X}^*\right),$$

where we rewrite the working model as a more general trace regression model (Rohde & Tsybakov, 2011) by treating $\mathbf{X}^* = \mathbf{X}\boldsymbol{W}^\top$ as the new covariate vector.

For instance, Qi et al. (2020) considered minimizing the deviance loss of the logistic regression for binary outcomes,

$$\arg\min_{f \in \mathcal{F}} \mathbb{E}\left[-\frac{R \boldsymbol{W}^\top \boldsymbol{f}}{\pi_A(\mathbf{X})} + \frac{\log\{1+\exp(\boldsymbol{W}^\top \boldsymbol{f})\}}{\pi_A(\boldsymbol{X})} \right].$$

Take the partial derivative with respect to $\boldsymbol{f}$ and we get

$$\Pr(R = 1 \mid \mathbf{X}, A = k) = \frac{\exp\{\boldsymbol{W}_k^\top \boldsymbol{f}(\mathbf{X})\}}{1+\exp\{\boldsymbol{W}_k^\top \boldsymbol{f}(\mathbf{X})\}},$$

and thus

$$\begin{aligned} D^*(\boldsymbol{X}) &= \mathop{\arg\max}_{k \in \{1,\ldots,K\}} \Pr(R = 1 \mid \mathbf{X}, A = k) \\ &= \mathop{\arg\max}_{k \in \{1,\ldots,K\}} \boldsymbol{W}_k^\top \boldsymbol{f}(\mathbf{X}), \end{aligned}$$

which shows that the least angle decision rule maximizes the quality function as expected. For linear decision rules, we have

$$\Pr(R = 1 \mid \mathbf{X}, A = k) = \frac{\exp\{\mathrm{tr}(\boldsymbol{B}^\top \mathbf{X}^*)\}}{1+\exp\{\mathrm{tr}(\boldsymbol{B}^\top \mathbf{X}^*)\}},$$

and we may estimate the coefficient matrix $\boldsymbol{B}$ by minimizing the negative log-likelihood of the corresponding trace regression model with regularization,

$$\mathop{\arg\min}_{\boldsymbol{B} \in \mathbb{R}^{p \times (K-1)}} -\frac{1}{n} \sum_{i=1}^{n} \left(\frac{r_i \,\mathrm{tr}(\boldsymbol{B}^\top \boldsymbol{x}_i^*)}{\pi_{a_i}(\boldsymbol{x}_i^*)} - \frac{\log\left[1+\exp\{\mathrm{tr}(\boldsymbol{B}^\top \boldsymbol{x}_i^*)\}\right]}{\pi_{a_i}(\boldsymbol{x}_i^*)} \right) + \lambda J(\boldsymbol{B}),$$

where $\boldsymbol{x}_i^*$ is the realization of $\mathbf{X}_i^*$, $J()$ can be the group lasso penalty, the nuclear norm penalty, or any other penalty invariant to orthogonal transformation.

There are different types of censoring when it comes to the survival outcomes. The most common right-censored outcome consists of a pair of random variables, $R = (T, \delta)$, where T and δ are the survival time and the event indicator, respectively. The survival time is the actual event time of interest when $\delta = 1$. Otherwise, the event time is censored and not observed exactly. For simplicity, we assume that all the event times are distinct without ties, and the baseline hazard functions are the same for different treatments. As the second example, Qi et al. (2020) illustrated how to extend the AD-learning to the right-censored data by utilizing the negative partial log-likelihood of the Cox model (Cox, 1972, 1975),

$$\underset{f \in \mathcal{F}}{\arg\min} -\frac{1}{n} \sum_{i:\delta_i=1} \left[\frac{\boldsymbol{w}_i^\top \boldsymbol{f}}{\pi_{a_i}(\boldsymbol{x}_i)} - \frac{1}{\pi_{a_i}(\boldsymbol{x}_i)} \log \left\{ \sum_{j:t_j \geq t_i} \exp(\boldsymbol{w}_i^\top \boldsymbol{f}) \right\} \right] + \lambda J(\boldsymbol{f}),$$

where t_i's are realizations of T_i's. For linear decision rules, we have

$$\underset{\boldsymbol{B} \in \mathbb{R}^{p \times (K-1)}}{\arg\min} \quad -\frac{1}{n} \sum_{i:\delta_i=1} \left(\frac{\operatorname{tr}(\boldsymbol{B}^\top \boldsymbol{x}_i^*)}{\pi_{a_i}(\boldsymbol{x}_i)} - \frac{1}{\pi_{a_i}(\boldsymbol{x}_i)} \log \left[\sum_{j:t_j \geq t_i} \exp\{\operatorname{tr}(\boldsymbol{B}^\top \boldsymbol{x}_i^*)\} \right] \right)$$
$$+ \lambda J(\boldsymbol{B}),$$

where $\boldsymbol{x}_i^*$'s are the new covariate vector. Qi et al. (2020) showed that the optimal decision rule for the survival outcomes can be written as

$$D^*(\mathbf{X}) = \underset{k \in \{1,\ldots,K\}}{\arg\max} \ \boldsymbol{W}_k^\top (-\boldsymbol{f}^*).$$

2.4 *Multi-Label Outcome-Weighted Deep Learning Algorithms*

Under the scenario that we mainly focus on in this chapter, patients are allowed to take one particular treatment among multiple options at one time. However, in practice, physicians may suggest patients take multiple treatments simultaneously for a potentially better treatment effect, which is often the case for patients with chronic diseases such as type 2 diabetes. Therefore, it is of great interest to obtain the optimal treatment strategy that can combine multiple treatments given the patient's characteristics. The possibility of recommending more than one treatment at the same time turns the problem into a multi-label classification problem thanks to the OWL framework that we reviewed in Sect. 1.1. Among others, Liang et al. (2018) proposed to utilize deep neural networks with the Hamming loss under the OWL framework to estimate the optimal treatment combination. The main idea is straightforward and the Fisher consistency of the proposed loss function was shown under certain conditions. In this section, we provide a review of their work.

We start with an introduction to the multi-label classification problem and its connection to the multi-category classification problem (see Sect. 1.4). The key characteristic of the multi-label classification problem is that the response variable can consist of a varying number of nonexclusive classes or labels. One may convert a multi-label classification problem to a multi-category classification problem by considering all possible label combinations, which is called the *label powerset transformation*. Nevertheless, such conversion treats each combination individually without taking into account the underlying correlation structure among labels. As a result, it is not scalable to the number of labels as the total number of label combinations is 2^K for K labels. For a multi-category classification problem with K labels, we need at least $K - 1$ classifiers. Thus, a multi-label classification problem transformed by label powerset requires at least $2^K - 1$ classifiers.

For better scalability, two strategies based on problem transformation are commonly used in practice. The first approach is called the *binary relevance* (Luaces et al., 2012), which converts a multi-label problem to a sequence of independent binary classification tasks. Each binary classifier corresponds to a particular label and decides if it should be assigned. A multi-label classification problem with K labels requires only K independent binary classifiers by using the binary relevance strategy, which greatly reduces the dimensionality of the estimation and increases computational efficiency. However, the independence assumption on the labels that the binary relevance strategy requires hardly holds in many applications. The second strategy is to ensemble the predicted labels from a collection of multi-category classifiers, each of which predicts a single label for one sample at one time. A simple way to ensemble the predictions is to adopt a voting scheme, where each predicted label that receives a sufficient percentage of votes is included in the final prediction.

Various classification methods have been adapted and proposed for the multi-label learning tasks in the literature. For instance, Tsoumakas and Katakis (2007) and Madjarov et al. (2012) provided a recent review and comparison of popular methods intended for the multi-label classification problem. A straightforward adaption is to utilize a loss function that is appropriate for the multi-label learning task. One such example is the Hamming loss, which measures the fraction of misclassified labels to the total number of labels. Let $\boldsymbol{A} = (A_1, \ldots, A_K)$ represent the treatment vector, where $A_k \in \{-1, 1\}$, $k \in \{1, \ldots, K\}$. We define $A_k = 1$ to indicate that the treatment A_k is adopted in the treatment strategy, and $A_k = -1$ otherwise. Given the sample $(\mathbf{X}, \boldsymbol{A})$, the Hamming loss of a decision rule $\boldsymbol{D}(\mathbf{X})$ is defined as follows:

$$\frac{1}{K}\sum_{k=1}^{K} \mathbf{1}\{A_k \neq D_k(\mathbf{X})\},$$

where $D_k(\mathbf{X}) \in \{-1, 1\}$ denotes the individual decision rule corresponding to the kth treatment, and the decision rule satisfies $\boldsymbol{D}(\mathbf{X}) = (D_1(\mathbf{X}), \ldots, D_K(\mathbf{X}))$. In contrast, we define the 0–1 loss as

$$1 - \prod_{k=1}^{K} \mathbf{1}\{A_k = D_k(\mathbf{X})\},$$

which is zero if and only if the treatment vector $\boldsymbol{A}$ coincides with the recommendation from the treatment rule $\boldsymbol{D}(\mathbf{X})$ completely. For example, we visualize the 0–1 loss and Hamming loss function for three labels in the left and right panel of Fig. 3, respectively. Although the number of misclassified labels given in the horizontal axis takes discrete values, we connect the loss values by an imaginary step function,

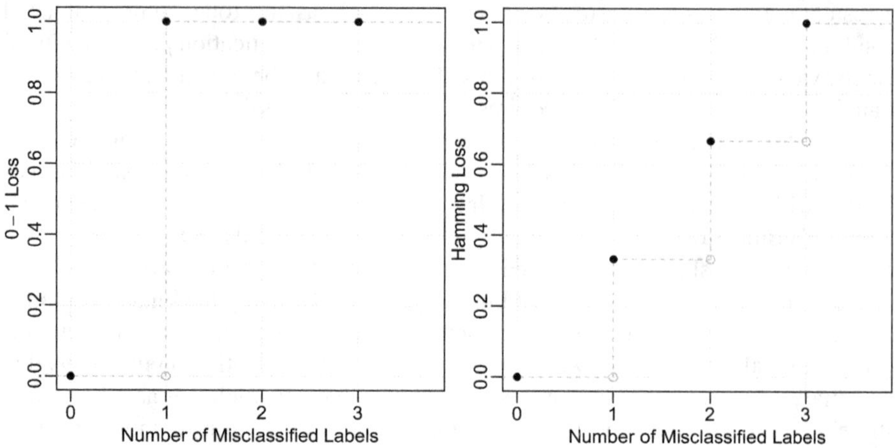

Fig. 3 Comparison between the 0–1 loss (on the left panel) and Hamming loss (on the right panel) for three labels. The loss values are indicated by solid black dots and connected by imaginary step functions (dashed gray lines) to better visualize the loss increment, where the jump points are indicated by hollow gray dots

respectively, for the 0–1 loss and the Hamming loss function, which visualizes the loss increment as the number of misclassified labels increases.

Liang et al. (2018) incorporated the Hamming loss into the OWL framework and proposed to estimate the optimal rule for recommending treatment combination by minimizing the *outcome weighted Hamming loss* given below.

$$\widehat{\boldsymbol{D}}^* \in \arg\min_{\boldsymbol{D}} \frac{1}{n}\sum_{i=1}^{n} \frac{r_i}{\pi_{\boldsymbol{a}_i}(\boldsymbol{x}_i)} \left[\frac{1}{K} \sum_{k=1}^{K} \mathbf{1}\{a_{ik} \neq D_k(\boldsymbol{x}_i)\} \right], \tag{13}$$

where $\boldsymbol{a}_i = (a_{i1}, \dots, a_{iK})$ is the assigned treatment vector for patient i satisfying $a_{ik} \in \{-1, 1\}$ for $k \in \{1, \dots, K\}$. For simplicity, Liang et al. (2018) assumed positive and real-valued rewards. The proposed loss in (13) is bounded from above by the OWL with the 0–1 loss, and the equivalency is attained for multi-category ITR problems. The problem of estimating the optimal ITR is then converted to a weighted multi-label classification problem under the OWL framework. Nevertheless, it is hard to directly minimize the loss in (13) as the indicator function is neither continuous nor convex. Similar to the OWL for binary treatments, Liang et al. (2018) adopted the hinge-loss as the surrogate loss function,

$$\widehat{\boldsymbol{D}}^* \in \arg\min_{\boldsymbol{D}} \frac{1}{n}\sum_{i=1}^{n} \frac{r_i}{\pi_{\boldsymbol{a}_i}(\boldsymbol{x}_i)} \left[\frac{1}{K} \sum_{k=1}^{K} \{1 - a_{ik} \neq D_k(\boldsymbol{x}_i)\}_+ \right], \tag{14}$$

where the decision function D_k can be estimated by linear models or the kernel methods (e.g., Bishop, 2006, Chapter 6). In particular, Liang et al. (2018) proposed classifiers based on the neural network models because of their great flexibility with the hidden layers and scalability for sample size. In addition, the neural network models have the *shared subspace* property (Yan et al., 2007), which means that all the decision functions depend on some common subspace of the predictors and thus efficiency can be borrowed from each other for estimation. The neural networks with a large number of hidden layers are often called *deep neural networks* or *deep learning*, which have been widely used to model complex non-linear relationships between the predictors and the response. For a general introduction to the deep neural networks, we refer readers to Goodfellow et al. (2016, Chapter 6) or Hastie et al. (2009, Chapter 11) among others.

For example, we visualize a neural network with three layers in Fig. 4. The first layer is the *input layer* that consists of all the covariates, $\boldsymbol{x} \in \mathbb{R}^p$. The middle layer is the *hidden layer* denoted by $\boldsymbol{h} = (h_1, \ldots, h_d)$ satisfying

$$h_j = \sigma(\boldsymbol{\theta}_{1j}^\top \boldsymbol{x} + \alpha_{1j}), \tag{15}$$

where $j \in \{1, \ldots, d\}$, σ is named the *activation function*, $\boldsymbol{\theta}_{1j}$ and α_{1j} represent the *weights* (or the coefficient vector) and the *bias* (or the intercept), respectively, for the jth *hidden unit*. The last layer is the *output layer* denoted by $\boldsymbol{D} = (D_1, \ldots, D_K)$ satisfying

$$D_k = \sigma(\boldsymbol{\theta}_{2k}^\top \boldsymbol{h} + \alpha_{2k}), \tag{16}$$

where $\boldsymbol{\theta}_{2j}$ and α_{2j} represent the weights and the bias for the kth *output unit* or the kth decision function in our context, respectively. The popular choices of the activation function include the *sigmoid* function, $\sigma(v) = 1/(1 + \exp(-v))$, and the *Rectified Linear Unit* (RelU), $\sigma(v) = v\mathbf{1}(v > 0)$. Although we used the same notation σ for the activation functions for the hidden layer and output layer in (15) and (16), we are free to utilize different activation functions at different layers. For the neural network model given in Fig. 4, we can rewrite D_k by

$$D_k(\boldsymbol{x}) = \sigma\{\boldsymbol{\theta}_{2k}^\top \sigma(\boldsymbol{\theta}_1 \boldsymbol{x} + \boldsymbol{\alpha}_1) + \alpha_{2k}\},$$

where $\boldsymbol{\theta}_1 = (\boldsymbol{\theta}_{11}, \ldots, \boldsymbol{\theta}_{1d})^\top$, $\boldsymbol{\alpha}_1 = (\alpha_{11}, \ldots, \alpha_{1d})^\top$, and the inner activation function σ performs the element-wise mapping with some abuse of notation.

Estimating the weights and biases of a neural network by gradient descent method is often called the *back-propagation*, which can be implemented in a paralleled fashion. For data with a large sample size, the *stochastic gradient descent* (SGD) proposed by Bottou (2010) is one of the widely used optimization algorithms for neural networks in practice. In a nutshell, the SGD suggests updating the gradient by using a small fraction of the random samples from the original data with or without replacement. Thus, it is scalable to the sample size.

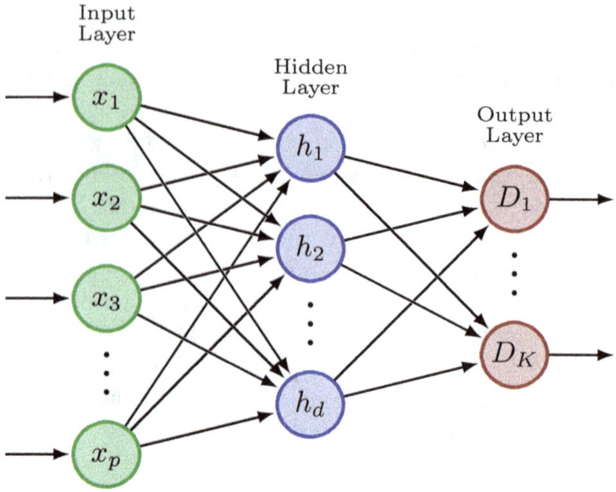

Fig. 4 Visualization of a neural network with one hidden layer for K treatments

The neural network models can be over-parameterized easily. The number of layers and units per layer plays an important role in determining the flexibility of the model. To prevent overfitting, Liang et al. (2018) considered imposing regularization to the loss function in (14) as follows:

$$\underset{\boldsymbol{\beta}}{\text{minimize}}\ \frac{1}{n}\sum_{i=1}^{n}\frac{r_i}{\pi_{\boldsymbol{a}_i}(\boldsymbol{x}_i)}\left[\frac{1}{K}\sum_{k=1}^{K}\{1-a_{ik}\neq D_k(\boldsymbol{x}_i)\}_+\right]+\lambda J(\boldsymbol{\beta}),$$

where $\boldsymbol{\beta}$ consists of the weight and the bias parameters of the neural network model, $J(\cdot)$ is the penalty function, and λ is the tuning parameter. The common choices of $J(\boldsymbol{\beta})$ include the ridge penalty and the lasso penalty for the weights. Another simple but effective method for mitigating overfitting is the *dropout* (Srivastava et al., 2014), which randomly drops some of the units from the input or the hidden layers, and their associated connections from the neural network in the training process. When making predictions, we use the complete neural network instead of the thinned network. However, the weight parameters associated with each unit are discounted by multiplying the probability of the unit being present during the training.

Liang et al. (2018) showed Fisher consistency of the proposed outcome-weighted Hamming loss under the condition that

$$D_k^*(\mathbf{X}) = \text{sign}\left\{\sum_{\boldsymbol{a}:a_k=1}\mathbb{E}(R \mid A=a,\mathbf{X}) - \sum_{\boldsymbol{a}:a_k=-1}\mathbb{E}(R \mid A=a,\mathbf{X})\right\},$$

which holds when there are no or small interactions between treatments. If strong interactions between treatments exist, it will be hard to borrow information among treatments, and thus the proposed method can fail to improve the estimation efficiency and give suboptimal results.

3 Summary

This chapter provides a selective review of various methods proposed in the recent literature for estimating optimal treatment rules in scenarios where multiple treatment options are available to patients. The reviewed methods mainly fall into two categories. The first approach follows the seminal outcome-weighted learning framework and transforms the problem into a classification task to minimize regret or loss caused by suboptimal treatment assignments. The second approach aims to learn decision functions directly through regression methods. Both approaches avoid the need to model conditional clinical outcomes, which makes them attractive for many applications in practice.

The methods for deriving the ITR have broad application areas not limited to healthcare. As a result, the literature on estimating the ITR has been blooming in recent decades with applications to various fields. Our review provided in this chapter is by no means comprehensive. For example, we restricted our scope to randomized clinical trials, where the propensity score given the pretreatment covariates is fixed and known. In general, one needs to estimate the propensity score additionally if the mechanism of the treatment assignment is unknown.

References

Bishop, C. M. (2006). *Pattern recognition and machine learning*. Springer.

Blatt, D., Murphy, S. A., & Zhu, J. (2004). A-learning for approximate planning. Technical Report 04-63, The Methodology Center, Pennsylvania State University, State College, PA.

Bottou, L. (2010). Large-scale machine learning with stochastic gradient descent. In *Proceedings of COMPSTAT'2010: 19th International Conference on Computational StatisticsParis France, August 22–27, 2010 Keynote, Invited and Contributed Papers* (pp. 177–186). Springer.

Chen, J., Fu, H., He, X., Kosorok, M. R., & Liu, Y. (2018). Estimating individualized treatment rules for ordinal treatments. *Biometrics, 74*(3), 924–933.

Cox, D. R. (1972). Regression models and life-tables. *Journal of the Royal Statistical Society. Series B (Methodological), 34*(2), 187–220.

Cox, D. R. (1975). Partial likelihood. *Biometrika, 62*(2), 269–276.

Cristianini, N., & Shawe-Taylor, J. (2000). *An introduction to support vector machines and other Kernel-based learning methods*. Cambridge University Press.

Goodfellow, I., Bengio, Y., & Courville, A. (2016). *Deep learning*. MIT Press.

Hastie, T., Rosset, S., Zhu, J., & Zou, H. (2009). Multi-class adaboost. *Statistics and Its Interface, 2*(3), 349–360.

Hastie, T., Tibshirani, R., & Friedman, J. (2009). *The elements of statistical learning: Data mining, inference, and prediction* (2nd ed.), Springer.

Lange, K., & Tong Wu, T. (2008). An MM algorithm for multicategory vertex discriminant analysis. *Journal of Computational and Graphical Statistics, 17*(3), 527–544.

Lee, Y., Lin, Y., & Wahba, G. (2004). Multicategory support vector machines: Theory and application to the classification of microarray data and satellite radiance data. *Journal of the American Statistical Association, 99*(465), 67–81.

Liang, M., Ye, T., & Fu, H. (2018). Estimating individualized optimal combination therapies through outcome weighted deep learning algorithms. *Statistics in Medicine, 37*(27), 3869–3886.

Lin, Y., & Zhang, H. H. (2006). Component selection and smoothing in multivariate nonparametric regression. *The Annals of Statistics, 34*(5), 2272–2297.

Liu, Y., Wang, Y., Kosorok, M. R., Zhao, Y., & Zeng, D. (2018). Augmented outcome-weighted learning for estimating optimal dynamic treatment regimens. *Statistics in Medicine, 37*(26), 3776–3788.

Liu, Y., Zhang, H. H., & Wu, Y. (2011). Hard or soft classification? Large-margin unified machines. *Journal of the American Statistical Association, 106*(493), 166–177.

Luaces, O., Díez, J., Barranquero, J., del Coz, J. J., & Bahamonde, A. (2012). Binary relevance efficacy for multilabel classification. *Progress in Artificial Intelligence, 1*(4), 303–313.

Madjarov, G., Kocev, D., Gjorgjevikj, D., & Džeroski, S. (2012). An extensive experimental comparison of methods for multi-label learning. *Pattern Recognition, 45*(9), 3084–3104.

Moodie, E. E. M., Platt, R. W., & Kramer, M. S. (2009). Estimating response-maximized decision rules with applications to breastfeeding. *Journal of the American Statistical Association, 104*(485), 155–165.

Murphy, S. A. (2003). Optimal dynamic treatment regimes. *Journal of the Royal Statistical Society: Series B (Statistical Methodology), 65*(2), 331–355.

Qi, Z., Liu, D., Fu, H., & Liu, Y. (2020). Multi-armed angle-based direct learning for estimating optimal individualized treatment rules with various outcomes. *Journal of the American Statistical Association, 115*(530), 678–691.

Qi, Z., & Liu, Y. (2018). D-learning to estimate optimal individual treatment rules. *Electronic Journal of Statistics, 12*(2), 3601–3638.

Qian, M., & Murphy, S. A. (2011). Performance guarantees for individualized treatment rules. *Annals of Statistics, 39*(2), 1180–1210.

Robins, J. M. (2004). Optimal structural nested models for optimal sequential decisions. In *Proceedings of the Second Seattle Symposium in Biostatistics* (pp. 189–326). Springer.

Rohde, A., & Tsybakov, A. B. (2011). Estimation of high-dimensional low-rank matrices. *The Annals of Statistics, 39*(2), 887–930.

Srivastava, N., Hinton, G., Krizhevsky, A., Sutskever, I., & Salakhutdinov, R. (2014). Dropout: A simple way to prevent neural networks from overfitting. *The Journal of Machine Learning Research, 15*(1), 1929–1958.

Tian, L., Alizadeh, A. A., Gentles, A. J., & Tibshirani, R. (2014). A simple method for estimating interactions between a treatment and a large number of covariates. *Journal of the American Statistical Association, 109*(508), 1517–1532.

Tibshirani, R. (1996). Regression shrinkage and selection via the Lasso. *Journal of the Royal Statistical Society. Series B (Methodological), 58*(1), 267–288.

Tsoumakas, G., & Katakis, I. (2007). Multi-label classification: An overview. *International Journal of Data Warehousing and Mining (IJDWM), 3*(3), 1–13.

Yan, R., Tesic, J., & Smith, J. R. (2007). Model-shared subspace boosting for multi-label classification. In *Proceedings of the 13th ACM SIGKDD International Conference on Knowledge Discovery and Data Mining* (pp. 834–843).

Yuan, M., & Lin, Y. (2006). Model selection and estimation in regression with grouped variables. *Journal of the Royal Statistical Society: Series B (Statistical Methodology), 68*(1), 49–67.

Zahid, F. M., & Tutz, G. (2013). Ridge estimation for multinomial logit models with symmetric side constraints. *Computational Statistics, 28*(3), 1017–1034.

Zhang, C., Chen, J., Fu, H., He, X., Zhao, Y.-Q., & Liu, Y. (2020). Multicategory outcome weighted margin-based learning for estimating individualized treatment rules. *Statistica Sinica, 30*, 1857–1879.

Zhang, C., & Liu, Y. (2014). Multicategory angle-based large-margin classification. *Biometrika, 101*(3), 625–640.

Zhang, C., Liu, Y., Wang, J., & Zhu, H. (2016). Reinforced angle-based multicategory support vector machines. *Journal of Computational and Graphical Statistics, 25*(3), 806–825.

Zhao, Y., Zeng, D., Rush, A. J., & Kosorok, M. R. (2012). Estimating individualized treatment rules using outcome weighted learning. *Journal of the American Statistical Association, 107*(499), 1106–1118.

Zhou, X., Wang, Y., & Zeng, D. (2018). Outcome-weighted learning for personalized medicine with multiple treatment options. In *2018 IEEE 5th International Conference on Data Science and Advanced Analytics (DSAA)* (pp. 565–574). IEEE.

Zou, H., & Hastie, T. (2005). Regularization and variable selection via the elastic net. *Journal of the Royal Statistical Society: Series B (Statistical Methodology), 67*(2), 301–320.

Statistical Reinforcement Learning and Dynamic Treatment Regimes

Tao Shen and Yifan Cui

Abstract This chapter introduces to readers the concept and methodology of reinforcement learning and modern-day dynamic treatment regimes in statistics. This discussion should be of interest to those who wish to go into the depth of statistical reinforcement learning. We start with introducing the Markov decision process. Then, several methods such as policy iteration, value iteration, temporal difference learning, and policy gradient are presented. Next, dynamic treatment regimes in both point exposure and time-varying settings are introduced. We also include a discussion of causal reinforcement learning which is an emerging field of policy learning. Several illustrative examples are provided for practitioners based on simulated data. This review aims to provide an overview of various state-of-the-art methods for reinforcement learning and dynamic treatment regimes. The sample codes provided in this chapter are available at https://github.com/taoshen2022/Statistical-Reinforcement-Learning-and-Dynamic-Treatment-Regimes-Supplementary-Material.

Keywords Causality · Dynamic treatment regimes · Markov decision process · Precision medicine · Reinforcement learning

1 Introduction

Reinforcement learning (RL) is a machine learning technique that enables an agent to learn in the environment by optimizing the accumulative reward (Sutton & Barto, 2018). By applying RL techniques to sequential decision-making, feedback from

---

T. Shen
Department of Statistics and Data Science, National University of Singapore, Singapore, Singapore
e-mail: taoshen@u.nus.edu

Y. Cui (✉)
Center for Data Science, Zhejiang University, Hangzhou, China
e-mail: cuiyf@zju.edu.cn

Y. Zhao, D.-G. Chen (eds.), *Statistics in Precision Health*, ICSA Book Series in Statistics, https://doi.org/10.1007/978-3-031-50690-1_7

experience is used to estimate optimal decision rules. Major applications of RL include gaming, personalized recommendations, robotics, discrete and continuous control, and precision medicine, etc (Chakraborty & Moodie, 2013; Kober et al., 2013; Neunert et al., 2020; Riedmiller & Gabel, 2007; Soliman, 2014). In this chapter, we introduce some core concepts of offline RL and briefly discuss several state-of-the-art methods without providing an in-depth introduction to RL.

The environment of the infinite horizon RL is typically stated in the form of a Markov decision process (MDP) (Howard, 1960; Murphy, 2005). Several classic RL techniques and algorithms including policy iteration, value iteration, temporal difference learning, and policy gradient are introduced (Agarwal et al., 2021; Clifton & Laber, 2020; Ernst et al., 2005; Howard, 1960; Lange et al., 2012; Levine et al., 2020; Li et al., 2023; Murphy, 2005; Neumann & Peters, 2008; Precup, 2000). Moreover, we review some pessimism algorithms (Chen & Jiang, 2019; Jin et al., 2021; Levine et al., 2020; Liu et al., 2020; Munos, 2003, 2005; Uehara & Sun, 2022; Xie et al., 2021) of the aforementioned RL methods, which have been recently developed to weaken the assumption regarding the sufficient coverage of the dataset that possibly fails to hold in practice. In contrast to being optimistic to promote exploration in online RL such as in the case of the well-known upper confidence bound algorithm (Garivier & Moulines, 2011), being pessimistic in offline RL guarantees that the suboptimality of the estimated policy only depends on how well the offline dataset covers the trajectory induced by the optimal policy.

Over the past few decades, there has been a growing interest in so-called dynamic treatment regimes (DTRs), particularly in medical and epidemiological settings. In contrast to static treatment regimes, a DTR consisting of a sequence of decision rules, one per stage of intervention, individualizes treatments to patients based on their evolving covariate and treatment history. The origin of DTR can be traced back to Robins and Greenland (1989); Robins (1986, 1997); Mark and Robins (1993) who invented inferential tools for DTRs in sequentially randomized trials. Since the 2000s, a prevailing strand of work has been devoted to estimating optimal DTR from a machine learning perspective (Chakraborty & Moodie 2013; Cui et al. 2017; Liu et al. 2018; Murphy 2003; Schulte et al. 2014; Wallace & Moodie 2015; Zhang et al. 2012; Zhang et al. 2012; Zhang & Zhang 2018; Zhao et al. 2012, 2015; Zhou et al. 2017; Kosorok & Laber 2019; Athey & Wager 2021). We briefly overview some important concepts and subsequently focus on some recent developments in DTRs.

Advanced developments of reinforcement learning and causal inference in healthcare applications have inspired groups of researchers to integrate both to tackle decision-making problems from a causal perspective. Recently, the integration has been seen in many works (Madumal et al., 2020; Zhu et al., 2019; Zhang et al., 2020; Zhang & Bareinboim, 2016; Lu et al., 2021; Cohen et al., 2020) and studied under the rubric of causal reinforcement learning (Causal RL, Elias (2023)). We use examples of confounded bandit and MDP to illustrate recent progress in this area.

Table 1 Different examples of three elements in MDP

State	Action	Reward	Reference
Blood glucose, insulin, etc.	Dose of insulin	Risk function of BG	Zhu et al. (2021)
Tumor size and toxicity	Drug level	Function of toxicity	Zhao et al. (2009)
Arterial carbon dioxide level, etc.	Heparin values	Fraction of time within range	Nemati et al. (2016)
Mean tumor diameters trajectory	Dosing options	Overall MTD reduction	Yauney and Shah (2018)

2 Reinforcement Learning and Markov Decision Process

A subfield of machine learning, RL is aimed at learning an optimal strategy, including a set of methods for learning how to behave "well" in an environment. In this section, we focus on the introduction of an environment described by a MDP, which is a discrete-time stochastic control process (Bellman, 1958) and can be traced back to the 1950s (Bellman, 1957). The environment setting bears a strong resemblance to a multi-stage decision-making process. In each time step, for a learning agent (e.g., robot, patient), an **action** is selected from a set of possible movements based on the current **state**, and then the agent receives a **reward** from the environment.

In most medical and epidemiological settings, state is related to the physical features of patients, whereas action is linked with treatment and reward is connected to an outcome of interest. Examples of state, action, and reward in a variety of medical applications described by MDP are listed in Table 1. A toy example[1] is provided in Fig. 1 and Table 2. As shown in the figure, different transition diagrams can be traced after taking disparate actions in an environment consisting of three states and three kinds of treatments. If we regard S_L as a standard state where patients are expected to stay, S_M as a state of serious illness, and S_H as a state representing the most severe case, the three treatments would have different curative effects on patients in each state. In detail, Treatment 1 keeps patients in S_L stable, and makes patients in S_H return to S_M, however, it cannot effectively return patients from S_M to S_L. Unlike Treatment 1, Treatment 2 is targeted on patients in S_M and S_H, allowing a possibility of returning to S_L. Treatment 3 is primarily used to assist patients in S_M to return to S_L, while it has little effect on patients in S_H. Therefore, if treatment is appropriately selected based on individualized state information, symptoms are expected to be relieved. An implementable pseudo-code for programming is proposed in Algorithm 1, and we use the function `env()` to indicate the environmental setup in Sect. 3.

Next, we will provide formal mathematical definitions and expressions.

[1] The sample codes provided in Sect. 3 are based on this toy example.

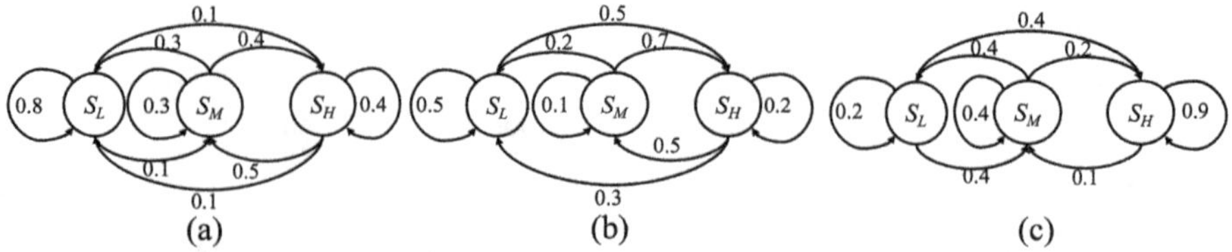

Fig. 1 The transition diagram of MDP: (**a**) if the patient is assigned with Treatment 1; (**b**) if the patient is assigned with Treatment 2; (**c**) if the patient is assigned with Treatment 3

Table 2 The reward function of the MDP setting used in sample code

State	Action	Reward
S_L	1	1.0
S_L	2	0.0
S_L	3	0.0
S_M	1	0.5
S_M	2	0.0
S_M	3	0.5
S_H	1	0.5
S_H	2	0.5
S_H	3	0.0

Definition 1 A MDP is defined as a tuple $M =< \mathcal{S}, \mathcal{A}, P, \rho, r, \gamma >$, where $\mathcal{S}$ is the state space that may be either discrete or continuous, $\mathcal{A}$ is the action space that can similarly be discrete or continuous, $P : \mathcal{S} \times \mathcal{A} \to \Delta(\mathcal{S})$ denote the transition function with $\Delta(\cdot)$ being the probability simplex, $\rho(\cdot)$ defines the initial state distribution, $r : \mathcal{S} \times \mathcal{A} \to \mathbb{R}$ defines a reward function, and $\gamma \in (0, 1)$ is a discount factor. If the elements in the tuple are all indexed by t, we define the tuple as M_t to represent the MDP setting for the t-th time step.

In Sects. 2 and 3, we focus on time-homogeneous MDP of an infinite time horizon, equipped with discrete state space and action space (see Jin et al. (2021); Li et al. (2022) for non-stationary settings), i.e., we assume

(i) The state space and action space are constant, i.e., $s_t \in \mathcal{S}_t \equiv \mathcal{S}$, $a_t \in \mathcal{A}_t \equiv \mathcal{A}$, with reward function denoted by $r_t = r(s_t, a_t)$ for any $t \geq 1$.
(ii) The data generating process satisfies the so-called Markovian property, i.e., $S_{t+1} \perp \{S_j, A_j\}_{1 \leq j < t} | S_t, A_t$ for any $t \geq 1$. Moreover, the transition dynamics are time-homogeneous, i.e., for any $t \geq 1$, $s \in \mathcal{S}$, $s' \in \mathcal{S}$ and $a \in \mathcal{A}$, we have $P_t \equiv P$ and $P(S_{t+1} = s' | S_t = s, A_t = a) = P(S_2 = s' | S_1 = s, A_1 = a) = P(s' | s, a) > 0$.

In addition, the agent often designates an upcoming action based on a policy $\pi \in \Pi$, which is a mapping from state space to action space, where $\Pi \subset (\mathcal{S} \to \Delta(\mathcal{A}))$ denotes the policy class. Note that a policy can be deterministic with its mapping from the state space to the action space, as well as stochastic with its mapping from the state space to a space of probabilistic distributions over the action

Algorithm 1: Environment of the toy example

1 $nS \leftarrow 3$

2 $nA \leftarrow 3$

3 **Construct reward matrix** : $R \leftarrow \begin{bmatrix} 1 & 0 & 0 \\ 0.5 & 0 & 0.5 \\ 0.5 & 0.5 & 0 \end{bmatrix}$

4 **Construct three transition matrices with the superscript denoting the action taken** :

$$P^{(1)} \leftarrow \begin{bmatrix} 0.8 & 0.1 & 0.1 \\ 0.5 & 0 & 0.5 \\ 0.2 & 0.4 & 0.4 \end{bmatrix}, P^{(2)} \leftarrow \begin{bmatrix} 0.3 & 0.3 & 0.4 \\ 0.2 & 0.1 & 0.7 \\ 0.4 & 0.4 & 0.2 \end{bmatrix}, P^{(3)} \leftarrow \begin{bmatrix} 0.1 & 0.5 & 0.4 \\ 0.3 & 0.5 & 0.2 \\ 0 & 0.1 & 0.9 \end{bmatrix}$$

space. By transitioning to a new state, following a policy to select an action and repeating the transition steps, the experience of the agent can be tracked to create trajectories. Formally, we define a trajectory as a sequence of states and actions, given by $\tau = (s_1, a_1, \cdots)$. The distribution of τ (Levine et al., 2020) for a given MDP and policy π can be represented as

$$p_\pi(\tau) = \rho(s_1) \prod_{t=1}^{\infty} \pi(a_t|s_t) P(s_{t+1}|s_t, a_t). \tag{1}$$

The goal of learning is to maximize the accumulative (future) reward following a specific policy (i.e., $J(\pi) = \mathbb{E}_{\tau \sim p_\pi(\cdot)}[\sum_{t=1}^{\infty} \gamma^t r_t]$). Two important quantities can be introduced namely, action-value function and value function. These two functions appear in almost every algorithm throughout Sect. 3.

Definition 2 Given the policy π, the action-value function Q is defined as

$$Q^\pi(s, a) = \mathbb{E}_{S_t, A_t, r_t, t \geq 1} \left[\sum_{t=1}^{\infty} \gamma^t r_t \middle| S_1 = s, A_1 = a \right], \tag{2}$$

where S_1, A_1 denote the initial state and action respectively.

Definition 3 Given the policy π, the value function V is defined as

$$V^\pi(s) = \mathbb{E}_{a \sim \pi(\cdot|s)} \left[Q^\pi(s, a) \right]. \tag{3}$$

The recursive relations between the value functions and action-value functions can be expressed by the following formulae:

$$Q^\pi(s, a) = r(s, a) + \gamma \mathbb{E}_{s' \sim P(\cdot|s,a)}[V^\pi(s')],$$

$$Q^\pi(s, a) = r(s, a) + \gamma \mathbb{E}_{s' \sim P(\cdot|s,a), a' \sim \pi(\cdot|s')}[Q^\pi(s', a')].$$

These formulae are known as Bellman equations (Bellman, 1957), which connect the function of the current state-action pair with that of the next state. The exploration of different policies leads to different value functions and action-value functions, and the agent aims to find the maximum value function (or action-value function) over the policy class. We provide the definitions of the optimal action-value function and value function below.

Definition 4 Given a policy class Π, the optimal action-value function Q^* and optimal value function V^* are defined as

$$Q^*(s, a) = \max_{\pi \in \Pi} Q^\pi(s, a), \tag{4}$$

$$V^*(s) = \max_{\pi \in \Pi} V^\pi(s), \tag{5}$$

respectively.

We also have the following Bellman optimality equations (Bellman, 1957) if Π consists of all measurable functions from $\mathcal{S}$ to $\mathcal{A}$:

$$Q^*(s, a) = r(s, a) + \gamma \sum_{s' \in \mathcal{S}} P(s'|s, a) \max_{a' \in \mathcal{A}} Q^*(s', a'), \tag{6}$$

$$V^*(s) = \max_{a \in \mathcal{A}} \left[r(s, a) + \gamma \sum_{s' \in \mathcal{S}} P(s'|s, a) V^*(s') \right]. \tag{7}$$

For a more detailed discussion, we refer to Sutton and Barto (2018).

3 Offline (Batch) Reinforcement Learning

The offline reinforcement learning problem (Ernst et al., 2005; Lange et al., 2012; Riedmiller, 2005) is composed of a data-driven framework. The agent cannot interact with the environment to obtain additional experience to improve its behavior, and only a static dataset $\mathcal{D} = \{(s_{i,t}, a_{i,t}, r_{i,t})\}_{1 \le i \le N, 1 \le t \le T_i}$ is provided to learn optimal value function and policy, with state transitions following P, action decisions based on behavior policy π_b. We also assume that the state-action tuples (s, a) are sampled according to $\mu(s, a)$. For many practical applications, conducting a new policy may be costly, and, in addition, a bad policy could be risky and harmful. Hence, the offline setting is often preferred, to reduce cost and avoid severe consequences. Therefore, it is of great importance for an offline RL system to have the ability to excavate the dynamics behind the environment through an offline dataset only. Prior to formally introducing the specific algorithms and methods, several basic topics need to be discussed.

Optimal Policy The optimal policy stands for a strategy that must be followed to attain maximum accumulative reward. If no constrains are specified, this policy can be derived directly from (7) according to

$$\pi^*(s) = \arg\max_{a \in \mathcal{A}} \left[r(s,a) + \gamma \sum_{s' \in \mathcal{S}} P(s'|s,a) V^*(s') \right]. \tag{8}$$

From another aspect $V^*(s) = \max_{a \in \mathcal{A}} Q^*(s,a)$, the optimal policy can be expressed according to

$$\pi^*(s) = \arg\max_{a \in \mathcal{A}} Q^*(s,a). \tag{9}$$

Bellman Operators There are several operators that are widely applied in the offline RL problems. To start with, a linear operator $T^\pi : \mathcal{S} \times \mathcal{A} \to \mathcal{S} \times \mathcal{A}$, representing the dynamics and policy backup, is set according to

$$(T^\pi f)(s,a) = \mathbb{E}_{s' \sim P(\cdot|s,a),\ a' \sim \pi(\cdot|s')}[f(s',a')], \forall f. \tag{10}$$

This operator is applied in the function approximation for action-value functions. In addition, operators derived from the Bellman equations (and the Bellman optimality equations) are also required. The Bellman backup operator $B^\pi : \mathcal{S} \times \mathcal{A} \to \mathcal{S} \times \mathcal{A}$ is set as

$$(B^\pi f)(s,a) = \left\{ r(s,a) + \gamma \mathbb{E}_{s' \sim P(\cdot|s,a), a' \sim \pi(\cdot|s')}[f(s',a')] \right\}, \forall f. \tag{11}$$

There are two useful Bellman optimality backup operators $B^* : \mathcal{S} \to \mathcal{S}$, $H^* : \mathcal{S} \times \mathcal{A} \to \mathcal{S} \times \mathcal{A}$, and they are set as follows:

$$(H^* v)(s) = \max_{a \in \mathcal{A}} \left[r(s,a) + \gamma \mathbb{E}_{s' \sim P(\cdot|s,a)} v(s') \right], \forall v, \tag{12}$$

$$(B^* f)(s,a) = \max_{a' \in \mathcal{A}} \left\{ r(s,a) + \gamma \mathbb{E}_{s' \sim P(\cdot|s,a)}[f(s',a')] \right\}, \forall f. \tag{13}$$

Function Approximation Throughout this section, several algorithms are provided in tabular-versions, where value estimates are stored in tables. In the pseudo-code, we denote $[V(s)]_{s \in \mathcal{S}}$ as a lookup table where each state has an entry to display its value function. Similarly, $[Q(s,a))]_{s \in \mathcal{S}, a \in \mathcal{A}}$ is the Q-value table with the (s,a)-th entry being the corresponding action-value function. However, generalizing this approach may not be possible for problems with very large state and action spaces. To address this, we introduce the function approximation into the offline RL problems, with the value functions (or action-value functions) estimated by different approximators, including linear combinations of features (Sard, 1963), nearest neighbors (Cover & Hart, 1967), decision trees (Breiman, 1996), random forests (Breiman, 2001), and neural networks (Wang, 2003; Lange & Riedmiller, 2010). Function approximation effectively reduces memory and computational complexity, and improves generalization of RL methods.

In the following, we use linear approximation for the action-value function as an example to help readers understand its implementation, though the idea could be extended to any type of approximations. If Q-function is a linear function on a feature vector $\mathbf{f}(s, a)$ (i.e., $Q_\phi = \mathbf{F}\phi$ where $\mathbf{F} \in \mathbb{R}^{|\mathcal{S}||\mathcal{A}|\times d}$ is the tabular form of d-dimensional $\mathbf{f}$, and ϕ is the parameter vector), ϕ can be selected using the Bellman residual minimization (Antos et al., 2008) or least-squares fixed-point approximation (Lagoudakis et al., 2002) to approximate Q^π for a given policy. Through the Bellman residual minimization, ϕ is selected such that the Q-function satisfies the Bellman equation as closely as possible in terms of the squared error. The normal equation (Levine et al., 2020) $\mathbf{F}\phi \approx B^\pi\mathbf{F}\phi = \mathbf{r} + \gamma T^\pi\mathbf{F}\phi$ ($\mathbf{r}$ is the reward vector) induces the least squares solution

$$\phi = [(\mathbf{F} - \gamma T^\pi\mathbf{F})^T(\mathbf{F} - \gamma T^\pi\mathbf{F})]^{-1}(\mathbf{F} - \gamma T^\pi\mathbf{F})^T\mathbf{r}, \tag{14}$$

provided the inverse exists. This expression minimizes the Bellman residual and is known as the Bellman residual minimizing solution. Regarding the least-squares fixed-point approximation, the solution of ϕ is extracted via repeatedly applying the following recurrence

$$\phi_{k+1} = (\mathbf{F}^T\mathbf{F})^{-1}\mathbf{F}^T(\mathbf{r} + \gamma T^\pi\mathbf{F}\phi_k),$$

until convergence. As $k \to \infty$, the fixed point of the projected Bellman operator can be obtained. Additional details are presented in Antos et al. (2008); Lagoudakis et al. (2002); Sutton and Barto (2018); Levine et al. (2020), and a congruous idea also leverages the approximation of a value function.

Distribution Shift In the offline RL problems, a new policy is learned, and a state-action distribution may be induced that is different from the provided data distribution μ. With the existence of the offline dataset $\mathcal{D}$, a useful assumption known as the full coverage assumption (Xie et al., 2021; Uehara & Sun, 2022) is adopted in several scenarios to ensure the stability of the estimation and a restricted distributional shift. The assumption mainly states that $\sup_{(s,a)} v(s, a)/\mu(s, a) < \infty$ for all admissible v(s) that are reachable for a policy. As such, the behavior policy is supposed to be sufficiently exploratory, engendering the offline dataset containing possible state-action pairs. A stronger and commonly used assumption known as global coverage (Xie et al., 2021; Uehara & Sun, 2022; Liu et al., 2020) states that $\sup_{(s,a)} v(s, a)/\mu(s, a)$ is bounded by a constant. In practice, sufficient coverage cannot be guaranteed throughout. As a consequence, we discuss cases with $\sup_{(s,a)} v^*(s, a)/\mu(s, a) < \infty$ (or bounded by a constant), where v^* represents the state-action occupancy distribution of some comparator policies (or the optimal policy), and we can refer it as partial coverage assumption (Uehara & Sun, 2022). This assumption implies that the offline data distribution only covers state-action distribution reachable for some high-quality comparator policies (Uehara & Sun, 2022).

When the estimators of the value functions or transitions are trained under a certain distribution, changes in the visited states will lead to an evaluation over a different distribution. Therefore, incomplete information encoded in the offline dataset can result in an undesirable policy. By introducing the idea of pessimism under uncertainty, penalties are imposed onto policies or value functions in certain algorithms, thereby controlling the distributional shift and further yielding policies robust to statistical errors (Zanette et al., 2021).

The remainder of this section is organized as follows. In Sects. 3.1 and 3.2, the methods and algorithms are aimed at finding the optimal policy. As many model-based methods are included, the discussions are conducted for a known model first. Subsequently, we provide techniques to tackle the issues stemming from restricted knowledge of the model. In addition, some pessimism algorithms are also briefly encapsulated. In Sects. 3.3 and 3.4, the analysis targets at a certain policy first, and the subsequent discussion focuses on policy learning.

3.1 Policy Iteration

In RL problems, iterative methods are widely implemented. One of the leading planning methods is dynamic programming (Howard, 1960), in which iterations pass through the index with non-stationary value estimates. Policy iteration (PI, Sutton and Barto (2018)) is one of the most classical approaches using the dynamic programming techniques. The classical policy iteration assumes that the transition probabilities and reward functions are known. Under this assumption, the method iterates over the whole state and action space to find a deterministic stationary policy that is no worse than the initial policy. The two steps forming the basis of the whole algorithm are: policy evaluation and policy improvement, which are straightforward applications of Bellman operators. They are presented in Algorithms 2 and 3, respectively.

As given in Algorithm 4, PI is an efficient algorithm to determine the optimal deterministic policy by combining the phases of policy evaluation and policy improvement. In particular, the algorithm eventually terminates in an environment with finite state and action space, as well as deterministic policy space. Moreover, the updated value functions at each iteration consist of a non-decreasing sequence (Sutton & Barto, 2018).

The reliability on $\mathcal{D}$ of the classical PI is weak as we have assumed a fully-known MDP model, however, such an assumption does not apply to many practical situations. If the knowledge of the front environment is poor, can we still identify a near-optimal policy via policy iterations? An approach is to directly replace

Algorithm 2: Policy evaluation

Input: π, ϵ
Initialize $\Delta > \epsilon$ and $V^{\pi} = [V^{\pi}(s)]_{s \in \mathcal{S}}$
while $\Delta > \epsilon$ **do**
    $\Delta = 0$
    for $s \in \mathcal{S}$ **do**
        $k = V^{\pi}(s)$
        $V^{\pi}(s) = \sum_a \pi(a|s) \sum_{s'} P(s'|s,a)[r(s,a) + \gamma V^{\pi}(s')]$
        $\Delta = \max(\Delta, |k - V^{\pi}(s)|)$
    end
end
output: $V^{\pi}(s)$ for all $s \in \mathcal{S}$

Algorithm 3: Policy improvement

Input: V^{π}
$\pi(s) \leftarrow \arg\max_{a \in \mathcal{A}} \left[r(s,a) + \gamma \sum_{s' \in \mathcal{S}} P(s'|s,a) V^{\pi}(s') \right], \forall s \in \mathcal{S}$
Output: $\pi(s)$ for all $s \in \mathcal{S}$

Algorithm 4: Policy iteration

Input: Π, ϵ
π randomly choose from Π
while *true* **do**
    $V^{\pi} \leftarrow$ POLICY EVALUATION (π, ϵ)
    $\pi^* \leftarrow$ POLICY IMPROVEMENT (V^{π})
    if $\pi^* = \pi$ **then**
        break
    else
        $\pi \leftarrow \pi^*$
    end
end
$V^* \leftarrow V^{\pi}$
Output: $V^*(s), \pi^*(s)$ for all $s \in \mathcal{S}$

the MDP components in Algorithm 4 by estimators trained from $\mathcal{D}$. The most commonly used estimators are Monte Carlo estimates (Sutton & Barto, 2018):

$$\hat{P}(s'|s,a) = \frac{\sum_{i=1}^{N} \sum_{j=1}^{T_i - 1} I(S_{i,j} = s, A_{i,j} = a, S_{i,j+1} = s')}{\sum_{i=1}^{N} (T_i - 1)}, \tag{15}$$

$$\hat{r}(s,a) = \frac{\sum_{i=1}^{N} \sum_{j=1}^{T_i} I(S_{i,j} = s, A_{i,j} = a) r_{i,j}}{\sum_{i=1}^{N} \sum_{j=1}^{T_i} I(S_{i,j} = s, A_{i,j} = a)}, \tag{16}$$

Algorithm 5: Least squares policy iteration

Input: $\mathcal{D}, \gamma, \mathbf{F}, \Pi$
Initialize $\pi_0 \in \Pi$
$\pi \leftarrow \pi_0$
$i \leftarrow 0$
$\tilde{\mathbf{A}} \leftarrow 0$ where $\tilde{\mathbf{A}}$ denote the estimate of $\mathbf{F}^T(\mathbf{F} - \gamma T^{\pi}\mathbf{F})$
$\tilde{b} \leftarrow 0$ where $\tilde{b}$ denote the estimate of $\mathbf{F}^T\mathbf{r}$
while $\pi \neq \pi'$ **do**
  $\pi \leftarrow \pi'$
  $i \leftarrow i + 1$
  for $t = 1, 2, \ldots, T_i - 1$ **do**
    $\tilde{\mathbf{A}} \leftarrow \tilde{\mathbf{A}} + \mathbf{f}(s_{i,t}, a_{i,t})[\mathbf{f}(s_{i,t}, a_{i,t}) - \gamma\mathbf{f}(s_{i,t+1}, \pi(s_{i,t+1}))]^T$
    $\tilde{b} \leftarrow \tilde{b} + \mathbf{f}(s_{i,t}, a_{i,t})r_{i,t}$
  end
  $\hat{\phi}^{\pi} \leftarrow \tilde{\mathbf{A}}^{-1}\tilde{b}$
  $\hat{Q}^{\pi} \leftarrow \mathbf{F}\hat{\phi}^{\pi}$
  $\pi'(s) \leftarrow \arg\max_{a\in\mathcal{A}} \hat{Q}^{\pi}(s, a), \forall s \in \mathcal{S}$
end
Output: $\pi(s)$ for all $s \in \mathcal{S}$

that use observed values to create the MDP model. The transition probability is the fraction of the times that the transition (s, a, s') is observed after taking action a at state s in the batch, whereas the reward $\hat{r}(s, a)$ is the average reward experienced after taking action a at state s. Other estimation methods such as Gaussian processes (Deisenroth & Rasmussen, 2011), Bayesian neural networks (Kononenko, 1989), and bootstrap ensembles (Chua et al., 2018) are also available.

Instead of estimating components of the environment to find reliable estimates of value functions, policy iteration algorithms use powerful function approximations in a variety of application problems. A standard offline algorithm known as the least-squares policy iteration (LSPI, Lagoudakis and Parr (2003)), works solely based on approximating the action-value functions. At each step of the policy evaluation, LSPI approximates $\hat{Q}^{\pi} = \mathbf{F}\hat{\phi}^{\pi}$ based on the selected π and batch data from $\mathcal{D}$. At each step of policy improvement, the policy is selected based solely on the current updated approximation. Details are presented in Algorithm 5. Other types of estimations on action-value functions are also applicable, and these algorithms can be generalized as a class known as fitted policy iteration (Antos et al., 2008) or approximate policy iteration (API, Munos (2003)).

The aforementioned methods are efficient and can produce estimated optimal policies when the global or full coverage assumption is satisfied. However, sufficient coverage is not always guaranteed in practice. Recently, researchers have proposed the idea of pessimism in offline reinforcement learning to tackle uncertainties (Jin et al., 2021; Liu et al., 2020; Agarwal et al., 2020; Buckman, 2020), especially when there are policies without sufficient support or infrequently visited states in the dataset. For a realizable function class and offline distribution that only

Algorithm 6: Constrained pessimistic policy optimization

1 **Input**: $\mathcal{M}$, $\mathcal{D}$, ϵ, Π
2 Obtain estimator $\hat{P}$ based on $\mathcal{D}$
3 Find the approximated value function $V_{P'}^{\pi}$ for each $P' \in \mathcal{M}, \pi \in \Pi$
4 Constrained policy optimization:

$$\hat{\pi}^* \leftarrow \arg\max_{\pi \in \Pi} \min_{P' \in \mathcal{M}} V_{P'}^{\pi} \ s.t. \mathbb{E}_{(s,a,r,s'),\mathcal{D}}[\delta(\hat{P}(\cdot|s,a), P'(\cdot|s,a))^2] \leq \epsilon,$$

where $\delta(\hat{P}(\cdot|s,a), P'(\cdot|s,a)) = \sup_{s' \in \mathcal{S}} |\hat{P}(s'|s,a) - P'(s'|s,a)|$ and $\mathbb{E}_{(s,a,r,s'),\mathcal{D}}[g(s,a,r,s')^2]$ denote $\frac{1}{\sum_i (T_i - 1)} \sum_{i=1}^{N} \sum_{t=1}^{T_i - 1} g(s_{i,t}, a_{i,t}, r_{i,t}, s_{i,t+1})^2$
5 **Output**: $\hat{\pi}^*$

provides partial coverage, a high-quality comparator policy can be learned by pessimistic methods. Constrained pessimistic policy optimization (CPPO, Uehara and Sun (2022)) is an algorithm that uses a realizable hypothesis class $\mathcal{M}$ of transition kernels, as well as an estimator $\hat{p}$ by seeking the least favorable model regarding constraint on total variation distance between P' and $\hat{P}$ (P' is a transition within class $\mathcal{M}$). As for $\hat{p}$, a general choice is to use $\hat{P} = \arg\max_{P' \in \mathcal{M}} \mathbb{E}_{(s,a,r,s'),\mathcal{D}}[\ln P'(s'|s,a)]$, where $\mathbb{E}_{(s,a,r,s'),\mathcal{D}}[\cdot]$ is introduced in line 4 of Algorithm 6. The algorithm is illustrated in Algorithm 6, in which lines 3 and 4 can be expanded into iterative processes, akin to the policy evaluation and policy improvement steps. More details are left for readers to explore in Uehara and Sun (2022).

Besides such the typical model-based method given in Algorithm 6, a model-free pessimistic PI algorithm is also introduced. Marginalized behavior supported policy iteration (MBS-PI, Liu et al. (2020)), shows its robustness by constraining the Bellman update that is only sufficiently covered by μ using a threshold b and estimated $\hat{\mu}$, and the corresponding pseudo-code is presented in Algorithm 7, with $\tilde{\mathcal{L}}_{\mathcal{D}}(f, f', \pi) = \mathbb{E}_{(s,a,r,s'),\mathcal{D},a' \sim \pi(\cdot|s')}[f(s,a) - r - \gamma I(\hat{\mu}(s,a) \geq b) f'(s',a')]^2$ representing the empirical loss of f given f' and policy π. Note that J is the number of policy evaluation steps, and I is the number of policy improvement steps. The output policy is guaranteed to converge to the approximate best decision policy within a set of policies that satisfy the bounded density ratio assumption. If the behavior data provides sufficient coverage of states and actions visited under the optimal policy, the algorithm will output a near-optimal policy. More details can be found in Liu et al. (2020).

3.2 *Value Iteration and Q Iteration*

The classical value iteration (VI) algorithm employs a theorem as follows

Algorithm 7: Marginalized behavior supported policy iteration

Input: $\mathcal{F}, \mathcal{D}, \hat{\mu}, b, \Pi, I, J$
Initialize $\pi_1 \in \Pi$
for $i = 1, \ldots, I$ **do**
    Initialize $Q_{i,1} \in \mathcal{F}$
    for $j = 1, \ldots, J$ **do**
        $Q_{i,j+1} \leftarrow \arg\min_{Q \in \mathcal{F}} \tilde{\mathcal{L}}_{\mathcal{D}}(Q, Q_{i,j}, \hat{\pi}_i)$
    end
    $\hat{\pi}_{i+1} \leftarrow \arg\max_{\pi \in \Pi} \mathbb{E}_{\mathcal{D}}[\mathbb{E}_{\pi}[I(\hat{\mu}(s,a) \geq b) Q_{i,J+1}]]$, where
    $\mathbb{E}_{\mathcal{D}}[\mathbb{E}_{\pi}[f]] := \frac{1}{\sum_i T_i} \sum_{i=1}^{N} \sum_{t=1}^{T_i} \pi(a_{i,t}|s_{i,t}) f(s_{i,t}, a_{i,t})$
end
Output: $\hat{\pi}_{I+1}$

Algorithm 8: Value iteration

Input: ϵ
Initialize $V = [V(s)]_{s \in \mathcal{S}}$, $V' = [V'(s)]_{s \in \mathcal{S}}$, $V'(s) \leftarrow 0$, $V(s) \leftarrow \infty$ for $\forall s \in \mathcal{S}$
while $||V - V'||_{\infty} > \epsilon$ **do**
    $V \leftarrow V'$
    $V'(s) \leftarrow (H^* V)(s), \forall s \in \mathcal{S}$
end
$V^* \leftarrow V$
$\pi^* \leftarrow \arg\max_{a \in \mathcal{A}} \left[r(s,a) + \gamma \sum_{s' \in \mathcal{S}} P(s'|s,a) V^*(s') \right], \forall s \in \mathcal{S}$
Output: $V^*(s), \pi^*(s)$ for all $s \in \mathcal{S}$

Theorem 1 (Sutton & Barto 2018) *For a MDP, let the fixed point obtained from the Bellman optimality backup operator H^* be denoted by V^*. Then the policy given by (8) is a stationary deterministic policy. The value function under this policy V^{π^*} satisfies $V^{\pi^*} = V^*$, which implies that there exists a stationary deterministic policy π^* whose value function is the fixed point of H^*.*

For a fully known environment (including transition probabilities and reward functions), the theorem suggests the basic logic of VI: finding the fixed point via repeatedly applying the recurrence relation of the Bellman optimality equation. The pseudo-code is provided in Algorithm 8.

Although the estimators of P and r can be directly applied to VI, if less knowledge of transition probabilities and reward functions is present, the estimation is of statistical concern regarding the suboptimality $V^*(s) - \hat{V}(s)$, where $\hat{V}(s)$ represents the estimated value function, especially for an underlying environment with large state and action spaces. Therefore, researchers have shifted their focus to the approximate value iteration (AVI, Munos (2005)), which usually iterates over the approximated action-value (value) functions.

A celebrated AVI algorithm, known as fitted Q iteration (FQI, Ernst et al. (2005)), updates the Q-values according to $Q_{k+1} = \arg\min_{Q \in \mathcal{F}} \mathbb{E}_{(s,a,r,s'),\mathcal{D}}[Q - \hat{B}^* Q_k]^2$ in the k-th iteration and returns the policy with respect to Q-values upon

Algorithm 9: Fitted Q iteration

Input: $\mathcal{D}, \gamma, \epsilon$
Initialize $\hat{Q}' = [\hat{Q}'(s,a)]_{s\in\mathcal{S},a\in\mathcal{A}}$, $\hat{Q} = [\hat{Q}(s,a)]_{s\in\mathcal{S},a\in\mathcal{A}}$, $\hat{Q}'(s,a) \leftarrow 0$, $\hat{Q}(s,a) \leftarrow \infty$ for $\forall (s,a) \in \mathcal{S} \times \mathcal{A}$
$i \leftarrow 0$
while $||\hat{Q} - \hat{Q}'||_\infty > \epsilon$ **do**
  $\hat{Q} \leftarrow \hat{Q}'$
  $i \leftarrow i+1$
  Produce a set $\{(s_{i,t}, a_{i,t}, r_{i,t} + \gamma \max_{a'\in\mathcal{A}} \hat{Q}(s_{i,t+1}, a'))\}_{1\le t\le T_i-1}$
  Use approximation method based on the set to obtain $\hat{Q}'$
end
$\hat{Q} \leftarrow \hat{Q}'$
$\hat{\pi}(s) \leftarrow \arg\max_{a\in\mathcal{A}} \hat{Q}(s,a) \forall s \in \mathcal{S}$
Output: $\hat{\pi}(s)$ for all $s \in \mathcal{S}$

termination, where $\mathcal{F}$ is a realizable Q-function class. A tabular version of FQI is provided in Algorithm 9. The entire procedure can be regarded as the Q-version value iteration combined with additional approximators. Within each iteration, the trajectory gathered from $\mathcal{D}$ and the Q-function computed in the previous step are put together to determine a new set used for computing the succeeding Q-function. Unlike directly finding value functions using the estimated transition kernels and reward functions, FQI provides a flexible way for users to construct approximation architectures (Clifton & Laber, 2020). As a consequence, to achieve the best bias-variance trade-off, the resolution of the learned architecture is free to be adjusted at each step. Moreover, the convergence of Algorithm 9 relies partially on the properties of the architecture selected (Kosorok & Moodie, 2015).

A tree-based fitted Q iteration (Ernst et al., 2005; Kosorok & Moodie, 2015) is used as an example to further provide insight into the algorithm. In tree-based methods, an estimated operator is formed based on built trees, and subsequently, the approximation at the i-th step can be performed according to $\hat{Q}' = \hat{B}_i^* \hat{Q}$, where

$$(\hat{B}_i^* f)(s,a) = \sum_{t=1}^{T_i-1} U((s_{i,t}, a_{i,t}), (s,a))[r(s_{i,t}, a_{i,t}) + \gamma \max_{a'\in\mathcal{A}} f(s_{i,t+1}, a')]. \tag{17}$$

Under the cases with p built regression trees, let $R_m(s,a)$ represent the region assigned by the m-th tree for input (s,a), and $I_R(s',a')$ be the indicator function indicating whether (s',a') lies in the region R. The kernel function U (Ernst et al., 2005) in (17) is then defined as

$$U((s_{i,t}, a_{i,t}), (s,a)) = \frac{1}{p}\sum_{m=1}^{p} \frac{I_{R_m(s,a)}(s_{i,t}, a_{i,t})}{\sum_{(s',a')\in\{(s_{i,t},a_{i,t})\}_{1\le t\le T_i}} I_{R_m(s,a)}(s',a')}.$$

The types of trees can be chosen from extra tree (Geurts et al., 2006), totally randomized tree (Geurts et al., 2006), and K-dimensional(KD)-tree (Bentley, 1979). If the normalizing conditions are satisfied (i.e., for each $1 \le i \le N$, $\sum_{t=1}^{T_i-1} |U((s_{i,t}, a_{i,t}), (s,a))| = 1, \forall (s,a)$), the tree-based FQI leads to convergence of the sequence of updated $\hat{Q}$ upon a certain time step j (Ernst et al., 2005):

$$||\hat{H}_j \hat{Q} - \hat{H}_j \hat{Q}'|| \le \gamma \, ||\hat{Q} - \hat{Q}'||_\infty \le ||\hat{Q} - \hat{Q}'||_\infty . \tag{18}$$

If practitioners take interest in the above discussion on tree-based FQI, they can run the following sample demo. The selected MDP environment is the same as the example in Fig. 1 with the reward function defined in Table 2.

```
#Set up the environment
env1 = env()
#Create offline dataset
data = env1.offline_set(10,500)
#Fitted Q iteration
opt_policy, Q =
Fitted_Q_iteration(data = data, nS = env1.nS, nA = env1.nA, N =
    100)
```

However, AVI suffers from bootstrapping errors in less-visited regions of the MDP. By introducing constraints or penalties into the value iteration, there are several pessimism approaches available. Similar to the discussion about pessimism algorithms in Sect. 3.1, we also summarize a model-based approach and a model-free approach in the subsequent narratives. Value iteration with lower concentration bound (VI-LCB, Rashidinejad et al. (2021); Li et al. (2022)), as a new-born model-based pessimistic approach, iterates over approximated action-value functions, incorporating with the estimated transition kernel and the Bernstein-style lower concentration bounds (Jin et al., 2018; Li et al., 2022; Rashidinejad et al., 2021). With the pseudo-code in Algorithm 10, it is noteworthy that action-value functions and value functions are approximated based on an estimated $\hat{P}$. Moreover, a Bernstein-style penalty term $\kappa(s, a; \hat{V})$, which is associated with the variance of the current estimate of the value function and the number of visitations of (s, a) in the offline data, is introduced to enforce pessimism. Under a partial coverage, Li et al. (2022) substantiates that the algorithm yields ϵ-accuracy with a certain sample complexity. Details about the Bernstein-style lower confidence bound and performance guarantees of VI-LCB can be found in Rashidinejad et al. (2021); Li et al. (2022).

Analogous to MBS-PI, marginalized behavior supported Q iteration (MBS-QI) is developed in Liu et al. (2020) to address constraints on the Bellman backups to have sufficient support from $\mathcal{D}$ in a Q-iterated style algorithm. The corresponding pseudo-code is shown in Algorithm 11, with $\tilde{\mathcal{L}}_{\mathcal{D}}(f, f') = \mathbb{E}_{(s,a,r,s'),\mathcal{D}}[f(s,a) - r - \gamma \max_{a' \in \mathcal{A}} I(\hat{\mu}(s,a) \ge b) f'(s', a')]^2$ denoting the empirical loss of f given f'. In practice, the minimization of empirical loss function is carried out

Algorithm 10: Value iteration with lower concentration bound

Input: $\mathcal{D}$, I
Initialize $\hat{Q}_1 \leftarrow 0$, $\hat{V}_1 \leftarrow 0$
Find an estimated transition kernel $\hat{P}$ based on $\mathcal{D}$
for $i = 1, \ldots, I$ **do**
  Compute the penalty term $\kappa(s, a; \hat{V}_i)$ for all $(s, a) \in \mathcal{S} \times \mathcal{A}$
  $\hat{Q}_{i+1}(s, a) \leftarrow \max\{r(s, a) + \gamma \sum_{s' \in \mathcal{S}} \hat{P}(s'|s, a)\hat{V}_i(s') - \kappa(s, a; \hat{V}_i), 0\}$ for all $(s, a) \in \mathcal{S} \times \mathcal{A}$
  $\hat{V}_{i+1}(s) \leftarrow \max_{a \in \mathcal{A}} \hat{Q}_{i+1}(s, a)$ for all $s \in \mathcal{S}$
end
$\hat{\pi}(s) \leftarrow \arg\max_{a \in \mathcal{A}} \hat{Q}_{I+1}(s, a)$ for all $s \in \mathcal{S}$
Output: $\hat{\pi}$

Algorithm 11: Marginalized behavior supported Q iteration

Input: $\mathcal{F}$, $\mathcal{D}$, $\hat{\mu}$, b, I
Initialize $Q_1 \in \mathcal{F}$
for $i = 1, \ldots, I$ **do**
  $Q_{i+1} \leftarrow \arg\min_{Q \in \mathcal{F}} \tilde{\mathcal{L}}_{\mathcal{D}}(Q, Q_i)$
  $\hat{\pi}_{i+1}(s) \leftarrow \arg\max_{a \in \mathcal{A}} I(\hat{\mu}(s, a) \geq b) Q_{i+1}(s, a)$
end
Output: $\hat{\pi}_{I+1}(s)$ for all $s \in \mathcal{S}$

via a series of optimization methods, including approximating different operators (Smola & Schölkopf, 1998), setting normal equations (Jefferys, 1980), introducing a variational auto-encoder (VAE, Doersch (2016)). Compared to FQI, MBS-QI concentrates on policies that satisfy a bounded density ratio assumption. For a complete discussion on this method, we refer readers to prior work (Liu et al., 2020) and their codes posted online.

Policy iteration v.s. Value iteration The structures of PI and VI are different: the output policy is explicitly extracted from one iteration and treated as an input for the next iteration in PI, while we finalize the output policy after completion of all iterations in VI. For the classical PI and VI algorithms, PI has a clear criterion, i.e., stability of updated policies to stop the iteration, but at the cost of complexity per iteration. VI requires the convergence of value functions, with more iterations but less complexity most of the time compared to PI. To deepen the understanding of these two algorithms, we provide sample code for practitioners to implement via Python to exhibit the methodology and properties. The environment setting is the same as before.

(continued)

```
#Set up the environment
env1 = env()
#Policy iteration
op_policy, V
= policy_improvement(env = env1, discount_factor = 0.5)
#Value iteration
op_policy, V
= value_iteration(env = env1, discount_factor = 0.5)
```

As for API and AVI, API usually requires an approximation on B^{π}, while AVI leverages approximations on B^* and H^*. Moreover, AVI is more flexible to have penalty terms within its estimation.

3.3 Temporal Difference Learning

Policy optimization can be solved by implementing various dynamic programming methods, to estimate a near-optimal policy and improve corresponding values. Meanwhile, considerable attention is also gained in evaluating different strategies from the offline data, particularly for applications like recommender ranking systems (Gauci et al., 2018), credit scoring (Creager et al., 2020) and clinical sedation regulation (Prasad et al., 2017). In the previous discussion, methods like model-based policy evaluation can be applied, however, learning a model (or a value) can be expensive and risky, regarding sample efficiency and model misspecification. Therefore, in this subsection, we introduce a more tractable approach, namely temporal difference (TD, Tesauro et al. (1995)) learning, to deal with both policy evaluation and optimization problems. First, a model-free policy evaluation based on TD learning is introduced. Next, we discuss the methods to identify the optimal policy and corresponding value function jointly via optimizing over a TD-related objective function.

By combining bootstrapping with sampling, TD learning algorithms can perform policy evaluation with an unknown transition kernel. The classical algorithms exploit an update rule, namely TD learning update (Tesauro et al., 1995), that is, for the t-th update,

$$V(s_t) \leftarrow V(s_t) + \alpha[r_t + \gamma V(s_{t+1}) - V(s_t)],$$

Algorithm 12: Temporal difference learning (batch version)

Input: $\mathcal{D}, \alpha, \gamma, \epsilon$
$V = [V(s)]_{s \in \mathcal{S}}, V(s) \leftarrow \infty$ for $\forall s \in \mathcal{S}$
$V' = [V'(s)]_{s \in \mathcal{S}} V'(s) \leftarrow 0$ for $\forall s \in \mathcal{S}$
$j \leftarrow 0$
while $||V - V'||_\infty > \epsilon$ **do**
  if $j < N$ **then**
    $j \leftarrow j + 1$
  end
  $V \leftarrow V'$
  for $i = 1, \ldots j$ **do**
    for $t = 1, \ldots T_i - 1$ **do**
      $V'(s_{i,t}) \leftarrow V(s_{i,t}) + \alpha[r_{i,t} + \gamma V(s_{i,t+1}) - V(s_{i,t})]$
    end
  end
end
Output: $V(s)$ for all $s \in \mathcal{S}$

where α is the step size, and $\delta_t = r_t + \gamma V(s_{t+1}) - V(s_t)$ is generally known as the TD error (Tesauro et al., 1995; Sutton & Barto, 2018), which measures the discrepancy between the true value function and its estimated counterpart. The classical TD learning method (Sutton & Barto, 2018) requires an online environment to receive immediate feedback, where transition tuples can be observed, and the value of the current state is updated with the value of the next state and obtained reward. The batch version of the TD learning algorithm is developed in Kalyanakrishnan and Stone (2007), which can be implemented with assistance of the offline dataset $\mathcal{D}$ to overcome the limitations of the offline setting. By feeding the histories/batches from $\mathcal{D}$, the value of the current state can be updated with the observed subsequent recordings, and finally the values of all states are amended. The algorithm aims at evaluating the policy behind observable trajectories, such as the behavior policy π_b. The converged estimate of value function is the same as that derived from the policy evaluation using the maximum likelihood MDP model (i.e., Eqs. (15) and (16)), and is referred to as the certainty equivalence estimate (Sutton & Barto, 2018). The pseudo-code is presented in Algorithm 12 for reference.

If a practitioner is interested in the aforementioned analysis, one can obtain TD-style model-free control by executing the following sample demo. The environment setting is the same as before. It is worthwhile noting that a sufficiently small α can lead to a stably converged result.

```
#Set up the environment
env1 = env()
#Create offline dataset
data = env1.offline_set(10,1000)
#Policy evaluation
b_policy = np.array([[1/3,1/3,1/3],
↪   [1/3,1/3,1/3],[1/3,1/3,1/3]])
```

```
V = policy_eval(b_policy, env1)
#TD learning
V = TD_batch(data, 10, env1, alpha = 0.05)
```

Because bootstrapping works best when significant and recognizable variations occur over the state and action for a certain duration, applying the update rule $V(s_t) \leftarrow V(s_t) + \alpha[r_t + \gamma V(s_{t+1}) - V(s_t)]$ is not as effective as expected (Sutton & Barto, 2018), especially when s_t is very often tantamount to s_{t+1}. Therefore, a n-step update rule (Sutton & Barto, 2018) is set as follows,

$$V(s_t) \leftarrow V(s_t) + \alpha[G_t^{(n)} + \gamma^n V(s_{t+n}) - V(s_t)],$$

where $G_t^{(n)} = \sum_{k=0}^{n-1} \gamma^k r_{t+k}$. Furthermore, backups can be conducted not only toward a n-step return, but also toward an average value of a set of multi-step returns, which induces the idea of TD(λ) algorithm (Seijen & Sutton, 2014). Based on a quantity, namely λ-return, which is defined by $G_t^{\lambda} = (1-\lambda)\sum_{n=1}^{T-t} \lambda^{n-1} G_t^{(n)} + \lambda^{T-t}(r_t + \cdots + \gamma^{T-t} r_T)$ for a trajectory with T steps and λ referring to the trace decay parameter with $0 \leq \lambda \leq 1$, TD(λ) can update values more efficiently, using the weighted average values with weights proportional to λ^{n-1} to conduct backups. Details for a variant of offline TD(λ) (Sutton & Barto, 2018) is provided in Algorithm 13, with a parameterized function approximator added.

To identify the estimated value of a policy different from the one embedded in the observations, can we apply the concept of TD learning to evaluate it? In off-policy evaluation, importance sampling (Glynn & Iglehart, 1989) is employed to approximately evaluate the expected cumulative (future) reward under a proposed policy π using the data following a known π_b. For example, combining importance sampling with the expected reward function (Glynn & Iglehart, 1989; Precup, 2000), it follows that

$$\mathbb{E}_\pi[r(s,a)] = \sum_a \pi(a|s) r(s,a) = \mathbb{E}_{\pi_b}\left[\frac{\pi(a|s)}{\pi_b(a|s)} r(s,a)\right], \tag{19}$$

Furthermore, regarding the learning objective $J(\pi)$ (Precup, 2000), similarly, it can be expressed as

$$J(\pi) = \mathbb{E}_{\tau \sim p_{\pi_b}(\cdot)}\left[\left(\prod_t \frac{\pi(a_t|s_t)}{\pi_b(a_t|s_t)}\right)\sum_t \gamma^t r(s_t, a_t)\right] \approx \sum_{i=1}^{N} w_{i,T_i} \sum_{t=1}^{T_i} \gamma^{t-1} r_{i,t}, \tag{20}$$

Algorithm 13: Offline TD(λ)

Input: $\mathcal{D}, \alpha, \gamma, \epsilon, \lambda, \hat{V}$, where $\hat{V}$ is a differentiable function with $\hat{V}(s, \mathbf{w})$ representing the approximate value of s given vector $\mathbf{w}$
Initialize $\mathbf{w}_1 > \mathbf{w}_0$
$j \leftarrow 0$
while $||\mathbf{w}_{j+1} - \mathbf{w}_j||_\infty > \epsilon$ **do**
    $\mathbf{w}_{1,1} \leftarrow \mathbf{w}_j$
    if $j < N$ **then**
        $j \leftarrow j + 1$
    end
    for $i = 1, \dots j$ **do**
        for $t = 1, \dots T_i - 2$ **do**
            $G_{i,t}^{(n)} \leftarrow \sum_{k=0}^{n-1} \gamma^k r_{i,t+k}$ for $n = 1, \dots, T_i - t$
            $G_{i,t} \leftarrow \sum_{k=0}^{T_i - t} \gamma^k r_{i,t+k}$
            $G_{i,t}^{\lambda} = (1 - \lambda) \sum_{n=1}^{T_i - t} \lambda^{n-1} G_{i,t}^{(n)} + \lambda^{T_i - t} G_{i,t}$
            Compute $\hat{V}(S_{i,t}, \mathbf{w}_{i,t})$ and its gradient $\nabla \hat{V}(S_{i,t}, \mathbf{w}_{i,t}$
            $\mathbf{w}_{i,t+1} \leftarrow \mathbf{w}_{i,t} + \alpha[G_t^{\lambda} - \hat{V}(S_{i,t}, \mathbf{w}_{i,t})] \nabla \hat{V}(S_{i,t}, \mathbf{w}_{i,t})$
        end
    end
    $\mathbf{w}_j \leftarrow \mathbf{w}_{j, T_i - 1}$
end
Finalize the converged vector as $\mathbf{w}$
Output: $\mathbf{w}$

where $w_{i,T_i} = \frac{1}{N} \prod_{t=1}^{T_i} \frac{\pi(a_{i,t}|s_{i,t})}{\pi_b(a_{i,t}|s_{i,t})}$. Due to high variance induced by the estimator described in (20), the importance weights from future time steps can be dropped out, leading to a new estimator as follows,

$$J(\pi) \approx \sum_{i=1}^{N} \sum_{t=1}^{T_i} w_{i,t} \gamma^{t-1} r_{i,t}. \tag{21}$$

If the behavior policy is unknown, we can replace the weight $w_{i,T}$ by its estimate $\hat{w}_{i,T_i} = \frac{1}{N} \prod_{t=1}^{T_i} \frac{\pi(a_{i,t}|s_{i,t})}{\hat{\pi}_b(a_{i,t}|s_{i,t})}$. Such series of derivation can also be conducted with respect to the value function. Inspired by this formulation, one can weight the target in Algorithm 12 by the ratio of the probabilities of observing the sample in the behavior policy and the evaluated policy, in order to evaluate the value corresponding to policy π. A modified updated rule (Sutton & Barto, 2018; Mahmood et al., 2014) takes the following form

$$V(s_t) \leftarrow V(s_t) + \alpha \left\{ \frac{\pi(a_t|s_t)}{\pi_b(a_t|s_t)} [r_t + \gamma V(s_{t+1}) - V(s_t)] \right\}.$$

Aside from policy evaluation, we might be interested in how to exploit the idea of TD learning to find an estimated optimal policy. A naive approach is to enumerate the policies within the policy class and evaluate corresponding values, which is not practical for an environment with large state-action space or large policy class. Two recent methods are summarized in the rest of this subsection, making use of TD error to offer some inspirational guidance for policy optimization. Greedy Gradient Q-learning (GGQ, Ertefaie and Strawderman (2018)) works by utilizing the temporal difference error of action-value functions. In details, if we define

$$e_{i,t+1} = r_{i,t} + \gamma \max_{a' \in \mathcal{A}} Q(s_{i,t+1}, a') - Q(s_{i,t}, a_{i,t}) \tag{22}$$

as the Q-version TD error for $i = 1, 2, \ldots, N, t = 1, 2, \ldots, T_i - 1$, then with Q-function approximated by linear features, the summation of trajectory loss is expected to satisfy

$$\sum_{i=1}^{N} \sum_{t=1}^{T_i - 1} e_{i,t+1} \nabla_{\phi} Q(s_{i,t}, a_{i,t}) = 0, \tag{23}$$

which conduces to an unbiased estimator of ϕ (Li et al., 2023). Furthermore, the corresponding estimated optimal policy (Ertefaie & Strawderman, 2018) is found via $\pi^*(s) = \arg\max_{a \in \mathcal{A}} Q_{\phi}(s, a)$ for $s \in \mathcal{S}$.

In addition, Zhou et al. (2022) has considered a novel proximal temporal consistency learning (pT-Learning) framework to estimate the optimal policy. A Lebesgue measurable function class is embedded into the TD error, which induces a modified TD error taking the form as

$$\tilde{e}_{t+1} = z(s_t, a_t)[r_t + \gamma V(s_{t+1}) - V(s_t)], \tag{24}$$

where $z(s, a)$ is a critic function belonging to a class of continuous functions under several regularity conditions. The problem is then constructed as an optimization problem

$$\min_{V,\pi} \max_{z} \mathbb{E}_{s_t, a_t \sim \pi}[z(s_t, a_t)[r_t + \gamma V(s_{t+1}) - V(s_t)]], \tag{25}$$

which can further be transformed into a single-stage minimization problem by introducing the proximal Bellman operator (Li et al., 2023). Thereafter, the optimization is solved by introducing primary and auxiliary parameters, constructing mini-batches, as well as applying the stochastic gradient descent method step by step. For details of pT-Learning and sample codes, we refer to Li et al. (2023); Zhou et al. (2022).

3.4 Policy Gradient

Up to now, we have discussed a variety of methods that follow a typical paradigm: learning value functions (or action-value functions), updating them, and learning a policy based on these values. These methods are classified into value-based methods (Sutton & Barto, 2018). Another class of methods, namely policy-based methods (Sutton & Barto, 2018) that are widely applied in robotic control (Peters & Schaal, 2006) and auto-driving system (Aradi et al., 2018), improve the policies directly, with policies being parametrized. A usual method of modeling $\pi(a|s,\theta)$ in a discrete action space is to form state-action preferences $b(s,a,\theta)$ for each (s,a) pair using softmax distribution (Gold et al., 1996):

$$\pi(a|s,\theta) = \frac{\exp(b(s,a,\theta))}{\sum_{a'\in\mathcal{A}}\exp(b(s,a',\theta))}, \tag{26}$$

where $b(s,a,\theta)$ can be parameterized arbitrarily (e.g., linear combinations of features (Sutton & Barto, 2018)). The parametrized policy function can be deterministic or stochastic in the modeling scheme, under which the action corresponding to the highest preference of each state will have the highest chance of being selected. If the optimal policy is deterministic, the preferences of the corresponding action will tend to be infinitely higher than those of other actions.

Policy gradient (Sutton et al., 1999) is a canonical method that attempts to find the gradient of an objective function with respect to the policy parameter and then carry out optimization towards the objective to find a solution for the parameter. Taking the learning objective $J(\pi)$ as an example, for some baseline function $c(s)$, the gradient of J (Sutton & Barto, 2018) with respect to θ is expressed as

$$\nabla_\theta J(\pi) = \mathbb{E}_{\tau\sim p_\pi(\cdot)}\left[\sum_{t=1}^{\infty}\gamma^t\nabla_\theta\log\pi(a_t|s_t,\theta)\left(\sum_{t'=t}^{\infty}\gamma^{t-t'}r(s_{t'},a_{t'})-c(s_t)\right)\right], \tag{27}$$

which can be further deduced to

$$\frac{1}{1-\gamma}\mathbb{E}_{s\sim\rho(\cdot),a\sim\pi(\cdot|s)}\left[\nabla_\theta\log\pi(a|s,\theta)\bar{A}(s,a)\right], \tag{28}$$

where $\bar{A}(s_t,a_t) = \left(\sum_{t'=t}^{\infty}\gamma^{t-t'}r(s_{t'},a_{t'})-c(s_t)\right)$. Combined with importance sampling, the gradient can be re-expressed (Levine et al., 2020) as

$$\nabla_\theta J(\pi) = \mathbb{E}_{s\sim\rho(\cdot),a\sim\pi_b(\cdot|s)}\left[\left(\prod_{t=1}^{\infty}\frac{\pi(a_t|s_t,\theta)}{\pi_b(a_t|s_t,\beta)}\right)(1-\gamma)^{-1}\nabla_\theta\log\pi(a|s,\theta)\bar{A}(s,a)\right], \tag{29}$$

where β is the parameter of π_b, with an estimator taking the form as follows

$$\nabla_\theta J(\pi) \approx (1-\gamma)^{-1} \sum_{i=1}^{N} \sum_{t=1}^{T_i} w_{i,t} \hat{A}(s_{i,t}, a_{i,t}) \nabla_\theta \log \pi(a_{i,t}|s_{i,t}, \theta), \tag{30}$$

with the assistance of dataset $\mathcal{D}$ and an estimator of $\bar{A}(s, a)$ denoted by $\hat{A}(s, a)$.

The classical policy gradient algorithms (Sutton & Barto, 2018) require an online environment to sample trajectories following a current policy. One can then leverage the sampled trajectories to estimate $\nabla_\theta J(\pi)$, and update θ by $\theta_{t+1} \leftarrow \theta_t + \alpha \nabla_{\theta_t} J(\pi)$ at the t-th time step with the learning rate α. Restricted by the offline setup, a naive way is to associate the updated rule with importance sampling (Levine et al., 2020). However, the behavior policy cannot always be captured, and even if π_b is provided, the resulting estimates of θ may have a large variance and the convergence is not always promised. Therefore, by combining approximate dynamic programming and policy gradient, a more reliable estimation with a much lower variance can be obtained. With the bootstrap of dynamic programming, function approximation, and TD learning, we introduce a method known as actor-critic (Konda & Tsitsiklis, 1999), which efficiently mitigates some deficiencies of purely value-based or policy-based methods.

The actor-critic method consists of two tasks: the critic updates the value function parameters, while the actor updates the policy parameters in the direction suggested by the critic. Diverse variants of the actor-critic methods have emerged in the last 20 years, however, here we focus on an off-policy variant with an approximation of the action-value function (Degris et al., 2012; Haarnoja et al., 2018; Heess et al., 2015). At each iteration, the actor-critic method aims to learn the Q-function corresponding to a current parameterized policy, and the policy parameter is updated subsequently based on the approximated policy gradient. The algorithm is similar to PI, with the policy improved using a gradient-based update procedure. Details (Degris et al., 2012; Levine et al., 2020) can be found in Algorithm 14. While in Algorithm 14, at the beginning of each iteration, additional interactions with the environment are required, the algorithm can be extended to a purely offline setting. For example, an offline actor-critic method can be obtained using $\mathcal{D}$ to pre-populate the data buffer. Additional details can be found in Levine et al. (2020).

If the optimal policy cannot be covered by the offline data, pessimism can be incorporated into the actor-critic method in the face of uncertainty. For instance, an algorithm, namely the pessimistic actor-critic for learning without exploration (PACLE, Zanette et al. (2021)), optimizes the lower bounds of value functions iteratively. More discussions are available in Zanette et al. (2021).

Algorithm 14: Off-policy actor-critic

Input: $I, J, G, \alpha_1, \alpha_2$
Initialize the value function parameter ϕ_1
Initialize the policy parameter θ_1
Initialize replay buffer O
for $i = 1, 2, \ldots, I-1$ **do**
  Repeatedly sample action from current policy π_{θ_i} and sample next state from MDP to append to buffer O
  $\phi_{i,1} \leftarrow \phi_i$
  for $j = 1, 2, \ldots, J-1$ **do**
    Sample batch O from O
    Estimate error from O:
    $\mathbf{e} = \sum_t (Q_{\phi_{i,j}}(s_t, a_t) - (r_t + \gamma \mathbb{E}_{a' \sim \pi_{\theta_i}(\cdot|s_{t+1})} Q_{\phi_i}(s_{t+1}, a')))^2$
    $\phi_{i,j+1} \leftarrow \phi_{i,j} - \alpha_1 \nabla_{\phi_{i,j}} \mathbf{e}$
  end
  $\phi_{i+1} \leftarrow \phi_{i,J}$
  $\theta_{i,1} \leftarrow \theta_i$
  for $g = 1, 2, \ldots, G-1$ **do**
    Sample batch of states (length n) from O
    For each state, sample corresponding action following $\pi_{\theta_{i,g}}$
    For each (s_t, a_t), compute $\hat{A}(s_t, a_t) = Q_{\phi_{i+1}}(s_t, a_t) - \mathbb{E}_{a \sim \pi_{\theta_{i,g}}(\cdot|s_t)}[Q_{\phi_{i+1}}(s_t, a)]$
    $\nabla_{\theta_{i,g}} J(\pi_{\theta_{i,g}}) \approx \frac{1}{n} \nabla_{\theta_{i,g}} \log \pi_{\theta_{i,g}}(a_t|s_t) \hat{A}(s_t, a_t)$
    $\theta_{i,g+1} \leftarrow \theta_{i,g} + \alpha_2 \nabla_{\theta_{i,g}} J(\pi_{\theta_{i,g}})$
  end
  $\theta_{i+1} \leftarrow \theta_{i,G}$
end
Output: ϕ_I, θ_I

4 Dynamic Treatment Regimes

Ideas of statistical reinforcement learning are also applicable to estimate the optimal treatment regime in sequentially randomized trials, as well as observational studies. In medical settings, actions, states, and rewards can be regarded as treatments, patients' specific covariates, and clinical outcomes, respectively (Clifton & Laber, 2020). For the problems consisting of an infinite time horizon, time-homogeneous MDP is a good setup for researchers to study optimal treatment regimes. In contrast, for the problems containing a finite time horizon, each trajectory from the decision process can be considered the complete history of a single patient (Clifton & Laber, 2020; Murphy, 2005) and the environment described by the MDP is no longer apropos. Therefore, we summarize several modern approaches to tackle this issue. To streamline our presentation, we focus on two types of statistical problems, namely individualized treatment regimes (ITRs) and multi-stage dynamic treatment regimes (DTRs).

An ITR is a rule in which the treatment is assigned to each patient based on the individual's characteristics among a set of possible treatments. Estimating optimal ITRs bears a strong resemblance to a single-stage problem in RL with discount factor $\gamma = 0$. To illustrate the idea using a mathematical framework, some general notations are introduced first. Similar to previous sections, we denote r as the outcome of interest, $A \in \{1, -1\}$ as the binary treatment indicator, and $S \in \mathcal{S}$ as a set of observed covariates. Our goal is to identify a treatment regime $\pi(\cdot)$ mapping from the covariate space $\mathcal{S}$ to the treatment space, under which the expected outcome for the population can be maximized. The optimal treatment regime is given by

$$\pi^*(S) = 2I\{\mathbb{E}(r(1) - r(-1)|S) > 0\} - 1, \tag{31}$$

where $r(1), r(-1)$ represent the potential outcomes of a person under a hypothetical intervention 1 and -1, respectively. Moreover, we define the potential outcome $r\{\pi(S)\}$ as

$$r\{\pi(S)\} = r(1)I\{\pi(S) = 1\} + r(-1)I\{\pi(S) = -1\}, \tag{32}$$

where $r\{\pi(S)\}$ indicates the potential outcome under an intervention with treatment assignment based on a regime $\pi(S)$. We make two standard causal assumptions, consistency and positivity, following Cui and Tchetgen (2021).

Additionally with an unconfoundedness assumption: $r(a) \perp A|S$ (for $a = 1, -1$), one can identify the counterfactual mean $\mathbb{E}[r\{\pi(S)\}]$ (Qian & Murphy, 2011), and the optimal treatment regime can be identified by

$$\pi^*(S) = 2I\{\mathbb{E}(r|S, A = 1) - \mathbb{E}(r|S, A = -1) > 0\} - 1.$$

An alternative formulation (Qian & Murphy, 2011; Cui & Tchetgen, 2021) for learning the optimal treatment regime is given by

$$\pi^* = \arg\max_{\pi} \mathbb{E}\left[\frac{I\{A = \pi(S)\}r}{f(A|S)}\right], \tag{33}$$

where $f(A|S)$ denotes the known randomization distribution of A given S. Furthermore, it is proposed that $\mathbb{E}[r(\pi(S))]$ can be maximized over a restricted set of functions (Zhang et al., 2012). Instead of treating the above learning problem as a maximization problem, one can form a weighted classification problem (Zhang et al., 2012; Zhao et al., 2012) with a 0–1 loss function as

$$\pi^* = \arg\min_{\pi} \mathbb{E}\left[\frac{r}{f(A|S)} I\{A \neq \pi(S)\}\right]. \tag{34}$$

Given the computational burden under this formulation, the hinge loss function can be applied to replace the 0–1 loss function, and then one can solve the classification problem via efficient optimization (e.g., using support vector machines) (Zhao et al., 2012). The ensuing work attempts to extend and refine the classification problem (Chen et al. 2016; Cui et al. 2017; Liu et al. 2018; Zhang et al. 2012; Zhang & Zhang 2018; Zhao et al. 2015; Zhao et al. 2015; Zhou et al. 2017). Notably, the methods aimed at learning ITR heavily rely on the unconfoundedness assumption, however, such an assumption is not always guaranteed in observational studies or randomized trials with non-compliance.

Multi-stage DTRs, which are also known as adaptive strategies (Lavori & Dawson, 2000), adaptive interventions, or adaptive treatment strategies (Murphy, 2005), are usually represented as a sequence of decision rules constructed based on patients' data (Murphy, 2003). We first introduce some general notations. Suppose that there is a sequence of K intermediate states and treatments, as well as the final outcome of interest r. The state S_j at time step j is a time-dependent covariate, whereas the action A_j is the treatment taken at the j-th time step. The order of occurrence is $S_1, A_1, S_2, A_2, \ldots, A_K, r$, and $\bar{S}_j = (S_1, \ldots, S_j)$ and $\bar{A}_j = (A_1, \ldots, A_j)$ are used to denote historical data (up to the j-th time step) of state and treatment, respectively. The potential outcomes are represented as $(S_1, S_2(\bar{a}_1), \cdots, S_K(\bar{a}_{K-1}), r(\bar{a}_K))$. Our goal is to maximize the counterfactual means for regime $(\pi_1, \ldots, \pi_K)$,

$$\mathbb{E}[r(\bar{\pi}_K)] = \mathbb{E}[r(\bar{a}_K)]_{a_1=\pi_1(S_1),\ldots,a_K=\pi_K(\bar{S}_K(\bar{a}_{K-1}),\bar{a}_{K-1})}, \tag{35}$$

where $\bar{S}_K(\bar{a}_{K-1}) = (S_1, S_2(a_1), \ldots, S_K(\bar{a}_{K-1}))$. Causal identifying assumptions are given below, and they are longitudinal counterparts of assumptions made in ITRs (Murphy, 2003; Robins, 1997, 2000).

1. Consistency: The potential outcomes are connected with the observed data by: $r = r(\bar{A}_K)$, $S_K = S_K(\bar{A}_{K-1}), \cdots, S_2 = S_2(A_1)$.
2. Positivity: For each treatment level a, $P(A_j = a|\bar{S}_j, \bar{A}_{j-1}) > 0$, $j = 1, 2, \ldots, K$. All treatments are possibly to be provided at each time step.
3. Unconfoundedness: $A_j \perp \{S_2(a_1), \ldots, S_{K-1}(\bar{a}_{K-2}), r(\bar{a}_K)\}$ for $j = 1, 2, \ldots, K$, given $\{S_1, A_1, \ldots, S_j\}$.

Under such assumptions, formula (35) can be rewritten as:

$$\mathbb{E}[r(\bar{\pi}_K)] = \mathbb{E}[\mathbb{E}[\ldots\mathbb{E}[r|\bar{S}_K, \bar{A}_{K-1}, A_K = \pi_K(\bar{S}_K, \bar{A}_{K-1})]\ldots|S_1, A_1 = \pi_1(S_1)]]. \tag{36}$$

Formula (36) is also known as Robins's G-computation formula (Gill & Robins, 2001; Robins, 1986, 1997) which treats the mean response as a function of the multivariate distribution of the longitudinal data. The optimal dynamic treatment regimes can be estimated via iterations (Murphy, 2003), i.e.,

$$V_0(\bar{S}_K, \bar{A}_{K-1}) = \sup_{a_K} \mathbb{E}[r|\bar{S}_K, \bar{A}_{K-1}, A_K = a_K],$$

$$\ldots$$

$$V_{K-j}(\bar{S}_j, \bar{A}_{j-1}) = \sup_{a_j} \mathbb{E}[V_{K-j-1}(\bar{S}_{j+1}, \bar{A}_j)|\bar{S}_j, \bar{A}_{j-1}, A_j = a_j],$$

for $j = 0, \cdots, K-1$. If $\mathbb{E}[|r||\bar{S}_K, \bar{A}_K]$ is bounded, the supremum of formula (36) over a regime class is equal to $\mathbb{E}[V_{K-1}(S_1)]$ (Murphy, 2003). To find $\mathbb{E}[V_{K-1}(S_1)]$, an intuitive solution is to capture the distribution of $(\bar{S}_K, \bar{A}_K, r)$ (e.g., estimating via a parametric, semiparametric or nonparametric model), following which the dynamic programming algorithms can be applied.

One classical method is linked with the regret function

$$\begin{aligned} \upsilon_j(a_j|\bar{S}_j, \bar{A}_{j-1}) = &\sup_{a_j} \mathbb{E}[V_{K-j-1}(\bar{S}_{j+1}, \bar{A}_j)|\bar{S}_j, \bar{A}_{j-1}, A_j = a_j] \\ &- \mathbb{E}[V_{K-j-1}(\bar{S}_{j+1}, \bar{A}_j)|\bar{S}_j, \bar{A}_{j-1}, A_j = a_j], \end{aligned}$$

and Murphy (2003) then shows that

$$\mathbb{E}[r|\bar{S}_K, \bar{A}_K] = \mathbb{E}[V_{K-1}(S_1)] + \sum_{j=1}^{K} \varphi_j(\bar{S}_j, \bar{A}_{j-1}) - \sum_{j=1}^{K} \upsilon_j(A_j|\bar{S}_j, \bar{A}_{j-1}), \quad (37)$$

where $\varphi_j(\bar{S}_j, \bar{A}_{j-1}) = V_{K-j}(\bar{S}_j, \bar{A}_{j-1}) - \mathbb{E}[V_{K-j}(\bar{S}_j, \bar{A}_{j-1})|\bar{S}_{j-1}, \bar{A}_{j-1}]$. It indicates that the derived response r is influenced by the initial state, the chance development over time of the states through φ_j, and the treatments assigned (Murphy, 2003; Henderson et al., 2010). As for estimation, one can introduce a parameter ϑ to parameterize the regret function $\upsilon_j(a_j|\bar{S}_j, \bar{A}_{j-1}; \vartheta)$ and follow a nonstandard iterative method designated by Murphy (2003); Moodie et al. (2007) to estimate ϑ without specifying the form of $\mathbb{E}[V_{K-1}(S_1)]$ and $\{\varphi_j(\bar{S}_j, \bar{A}_{j-1})\}$ explicitly.

Furthermore, the last two decades have witnessed tremendous progress in estimating the optimal DTR, and there are two main directions of statistical learning approaches. The first type is advantage learning (Schulte et al., 2014), which posits models for the contrast (or the regret) of the outcome model and the probability of the observed treatment assignment at each decision point. The classical advantage learning methods are created by Murphy (2003); Robins (1997) and these methods can be implemented through a backward recursive fitting procedure (Murphy, 2003; Robins, 2004). In addition, other advantage learning methods have also been developed (e.g., the optimal double regime structural nested mean model (Robins, 2004), regression on regret functions (Henderson et al., 2010; Moodie et al., 2007)). Another direction leads to a class of methods called quality learning (Schulte et al., 2014; Moodie et al., 2014) which involves the same recursive strategy and is based on posited regression models for the outcome of interest given patient information at each decision point. Owing to its regression-based foundations, the idea of quality learning is widely used (Chakraborty & Moodie, 2013; Chakraborty et al., 2010; Huang & Ning, 2012; Moodie et al., 2014; Wallace & Moodie, 2015; Zhao et al., 2009), despite the possibility of model misspecification.

In the previous discussion, finding optimal DTRs targets mainly at quantities with respect to the final outcome r. In fact, many researchers have also turned their attention toward optimizing the overall reward (Neumann & Peters, 2008; Zhao et al., 2015, 2009) to obtain DTRs by introducing historical outcomes $r_1, \cdots, r_{K-1}, r_K$ into the setup, noting that r_j can be part of S_j. The decision-making process is analogous to finding an optimal policy in multi-stage finite RL problems with $\gamma = 1$. For example, Zhao et al. (2015) formulates the value function as

$$V^\pi = \mathbb{E}\left[\frac{(\sum_{j=1}^K r_j)\prod_{j=1}^K I(A_j = \pi_j(\bar{S}_j, \bar{A}_{j-1}))}{\prod_{j=1}^K \omega_j(A_j, \bar{S}_j, \bar{A}_{j-1})}\right], \tag{38}$$

where ω_js are assumed to be known in randomized trials with $\omega_j(a_j, \bar{S}_j, \bar{A}_{j-1}) = P(A_j = a_j|\bar{S}_j, \bar{A}_{j-1}), j = 1, 2, \ldots, K$. Instead of treating the above learning problem as a maximization problem, a weighted classification problem (Zhao et al., 2015) can be formed to minimize

$$\mathbb{E}\left[\frac{(\sum_{j=t}^K r_j)\prod_{j=t+1}^K I(A_j = \pi_j^*(\bar{S}_j, \bar{A}_{j-1}))}{\prod_{j=t}^K \omega_j(A_j, \bar{S}_j, \bar{A}_{j-1})} I(A_t \neq \pi_t(\bar{S}_t, \bar{A}_{t-1}))\right], \tag{39}$$

for each t. Example codes for implementation are available at Wallace et al. (2020); Chen et al. (2020).

Recently, there has been a rapid growth in the literature on the estimation of optimal ITRs based on randomized trials subject to non-compliance and observational studies which is beyond point identification, e.g., see Han (2021); Cui and Tchetgen (2021) for sign identification and Cui (2021) for partial identification. In the future, it would be beneficial to extend the study beyond point identification in DTR settings. Another potential direction is to incorporate machine learning methods (e.g., random forests (Tao et al., 2018; Zhu et al., 2015; Athey et al., 2019; Cui et al., 2020), neural networks (Wang et al., 2018), decision lists (Zhang et al., 2018)) to find optimal DTRs, which gives rise to deal with different types of datasets, including high-dimensional covariates (Lu et al., 2013; Gunter et al., 2011; Bian et al., 2023; Liu et al., 2017, 2019), continuous-valued treatments (Rich et al., 2014; Kallus & Zhou, 2018), multi-dimensional outcomes (Lizotte et al., 2010), censored outcomes (Zhao et al., 2011) and discrete outcomes (Moodie et al., 2014; Bian et al., 2022). Moreover, instead of the mean criterion, other criteria (e.g., median or other quantiles) are also of interest in certain applications (Wang et al., 2018; Linn et al., 2017). In addition, some researchers have had success estimating ITRs, as seen in Leqi and Kennedy (2021) for optimizing conditional median outcome and Qi et al. (2019) for optimizing conditional-value-at-risk. Using those methods to formulate multi-stage optimal decision rules may lead to important scientific discoveries.

5 Causality and RL

Equipped with tools of causal inference, an agent could explore the dynamics with the presence of unobserved variables, and conduct counterfactual reasoning, which effectively enhances the learning ability in the RL framework, especially when applied in the medicine or epidemiology domain where exploration and randomization are usually restricted. Moreover, causal RL makes good use of causal mechanisms to handle prototypical problems that appear frequently in RL, e.g., sample efficiency (Wang et al., 2021), distributional shift (Rojas-Carulla et al., 2018), and partial observations (Shi et al., 2022). For the remainder of the section, we will summarize a few advanced results established in the applied work of multi-armed bandits (MABs) and MDP.

MAB is a simple model, that can be regarded as a special case of the MDP, describing the interaction between an agent and an environment with only one state element, a finite number of actions, and a finite time horizon. The formal definition is provided below.

Definition 5 A Multi-armed bandit problem is defined as a tuple $(\mathcal{A}, T, r)$, where $\mathcal{A} = \{a_1, \ldots, a_m\}$ is the action space consisting of m arms, T denotes length of time horizon, and for $t \in \{1, \cdots, T\}$, the reward function at the t-th time step is denoted by $r_t = r(a_t, u_t)$, where u_t represents value of the unobserved variable U_t.

Graphical representations of standard environments are displayed in Fig. 2a and b, where for each time step, $U_t \equiv 0$ or U_t only affects the reward r_t demonstrating unconfoundedness. In particular, for the offline case, the actions chosen at these time steps are guided by behavior policies π_b, while intervening strategies π are adopted to control the selected arms in the online case. Interestingly, for any action a taken at the t-th time step, we have identical $P(r_t|a)$ regardless of the action being the consequence of intervention π or predilection π_b. Concerning maximizing the cumulative reward in a long run, such a finding can then assist practitioners in adjusting their strategies in online experiments, in order to find an optimal solution.

The standard MAB problem has gained considerable attention in various application domains, such as mobile health, marketing, and gaming. Meanwhile, classical algorithms are put forward to balance exploration and exploitation, to conduct sequential decision-making, such as the greedy algorithm (Lattimore & Szepesvári,

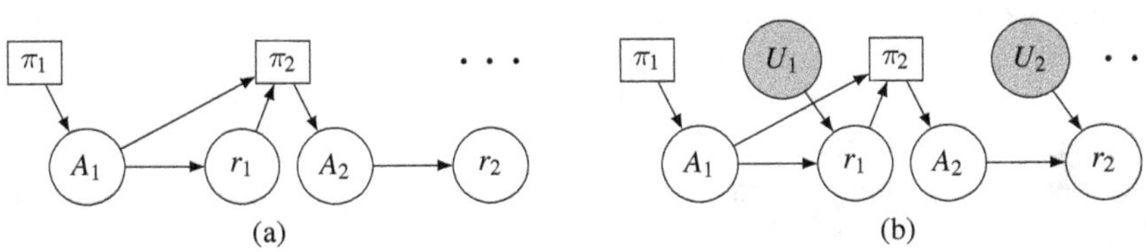

Fig. 2 Graphical representation of the standard MAB problem ($\pi_1, \pi_2, \cdots$ denote the behavior policy at each time step)

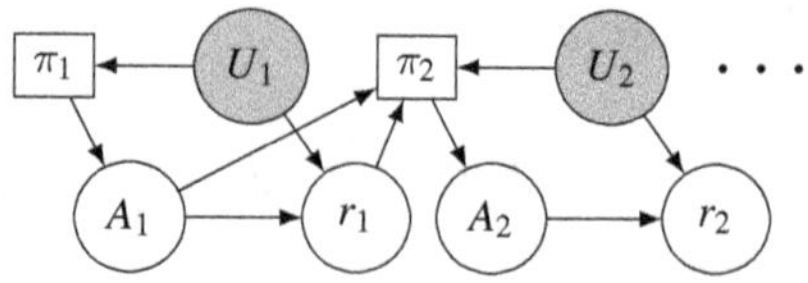

Fig. 3 Graphical representation of the MAB problem with the existence of unmeasured confounders

2020), the explore-then-commit algorithm (Garivier et al., 2016), the upper confidence bounds algorithm (Auer & Ortner, 2010), the Thompson sampling algorithm (Agrawal & Goyal, 2012), and their ongoing works (Garivier & Cappé, 2011; Agrawal & Goyal, 2017; Kandasamy et al., 2018).

However, the unmeasured variables, namely $U_1, \cdots, U_T$, affect both actions and reward, whose presence suggests the inconsistency between offline and online data-generation modes and imposes difficulties on further identification and optimization. One of the possible graphical representations is shown in Fig. 3, and we use it as an example case to introduce a causal RL algorithm as below. Let $r(a)$ represent the potential outcome with action A taking value a. We can find that instead of considering

$$\arg\max_a \mathbb{E}[r|A = a], \tag{40}$$

where a is selected according to π, an objective with a comparison between the expected reward obtained by agents taking actions in favor or against their predilection should be considered, i.e.,

$$\arg\max_{a'} \mathbb{E}[r(a')|A = a], \tag{41}$$

where a is selected according to π_b. The consistency axiom suggests that $\mathbb{E}[r|A = a] = \mathbb{E}[r(a)|A = a]$ for any a following π_b, which is accessible from the offline data. Therefore, it is of interest to estimate counterfactuals (i.e., $\mathbb{E}[r(a')|A = a]$ with $a' \neq a$), and seek a more payable strategy. Considering a simple case with binary reward and $\mathcal{A} = \{1, 2\}$, Bareinboim et al. (2015) introduce Causal Thompson Sampling (CTS), which follows logic from a traditional bandit algorithm, namely Thompson Sampling. The core idea is to intervene on any subject before they execute their behavior policy, take a counter-intuitive action, and make inference from the estimation of $\{\mathbb{E}[r(a')|A = a]\}_{a,a' \in \mathcal{A}}$. Details can be found in Algorithm 15. Moreover, example codes are accessible for practitioners referring to the function `CTS()` in the digital lab using the experimental setup in Bareinboim et al. (2015), which demonstrates a faster convergence rate and smaller regret, compared with traditional bandit algorithms.

Besides, much literature has also been devoted to different types of MAB problems with latent variables (Lee & Bareinboim, 2019; Forney et al., 2017), as well as conducting regret analysis and sample complexity evaluation in the confounded setting (Lattimore et al., 2016; Lu et al., 2020).

Algorithm 15: Causal Thompson sampling

Input: Offline data $\mathcal{D}$
Get an observational distribution P_{obs} from $\mathcal{D}$
Seed distribution $\mathbb{E}[r(a)|A] \leftarrow P_{obs}(r|A)$
for $t = 1, 2, \ldots, T$ **do**
  Get intuition a from behavior
  $Q_1 \leftarrow \mathbb{E}[r(a')|A = a]$ for $a' \neq a$
  $Q_2 \leftarrow \mathbb{E}[r(a)|A = a]$
  $w \leftarrow [1, 1]$
  $bias \leftarrow 1 - |Q_1 - Q_2|$
  if $Q_1 > Q_2$ **then**
    $w[a] \leftarrow bias$
  end
  else
    $w[a'] \leftarrow bias$
  end
  Choose arm $a' \leftarrow \arg\max_{a' \in \mathcal{A}} \beta(s(a', a), f(a', a))w[a']$, where $\beta(s(a', a), f(a', a))$ means sampling from a Beta distribution with two parameters: $s(a', a)$ denote times of $r = 1$ when choosing action a' for original predilection on a, and $f(a', a)$ denote times of $r = 0$ when choosing action a' for original predilection on a.
  Pull a' and get r
  Use r to update $\mathbb{E}[r(a')|A = a]$
end

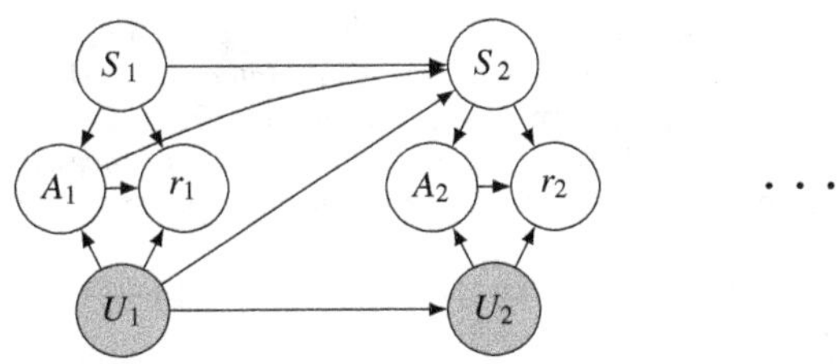

Fig. 4 Graphical representation of the MDP problem with the existence of unmeasured confounders

Inspired by the progress made in confounded MABs, studies on confounders also transfer to MDP. However, such extension is much more challenging, as the causal relations in MDP problems are more complex because the confounders affect not only actions and reward, but also mutually dependent states, with one of the possible graphical representations shown in Fig. 4. In fact, the area is fast growing, and we summarize some recent progress in the following.

Analogous to our observation in MABs, the existence of unmeasured confounders also leads to the unconformity between the offline and online cases in MDP problems, which causes suboptimality if one applies traditional RL algorithms (Zhang & Bareinboim, 2016). Therefore, there is a clear need to develop systematic RL methods from a causal perspective, for example, see Liao et al. (2021), Lu et al. (2022), Fu et al. (2022), Miao et al. (2022) for instrumental variable approaches (Imbens & Angrist, 1994; Angrist et al., 1996) and Bennett and Kallus (2023) for the application of recently proposed proximal causal inference (Miao et al., 2018; Tchetgen et al., 2020; Cui et al., 2023).

References

Agarwal, A., Kakade, S. M., Lee, J. D., & Mahajan, G. (2021). On the theory of policy gradient methods: Optimality, approximation, and distribution shift. *Journal of Machine Learning Research, 22*(98), 1–76.

Agarwal, R., Schuurmans, D., & Norouzi, M. (2020). An optimistic perspective on offline reinforcement learning. In *International conference on machine learning* (pp. 104–114). PMLR.

Agrawal, S., & Goyal, N. (2012). Analysis of Thompson sampling for the multi-armed bandit problem. In *Conference on learning theory* (pp. 39–1). JMLR Workshop and Conference Proceedings.

Agrawal, S., & Goyal, N. (2017). Near-optimal regret bounds for Thompson sampling. *Journal of the ACM (JACM), 64*(5), 1–24.

Angrist, J. D., Imbens, G. W., & Rubin, D. B. (1996). Identification of causal effects using instrumental variables. *Journal of the American Statistical Association, 91*(434), 444–455.

Antos, A., Szepesvári, C., & Munos, R. (2008). Learning near-optimal policies with bellman-residual minimization based fitted policy iteration and a single sample path. *Machine Learning, 71*(1), 89–129.

Aradi, S., Becsi, T., & Gaspar, P. (2018). Policy gradient based reinforcement learning approach for autonomous highway driving. In *2018 IEEE conference on control technology and applications (CCTA)* (pp. 670–675). IEEE.

Athey, S., Tibshirani, J., & Wager, S. (2019). Generalized random forests. *The Annals of Statistics, 47*(2), 1148–1178.

Athey, S., & Wager, S. (2021). Policy learning with observational data. *Econometrica, 89*(1), 133–161.

Auer, P., & Ortner, R. (2010). UCB revisited: Improved regret bounds for the stochastic multi-armed bandit problem. *Periodica Mathematica Hungarica, 61*(1–2), 55–65.

Bareinboim, E., Forney, A., & Pearl, J. (2015). Bandits with unobserved confounders: A causal approach. *Advances in Neural Information Processing Systems, 28*, 1342–1350.

Bellman, R. (1957). A Markovian decision process. *Journal of Mathematics and Mechanics, 6*(4), 679–684.

Bellman, R. (1958). Dynamic programming and stochastic control processes. *Information and Control, 1*(3), 228–239.

Bennett, A., & Kallus, N. (2023). Proximal reinforcement learning: Efficient off-policy evaluation in partially observed Markov decision processes. *Operations Research*. Preprint. arXiv:2110.15332.

Bentley, J. L. (1979). Multidimensional binary search trees in database applications. *IEEE Transactions on Software Engineering, SE-5*(4):333–340.

Bian, Z., Moodie, E. E., Shortreed, S. M., & Bhatnagar, S. (2023). Variable selection in regression-based estimation of dynamic treatment regimes. *Biometrics, 79*(2), 988–999.

Bian, Z., Moodie, E. E., Shortreed, S. M., Lambert, S. D., & Bhatnagar, S. (2023). Variable selection for individualized treatment rules with discrete outcomes. *Journal of the Royal Statistical Society Series C: Applied Statistics*. https://doi.org/10.1093/jrsssc/qlad096.

Breiman, L. (1996). Bagging predictors. *Machine Learning, 24*(2), 123–140.

Breiman, L. (2001). Random forests. *Machine Learning, 45*(1), 5–32.

Buckman, J., Gelada, C., & Bellemare, M. G. (2020). The importance of pessimism in fixed-dataset policy optimization. Preprint. arXiv:2009.06799.

Chakraborty, B., & Moodie, E. (2013). Statistical methods for dynamic treatment regimes. *Springer-Verlag. doi, 10*(978–1), 4–1.

Chakraborty, B., Murphy, S., & Strecher, V. (2010). Inference for non-regular parameters in optimal dynamic treatment regimes. *Statistical Methods in Medical Research, 19*(3), 317–343.

Chen, G., Zeng, D., & Kosorok, M. R. (2016). Personalized dose finding using outcome weighted learning. *Journal of the American Statistical Association, 111*(516), 1509–1521.

Chen, J. & Jiang, N. (2019). Information-theoretic considerations in batch reinforcement learning. In *International conference on machine learning* (pp. 1042–1051). PMLR.

Chen, Y., Liu, Y., Zeng, D., & Wang, Y. (2020). *DTRlearn2: Statistical learning methods for optimizing dynamic treatment regimes*. R package version 1.1.

Chua, K., Calandra, R., McAllister, R., & Levine, S. (2018). Deep reinforcement learning in a handful of trials using probabilistic dynamics models. *Advances in Neural Information Processing Systems, 31*, 4759–4770.

Clifton, J., & Laber, E. (2020). Q-learning: theory and applications. *Annual Review of Statistics and Its Application, 7*, 279–301.

Cohen, A. O., Nussenbaum, K., Dorfman, H. M., Gershman, S. J., & Hartley, C. A. (2020). The rational use of causal inference to guide reinforcement learning strengthens with age. *NPJ Science of Learning, 5*(1), 1–9.

Cover, T., & Hart, P. (1967). Nearest neighbor pattern classification. *IEEE Transactions on Information Theory, 13*(1), 21–27.

Creager, E., Madras, D., Pitassi, T., & Zemel, R. (2020). Causal modeling for fairness in dynamical systems. In *International conference on machine learning* (pp. 2185–2195). PMLR.

Cui, Y. (2021). Individualized decision-making under partial identification: Three perspectives, two optimality results, and one paradox. *Harvard Data Science Review, 3*(3). https://hdsr.mitpress.mit.edu/pub/1h4a86jh.

Cui, Y., Kosorok, M. R., Sverdrup, E., Wager, S., & Zhu, R. (2020). Estimating heterogeneous treatment effects with right-censored data via causal survival forests. *Journal of the Royal Statistical Society: Series B, 85*(2), 179–211.

Cui, Y., Pu, H., Shi, X., Miao, W., & Tchetgen, E. T. (2023). Semiparametric proximal causal inference. *Journal of the American Statistical Association*, 1–12. Preprint. arXiv:2011.08411.

Cui, Y., & Tchetgen, E. T. (2021). On a necessary and sufficient identification condition of optimal treatment regimes with an instrumental variable. *Statistics & Probability Letters, 178*, 109180.

Cui, Y., & Tchetgen, E. T. (2021). A semiparametric instrumental variable approach to optimal treatment regimes under endogeneity. *Journal of the American Statistical Association, 116*(533), 162–173.

Cui, Y., Zhu, R., & Kosorok, M. (2017). Tree based weighted learning for estimating individualized treatment rules with censored data. *Electronic Journal of Statistics, 11*(2), 3927.

Degris, T., White, M., & Sutton, R. S. (2012). Off-policy actor-critic. Preprint. arXiv:1205.4839.

Deisenroth, M., & Rasmussen, C. E. (2011). Pilco: A model-based and data-efficient approach to policy search. In *Proceedings of the 28th international conference on machine learning (ICML-11)* (pp. 465–472). Citeseer.

Doersch, C. (2016). Tutorial on variational autoencoders. Preprint. arXiv:1606.05908.

Elias, B. (2023). *Causal reinforcement learning*. https://crl.causalai.net. Last modified: 2020-12-28.

Ernst, D., Geurts, P., & Wehenkel, L. (2005). Tree-based batch mode reinforcement learning. *Journal of Machine Learning Research, 6*, 503–556.

Ertefaie, A., & Strawderman, R. L. (2018). Constructing dynamic treatment regimes over indefinite time horizons. *Biometrika, 105*(4), 963–977.

Forney, A., Pearl, J., & Bareinboim, E. (2017). Counterfactual data-fusion for online reinforcement learners. In *International conference on machine learning* (pp. 1156–1164). PMLR.

Fu, Z., Qi, Z., Wang, Z., Yang, Z., Xu, Y., & Kosorok, M. R. (2022). Offline reinforcement learning with instrumental variables in confounded markov decision processes. Preprint. arXiv:2209.08666.

Garivier, A., & Cappé, O. (2011). The KL-UCB algorithm for bounded stochastic bandits and beyond. In *Proceedings of the 24th annual conference on learning theory* (pp. 359–376). JMLR Workshop and Conference Proceedings.

Garivier, A., Lattimore, T., & Kaufmann, E. (2016). On explore-then-commit strategies. *Advances in Neural Information Processing Systems, 29*.

Garivier, A., & Moulines, E. (2011). On upper-confidence bound policies for switching bandit problems. In *International conference on algorithmic learning theory* (pp. 174–188). Springer.

Gauci, J., Conti, E., Liang, Y., Virochsiri, K., He, Y., Kaden, Z., Narayanan, V., Ye, X., Chen, Z., & Fujimoto, S. (2018). Horizon: Facebook's open source applied reinforcement learning platform. Preprint. arXiv:1811.00260.

Geurts, P., Ernst, D., & Wehenkel, L. (2006). *Extremely randomized trees. Machine Learning, 63*(1), 3–42.

Gill, R. D., & Robins, J. M. (2001). Causal inference for complex longitudinal data: The continuous case. *Annals of Statistics, 29*(6), 1785–1811.

Glynn, P. W., & Iglehart, D. L. (1989). Importance sampling for stochastic simulations. *Management Science, 35*(11), 1367–1392.

Gold, S., Rangarajan, A., et al. (1996). Softmax to softassign: Neural network algorithms for combinatorial optimization. *Journal of Artificial Neural Networks, 2*(4), 381–399.

Gunter, L., Zhu, J., & Murphy, S. (2011). Variable selection for qualitative interactions. *Statistical Methodology, 8*(1), 42–55.

Haarnoja, T., Zhou, A., Abbeel, P., & Levine, S. (2018). Soft actor-critic: Off-policy maximum entropy deep reinforcement learning with a stochastic actor. In *International conference on machine learning* ((pp. 1861–1870)). PMLR.

Han, S. (2021). Comment: Individualized treatment rules under endogeneity. *Journal of the American Statistical Association, 116*(533), 192–195.

Heess, N., Wayne, G., Silver, D., Lillicrap, T., Erez, T., & Tassa, Y. (2015). Learning continuous control policies by stochastic value gradients. *Advances in Neural Information Processing Systems, 28*, 2944–2952.

Henderson, R., Ansell, P., & Alshibani, D. (2010). Regret-regression for optimal dynamic treatment regimes. *Biometrics, 66*(4), 1192–1201.

Howard, R. A. (1960). *Dynamic programming and markov processes*. Wiley.

Huang, X., & Ning, J. (2012). Analysis of multi-stage treatments for recurrent diseases. *Statistics in Medicine, 31*(24), 2805–2821.

Imbens, G., & Angrist, J. (1994). Identification and estimation of local average treatment effects. *Econometrica, 62*(2), 467–475.

Jefferys, W. H. (1980). On the method of least-squares. *The Astronomical Journal, 85*, 177.

Jin, C., Allen-Zhu, Z., Bubeck, S., & Jordan, M. I. (2018). Is Q-learning provably efficient? *Advances in Neural Information Processing Systems, 31*, 4868–4878.

Jin, Y., Yang, Z., & Wang, Z. (2021). Is pessimism provably efficient for offline rl? In *International conference on machine learning* (pp. 5084–5096). PMLR.

Kallus, N., & Zhou, A. (2018). Policy evaluation and optimization with continuous treatments. In *International conference on artificial intelligence and statistics* (pp. 1243–1251). PMLR.

Kalyanakrishnan, S., & Stone, P. (2007). Batch reinforcement learning in a complex domain. In *Proceedings of the 6th international joint conference on Autonomous agents and multiagent systems* (pp. 1–8).

Kandasamy, K., Krishnamurthy, A., Schneider, J., & Póczos, B. (2018). Parallelised Bayesian optimisation via Thompson sampling. In *International conference on artificial intelligence and statistics* (pp. 133–142). PMLR.

Kober, J., Bagnell, J. A., & Peters, J. (2013). Reinforcement learning in robotics: A survey. *The International Journal of Robotics Research, 32*(11), 1238–1274.

Konda, V., & Tsitsiklis, J. (1999). Actor-critic algorithms. *Advances in Neural Information Processing Systems, 12*, 1008–1014.

Kononenko, I. (1989). Bayesian neural networks. *Biological Cybernetics, 61*(5), 361–370.

Kosorok, M. R., & Laber, E. B. (2019). Precision medicine. *Annual Review of Statistics and Its Application, 6*, 263.

Kosorok, M. R., & Moodie, E. E. (2015). *Adaptive treatment strategies in practice: Planning trials and analyzing data for personalized medicine*. SIAM.

Lagoudakis, M. G., & Parr, R. (2003). Least-squares policy iteration. *The Journal of Machine Learning Research, 4*, 1107–1149.

Lagoudakis, M. G., Parr, R., & Littman, M. L. (2002). Least-squares methods in reinforcement learning for control. In *Hellenic conference on artificial intelligence* (pp. 249–260). Springer.

Lange, S., Gabel, T., & Riedmiller, M. (2012). Batch reinforcement learning. In *Reinforcement learning* (pp. 45–73). Springer.

Lange, S., & Riedmiller, M. (2010). Deep auto-encoder neural networks in reinforcement learning. In *The 2010 international joint conference on neural networks (IJCNN)* (pp. 1–8). IEEE.

Lattimore, F., Lattimore, T., & Reid, M. D. (2016). Causal bandits: Learning good interventions via causal inference. *Advances in Neural Information Processing Systems, 29*, 1189–1197.

Lattimore, T., & Szepesvári, C. (2020). *Bandit algorithms*. Cambridge University.

Lavori, P. W., & Dawson, R. (2000). A design for testing clinical strategies: biased adaptive within-subject randomization. *Journal of the Royal Statistical Society: Series A (Statistics in Society), 163*(1), 29–38.

Lee, S., & Bareinboim, E. (2019). Structural causal bandits with non-manipulable variables. In *Proceedings of the AAAI conference on artificial intelligence* (Vol. 33, pp. 4164–4172).

Leqi, L., & Kennedy, E. H. (2021). Median optimal treatment regimes.

Levine, S., Kumar, A., Tucker, G., & Fu, J. (2020). Offline reinforcement learning: Tutorial, review, and perspectives on open problems. Preprint. arXiv:2005.01643.

Li, G., Shi, L., Chen, Y., Chi, Y., & Wei, Y. (2022). Settling the sample complexity of model-based offline reinforcement learning. Preprint. arXiv:2204.05275.

Li, M., Shi, C., Wu, Z., & Fryzlewicz, P. (2022). Reinforcement learning in possibly nonstationary environments.

Liao, L., Fu, Z., Yang, Z., Wang, Y., Kolar, M., & Wang, Z. (2021). Instrumental variable value iteration for causal offline reinforcement learning. Preprint. arXiv:2102.09907.

Linn, K. A., Laber, E. B., & Stefanski, L. A. (2017). Interactive q-learning for quantiles. *Journal of the American Statistical Association, 112*(518), 638–649.

Liu, N., Liu, Y., Logan, B., Xu, Z., Tang, J., & Wang, Y. (2019). Learning the dynamic treatment regimes from medical registry data through deep q-network. *Scientific reports, 9*(1), 1–10.

Liu, Y., Logan, B., Liu, N., Xu, Z., Tang, J., & Wang, Y. (2017). Deep reinforcement learning for dynamic treatment regimes on medical registry data. In *2017 IEEE international conference on healthcare informatics (ICHI)* (pp. 380–385). IEEE.

Liu, Y., Swaminathan, A., Agarwal, A., & Brunskill, E. (2020). Provably good batch off-policy reinforcement learning without great exploration. *Advances in neural information processing systems 33*, 1264–1274.

Liu, Y., Wang, Y., Kosorok, M. R., Zhao, Y., & Zeng, D. (2018). Augmented outcome-weighted learning for estimating optimal dynamic treatment regimens. *Statistics in medicine, 37*(26), 3776–3788.

Lizotte, D. J., Bowling, M. H., & Murphy, S. A. (2010). Efficient reinforcement learning with multiple reward functions for randomized controlled trial analysis. In *ICML*.

Lu, M., Yang, W., Zhang, L., & Zhang, Z. (2022). Statistical estimation of confounded linear MDPs: An instrumental variable approach. Preprint. arXiv:2209.05186.

Lu, W., Zhang, H. H., & Zeng, D. (2013). Variable selection for optimal treatment decision. *Statistical Methods in Medical Research, 22*(5), 493–504.

Lu, Y., Meisami, A., Tewari, A., & Yan, W. (2020). Regret analysis of bandit problems with causal background knowledge. In *Conference on uncertainty in artificial intelligence* (pp. 141–150). PMLR.

Lu, Y., Xu, Z., & Tewari, A. (2021). Bandit algorithms for precision medicine. Preprint. arXiv:2108.04782.

Madumal, P., Miller, T., Sonenberg, L., & Vetere, F. (2020). Explainable reinforcement learning through a causal lens. In *Proceedings of the AAAI conference on artificial intelligence* (Vol. 34, pp. 2493–2500).

Mahmood, A. R., Van Hasselt, H. P., & Sutton, R. S. (2014). Weighted importance sampling for off-policy learning with linear function approximation. *Advances in Neural Information Processing Systems, 27*, 3014–3022.

Mark, S. D., & Robins, J. M. (1993). A method for the analysis of randomized trials with compliance information: An application to the multiple risk factor intervention trial. *Controlled Clinical Trials, 14*(2), 79–97.

Miao, R., Qi, Z., & Zhang, X. (2022). Off-policy evaluation for episodic partially observable markov decision processes under non-parametric models. *Accepted at Advances in Neural Information Processing Systems(NeurIPS), 35*, 593–606.

Miao, W., Geng, Z., & Tchetgen Tchetgen, E. J. (2018). Identifying causal effects with proxy variables of an unmeasured confounder. *Biometrika, 105*(4), 987–993.

Moodie, E. E., Dean, N., & Sun, Y. R. (2014). Q-learning: Flexible learning about useful utilities. *Statistics in Biosciences, 6*(2), 223–243.

Moodie, E. E., Richardson, T. S., & Stephens, D. A. (2007). Demystifying optimal dynamic treatment regimes. *Biometrics, 63*(2), 447–455.

Munos, R. (2003). Error bounds for approximate policy iteration. In *ICML* (Vol. 3, pp. 560–567).

Munos, R. (2005). Error bounds for approximate value iteration. In *Proceedings of the National Conference on Artificial Intelligence* (Vol. 20, p. 1006). AAAI Press; MIT Press; 1999.

Murphy, S. A. (2003). Optimal dynamic treatment regimes. *Journal of the Royal Statistical Society: Series B (Statistical Methodology), 65*(2), 331–355.

Murphy, S. A. (2005). An experimental design for the development of adaptive treatment strategies. *Statistics in Medicine, 24*(10), 1455–1481.

Murphy, S. A. (2005). A generalization error for q-learning. *Journal of Machine Learning Research, 6*, 1073–1097.

Nemati, S., Ghassemi, M. M., & Clifford, G. D. (2016). Optimal medication dosing from suboptimal clinical examples: A deep reinforcement learning approach. In *2016 38th annual international conference of the IEEE engineering in medicine and biology society (EMBC)* (pp. 2978–2981). IEEE.

Neumann, G., & Peters, J. (2008). Fitted q-iteration by advantage weighted regression. *Advances in Neural Information Processing Systems, 21*, 1177–1184.

Neunert, M., Abdolmaleki, A., Wulfmeier, M., Lampe, T., Springenberg, T., Hafner, R., Romano, F., Buchli, J., Heess, N., & Riedmiller, M. (2020). Continuous-discrete reinforcement learning for hybrid control in robotics. In *Conference on Robot learning* (pp. 735–751). PMLR.

Peters, J., & Schaal, S. (2006). Policy gradient methods for robotics. In *2006 IEEE/RSJ international conference on intelligent robots and systems* (pp. 2219–2225). IEEE.

Prasad, N., Cheng, L.-F., Chivers, C., Draugelis, M., & Engelhardt, B. E. (2017). A reinforcement learning approach to weaning of mechanical ventilation in intensive care units. Preprint. arXiv:1704.06300.

Precup, D. (2000). Eligibility traces for off-policy policy evaluation. In *Computer science department faculty publication series* (p. 80).

Qi, Z., Cui, Y., Liu, Y., & Pang, J.-S. (2019). Estimation of individualized decision rules based on an optimized covariate-dependent equivalent of random outcomes. *SIAM Journal on Optimization, 29*(3), 2337–2362.

Qian, M., & Murphy, S. A. (2011). Performance guarantees for individualized treatment rules. *Annals of Statistics, 39*(2), 1180.

Rashidinejad, P., Zhu, B., Ma, C., Jiao, J., & Russell, S. (2021). Bridging offline reinforcement learning and imitation learning: A tale of pessimism. *Advances in Neural Information Processing Systems, 34*, 11702–11716.

Rich, B., Moodie, E., & Stephens, D. (2014). Adaptive individualized dosing in pharmacological studies: Generating candidate dynamic dosing strategies for warfarin treatment. *Report, McGill University, Montreal, QC, Canada.*

Riedmiller, M. (2005). Neural fitted q iteration–first experiences with a data efficient neural reinforcement learning method. In *European conference on machine learning* (pp. 317–328). Springer.

Riedmiller, M., & Gabel, T. (2007). On experiences in a complex and competitive gaming domain: Reinforcement learning meets robocup. In *2007 IEEE symposium on computational intelligence and games* (pp. 17–23). IEEE.

Robins, J. (1986). A new approach to causal inference in mortality studies with a sustained exposure period–application to control of the healthy worker survivor effect. *Mathematical Modelling, 7*(9–12), 1393–1512.

Robins, J. M. (1997). Causal inference from complex longitudinal data. In *Latent variable modeling and applications to causality* (pp. 69–117). Springer.
Robins, J. M. (2000). Marginal structural models versus structural nested models as tools for causal inference. In *Statistical models in epidemiology, the environment, and clinical trials* (pp. 95–133). Springer.
Robins, J. M. (2004). Optimal structural nested models for optimal sequential decisions. In *Proceedings of the second seattle symposium in biostatistics* (pp. 189–326). Springer.
Robins, J. M., & Greenland, S. (1989). Estimability and estimation of excess and etiologic fractions. *Statistics in Medicine, 8*(7), 845–859.
Rojas-Carulla, M., Schölkopf, B., Turner, R., & Peters, J. (2018). Invariant models for causal transfer learning. *The Journal of Machine Learning Research, 19*(1), 1309–1342.
Sard, A. (1963). *Linear approximation* (Vol. 9). American Mathematical Society.
Schulte, P. J., Tsiatis, A. A., Laber, E. B., & Davidian, M. (2014). Q-and a-learning methods for estimating optimal dynamic treatment regimes. *Statistical Science: A Review Journal of the Institute of Mathematical Statistics, 29*(4), 640.
Seijen, H., & Sutton, R. (2014). True online TD (lambda). In *International conference on machine learning* (pp. 692–700). PMLR.
Shi, C., Uehara, M., Huang, J., & Jiang, N. (2022). A minimax learning approach to off-policy evaluation in confounded partially observable markov decision processes. In *International Conference on Machine Learning* (pp. 20057–20094). PMLR.
Smola, A. J., & Schölkopf, B. (1998). On a kernel-based method for pattern recognition, regression, approximation, and operator inversion. *Algorithmica, 22*(1), 211–231.
Soliman, Y. M. (2014). Personalized medical treatments using novel reinforcement learning algorithms. Preprint. arXiv:1406.3922.
Sutton, R. S., & Barto, A. G. (2018). *Reinforcement learning: An introduction*. MIT Press.
Sutton, R. S., McAllester, D., Singh, S., & Mansour, Y. (1999). Policy gradient methods for reinforcement learning with function approximation. *Advances in Neural Information Processing Systems, 12*, 1057–1063.
Tao, Y., Wang, L., & Almirall, D. (2018). Tree-based reinforcement learning for estimating optimal dynamic treatment regimes. *The Annals of Applied Statistics, 12*(3), 1914.
Tchetgen Tchetgen, E. J., Ying, A., Cui, Y., Shi, X., & Miao, W. (2020). An introduction to proximal causal learning. Preprint. arXiv:2009.10982.
Tesauro, G., et al. (1995). Temporal difference learning and td-gammon. *Communications of the ACM, 38*(3), 58–68.
Uehara, M., & Sun, W. (2022). Pessimistic model-based offline reinforcement learning under partial coverage. In *International Conference on Learning Representations*.
Wallace, M., Moodie, E. E. M., Stephens, D. A., Simoneau, G., & Schulz, J. (2020). *DTRreg: DTR estimation and inference via G-Estimation, dynamic WOLS, Q-Learning, and dynamic weighted survival modeling (DWSurv)*. R package version 1.7.
Wallace, M. P., & Moodie, E. E. (2015). Doubly-robust dynamic treatment regimen estimation via weighted least squares. *Biometrics, 71*(3), 636–644.
Wang, L., Yang, Z., & Wang, Z. (2021). Provably efficient causal reinforcement learning with confounded observational data. *Advances in Neural Information Processing Systems, 34*, 21164–21175.
Wang, L., Zhang, W., He, X., & Zha, H. (2018). Supervised reinforcement learning with recurrent neural network for dynamic treatment recommendation. In *Proceedings of the 24th ACM SIGKDD international conference on knowledge discovery & data mining* (pp. 2447–2456).
Wang, L., Zhou, Y., Song, R., & Sherwood, B. (2018). Quantile-optimal treatment regimes. *Journal of the American Statistical Association, 113*(523), 1243–1254.
Wang, S.-C. (2003). Artificial neural network. In *Interdisciplinary computing in java programming* (pp. 81–100). Springer.
Xie, T., Cheng, C.-A., Jiang, N., Mineiro, P., & Agarwal, A. (2021). Bellman-consistent pessimism for offline reinforcement learning. *Advances in Neural Information Processing Systems, 34*, 6683–6694.

Yauney, G., & Shah, P. (2018). Reinforcement learning with action-derived rewards for chemotherapy and clinical trial dosing regimen selection. In *Machine learning for healthcare conference* (pp. 161–226). PMLR.

Zanette, A., Wainwright, M. J., & Brunskill, E. (2021). Provable benefits of actor-critic methods for offline reinforcement learning. *Advances in Neural Information Processing Systems, 34*, 13626–13640.

Zhang, B., Tsiatis, A., Laber, E., & Davidian, M. (2012). A robust method for estimating optimal treatment regimes. *Biometrics, 68*, 05.

Zhang, B., Tsiatis, A. A., Davidian, M., Zhang, M., & Laber, E. (2012). Estimating optimal treatment regimes from a classification perspective. *Stat, 1*(1), 103–114.

Zhang, B., & Zhang, M. (2018). C-learning: A new classification framework to estimate optimal dynamic treatment regimes. *Biometrics, 74*(3), 891–899.

Zhang, J., & Bareinboim, E. (2016). Markov decision processes with unobserved confounders: A causal approach. Technical report, Technical Report R-23, Purdue AI Lab.

Zhang, J., Kumor, D., & Bareinboim, E. (2020). Causal imitation learning with unobserved confounders. *Advances in Neural Information Processing Systems, 33*, 12263–12274.

Zhang, Y., Laber, E. B., Davidian, M., & Tsiatis, A. A. (2018). Interpretable dynamic treatment regimes. *Journal of the American Statistical Association, 113*(524), 1541–1549.

Zhao, Y., Kosorok, M. R., & Zeng, D. (2009). Reinforcement learning design for cancer clinical trials. *Statistics in Medicine, 28*(26), 3294–3315.

Zhao, Y., Zeng, D., Rush, A. J., & Kosorok, M. R. (2012). Estimating individualized treatment rules using outcome weighted learning. *Journal of the American Statistical Association, 107*(499), 1106–1118.

Zhao, Y., Zeng, D., Socinski, M. A., & Kosorok, M. R. (2011). Reinforcement learning strategies for clinical trials in nonsmall cell lung cancer. *Biometrics, 67*(4), 1422–1433.

Zhao, Y.-Q., Zeng, D., Laber, E. B., & Kosorok, M. R. (2015). New statistical learning methods for estimating optimal dynamic treatment regimes. *Journal of the American Statistical Association, 110*(510), 583–598.

Zhao, Y.-Q., Zeng, D., Laber, E. B., Song, R., Yuan, M., & Kosorok, M. R. (2015). Doubly robust learning for estimating individualized treatment with censored data. *Biometrika, 102*(1), 151–168.

Zhou, W., Zhu, R., & Qu, A. (2022). Estimating optimal infinite horizon dynamic treatment regimes via pt-learning. *Journal of the American Statistical Association* (pp. 1–14).

Zhou, X., Mayer-Hamblett, N., Khan, U., & Kosorok, M. R. (2017). Residual weighted learning for estimating individualized treatment rules. *Journal of the American Statistical Association, 112*(517), 169–187.

Zhou, W., Li, Y., & Zhu, R. (2023). Policy learning for individualized treatment regimes on infinite time horizon. Preprint. arXiv:2309.13458.

Zhu, J., Zhang, Y., Rao, W., Zhao, Q., Li, J., & Wang, C. (2021). Reinforcement learning for diabetes blood glucose control with meal information. In *International symposium on bioinformatics research and applications* (pp. 80–91). Springer.

Zhu, R., Zeng, D., & Kosorok, M. R. (2015). Reinforcement learning trees. *Journal of the American Statistical Association, 110*(512), 1770–1784.

Zhu, S., Ng, I., & Chen, Z. (2019). Causal discovery with reinforcement learning. Preprint. arXiv:1906.04477.

Part II
New Advances in Statistical Methods of Precision Medicine and the Applications

Integrative Learning to Combine Individualized Treatment Rules from Multiple Randomized Trials

Xin Qiu, Donglin Zeng, and Yuanjia Wang

Abstract Using individualized treatment rules (ITRs) to tailor treatment to individual patient's characteristics and intermediate responses holds promise to improve treatment response of chronic disorders. However, several barriers, in particular, lack of generalizability or reproducibility of ITRs derived from a single study and lack of power to detect treatment modifiers as tailoring variables, pose serious challenges for implementing ITRs in clinical practices. In this work, we propose a novel integrative learning method to combine evidence from multiple clinical trials to yield an integrative ITR that improves both precision and reproducibility. Since subject-specific covariates available in each trial may differ, ITRs learned from each study can depend on a different resolution of patient-specific characteristics. Our method does not require all studies to use the same set of covariates, and thus allows study-specific ITRs to be transferable across studies. Specifically, to transfer information we propose integrative learning to enhance a high-resolution ITR by borrowing information from coarsened ITRs or vice versa. We conduct extensive simulation studies to show that the integrative learning has improved performance compared to using single study data. We apply the developed method to multiple clinical trials of major depressive disorder (MDD) and other co-morbid mental disorders. We demonstrate that the integrative ITR yields a greater benefit and has improved precision compared to single-study ITRs or non-personalized rules for treating major depression.

X. Qiu
Janssen Research and Development, Raritan, NJ, USA

D. Zeng
Department of Biostatistics, University of Michigan, Ann Arbor, MI, USA
e-mail: dzeng@umich.edu

Y. Wang (✉)
Department of Biostatistics, Mailman School of Public Health, Columbia University, New York, NY, USA
e-mail: yw2016@cumc.columbia.edu

Y. Zhao, D.-G. Chen (eds.), *Statistics in Precision Health*, ICSA Book Series in Statistics, https://doi.org/10.1007/978-3-031-50690-1_8

Keywords Precision medicine · Interaction · Machine learning · Integrative data analysis · Generalizability · Data pooling · Randomized controlled trials

1 Introduction

Chronic disorders exert immense burden on healthcare. For example, mental disorders accounted for 7.4% of total disease burden in 2010 (Whiteford et al., 2013). Extensive efforts and resources have been devoted to developing pharmacological and behavioral treatments for mental disorders, but no universally effective strategy is discovered (Trivedi et al., 2006). Furthermore, treatment outcomes still remain inadequate or poor and considerable heterogeneity has been documented across mental disorders (Stroup et al., 2006; Trivedi et al., 2016). These facts call for personalized treatments that can sufficiently take into account of individual patient's psychosocial characteristics and/or biomarkers (Holmes et al., 2009).

Randomized controlled trials (RCTs) provide high quality data for learning individualized treatment rules (ITRs) to tailor treatments to patients. However, several barriers hamper the success of constructing ITRs and their implementation in clinical practices. First, the general patient population at large can be different from the trial patients due to differences in time, geographic location, and restricted trial eligibility criteria (Justice et al., 1999), contributing to non-reproducibility and limited generalizability of ITRs. Furthermore, a single trial is usually only powered to detect the average treatment effect instead of subgroup effects, let alone optimal ITRs. Therefore, the ITRs based on a single trial may perform well for the trial patients but not be useful for patients in other studies or in the community. A third challenge is to how to ensure the treatment rules are simple, transparent, and interpretable by clinicians so that they can be easily used in clinical settings.

A cost-effective way to remedy the limited sample size problem and improve reproducibility and generalizability is to combine evidence from multiple RCTs. One commonly used method is meta-analysis that produces a pooled estimate of treatment effects from multiple trials (Haidich, 2010; Cipriani et al., 2018). However, conducting meta-analysis for subgroups across multiple trials is difficult because subgroups are often identified specific to each study in an exploratory fashion, and there are very few well-designed studies available to examine the same subgroup effect. For example, a systematic review of subgroup analysis of RCTs for major depressive disorder (MDD) reveals that among 211 trials that met our inclusion criteria, only one common treatment modifier, baseline severity of depression, was identified among studies that reported subgroup effects. This suggests that the literature is incomplete and inconclusive given the substantial observed heterogeneity in treatment effects for MDD. Some studies use meta-regression to explore whether trial-level factors (e.g., gender proportion or intervention type) modify the treatment effect, but this approach is often largely underpowered due to a small number of trials and the use of effect sizes aggregated at the trial level instead of individual participants' data (Dagne et al., 2016).

Alternatively, integrative data analysis (IDA) (Curran & Hussong, 2009) pools individual-level participant data from multiple studies and analyze the pooled data as if it were from a single study. IDA has been proposed to synthesize evidence for treatments of MDD (Brown et al., 2018) or detect genetic risk factors from multiple cancer studies (Ma et al., 2011). One key requirement for IDA is that multiple studies share the same biological background and collect the same set of covariates. This may not be the case in practice since RCTs are often conducted under different research aims. When new evidence from biological research emerges, hypotheses regarding biomarkers as predictive variables for pharmacotherapy of MDD were proposed. For example, in a recent large, multi-site study of MDD, Establishing Moderators and Biosignatures of Antidepressant Response in clinical Care (EMBARC, Trivedi et al., 2016), biomarkers for predicting treatment responses are one of the foci of data collection. In addition to the usual clinical and neuropsychiatric measures, the study collected several novel baseline covariates including behavioral phenotyping (BP) and functional magnetic resonance imaging (fMRI) measures. In comparison, earlier studies such as Sequenced Treatment Alternatives to Relieve Depression (STAR*D, Rush et al., 2004) and Healing Emotion After Loss (HEAL, Shear et al., 2016) only collected clinical variables and assessments of depression symptomatology, quality of life, work and social adjustment measures. Thus, IDA cannot be directly applied to synthesize evidence from recent and earlier studies without substantial efforts for measure harmonization (e.g., Siddique et al., 2018).

Because baseline patient-specific covariates available in each trial differ, ITRs learned from each study may depend on a different resolution of patient-specific characteristics. For example, the ITR based on EMBARC may depend on both clinical variables and neuroimaging biomarkers, whereas ITRs from STAR*D or HEAL can only depend on clinical variables. The fitted ITRs may yield opposite treatment recommendations for the same patient. A conundrum is that although ITRs learned with a rich set of covariates (high resolution) may be expected to prescribe a more refined or accurate treatment for an individual, these ITRs estimated from a finite sample can be less reliable due to an increased number of tailoring variables involved in the learning process. Conversely, the coarsened ITRs learned with a smaller set of covariates (low resolution) are more robust but may not provide a truly optimal personalized treatment due to that some useful tailoring variables are not available. Therefore, the reconciliation and integration of ITRs learned from heterogeneous resolutions of patient characteristics are important when combining ITRs from multiple trials.

In this work, to tackle the challenges when combining evidence from multiple trials, we propose a novel integrative learning to synthesize evidence across trials and estimate an integrative ITR. Our method does not require all studies to collect a common set of covariates and thus allows study-specific ITRs to be transferable across studies. For example, the integrative learning allows evidence to be combined from ITRs identified in recent RCTs that collected emerging biomarkers with earlier RCTs that focused on clinical and psychiatric markers. The method will integrate information across different trials by a regularized value function and use a data-driven method to determine how much information each study contributes to the integrative ITR. Optimization of the regularized value function can be easily solved under the framework of existing methods (e.g., Zhao et al., 2012; Qiu et al., 2018) with an augmentation term related to clinical outcomes. We provide theoretical justification to show that the integrative learning optimizes treatment selection by evaluating the combined treatment effects from multiple trials. We demonstrate its superior performance by simulations and applications to integrate ITRs from EMBARC and HEAL. Lastly, we identify clinical, behavioral and neuroimaging measures to construct an ITR from EMBARC and show that by integrating HEAL ITRs, the performance is much improved compared to using EMBARC alone.

The rest of this paper is organized as follows. In Sect. 2, we introduce the rationale and algorithm of the integrative learning method. In the last part of this section, we derive the underlying theoretical Bayesian rules of the proposed method to justify integrative learning. In Sect. 3, we perform simulation studies to show improvement as compared to existing machine learning methods for single studies. In Sect. 4, we compare ITRs fitted from the integrative learning using EMBARC and HEAL data with existing approaches. We conclude the paper with discussion and several extensions in Sect. 5.

2 Methodology

We first give a brief review of learning ITRs from a single trial, and then describe the integrative learning in a multiple-trial setting.

2.1 *Learning ITRs from a Single Trial*

Consider a two-arm RCT with $A = 1$ or $A = -1$ indicating two treatment options. Let R denote a post-treatment outcome (e.g., reduction in depressive symptoms), and assume a large value of R to be desirable. Let $\boldsymbol{X}$ denote a vector of subject-specific baseline covariates of p-dimension. For the ith subject, we observe $(\boldsymbol{X}_i, A_i, R_i)$ with the randomization probability denoted as $\pi_i = P(A_i|\boldsymbol{X}_i)$. An ITR, denoted as $\mathcal{D}(\boldsymbol{X}) = \text{sign}(f(\boldsymbol{X}))$, is a binary decision function that maps patient's characteristics $\boldsymbol{X}$ onto the treatment decision domain $A = \{-1, 1\}$, and $f(\boldsymbol{X})$ is the ITR decision function, where its sign determines the choice of treatment 1 or -1 based on $\boldsymbol{X}$. The optimal ITR is the ITR that maximizes the value function defined as the expected post-treatment clinical outcome using the ITR to assign treatment for patients, which is equivalent to $V(\mathcal{D}) = E\big[RI(A = \mathcal{D}(\boldsymbol{X}))/\pi(A|\boldsymbol{X})\big]$. For RCT data, the empirical value function to be maximized can be expressed as

$$\frac{1}{n}\sum_{i=1}^{n}\frac{R_i}{\pi_i}I\{A_i = \text{sign}(f(\boldsymbol{X}_i))\} = \frac{1}{n}\sum_{i=1}^{n}\frac{R_i}{\pi_i}I\{A_i f(\boldsymbol{X}_i) > 0\}. \tag{1}$$

Many methods have been proposed to estimate the optimal DTR using data collected from an RCT, including parametric or semiparametric modeling of the conditional mean of R given $(A, \boldsymbol{X})$ (Qian & Murphy, 2011; Moodie et al., 2007) and directly maximizing the empirical value function based on machine learning algorithms (Zhao et al., 2012; Liu et al., 2018; Qiu et al., 2018). In the following sections, since we will focus on the methodology to integrate ITRs across studies, we assume that one particular method, which is not essential for this paper, is used to learn the optimal ITR decision function from individual studies.

Furthermore, to compare candidate ITRs including the estimated optimal one, we define the benefit function of $\mathcal{D}(\boldsymbol{X})$ for subjects with covariates $\boldsymbol{X}$ as

$$\delta(\boldsymbol{X}) = E\left[R|A = \text{sign}(f(\boldsymbol{X})), \boldsymbol{X}\right] - E\left[R|A \neq \text{sign}(f(\boldsymbol{X})), \boldsymbol{X}\right]. \tag{2}$$

The benefit function measures the difference in the expected conditional outcome between treatments assigned according to $\mathcal{D}(\boldsymbol{X})$ and according to its complementary rule, $-\mathcal{D}(\boldsymbol{X})$, given a patient's covariates $\boldsymbol{X}$. The benefit function can be estimated empirically, where the conditional mean of R given $A = a$ and $\boldsymbol{X}$ is obtained from fitting some parametric or nonparametric regression models (Qiu et al., 2018). This function provides a tool to assess optimality of any estimated ITR. The single-trial ITR learning approaches and benefit function serve as the basis for our integrative learning with multiple trials.

2.2 *Integrative ITRs for Multiple Studies*

The ITR learned from a single study optimizes its performance on the present study but does not guarantee good performance on a future sample. A model that uses information from multiple studies is expected to improve generalizability and thus has better reproducibility. Because not the same covariates are collected in each study (i.e., EMBARC collected neuroimaging and behavioral phenotyping measures but not HEAL), one cannot directly combine study data and perform a single analysis as in IDA. Taking weighted averages of the estimated decision functions also does not apply because each ITR is obtained by maximizing the empirical value function, which has a complex, nonlinear relationship with the decision function $f(\boldsymbol{X})$. In this section, we propose a novel integrative learning method to tackle these challenges and provide an integrative ITR.

2.2.1 Integrative Learning for High-Resolution ITRs Using Coarsened ITRs

First consider how to enhance learning the optimal high-resolution ITR that depends on a rich set of variables using coarsened ITRs learned from smaller subsets of variables. Let $(\boldsymbol{X}_i, A_i, R_i), i = 1, \ldots, n$, be a rich set of baseline covariates, assigned treatments, and treatment outcomes collected on the ith participant in a comprehensive trial, denoted as study 0. For the other K trials, only a subset of covariates are collected by design. The observed data from these trials consist of $(\boldsymbol{Z}_{jk}, A_{jk}, R_{jk})$ with $\boldsymbol{Z}_{jk}$ denotes available covariates for the jth patient in the kth trial for $j = 1, ..., n_k, k = 1, ..., K$.

The richest variable set $\boldsymbol{X}$ defines the optimal high-resolution ITR (denoted by f_0), in terms of the conditional outcome given $\boldsymbol{X}$ and treatments. Variables $\boldsymbol{Z}_k \subset \boldsymbol{X}$ as a subset of $\boldsymbol{X}$ define a coarsened, low-resolution ITR (denoted by f_k), in terms of the conditional outcome given $\boldsymbol{Z}_k$ and treatments. Our rationale is that when applied to patients in the comprehensive trial, each low-resolution f_k still maintains certain optimality when only using the subset of variables $\boldsymbol{Z}_k \subset \boldsymbol{X}$. Although f_k is obtained by maximizing the average value function of the conditional outcome given coarsened groups (e.g., strata formed by conditioning on $\boldsymbol{Z}_k$), such that it may not be optimal for every individual in the finest group stratified by $\boldsymbol{X}$, f_k is informative of f_0 in the following sense: randomly select a patient from subpopulations stratified by conditioning on $\boldsymbol{Z}_k$ and apply to this patient the finest optimal ITR f_0 based on $\boldsymbol{X}$; we should expect that most of times, the treatment recommendations from f_0 should be the same as those based on f_k, because f_k should maximize the value given $\boldsymbol{Z}_k$ when averaged over all subjects. Based on this rationale, we propose the following method to integrate evidence from trials with subsets of covariates to improve learning the optimal ITR from the trial with comprehensive covariates.

First, for $k = 1, ..., K$, we use some particular learning method, for instance, O-learning (Zhao et al., 2012; Liu et al., 2018) or its robust version (Qiu et al., 2018), to obtain the optimal rule that depends on the study variables in the kth trial. Denote the fitted decision function as $\widetilde{f}_k(\boldsymbol{Z}_k)$ and the corresponding benefit function estimated from study k as

$$\widetilde{\delta}_k(\boldsymbol{Z}_k) = \widehat{E}_k(R_k|A_k = \text{sign}(\widetilde{f}_k(\boldsymbol{Z}_k)), \boldsymbol{Z}_k) - \widehat{E}_k(R_k|A_k \neq \text{sign}(\widetilde{f}_k(\boldsymbol{Z}_k)), \boldsymbol{Z}_k),$$

where $\widehat{E}_k(R_k|A_k = a, \boldsymbol{Z}_k)$ is the estimated mean from a regression model using data from trial k. In our numeric implementation, a linear regression with assigned treatment, baseline covariates and their interactions as predictors is used in the regression. Thus, for any subject (including the subjects in the high-resolution study 0) whose covariates $\boldsymbol{Z}_k$ are observed to be z_k, the learned ITR based on the kth study data yields that the optimal treatment assignment should be 1 if $\widetilde{f}_k(z_k) > 0$, and -1 otherwise. Under this treatment assignment scheme, the estimated benefit is $\widetilde{\delta}_k(z_k)$ for these subjects.

Hence, for each subject i in study 0, with the ITR $\widetilde{f}_k$ learned from auxiliary study k in addition to his or her observed data $(A_i, \boldsymbol{X}_i, R_i)$, we expect that the optimal treatment should not be significantly different from $\text{sign}(\widetilde{f}_k(\boldsymbol{Z}_{ik}))$ and the expected benefit should not be far from $\widetilde{\delta}_k(\boldsymbol{Z}_{ik})$. Note that $\boldsymbol{Z}_{ik}$ is the set of the covariates designed to be collected in study k and is a subset of $\boldsymbol{X}_i$. Next, we integrate this external information to learn the optimal treatment rule using the data from study 0. Specifically, we introduce an additional regularization to incorporate $\widetilde{f}_k$ as auxiliary prior information in the learning approach that is in line with multi-task learning

(Liu et al., 2015). The regularized value function for learning an integrative ITR f_0 is

$$\frac{1}{n}\sum_{i=1}^{n}\frac{R_i}{\pi_i}I(A_i f_0(\boldsymbol{X}_i)>0)+\left\{\sum_{k=1}^{K}\frac{\lambda_k}{n}\sum_{i=1}^{n}\frac{\widetilde{\delta}_{ik}}{\pi_i}I(f_0(\boldsymbol{X}_i)\widetilde{f}_k(\mathbf{Z}_{ik})>0)\right\},\tag{3}$$

where λ_k is a tuning parameter to be chosen data-adaptively, and $\widetilde{\delta}_{ik} = \widetilde{\delta}_k(\mathbf{Z}_{ik})$ is the benefit of using $\widetilde{f}_k$ for study 0 subjects. More specifically, to obtain $\widetilde{\delta}_{ik}$, we use the estimated benefit function in the kth trial data and then predict the estimated benefit $\widetilde{\delta}_{ik}$ on subject i in study 0. Note that the first term in (3) is the empirical value function associated with f_0. The second term is a regularization term such that the more consistent the signs of f_0 and $\widetilde{f}_k$, the larger the value. Thus, the regularization encourages the ITR derived from study 0 to be consistent with the ITR derived from study k, and the amount of contribution from study k scales with the benefit associated with $\widetilde{f}_k$ evaluated on study 0 subjects. In other words, the studies with a larger benefit have a greater contribution.

Note that maximizing (3) is equivalent to maximizing

$$\frac{1}{n}\sum_{i=1}^{n}\left[\frac{R_i}{\pi_i}I(A_i f_0(\boldsymbol{X}_i)>0)\right.$$
$$\left.+\left\{\sum_{k=1}^{K}\frac{\lambda_k}{2}\frac{\widetilde{\delta}_{ik}}{\pi_i}\text{sign}(A_i f_0(\boldsymbol{X}_i))\text{sign}(A_i\widetilde{f}_k(\mathbf{Z}_{ik}))+\sum_{k=1}^{K}\frac{\lambda_k}{2}\frac{\widetilde{\delta}_{ik}}{\pi_i}\right\}\right],$$

which can be re-expressed (up to a constant) as $\frac{1}{n}\sum_{i=1}^{n}\frac{\widetilde{R}_i}{\pi_i}I(A_i f_0(\boldsymbol{X}_i)>0)$ after defining $\widetilde{R}_i = R_i + \sum_{k=1}^{K}\lambda_k\widetilde{\delta}_{ik}\left\{2I(A_i\widetilde{f}_k(\mathbf{Z}_{ik})>0)-1\right\}$. Thus, the objective function (3) is equivalent to another empirical value function, with the outcome for individual i augmented from R_i to $\widetilde{R}_i$. Hence, learning integrative high-resolution ITR by maximizing (3) can be transferred to another weighted classification problem with $\widetilde{R}_i$ as the weights, and the solution can be obtained by minimizing a weighted surrogate hinge loss or ramp loss function (Qiu et al., 2018). The tuning parameters λ_k can be selected by cross-validation. In practice, since $\widetilde{\delta}_{ik}$ will be estimated separately for each study k, using the same tuning parameter across studies will ease computational burden.

2.2.2 Integrative Learning for Coarsened ITRs Using High-Resolution ITRs

In clinical settings with limited resources (e.g., when cost constraint does not permit the collection of expensive neuroimaging biomarkers), an ITR that depends only on easily assessed clinical variables can be practically useful. Thus, we propose an integrative learning approach to enhance coarsened ITRs, say ITR for study k, using information generated from a study with high-resolution ITRs, i.e., study 0.

Our rationale is that the average benefit of the high-resolution ITR f_0 evaluated in a coarsened subgroup stratified by $\boldsymbol{Z}_k$ should be the same as the average benefit of the coarsened ITR f_k. Thus, we first obtain the optimal high-resolution treatment rule, denoted by $\text{sign}(\widetilde{f}_0(\boldsymbol{X}))$, and its benefit function, denoted by $\widetilde{\delta}_0(\boldsymbol{X})$, using study 0 data. To incorporate this information to improve the ITR learning in study k, consider estimating f_k by maximizing the following regularized value function similar to (3), but exchange the roles of f_0 and f_k as

$$\frac{1}{n_k}\sum_{j=1}^{n_k}\frac{R_{jk}}{\pi_{jk}}I(A_{jk}f_k(\boldsymbol{Z}_{jk})>0)+\frac{\lambda_k}{n}\sum_{i=1}^{n}\frac{\widetilde{\delta}_i}{\pi_i}I(f_k(\boldsymbol{Z}_{ik})\widetilde{f}_0(\boldsymbol{X}_i)>0),$$
$$k=1,\cdots,K, \tag{4}$$

where λ_k is a tuning parameter and weights $\widetilde{\delta}_i=\widetilde{\delta}_0(\boldsymbol{X}_i)$ are the benefits of using $\widetilde{f}_0$ to assign treatment for subject i. In (4), the first component is the empirical value function for each ITR f_k evaluated on the kth trial data. The second component encourages the low-resolution f_k to yield the same treatment recommendation as the high-resolution $\widetilde{f}_0$ as much as possible, especially for subjects with a large benefit $\widetilde{\delta}_i$, so the contribution of the high resolution study 0 scales with the benefit of estimated $\widetilde{f}_0$.

To learn the kth trial integrative ITR, re-express (4), up to a constant, as

$$\frac{1}{n_k}\sum_{j=1}^{n_k}\frac{R_{jk}}{\pi_{jk}}I(A_{jk}f_k(\boldsymbol{Z}_{jk})>0)$$
$$+\frac{\lambda_k}{n}\sum_{i=1}^{n}\frac{\widetilde{\delta}_i}{\pi_i}\left(2I(A_i\widetilde{f}_0(\boldsymbol{X}_i)>0)-1\right)I(A_if_k(\boldsymbol{Z}_{ik})>0).$$

The objective function can be viewed as augmenting the original data set by combining data from both study k and study 0; however, for subject i in study 0, instead of using R_i as the reward outcome in the augmentation term, we use $\widetilde{R}_i=\lambda_k n_k n^{-1}\widetilde{\delta}_i(2I(A_i\widetilde{f}_0(\boldsymbol{X}_i)>0)-1)$, which is less variable than R_i. For optimization, concatenate data from the kth study, $(\boldsymbol{Z}_{jk},A_{jk},R_{jk})$, with additional data from study 0, $(\boldsymbol{Z}_{ik},A_i,\widetilde{R}_i)$. The standard learning method can then be applied to the combined data to estimate the optimal integrative ITR f_k. Additionally, tuning parameter λ_k weighs the importance of the augmentation and will be chosen by cross-validation.

As a note, the integrative learning can be directly applied to handle single studies with blockwise data collection pattern (e.g., all subjects have clinical measures and a subgroup of subjects do not have any neuroimaging biomarkers) if block patterns are assumed to be completely random. In this case, we treat data of different block patterns as independent trials then can apply the proposed method to integrate optimal ITRs from each data. The details are given in Appendix "Special Case: Integrative Learning to Analyze a Trial with Block-Wise Collection of Covariates".

2.3 Rationale of Integrative Learning

In this section, we provide rationale to justify the proposed integrative leaning methods under conditions. To this end, we evaluate the expectation of the empirical objective function that the integrative ITR maximizes and derive the Bayesian rule for a given objective function. We assume that all the trials are obtained from subpopulations with similar conditional treatment effects given covariates. We show in Sect. 4 some preliminary analysis to check these assumptions. Since the learned ITRs will be applied to the same subpopulations, for the ease of presentation we remove the subscript in the conditional expectations $E_k[\cdot]$ for clarity. Furthermore, if letting $R(a)$ be the potential outcome when treated by a, $a \in \{-1, 1\}$, we assume that $R = \sum_a R(a)I(A = a)$. Lastly, by the randomization design of each trial, A_k is independent of $\{R(a)\}$ given $\boldsymbol{Z}_k$ for $k = 1, \ldots, K$ and A is independent of $\{R(a)\}$ given $\boldsymbol{X}$ in trial 0.

We first consider the ITR learned from the integrative learning method when the low-resolution trials are integrated to the high-resolution trial. Assume that the asymptotic limits of $\widetilde{f}_k(\boldsymbol{Z}_k)$ and $\widetilde{\delta}_k(\boldsymbol{Z}_k)$ in (3) are, $f_k(\boldsymbol{Z}_k)$ and $\delta_k(\boldsymbol{Z}_k)$, respectively. The Bayesian rule of the integrative learning is

$$\text{sign}\left(E[R(1)|\boldsymbol{X}] - E[R(-1)|\boldsymbol{X}] + \sum_{k=1}^{K} \lambda_k(E[R(1)|\boldsymbol{Z}_k] - E[R(-1)|\boldsymbol{Z}_k])\right).$$

The first term has the same sign as the theoretical optimal high-resolution ITR using the rich set of variables $\boldsymbol{X}$. The second term aggregates coarsened treatment recommendations and benefits that are fitted from the smaller sets of variables $\boldsymbol{Z}_k$. Therefore, the integrated ITR converges to an ITR that prescribes treatment with larger weighted conditional benefits across trials. In this sense, the integrated ITR borrows information from all trials.

When λ_k's are chosen sufficiently small and the benefit in trial 0 is assumed to be strictly positive, this rule is equivalent to sign $(E[R(1)|\boldsymbol{X}] - E[R(-1)|\boldsymbol{X}])$, which is the Bayesian ITR from trial 0 only. However, according to the expression in (3), the ITR from our integrative learning not only uses data $(R_i, A_i, \boldsymbol{X}_i)$ from trial 0, but also incorporates essentially independent data $(\widetilde{\delta}_{ik}, \widetilde{f}_k(\boldsymbol{Z}_i), \boldsymbol{Z}_i)$ based on trial k (conditional on $\boldsymbol{Z}_i$). Thus, the proposed ITR estimator is more efficient due to using observations from all trials. Hence, we obtain the following result:

Result 1 When λ_k's are chosen to be close to zero, the integrative high-resolution rule is equivalent to the optimal ITR from the high resolution trial only, but due to using additional independent observations, tends to have a higher accuracy in terms of a smaller excess risk bound of the value function given in Zhao et al. (2012) or Liu et al. (2018) .

Next, we study the ITR from the integrative learning when using the high-resolution trial to improve the ITR from the coarsened trial. The Bayesian rule for the integrative learning of low-resolution rule is

$$\begin{aligned}\text{sign}(f_k^*(\mathbf{Z}_k)) = \text{sign}\big(&E[R_k|A_k = 1, \mathbf{Z}_k] - E[R_k|A_k = -1, \mathbf{Z}_k]\\&+\lambda_k \frac{g_0(\mathbf{Z}_k)}{g_k(\mathbf{Z}_k)}(E[R|A = 1, \mathbf{Z}_k] - E[R|A = -1, \mathbf{Z}_k])\big),\end{aligned}$$

where $g_0(\cdot)$ and $g_k(\cdot)$ are the density functions for $\mathbf{Z}_k$ in study 0 and study k, respectively. The Bayesian rule is a weighted combination of the conditional treatment differences given $\mathbf{Z}_k$ evaluated on study k and evaluated on study 0. In particular, λ_k gives the weight to integrate evidence from study 0. If we further assume that A is independent of $\{R(a)\}$ given $\mathbf{Z}$ in trial 0, then $E[R|A = 1, \mathbf{Z}_k] - E[R|A = -1, \mathbf{Z}_k]) = E[R(1)|\mathbf{Z}_k] - E[R(-1)|\mathbf{Z}_k]$. So the Bayesian rule is $\text{sign}(f_k^*(\mathbf{Z}_k)) = \text{sign}\big(E[R(1)|\mathbf{Z}_k] - E[R(-1)|\mathbf{Z}_k]\big)$, which is the ITR learned using data in trial k only. Our estimator has a higher accuracy due to using independent observations from both trial k, $(R_{ik}, A_{ik}, \mathbf{Z}_{ik})$, and trial 0, $(\widetilde{\delta}_{i0}, \widetilde{f}_0(\mathbf{Z}_{ik}), \mathbf{Z}_{ik})$. In other words, we arrive at the following result:

Result 2 When the treatment rule in trial 0 only depends on $\mathbf{Z}$, the ITR from the integrative learning leads to the same optimal ITR as using trial k only, but due to using additional data from trial 0, has a higher accuracy in terms of a smaller excess risk bound of the value function given in Zhao et al. (2012) or Liu et al. (2018).

The derivations of Results 1 and 2 are provided in Appendix "Baysian Rules for Integrative Learning".

3 Simulation Study

3.1 Simulation Design

To mimic the real world applications with considerable between-patient heterogeneity, we simulated patients from four subgroups with probabilities denoted by (p_1, p_2, p_3, p_4). Depending on the subgroup, patients have different distribution of covariates $\boldsymbol{X}$ and outcome R given $\boldsymbol{X}$. Patient's subgroup membership is usually not observed in practice and thus not used in the analyses. Five covariates (denoted as X_1,..., X_5) informative of the optimal treatment and outcome were independently simulated. The first three covariates were normally distributed with variance one. Their means are the same within each latent group but differ across four groups with values $1, 0.5, -1$ and 0.5, respectively. The last two covariates follow standard normal distribution. The clinical outcome for subjects in the kth latent group was generated by

$$\begin{aligned} R = 1 &+ I(A = 1)(\beta_{1k} + 0.5X_4 + \alpha_{1k}X_5) \\ &+ I(A = -1)(\beta_{2k} + 0.5X_4 + \alpha_{2k}X_5) + \epsilon_k, \end{aligned} \tag{5}$$

where $\epsilon \sim N(0, 0.25 + \alpha_{1k}^2)$, and $\boldsymbol{\beta} = [\beta_{lk}]_{2\times 4} = \begin{bmatrix} 0.8 & 0.3 & 0 & 0 \\ 0 & 0 & 0.8 & 0.3 \end{bmatrix}$, $\boldsymbol{\alpha} = [\alpha_{lk}]_{2\times 4} = \begin{bmatrix} 1 & 0.6 & 0.5 & 0.3 \\ 0.5 & 0.3 & 1 & 0.6 \end{bmatrix}$. Thus, treatment $A = 1$ has a greater average effect for subjects in subgroups 1 and 2, and the alternative treatment $A = -1$ has a greater average effect in subgroups 3 and 4. In addition, a noise variable X_6 independent of the clinical outcome was generated from a standard normal distribution. Treatment $A = 1$ or -1 was randomly assigned to each subject with equal probability.

To illustrate our method, we generated the high-resolution data (study 0), including all the observed covariates $\boldsymbol{X} = (X_1, \cdots, X_6)$, and the low-resolution data (study 1) with only $\boldsymbol{Z}_1 = (X_1, \cdots, X_4)$. Furthermore, to mimic between-study heterogeneity in real world applications (Justice et al., 1999), for these two RCTs, patients were simulated from the latent subgroups with different probabilities. That is, we assumed that within each group the conditional distribution of R given $\boldsymbol{X}$ is the same as described in (5), but the mixing probabilities were different between the trials. We set the probabilities of patients belonging to each of the four subgroups as $(p_1, p_2, p_3, p_4) = (0.4, 0.1, 0.4, 0.1)$ for study 0 and $(0.2, 0.3, 0.2, 0.3)$ for study 1. This simulation setting can represent applications where a trial conducted at one location may consist of, for example, more elderly patients than another trial at a different location, but the distribution of clinical outcome given age and other covariates is the same.

In order to test the generalizability of the proposed integrative learning, we evaluated the estimated ITR on different target populations, aiming to simulate different settings of the target population where the ITR will be implemented. The evaluation population also consists of a mixture of four groups of patients with the same conditional distribution of R given $\boldsymbol{X}$ within group as described above. For settings 1 through 4, the target validation population has mixing proportions as $(p_1, p_2, p_3, p_4) = (0.25, 0.25, 0.25, 0.25)$, $(0.2, 0.2, 0.3, 0.3)$, $(0.5, 0, 0.5, 0)$, and $(0, 0.5, 0, 0.5)$, respectively. These probabilities are different from that of study 0 and study 1. Thus the distribution of covariates $\boldsymbol{X}$ is different for the data used to learn ITRs and the data used to evaluate ITRs. Validation settings 3 and 4 represent extreme cases when only subgroups with large or small benefits are considered. We simulated a large independent validation data ($N = 10{,}000$) under these four settings to compare methods.

Lastly, for the treatment rules to have a realistic chance of being applied broadly and consistently in real-world clinical setting, simple and transparent rules interpretable by clinicians are desirable (e.g., linear rules). Thus, in simulation studies we estimated the optimal ITRs as linear functions of covariates. Moreover, we adopted a recently developed method referred as asymptotically best linear O-learning (ABLO, Qiu et al., 2018), because it guarantees the optimal value when the treatment rules are restricted to the class of linear rules. This approach uses a more precise loss function, the ramp loss (McAllester & Keshet, 2011), instead of hinge-loss to replace the zero-one loss in the empirical value function in (3) and (4).

3.2 Simulations for Integrative Learning to Improve High-Resolution ITR

In this section, we evaluate the performance for improving high-resolution ITR by coarsened ITRs. In each simulation, study 1 data were used to learn a coarsened ITR f_1 with only 4 variables, namely $\boldsymbol{Z}_1 = (X_1, \cdots, X_4)$. The integrative ITR approach used all 6 covariates in $\boldsymbol{X}$ and included the coarsened ITR as auxiliary information as in (3). The predicted benefit $\widetilde{\delta}_{i1}$ under ITR f_1 for subject i was computed based on a linear regression model using study 1 data with treatment assignment, covariates and their interactions as predictors. Since in some real data applications estimating the benefit may not be possible (e.g., only the coarsened ITR is available but not the individual participant data in study 1), we will also consider letting $\widetilde{\delta}_{i1}$ be a constant of one. We compare the integrative learning weighted by the estimated benefit and weighted equally (denoted as Int-Ben and Int-Equal) to using only high-resolution study 0 with covariates $\boldsymbol{X}$ (denoted as ABLO-H) and using low-resolution study 1 with covariates $\boldsymbol{Z}_1$ (denoted as ABLO-L). The sample size was 200 for both studies. Cross validation was used to select the tuning parameter λ. We report the benefit, value function and treatment selection accuracy defined as the proportion of subjects for whom the estimated optimal treatment was the same as the true optimal treatment.

The results from 100 replications are summarized in Fig. 1 and Table 1. For the first three validation settings, integrative learning ITRs using both trials' data outperform single-study ITRs in terms of a higher optimal treatment selection accuracy, larger benefit and greater value function. The benefit function and value function estimated by integrative learning are also much more precise with their standard deviations reduced by more than 30% compared to single-study ITRs. For the last setting where the target population only consists of low benefit subjects, Int-Ben that incorporated predicted benefits outperforms ITR using study 0 only, but not Int-Equal. With more data used to estimate an ITR, the integrative learning is more likely to be generalized to a future sample with a higher expected outcome and accuracy in most scenarios of the target population considered in simulations. Integrative learning weighted by the estimated benefit performs better than using equal weights since the estimated benefits contain useful information in learning the ITR.

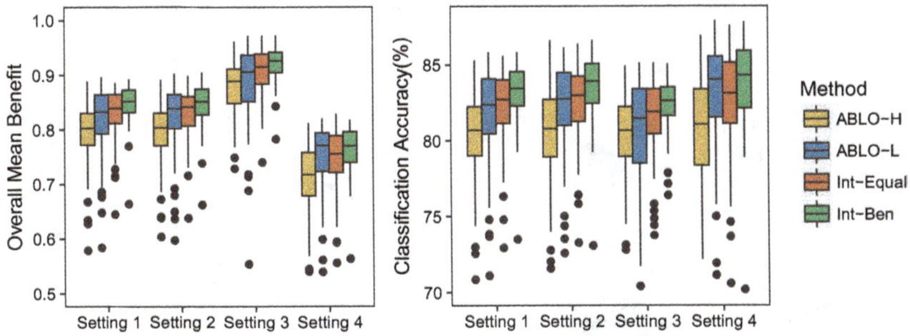

Fig. 1 ITR benefit and optimal treatment allocation accuracy for the four methods. ABLO-H is estimated by ABLO using high-resolution data only; ABLO-L is estimated by ABLO using low-resolution data only; Int-Equal is estimated by integrative learning for high-resolution ITR with equal weights for all subjects; Int-Ben is estimated by integrative learning for high-resolution ITR weighted by estimated benefit

Table 1 Mean and standard deviation of the benefit, value and accuracy using integrative learning for high-resolution ITRs comparing to ITRs by ABLO on single studies[a]

		ABLO-H	ABLO-L	Int-Equal	Int-Ben
Benefit	Setting 1	0.797(0.054)	0.821(0.059)	0.830(0.045)	0.848(0.032)
	Setting 2	0.795(0.057)	0.824(0.056)	0.830(0.044)	0.847(0.037)
	Setting 3	0.879(0.047)	0.887(0.069)	0.905(0.044)	0.920(0.031)
	Setting 4	0.708(0.065)	0.751(0.058)	0.744(0.056)	0.765(0.038)
Value	Setting 1	0.664(0.028)	0.676(0.031)	0.681(0.023)	0.689(0.017)
	Setting 2	0.659(0.028)	0.673(0.028)	0.676(0.022)	0.685(0.018)
	Setting 3	0.824(0.024)	0.828(0.035)	0.838(0.023)	0.846(0.017)
	Setting 4	0.500(0.033)	0.521(0.029)	0.518(0.028)	0.528(0.019)
Accuracy	Setting 1	0.805(0.028)	0.819(0.030)	0.822(0.025)	0.832(0.019)
	Setting 2	0.806(0.030)	0.822(0.031)	0.825(0.025)	0.836(0.021)
	Setting 3	0.804(0.025)	0.806(0.035)	0.816(0.023)	0.824(0.016)
	Setting 4	0.805(0.038)	0.830(0.036)	0.826(0.035)	0.840(0.026)

[a] See Fig. 1

3.3 Simulations for Integrative Learning of Coarsened ITR

Here we evaluate the performance for enhancing the coarsened ITR using the high-resolution ITR. The simulation settings are similar to the previous section. The only difference is that we assume two variables X_5 and X_6 are not collected in the target population due to resource constraints (e.g., no neuroimaging measure is collected). Therefore, the validation data used to evaluate ITRs consists of $(\boldsymbol{Z}_1, R)$ and the high-resolution ITR cannot be applied. For each simulated data, a high-resolution ITR is learned from study 0 and incorporated as auxiliary information to learn integrative coarsened ITR as in (4). We compare the integrative coarsened

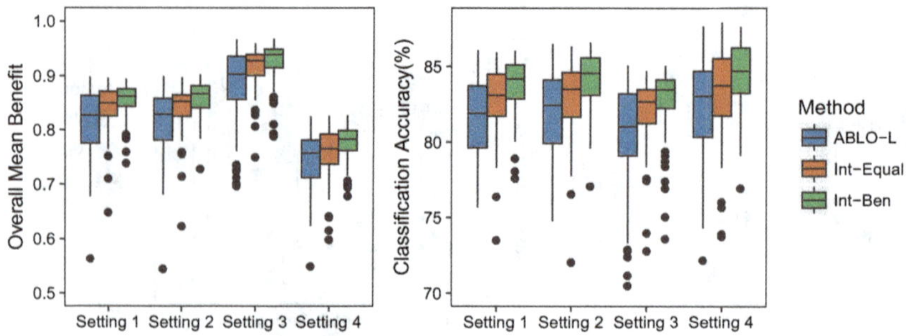

Fig. 2 Overall ITR benefit and optimal treatment allocation accuracy for the three methods. ABLO-L is estimated by ABLO using low-resolution data only; Int-Equal is estimated by integrative learning for low-resolution ITR with equal weights for all subjects; Int-Ben is estimated by integrative learning for low-resolution ITR weighted by estimated benefit

Table 2 Mean and standard deviation of overall benefit, value and accuracy rates using integrative learning for low-resolution ITRs comparing to ITRs by ABLO on single studies[a]

		ABLO-L	Int-Equal	Int-Ben
Benefit	Setting 1	0.814(0.060)	0.841(0.040)	0.855(0.031)
	Setting 2	0.817(0.057)	0.840(0.040)	0.857(0.032)
	Setting 3	0.885(0.064)	0.916(0.036)	0.927(0.035)
	Setting 4	0.742(0.058)	0.756(0.048)	0.774(0.032)
Value	Setting 1	0.672(0.031)	0.686(0.021)	0.693(0.016)
	Setting 2	0.669(0.029)	0.681(0.020)	0.689(0.016)
	Setting 3	0.828(0.033)	0.844(0.019)	0.849(0.018)
	Setting 4	0.517(0.029)	0.524(0.024)	0.533(0.016)
Accuracy	Setting 1	0.814(0.030)	0.828(0.021)	0.837(0.018)
	Setting 2	0.818(0.031)	0.831(0.023)	0.841(0.018)
	Setting 3	0.805(0.035)	0.822(0.020)	0.828(0.021)
	Setting 4	0.823(0.037)	0.833(0.031)	0.846(0.023)

[a] See Fig. 2

ITR (both weighted by estimated benefit and weighted equally) to the single-study coarsened ITR using low-resolution covariates $\boldsymbol{Z}_1$ in study 1.

Simulation results from 100 replications are summarized in Fig. 2 and Table 2. Similar to before, for all validation settings, integrative coarsened ITRs (Int-Equal and Int-Ben) using data from both trials outperform the single-study ITRs (ABLO-L) using study 1 data alone. Integrative learning weighted by estimated benefit performs better than using equal weights. The integrative learning is also more efficient compared to ABLO-L, reducing the standard deviations of the benefit and value function by about 40%. Thus, the integrative ITR retains the simplicity by including only four covariates, but improves the performance by incorporating additional information from the high resolution rule fitted from study 0 with more covariates.

Additional simulation studies were performed to examine robustness of our method and shown in Appendix section "Simulation Studies of Sensitivity Analysis". In these sensitivity analyses, in addition to mixing proportions, the conditional outcome models of R given X in four subgroups differ across study 0, study 1, and the validation populations. The simulation results show that depending on the nature of difference between studies, the low-resolution rule may perform better than the high-resolution rule. However, integrative ITRs remain as the best method. All methods become more variable when the validation population becomes substantially different from the training data population.

4 Real Data Application

Our high-resolution study (study 0) is EMBARC, a two-stage, multi-site randomized trial designed to discover biosignatures associated with response to treatment for MDD (Trivedi et al., 2016). In the first stage, participants were randomized to an 8-week treatment of sertraline (a Selective Serotonin Reuptake Inhibitor, SSRI) or placebo. We will focus on the first stage in our analysis. The primary outcome, Hamilton Depression Rating Scale (HAM-D) was measured at week 8 to assess the severity of depression. The clinical outcome was the change in HAM-D score from baseline to week 8, where a larger value of the change corresponds to better treatment response (a greater reduction in depressive symptoms). There were 242 participants with complete baseline demographics, clinical variables and outcomes, including gender, race, age, baseline HAM-D score, and Quick Inventory of Depressive Symptomatology (QIDS) score. Among all 242 patients, only 138 had neuroimaging and behavioral phenotyping measures. For learning the high-resolution ITR, 138 patients with complete clinical measures, behavioral phenotyping and neuroimaging measures were considered as study 0. The covariates were divided into tiers based on prior studies and hypothesis of their roles in treating depression (Trivedi et al., 2016). We considered tier 1 variables and performed univariate screening by a linear regression model including the covariates, assigned treatment and their interaction to assess all tier 1 behavioral phenotyping and brain imaging measures. The most informative measures were "Effect of Flanker interference on accuracy" and functional magnetic resonance imaging (fMRI) measure "Pregenual cingulate (seed) to right amygdala conflict adaptation coupling". Flanker accuracy effect is a measure of interference effects, with a higher score indicating an increased interference effects or reduced cognitive control. The fMRI measure is the difference of the connectivities between Pregenual cingulate and Right amygdala during incongruent minus congruent frames, during the emotion recognition task (ERT), which is a standard neuropsychiatric test. Both variables were included when estimating a high-resolution ITR using EMBARC data.

Our low-resolution study (study 1) is HEAL (Shear et al., 2016), a multi-site RCT where patients affected by complicated grief were randomized to receive citalopram (an SSRI), placebo, citalopram+psychotherapy, or placebo+psychotherapy. Since a large proportion (about 70%) of patients with complicated grief are also suffering from MDD (Shear et al., 2016), we included the subgroup of patients with current MDD and were randomized to either citalopram or placebo in the analysis. There were 74 patients and the outcome was the change of QIDS score from baseline to week 12. To learn a coarsened ITR, we used age, gender and baseline QIDS score in the HEAL study as the covariates.

Since our approach does not require the marginal distribution of baseline covariates to be the same across studies but only the direction of conditional treatment effects, we perform several preliminary studies to examine the conditional treatment effects for several covariates available in both EMBARC and HEAL. The treatment effect for female is 2.83 and 1.66 for EMBARC and HEAL, respectively. While the conditional treatment effects for Male is -0.83 and -0.54. The directions of treatment effects in both groups are the same. For baseline depression severity, the conditional treatment effects for the severe group (QIDS $\geq$ 16) is 2.19 in EMBARC and 2.57 in HEAL, while they are both near 0 for the non-severe group (QIDS $<$ 16). For age, most of HEAL subjects are more than 50 years old, for the subgroup older than age 50 in both trials, the directions of the treatment effect are also the same in two studies.

We applied the same four methods as the simulation studies to EMBARC and HEAL. For all methods, we randomly split the EMBARC sample into a training and testing set with a 2:1 ratio and repeated the procedure 100 times. The value function and ITR benefits were evaluated on the testing set. For integrative learning methods, we bootstrapped the HEAL sample to learn a coarsened ITR for every repetition. Results for low-resolution rule using the HEAL data (ABLO-L), high-resolution rule using EMBARC complete case only (138 subjects; ABLO-H), integrative learning with equal weight (Int-Equal) and integrative learning with estimated benefit (Int-Ben) are compared in Table 3 and Fig. 3. The non-personalized rules yield a change in HAM-D score of 7.48 points for SSRI and 5.84 for placebo, with a difference of 1.64. The ITR estimated by Int-Ben yields a HAM-D change score of 7.71, which is larger than Int-Equal (7.64), ABLO-H (7.63), and ABLO-L (7.24). The ITR benefit of Int-Ben (2.24) is also larger than Int-Equal (2.04), ABLO-H (2.04), and ABLO-L (1.33); and much larger than the non-personalized rule (1.64). The estimation of value function is more efficient by using Int-Ben (sd = 1.27) compared to ABLO-H (sd = 1.40), while the estimation of ITR benefit is equally efficient for Int-Ben and ABLO-H.

In two studies, we focus on patients with MDD treated by either SSRI or placebo. HAM-D and QIDS are both common outcome measures of the severity of depression symptoms. Given the direction of treatment effect stratified by gender, age groups and baseline severity of MDD are quite similar, integrative learning

Table 3 Overall performance of the four methods in EMBARC study using HEAL ITR

Method	ABLO-L	ABLO-H	Int-Equal	Int-Ben
Value	7.24(1.60)	7.63(1.40)	7.64(1.36)	7.71(1.27)
Benefit	1.33(2.49)	2.04(1.98)	2.04(2.18)	2.24(1.98)

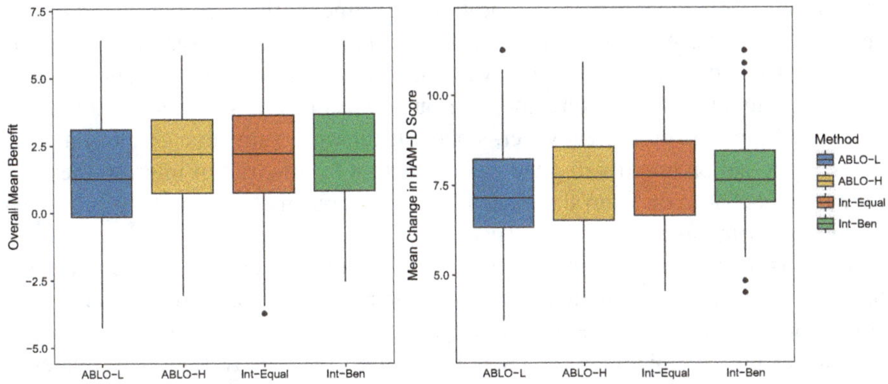

Fig. 3 Overall performance of the four methods in EMBARC study using HEAL ITR. Distribution of estimated ITR benefit (the higher the better) and change in HAM-D score (the larger the better) at the end of stage 1 treatment for four methods (based on 100 cross-validation)

improves the predictive performance by combining more information from two studies compared to using EMBARC only.

In Appendix "EMBARC Analysis with Blockwise Data Collection", we present an extension to integrate ITRs fitted from subjects with and without neuroimaging biomarkers in EMBARC. The results show that integrated high-resolution rule outperforms ITR using only subjects with complete imaging measures.

5 Discussion

In this work, we propose machine learning methods to estimate integrative ITRs that combine information across multiple trials. Our integrative methods improve the efficiency and reproducibility not only for high-resolution ITRs but also for coarsened ITRs by using a regularized value function to incorporate information from other related studies. A data-driven method determines how much evidence each study contributes to the integrative ITR. In practice, it is rare for RCTs to collect the same set of baseline covariates, given different hypotheses and goals across studies. The integrative learning accommodates heterogeneous sets of baseline measures and allows more flexibility compared to IDA that combines multiple data sets into a single data set.

We derive the underlying Bayesian rules for the two integrative schemes, which show that the integrative rule maximizes a weighted combination of conditional treatment effects given different subsets of covariates. When the direction (sign) of the true conditional treatment effects given subsets of covariates is the same, the integrative rule converges to the theoretical optimal rule. This is a relatively weak assumption on integration since we do not place assumptions on the marginal distributions nor the magnitude of conditional treatment effects between studies and subsets of patients. As shown in our real data application, this assumption can be examined from data for covariates available across studies. The benefit function requires to model the conditional mean of outcomes given covariates and treatment which can be misspecified. However, since our tuning parameters are chosen data-adaptively based on maximizing the value function for the trial of interest, they can govern the contribution from the other trials. Thus, our method can be robust even if the posited outcome model is misspecified.

We considered the baseline covariates of smaller trials are nested within a comprehensive trial. When there are overlapping but non-nested covariates, the "comprehensive trial" will be the study with the largest number of covariates and the low resolution covariate sets will be defined by the overlap between each smaller study and the "comprehensive study". We expect that having all studies with comparable sample sizes would facilitate the integrative learning so that the final rule is not driven by a dominating study.

Although here we focus on linear decision rules, our method can be generalized to any nonparametric rules or using other loss functions to replace zero-one loss for optimization. Here a linear regression is used to estimate the benefit of ITRs estimated from other studies, while other nonparametric or machine learning methods can be considered. Another extension is to apply integrative learning to observational studies, where the difficulty is how to properly weight the evidence of RCTs versus observational studies when both types of studies are present. Due to the virtue of randomization of RCTs, our integrative learning methods are consistent no matter which subset of variables is included in the integration. However, this may not hold for observational studies depending on whether all confounders are included in the subset of variables, which complicates how to weight study-specific ITRs in observational studies.

In general, performing integration does not always improve performance especially when f_0 and f_k yield different treatment recommendations for a large number of subjects. However, our tuning parameters are chosen adaptively based on data to prevent integrative learning from deteriorating performance in this case. For example, when the validation population is similar to the first trial patients, but has substantially different distribution from the second trial patients, we expect that the tuning parameter will be chosen close to zero (to ensure a higher value is achieved), and therefore the second trial does not assist in improving the performance of the ITR. When the validation population is different from all trials and no data is collected from the validation population, the performance gain or loss depends on the nature of differences. We performed a few sensitivity analyses to examine this phenomena in Appendix (e.g., Tables 4 and 5). Our integrative learning requires treatments being compared to be the same across trials. It would be of interest to extend our methods to handle multiple treatments using evidence from direct and indirect comparisons, which is similar to the network meta-analysis (Tonin et al., 2017).

Some alternative methods may be considered, e.g., imputation-based methods to impute missing covariates in the high-resolution studies. However, these methods heavily rely on the correct specification of the imputation models. One may consider to the value function instead of benefit function to combine ITRs in the integrated learning. However, benefit function, which is the contrast of the value function under the estimated optimal treatment rule and its complementary rule, tends to be more interpretable based on the expression of the Bayesian rule (e.g., Sect. 2.3). It would be desirable to extend current methods to the cases where individual level data are not available but the summary statistics or treatment rules are available. For example, consider an objective function that combines information and encourages consistency of recommendations between studies.

Appendix

Special Case: Integrative Learning to Analyze a Trial with Block-Wise Collection of Covariates

With the development of new biomarker collection techniques, multiple sources of data may be collected from different domains. In practice, it is common that subjects entail block-wise patterns of baseline covariates (Xiang et al., 2013). In EMBARC, some subjects have complete demographics, clinical, neuroimaging (fMRI) and behavioral phenotyping (BP) measures, while others lack the entire domain of fMRI and BP measures (e.g., Fig. 4).

Assume that the blockwise variable collection pattern is completely random and noninformative. The data collected from a single study can be treated as collected from multiple sub-studies based on the domains of covariates that are available. More specifically, in this context study 0 consists of subjects with complete information on all baseline variables, and subjects who have the same block of

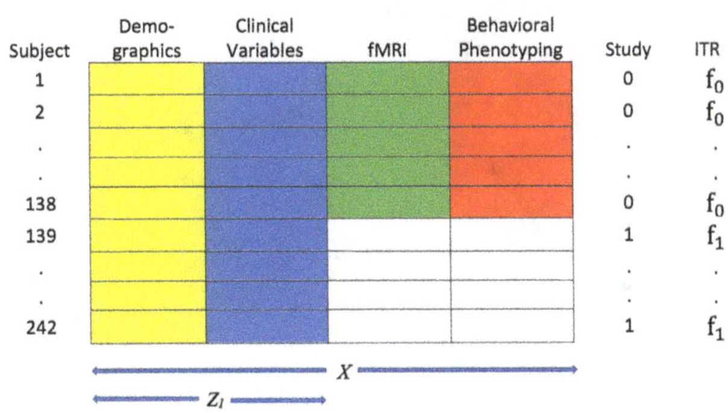

Fig. 4 An example of data collected with blockwise variable domains in EMBARC. White color indicates the measures for that domain are not collected, other colors indicate the measures that are complete. In this example, $\boldsymbol{X}$ includes covariates from all four domains; $\boldsymbol{Z}_1$ includes variables from demographics and clinical variables

variables will be considered as form a sub-study. The total number of sub-studies, K, is the total number of block patterns of variables. For example, in EMBARC study presented in Fig. 4, there are 2 sub-studies. Standard approach will be a complete-case analysis where only study 0 subjects ($n = 138$) will be included to learn a high-resolution ITR. In our approach, study 1 ($n = 104$) provides a coarsened ITR $\widetilde{f}_k(\boldsymbol{Z}_1)$ based on the available covariates $\boldsymbol{Z}_1$; using study 0 subjects with complete covariates $\boldsymbol{X}$ and incorporating $\widetilde{f}_k(\boldsymbol{Z}_1)$ as auxiliary information, methods in Sect. 2.2.2 based on the regularized value function (3) in the main paper provide an integrative high-resolution ITR. Compared to standard complete-case analysis using study 0 subjects only, the integrative learning offers an opportunity to use all collected data efficiently. Similarly, methods in Sect. 2.2.2 can be used to learn an integrative low-resolution rule if constructing simple ITRs without fMRI and BP measures is desirable.

Baysian Rules for Integrative Learning

In this section, we provide details to derive Baysian Rules in Sect. 2.3.

We first consider the ITR learned from the integrative learning method when the low-resolution trials are integrated to the high-resolution trial. For given λ_k's, it is clear that the Bayesian rule for learning integrative high-resolution ITR maximizes

$$E\left[\left\{R+\sum_{k=1}^{K}\lambda_k\delta_k(\boldsymbol{Z}_k)\left(2I(Af_k(\boldsymbol{Z}_k)>0)-1\right)\right\}\frac{I(Af_0(\boldsymbol{X})>0)}{\pi(A|\boldsymbol{X})}\right]$$

over $f_0(\boldsymbol{X})$, where $E_0(\cdot)$ denotes the expectation under the joint distribution of $(R, \boldsymbol{X})$ in study 0, and $\pi(a|\boldsymbol{X}) = P(A = a|\boldsymbol{X})$ in study 0. Using the results from the outcome-weighted learning (Zhao et al., 2012), we immediately obtain that the Bayesian rule, denoted by $f_0^*(\boldsymbol{X})$, satisfies

$$\begin{aligned}
\operatorname{sign}(f_0^*(\boldsymbol{X})) &= \operatorname{sign}\left(E\left\{R+\sum_{k=1}^{K}\lambda_k\delta_k(\boldsymbol{Z}_k)\left(2I(Af_k(\boldsymbol{Z}_k)>0)-1\right)\Big|A=1,\boldsymbol{X}\right\}\right. \\
&\qquad \left.-E\left\{R+\sum_{k=1}^{K}\lambda_k\delta_k(\boldsymbol{Z}_k)\left(2I(Af_k(\boldsymbol{Z}_k)>0)-1\right)\Big|A=-1,\boldsymbol{X}\right\}\right) \\
&= \operatorname{sign}\left(E[R|A=1,\boldsymbol{X}]-E[R|A=-1,\boldsymbol{X}]\right. \\
&\qquad \left.+\sum_{k=1}^{K}\lambda_k\delta_k(\boldsymbol{Z}_k)\operatorname{sign}(f_k(\boldsymbol{Z}_k))\right).
\end{aligned}$$

Note that $\delta_k(\mathbf{Z}_k)\text{sign}(f_k(\mathbf{Z}_k)) = E_k[R_k|A_k = 1, \mathbf{Z}_k] - E_k[R_k|A_k = -1, \mathbf{Z}_k]$. Thus, the Bayesian rule is

$$\text{sign}\left(E[R|A=1,\boldsymbol{X}] - E[R|A=-1,\boldsymbol{X}]\right.$$
$$\left. + \sum_{k=1}^{K} \lambda_k (E[R_k|A_k = 1, \mathbf{Z}_k] - E[R|A_k = -1, \mathbf{Z}_k])\right),$$

which is equivalent to

$$\text{sign}\left(E[R(1)|\boldsymbol{X}] - E[R(-1)|\boldsymbol{X}] + \sum_{k=1}^{K} \lambda_k (E[R(1)|\mathbf{Z}_k] - E[R(-1)|\mathbf{Z}_k])\right).$$

We next study the ITR from the integrative learning when using the high-resolution trial to improve the ITR from the coarsened trial. Using the same argument as before, we obtain that the Bayesian rule for the low-resolution study k maximizes

$$E\left[\frac{R_k}{\pi_k(A_k|\mathbf{Z}_k)} I(A_k f_k(\mathbf{Z}_k) > 0)\right]$$
$$+\lambda_k E\left[\frac{\delta_0(\boldsymbol{X})}{\pi(A|\boldsymbol{X})}\text{sign}(Af_0(\boldsymbol{X}))I(Af_k(\mathbf{Z}_k) > 0)\right]$$
$$= E\left[\frac{R_k}{\pi_k(A_k|\mathbf{Z}_k)} I(A_k f_k(\mathbf{Z}_k) > 0)\right]$$
$$+\lambda_k E\left[\frac{\delta_0(\boldsymbol{X})}{\pi(A|\boldsymbol{X})}\frac{g_0(\mathbf{Z}_k)}{g_k(\mathbf{Z}_k)}\text{sign}(Af_0(\boldsymbol{X}))I(Af_k(\mathbf{Z}_k) > 0)\right],$$

where $g_0(\cdot)$ and $g_k(\cdot)$ are the density functions for $\mathbf{Z}_k$ in study 0 and study k, respectively. This expression, up to a constant independent of f_k, is equivalent to

$$E_k\left[\{E[R_k|A_k = 1, \mathbf{Z}_k] - E[R_k|A_k = -1, \mathbf{Z}_k]\}\,\text{sign}(f_k(\mathbf{Z}_k))\right]$$
$$+ \lambda_k E\left[\delta_0(\boldsymbol{X})\frac{g_0(\mathbf{Z}_k)}{g_k(\mathbf{Z}_k)}\text{sign}(f_0(\boldsymbol{X}))\text{sign}(f_k(\mathbf{Z}_k))\right].$$

Therefore, the Bayesian rule for the integrative learning of low-resolution rule is

$$\text{sign}(f_k^*(\mathbf{Z}_k)) = \text{sign}\big(E[R_k|A_k = 1, \mathbf{Z}_k] - E[R_k|A_k = -1, \mathbf{Z}_k]$$
$$+ \lambda_k \frac{g_0(\mathbf{Z}_k)}{g_k(\mathbf{Z}_k)}(E[R|A = 1, \mathbf{Z}_k] - E[R|A = -1, \mathbf{Z}_k])\big).$$

Simulation Studies of Sensitivity Analysis

In this setting, we perform sensitivity analysis where the simulation models of R given $\boldsymbol{X}$ is different across trials and the validation population. The clinical outcome R is generated by

$$R = \phi(\boldsymbol{X}) * A + 0.5 * X_1 + \epsilon, \quad \epsilon \sim N(0, 0.25)$$
$$\phi(\boldsymbol{X}) = X_1 - 0.5X_2 - \alpha X_3 - \beta X_4 + 0.3(X_5^2 > 0.64) - 0.3(X_6 > 0). \qquad (6)$$

In the high-resolution study 0, $X_1, \cdots, X_5$ are observed, and X_6 is an unobserved latent variable. The interaction effects of X_3 and X_4 were both $\alpha = \beta = 0.8$. In the low-resolution study 1, only X_1, X_2 and X_3 are collected and used to estimate a coarsened ITR. The interaction effects of X_3 and X_4 were set to $\alpha = \beta = 0.1$, reflecting that both X_1 and X_4 contribute less to the selection of optimal treatment. All covariates $X_1, \cdots, X_6$ were generated independently from a standard normal distribution. Treatments were randomly assigned for each subject with equal probability.

In the validation data, we simulated a more general population with different treatment interaction effects than the training data collected from study 0 and study 1. For validation settings 1 to 4, we set $\alpha = \beta = 0.3, 0.4, 0.5$, or 0.6. Similar to Sect. 3, we first learn integrative high-resolution ITR from study 0 using coarsened ITR estimated from study 1, and computed performance measures using validation data. Results from 100 replications are shown in Table 4 and Fig. 5. In validation setting 1, ABLO-L performs better than ABLO-H with a larger benefit and accuracy, but worse than Int-Ben and Int-Equal. This is because the interaction effect of X_3 and X_4 in the validation population is 0.3, which is closer to the sample collected

Table 4 Mean and standard deviation of overall benefit, value and accuracy rates for integrative learning of high-resolution ITR comparing to ITRs by ABLO on single studies

		ABLO-H	ABLO-L	Int-Equal	Int-Ben
Benefit	Setting 1	1.757(0.057)	1.824(0.028)	1.849(0.063)	1.868(0.047)
	Setting 2	1.874(0.044)	1.816(0.037)	1.935(0.041)	1.951(0.029)
	Setting 3	2.014(0.032)	1.779(0.053)	2.042(0.028)	2.056(0.019)
	Setting 4	2.190(0.029)	1.821(0.059)	2.195(0.044)	2.209(0.027)
Value	Setting 1	0.887(0.027)	0.913(0.014)	0.932(0.030)	0.941(0.022)
	Setting 2	0.938(0.022)	0.901(0.019)	0.967(0.020)	0.975(0.014)
	Setting 3	1.015(0.018)	0.901(0.027)	1.031(0.014)	1.038(0.010)
	Setting 4	1.089(0.016)	0.905(0.029)	1.092(0.023)	1.098(0.014)
Accuracy	Setting 1	0.852(0.020)	0.880(0.012)	0.890(0.029)	0.898(0.022)
	Setting 2	0.875(0.018)	0.848(0.013)	0.904(0.020)	0.913(0.017)
	Setting 3	0.903(0.017)	0.820(0.015)	0.919(0.016)	0.928(0.012)
	Setting 4	0.921(0.014)	0.797(0.013)	0.923(0.021)	0.930(0.015)

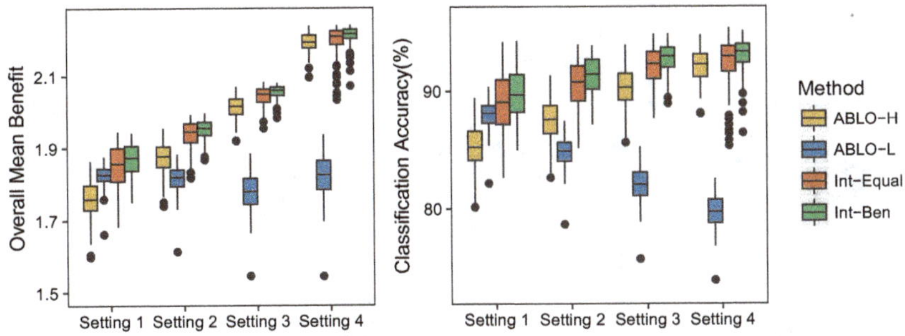

Fig. 5 ITR benefit and optimal treatment allocation accuracy for the four methods. See Fig. 1. ABLO-H is estimated by ABLO using high-resolution data only; ABLO-L is estimated by ABLO using low-resolution data only; Int-Equal is estimated by integrative learning for high-resolution ITR with equal weights for all subjects; Int-Ben is estimated by integrative learning for high-resolution ITR weighted by estimated benefit

Table 5 Mean and standard deviation of overall benefit, value and accuracy rates for integrative learning of low-resolution ITR comparing to ITRs by ABLO on single studies

		ABLO-L	Int-Equal	Int-Ben
Benefit	Setting 1	1.794(0.029)	1.816(0.020)	1.821(0.018)
	Setting 2	1.853(0.041)	1.895(0.029)	1.911(0.021)
	Setting 3	1.844(0.051)	1.903(0.042)	1.927(0.032)
	Setting 4	1.844(0.065)	1.922(0.056)	1.956(0.043)
Value	Setting 1	0.876(0.017)	0.890(0.011)	0.895(0.009)
	Setting 2	0.929(0.021)	0.950(0.014)	0.958(0.010)
	Setting 3	0.955(0.025)	0.983(0.021)	0.995(0.016)
	Setting 4	0.919(0.032)	0.958(0.028)	0.976(0.022)
Accuracy	Setting 1	0.881(0.012)	0.890(0.009)	0.891(0.008)
	Setting 2	0.857(0.014)	0.871(0.010)	0.876(0.008)
	Setting 3	0.825(0.013)	0.840(0.011)	0.847(0.009)
	Setting 4	0.806(0.015)	0.826(0.015)	0.835(0.012)

from study 1 than study 0. ABLO-L is more efficient compared to other methods with a smaller standard deviation in estimating benefit and value function. This is because the coarsened ITR is simpler than high-resolution ITR and with fewer model assumptions, and therefore providing less variable ITRs. However, when the treatment effect increases in setting 2 to 4, integrative learning Int-Ben performs the best with the largest benefit and accuracy, and with the smallest standard deviation.

Next, we learn integrative low-resolution ITR (study 1) using high-resolution ITR estimated from study 0. Results are shown in Table 5 and Fig. 6. In all validation settings, Int-Ben performs the best in terms of a larger estimated benefit and a smaller standard deviation by borrowing information from the high-resolution ITR. Overall, the ITR benefit becomes less efficient for all methods from setting 1 to 4 even though the treatment effect increases. This is because the validation data

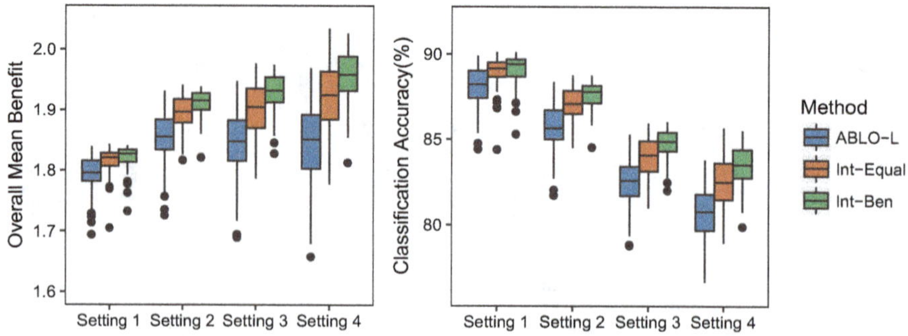

Fig. 6 ITR benefit and optimal treatment allocation accuracy for the three methods. See Fig. 4

Table 6 Overall performance of the four methods to handle blockwise variable collection in EMBARC study

Method	ABLO-L	ABLO-H	Int-Equal	Int-Ben
Value	7.22(1.44)	7.63(1.40)	7.67(1.35)	7.95(1.37)
Benefit	1.25(2.21)	2.04(1.98)	2.11(1.97)	2.68(2.02)

becomes more and more different to the study where the low-resolution ITR was estimated.

EMBARC Analysis with Blockwise Data Collection

Here, we present an application to handle covariates collected in block-wise fashion when learning an ITR. As seen in Fig. 4, 104 subjects in EMBARC had complete demographics and clinical measures but did not have neuroimaging and behavioral phenotyping measures. In the following, we will use these 104 subjects as study 1 with low-resolution variables to improve the performance of the high-resolution ITR learned from the remaining 138 subjects with complete covariates from all four domains (study 0). The coarsened ITR learned from study 1 includes all the clinical measures at baseline, i.e., age, gender, race, baseline QIDS score, and HAM-D score, while the high-resolution ITR includes the fMRI and behavioral phenotyping measures. We repeated the same procedure as before and report the results in Table 6 and Fig. 7. The ITR benefit of Int-Ben (2.68) is much larger than using complete case only (2.04) and the coarsened rule (1.25). The coarsened ITR learned from EMBARC included more covariates and more subjects than HEAL, which leads to a greater improvement in the performance of the integrative ITR than using HEAL coarsened ITR.

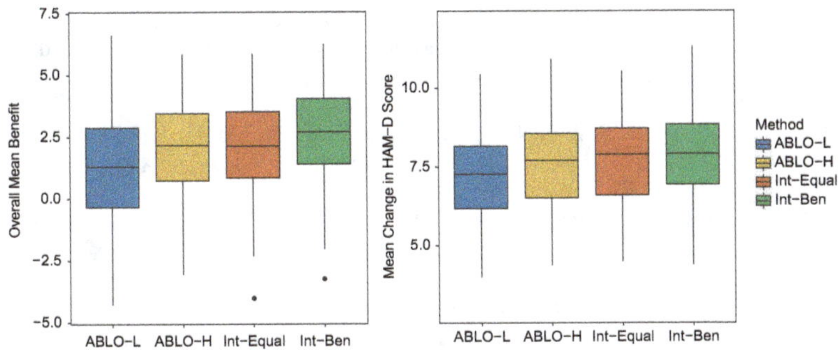

Fig. 7 Performance of the four methods to handle blockwise variable collection in EMBARC. Distribution of estimated ITR benefit (the higher the better) and change in HAM-D score (the larger the better) at the end of stage 1 treatment for four methods (based on 100 cross-validation)

References

Brown, C. H., Brincks, A., Huang, S., Perrino, T., Cruden, G., Pantin, H., Howe, G., Young, J. F., Beardslee, W., Montag, S., et al. (2018). Two-year impact of prevention programs on adolescent depression: An integrative data analysis approach. *Prevention Science, 19*(1), 74–94.

Cipriani, A., Furukawa, T. A., Salanti, G., Chaimani, A., Atkinson, L. Z., Ogawa, Y., Leucht, S., Ruhe, H. G., Turner, E. H., Higgins, J. P., et al. (2018). Comparative efficacy and acceptability of 21 antidepressant drugs for the acute treatment of adults with major depressive disorder: A systematic review and network meta-analysis. *The Lancet, 391*(10128), 1357–1366.

Curran, P. J., & Hussong, A. M. (2009). Integrative data analysis: The simultaneous analysis of multiple data sets. *Psychological Methods, 14*(2), 81–100.

Dagne, G. A., Brown, C. H., Howe, G., Kellam, S. G., & Liu, L. (2016). Testing moderation in network meta-analysis with individual participant data. *Statistics in Medicine, 35*(15), 2485–2502.

Haidich, A.-B. (2010). Meta-analysis in medical research. *Hippokratia, 14*(Suppl 1), 29–37.

Holmes, M. V., Shah, T., Vickery, C., Smeeth, L., Hingorani, A. D., & Casas, J. P. (2009). Fulfilling the promise of personalized medicine? Systematic review and field synopsis of pharmacogenetic studies. *PLoS One, 4*(12), e7960.

Justice, A. C., Covinsky, K. E., & Berlin, J. A. (1999). Assessing the generalizability of prognostic information. *Annals of Internal Medicine, 130*(6), 515–524.

Liu, C., Xu, X., & Hu, D. (2015). Multiobjective reinforcement learning: A comprehensive overview. *IEEE Transactions on Systems, Man, and Cybernetics: Systems, 45*(3), 385–398.

Liu, Y., Wang, Y., Kosorok, M. R., Zhao, Y., & Zeng, D. (2018). Augmented outcome-weighted learning for estimating optimal dynamic treatment regimens. *Statistics in Medicine, 37*(26), 3776–3788.

Ma, S., Huang, J., & Song, X. (2011). Integrative analysis and variable selection with multiple high-dimensional data sets. *Biostatistics, 12*(4), 763–775.

McAllester, D. A., & Keshet, J. (2011). Generalization bounds and consistency for latent structural probit and ramp loss. In *Neural Information Processing Systems* (pp. 2205–2212).

Moodie, E. E., Richardson, T. S., & Stephens, D. A. (2007). Demystifying optimal dynamic treatment regimes. *Biometrics, 63*(2), 447–455.

Qian, M., & Murphy, S. A. (2011). Performance guarantees for individualized treatment rules. *Annals of Statistics, 39*(2), 1180–1210.

Qiu, X., Zeng, D., & Wang, Y. (2018). Estimation and evaluation of linear individualized treatment rules to guarantee performance. *Biometrics, 72*(2), 517–528.

Rush, A. J., Fava, M., Wisniewski, S. R., Lavori, P. W., Trivedi, M. H., Sackeim, H. A., Thase, M. E., Nierenberg, A. A., Quitkin, F. M., Kashner, T. M., et al. (2004). Sequenced treatment alternatives to relieve depression (STAR*D): rationale and design. *Controlled Clinical Trials, 25*(1), 119–142.

Shear, M. K., Reynolds, C. F., Simon, N. M., Zisook, S., Wang, Y., Mauro, C., Duan, N., Lebowitz, B., & Skritskaya, N. (2016). Optimizing treatment of complicated grief: A randomized clinical trial. *JAMA Psychiatry, 73*(7), 685–694.

Siddique, J., de Chavez, P. J., Howe, G., Cruden, G., & Brown, C. H. (2018). Limitations in using multiple imputation to harmonize individual participant data for meta-analysis. *Prevention Science, 19*(1), 95–108.

Stroup, T. S., Lieberman, J. A., McEvoy, J. P., Swartz, M. S., Davis, S. M., Rosenheck, R. A., Perkins, D. O., Keefe, R. S., Davis, C. E., Severe, J., et al. (2006). Effectiveness of olanzapine, quetiapine, risperidone, and ziprasidone in patients with chronic schizophrenia following discontinuation of a previous atypical antipsychotic. *American Journal of Psychiatry, 163*(4), 611–622.

Tonin, F. S., Rotta, I., Mendes, A. M., & Pontarolo, R. (2017). Network meta-analysis: A technique to gather evidence from direct and indirect comparisons. *Pharmacy Practice (Granada), 15*(1), 943.

Trivedi, M. H., McGrath, P. J., Fava, M., Parsey, R. V., Kurian, B. T., Phillips, M. L., Oquendo, M. A., Bruder, G., Pizzagalli, D., Toups, M., et al. (2016). Establishing moderators and biosignatures of antidepressant response in clinical care (EMBARC): Rationale and design. *Journal of Psychiatric Research, 78*, 11–23.

Trivedi, M. H., Rush, A. J., Wisniewski, S. R., Nierenberg, A. A., Warden, D., Ritz, L., Norquist, G., Howland, R. H., Lebowitz, B., McGrath, P. J., et al. (2006). Evaluation of outcomes with citalopram for depression using measurement-based care in STAR*D: Implications for clinical practice. *American Journal of Psychiatry, 163*(1), 28–40.

Whiteford, H. A., Degenhardt, L., Rehm, J., Baxter, A. J., Ferrari, A. J., Erskine, H. E., Charlson, F. J., Norman, R. E., Flaxman, A. D., Johns, N., et al. (2013). Global burden of disease attributable to mental and substance use disorders: Findings from the global burden of disease study 2010. *The Lancet, 382*(9904), 1575–1586.

Xiang, S., Yuan, L., Fan, W., Wang, Y., Thompson, P. M., & Ye, J. (2013). Multi-source learning with block-wise missing data for alzheimer's disease prediction. In *Proceedings of the 19th ACM SIGKDD International Conference on Knowledge Discovery and Data Mining* (pp. 185–193). ACM.

Zhao, Y., Zeng, D., Rush, A. J., & Kosorok, M. R. (2012). Estimating individualized treatment rules using outcome weighted learning. *Journal of the American Statistical Association, 107*(499), 1106–1118.

Adaptive Semi-supervised Inference for Optimal Treatment Decisions with Electronic Medical Record Data

Kevin Gunn, Wenbin Lu, and Rui Song

Abstract A treatment regime is a rule that assigns a treatment to patients based on their covariate information. Recently, estimation of the optimal treatment regime that yields the greatest overall expected clinical outcome of interest has attracted a lot of attention. In this work, we consider estimation of the optimal treatment regime with electronic medical record data under a semi-supervised setting. Here, data consist of two parts: a set of 'labeled' patients for whom we have the covariate, treatment and outcome information, and a much larger set of 'unlabeled' patients for whom we only have the covariate information. We proposes an imputation-based semi-supervised method, utilizing 'unlabeled' individuals to obtain a more efficient estimator of the optimal treatment regime. The asymptotic properties of the proposed estimators and their associated inference procedure are provided. Simulation studies are conducted to assess the empirical performance of the proposed method and to compare with a fully supervised method using only the labeled data. An application to an electronic medical record data set on the treatment of hypotensive episodes during intensive care unit (ICU) stays is also given for further illustration.

Keywords Electronic medical record data · Kernel regression · Optimal treatment decision · Semi-supervised learning

1 Introduction

Precision medicine, which is focused on developing individualized treatment decision rules by utilizing patients' covariate information, has earned considerable interest. A treatment decision rule is a deterministic function that assigns a treatment to patients according to their covariates. One main objective of precision medicine

K. Gunn · W. Lu (✉) · R. Song
Department of Statistics, North Carolina State University, Raleigh, NC, USA
e-mail: kpgunn@ncsu.edu; wlu4@ncsu.edu

Y. Zhao, D.-G. Chen (eds.), *Statistics in Precision Health*, ICSA Book Series in Statistics, https://doi.org/10.1007/978-3-031-50690-1_9

is to find the optimal treatment rule (OTR) that maximizes patients' expected outcomes of interest. There is a great deal of work on estimation of optimal treatment rules from randomized clinical trials or observational studies (e.g. Watkins & Dayan, 1992; Murphy, 2003; Robins, 2004; Moodie et al., 2007; Zhao et al., 2012; Zhang et al., 2012; Fan et al., 2017).

A recent area of interest in estimation of optimal treatment rules has been the utilization of Electronic Medical Record (EMR) data. For example, individualized treatment decisions are particularly suited to the ICU due to patients' heterogeneity for treating diseases like sepsis and acute kidney injury (Maslove et al., 2017). EMR data allow researchers to explore optimal treatment rules in specific clinical scenarios not feasible in randomized clinical trials (RCT) such as patients suffering from sepsis (Raghu et al., 2017). They can also provide guidance where there is little previous research conducted, such as second line treatment choices for type 2 diabetes episodes (Wang et al., 2016). The recent shift in hospitals and other healthcare organizations to store clinical information in EMRs provide investigators with an immense amount of detailed information on the health status and clinical process of patients during interactions with their healthcare system. For example, MIMIC-III (Johnson et al., 2016) is an openly available critical care database that encompasses medical record chart data recorded by clinicians, such as demographics, medical diagnoses, medications, vital signs, laboratory measurements (obtained in and out of the hospital), progress notes, imaging results, and billing information. The wealth of detailed clinical data provides the opportunity to enhance clinical decisions for accurate personalized medicine.

Although secondary analysis of EMR data can serve as an important tool in improving patient outcomes, a major challenge in utilizing EMR data for estimation of OTR arises from the missingness and measurement errors in patients' records since EMR data are usually noisy and not recorded for research purposes (Weiskopf et al., 2013; Beaulieu-Jones et al., 2018). In this paper, we consider the estimation of OTR under a semi-supervised setting using EMR data. Specifically, data consist of two parts: a set of 'labeled' patients for whom we have the covariate, treatment and outcome information, and a much larger set of 'unlabeled' patients for whom we only have the covariate information.

In classical semi-supervised learning, the knowledge gained through $\mathbb{P}_{\mathbf{X}}$ (distribution of covariates) from 'unlabeled' individuals is incorporated to improve inference on $\mathbb{P}_{\mathbf{Y}|\mathbf{X}}$ (Chapelle et al., 2006; Zhu & Goldberg, 2009). When the amount of missing information far exceeds the amount of available information, semi-supervised learning techniques have been used for improving the prediction with EMR data. For example, Chakrabortty and Cai (2018) proposed semi-supervised learning techniques to improve estimation efficiency of regression parameters in a linear working model using EMR data for studying the association between biomarkers and rheumatoid arthritis. Beaulieu-Jones et al. (2016), Garla et al. (2013) and Hong et al. (2019) developed semi-supervised methods for more accurate phenotype prediction when only a small subset of the patients underwent a manual chart review by a physician.

When estimating the OTR under the semi-supervised setting, we propose to use the covariate information in unlabeled data to derive the estimation of the contrast function, i.e. $E\{\mathbf{Y}|\mathbf{X}, \mathbf{A} = 1\} - E\{\mathbf{Y}|\mathbf{X}, \mathbf{A} = 0\}$, which in turn improves the estimation of the linear decision rule based on a semiparametric working model. Specifically, the proposed semi-supervised (SS) method is a two-step estimation procedure. In the first step, we develop a kernel estimator of the contrast function based on the labeled data. In the second step, the contrast function is imputed for unlabeled data based on their covariate information, and then a linear decision rule is estimated based on the imputed contrast function.

The article is organized as follows. Section 2 provides an overview for optimal treatment decision using EMR data. The framework and assumptions are also presented here. An estimation procedure for the optimal linear treatment regime using the labeled data only is introduced in Sect. 3. The proposed semi-supervised learning method using both labeled and unlabeled data is introduced in Sect. 4. Simulation studies are conducted in Sect. 5 to compare the empirical performance of the methods introduced in Sects. 3 and 4. An application to an EMR data set for studying patients in the intensive care unit (ICU) undergoing a hypotensive episode is given in Sect. 6. The paper is concluded with discussions. All the proofs are provided in the Appendix.

2 Optimal Treatment Decision with EMR Data

When estimating optimal treatment regime using EMR data, a challenge arises from that a large portion of outcomes of interest and/or treatment information may not be available. For example, Lee et al. (2012) performed a retrospective study on patients in MIMIC-III that underwent a single hypotensive episode during their ICU stay. The goal is to examine outcomes of patients receiving the IV fluid resuscitation or vasoactive therapy. In this study, the treatment information is not recorded for many subjects, despite one of the treatments was typically used as a first line of therapy against hypotension (Kellum et al., 2012). In addition to the missingness in the treatment information, one primary outcome of interest, serum creatinine, may also be missing, even though it is a marker for kidney damage, a concern for patients with hypotension. In this work, we will develop a semi-supervised learning method for estimating optimal treatment regime when missingness occurs in the response and/or treatment for a large proportion of subjects.

Next, we present the framework and assumptions for the considered semi-supervised optimal treatment decision problem. Let $A \in \mathcal{A}$ denote the treatments received by a patient, where $\mathcal{A}$ is the set of possible treatments. Here, we only consider binary treatment options, i.e. $\mathcal{A} = \{0, 1\}$. Let $\mathbf{X} \in \mathcal{X} \subseteq \mathbb{R}^p$ be a vector of covariates ascertained prior to treatment. Let $Y \in \mathbb{R}$ denote the outcome of interest. Here, a larger value of Y indicates a better outcome. The observed data consist two parts: a set of fully observed data, namely the labeled data,

$\mathcal{O} = \left\{\mathcal{O}_i = (Y_i, A_i, \mathbf{X}_i) : i = 1, \ldots, n\right\}$; and a set of partially observed data, namely the unlabeled data, $\mathcal{U} = \left\{\mathbf{X}_j : j = n+1, \ldots, n+N\right\}$. Here, N can be much larger than n. In this work, we assume that the response and treatment information in $\mathcal{U}$ are missing completely at random (MCAR).

Let $Y^*(a)$ denote the potential outcome that would be observed if a patient were given treatment a, where $a = 1$ or 0. We make the following standard assumptions in causal inference literature (Rubin, 1978; Rosenbaum & Rubin, 1983): (i) *stable unit treatment value assumption (SUTVA)* $Y_i = A_i Y^*(1) + (1 - A_i) Y^*(0)$; (ii) *no unmeasured confounders* $\{Y^*(1), Y^*(0)\} \perp A \mid \mathbf{X}$; (iii) *positivity assumption* $0 < \pi(\mathbf{X}) < 1$ for any $\mathbf{X}$, where $\pi(\mathbf{X}) = P(A = 1 \mid \mathbf{X})$ denotes the propensity score.

A treatment regime is a function that maps the covariates onto the treatment space, $\delta(\mathbf{X}) : \mathcal{X} \to \mathcal{A}$. Define $Y^*(\delta) = \delta(\mathbf{X})Y^*(1) + \{1 - \delta(\mathbf{X})\}Y^*(0)$. The optimal treatment regime is defined as $\delta^{opt} = \arg\max_{\delta \in \Delta} E\{Y^*(\delta)\}$, where Δ is the class of interested treatment regimes.

3 Optimal Treatment Decision Using Labeled Data Only

We assume the following model for the Q-function, i.e.

$$Q(A, \mathbf{X}) = E(Y \mid \mathbf{X}, A) = \mu(\mathbf{X}) + AC(\mathbf{X}),$$

where $\mu(\mathbf{X})$ is the baseline mean function and $C(\mathbf{X})$ is the contrast function. Here, both functions $\mu(\mathbf{X})$ and $C(\mathbf{X})$ are unspecified. Under the assumptions (i) and (ii), the optimal treatment regime is given by $\delta^{opt}(\mathbf{X}) = I\{C(\mathbf{X}) > 0\}$.

In this work, we are interested in a class of linear treatment regime $\delta(\mathbf{X}) = I(\boldsymbol{\beta}'\widetilde{\mathbf{X}} > 0)$, where $\widetilde{\mathbf{X}} = \left(1, \mathbf{X}'\right)'$, because they are easily interpretable and implementable in practical applications. In other words, we posit the following working model: $E(Y \mid \mathbf{X}, A) = \mu(\mathbf{X}) + A(\boldsymbol{\beta}'\widetilde{\mathbf{X}})$. To estimate the parameters $\boldsymbol{\beta}$ in the working model, we adopt a similar approach in Song et al. (2017) based on transformed responses. Specifically, define

$$\widetilde{Y}_i = \frac{Y_i\{A_i - \hat{\pi}(\mathbf{X}_i)\}}{\hat{\pi}(\mathbf{X}_i)\{1 - \hat{\pi}(\mathbf{X}_i)\}}, \qquad i = 1, \cdots, n,$$

where $\hat{\pi}(\cdot)$ is a consistent estimate for the propensity score. For example, we can posit a logistic regression model for the propensity score and $\hat{\pi}$ can be obtained by the maximum likelihood estimation using the labeled data.

When $\hat{\pi}$ is replaced by the true propensity score in $\tilde{Y}_i$, it can be shown that $E(\tilde{Y}_i|\mathbf{X}_i) = C(\mathbf{X}_i)$. Therefore, an estimator for $\boldsymbol{\beta}$ in the working model can be obtained by

$$\widehat{\boldsymbol{\beta}}_{TR} = \arg\min_{\beta} \sum_{i=1}^{n} \left(\widetilde{Y}_i - \boldsymbol{\beta}'\widetilde{\mathbf{X}}_i\right)^2. \tag{1}$$

Define $\boldsymbol{\beta}^* = \arg\min_{\boldsymbol{\beta}} E\{C(\mathbf{X}_i) - \boldsymbol{\beta}'\widetilde{\mathbf{X}}_i\}^2$. Following the similar arguments in Lu et al. (2013), it can be shown that

$$\sqrt{n}\left(\widehat{\boldsymbol{\beta}}_{TR} - \boldsymbol{\beta}^*\right) = n^{-1/2}\sum_{i=1}^{n} \Psi_{TR}\left(\mathcal{O}_i\right) + o_p(1), \tag{2}$$

where $\Psi_{TR}(\mathcal{O}_i)$ are i.i.d. random vectors with mean zero. It follows that $\sqrt{n}\left(\widehat{\boldsymbol{\beta}}_{TR} - \boldsymbol{\beta}^*\right)$ converges in distribution to a normal random vector with mean zero and variance $\mathcal{V}_{\boldsymbol{\beta}_{TR}} = E\left\{\Psi_{TR}\left(\mathcal{O}_i\right)\Psi_{TR}'\left(\mathcal{O}_i\right)\right\}$. Here, $\mathcal{V}_{\boldsymbol{\beta}_{TR}}$ can be consistently estimated by

$$\widehat{\mathcal{V}}_{\boldsymbol{\beta}_{TR}} = n^{-1}\sum_{i=1}^{n} \widehat{\Psi}_{TR}\left(\mathcal{O}_i\right)\widehat{\Psi}_{TR}'\left(\mathcal{O}_i\right),$$

where $\widehat{\Psi}_{TR}(\mathcal{O}_i)$ is a consistent estimate of $\Psi_{TR}(\mathcal{O}_i)$ that can be obtained by the plug-in method.

4 Semi-supervised Learning for Optimal Treatment Decision

The proposed semi-supervised learning consists of two steps: (i) compute the kernel estimators of $Q(\mathbf{X}, a)$ for $a = 0, 1$ based on labeled data; (ii) impute contrast functions, $C(\mathbf{X}_j) = Q\left(\mathbf{X}_j, 1\right) - Q\left(\mathbf{X}_j, 0\right)$, $j = n+1, \ldots, N$, for unlabeled data, and regress the imputed contrast functions on $\mathbf{X}_j$'s to obtain a semi-supervised estimator for $\boldsymbol{\beta}$.

4.1 Nonparametric Imputation of the Q-Functions

The kernel estimators of $Q(\mathbf{X}, a)$ are given by

$$\widehat{Q}^{(np)}(\mathbf{x}, 1) = \frac{h^{-p}\sum_{i=1}^{n} W\left(\frac{\mathbf{x}-\mathbf{X_i}}{h}\right) A_i Y_i}{h^{-p}\sum_{i=1}^{n} W\left(\frac{\mathbf{x}-\mathbf{X_i}}{h}\right) A_i} \tag{3}$$

and,

$$\widehat{Q}^{(np)}(\mathbf{x}, 0) = \frac{h^{-p}\sum_{i=1}^{n} W\left(\frac{\mathbf{x}-\mathbf{X_i}}{h}\right)(1-A_i)Y_i}{h^{-p}\sum_{i=1}^{n} W\left(\frac{\mathbf{x}-\mathbf{X_i}}{h}\right)(1-A_i)}, \tag{4}$$

where $W(\cdot) : \mathbb{R}^p \to \mathbb{R}$ is a suitable kernel function and h is a bandwidth that converges to zero as n goes to infinity. Define $\widehat{C}^{(np)}(\mathbf{x}) = \widehat{Q}^{(np)}(\mathbf{x}, 1) - \widehat{Q}^{(np)}(\mathbf{x}, 0)$. The estimate of $\boldsymbol{\beta}$ based on imputed contrast functions for unlabeled data is given by

$$\widehat{\boldsymbol{\beta}}_{np} = \arg\min_{\beta} \sum_{j=n+1}^{n+N} \left\{\widehat{C}^{(np)}(\mathbf{X}_j) - \boldsymbol{\beta}'\widetilde{\mathbf{X}}_j\right\}^2. \tag{5}$$

The corresponding linear treatment regime is $\widehat{\delta}_{np}(\mathbf{X}) = I(\widehat{\boldsymbol{\beta}}'_{np}\widetilde{\mathbf{X}} > 0)$.

Next, we establish the asymptotic distribution of $\widehat{\boldsymbol{\beta}}_{np}$. The subsequent assumptions are needed for Theorem 1 given below. As in the literature for kernel estimation (Fan, 1992; Hansen, 2008; Newey, 1994; Chakrabortty & Cai, 2018), we assume the following conditions.

(C1) The Q-functions $Q(\mathbf{x}, a)$, the density function $f(\mathbf{X})$ of $\mathbf{X}$ and the propensity score $\pi(\mathbf{x})$ are r times continuously differentiable with bounded rth derivatives on $\mathbf{X} \in \mathcal{X}$.

(C2) The function $W(\cdot)$ is a symmetric r-th order kernel for some integer $r \geq 2$. In addition, $W(\cdot)$ is Lipschitz continuous and has bounded support.

(C3) $E(|AY|^s) < \infty$ and $E(|(1-A)Y|^s) < \infty$ for some $s > 2$.

(C4) $E(|AY|^s \mid \mathbf{X} = \mathbf{x})f(\mathbf{x}) = E(|Y|^s \mid \mathbf{X} = \mathbf{x})\pi(\mathbf{x})f(\mathbf{x})$ and $f(\mathbf{x})$ are bounded on $\mathcal{X}$ and $\inf_{\mathbf{x}\in\mathcal{X}} f(\mathbf{x}) > 0$.

These conditions are assumed to ensure the desired convergence rate of the kernel estimator of the contrast function such that the asymptotic normality of $\widehat{\boldsymbol{\beta}}_{np}$ can be established.

Theorem 1 *Suppose $n^{1/2}h^r \to 0$ and $\sqrt{\frac{\log n}{nh^p}} \to 0$ as $h \to 0$, $n, N \to \infty$ and $n/N \to 0$. Then, under Conditions (C1)-(C4), we have*

$$n^{1/2}\left(\widehat{\boldsymbol{\beta}}_{np} - \boldsymbol{\beta}^*\right) = n^{-1/2}\sum_{i=1}^{n} \Psi_{np}(\mathcal{O}_i) + o_p(1),$$

where $\Psi_{np}(\mathcal{O}_i) = \left\{\frac{A_i}{\pi(\mathbf{X}_i)} - \frac{1-A_i}{1-\pi(\mathbf{X}_i)}\right\}\Lambda^{-1}\widetilde{\mathbf{X}}_i\{Y_i - Q(\mathbf{X}_i, A_i)\}$ and $\Lambda = E(\widetilde{\mathbf{X}}_j\widetilde{\mathbf{X}}'_j)$.

Theorem 1 implies that $n^{1/2}(\widehat{\boldsymbol{\beta}}_{np} - \boldsymbol{\beta}^*)$ converges in distribution to a multivariate normal $N_{p+1}(0, \mathcal{V}_{\boldsymbol{\beta}_{np}})$ with $\mathcal{V}_{\boldsymbol{\beta}_{np}} = E\left\{\Psi_{np}(\mathcal{O}_i)\Psi'_{np}(\mathcal{O}_i)\right\}$. The asymptotic variance matrix $\mathcal{V}_{\boldsymbol{\beta}_{np}}$ can be consistently estimated by the usual plug-in method.

As discussed in Fan (1992), kernel estimates may suffer finite sample bias especially when sample size is small. This in turn will cause relatively large finite sample bias in $\widehat{\boldsymbol{\beta}}_{np}$ based on imputed contrast functions. In the next section, we develop a similar cross-validation technique to correct the finite sample bias in kernel estimation as in Chakrabortty and Cai (2018). In addition, when the number of covariates is relatively large, some other nonparametric regression methods, such as random forest, may be used for estimating the Q functions since kernel estimation may suffer the curse of dimensionality.

4.2 Semi-nonparametric Imputation of the Q-Functions

Consider a $\mathcal{K}$-fold cross-validation. Let $\mathcal{O}^{(k)}$, $k = 1, .., \mathcal{K}$, denote the corresponding partition. First, we compute the kernel estimators of the Q-functions based on data excluding the k-th fold, denoted by $\widehat{Q}^{np}_{-k}(\mathbf{x}, a)$. Next, we conduct a linear refitting step to correct the bias of $\widehat{Q}^{(np)}(\mathbf{x}, a)$. Specifically, define

$$\widehat{\boldsymbol{\theta}}_1 = \arg\min_{\boldsymbol{\theta}_1} \sum_{k=1}^{\mathcal{K}} \sum_{i \in \mathcal{O}^{(k)}} \frac{A_i}{\hat{\pi}(\mathbf{X}_i)} \left\{Y_i - \widehat{Q}^{(np)}_{-k}(\mathbf{X}_i, A_i) - \boldsymbol{\theta}'_1 \widetilde{\mathbf{X}}_i\right\}^2, \tag{6}$$

and

$$\widehat{\boldsymbol{\theta}}_0 = \arg\min_{\boldsymbol{\theta}_0} \sum_{k=1}^{\mathcal{K}} \sum_{i \in \mathcal{O}^{(k)}} \frac{1 - A_i}{1 - \hat{\pi}(\mathbf{X}_i)} \left\{Y_i - \widehat{Q}^{(np)}_{-k}(\mathbf{X}_i, A_i) - \boldsymbol{\theta}'_0 \widetilde{\mathbf{X}}_i\right\}^2. \tag{7}$$

The semi-nonparametric estimators of the Q-functions are given by

$$\widehat{Q}^{(SS)}(\mathbf{x}, a, \widehat{\boldsymbol{\theta}}_a) = \frac{1}{\mathcal{K}} \sum_{k=1}^{\mathcal{K}} \widehat{Q}^{(np)}_{-k}(\mathbf{x}, a) + \widehat{\boldsymbol{\theta}}'_a \widetilde{\mathbf{x}}, \tag{8}$$

for $a \in \{0, 1\}$. Define $\widehat{C}^{(SS)}(\mathbf{x}) = \widehat{Q}^{(SS)}(\mathbf{x}, 1, \widehat{\boldsymbol{\theta}}_1) - \widehat{Q}^{(SS)}(\mathbf{x}, 0, \widehat{\boldsymbol{\theta}}_0)$. A least squares estimate $\widehat{\boldsymbol{\beta}}_{SS}$ can be obtained as in (5) based on imputed contrast functions $\widehat{C}^{(SS)}(\mathbf{X}_j)$, $j = n + 1, \ldots, N$. The corresponding linear treatment regime is $\widehat{\delta}_{SS}(\mathbf{X}) = I(\widehat{\boldsymbol{\beta}}'_{SS} \widetilde{\mathbf{X}} > 0)$.

Theorem 2 *Under the same conditions assumed in Theorem 1, we have*

$$n^{1/2}\left(\widehat{\boldsymbol{\beta}}_{SS} - \boldsymbol{\beta}^*\right) = n^{-1/2}\sum_{i=1}^{n}\Psi_{SS}(\mathcal{O}_i) + o_p(1),$$

where $\Psi_{SS}(\mathcal{O}_i) = \left\{\frac{A_i}{\pi(\mathbf{X}_i)} - \frac{1-A_i}{1-\pi(\mathbf{X}_i)}\right\}\Lambda^{-1}\widetilde{\mathbf{X}}_i\left\{Y_i - Q^{(SS)}(\mathbf{X}_i, A_i, \boldsymbol{\theta}_A)\right\}$ *and* $Q^{(SS)}(\mathbf{x}, a, \boldsymbol{\theta}_a) = Q(\mathbf{x}, a) + \boldsymbol{\theta}_a'\mathbf{x}$. *Here* $\boldsymbol{\theta}_a$ *is the limit of* $\widehat{\boldsymbol{\theta}}_a$, $a = 0, 1$.

The asymptotic variance of $\widehat{\boldsymbol{\beta}}_{SS}$ can be estimated by the usual plug-in method. In general, it can be shown that $\widehat{\boldsymbol{\theta}}_a = 0$, $a = 0, 1$. Therefore, the estimators $\widehat{\boldsymbol{\beta}}_{NP}$ and $\widehat{\boldsymbol{\beta}}_{SS}$ have the same asymptotic variance. To reduce the bias in the variance estimation, a double cross-validation method as proposed in Chakrabortty and Cai (2018) can be adopted. The proof of Theorems 1 and 2 are outlined in the Appendix.

5 Simulation Analysis

We conducted simulation studies to examine the finite-sample performance of the estimators introduced in Sect. 3 (denoted by TR) and in Sect. 4.2 (denoted by SS). Our simulation study design considers varying degrees of model misspecification in $C(\mathbf{X})$ to better understand the advantages of SS when estimating $\delta_{\boldsymbol{\beta}}^{opt}$. During each simulation replication, the propensity score will be estimated to better reflect the analysis of an observation study with EMR data. In the implementation of our proposed SS estimator, a Gaussian kernel was used for $W(\cdot)$ and the bandwidth parameter h was chosen using cross-validation. In addition we set $\mathcal{K} = 5$. We present the percent of correct decisions (PCD) and value function to illustrate the benefits of semi-supervised prediction on assessing $\delta_{\boldsymbol{\beta}}^{opt}$. Furthermore, we study the bias, empirical standard deviation (SD), estimated standard error (SE), and the relative efficiency of $\widehat{\boldsymbol{\beta}}_{SS}$ with respect to $\widehat{\boldsymbol{\beta}}_{TR}$.

5.1 Data Generation

In this simulation, n is the amount of subjects in $\mathcal{O}$ and N is the amount of subjects in $\mathcal{U}$. The N unobserved responses were simulated to be missing completely at random (MCAR), where the response and treatment were not observed. Let $\mathbf{X} \sim N\left(0_p, I_p\right)$, where $\mathbf{X}$ is contained in $[-5, 5]^p$ to ensure it is in a compact set, and $\mathcal{A} \in \{0, 1\}$. The following models were tested with $p = 2$, $n = 500$ and $N = 5000$. We examined three choices for $C(\mathbf{X})$,

- *Model 1 (Linear)*: $\mathbf{Y} = \boldsymbol{\mu}(\mathbf{X}) + \mathbf{A}\left(\boldsymbol{\eta}'\mathbf{X}\right) + \epsilon$
- *Model 2 (Cubic)*: $\mathbf{Y} = \mu(\mathbf{X}) + \mathbf{A}\left(\boldsymbol{\eta}'\mathbf{X}\right)^3 + \epsilon$
- *Model 3 (Sine)*: $\mathbf{Y} = \mu(\mathbf{X}) + \mathbf{A}\left(\sin(\boldsymbol{\eta}'\mathbf{X})\right) + \epsilon$,

where $\boldsymbol{\eta} = \mathbf{1}_2$ for Models 1 and 3, $\boldsymbol{\eta} = (0.3, 0.6)'$ for Model 2, and $\boldsymbol{\epsilon} \sim \mathcal{N}(0, 1)$. Two baseline functions were studied. The first one considered a situation where the baseline effect for subjects is small and the second function considered a larger baseline effect,

1. $\mu(\mathbf{X}) = \left(\boldsymbol{\omega}'\mathbf{X}\right)^3$,
2. $\mu(\mathbf{X}) = \left(\boldsymbol{\alpha}'\mathbf{X}\right)\left(1 + \boldsymbol{\omega}'\mathbf{X}\right)$.

We studied two different baseline functions to understand the flexibility of SS in estimating $C(\mathbf{X})$. The models were studied with $\boldsymbol{\omega} = (0.5, 0.5)'$ and $\boldsymbol{\alpha} = (0.75, 0.75)'$. We simulated a propensity score of $\pi(\mathbf{X}) = logit(0.5X_1 - 0.5X_2)$, and an allocation of treatment as $\mathbf{A} \sim bernoulli(p = \pi(\mathbf{X}))$. The propensity scores were generated to simulate observational data where the allocation of treatment depended on a patient's health status information. The true values, $\boldsymbol{\beta}^*$, were estimated by simulating a fully observed Monte Carlo data set of size 500,000.

5.2 Simulation Results

The results of interest were percent of correct decisions (PCD), value function (V), and component-wise relative efficiency (RE) over 500 replications. The component-wise RE for each estimator was calculated as $\sum_{i=1}^{500} \|\widehat{\boldsymbol{\beta}}_{TR,i,j} - \boldsymbol{\beta}_j^*\|^2 / \sum_{k=1}^{500} \|\widehat{\boldsymbol{\beta}}_{SS,j} - \boldsymbol{\beta}_j^*\|^2$, where $j = 1, \cdots, p+1$. During each replication, new vectors $\mathbf{Y}$, $\mathbf{A}$ and matrix $\mathbf{X}$ were simulated. The percent of correct decisions (PCD) was calculated for each method,

$$PCD_i = 1 - \frac{1}{5500} \sum_{k=1}^{5500} \left| I(\widehat{\boldsymbol{\beta}}'\widetilde{\mathbf{X}}_k > 0) - I(\boldsymbol{\beta}^{*\prime}\widetilde{\mathbf{X}}_k > 0) \right|,$$

and then averaged over all 500 simulations, $PCD = (1/500) \sum_{i=1}^{500} PCD_i$. The true value function was calculated with the Monte Carlo data set of sample size 500,000 as,

$$V_0 = \frac{1}{500{,}000} \sum_{m=1}^{500{,}000} \{\mu(\mathbf{X}_m) + I(C(\mathbf{X}_m) > 0)C(\mathbf{X}_m)\}.$$

Estimation of the value function using TR or SS was calculated as,

$$\widehat{V}_i = \frac{1}{500{,}000} \sum_{m=1}^{500{,}000} \left\{\mu(\mathbf{X}_m) + \widehat{\delta}_{\boldsymbol{\beta}}^{opt} C(\mathbf{X}_m)\right\},$$

on the same Monte Carlo data set and then averaged over 500 replications.

Table 1 Treatment decision results: V_0, the value of the true optimal treatment rule; V, the average of the values of estimated optimal linear treatment rules over 500 replications; PCD, average of percent of correct decisions over 500 replications. Empirical standard deviations are provided in parentheses

			TR		SS	
$\mu(\mathbf{X})$	Model	V_0	V	PCD	V	PCD
$(\omega'\mathbf{X})^3$	Linear	0.56	0.54 (0.04)	0.92 (0.06)	0.56 (0.01)	0.96 (0.02)
	Cubic	0.24	0.22 (0.06)	0.87 (0.12)	0.24 (0.01)	0.93 (0.05)
	Sine	0.32	0.21 (0.12)	0.80 (0.17)	0.26 (0.06)	0.88 (0.09)
$(\alpha'\mathbf{X})(1+\omega'\mathbf{X})$	Linear	1.31	1.29 (0.05)	0.91 (0.06)	1.31 (0.01)	0.96 (0.02)
	Cubic	0.99	0.98 (0.05)	0.86 (0.11)	0.99 (<0.01)	0.94 (0.04)
	Sine	1.06	0.96 (0.11)	0.80 (0.16)	1.02 (0.04)	0.89 (0.06)

Table 1 summarized the value function and PCD results under all 6 settings. Across all settings, we observed an improvement in the average value function, average PCD, and their respective SD while using the SS method compared to TR. Table 2 presented the mean bias, standatrd deviation (SD), average of estimated standard errors (SE), and coverage probabilities of the 95% Wald-type confidence intervals of the estimates over 500 replications. In addition, the relative efficiency (RE) of the SS estimators compared with the TR estimator was also reported. Based on the results, we observed that both estimators are nearly unbiased, but the SS estimators are more efficient that the TR estimator. The RE ranged from 2.13 to 5.37.

6 Application to an EMR Study

We apply our proposed method, SS, to an EMR study on patients undergoing a hypotensive episode in the ICU within the MIMIC-III database. It is important to treat hypotensive episodes (HE) in ICU patients to minimize end-organ damage. A marker of end-organ damage is an increase in serum creatinine post hypotensive episode (Lehman et al., 2010). Specifically, the outcome of interest considered in our study is the difference between a post-episode serum creatinine measurement taken no more than 72 hours after the HE and a baseline serum creatinine measurement taken no earlier than 24 hours before the HE. Two available treatments for HE's include IV fluid resuscitation and vasoactive therapy (Lee et al., 2012). Our objective is to determine the optimal initial treatment for each individual between IV fluid resuscitation and vasopressor interventions to minimize any increase in serum creatinine 72 hours post-hypotensive episode.

Lee et al. (2012) defines the initiation of an HE as the occurrence of two successive mean arterial pressure (MAP) measurement values > 60 *mm Hg*, followed by two MAP measurements ≤ 60 *mm Hg*. The end of an HE is characterized by two successive MAP measurements ≤ 60 *mm Hg*, followed by two successive MAP

Table 2 Parameters estimation results: Bias, the bias of the estimates over 500 replications; SD, empirical standard deviation of the estimates; SE, average of estimated standard errors; CP, empirical coverage probability of Wald-type 95% confidence intervals; RE, relative efficiency of the SS estimates compared with the TR estimates

(a) $\mu(\mathbf{X}) = (\boldsymbol{\omega}'\mathbf{X})^3$

		TR				SS				
Model	$\boldsymbol{\beta}$	Bias	SD	SE	CP	Bias	SD	SE	CP	RE
Linear	0	−0.010	0.220	0.209	0.94	−0.005	0.123	0.122	0.95	3.21
	1	−0.021	0.329	0.347	0.98	−0.008	0.182	0.173	0.93	3.28
	1	−0.020	0.355	0.347	0.97	−0.004	0.198	0.175	0.92	3.22
Cubic	0	−0.008	0.206	0.205	0.94	0.001	0.123	0.132	0.95	2.80
	0.41	0.002	0.352	0.338	0.95	−0.002	0.212	0.193	0.95	2.74
	0.81	0.002	0.423	0.390	0.94	−0.010	0.239	0.212	0.91	3.14
Sine	0	−0.006	0.171	0.168	0.96	0.004	0.118	0.116	0.95	2.13
	0.37	−0.007	0.282	0.270	0.94	−0.011	0.170	0.158	0.91	2.74
	0.37	0.011	0.296	0.272	0.94	0.011	0.176	0.161	0.92	2.82

(b) $\mu(\mathbf{X}) = (\boldsymbol{\alpha}'\mathbf{X})(1 + \boldsymbol{\omega}'\mathbf{X})$

		TR				SS				
Model	$\boldsymbol{\beta}$	Bias	SD	SE	CP	Bias	SD	SE	CP	RE
Linear	0	−0.017	0.242	0.230	0.94	−0.007	0.114	0.116	0.95	4.53
	1	−0.012	0.348	0.326	0.94	0.011	0.150	0.151	0.93	5.37
	1	−0.021	0.352	0.347	0.94	−0.011	0.163	0.154	0.91	4.66
Cubic	0	−0.005	0.236	0.223	0.93	−0.026	0.121	0.122	0.95	3.77
	0.41	−0.014	0.341	0.313	0.93	0.003	0.155	0.154	0.92	4.86
	0.81	−0.004	0.432	0.392	0.91	−0.008	0.193	0.186	0.92	4.99
Sine	0	0.003	0.210	0.202	0.94	0.009	0.117	0.113	0.95	3.14
	0.37	0.002	0.296	0.289	0.94	−0.004	0.146	0.146	0.94	4.12
	0.37	−0.005	0.300	0.290	0.93	0.001	0.136	0.140	0.95	4.90

measurements > 60 *mm Hg*. The study cohort includes subjects greater than or equal to 15 years old who experienced a single hypotensive episodes during their ICU stay, which amounts to 3,372 subjects.

To formulate this problem into the OTD framework, we must define Y as the negative of the difference between post-HE and pre-HE serum creatinine measurements. The treatment space is defined as $A = 1$ if the patient was given vasopressor treatment, and $A = 0$ if they received IV fluid resuscitation. Vasopressor treatments include dobutamine, dopamine, epinephrine, norepinephrine, phenylephrine, milrinone or vasopressin during the HE. IV fluid treatment is defined as receiving at least one infusion of colloids, or isotonic crystalloids of at least 250 ml (Lee et al., 2012). The predictors, $\mathbf{X}$, include normalized versions of mean MAP and mean heart rate in the 3 hour window immediately before the HE onset. A total of 945 subjects received either IV fluid resuscitation or vasopressor treatment as their sole treatment. The number of patients untreated with solely IV fluid resuscitation or vasoactive therapy and/or missing response information resulted

Table 3 Estimated parameters and their associated standard errors (SE) and p-values

	TR			SS		
Predictors	β_{TR}	SE	p-value	β_{SS}	SE	p-value
Intercept	0.020	0.136	0.883	−0.050	0.126	0.695
Mean MAP	0.043	0.094	0.640	0.037	0.066	0.579
Mean heart rate	−0.096	0.057	0.089	−0.104	0.050	0.040

Table 4 Treatment allocation given by SS and TR

		SS	
	Treatment	IV fluid	Vasopressors
TR	IV fluid	1430	29
	Vasopressors	459	1454

in 2,427 subjects. For this data analysis, we are focusing on the semi-supervised setting with a small set of labeled data. Therefore we randomly choose $n = 300$ out of 945 subjects to create the fully observed set, $\mathcal{O}$, and place the remaining 3,072 subjects into $\mathcal{U}$. Under such a sampling scheme, the missing completely at random assumption is held.

Propensity scores were modeled with a logistic regression model that included the following covariates: baseline creatinine, age, gender, indicator function for patients that received surgical services, log transformation of the elixhauser comorbidity score (method of categorizing comorbidity of patients based on their ICD-9 codes) (van Walraven et al., 2009), log transformation of simplified acute physiologic score (SAPS) (scoring system to reflect the risk of death upon admission to ICU) (Le Gall et al., 1984), total urine output in the 3 hour window prior to the HE, average mean heart rate in the 3 hour window prior to the HE, and average mean arterial pressure (MAP) in the 3 hour window prior to the HE. We perform TR and SS on the data set to obtain estimates and standard errors for $\boldsymbol{\beta}$. A Gaussian kernel was chosen for $W(\cdot)$ in Eqs. 3 and 4 and $\mathcal{K} = 5$ similar to Sect. 5. Table 3 presents the point estimates, estimated SE, and p-values for testing null effects. SS and TR estimates are relatively close, but SS estimates have smaller SE than TR estimates suggesting they are more efficient as demonstrated in simulations.

In Table 4, we demonstrate treatment allocation given by both methods on all subjects in the data set. The OTD estimated by SS and TR produce similar treatment decisions for the vast majority of subjects. SS assigns more patients to treatment with IV fluids than TR and is less likely to assign vasoactive therapy.

7 Discussion

We proposed a new semi-supervised learning method for estimating optimal treatment regime using both labeled and unlabeled data from EMR. In this work, we focused on binary treatments, but the proposed method can be easily extended to accommodate multiple treatments. In addition, we considered a semi-parametric working model with linear treatment effects for its simplicity. The estimated linear treatment regime may not give the highest value function among the class of linear decision rules since it is obtained based on a regression method. Alternatively, we may consider an estimated linear treatment regime that maximizes the estimated value function or equivalently the estimated contrast function. However, the resulting estimator of $\boldsymbol{\beta}$ has a slower cubic convergence rate and its associated inference will become more challenging. Moreover, the proposed method can be generalized to accommodate nonlinear treatment effects using spline representation techniques.

The proposed method is built on kernel estimation for imputing the contrast functions of unlabeled data, which can not handle high-dimensional predictors. To incorporate high-dimensional predictors, some dimension reduction techniques, such as principal component analysis or sliced inverse regression, can be utilized to reduce the dimension of covariates into a lower-dimensional subspace, and then kernel estimators can be applied as in Chakrabortty and Cai (2018). Moreover, in this paper we only consider the MCAR setting, but the proposed semi-supervised learning method can be easily extended to the case when Y and A are missing at random (MAR) by using the inverse missing probability weighted estimation.

The proposed semi-supervised learning method only uses variables obtained from structured EMR data. MIMIC-III also contains unstructured data through progress reports and discharge summaries recorded during a patient's hospital stay. Clinical text data holds a great deal of valuable patient information that can be used for building the optimal treatment regime. Clinical natural language processing tools, such as cTakes, can be used to extract information from clinical texts, such as patients' disease status, symptoms, prescribed treatments, and family history (Savova et al., 2010). How to incorporate these information in the optimal treatment decision is an interesting question that warrants future investigation.

Appendix 1: Proof of Theorem 1

Let $\widehat{C}(\mathbf{X}) = \widehat{Q}^{(np)}(\mathbf{X}, 1) - \widehat{Q}^{(np)}(\mathbf{X}, 0)$. We have

$$
\begin{aligned}
\widehat{\boldsymbol{\beta}}_{np} - \boldsymbol{\beta}^* &= \Lambda^{-1}\left[N^{-1}\sum_{j=n+1}^{n+N}\widetilde{\mathbf{X}}_j\left\{\widehat{C}(\mathbf{X}_j) - \boldsymbol{\beta}^{*\prime}\widetilde{\mathbf{X}}_j\right\}\right] \\
&= \Lambda^{-1}\left[N^{-1}\sum_{j=n+1}^{n+N}\widetilde{\mathbf{X}}_j\left\{\widehat{C}(\mathbf{X}_j) - C(\mathbf{X}_j)\right\}\right] \\
&\quad + \Lambda^{-1}\left[N^{-1}\sum_{j=n+1}^{n+N}\widetilde{\mathbf{X}}_j\left\{C(\mathbf{X}_j) - \boldsymbol{\beta}'\widetilde{\mathbf{X}}_j\right\}\right] \\
&= \Lambda^{-1}E\left[\widetilde{\mathbf{X}}\left\{\widehat{C}(\mathbf{X}) - C(\mathbf{X})\right\}\right] + O_p(N^{-1/2}).
\end{aligned}
$$

The first step follows from the normal equations. The last step is due to the fact

$$\Lambda^{-1}\left[N^{-1}\sum_{j=n+1}^{n+N}\widetilde{\mathbf{X}}_j\left\{\widehat{C}(\mathbf{X}_j)-C(\mathbf{X}_j)\right\}\right]=\Lambda^{-1}E\left[\widetilde{\mathbf{X}}\left\{\widehat{C}(\mathbf{X})-C(\mathbf{X})\right\}\right]+o_p(1)$$

by standard arguments involving the weak law of large numbers. According to the central limit theorem $N^{-1/2}\left[N^{-1/2}\sum_{j=n+1}^{N}\Lambda^{-1}\left\{C(\mathbf{X})_j-\boldsymbol{\beta}'\widetilde{\mathbf{X}}_j\right\}\right]=O_p(N^{-1/2})$. Multiplying both sides by $n^{1/2}$ we have

$$n^{1/2}\left(\widehat{\boldsymbol{\beta}}_{np}-\boldsymbol{\beta}\right)=n^{1/2}\Lambda^{-1}E\left[\widetilde{\mathbf{X}}\left\{\widehat{C}(\mathbf{X})-C(\mathbf{X})\right\}\right]+O_p\left((n/N)^{\frac{1}{2}}\right) \tag{9}$$

$$=n^{1/2}\Lambda^{-1}E\left[\widetilde{\mathbf{X}}\left\{\widehat{Q}^{(np)}(\mathbf{X},1)-Q^{(np)}(\mathbf{X},1)\right\}\right] \tag{10}$$

$$-n^{1/2}\Lambda^{-1}E\left[\widetilde{\mathbf{X}}\left\{\widehat{Q}^{(np)}(\mathbf{X},0)-Q^{(np)}(\mathbf{X},0)\right\}\right]$$

$$+O_p\left((n/N)^{\frac{1}{2}}\right). \tag{11}$$

Note that $n/N\to 0$ implying $O_p\left((n/N)^{\frac{1}{2}}\right)\equiv o_p(1)$. Next, let $\tau(\mathbf{X})=\pi(\mathbf{X})f(\mathbf{X})$ and $\widehat{\tau}(\mathbf{X})=\frac{1}{nh^p}\sum_{i=1}^{n}A_iW_h(\mathbf{X}_i-\mathbf{X})$, where $W_h(\mathbf{X}_i-\mathbf{X})=W(\frac{\mathbf{X}_i-\mathbf{X}}{h})$. Let's rewrite $E\left[\widetilde{\mathbf{X}}\left\{\widehat{Q}^{(np)}(\mathbf{X},1)-Q^{(np)}(\mathbf{X},1)\right\}\right]$ as

$$=E\left\{\widetilde{\mathbf{X}}\left(\frac{\frac{1}{nh^p}\sum_{i=1}^{n}A_iW_h(\mathbf{X}_i-\mathbf{X})\left\{Y_i-Q^{(np)}(\mathbf{X},1)\right\}}{\tau(\mathbf{X})}\right)\right\} \tag{12}$$

$$+E\left\{\widetilde{\mathbf{X}}\left(\widehat{Q}^{(np)}(\mathbf{X},1)-Q^{(np)}(\mathbf{X},1)\right)\left\{\frac{\tau(\mathbf{X})-\widehat{\tau}(\mathbf{X})}{\tau(\mathbf{X})}\right\}\right\}=H_{n,1}^{(1)}+H_{n,2}^{(1)}. \tag{13}$$

Then $H_{n,1}^{(1)}$ is equivalent to:

$$=\frac{1}{nh^p}\sum_{i=1}^{n}A_i\int\widetilde{\mathbf{X}}_i\left\{Y_i-Q^{(np)}(\mathbf{X},1)\right\}\frac{W_h(\mathbf{X}-\mathbf{X}_i)}{\tau(\mathbf{X})}f(\mathbf{X})d\mathbf{X}$$

$$=\frac{1}{nh^p}\sum_{i=1}^{n}A_i\int\widetilde{\mathbf{X}}_i\left\{Y_i-Q^{(np)}(\mathbf{X},1)\right\}\frac{W_h(\mathbf{X}-\mathbf{X}_i)}{\pi(\mathbf{X})}d\mathbf{X}$$

$$=\frac{1}{n}\sum_{i=1}^{n}A_i\int\left(\widetilde{\mathbf{X}}_i+ht_i\right)\left\{Y_i-Q^{(np)}(\mathbf{X}_i+ht_i,1)\right\}\frac{W(t_i)}{\pi\left(\mathbf{X}_i+ht_i\right)}dt_i.$$

By assumptions (C1) and (C2), and Taylor expansion of $H_{n,1}^{(1)}$ in ht_i for sufficiently small h, it leads to

$$H_{n,1}^{(1)} = \frac{1}{n}\sum_{i=1}^{n}\frac{A_i}{\pi\left(\mathbf{X}_i\right)}\widetilde{\mathbf{X}}_i\left\{Y_i - Q^{(np)}(\mathbf{X}_i, 1)\right\} + O_p(h^r).$$

Since $n^{1/2}h^r \to 0$ as $n \to \infty$,

$$n^{1/2}\Lambda^{-1}H_{n,1}^{(1)} = \widetilde{H}_{n,1}^{(1)} = n^{-1/2}\sum_{i=1}^{n}\frac{A_i}{\pi\left(\mathbf{X}_i\right)}\Lambda^{-1}\widetilde{\mathbf{X}}_i\left\{Y_i - Q^{(np)}(\mathbf{X}_i, 1)\right\} + o_p(1). \tag{14}$$

Let $q(\mathbf{X}) = \widehat{Q}^{(np)}(\mathbf{X}, 1) - Q^{(np)}(\mathbf{X}, 1)$, $l(\mathbf{X}) = \dfrac{\tau(\mathbf{X}) - \widehat{\tau}(\mathbf{X})}{\tau(\mathbf{X})} = 1 - \dfrac{\widehat{\tau}(\mathbf{X})}{\tau(\mathbf{X})}$. Given that $\mathbf{X}$ is bounded, $\sqrt{\frac{\ln n}{nh^p}} \to 0$ as $n \to \infty$, as well as assumptions (C1), (C3), and (C4), *Lemma B.1* in Newey (1994), it can be shown that $\sup_{\mathbf{x}\in\mathcal{X}}|\widehat{\tau}(\mathbf{X}) - \tau(\mathbf{X})| = o_p(1)$. It also holds that $\sup_{\mathbf{x}\in\mathcal{X}}|q(\mathbf{X})| = o_p(1)$ through *Theorem 8* of Hansen (2008). It follows from these results that $H_{n,2}^{(1)} = E\left\{\widetilde{\mathbf{X}}q(\mathbf{X})l(\mathbf{X})\right\}$ and,

$$n^{1/2}\Lambda^{-1}H_{n,2}^{(1)} \le n^{1/2}\Lambda^{-1}\sup_{\mathbf{x}\in\mathcal{X}}\left\{\|\widetilde{\mathbf{X}}\|\ |q(\mathbf{X})|\ |l(\mathbf{X})|\right\} = o_p(1). \tag{15}$$

Moreover, using similar techniques we can prove,

$$n^{1/2}\Lambda^{-1}E\left[\widetilde{\mathbf{X}}\left\{\widehat{Q}^{(np)}(\mathbf{X}, 0) - Q^{(np)}(\mathbf{X}, 0)\right\}\right] = \widetilde{H}_{n,1}^{(0)} + o_p(1). \tag{16}$$

This leads to

$$\begin{aligned} n^{1/2}\Lambda^{-1}E\left[\widetilde{\mathbf{X}}\left\{\widehat{C}(\mathbf{X}) - C(\mathbf{X})\right\}\right] &= \widetilde{H}_{n,1}^{(1)} - \widetilde{H}_{n,1}^{(0)} + o_p(1) \\ &= n^{-1/2}\sum_{i=1}^{n}\Psi_{np}(\mathbf{O}_i) + o_p(1). \end{aligned}$$

It approves the asymptotic results established in Theorem 1.

Appendix 2: Proof of Theorem 2

The proof of Theorem 2 follows similarly to that of Theorem 1. Let $\widehat{C}^{(SS)}(\mathbf{X}) = \widehat{Q}^{(SS)}(\mathbf{x}, 1, \widehat{\boldsymbol{\theta}}_1) - \widehat{Q}^{(SS)}(\mathbf{X}, 0, \widehat{\boldsymbol{\theta}}_0)$. We have

$$n^{1/2}\left(\widehat{\boldsymbol{\beta}}_{SS} - \boldsymbol{\beta}\right) = n^{1/2}\Lambda^{-1}E\left[\widetilde{\mathbf{X}}\left\{\widehat{C}(\mathbf{X}) - C(\mathbf{X})\right\}\right] + o_p(1) \tag{17}$$

$$= n^{1/2}\Lambda^{-1}E\left[\widetilde{\mathbf{X}}\left\{\widehat{Q}^{(SS)}(\mathbf{X}, 1, \widehat{\boldsymbol{\theta}}_1) - Q^{(SS)}(\mathbf{X}, 1, \boldsymbol{\theta}_1)\right\}\right] \tag{18}$$

$$-n^{1/2}\Lambda^{-1}E\left[\widetilde{\mathbf{X}}\left\{\widehat{Q}^{(SS)}(\mathbf{X}, 0, \widehat{\boldsymbol{\theta}}_0) - Q^{(SS)}(\mathbf{X}, 0, \boldsymbol{\theta}_0)\right\}\right] + o_p(1). \tag{19}$$

Next, recall that

$$\widehat{Q}^{(SS)}(\mathbf{x}, a, \widehat{\boldsymbol{\theta}}_a) = \frac{1}{\mathcal{K}}\sum_{k=1}^{\mathcal{K}} \widehat{Q}_k^{(np)}(\mathbf{x}, a) + \widehat{\boldsymbol{\theta}}_a'\widetilde{\mathbf{x}},$$

and

$$Q^{(SS)}(\mathbf{x}, a, \boldsymbol{\theta}_a) = Q^{(np)}(\mathbf{x}, a) + \boldsymbol{\theta}_a'\widetilde{\mathbf{x}}.$$

This allows us to rewrite Equation (18) in the following way

$$n^{1/2}\Lambda^{-1}\left\{\frac{1}{N}\sum_{j=n+1}^{N} \widetilde{\mathbf{X}}_j\left\{\frac{1}{\mathcal{K}}\sum_{k=1}^{\mathcal{K}} \widehat{Q}_k^{(np)}\left(\mathbf{X}_j, a\right) + \widehat{\boldsymbol{\theta}}_a'\widetilde{\mathbf{X}}_j\right\} - Q^{(np)}\left(\mathbf{X}_j, a\right) + \boldsymbol{\theta}_a'\widetilde{\mathbf{X}}_j\right\} + o_p(1). \tag{20}$$

Equation (19) can be rewritten analogously. Define

$$\widehat{S}_a^{\mathcal{K}} = \frac{1}{\mathcal{K}}\sum_{k=1}^{\mathcal{K}}\left\{\frac{1}{N}\sum_{j=n+1}^{n+N}\widetilde{\mathbf{X}}_j\left\{\widehat{Q}_k^{(np)}(\mathbf{X}_j, a) - Q^{(np)}(\mathbf{X}_j, a)\right\}\right\}.$$

Rearranging the summands in Eq. (20), we can obtain

$$n^{1/2}\left(\widehat{\boldsymbol{\beta}}_{SS} - \boldsymbol{\beta}\right) = n^{1/2}\Lambda^{-1}\left\{\Lambda(\widehat{\boldsymbol{\theta}}_1 - \boldsymbol{\theta}_1) + \widehat{S}_1^{\mathcal{K}}\right\} - n^{1/2}\Lambda^{-1}\left\{\Lambda(\widehat{\boldsymbol{\theta}}_0 - \boldsymbol{\theta}_0) + \widehat{S}_0^{\mathcal{K}}\right\} + o_p(1). \tag{21}$$

Following similar arguments in the proof of *Theorem 3.2* in Chakrabortty and Cai (2018), we can approve the asymptotic results established in Theorem 2.

References

Beaulieu-Jones, B. K., Greene, C. S., et al. (2016). Semi-supervised learning of the electronic health record for phenotype stratification. *Journal of Biomedical Informatics, 64*, 168–178.

Beaulieu-Jones, B. K., Lavage, D. R., Snyder, J. W., Moore, J. H., Pendergrass, S. A., & Bauer, C. R. (2018). "Characterizing and managing missing structured data in electronic health records: Data analysis. *JMIR Medical Informatics, 6*, e11.

Chakrabortty, A., & Cai, T. (2018). Efficient and adaptive linear regression in semi-supervised settings. *The Annals of Statistics, 46*, 1541–1572.

Chapelle, O., Scholkopf, B., & Zien, A. (2006), *Semi-supervised learning*. MIT Press.

Fan, C., Lu, W., Song, R., & Zhou, Y. (2017). Concordance-assisted learning for estimating optimal individualized treatment regimes. *Journal of the Royal Statistical Society: Series B (Statistical Methodology), 79*(5), 1565–1582.

Fan, J. (1992). Design-adaptive nonparametric regression. *Journal of the American statistical Association, 87*, 998–1004.

Garla, V., Taylor, C., & Brandt, C. (2013). Semi-supervised clinical text classification with Laplacian SVMs: An application to cancer case management. *Journal of Biomedical Informatics, 46*, 869–875.

Hansen, B. E. (2008). Uniform convergence rates for kernel estimation with dependent data. *Econometric Theory, 24*, 726–748.

Hong, C., Liao, K. P., & Cai, T. (2019). Semi-supervised validation of multiple surrogate outcomes with application to electronic medical records phenotyping. *Biometrics, 75*, 78–89.

Johnson, A. E., Pollard, T. J., Shen, L., Li-wei, H. L., Feng, M., Ghassemi, M., Moody, B., Szolovits, P., Celi, L. A., & Mark, R. G. (2016). MIMIC-III, a freely accessible critical care database. *Scientific Data, 3*, 160035.

Kellum, J. A., Lameire, N., Aspelin, P., Barsoum, R. S., Burdmann, E. A., Goldstein, S. L., Herzog, C. A., Joannidis, M., Kribben, A., Levey, A. S., et al. (2012). Kidney disease: improving global outcomes (KDIGO) acute kidney injury work group. KDIGO clinical practice guideline for acute kidney injury. *Kidney International Supplements, 2*, 1–138.

Le Gall, J., Loirat, P., Alperovitch, A., Glaser, P., Granthil, C., Mathieu, D., Mercier, P., Thomas, R., & Villers, D. (1984). A simplified acute physiology score for ICU patients. *Critical Care Medicine, 12*, 975–977.

Lee, J., Kothari, R., Ladapo, J. A., Scott, D. J., & Celi, L. A. (2012). Interrogating a clinical database to study treatment of hypotension in the critically ill. *BMJ Open, 2*, e000916.

Lehman, L. w., Saeed, M., Moody, G., & Mark, R. (2010). Hypotension as a risk factor for acute kidney injury in ICU patients. *Computing in Cardiology, 37*, 1095.

Lu, W., Zhang, H. H., & Zeng, D. (2013). Variable selection for optimal treatment decision. *Statistical Methods in Medical Research, 22*, 493–504.

Maslove, D. M., Lamontagne, F., Marshall, J. C., & Heyland, D. K. (2017). A path to precision in the ICU. *Critical Care, 21*, 79.

Moodie, E. E., Richardson, T. S., & Stephens, D. A. (2007). Demystifying optimal dynamic treatment regimes. *Biometrics, 63*, 447–455.

Murphy, S. A. (2003). Optimal dynamic treatment regimes. *Journal of the Royal Statistical Society: Series B (Statistical Methodology), 65*, 331–355.

Newey, W. K. (1994). Kernel estimation of partial means and a general variance estimator. *Econometric Theory, 10*, 1–21.

Raghu, A., Komorowski, M., Celi, L. A., Szolovits, P., & Ghassemi, M. (2017). Continuous state-space models for optimal sepsis treatment: A deep reinforcement learning approach. In *Machine Learning for Healthcare Conference* (pp. 147–163).

Robins, J. M. (2004). Optimal structural nested models for optimal sequential decisions. In *Proceedings of the Second Seattle Symposium in Biostatistics* (pp. 189–326). Springer.

Rosenbaum, P. R., & Rubin, D. B. (1983). The central role of the propensity score in observational studies for causal effects. *Biometrika, 70*, 41–55.

Rubin, D. B. (1978). Bayesian inference for causal effects: The role of randomization. *The Annals of Statistics, 6*, 34–58.

Savova, G. K., Masanz, J. J., Ogren, P. V., Zheng, J., Sohn, S., Kipper-Schuler, K. C., & Chute, C. G. (2010). Mayo clinical text analysis and knowledge extraction system (cTAKES): Architecture, component evaluation and applications. *Journal of the American Medical Informatics Association, 17*, 507–513.

Song, R., Luo, S., Zeng, D., Zhang, H. H., Lu, W., & Li, Z. (2017). Semiparametric single-index model for estimating optimal individualized treatment strategy. *Electronical Journal of Statistics, 11*, 364–384.

van Walraven, C., Austin, P. C., Jennings, A., Quan, H., & Forster, A. J. (2009). A modification of the Elixhauser comorbidity measures into a point system for hospital death using administrative data. *Medical Care, 47*, 626–633.

Wang, Y., Wu, P., Liu, Y., Weng, C., & Zeng, D. (2016). Learning optimal individualized treatment rules from electronic health record data. In *2016 IEEE International Conference on Healthcare Informatics (ICHI)* (pp. 65–71).

Watkins, C. J., & Dayan, P. (1992). Q-learning. *Machine Learning, 8*, 279–292.

Weiskopf, N. G., Hripcsak, G., Swaminathan, S., & Weng, C. (2013). Defining and measuring completeness of electronic health records for secondary use. *Journal of Biomedical Informatics, 46*, 830–836.

Zhang, B., Tsiatis, A. A., Laber, E. B., & Davidian, M. (2012). A robust method for estimating optimal treatment regimes. *Biometrics, 68*, 1010–1018.

Zhao, Y., Zeng, D., Rush, A. J., & Kosorok, M. R. (2012). Estimating individualized treatment rules using outcome weighted learning. *Journal of the American Statistical Association, 107*, 1106–1118.

Zhu, X., & Goldberg, A. B. (2009). Introduction to semi-supervised learning. *Synthesis Lectures on Artificial Intelligence and Machine Learning, 3*, 1–130.

Estimation and Inference for Individualized Treatment Rules Using Efficient Augmentation and Relaxation Learning

Muxuan Liang and Ying-Qi Zhao

Abstract Individualized treatment rules aim to identify the best strategy—if, when, which, and to whom treatment should be applied, to optimize the health outcomes. One type of approaches to learn such rules estimates individualized treatment rules through a weighted classification framework. Particularly, in this chapter, we introduce doubly-robust learning methods called efficient augmentation and relaxation learning (EARL) and its extensions to high-dimensional data such as genetic profiles. These methods can consistently estimate the optimal individualized treatment rule as long as model specifications of either a propensity or outcome models are correct. More importantly, we also provide an inferential tool for EARL and its extensions under a linear decision rule. These inferential tools provide p-values and confidence intervals to characterize how predictors affect the best treatment decision. With these inferential tools, EARL and its extension can inform the practical use of the treatment and facilitate scientific discovery.

Keywords Individualized treatment rule · Double-robustness · High-dimensional inference · Semiparametric inference · Non-differentiable surrogate loss

1 Introduction

Individualized treatment rules (ITRs) are decision rules that map the patient-level information $X \in \mathcal{X}$ into the space of intervention, $A \in \mathcal{A}$, where $\mathcal{X} \subset \mathbb{R}^p$, p is the number of the covariates, and $\mathcal{A}$ is the set of available interventions. Among ITRs, the decision rule which maximizes the mean outcome if the rule were applied to a target population is called the *optimal ITR* (Qian and Murphy, 2011).

M. Liang (✉)
Department of Biostatistics, University of Florida, Gainesville, FL, USA
e-mail: muxuan.liang@ufl.edu

Y.-Q. Zhao
Public Health Sciences Divisions, Fred Hutchinson Cancer Research Center, Seattle, WA, USA
e-mail: yqzhao@fredhutch.org

Y. Zhao, D.-G. Chen (eds.), *Statistics in Precision Health*, ICSA Book Series in Statistics, https://doi.org/10.1007/978-3-031-50690-1_10

Many methods have been proposed to estimate the optimal ITR. These methods can be categorized into two categories—indirect and direct learning. Indirect learning of the optimal ITR estimates the *conditional average treatment effect (CATE)* and transforms the resulting estimates into a decision rule using an indicator function. Specifically, let $\Delta(X)$ be the CATE and $\widehat{\Delta}(X)$ be an estimate. Then an estimate of the optimal ITR can be defined as $D_{\text{opt}}(X) = 1\left\{\widehat{\Delta}(X) \geq 0\right\}$. In this procedure, the estimated ITR closely depends on the estimate of CATE, and many indirect learning approaches focus on how to model CATE to avoid model misspecification. Q-learning approaches (Watkins and Dayan, 1992; Chakraborty et al., 2010; Qian and Murphy, 2011; Laber et al., 2014a) implicitly model CATE via positing a fully specified model on the means of the outcomes conditional on covariates and interventions. A-learning methods utilizing the propensity score posit a model assumption on CATE (Murphy, 2003; Lu et al., 2013; Shi et al., 2016, 2018; Liang and Yu, 2022). Especially, Liang and Yu (2022) considered a dimension reduction strategy that is flexible and efficient. Direct learning approaches directly search for the optimal ITR over a pre-specified class of ITRs to maximize an estimate of the mean outcome (Laber et al., 2014b; Athey and Wager, 2017; Davidian et al., 2014; Zhang et al., 2012b; Qi and Liu, 2018; Zhang and Zhang, 2018; Jiang et al., 2019). Zhao et al. (2012a) proposed to estimate the optimal ITR within the weighted classification framework, with weights proportional to the observed outcomes.

Another important topic is the inference of the estimated ITR. The inference of an ITR can be separated into two related questions. First, how should we quantify the performance of an estimated ITR? Second, what are the driving factors of the estimated ITR? In order to answer these questions, we need some inferential tools. To answer the first question, we need to provide interval estimations for the optimal value or the value under the estimated ITR. The inference of the optimal value has been shown to be a challenging problem, especially at exception laws where there is a subgroup of the patients whose treatment effect vanishes (Chakraborty et al., 2010; Laber et al., 2014b; Goldberg et al., 2013). Chakraborty et al. (2014) proposed a bootstrap procedure to construct a confidence interval for the value; Luedtke and Van Der Laan (2016) proposed an online one-step estimator for the value. Shi et al. (2020) proposed to use a subagging procedure to aggregate value estimates obtained by repeated sample-splitting. We refer to these methods as the inferential tools for the optimal value. To answer the second question, we need to provide marginal/joint interval estimations for the contribution of each coefficient. There is relatively less literature covering this topic.

In this chapter, based on the direct learning framework, we introduce a new type of learning approach—Efficiency Augmentation and Relaxation Learning (EARL) (Zhao et al., 2019). The EARL enjoys the advantages of the direct approaches and combines the information from both the propensity score and outcome regression models. Due to this, the EARL is doubly robust in the sense that either the propensity model or the outcome regression models are correctly specified, the resulting estimator is consistent. We further introduce an inference procedure to identify the driving factors when focusing on linear decision rules.

2 A Novel Learning Approach to Optimal Individualized Treatment Rules

2.1 Setup and Notation

In this section, we first introduce some notations and causal assumptions. Let X be a p-dimensional vector representing covariate information, A be a binary treatment ($\mathcal{A} = \{-1, 1\}$), and Y be the outcome of interest. We adopt the potential outcome framework (Rubin, 1974, 2005). Denote the potential outcomes under treatment $A = a$ as $Y(a)$. Then the observed outcome is $Y = \sum_a Y(a)1\{A = a\}$. This implies that for each patient, only one of the potential outcomes is observed. To ensure the identifiability of the causal effect, we assume the following conditions:

1. The stable unit treatment value assumption (SUTVA) (Imbens and Rubin, 2015);
2. No unmeasured confounding, i.e., $(Y(1), Y(-1)) \perp A \mid X$, where $\perp$ denotes independence;
3. Consistency $Y = Y(A)$.

The SUTVA condition guarantees that one patient's treatment assignment does not affect the potential outcomes of another patient. It also assumes that there are no different versions of the treatment. It is satisfied when the data triplets (X, A, Y)'s are independently identically distributed. The assumption of no unmeasured confounding is also called the strong ignorability condition. It assumes that the covariate information X captures all the correlations between the treatment assignment mechanism and the potential outcomes. These conditions are automatically satisfied in a randomized clinical trial, where the patients in the treatment arm share similar characteristics as those in the control arm due to randomization. If data are collected from an observational data source, these conditions are unverifiable. In practice, a large number of covariates are usually collected in the study in the hope that the confounding effect of treatment could be controlled.

An ITR denoted by $D(X)$ is a mapping from covariate space $\mathcal{X}$ to the treatment space $\mathcal{A}$. Given an ITR $D(X)$, the outcome is $Y(D(X))$, and the average outcome under $D(X)$ is

$$V(D) = E[Y(D(X))],$$

which is also called the value function. The optimal ITR is defined as the maximizer of the value function over all possible ITRs, i.e., $D_{\text{opt}} = \arg\max_D\{V(D)\}$.

Let $\pi_a(x) = P(A = a \mid X = x)$ and $Q_a(x) = E(Y \mid X = x, A = a)$. Then,

$$
\begin{aligned}
V(D) &= E[Y(1)1\{D(X) = 1\} + Y(-1)1\{D(X) = -1\}] \\
&= E[Q_1(X)1\{D(X) = 1\} + Q_{-1}(X)1\{D(X) = -1\}].
\end{aligned}
$$

It can be seen that $D_{\text{opt}}(x) = \text{sgn}\{Q_1(x) - Q_{-1}(x)\}$. One approach to estimate the optimal ITR is estimating $Q_a(x)$'s, which are referred to as the Q-function. This type of approach posits a model for $Q_a(x)$ and uses the prediction of Q-functions to estimate the optimal ITR. This framework is also called the indirect learning because it estimates ITR through estimating $Q_a(x)$'s. However, because the tight relationship between $Q_a(x)$'s and the estimator of the optimal ITR, if the model for $Q_a(x)$'s is not correctly specified, the estimation for the optimal ITR can be inconsistent even if $Q_1(x) - Q_{-1}(x)$ enjoys relative simple structure. To illustrate, we consider the following example provided in Qian and Murphy (2011). Suppose that X follows a uniform distribution in $[-1, 1]$, and A takes values in $\{-1, 1\}$ with equal probability. The outcome Y is generated from a standard normal distribution with mean $(X - 1/3)^2 A$. Given the data generation process, the optimal ITR $D_{\text{opt}}(x) \equiv 1$ because the CATE $(X - 1/3)^2$ is always positive. Consider linear regression models for $Q_a(x)$'s,

$$
Q_a(X) = \beta_0 + \beta_1 X + \psi_0 A + \psi_1 X A.
$$

The $Q_1(X) - Q_{-1}(X) = 2\psi_0 + 2\psi_1 X$ is linear and the optimal ITR enjoys a linear form $D_{\text{opt}}(X) = \text{sgn}\{\psi_0 + \psi_1 X\}$. Although the optimal ITR given the data generation process is indeed linear, we will see that the ITR estimated under linear models for $Q_a(x)$'s is biased. Specifically, minimizing the prediction error yields the $\widetilde{Q}_a(X) = (4/9 - 2/3X)A$. The estimated ITR $\widehat{D}(X) = \text{sgn}\{4/9 - 2/3X > 0\}$. It is sub-optimal because for patients satisfying that $4/9 - 2/3X < 0$, where the estimated ITR assigns $A = -1$ to the patients, and the optimal ITR assigns $A = 1$. Further, if we calculate the value function, $V(D_{\text{opt}}) = 4/9 > 29/81 = V(\widehat{D})$.

2.2 *Recap: Outcome Weighted Learning*

In this section, we introduce the framework of the outcome weighted learning, which is closely related to the value function and the proposed approach. To start with, we introduce an estimator for the value function $V(D)$ for any fixed D. Notice that

$$
\widehat{V}_{\text{OWL}}(D) = E_n[\widehat{W}_A 1\{D(X) = A\}],
$$

where

$$\widehat{W}_a = \frac{YI\{A = a\}}{\widehat{\pi}_a(X)},$$

where $\widehat{\pi}_a$ be the estimator of π_a. Equivalently, maximizing the $\widehat{V}_{\text{OWL}}(D)$ is equivalent to minimizing

$$E_n[\widehat{W}_A 1\{D(X) \neq A\}].$$

However, the space of all possible ITRs is huge. Furthermore, the involved nonsmooth and discontinuous indicator function makes the objective hard to optimize directly. One common solution in the machine learning literature is to replace the indicator function with a convex surrogate loss (Boser et al., 1992; Cortes and Vapnik, 1995; Freund and Schapire, 1997; Friedman et al., 2000; Bartlett et al., 2006). In addition, to reduce the complexity of searching over all possible ITRs, we restrict the class of ITRs and impose a penalty to avoid overfitting. Specifically, the outcome weighted learning aims to minimize a penalized empirical risk, i.e.

$$E_n[\widehat{W}_A \phi(Af)] + \lambda_n \|f\|_{\mathcal{F}}.$$

where $\mathcal{F}$ is a pre-specified class of the ITRs, $\phi(\cdot)$ is a convex surrogate loss, and $\|\cdot\|_{\mathcal{F}}$ is an associated norm of $\mathcal{F}$. The outcome weighted learning can be implemented easily using the developed R package. For example, the *glmnet* Package can be sued to solve this optimization when f is a linear decision rule and $\phi(\cdot)$ is chosen as a logistic loss. However, when Y is negative, the objective function may not be convex; in addition, to achieve fisher consistency, the propensity score model needs to be correctly specified.

Another type of approach aims at recasting the maximization of the value function into a weighted classification problem and directly searching for the optimal ITR (Zhang et al., 2012a,b). For example, Zhang et al. (2012b) directly optimized the value function given a class of ITRs. Zhang et al. (2012a) shown that the maximizer of the value function is equivalent to

$$\min_g E\left[|C(X)|[1\{C(X)\} - g(X)]^2\right],$$

where $C(X) = Q_1(X) - Q_{-1}(X)$, and $g(X)$ takes value only in $\{0, 1\}$. The $g(X) = 1$ represents that the treatment should be assigned; otherwise, the control should be assigned. Further, Zhang and Zhang (2018) shows that the optimal ITR can also be defined as

$$E[|C(X)|1\{D(X) = A\}].$$

They extend this idea to accommodate dynamic treatment regimes where there are multiple stages of the decisions to make.

2.3 Efficiency Augmentation and Relaxation Learning

In this section, we introduce a novel learning approach that incorporates both the propensity score and the outcome regression. The proposed method achieves fisher consistency either the propensity score model or the outcome regression model is correctly specified (double robustness). To start with, we introduce a doubly robust estimator for the value function $V(D)$ for any fixed D. Let $\widehat{Q}_a$'s be the estimator of the Q_a's, and $\widehat{\pi}_a$ be the estimator of π_a. An augmented inverse propensity weighting (AIPW) estimator of $V(D)$ can be formulated as

$$\widehat{V}_{\mathrm{DR}}(D) = E_n[\widehat{W}_1 1\{D(X) = 1\} + \widehat{W}_{-1} 1\{D(X) = -1\}],$$

where

$$\widehat{W}_a = W_a(Y, X, A, \widehat{\pi}, \widehat{Q}) = \frac{YI\{A = a\}}{\widehat{\pi}_a(X)} - \frac{[I\{A = a\} - \widehat{\pi}_a(X)]\,\widehat{Q}_a(X)}{\widehat{\pi}_a(X)}.$$

for $a \in \{-1, 1\}$, and $E_n[\cdot]$ is the empirical mean over the observed data. Comparing $\widehat{V}_{\mathrm{DR}}(D)$ and $\widehat{V}_{\mathrm{OWL}}(D)$, the $\widehat{V}_{\mathrm{OWL}}(D)$ can be considered as a special case of $\widehat{V}_{\mathrm{DR}}(D)$ where $\widehat{Q}_a(X)$ is mis-specified as 0.

Next, in order to estimate the optimal ITR, we maximize the estimated value function over the all possible ITRs,

$$\max_D \widehat{V}_{\mathrm{DR}}(D).$$

Equivalently, we can consider the following minimization problem over all possible ITRs

$$\min_D E_n[\widehat{W}_1 1\{D(X) \neq 1\} + \widehat{W}_{-1} 1\{D(X) \neq -1\}].$$

However, the space of all possible ITRs is huge. Similar to the outcome weighted learning, we consider the penalized empirical risk minimization

$$\widehat{L}^{\lambda_n}_{\phi,\mathcal{F}}(f) = E_n\left[\widehat{W}_1\phi(f) + \widehat{W}_{-1}\phi(-f)\right] + \lambda_n \|f\|_{\mathcal{F}}.$$

This method, called the Efficiency Augmentation and Relaxation Learning (EARL) method, was first proposed in Zhao et al. (2019).

Example 1 In this example, we focus on the linear function class. Specifically, we consider that $\mathcal{F} = \{f : f(X) = X^\top\beta\}$. The corresponding ITR will assign $A = 1$ to a patient if his or her covariate information X satisfies that $X^\top\beta > 0$. In this case, we may choose that $\|f\|_{\mathcal{F}} = \|\beta\|_1$. Then the proposed optimization problem is

$$\min_\beta \widehat{L}^{\lambda_n}_{\phi,\mathcal{F}}(\beta) = E_n\left[\widehat{W}_1\phi(X^\top\beta) + \widehat{W}_{-1}\phi(-X^\top\beta)\right] + \lambda_n\|\beta\|_1.$$

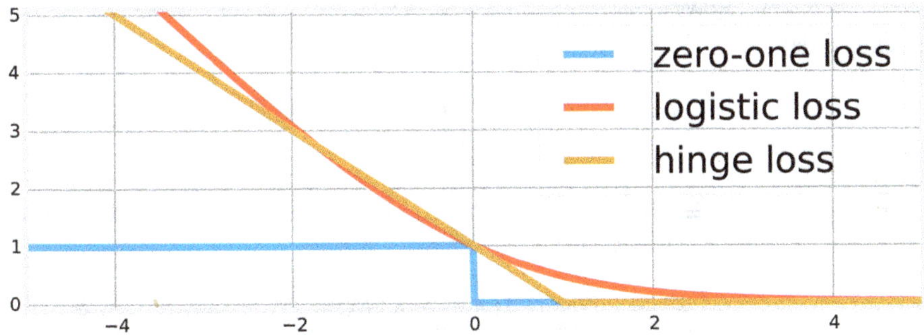

Fig. 1 Different choices of the surrogate loss functions

Suppose that $\widehat{\beta}$ is the minimizer of the proposed optimization problem, the estimated ITR is $\widehat{D} = \text{sgn}\left\{X^\top \widehat{\beta}\right\}$.

There are multiple options of the surrogate loss functions. One of the common choices is a logistic loss function, which is strictly convex and differentiable. This type of loss is often referred to as soft-margin classifiers (Wahba et al., 1999; Wahba, 2002). Another popular choice is a hinge loss function. The hinge loss function is piece-wise linear and thus non-differentiable, which is often referred to as hard-margin classifiers (Wahba et al., 1999; Wahba, 2002). Figure 1 presents these commonly selected loss functions. In a classification problem, when the class probabilities are complex, a hard-margin classifier may have better performance because it does not over-penalize the misclassified samples compared with soft-margin classifiers. Likewise, when the $Q_1(X) - Q_{-1}(X)$ is complex, using hard-margin classifiers may lead to improved performance.

In the implementation of the EARL approach, one crucial step is how to fit the nuisance parameters. The nuisance parameters include π_a and Q_a's. In some scenarios, we may have information on how to construct the nuisance parameters. For example, in a randomized clinical trial, the propensity π_a is usually known. However, in an observational study, we need to estimate these nuisance parameters using the observed samples. For example, a logistic regression treating A as the output and X as the predictor is commonly used in estimating π_a; generalized linear models for Q_a's can be applied for the treatment and the control arm separately. Besides these parametric approaches, other semiparametric or nonparametric methods can also be applied. One of the nonparametric methods to consider is the Nadaraya–Watson estimator (Nadaraya, 1964; Watson, 1964; Bierens, 1988). The Nadaraya–Watson estimator uses a kernel function $K(\cdot)$ satisfying

$$K(x) \geq 0, \qquad \int K(x)\text{d}x = 1,$$

$$\int xK(x)\text{d}x = 0, \qquad \text{and} \int x^2 K(x)\text{d}x > 0.$$

Then, the Nadaraya–Watson estimator for $\pi_a(x) = P(A = a|X) = E[1\{A = a\} \mid X]$ can be defined as

$$\widehat{\pi}_a(x) = \frac{E_n\left[1\{A = a\}K_h(X - x)\right]}{E_n\left[K(X - x)\right]},$$

where $K_h(X - x) = \frac{1}{h^p}K((X - x)/h)$. Similarly, we can estimate $Q_a(x) = E[Y(a) \mid X] = E[Y \mid A = a, X]$ by

$$\widehat{Q}_a(x) = \frac{E_n\left[Y1\{A = a\}K_h(X - x)\right]}{E_n\left[1\{A = a\}K_h(X - x)\right]}.$$

The bandwidth h can be selected by the cross-validation. When the number of the covariate is large (p is large), the performance of the Nadaraya–Watson estimators may not be ideal due to the curse of the dimensionality. To accommodate more covariates, we may consider semiparametric models. Specifically, we can first consider dimension reduction approach such as the sliced inverse regression (Li, 1991) and the minimum average variation estimation (Yin and Li, 2011). Then, based on the reduced dimension, we can use the Nadaraya–Watson kernel estimation. In addition to the statistical methods, machine learning methods can also be applied. For example, we could use the gradient boosted trees and deep neural networks to fit the nuisance parameters. Wager and Athey (2018) proposed causal trees that can be used to estimate $Q_a(x)$; Shalit et al. (2017) proposed a neural network approach to estimate $Q_a(x)$ and provided a generalization bound. Although we can accommodate any nonparametric or machine learning methods, in theory, we require that the product of the convergence rates of nuisance parameter estimations is faster than $n^{-1/2}$. Specifically, assume that $\sup_x \|\widehat{\pi}_a(x) - \pi_a(x)\| = O_p(n^{-\alpha})$ and $\sup_x \|\widehat{Q}_a(x) - Q_a(x)\| = O_p(n^{-\gamma})$, we require that $\alpha + \gamma > 1/2$. This can be attained if either the convergence rate of $\widehat{\pi}_a$ or $\widehat{Q}_a$ is sufficiently fast. For example, if π_a is estimated by a regression spline estimator and is known to be p_π-dimensional (low dimension) by design, we have $\sup_x |\widehat{\pi}_a(x) - \pi_a(x)| = O_p\left(n^{-1/3}\right)$, where π_a is assumed to belong to the Hölder class with a smoothness parameter greater than $5p_\pi$ (Newey, 1997). Then $n^{-\alpha-\gamma} \ll n^{-1/2}$ is satisfied when $n^{-\gamma} \ll n^{-1/6}$.

However, directly using plug-in weights $\widehat{W}_a$'s where the nuisance parameters are fitted on the entire dataset may lead to deteriorating performances due to potentially significant estimation errors of the nuisance parameters. To reduce the impact of the estimation error of the nuisance parameters on learning the ITR, we adopt an algorithm called cross-fitting (Chernozhukov et al., 2018). In this algorithm, we fit the nuisance parameters on only part of the samples and then estimate the ITRs using the other part of the dataset with the estimated nuisance parameters plugged in. This procedure ensures that the estimators for the nuisance parameters are independent of the data used to construct the proposed minimization problem.

Suppose we split the entire dataset into K equally-sized parts denoted as I_k's. Let $T(\theta)$ be a statistic depending on some parameters θ. We use the superscript to represent which part of the data is used to construct the statistic/parameter. For example, $T^{(k)}(\theta)$ represents the statistic is constructed using the data indexed in I_k. In addition, we use the subscript $(-k)$ to represent a parameter estimated using the data indexed in I_k^c. For example, $T^{(k)}(\theta^{(-k)})$ represents a statistic constructed using the data indexed in I_k with the parameter estimated using the data indexed in I_k^c. Similarly, we use $E_n^{(k)}[\cdot]$ to represent the empirical mean on data indexed in I_k; and $E_n^{(-k)}[\cdot]$ to represent the empirical mean on data indexed in I_k^c.

We first randomly split the entire dataset into K parts with equal sample sizes. Then, for each $k = 1, \ldots, K$, we use the sample indexed in I_k^c to esitmate the nusiance paramters and we denote the estimated nuisance parameters as $\widehat{p_a}^{(-k)}$ and $\widehat{Q_a}^{(-k)}$'s. Using the estimated nuisance parameters, we can form the weights $\widehat{W}_a^{(-k)}$'s and solve

$$\min_{f_{\mathcal{F}}} \widehat{L}^{\lambda_n}_{\phi,\mathcal{F}}(f) = E_n^{(k)}\left[\widehat{W}_1^{(-k)}\phi(f) + \widehat{W}_{-1}^{(-k)}\phi(-f)\right] + \lambda_n \|f\|_{\mathcal{F}}.$$

We denote the minimizer as $\widehat{f}^{(k)}$. Then we can go through the same procedure with different choices of k. We denote the minimizer as $\widehat{f}^{(k)}$. To compensate on the efficiency loss in the sample splitting, we average the two minimizers and obtain the final estimator

$$\widehat{f} = \frac{1}{K}\sum_{k=1}^{K} \widehat{f}^{(k)}.$$

The following figure (Fig. 2) summarizes the entire procedure with $K = 2$.

The proposed method can be implemented by using established software. For example, *R* Package, *glmnet*, can be used when the logistic loss is employed as the surrogate loss; and quadratic programming or gradient descent algorithm can be carried out when using the hinge loss (Zhao et al., 2012b).

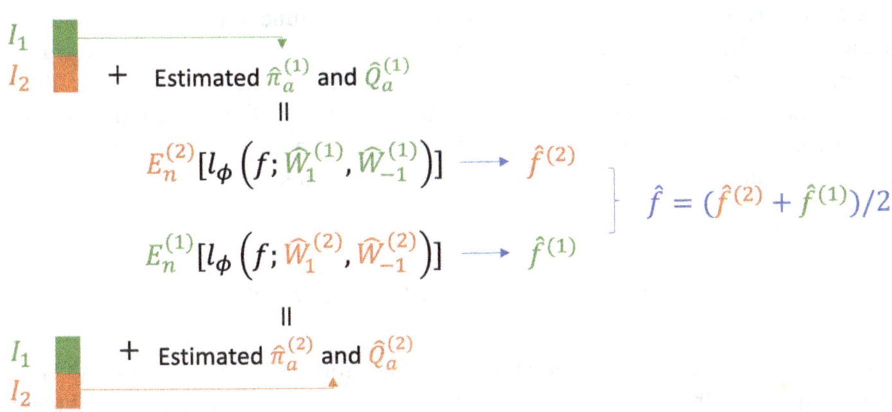

Fig. 2 Cross-fitting algorithm for EARL approach

3 Inference of the Estimated Linear ITR

In this section, we aim to provide an inference procedure to the linear ITR estimated by the EARL approach. In EARL procedure, there are many possible choices of the class of decision rules $\mathcal{F}$. In practice, for the interpretability, we may choose a linear decision rule. To interpret the estimated linear ITR, an inference procedure which produces p-values or interval estimations for the coefficients in the linear decision rule is important. Specifically, we minimize the following (population) risk

$$\min_{\beta} L_{\phi} = E\left[W_1\phi(X^\top\beta) + W_{-1}\phi(-X^\top\beta)\right],$$

where $W_a = E[Y(a) \mid X]$. Denote the minimizer as β^*. Our inference procedure aims to provide p-values or interval estimations for the coordinates of β^*. To illustrate the procedure, in this section, we consider the hypothesis testing for

$$H_0 : \beta_1^* = 0, \quad H_1 : \beta_1^* \neq 0,$$

where β_j^* represents the j-coordinate of the β^*.

To estimate β^*, we use the EARL approach with $\mathcal{F} = \{f(X) = X^\top\beta\}$. When the number of the covariate is much smaller than the sample size, we can choose $\lambda_n = 0$ to avoid any bias induced by the penalization. When the number of the covariate is much larger than the sample size, we have to impose a penalization which induces the additional bias. In this case, we need to adjust the estimate to remove the induced bias. This procedure is related to the methods in high-dimensional inference literature and will be discussed in Sect. 4. In this section, we focus on a low-dimensional setting. In our discussion, we consider the case where the surrogate loss is differentiable and the case where the surrogate loss is non-differentiable. The key difference is that when the surrogate loss is non-differentiable, the hessian of the empirical risk does not exist. Consequently, the hessian of the (population) risk L_{ϕ} involves non-regular functions such as a Dirac function. The Dirac function achieves infinity at a single point and vanishes at all other points; the integral of a Dirac function over the real line is 1. To estimate the hessian involving these non-regular functions, special treatment is needed. In the following, we first discuss the inference procedure under a differentiable surrogate loss and then highlight the difference when we use a non-differentiable surrogate loss.

3.1 *Inference Under a Differentiable Surrogate Loss*

When the surrogate loss is differentiable, one possible approach is to use the Wald test. To illustrate the idea, we first consider the case where π_a and Q_a's are known and then extend to the case where these parameters are unknown. When the π_a and

Q_a's are known, the estimator $\widehat{\beta}$ is obtained via minimizing

$$\widehat{L}^0_{\phi,\mathcal{F}}(\beta) = E_n\left[W_1\phi(X^\top\beta) + W_{-1}\phi(-X^\top\beta)\right].$$

The score function for β^* is

$$S_n(\beta) = E_n\left[\left\{W_1\phi'(X^\top\beta) - W_{-1}\phi'(-X^\top\beta)\right\}X\right].$$

By the definition of $\widehat{\beta}$, we have

$$\begin{aligned} -\sqrt{n}S_n(\beta^*) &= \sqrt{n}\left\{S_n(\widehat{\beta}) - S_n(\beta^*)\right\} \\ &= \sqrt{n}\nabla S_n(\beta^*)(\widehat{\beta} - \beta^*) + o_p(1), \end{aligned}$$

where $\nabla S_n(\beta^*)$ is the gradient of $S_n(\beta)$ with respect to β evaluated at $\beta = \beta^*$. By the central limit theorem, under mild conditions, we have

$$\sqrt{n}S_n(\beta^*) \to N(0, \Sigma),$$

where $\Sigma = E\left[\left\{W_1\phi'(X^\top\beta^*) - W_{-1}\phi'(-X^\top\beta^*)\right\}^2 XX^\top\right]$. In addition, we have that

$$\nabla S_n(\beta^*) \to \nabla S(\beta^*)$$

in probability, where $S(\beta^*) = E[S_n(\beta^*)]$ and $\nabla S(\beta^*)$ is the gradient of $S(\beta)$ with respect to β evaluated at $\beta = \beta^*$. Combining these results, we have

$$\sqrt{n}(\widehat{\beta} - \beta^*) \to N(0, \left\{\nabla S(\beta^*)\right\}^{-1}\Sigma\left\{\nabla S(\beta^*)\right\}^{-1}).$$

Let σ^2 be the (1, 1)-entry of the matrix $\{\nabla S(\beta^*)\}^{-1}\Sigma\{\nabla S(\beta^*)\}^{-1}$. In addition, we estimate σ^2 by replacing β^* with $\widehat{\beta}$ in the formula and denote the estimate as $\widehat{\sigma}^2$. We can construct the Wald statistic

$$T_{\mathrm{w}} = \frac{\widehat{\beta}_1^2}{\widehat{\sigma}^2/n}.$$

Under the null hypothesis, the Wald statistic T_{w} should converge to a χ^2-distribution with one degree of freedom.

Another strategy is to use the score test. In the score test, we focus on the score related to β_1^*, i.e.

$$S_n(\beta_1; \beta_{-1}) = E_n\left[\left\{W_1\phi'(X^\top\beta) - W_{-1}\phi'(-X^\top\beta)\right\}X_1\right].$$

Under the null hypothesis, we construct the score test statistic

$$T_{\mathrm{s}} = \sqrt{n} S_n(0; \widehat{\beta}_{-1}).$$

In order to obtain the asymptotic distribution of T_{s}, we consider the Taylor's expansion

$$\begin{aligned}
&\sqrt{n} T_{\mathrm{s}}(0; \widehat{\beta}_{-1}) \\
&= \sqrt{n} T_{\mathrm{s}}(0; \beta^*_{-1}) + \sqrt{n} \left\{ T_{\mathrm{s}}(0; \widehat{\beta}_{-1}) - T_{\mathrm{s}}(0; \beta^*_{-1}) \right\} \\
&= \sqrt{n} T_{\mathrm{s}}(0; \beta^*_{-1}) + \sqrt{n} \nabla^{\top}_{\beta_{-1}} T_{\mathrm{s}}(0; \beta^*_{-1}) \left(\widehat{\beta}_{-1} - \beta^*_{-1} \right) + o_p(1),
\end{aligned}$$

where $\nabla_{\beta_{-1}} T_{\mathrm{s}}(0; \beta^*_{-1})$ is the gradient of $T_{\mathrm{s}}(\beta_1; \beta_{-1})$ with respect to β_{-1} evaluated at $\beta_{-1} = \beta^*_{-1}$. Under the null hypothesis, the first two terms are asymptotically normal. By calculating the asymptotic covariance, we can derive that under the null hypothesis, the score test statistic is asymptotic normal.

When π_a and Q_a's are unknown, we can apply the cross-fitting procedure to obtain a valid test statistic. We split the entire dataset into K parts with equal sample sizes. For each k, we use the samples indexed in I_k^c to fit these nuisance parameters and then use the samples indexed in I_k to calculate the test statistic. Finally, we average the test statistics derived from each part of the data. For example, if we use the Wald test statistic, we can derive $\widehat{\beta}_1^{(k)}$ and $\widehat{\sigma}^{(k)}$, the estimators for β_1^* and σ^2, estimated using the samples indexed in I_k with nuisance parameters fitted using samples indexed in I_k^c. Then, we can calculate the Wald test statistic for the k-th part.

$$T_{\mathrm{w}}^{(k)} = \frac{\left(\widehat{\beta}_1^{(k)} \right)^2}{K \left(\widehat{\sigma}^{(k)} \right)^2 / n}.$$

We average these test statistics by

$$T_{\mathrm{w}} = \frac{1}{K} \sum_{k=1}^{K} T_{\mathrm{w}}^{(k)}.$$

For the score test, we average the test statistics by

$$T_{\mathrm{s}} = \frac{1}{K} \sum_{k=1}^{K} T_{\mathrm{s}}^{(k)},$$

where $T_{\mathrm{s}}^{(k)}$ is the score test statistic calculated using the samples indexed in I_k with nuisance parameters fitted using the samples indexed in I_k^c.

3.2 Inference Under a Non-differentiable Surrogate Loss

When the surrogate loss is non-differentiable, the regular Hessian of the surrogate loss does not exists. In this case, the asymptotic variance of T_{w}/ the Hessian of the risk L_ϕ is non-regular because it may involve non-regular function such as a Dirac function. Thus, we need additional efforts to estimate the asymptotic variance. For simplicity, we focus on the hinge loss as the surrogate loss. If the π_a and Q_a's are known, the estimator $\widehat{\beta}$ is obtained via minimizing

$$\widehat{L}^{\lambda}_{\phi,\mathcal{F}}(\beta) = E_n\left[W_1(1 - X^\top\beta)_+ + W_{-1}(1 + X^\top\beta)_+\right] + \lambda\|\beta\|^2,$$

where $(z)_+ = \max\{z, 0\}$. To construct the Wald test statistic, we need to take the gradient and the hessian of the surrogate loss. However, the gradient and the hessian for the hinge loss/empricial risk do not exist. To overcome the non-differentiability, the key is to approximate the gradient and the hessian of the risk $L_\phi = E\left[\widehat{L}^{0}_{\phi,\mathcal{F}}(\beta)\right]$. Koo et al. (2008) show that using the subgradient and a global kernel function, we can derive a valid Wald-type test statistic for the empirical risk minimization using the hinge loss. Following this idea, we can show that

$$\begin{aligned}&\sqrt{n}(\widehat{\beta} - \beta^*)\\ &= -\sqrt{n}H(\beta^*)^{-1}E_n\left[\left\{W_1 1\{X^\top\beta^* \le 1\} - W_{-1}1\{X^\top\beta^* \ge -1\}\right\}X\right] + o_p(1),\end{aligned}$$

where $H(\beta) = E\left[\left\{W_1\delta(1 - X^\top\beta) + W_1\delta(1 + X^\top\beta)\right\}XX^\top\right]$. The $\delta(\cdot)$ is known as the Dirac function which obtains $+\infty$ at point 0 and vanished at other points. By the central limit theorem, we have that

$$\sqrt{n}E_n\left[\left\{W_1 1\{X^\top\beta^* \le 1\} - W_{-1}1\{X^\top\beta^* \ge -1\}\right\}X\right] \to N(0, \Sigma),$$

where $\Sigma = E\left[\left\{W_1 1\{X^\top\beta^* \le 1\} - W_{-1}1\{X^\top\beta^* \ge -1\}\right\}^2 XX^\top\right]$. Thus, the $\sqrt{n}(\widehat{\beta} - \beta^*)$ converges to a Gaussian random variable with the asymptotic variance $H(\beta^*)^{-1}\Sigma H(\beta^*)^{-1}$ when $n \to +\infty$. In order to form a Wald test statistic, we need to estimate $H(\beta^*)$ and Σ. It can be shown that

$$\begin{aligned}H(\beta) = {}& f_{X^\top\beta}(1)E\left[W_1XX^\top \mid X^\top\beta = 1\right]\\ &+ f_{X^\top\beta}(-1)E\left[W_{-1}XX^\top \mid X^\top\beta = -1\right],\end{aligned}$$

which is related to the density of $X^\top\beta$ and the conditional expectations of $W_aXX^\top$'s given $X^\top\beta = a$.

We can use the Nadaraya–Watson estimators to estimate $f_{X^\top\beta}(a)$ and $E\left[W_a XX^\top \mid X^\top\beta = a\right]$ separately

$$\widehat{f}_{X^\top\beta}(a) = E_n\left[K_h(X^\top\beta - a)\right],$$

$$\widehat{E}\left[W_a X_{j_1} X_{j_2} \mid X^\top\beta = a\right] = \frac{E_n\left[W_a X_{j_1} X_{j_2} K_h(X^\top\beta - a)\right]}{E_n\left[K_h(X^\top\beta - a)\right]}.$$

Then we can form a plug-in estimator for $H(\beta)$, i.e.,

$$\widehat{H}(\beta) = E_n\left[\left\{W_1 K_h(X^\top\beta - 1) + W_1 K_h(1 + X^\top\beta)\right\} XX^\top\right].$$

We can also view the above estimator by using the kernel function to approximate the Dirac function in $H(\beta)$. Figure 3 shows that with a smaller bandwidth h, the kernel function would gradually approximate the Dirac function.

As shown in Koo et al. (2008), the $H(\beta^*)$ can be estimated by $\widehat{H}(\widehat{\beta})$ under mild conditions. In addition, the Σ can also be estimated by a plug-in estimator

$$\widehat{\Sigma} = E_n\left[\left\{W_1 1\{X^\top\widehat{\beta} \le 1\} - W_{-1} 1\{X^\top\widehat{\beta} \ge -1\}\right\}^2 XX^\top\right].$$

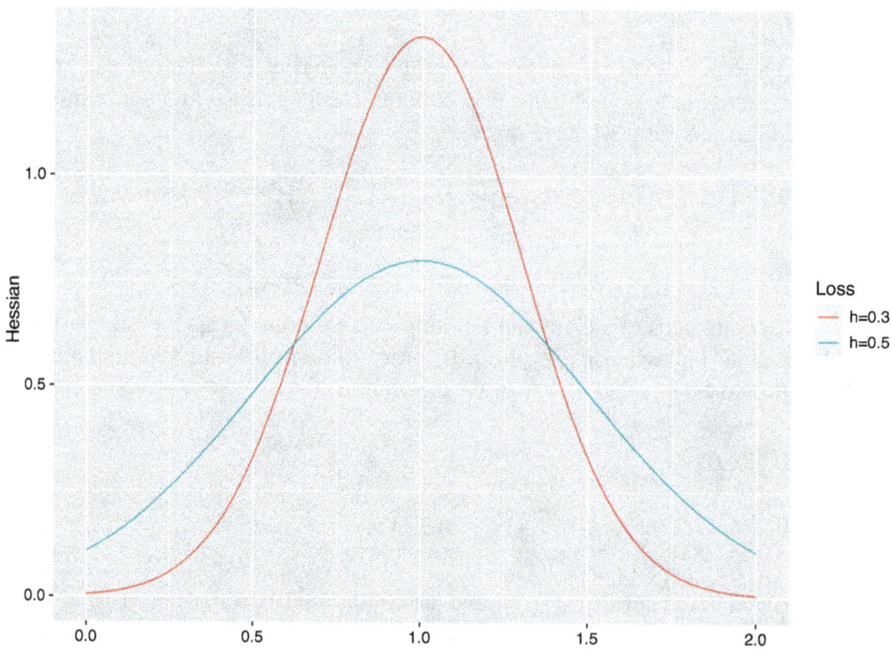

Fig. 3 With a smaller bandwidth, the kernel function approximates the Dirac function which obtains ∞ at 1 and vanishes at all other points

The details of the required conditions can be found in Koo et al. (2008). Among these conditions, one of the important implications is that the samples cannot concentrate on the boundary $X^\top \beta^* = 1$ and $X^\top \beta^* = -1$. This implication ensures the consistency of the plug-in estimators. We suggest choosing the bandwidth by cross-validation or using the optimal bandwidth $h_{\text{opt}} = Cn^{-1/5}$, which minimizes the squared bias and variance when using a Gaussian kernel function as a kernel approximation in estimating $H(\beta)$ for any fixed β. This order of the optimal bandwidth is also consistent with the optimal bandwidth selection in kernel regressions using the Gaussian kernel (Nadaraya, 1964). When the nuisance parameters are unknown, we could also combine with the cross-fitting algorithm to construct a valid inference procedure.

4 Inference Under High-Dimensional Covariate

In this section, we introduce the estimation and inference of the linear ITRs in a high-dimensional setting. The high-dimensional setting applies to the cases where the total number of covariates is equal to or larger than the sample size. We consider the following penalized empirical risk minimization problem, i.e.

$$\widehat{L}^{\lambda_n}_{\phi,\mathcal{F}}(\beta; W_1, W_{-1}) = E_n\left[W_1\phi(X^\top\beta) + W_{-1}\phi(-X^\top\beta)\right] + \lambda_n\|\beta\|_1,$$

where λ is a tuning parameter which can be tuned by cross-validation.

Although the estimation procedure with high-dimensional covariates is straightforward by incorporating a lasso penalty, the inference procedure is not easy due to the bias induced by $\lambda_n\|\beta\|_1$. For the Wald test, due to the bias, the $\sqrt{n}(\widehat{\beta} - \beta^*)$ is not asymptotic normal. For the score test, by Taylor expansion, we have

$$\begin{aligned}
&\sqrt{n}T_{\text{s}}(0; \widehat{\beta}_{-1}) \\
&= \sqrt{n}T_{\text{s}}(0; \beta^*_{-1}) + \sqrt{n}\left\{T_{\text{s}}(0; \widehat{\beta}_{-1}) - T_{\text{s}}(0; \beta^*_{-1})\right\} \\
&= \sqrt{n}T_{\text{s}}(0; \beta^*_{-1}) + \sqrt{n}\nabla^\top_{\beta_{-1}} T_{\text{s}}(0; \beta^*_{-1})\left(\widehat{\beta}_{-1} - \beta^*_{-1}\right) + o_p(1),
\end{aligned}$$

Because $\sqrt{n}\nabla^\top_{\beta_{-1}} T_{\text{s}}(0; \beta^*_{-1})\left(\widehat{\beta}_{-1} - \beta^*_{-1}\right)$ may not be asymptotically normal due to the bias induced by the penalization, the score test statistic $\sqrt{n}T_{\text{s}}(0; \widehat{\beta}_{-1})$ may not be asymptotically normal. Thus, special treatments are needed to infer β^* in the high-dimensional settings.

4.1 Split-and-Pooled De-correlated Score Test for Differentiable Surrogate Loss

In this section, we introduce the split-and-pooled de-correlated score test to accommodate the high-dimensional covariates when the surrogate loss is differentiable (Liang et al., 2020). This is generalized from the de-correlated score developed in Ning and Liu (2017). The proposed algorithm can also deal with nuisance parameters fitted by nonparametric methods due to the intrinsic spirit of the cross-fitting algorithm. The key idea is to project the score function in a certain direction such that the constructed score is de-correlated with the estimation error of $\widehat{\beta}_{-1}$. Specifically, we consider

$$T_s(\beta, w^*; W_1, W_{-1}) = E_n\left[\left\{W_1\phi^{'}(X^\top\beta) - W_{-1}\phi^{'}(-X^\top\beta)\right\}(X_1 - X_{-1}^\top w^*)\right],$$

where w^* is a $p-1$ dimensional vector, defined as the minimizer of

$$\min_w E\left[\left\{W_1\phi^{''}(X^\top\beta) + W_{-1}\phi^{''}(-X^\top\beta)\right\}(X_1 - X_{-1}^\top w)^2\right].$$

Let $\widehat{\beta}_{\text{null}} = (0, \widehat{\beta}_{-1})^\top$. To see the key idea, by the Taylor's expansion, under mild conditions, we have that

$$\begin{aligned}&\sqrt{n}T_s(\widehat{\beta}_{\text{null}}, w^*; W_1, W_{-1})\\ &= \sqrt{n}T_s(\beta^*, w^*; W_1, W_{-1}) + \sqrt{n}\nabla T_s(\beta^*, w^*; W_1, W_{-1})^\top(\widehat{\beta}_{\text{null}} - \beta^*) + o_p(1).\end{aligned}$$

The first term $T_s(\beta^*, w^*; W_1, W_{-1})$ is asymptotically normal. The second term

$$\begin{aligned}&\sqrt{n}\nabla T_s(\beta^*, w^*; W_1, W_{-1})^\top(\widehat{\beta}_{\text{null}} - \beta^*)\\ &= \sqrt{n}E_n\left[\left\{W_1\phi^{''}(X^\top\beta^*) + W_{-1}\phi^{''}(-X^\top\beta^*)\right\}(X_1 - X_{-1}^\top w^*)X_{-1}^\top\right]\\ &\quad\times(\widehat{\beta}_{-1} - \beta_{-1}^*)\\ &= o_p(1)\end{aligned}$$

due to the fact that w^* is the solution of

$$E\left[\left\{W_1\phi^{''}(X^\top\beta^*) + W_{-1}\phi^{''}(-X^\top\beta^*)\right\}(X_1 - X_{-1}^\top w)X_{-1}\right] = 0.$$

Thus, we could use $\sqrt{n}T_s(\widehat{\beta}_{\text{null}}, w^*; W_1, W_{-1})$ as a test statistic if w^*, W_1, and W_{-1} are known.

To estimate w^*, we can optimize

$$\min_w E_n\left[\left\{W_1\phi^{''}(X^\top\widehat{\beta}) + W_{-1}\phi^{''}(-X^\top\widehat{\beta})\right\}(X_1 - X_{-1}^\top w)^2\right] + \mu_n\|w\|_1,$$

where μ_n is a tuning parameter and can be tuned by cross-validation.

The entire algorithm can be separated into three steps. First, we estimate the nuisance parameters W_1 and W_{-1}; then we estimate β^* and use the estimated coefficients to estimate w^*; finally, we can plug in the de-correlated score. To accommodate the nuisance parameters fitted by nonparametric methods, we can employ the cross-fitting algorithm. The entire procedure is summarized in the following algorithm.

Algorithm 1: Inference of β^* using a sample-split procedure

Input: A random seed; n samples; a positive integer K.
Output: $\widehat{\beta}$ and a p-value for $\mathcal{H}_0 : \beta_1^* = 0$.
Randomly split data into K parts I_k's with equal size, and set $k = 1$;
Estimate π_a and Q_a on I_1^c and denote the estimator as $\widehat{\pi}_a^{(-k)}$ and $\widehat{Q}_a^{(-k)}$;
Obtain a data-split estimator $\widehat{\beta}^{(k)}$ on I_k by

$$\min_{\beta} \widehat{L}_{\phi,\mathcal{F}}^{\lambda_{n,k},(k)}(\beta; W_1^{(-k)}, W_{-1}^{(-k)}),$$

where $\lambda_{n,k}$ is tuned by cross-validation ;
Obtain an estimator $\widehat{w}^{(k)}$ for w^* by

$$\min_{w} E_n^{(k)}\left[\left\{W_1^{(-k)}\phi''(X^\top\widehat{\beta}^{(k)}) + W_{-1}^{(-k)}\phi''(-X^\top\widehat{\beta}^{(k)})\right\}(X_1 - X_{-1}^\top w)^2\right] + \mu_{n,k}\|w\|_1,$$

where $\mu_{n,k}$ is tuned by cross-validation ;
Construct the data-split de-correlated score test statistic $S^{(k)}(\widehat{\beta}_{\text{null}}^{(k)}, \widehat{w}^{(k)})$, and the estimator of the variance
$\widehat{\sigma}_k^2 = E_n^{(k)}\left[\left\{\nabla l_\phi\left(\widehat{\beta}^{(k)}; \widehat{\Omega}_+^{(-k)}, \widehat{\Omega}_-^{(-k)}\right)\right\}^2 (X_1 - X_{-1}^\top\widehat{w}^{(k)})^2\right]$;
Set $k = 2, 3, \ldots, K$, and repeat Step 2 and 5. Obtain $\{\widehat{\beta}^{(k)}\}_{k=1}^K$ and $\left\{T_s^{(k)}\left(\widehat{\beta}_{\text{null}}^{(k)}, \widehat{w}^{(k)}\right)\right\}_{k=1}^K$ as well as $\{\widehat{\sigma}_k^2\}_{k=1}^K$. Aggregate them by

$$\widehat{\beta} = K^{-1}\sum_{k=1}^K \widehat{\beta}^{(k)},\ T_s = K^{-1}\sum_{k=1}^K T_s^{(k)}\left(\widehat{\beta}_{\text{null}}^{(k)}, \widehat{w}^{(k)}\right), \widehat{\sigma}^2 = K^{-1}\sum_{k=1}^K \widehat{\sigma}_k^2.$$

Calculate the p-value by $2\left(1 - \Phi(n^{1/2}|S|/\widehat{\sigma})\right)$, where $\Phi(\cdot)$ is the cumulative distribution function of a standard normal distribution.

4.2 *Kernel-Smoothened De-correlated Score Test for Non-differentiable Loss*

When the surrogate loss is non-differentiable, the split-and-pooled de-correlated score cannot be directly applied because the hessian of the surrogate loss does not exist. Similar to the low-dimensional situation, we consider to approximate the hessian of the risk L_ϕ by a kernel function (Liang et al., 2021+). Specifically, we define the w^* as the solution to

$$\min_w E\left[\left\{W_1\delta(1-X^\top\beta^*)+W_1\delta(1+X^\top\beta^*)\right\}(X_1-X_{-1}^\top w)^2\right].$$

and estimate w^* by

$$\min_w E_n\left[\left\{W_1K_h(1-X^\top\beta^*)+W_1K_h(1+X^\top\beta^*)\right\}(X_1-X_{-1}^\top w)^2\right]+\mu_n\|w\|_1.$$

Because the $K_h(\cdot)$ is non-negative, the proposed optimization is strictly convex in w. However, due to the kernel approximation, the inference procedure is much more involved. Assume that W_a's are known. We can split the entire sample into K folds. We leave the k-th fold to estimate w and use the rest folds to fit β^*. Specifically, for each k, we use the sample indexed in I_k^c to estimate β^*. Then we optimize a cross-averaged loss function to estimate w. Finally, we construct the kernel-smoothened de-correlated score. The cross-fitting algorithm can be utilized for a valid inference procedure when W_a's need to be estimated. Details can be found in Liang et al. (2021+).

5 Simulations

In this section, we conduct simulations to show (1) the advantage of the EARL approach vs. the Q-learning approach; (2) the advantage of the EARL approach using a hinge loss vs. a logistic loss.

5.1 *The Advantage of the EARL vs. Q-Learning Approach*

To show the advantage of the EARL approach vs. Q-learning approach, we test our estimation and inference procedure under various simulation scenarios. Let

$$\Delta(X)=\{Q(1;X)-Q(-1;X)\}/2$$

and

$$S(X)=\{Q(1;X)+Q(-1;X)\}/2.$$

We generate $X\sim N\left(0,I_{p\times p}\right)$, and $Y=A\Delta(X)+S(X)+\epsilon$, $\epsilon\sim N(0,1)$. Let $\beta^{\text{opt}}=(1,1,-1,-1,0,\ldots,0)^\top$, $\beta_S^*=(-1,-1,1,-1,0,\ldots,0)^\top$, and $\beta_\pi^*=(1,-1,0,\ldots,0)^\top$. The following scenarios are considered:

(I). $\Delta(X) = \xi X^\top \beta^{\mathrm{opt}}$, $S(X) = 0.4X^\top \beta_S^*$, and

$$\pi(1; X) = \exp(0.4X^\top \beta_\pi^*) / \left\{1 + \exp(0.4X^\top \beta_\pi^*)\right\};$$

(II). $\Delta(X) = \left\{\Phi\left(\xi X^\top \beta^{\mathrm{opt}}\right) - 0.5\right\} \times \widetilde{\Delta}(X)$, $S(X) = \exp\left(0.4X^\top \beta_S^*\right)$, $\pi(1; X) = \exp\{(X_1^2 + X_2^2 + X_1 X_2)/4\} / \left[1 + \exp\{(X_1^2 + X_2^2 + X_1 X_2)/4\}\right]$, where $\widetilde{\Delta}(X) = 2(\sum_{j=1}^4 X_j)^2 + 2\xi$ and $\Phi(\cdot)$ is the cdf of the standard normal distribution.

Under these settings, the magnitude of the treatment effect $\Delta(X)$ changes with ξ, which ranges from 0.1 to 1. Scenario (I) features a linear outcome model $Q(a; X)$ for both $a = 1$ and $a = -1$, and a logistic model for the propensity. Scenario (II) has a nonlinear treatment effect $\Delta(X)$, though the decision boundary is still linear. The treatment assignment mechanism is also complex.

We compare the pooled estimator with Q-learning (Qian and Murphy, 2011). With high-dimensional covariates, we fit a linear regression with a lasso penalty in Q-learning for all scenarios. In our simulation experiments, we will test and construct confidence intervals for β_j^*'s, $j = 1, \ldots, 8$, the j-th coordinate of β^*, which by abuse of notations, denote the limits of estimates under either method. We conduct the hypothesis testing for Q-learning using the decorrelated score test proposed in Ning and Liu (2017). An R package called `ITRInference` is coded to implement the proposed method and Q-learning approach, which is open source at https://github.com/muxuanliang/ITRInference. For the proposed method, the user can specify the method or select from a list of candidates to estimate nuisance parameters. In our implementation, we estimate π and Q functions nonparametrically for all scenarios. To be more specific, we first implement a distance correlation-based variable screening procedure (Li et al., 2012). We then fit a kernel regression using the selected variables after screening. When estimating π, we set caps at 0.1 and 0.9 to trim extreme values.

In all scenarios, the sample size n and the dimension p range from 350, 500, 800, 1600, 2500 to 8000. We set the nominal significance level at 0.05, and the nominal coverage at 95%. We report the type I errors, the powers of the hypothesis tests, and the value functions under the estimated decision rules out of 500 replications. In particular, we present the type I errors for testing β_5^* to β_8^*, and the powers for testing β_1^* to β_4^*.

Figures 4 and 5 show the simulation results for different scenarios, with the sample size n varied and the p and ξ fixed. As expected, in Scenarios (I) (Fig. 4) where the regression model is correctly specified for Q-learning, Q-learning yields a better value function. Conversely, the proposed method outperforms the Q-learning method in Scenario (II) (Fig. 5). In terms of the type I error and power, the proposed method is comparable to the Q-learning approach in Scenario (I) (Fig. 4). For Scenarios (II) (Fig. 5), our method is more powerful, and the type I errors are well controlled.

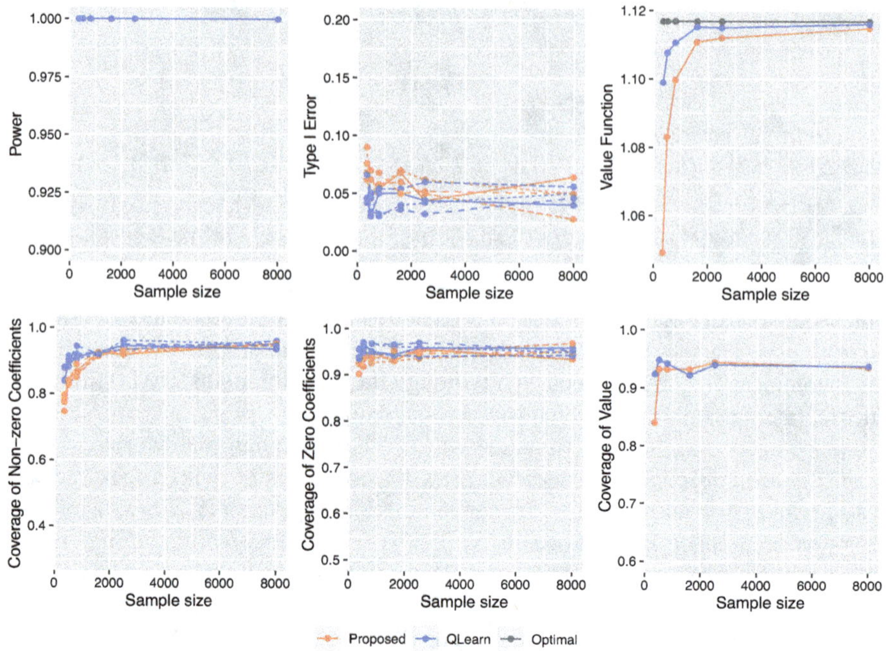

Fig. 4 Simulation results for Scenario (I) with the change of sample size when $\xi = 0.7$ and $p = 2500$. Types of the line represent different coefficients

5.2 *The Advantage of the EARL Using a Hinge Loss vs. a Logistic Loss*

To show the advantage of the EARL approach using a hinge loss vs. a logistic loss, we generate data by the following models.

(III). The population is a mixture of two subgroups with a probability 0.4 from Group I and probability 0.6 from Group II. For Group I, the covariate vector is generated from $N(\xi\mu_0, I_{p\times p} - 0.1e_1e_1^\top)$; for Group II, the covariate vector is generated from $N(\xi\mu_1, I_{p\times p})$. For each group, $Y(-1) = (X^\top\gamma)^2 + \epsilon$, where $\gamma = (-0.4, -0.4, 0.4, -0.4, 0, \cdots, 0)^\top$ and ϵ follows a standard normal distribution. For Group I, the treatment effect is $E[Y(1) - Y(-1)|X] = |X_1| + 0.5$; for Group II, the treatment effect is $E[Y(1) - Y(-1)|X] = -(|X_1| + 0.5)$. The treatment assignment mechanism follows $P(A = 1 \mid X) = \exp(0.25 \times (X_1^2 + X_2^2 + X_1X_2)/(1 + 0.25 \times (X_1^2 + X_2^2 + X_1X_2))$. The observed outcome $Y = Y(a)$ if the treatment a is assigned to the patient.

The parameter ξ controls the magnitude of the overlapping of two subgroups. With a larger ξ, the two subgroups become easier to be separated by a linear decision rule. In our simulations, we gradually increase ξ from 0 to 1 (by 0.1), sample size

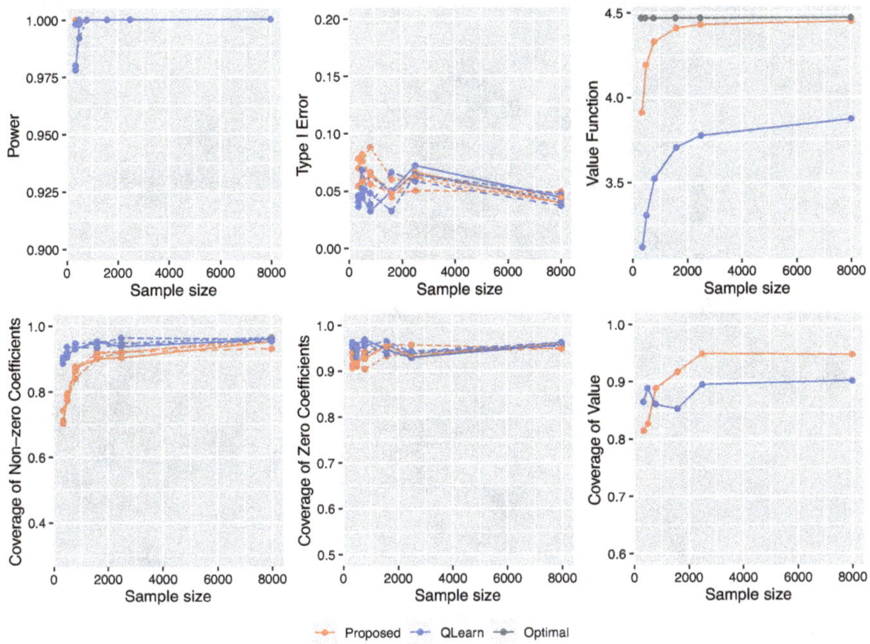

Fig. 5 Simulation results for Scenario (II) with the change of sample size when $\xi = 0.8$ and $p = 2500$. Types of the line represent different coefficients

n from 500 to 1600, and the dimension of the covariate p from 500 to 1600, which lead to 120 settings in total. We use kernel regression after the variable screening to estimate both nuisance parameters.

To evaluate the performance of the estimated ITR, we generate a testing dataset with a sample size of 10^4. On this testing dataset, we calculate the value defined as $\widehat{E}_{\text{test}}[Y(\text{sgn}(X^\top\widehat{\beta}))]$ under the estimated ITR, where $\widehat{E}_{\text{test}}[\cdot]$ is the empirical average over all testing samples. To evaluate the inference procedure, we report the Type I error and power for the first eight coordinates of β^*. Each simulation setting is repeated for 500 times. The true value of β^* is estimated by the average over these 500 repeats with $n = 2500$.

Figures 6 and 7 show the simulation results. We can see that the hinge loss tends to perform better when ξ are extremes (smaller than 0.2; larger than 0.8). From the power and Type I error, the proposed inference method is valid for the hinge loss.

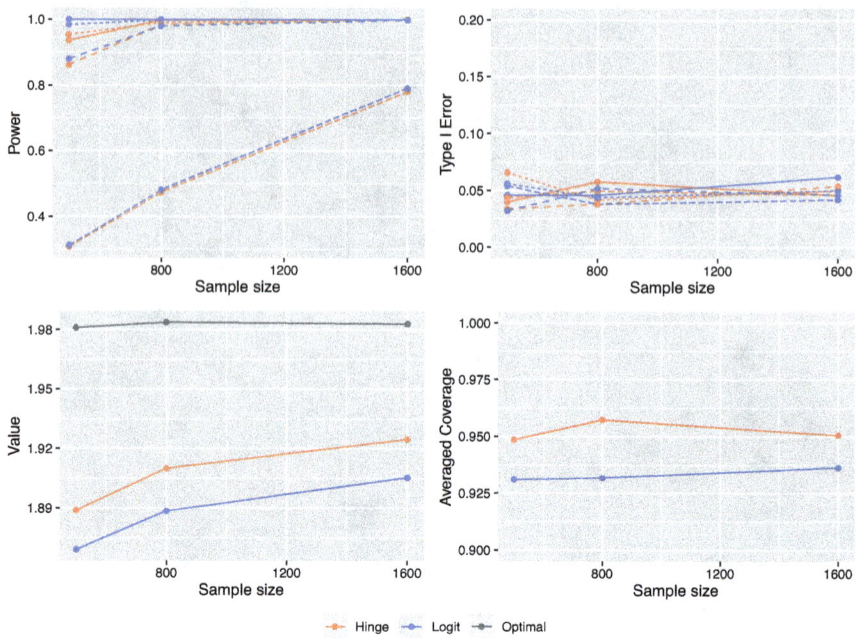

Fig. 6 Simulation results for Scenario (III) with the change of sample size when $\xi = 0.8$ and $p = 800$. Types of the line represent different coefficients

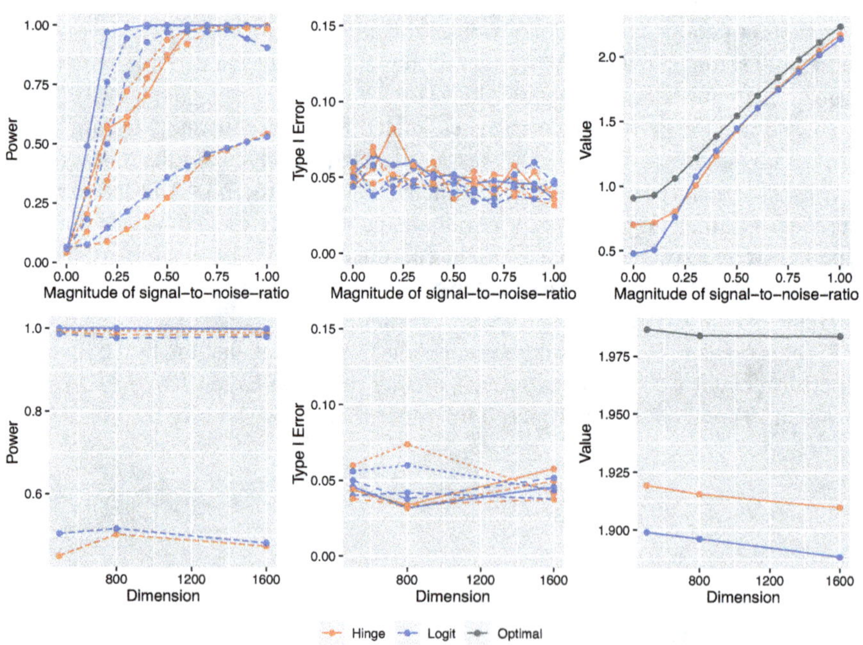

Fig. 7 Simulation results for Scenario (III) with the change of ξ ($n = 800$; $p = 1500$) and the change of p ($n = 800$; $\xi = 0.8$). Types of the line represent different coefficients

6 Real Data Example

In this section, we apply our proposed inference method to identify the driving factors of the estimated ITR aiming at controlling HbA1c for Type-II diabetic Medicare patients. The data source is the electronic medical records collected through the Heath Innovation Program at the University of Wisconsin, linked with medicare claims data. The dataset is not de-identified, and thus cannot be shared publicly. The dataset contains $n = 9101$ patients and 40 covariates, including socio-demographic variables, disease history, baseline HbA1c levels, etc. The outcome of interest is whether the patient successfully controls the HbA1c below 8% after a year. The task is to select the optimal treatment to maximize the success rate. The possible treatment options include $A = 1$ if the patient received any hypoglycemic agent and $A = -1$ otherwise. We also consider both raw covariate and their first-order interactions. We rank these covariates by their variances and select $p = 120$ covariates with top variances.

Using this dataset, we consider two analyses with different goals. In the first analysis, our goal is to compare the performance of using a hinge loss with other methods; in the second analysis, our goal is to identify the driving factors of the learned ITR under a hinge loss using the proposed inference method.

To achieve the first goal, we split the entire dataset into a training (80%) and testing (20%) dataset. We consider the weighted classification framework using a hinge loss and a logistic loss as well as the Q-learning approach. In the Q-learning approach, the outcome regression model is fully parametric and assumed to be linear. To compare different methods, we estimate the propensity score non-parametrically on the testing dataset and estimate the value function of the ITRs learned by each method using the estimated propensity. The entire procedure is repeated 100 times. Table 1 shows that the weighted classification framework using a hinge loss leads to the largest value compared with other methods.

To identify the driving factor of the learned ITR under a hinge loss, we use the entire dataset and implement the proposed inference procedure. Table 2 shows that the three significant covariates under a hinge loss after controlling for false discovery rate, FDR < 0.05. To cross-validate our findings, we also implement the inference procedure for the logistic loss and the Q-learning. To compare the selected variable between different methods, the logistic loss selects only one covariate, which is also selected by the hinge loss, and the Q-Learning selects none after controlling the FDR.

Table 1 Results for comparisons on value functions

Method	Mean	Sd
Observed	0.860	0.008
Hinge	0.871	0.013
Logistic	0.867	0.011
Q-learning	0.870	0.014

Table 2 Coefficients and p-value for the identified significant covariate of the estimated optimal ITR. Special chronic conditions refer to chronic conditions including amputation, chronic blood loss, drug abuse, lymphoma, metastatistic cancer, and peptic ulcer disease. Bucketized age refers to a variable created by bucketizing the raw age by its observed quartiles

Covariate	Coef	P-value
Hypertension (Yes)	−0.0451	8.32×10^{-4}
Chronic complications (Yes): Female	0.1244	4.51×10^{-6}
Hypertension (Yes): baseline A1c	0.0627	6.95×10^{-7}

7 Summary

The EARL method estimates the optimal ITR via optimizing a doubly robust estimator for the value function with a surrogate loss function. It is doubly robust in the sense that when either the propensity estimate or the outcome model estimates are correct, the EARL is consistent. In addition, we can construct a valid inference procedure by using the cross-fitting algorithm. When using a non-differentiable surrogate loss function whose hessian does not exist, we can approximate the hessian of the risk by incorporating a kernel function. Further, combined with the inference technique in high-dimensional literature, the proposed inference procedure is extended to accommodate high-dimensional covariates.

We note that the proposed inference procedure is for β^*. However, there is no guarantee that the corresponding decision rule $D^* = \text{sgn}\{X^\top \beta^*\}$ is the optimal ITR. When the $\mathcal{F}$ contains arbitrary functions, Fisher consistency shows that minimizing the surrogate objective leads to a minimizer of the original objective; but such consistency is not guaranteed when the $\mathcal{F}$ is constrained in the class of linear functions. Liang et al. (2020) provided a sufficient condition such that $D^* = \text{sgn}\{X^\top \beta^*\}$ is indeed the optimal ITR. In addition, when the estimated ITR performs well, the driving factors of the estimated ITR may also be of interest. In this case, the proposed inference procedure can still be used to identify the important driving factors of the decision. Thus, it is a powerful and swift tool to detect the source of the heterogeneous treatment effect in a complex setting when many covariates exist.

References

Athey, S., & Wager, S. (2017). Efficient policy learning. *arXiv preprint arXiv:1702.02896*.

Bartlett, P. L., Jordan, M. I., & Mcauliffe, J. D. (2006). Convexity, classification, and risk bounds. *Journal of the American Statistical Association, 101*(473), 138–156. ISSN 01621459. http://www.jstor.org/stable/30047445

Bierens, H. J. (1988). *The Nadaraya-Watson kernel regression function estimator*.

Boser, B. E., Guyon, I. M., & Vapnik, V. N. (1992). A training algorithm for optimal margin classifiers. In *Proceedings of the Fifth Annual Workshop on Computational Learning Theory* (pp. 144–152).

Chakraborty, B., Laber, E. B., & Zhao, Y.-Q. (2014). Inference about the expected performance of a data-driven dynamic treatment regime. *Clinical Trials, 11*(4), 408–417.

Chakraborty, B., Murphy, S., & Strecher, V. (2010). Inference for non-regular parameters in optimal dynamic treatment regimes. *Statistical Methods in Medical Research, 19*(3), 317–343.

Chernozhukov, V., Chetverikov, D., Demirer, M., Duflo, E., Hansen, C., Newey, W., & Robins, J. (2018). Double/debiased machine learning for treatment and structural parameters. *The Economic Journal, 21*(1), C1–C68. https://doi.org/10.1111/ectj.12097. https://onlinelibrary.wiley.com/doi/abs/10.1111/ectj.12097

Cortes, C., & Vapnik, V. (1995). Support-vector networks. *Machine Learning, 20*(3), 273–297.

Davidian, M., Tsiatis, A., & Laber, E. (2014). Value search estimators. In *Dynamic treatment regimes* (pp. 1–40). Springer.

Freund, Y., & Schapire, R. E. (1997). A decision-theoretic generalization of on-line learning and an application to boosting. *Journal of Computer and System Sciences, 55*(1), 119–139.

Friedman, J., Hastie, T., & Tibshirani, R. (2000). Additive logistic regression: A statistical view of boosting (with discussion and a rejoinder by the authors). *The Annals of Statistics, 28*(2), 337–407.

Goldberg, Y., Song, R., & Kosorok, M. R. (2013). *Adaptive Q-learning*. Collections (Vol. 9, pp. 150–162). Institute of Mathematical Statistics . https://doi.org/10.1214/12-IMSCOLL911.

Imbens, G. W., & Rubin, D. B. (2015). *Causal Inference for Statistics, Social, and Biomedical Sciences: An Introduction.* Cambridge University Press. https://doi.org/10.1017/CBO9781139025751

Jiang, B., Song, R., Li, J., & Zeng, D. (2019). Entropy learning for dynamic treatment regimes. *Statistica Sinica, 29*(4), 1633.

Koo, J.-Y., Lee, Y., Kim, Y., & Park, C. (2008). A bahadur representation of the linear support vector machine. *The Journal of Machine Learning Research, 9*, 1343–1368.

Laber, E. B., Linn, K. A., & Stefanski, L. A. (2014a). Interactive model building for Q-learning. *Biometrika, 101*(4), 831–847. ISSN 0006-3444. https://doi.org/10.1093/biomet/asu043

Laber, E. B., Lizotte, D. J., Qian, M., Pelham, W. E., & Murphy, S. A. (2014b). Dynamic treatment regimes: Technical challenges and applications. *Electronic Journal of Statistics, 8*(1), 1225.

Li, K.-C. (1991). Sliced inverse regression for dimension reduction. *Journal of the American Statistical Association, 86*(414), 316–327.

Li, R., Zhong, W., & Zhu, L. (2012). Feature screening via distance correlation learning. *Journal of the American Statistical Association, 107*(499), 1129–1139. PMID: 25249709. https://doi.org/10.1080/01621459.2012.695654.

Liang, M., Choi, Y.-G., Ning, Y., Smith, M. A., & Zhao, Y.-Q. (2020). Estimation and inference on high-dimensional individualized treatment rule in observational data using split-and-pooled de-correlated score. *arXiv preprint arXiv:2007.04445.*

Liang, M., Ning, Y., Smith, M., & Zhao, Y.-Q. (2021+). Statistical inference of decision rules under a non-differentiable surrogate loss in a general classification framework. *Working paper.*

Liang, M., & Yu, M. (2022). A semiparametric approach to model effect modification. *Journal of the American Statistical Association, 117*(538), 752–764. https://doi.org/10.1080/01621459.2020.1811099.

Lu, W., Zhang, H. H., & Zeng, D. (2013). Variable selection for optimal treatment decision. *Statistical Methods in Medical Research, 22*(5), 493–504.

Luedtke, A. R., & Van Der Laan, M. J. (2016). Statistical inference for the mean outcome under a possibly non-unique optimal treatment strategy. *Annals of Statistics, 44*(2), 713.

Murphy, S. A. (2003). Optimal dynamic treatment regimes. *Journal of the Royal Statistical Society Series B, 65*(2), 331–355.

Nadaraya, E. A. (1964). On estimating regression. *Theory of Probability & Its Applications, 9*(1), 141–142.

Newey, W. K. (1997). Convergence rates and asymptotic normality for series estimators. *Journal of Economics, 79*(1), 147–168. ISSN 0304-4076. https://doi.org/10.1016/S0304-4076(97)00011-0. http://www.sciencedirect.com/science/article/pii/S0304407697000110

Ning, Y., & Liu, H. (2017). A general theory of hypothesis tests and confidence regions for sparse high dimensional models. *Annals of Statistics, 45*(1), 158–195. https://doi.org/10.1214/16-AOS1448

Qi, Z., & Liu, Y. (2018). D-learning to estimate optimal individual treatment rules. *Electronic Journal of Statistics, 12*(2), 3601–3638.

Qian, M., & Murphy, S. A. (2011). Performance guarantees for individualized treatment rules. *Annals of Statistics, 39*(2), 1180.

Rubin, D. B. (1974). Estimating causal effects of treatments in randomized and nonrandomized studies. *Journal of Educational Psychology, 66*(5), 688.

Rubin, D. B. (2005). Causal inference using potential outcomes. *Journal of the American Statistical Association, 100*(469), 322–331. https://doi.org/10.1198/016214504000001880

Shalit, U., Johansson, F. D., & Sontag, D. (2017). Estimating individual treatment effect: Generalization bounds and algorithms. In *International Conference on Machine Learning* (pp. 3076–3085). PMLR.

Shi, C., Fan, A., Song, R., & Lu, W. (2018). High-dimensional a-learning for optimal dynamic treatment regimes. *Annals of Statistics, 46*(3), 925–957. https://doi.org/10.1214/17-AOS1570

Shi, C., Lu, W., & Song, R. (2020). Breaking the curse of nonregularity with subagging—inference of the mean outcome under optimal treatment regimes. *Journal of Machine Learning Research, 21*(176), 1–67.

Shi, C., Song, R., & Lu, W. (2016). Robust learning for optimal treatment decision with np-dimensionality. *Electronic Journal of Statistics, 10*(2), 2894–2921. https://doi.org/10.1214/16-EJS1178

Wager, S., & Athey, S. (2018). Estimation and inference of heterogeneous treatment effects using random forests. *Journal of the American Statistical Association, 113*(523), 1228–1242.

Wahba, G. (2002). Soft and hard classification by reproducing kernel hilbert space methods. *Proceedings of the National Academy of Sciences, 99*(26), 16524–16530.

Wahba, G., et al. (1999). Support vector machines, reproducing kernel hilbert spaces and the randomized gacv. *Advances in Kernel Methods-Support Vector Learning, 6*, 69–87.

Watkins, C. J. C. H., & Dayan, P. (1992). Q-learning. *Machine Learning, 8*(3), 279–292. https://doi.org/10.1007/BF00992698

Watson, G. S. (1964). Smooth regression analysis. *Sankhyā: The Indian Journal of Statistics, Series A, 26*, 359–372.

Yin, X., & Li, B. (2011). Sufficient dimension reduction based on an ensemble of minimum average variance estimators. *Annals of Statistics, 39*(6), 3392–3416. ISSN 00905364. https://doi.org/10.1214/11-AOS950

Zhang, B., Tsiatis, A. A., Davidian, M., Zhang, M., & Laber, E. (2012a). Estimating optimal treatment regimes from a classification perspective. *Statistical, 1*(1), 103–114.

Zhang, B., Tsiatis, A. A., Laber, E. B., & Davidian, M. (2012b). A robust method for estimating optimal treatment regimes. *Biometrics, 68*(4), 1010–1018.

Zhang, B., & Zhang, M. (2018). C-learning: A new classification framework to estimate optimal dynamic treatment regimes. *Biometrics, 74*(3), 891–899.

Zhao, Y., Zeng, D., Rush, A. J., & Kosorok, M. R. (2012a). Estimating individualized treatment rules using outcome weighted learning. *Journal of the American Statistical Association, 107*(499), 1106–1118. ISSN 0162-1459. https://doi.org/10.1080/01621459.2012.695674

Zhao, Y., Zeng, D., Rush, A. J., & Kosorok, M. R. (2012b). Estimating individualized treatment rules using outcome weighted learning. *Journal of the American Statistical Association, 107*(499), 1106–1118. PMID: 23630406. https://doi.org/10.1080/01621459.2012.695674.

Zhao, Y.-Q., Laber, E. B., Ning, Y., Saha, S., & Sands, B. E. (2019). Efficient augmentation and relaxation learning for individualized treatment rules using observational data. *Journal of Machine Learning Research, 20*(48), 1–23.

Subgroup Analysis Using Doubly Robust Semiparametric Procedures

Anqi Yin, Ao Yuan, and Ming T. Tan

Abstract Subgroup analysis is commonly performed in clinical trials and is an important means for precision medicine as the treatment improvement could well be different between two subgroups of patients. One goal is then to identify if there exists such a subgroup in the patient population. Since the treatment assignment in the study population may not be covariate balanced, and the naive estimate of treatment effect such as a group mean may be biased. To infer the true effect of a new treatment against a current one, and classify patients into the new treatment favorable subgroup if it exists, causal inference methods are needed. The doubly robust (DR) procedure for causal inference represents an important methodological advancement, where a propensity score model and a regression model need to be specified. As long as one of the two models is specified correctly, the DR causal estimate of the treatment will be unbiased asymptotically. In practice, DRE may fail and have large bias and variance, even when the propensity and/or outcome models only depart slightly from the true model. Recently we developed an enhanced DRE by utilizing more flexible semiparametric propensity and regression models and applied it to subgroup analysis. The key idea is to enhance and construct the DRE to achieve robustness and numerical stability, and then identify treatment favorable subgroup if it exists. In this chapter, we will explore an alternative construct of the DRE after a brief review of the DRE and enhanced DRE. Then we will use the enhanced doubly robust procedure in subgroups analysis, which is to identify subgroup if there exists groups of patients who benefit significantly more from the treatment over the control, and compare it with the previously developed enhanced procedure and the naive estimator in both simulation and in the analysis of a smoking cessation study. We found that the two enhanced DR estimators are both robust and overall give consistent results.

A. Yin (✉) · A. Yuan · M. T. Tan
Department of Biostatistics, Bioinformatics and Biomathematics, Georgetown University, Washington, DC, USA
e-mail: ay380@georgetown.edu; ay312@georgetown.edu; mtt34@geogetown.edu

Y. Zhao, D.-G. Chen (eds.), *Statistics in Precision Health*, ICSA Book Series in Statistics, https://doi.org/10.1007/978-3-031-50690-1_11

Keywords Causal effect · Clinical trial · Doubly robust · Semiparametric model · Sub-group

1 Introduction

Subgroup analysis is commonly performed in clinical trials and is an important means for precision medicine as the treatment improvement could well be different between two subgroups of patients. For example, Goldhrisch et al. (2002) reported that patients with ER-negative tumors benefited substantially from chemotherapy, while those with ER-positive tumors did not benefit; in the studies of biomarkers predictive of survival of advanced head and neck cancer patients following treatments in three randomized control trials (Cullen et al., 2009), the efficacy in patients with biomarker positive is clearly different from those who are biomarker negative. Thus identifying the treatment favorable subgroups if they exist, and being able to make confirmatory causal inference on the differential treatment effects are of important significance (Ruberg et al., 2010). This has been one of the major topics in recent research in precision medicine and biostatistics, leading to several advances in statistical methodology as in Shen and He (2015), Ma and Huang (2017), Fan et al. (2017), Egleston et al. (2017), Yuan et al. (2018) among others.

When the treatment assignments are not completely random or imbalanced in covariates, the naive group mean estimates of treatment effects are biased, and so the causal inference method should be used, which is an essential part in today's clinical studies (Rubin, 1974; Rosenbaum & Rubin, 1983; Holland, 1986). Pearl et al. (2016) gave a comprehensive review of development and methods in this field. The doubly robust estimator (DRE) is a more recent and popular approach for the causal inference (Robins et al., 1994; Robins, 1997; Lunceford & Davidian, 2004; Bang & Robins, 2005). It requires the specification of a propensity score model and an outcome regression model, and gives unbiased estimate of treatment difference when one of the two models is correctly specified, hence the name doubly robust. The ensemble learning methods (Breiman et al., 1996; Opitz & Maclin, 1999; Rani & Muneeswaran, 2018; Bonab & Can, 2019) are also very robust and flexible, but the semiparametric model may be more preferable in characterizing the relationship between response and covariates, via the regression coefficients. Also the former is computationally intensive and is not as convenient to use as the latter. In practice, any subjectively specified model more or less deviates from the true one. So it is still likely that in the doubly robust procedure both models are incorrectly specified and the inference result may not be reliable. To overcome this issue, recently Yuan et al. (2021) proposed a flexible robust causal inference method for testing and estimating treatment effect, and investigated its asymptotic properties. Yuan et al. (2022) used this method for subgroup analysis, and proposed a method to classify subjects into significant treatment favorable or non-significant groups according to the observed covariates.

This chapter focuses on the existence and identification of subgroups between which the treatment difference is different. We will review methods using the DRE and enhanced DRE, and then utilize an alternative construct of the DRE. Then we develop the corresponding enhanced doubly robust procedure and the resulting in method for subgroups analysis, namely, to determine whether there exists groups of subjects who benefit significantly more from treatment than the control group, and compare it with the previously developed enhanced procedure (Yuan et al., 2022) and the naive estimator in both simulation and the analysis of a smoke cessation study.

The rest of the chapter is organized as follows. In Sect. 2, we describe the method of Yuan et al. (2022). Section 3.1 reports simulation results comparing the two commonly used DRE constructs using the framework in Yuan et al. (2022) and identify treatment favorable subgroup. The methods are then used to analyze a smoking cessation data in Sect. 3.2. We conclude with a discussion and remarks about the far reaching implications of the proposed method.

2 The Method

We first briefly describe the background, data and notations. Denote the observed data as $D_n = \{(y_i, t_i, x_i) : i = 1, \ldots, n\}$, where for given individual i, $y_i \in R$ is the response, t_i is the treatment indicator, $t_i = 1($ or $0)$ if treatment (or control), and $x_i = (x_{i1}, \ldots, x_{id})' \in R^d$ is the covariate matrix. Adopting common notations in causal inference, the response is written as

$$y_i = t_i y_{i1} + (1 - t_i) y_{i0},$$

where y_{i1} is response of subject i in the treatment group, and y_{i0} is that in the control group, only one of them can be observed, the unobserved one is called *counterfactual outcome*. Before describing our method, we first give a brief review of the doubly robust procedure.

2.1 *Brief Review of the Doubly Robust Procedures*

If the treatment assignment is independent among the participants, or in other words if the unobserved y_{ij}'s are missing completely at random (MCAR), then the causal effect is unbiasedly estimated by

$$\Delta = E(Y_{i1}|T_i = 1) - E(Y_{i0}|T_i = 0) = E(Y_i|T_i = 1) - E(Y_i|T_i = 0) := \mu_1 - \mu_0.$$

The μ_j's are easily estimated by their treatment group empirical means (naive estimates) and are unbiased. However, in practice the MCAR assumption is often violated, and it is known that the above naive estimates of μ_j's (and hence of Δ) are biased. To overcome this problem, the early method is to use the propensity score and the inverse probability weighting (IPW) approach (Rosenbaum & Rubin, 1983), however the method fails if the propensity score is incorrectly specified. The doubly robust procedure, based on Robins et al. (1994), Robins (1997), Lunceford and Davidian (2004) is an improvement of the IPW method.

Under assumption of missing at random (MAR), there are several commonly used estimates of the μ_j's. Denote $P(T = 1|X) := \pi(X)$, the *propensity score*. If $\pi(\cdot)$ is correctly specified, the IPW estimate $\sum_{i=1}^{n} t_i y_i/\pi(x_i)$ is unbiased estimate of μ_1 (Horvitz & Thompson, 1952), but it is still biased if $\pi(\cdot)$ is incorrectly specified. Denote $E(Y|X, T = j) = m_j(X, \beta_j)$, $(j = 0, 1)$, the *outcome model*. Under MAR, the commonly used DRE is the following (Robins et al., 1994) estimate of $\Delta = \mu_1 - \mu_0$ is doubly unbiased

$$\hat{\Delta}^R = \frac{1}{n}\sum_{i=1}^{n}\Big(\frac{t_i\big(y_i - m_1(x_i, \hat{\beta}_1)\big)}{\pi(x_i, \hat{\gamma})} + m_1(x_i, \hat{\beta}_1)$$

$$- \frac{(1-t_i)\big(y_i - m_0(x_i, \hat{\beta}_0)\big)}{1-\pi(x_i, \hat{\gamma})} - m_0(x_i, \hat{\beta}_0)\Big),$$

in that it is an unbiased estimate of Δ, as long as either $\pi(\cdot,\cdot)$ or $m_j(\cdot,\cdot)$ $(j = 0, 1)$ are correctly specified, where γ and β are the coefficients in the specified propensity score and regression models. So compared to the IPW estimator, it provides double protection against model misspecifications. However, there are different constructions of DRE's. For example in Kang et al. (2007) the following construction of DRE for μ_1 is described

$$\hat{\mu}_1 = \frac{1}{n}\sum_{i=1}^{n} m_1(x_i) + \frac{\sum_{i=1}^{n} t_i\big(y_i - m_1(x_i)\big)/\pi(x_i)}{\sum_{i=1}^{n} t_i/\pi(x_i)}.$$

The corresponding DRE $\hat{\Delta}^K$ is constructed accordingly.

In this report we compare the performances of the two DREs $\hat{\Delta}^R$, $\hat{\Delta}^K$ and the naive estimator, and derive the enhanced DRE based on the new construct, using the same semiparametric robust propensity score and outcome models in Yuan et al. (2021, 2022).

2.2 The Semiparametric Propensity Score and Outcome Models

As mentioned in the Introduction, most of the existing doubly robust method are parametric and may still be non-robust, and so we use the more flexible and robust semiparametric method in Yuan et al. (2021, 2022). It has two steps, estimating

Δ via enhanced robust procedure, and then classifying the subjects. To use this enhanced robust procedure to estimate Δ, we need to specify the propensity score and the outcome models, as described below.

The Propensity Score Model The classical propensity score model is the logistic model

$$P(T = 1|x) = \frac{\exp(x^\top \gamma)}{1 + \exp(x^\top \gamma)}.$$

Although this model is a good approximation in many cases, it is not flexible and robust enough. In Yuan et al. (2021, 2022) the propensity score model is specified as

$$P(T = 1|x) = \pi(x^\top \gamma), \quad \pi(\cdot) \in \Pi, \quad \|\gamma\| = 1 \tag{1}$$

where Π is a collection of non-negative monotone increasing functions upper bounded by 1, in particular the collection of all distribution functions. Monotone link function has been used in a number of literatures (e.g. Friedman and Tibshirani (1984), Hawkins (1994), Mukherjee (1988), Ramsay (1988), He and Shi (1998), Kelly and Rice (1990), Mammen (1991), Diggle et al. (1999), Qin et al. (2014, 2022)). For general single index model, the constraint $\|\gamma\| = 1$ together with the fact that first component of γ having fixed sign is for the semiparametric model identifiability (e.g. Balabdaoui et al. 2018). For monotone link function, the latter condition is not needed. Here we assume $dim(x) > 1$, otherwise γ is absorbed into $\pi(\cdot)$. Under model (1), the log-likelihood for data $\{(t_i, x_i) : i = 1, ..., n\}$ is

$$\ell_T(\gamma, \pi|D_n) = \sum_{i=1}^{n} \Big(t_i \log \pi(x_i^\top \gamma) + (1 - t_i) \log \big(1 - \pi(x_i^\top \gamma)\big)\Big). \tag{2}$$

We estimate (γ, π) by the MLE $(\hat{\gamma}, \hat{\pi})$,

$$(\hat{\gamma}, \hat{\pi}) = \arg \max_{(\gamma,\pi)\in(\Gamma,\Pi)} \ell_T(\gamma, \pi|D_n).$$

In particular, $\hat{\gamma}$ can be computed from the profile likelihood $\tilde{\ell}_T(\gamma|D_n) = \ell(\gamma, \hat{\pi}(\gamma)|D_n)$, find the maximizer $\tilde{\gamma}$ of $\tilde{\ell}_T(\gamma|D_n)$, then normalize it to get $\hat{\gamma} = \tilde{\gamma}/\|\tilde{\gamma}\|$. The computation of $\hat{\pi}(\cdot)$ can be carried out using the well known pool adjacent violator algorithm (PAVA). This algorithm is adopted by a number of researchers, such as in Qin et al. (2014), Qin et al. (2022). For the details of the computation of $(\hat{\gamma}, \hat{\pi})$ see Yuan et al. (2021).

The Outcome Model In many existing doubly robust procedure, $m_j(x, \beta_j)$ are either specified parametric or by a known form of $m_j(\cdot, \cdot)$, mostly the linear model.

For flexibility and robustness, in Yuan et al. (2021, 2022), the outcome model is specified as

$$\begin{aligned} y_{ij} &= m_j(x_i^\mathsf{T}\beta_j) + \epsilon_i, \quad \epsilon_i \perp (y_i, x_i), \quad E(\epsilon_i) = 0, \quad m_j \in \mathscr{M}, \\ i &= 1, ..., n;\ j = 0, 1. \end{aligned} \tag{3}$$

where β_j is the regression coefficients satisfying $\beta_j \in \mathscr{B} = \{\beta : ||\beta|| = 1\}$, in which the constraint $||\beta|| = 1$ is for model identifiability; ϵ_i is the model residual error, the notation $\epsilon_i \perp (y_i, x_i)$ means ϵ_i and (y_i, x_i) are independent, and $\mathscr{M}$ is the collection of all bounded monotone increasing functions on R. For further robustness we estimate (β_j, m_j) by the least squares estimator (LSE), let

$$\begin{aligned} \ell(\beta_0, m_0|D_n) &= \sum_{i=1}^{n}(1-t_i)\big(y_i - m_0(x_i^\mathsf{T}\beta_0)\big)^2, \\ \ell(\beta_1, m_1|D_n) &= \sum_{i=1}^{n} t_i\big(y_i - m_1(x_i^\mathsf{T}\beta_1)\big)^2 \end{aligned} \tag{4}$$

and the parameters are estimated by

$$(\hat{\beta}_j, \hat{m}_j) = \arg \min_{(\beta_j, m_j)\in(\mathscr{B},\mathscr{M})} \ell(\beta_j, m_j|D_n), \qquad (j = 0, 1). \tag{5}$$

In particular, $(\hat{\beta}_j, \hat{m}_j)$ can be obtained using an iterative method, for details see Yuan et al. (2021), and the asymptotic properties of $\hat{\gamma}$, $\hat{\pi}(\cdot)$, $\hat{\beta}$ and $\hat{m}_j(\cdot)$ are given there.

The Doubly Robust Estimators of Δ The treatment effect μ_j and Δ are jointly determined by $(\pi(\cdot), \gamma, m_j(\cdot), \beta_j : j = 0, 1)$. After obtaining $(\hat{\gamma}, \hat{\pi})$ and $(\hat{\beta}_j, \hat{m}_j : j = 0, 1)$, the two commonly used DREs of Δ we consider are

$$\begin{aligned} \hat{\Delta}^R = \hat{\mu}_1 - \hat{\mu}_0 &:= \frac{1}{n}\sum_{i=1}^{n}\hat{\Delta}_i \\ = \frac{1}{n}\sum_{i=1}^{n}\Big(&\frac{t_i\big(y_i - m_1(x_i, \hat{\beta}_1)\big)}{\pi(x_i, \hat{\gamma})} + m_1(x_i, \hat{\beta}_1) \\ &- \frac{(1-t_i)\big(y_i - m_0(x_i, \hat{\beta}_0)\big)}{1-\pi(x_i, \hat{\gamma})} - m_0(x_i, \hat{\beta}_0)\Big) \end{aligned} \tag{6}$$

and

$$\hat{\Delta}^K = \frac{1}{n}\sum_{i=1}^{n}\Big(\hat{m}_1(x_i,\hat{\beta}_1) - \hat{m}_0(x_i,\hat{\beta}_0)\Big) + \frac{\sum_{i=1}^{n}\big(y_i - \hat{m}_1(x_i,\hat{\beta}_1)\big)t_i/\hat{\pi}(x_i,\hat{\gamma})}{\sum_{i=1}^{n} t_i/\hat{\pi}(x_i,\hat{\gamma})}$$

$$- \frac{\sum_{i=1}^{n}\big(y_i - \hat{m}_0(x_i,\hat{\beta}_0)\big)(1-t_i)/\big(1-\hat{\pi}(x_i,\hat{\gamma})\big)}{\sum_{i=1}^{n}(1-t_i)/\big(1-\hat{\pi}(x_i,\hat{\gamma})\big)}. \tag{7}$$

In (7), the sum of inverse propensity scores is used to adjust the estimate, and intuitively it is expected to be more stable than (6), in which each individual inverse propensity score is used in the adjustment.

The naive estimator is given by, which is known to be biased when the treatment assignment is not completely random

$$\tilde{\Delta} = \frac{\sum_{i=1}^{n} t_i y_i}{\sum_{i=1}^{n} t_i} - \frac{\sum_{i=1}^{n}(1-t_i)y_i}{\sum_{i=1}^{n}(1-t_i)}.$$

Test the Significance Let $\hat{\Delta}$ be either $\hat{\Delta}^R$, $\hat{\Delta}^K$ or $\tilde{\Delta}$. In Yuan et al. (2022), to test the null hypothesis $H_0 : \Delta = 0$ vs $H_1 : \Delta > 0$. The Wald test is convenient to use, under H_0,

$$T_n := \frac{\hat{\Delta}}{\sqrt{Var(\hat{\Delta})}} \xrightarrow{D} N(0,1).$$

Denote $T_{n,R}$ or $T_{n,K}$ if $\hat{\Delta}^R$ or $\hat{\Delta}^K$ is used. For $\hat{R} = \hat{\Delta}^R$, $Var(\hat{\Delta})$ can be estimated as,

$$Var(\hat{\Delta}) \approx \frac{1}{n^2}\sum_{i=1}^{n}(\hat{\Delta}_i - \hat{\Delta})^2.$$

For $\hat{\Delta} = \hat{\Delta}^K$ or $\tilde{\Delta}$, $Var(\hat{\Delta})$ can be estimated using repetition or bootstrap.

For given significance level α, H_0 is rejected if $T_n > \Phi^{-1}(1-\alpha)$, where $\Phi^{-1}(\cdot)$ is the quantile function of $N(0,1)$.

Subgroup Identification and Subjects Classification If the null hypothesis H_0 is rejected, it suggests that there exist subgroups with different treatment improvements. We then identify the groups by classifying the subjects into two subgroups according to their covariate values, one group is significantly more favorable to the treatment, and the other not so. In Yuan et al. (2022), subject i is classified into the treatment favorable sub-group if

$$\hat{m}_1(x^\top\hat{\beta}_1) - \hat{m}_0(x^\top\hat{\beta}_0) \geq \hat{\Delta}(1+\eta), \tag{8}$$

for some pre-specified η, such as $\eta = 0.05, ..., 0.1$ as in Foster et al. (2011).

In the virtual-twin method of Foster et al. (2011) with their estimate $\hat{\Delta}$, they take the cut-off point $C = \hat{\Delta} + 0.1$ or $C = \hat{\Delta} + 0.05$. We think it is more reasonable to take $C = \hat{\Delta}(1 + \eta)$. Our classifying method is different from the linear method (Foster et al., 2011) in which a subject with covariates x is classified to the treatment favorable group if $x^\top \hat{\beta} > 0$. Here our classification rule is non-linear.

3 Simulation Study and Application

3.1 Simulation Study

For ease of comparison, in assessing the performances of $\hat{\Delta}^R$, $\hat{\Delta}^K$ and $\tilde{\Delta}$, we use the same data and parameter estimates in the simulation studies in Yuan et al. (2022), in which several data sets are simulated with sample size $n = 500$ each. Different data sets have different dimensions of covariates and parameters values, as described again below. For the propensity score model, the t_i's are generated from model (1), under five different $\pi(\cdot)$ distribution functions: (1) N(1,1.5^2); (2) N(2,1); (3) N(1,1.5^2); (4) logistic(9,3); (5) logistic(6,2).

Five different combinations of constant times distribution functions for $m_j(\cdot)$'s are selected: (1): m_1: 8×Gamma(5.5, 1), m_0: 6×Gamma(5.5,1), (2): m_1: 10×N(0,4), m_0: 7×N(0, 4), (3): m_1: 10×N(0, 1), m_0: 7×N(0, 1), (4): m_1: 10×N(2, 4), m_0: 7×Gamma(2, 2), (5): m_1: 9×N(1.5, =1), m_0: 7×Gamma(3, 2).

In Yuan et al. (2022), the estimation results are presented with different choices of γ based on the proposed semiparametric propensity score model and compared with the results from standard logistic model; and estimation results are shown with different choices of β_j based on the proposed outcome model and compare the results from the commonly used linear model. After obtaining $\hat{\gamma}$ and $(\hat{\beta}, \hat{m})$, we compute the DREs $\hat{\Delta}^R$ and $\hat{\Delta}^K$ and then test the null hypothesis $H_0 : \Delta = 0$ vs $H_1 : \Delta > 0$ based on the two estimators, and compare the results with those from the naive estimator $\tilde{\Delta}$.

Then the subjects are classified into the treatment favorable subgroup if $\hat{m}_1(x^\top \hat{\beta}_1) - \hat{m}_0(x^\top \hat{\beta}_0) \geq C$ for cut-off point $C_1 = \hat{\Delta} + 0.05$ and $C_2 = \hat{\Delta} + 0.1$. Ten different scenarios are considered, and π, m_1 and m_0 are the distribution functions of the following: (1): π: N(0, 1), m_1: 13×Gamma(4, 1), m_0: 7×Gamma(4, 1), (2): π: N(1, 1.5^2), m_1: 9×Gamma(5.5, 1), m_0: 2×Gamma(5.5,1), (3): π: N(2, 1), m_1: 10×N(0, 4), m_0: 6×N(0, 4), (4): π: logistic(2, 1), 10×Gamma(6.5, 1), m_0: 7×Gamma(6.5,1), (5): π: logistic(5, 2), m_1: 10×Gamma(2, 2), m_0: 3×Gamma(2, 2), (6): π: N(1, 1.5^2), m_1: 10×Normal(0,1), m_0: 7×N(0,1), (7): π: logistic(9, 3), m_1: 12×N(2, 4), m_0: 3×Gamma(2, 2), (8): π: logistic(6, 2), m_1: 10×N(1.5,1), m_0: 7×Gamma(3, 2), (9): π: Exp(1), m_1: 13×Gamma(4, 1), m_0: 7×Gamma(4, 1), (10): π: N(0, 1), m_1: 13×Gamma(4,1), m_0: 13×Gamma(4,1). The true causal effects Δ_0, the estimated causal effects $\hat{\Delta}^R$ given in (6), $\hat{\Delta}^K$ given in (7), the naive estimate $\tilde{\Delta}$, the values of the test statistic T_n, and classification results are shown in Table 1;

Table 1 Classification results for simulated data

Scenario	Δ_0	$\hat{\Delta}^R$	T_{nR}	C_{1R}	$Prop_{1R}$	C_{2R}	$Prop_{2R}$	$\hat{\Delta}^K$	T_{nK}	C_{1K}	$Prop_{1K}$	C_{2K}	$Prop_{2K}$	$\tilde{\Delta}$
1	2.962	3.300	12.303	3.465	0.456	3.630	0.416	3.323	10.813	3.489	0.422	3.655	0.416	3.735
2	0.702	1.112	6.939	1.168	0.296	1.223	0.296	1.065	7.884	1.118	0.296	1.171	0.296	1.148
3	3.240	3.061	12.982	3.214	0.576	3.367	0.554	2.997	14.596	3.147	0.580	3.297	0.556	4.133
4	0.598	0.098	0.606	0.103	0.774	0.108	0.774	0.138	0.761	0.144	0.768	0.151	0.768	−2.325
5	2.106	3.192	6.182	3.351	0.402	3.511	0.396	2.760	10.421	2.898	0.432	3.036	0.430	5.464
6	2.740	4.506	24.498	4.731	0.302	4.956	0.282	4.606	19.288	4.836	0.298	5.066	0.280	6.510
7	5.923	6.966	25.693	7.314	0.570	7.663	0.536	6.877	34.572	7.221	0.574	7.565	0.544	10.885
8	4.684	5.764	28.379	6.053	0.418	6.341	0.416	5.739	31.414	6.026	0.418	6.313	0.418	8.424
9	2.962	3.105	9.241	3.261	0.450	3.416	0.396	3.189	7.626	3.348	0.410	3.508	0.390	3.997
10	0.003	0.235	0.725	0.247	0.480	0.259	0.478	0.263	0.710	0.276	0.474	0.289	0.474	1.003

$Prop_1$ and $Prop_2$ are the proportions of subjects classified into subgroup A using cut-off C_1 and C_2 respectively. The superscript R and K represents corresponding values of the methods in Robins et al. (1994) and Kang et al. (2007) respectively. We see that the estimated causal effects are similar, but they are overall quite different from the naive estimates. This suggests potentially serious consequence if the naive estimate is used without proper adjustment. For Scenario 4, the estimated causal effect is around 0.1, while the naive estimate is -2.325, which is seriously biased. Also, for Scenario 5–8, the naive estimates deviate a lot from the true causal effects. For Scenario 10, we considered the null model ($m_1 = m_0$), and the classification results are close to 0.5, which suggest that the subgroup does not exist and it's reasonable. For these data sets, the classification results are not sensitive to the cut-off C_1 or C_2 to be used. The classification proportions are determined by the covariates via (8), they have close relationship to the value of T_n, but not in a simple monotonic manner. For example, in Scenario 1, based on the cut-off C_{1R}, the $Prop_{1R}$ is 0.456, which means 45.6% subjects are significantly favorable to the treatment group, and individuals in this subgroup will have an enhanced treatment effect.

From Table 1, the two DREs $\hat{\Delta}^R$ and $\hat{\Delta}^K$ are fairly close to each other and to the true value across all scenarios, while the naive estimator $\tilde{\Delta}$ is far from the true value Δ_0, as expected. It seems that overall the alternative estimate $\hat{\Delta}^K$ performs slightly better than $\hat{\Delta}^R$.

Table 2 shows the classification results for different sample size based on Scenario 3. When the sample size reaches 500, the two DREs $\hat{\Delta}^R$ and $\hat{\Delta}^K$ are quite stable and close to the true values, while the naive estimator deviates a lot.

3.2 Application to a Smoking Cessation Study

We use our method to analyze a subset of the data from the NHEFS (wwwn.cdc.gov/nchs/nhanes/nhefs/, National Health and Nutrition Examination Survey Data I Epidemiologic Follow-up Study). We performed the analyses based on the data from 1566 cigarette smokers aged 25–74 years, and estimated the average causal effect of smoking cessation (treatment/intervention) on weight gain (outcome) in Yuan et al. (2022). We considered treatment 1 (quit smoking = yes) and treatment 2 (quit smoking = no), with sample sizes 403 and 1163 respectively. Eight covariates were included: age (years), sex (0 = male, 1 = female), race (0 = white, 1 = black or other), smoking intensity (number of cigarettes smoked per day), education (1 = 8th grade or less, 2 = HS dropout, 3 = HS, 4 = College dropout, 5 = College or more), active life (0 = very active, 1 = moderately active, 2 = inactive), exercise (0 := much exercise, 1 = moderate exercise, 2 = little or no exercise) and baseline weight (kg). The estimated causal effects $\hat{\Delta}$ given in (6) of the proposed semiparametric model and the naive estimate $\tilde{\Delta}$ is 2.795 and 2.541 respectively, which is quite different. The results of Robins et al. (1994) and Kang et al. (2007) are quite similar. In this study, we classify the subjects to the subgroup with/without an enhanced treatment effect and the results can be found in Table 3. Based on the criterion given

Table 2 Classification results for different sample size based on Scenario 3 for simulated data

N	Δ_0	$\hat{\Delta}^R$	T_{nR}	C_{1R}	$Prop_{1R}$	C_{2R}	$Prop_{2R}$	$\hat{\Delta}^K$	T_{nK}	C_{1K}	$Prop_{1K}$	C_{2K}	$Prop_{2K}$	$\tilde{\Delta}$
300	3.177	2.713	6.698	2.849	0.600	2.985	0.600	2.581	9.428	2.711	0.603	2.840	0.600	4.270
500	3.240	3.061	12.982	3.214	0.576	3.367	0.554	2.997	14.596	3.147	0.580	3.297	0.556	4.133
1000	3.252	3.076	19.934	3.230	0.571	3.384	0.567	3.048	19.663	3.201	0.571	3.353	0.569	4.346

Table 3 Classification results (real data)

	$\hat{\Delta}$	T_n	C_1	$Prop_1$	C_2	$Prop_2$	$\widetilde{\Delta}$
$\hat{\Delta}^R$	2.795	5.752	2.934	95.59%	3.074	95.59%	2.541
$\hat{\Delta}^K$	2.793	5.918	2.933	95.59%	3.072	95.59%	2.541

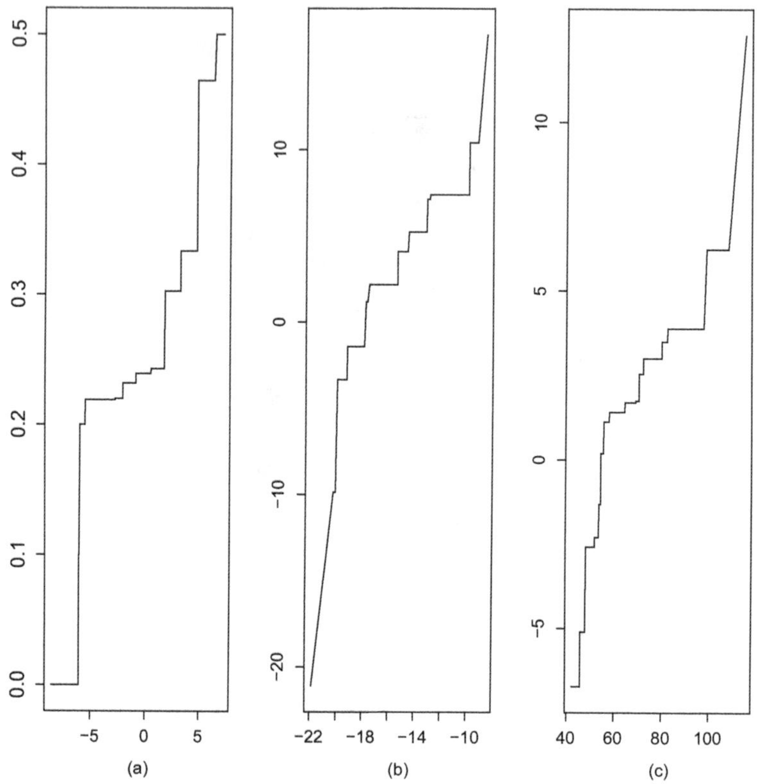

Fig. 1 Propensity score model and regression model (real data). (**a**), (**b**) and (**c**): $\hat{\pi}$, $\hat{m}_1$ and $\hat{m}_0$

in (8) and the cut-off C_1, 1497 (95.59%) subjects are classified into the smoking cessation favorable subgroup, which means that the individuals in the subgoup will gain weight if they quit smoking. Also, 379 out of 403 of them had actually quit smoking, while 1118 out of 1163 had not quit smoking. Although the cut-off C_2 is different from C_1, we still get the same estimated proportion of subgroup favorable to treatment 1.

From Table 3, both the subgroup analysis methods, one based on $\hat{\Delta}^K$, the alternative DRE developed here, and the other based on our previous $\hat{\Delta}^R$, give similar results in treatment difference estimate and classification, although the estimated regression functions differ to some extent, shown in Fig. 1 along with the estimated propensity score.

The figures show non-linearity of the estimated propensity score function (a) and estimated regression functions (b) and (c), and different regression patterns for the two treatments.

We attempted to set part of the data as training data and testing data, and apply the methods in the training data to get the subgroup and the estimates, and then show whether the subgroup makes sense in the testing data. However, the approach is not feasible in this particular case, because the sample size of one of the subgroups is very small (less than 5% patients), splitting the data making it even smaller. Here the point is to examine in detail the identified 5% patients to see if there's anything unique about them, and to obtain insights for planning future studies.

Concluding Remarks We have developed an alternative enhanced DRE and subgroup analysis method for identifying and classifying subjects into treatment favorable and un-favorable subgroups, and compared it with the previous method with a different enhanced DRE. The approach is based on a pair of semi-parametric specifications of the propensity score and outcome models, and the doubly robust correction procedure. The method is easy to implement and use. Simulation studies have shown that the new enhanced DRE is slightly more closer to the true value and stable but both of the two versions of DREs perform well, and outperform the naive estimator. The analysis of the smoking cessation data in different subgroups using the two doubly robust estimators give consistent results. It seems the semiparametric modeling provides the needed additional robustness to DREs. It would be of interest to use the approach to more real studies to gain more insights on how the method performs. There is one limitation of our proposed method. Currently we use the monotone increasing functions for propensity score and outcome model. However, in practice they may not be monotone functions, and convex or concave functions can also be considered for the propensity score or outcome model.

Data Availability Statement
A zip file with R code and datasets used for the simulations and applications are included in this published chapter and its supplementary information files.

References

Balabdaoui, F., Groeneboom, P., & Hendrickx, K. (2018). Score estimation in the monotone single index model. *Scandinavian Journal of Statistics, 46*, 517–544.

Bang, H., & Robins, J. M. (2005). Doubly robust estimation in missing data and causal inference models. *Biometrics, 61*, 962–972.

Bonab, H., & Can, F. (2019). *Less is more: a comprehensive framework for the number of components of ensemble classifiers*. TNNLS. USA: IEEE. arXiv:1709.02925.

Breiman, L. (1996). Stacked regression. *Machine Learning, 24*. https://doi.org/10.1007/BF00117832

Cullen, K. J., Schumaker, L., Nikitakis, N., et al. (2009). Beta-Tubulin-II expression strongly predicts outcome in patients receiving induction chemotherapy for locally advanced squamous carcinoma of the head and neck: a companion analysis of the TAX 324 trial. *Journal of Clinical Oncology, 27*(36), 622–628.

Diggle, P., Morris, S., & Morton-Jones, T. (1999). Case-control isotonic regression for investigation of elevation in risk around a risk source. *Statistics in Medicine, 18*, 1605–1613.

Egleston, B. L., Uzzo, R. G., & Wong, Y.-N. (2017). Latent class survival models linked by principal stratification to investigate heterogeneous survival subgroups among individuals with early stage kidney. *Journal of the American Statistical Association, 112*, 534–546.

Fan, A., Song, R., & Lu, W. (2017). Change-plane analysis for subgroup detection and sample size calculation. *Journal of the American Statistical Association, 112*, 769–778.

Foster, J. C., Taylor, J. M. C., & Ruberg, S. J. (2011). Subgroup identification from randomized clinical trial data. *Statistics in Medicine, 30*, 2867–2880.

Friedman, J. H., & Tibshirani, R. (1984). The monotone smoothing of scatterplots. *Technometrics, 26*, 243–250.

Goldhrisch, A. S. C., Collieoni, M., Nasi, M. L., et al. (2002). Endocrine responsiveness and tailoring adjuvant therapy for postmenopausal lymph node-negative breast cancer: a randomized trial. *Journal of the National Cancer Institute, 94*, 1054–1065.

Hawkins, D. M. (1994). Fitting monotonic polynomials to data. *Computational Statistics Quarterly, 9*, 233–247.

He, X., & Shi, P. (1998). Monotone B-spline smoothing. *Journal of the American Statistical Association, 93*, 643–650.

Kang, J. D. Y., & Schafer, J. L. (2007). Demystifying double robustness: a comparison of alternative strategies for estimating a population mean from incomplete data. *Statistical Sceince, 22*(4), 523–539.

Kelly, C., & Rice, J. (1990). Monotone smoothing with application to dose-response curves and the assessment of Synergism, *Biometrics, 46*, 1071–1085.

Holland, P. W. (1986). Statistics and causal inference. *Journal of the American Statistical Association, 81* (396), 945–960.

Horvitz, D. G., & Thompson, D. J. A (1952). Generalization of sampling without replacement from a finite universe. *Journal of the American Statistical Association, 47*, 663–685.

Lunceford, J. K., & Davidian, M. (2004). Stratification and weighting via the propensity score in estimation of causal treatment effects: a comparative study. *Statistics in Medicine, 23*(19), 2937–2960.

Ma, S., & Huang, J. (2017). A concave pairwise fusion approach to subgroup analysis. *Journal of the American Statistical Association, 112*, 410–423.

Mammen, E. (1991). Estimating a smooth monotone regression function. *Annals of Statistics, 19*, 724–740.

Mukherjee, H. (1988). Monotone nonparametric regression. *Annals of Statistics, 16*, 741–750.

Opitz, D., & Maclin, R. (1999). Popular ensemble methods: An empirical study. *Journal of Artificial Intelligence Research, 11*, 169–198

Pearl, J., Glymour, M., & Jewell, N. P. (2016). *Causal inference in statistics: A primer*. John Wiley & Sons.

Qin, J., Garcia, T. P., Ma, Y., Tang, M.-X., Marder, K., & Wang, Y. (2014). Combining isotonic regerssion and EM algorithm to predict risk under monotonicity constraint. *Annals of Applied Statistics, 8*, 1182–1208.

Qin, J., Sun, Y., Yuan, A., & Huang, C-Y. (2022). Maximum likelihood estimation for shape-resrticted single-index hazard models. *Journal of data science, 21*(4), 681–695. https://doi.org/10.6339/22-JDS1061

Ramsay, J. (1988). Monotone regression splines in action. *Statistical Science, 3*, 425–441.

Rani, P., & Muneeswaran, K. (2018). Emotion recognition based on facial components. *Sadhana, 43*(3). https://doi.org/10.1007/s12046-018-0801-6

Robins, J., Rotnitzky, A., & Zhao, L. (1994). Estimation of regression coefficients when some of the regressors are not always observed. *Journal of the American Statistical Association, 89*, 846–866.

Robins, J. M. (1997). Marginal structural models. In *Proceedings of the Section on Bayesian Statistical Science* (pp. 1–10), Alexandria, VA: American Statistical Association.

Rosenbaum, P., & Rubin, D. B. (1983). The central role of the propensity score in observational studies for causal effects. *Biometrika, 70*, 41–55.

Ruberg, S. J., Chen, L., & Wang, Y. (2010). The mean doesn't mean as much any more: finding sub-groups for tailored therapeutics. *Clinical Trials, 7*, 574–583.

Rubin, D. (1974). Estimating causal effects of treatments in randomized and nonrandomized studies. *Journal of Educational Psycology, 66*(5), 688–701.

Shen, J., & He, X. (2015). Inference for subgroup analysis with a structured logistic-normal mixture model. *Journal of the American Statistical Association, 110*, 303–312.

Yuan, A., Chen, X., Zhou, Y., & Tan, M. T. (2018). Subgroup analysis with semiparametric models toward precision medicine. *Statistics in Medicine, 37*, 1830–1845.

Yuan, A., Yin, A., & Tan, M. T. (2021). Enhanced doubly robust procedure for causal inference. *Statistics in Biosciences*. https://doi.org/10.1007/s12561-021-09300-y

Yuan, A., Yin, A., & Tan, M. T. (2022). Flexible and robust procedure for subgroup inference. *Biostatistics and Epidemiology, 6*(2), 314–328. https://doi.org/10.1080/24709360.2022.2127650

A Selective Overview of Fusion Penalized Learning in Latent Subgroup Analysis for Precision Medicine

Mingming Liu, Queen Ikhelowa, and Shujie Ma

Abstract Modern technology has facilitated the collection of large-scale data from electronic medical records, health insurance databases and other platforms. Big data along with rapid growth in computational power are creating unprecedented opportunities in precision medicine that is at the forefront of medical research. Understanding the disease heterogeneity is essential to the development of precision medicine that aims to tailor treatments to subgroups of patients with similar characteristics. One major challenge of achieving this goal is that the heterogeneity can be driven by latent factors which are unobservable in the datasets. This article gives a review on a pairwise fusion penalized learning method for latent subgroup analysis. This approach has received increasing attention in recent years. The problem of subgroup investigation can be formulated into a heterogeneous regression model that allows the parameters to be subject-dependent with unknown grouping information. We then apply the fusion penalized method that can automatically divide the patients into different groups as well as estimating the parameters simultaneously. This machine learning approach is flexible, can be adapted into different regression models and data settings, and requires mild assumptions. We will discuss in detail the computational algorithms for this method, and its applications to longitudinal data with different types of responses such as continuous, binary and count data. We will also illustrate the performance of this method through simulation studies and a biomedical data application.

For the Alzheimer's Disease Neuroimaging Initiative: Data used in preparation of this article were obtained from the ADNI database (adni.loni.usc.edu). As such, the investigators within the ADNI contributed to the design and implementation of ADNI and/or provided data but did not participate in analysis or writing of this report. A complete listing of ADNI investigators can be found at: http://adni.loni.usc.edu/wp-content/uploads/how_to_apply/ADNI_Acknowledgement_List.pdf.

M. Liu · Q. Ikhelowa · S. Ma (✉)
Department of Statistics, University of California, Riverside, CA, USA
e-mail: mliu034@ucr.edu; qikhe001@ucr.edu; shujie.ma@ucr.edu

Y. Zhao, D.-G. Chen (eds.), *Statistics in Precision Health*, ICSA Book Series in Statistics, https://doi.org/10.1007/978-3-031-50690-1_12

Keywords Fusion penalty · Heterogeneity · Longitudinal data · Precision medicine · Subgroup analysis

1 Introduction

Subgroup analysis plays an important role in precision medicine. It aims at developing individualized treatment strategies by taking into account individual variability in genes, environment, lifestyle and so forth, so that the targeted subgroups can benefit most from a specific treatment. The treatment effects are often heterogeneous (Kravitz et al., 2004; Sørensen, 1996), and failure to detect heterogeneity can yield misleading results. For example, radiation therapy can help control the growth of tumors for some patients with cancer, but it may be ineffective in treating other patients. Understanding the disease heterogeneity is essential to the development of precision medicine that aims to tailor treatments to subgroups of patients with similar characteristics.

Traditionally, a subgroup analysis can be performed by splitting the data into subgroups based on observed grouping variables. In practice, it is mostly infeasible to explore high-order interactions among many variables, and moreover the heterogeneity can be driven by latent factors which are unobservable in the datasets. A popular approach for latent subgroup analysis is to employ a mixture modelling strategy (Banfield & Raftery, 1993; Everitt & Hand, 1981; Shen & He, 2015). This method needs to pre-specify the number of subgroups, which is often unknown in practice, and also requires an underlying distribution for the data. To tackle this problem, Ma (2017) proposed a concave pairwise fusion penalized method to group the means of clinical outcomes, in which the heterogeneity is driven by unobserved latent factors, and is modeled through subject-specific intercepts. This fusion learning method can automatically divide the patients into different subgroups as well as estimating the parameters simultaneously. Ma et al. (2019) further extended this method to estimate heterogeneous treatment effects (HTE) in a regression model which allows the coefficients of treatment variables to be subject-dependent with unknown grouping information.

The heterogeneous regression model for HTE considered in Ma et al. (2019) is

$$y_i = \boldsymbol{x}_i^T \boldsymbol{\beta}_i + \boldsymbol{z}_i^T \boldsymbol{\eta} + \varepsilon_i, \quad i = 1, \ldots, n, \tag{1}$$

where y_i is an outcome variable, $\boldsymbol{x}_i$ is a vector of treatment variables, and $\boldsymbol{z}_i$ is a vector of observed baseline covariates. Moreover, $\boldsymbol{\eta}$ and $\boldsymbol{\beta}_i$ are unknown regression coefficients and the ε_i are i.i.d. random errors with $E(\varepsilon_i) = 0$. The first entry in each $\boldsymbol{x}_i$ is 1, so the intercept is included in $\boldsymbol{\beta}_i$. Under this model and the strongly ignorable treatment assignment assumption (Rosenbaum & Rubin, 1983), the subject-specific regression coefficient $\boldsymbol{\beta}_i$ is the causal effect of treatment $\boldsymbol{x}_i$ on the ith subject, and the distribution of $\boldsymbol{\beta}_i$ across i represents the true HTE. The heterogeneity in treatment effects, characterized by $\boldsymbol{\beta}_i$, can be driven by both observed and unobserved latent baseline covariates. A conventional way to identify and estimate model (1) is to consider $\boldsymbol{\beta}_i$ as a function of some baseline covariates $\boldsymbol{t}_i$, where $\boldsymbol{t}_i$ is a subset of z_i. This method ignores possible latent HTE. Even though the HTE is assumed to be completely determined by the observed covariates, it is unclear in what specific form the function should be. A linear form is typically assumed, but it may not be flexible enough to explore HTE. Moreover, higher-order interactions can be involved if many baseline variables are present, leading to a greater risk of false positive results (Wang et al., 2007). To solve the aforementioned problems, Ma (2017) and Ma et al. (2019) assume that the subjects in the same group share the same treatment coefficients, whereas the grouping information is unknown from *a priori* knowledge. A pairwise fusion penalized method is then applied to simultaneously recover the grouping structure and the parameters in the model.

Biomedical studies often involve longitudinal data, in which the outcome variable is repeatedly measured over time for each patient. With repeated measurements, a longitudinal study can increase the power to detect subgroups, as well as improving the precision of estimating treatment effects (Zeger & Liang, 1992). The fusion learning method provided in Ma (2017) and Ma et al. (2019) can be naturally extended to longitudinal settings, see, e.g., Liu et al. (2021), Liu et al. (2022), Zhu (2021), among others. Different from the existing works that focus on a particular type of responses, in this article we will provide an overview on a general framework of generalized heterogeneous regression models for latent subgroup analysis of longitudinal data. The setup can be applied to continuous, binary and count responses under mild regularity conditions. As an extension, we will present three specific models that are of practical use, including the subject-specific intercepts, coefficients and nonparametric regression models. The pairwise fusion penalized method is applied for simultaneously identifying the latent subgroups, recovering the memberships and estimating the parameters. Moreover, we will discuss an alternating direction method of multiplers (ADMM) algorithm (Boyd et al., 2011) for computing and implementing the fusion method.

The rest of the article is organized as follows. Section 2 introduces the generalized heterogeneous regression model. Section 3 presents three specific models as special cases of the proposed model. In Sect. 4, we describe a general estimation procedure and the ADMM algorithm. We then present some numerical results in Sect. 5. A real data application is included in Sect. 6. Some concluding remarks and further discussions are given in Sect. 7.

2 A Generalized Heterogeneous Regression Model

It is well known that the generalized linear models (GLMs) have been widely used to analyze the longitudinal data. We use GLMs with subject-dependent coefficients to investigate heterogeneous causal effects of some treatment variables. Suppose that the longitudinal data consist of

$$\left\{(Y_{ij}, \boldsymbol{X}_{ij}, \boldsymbol{Z}_{ij}), 1 \leq i \leq n, 1 \leq j \leq m_i\right\},$$

where Y_{ij} is the response for the ith subject at the jth measurement, $\boldsymbol{X}_{ij}$ is a p-vector of treatment variables associated with heterogeneous effects across the subjects, and $\boldsymbol{Z}_{ij}$ is a q-vector of covariates having homogeneous effects. The response variable here can be either continuous or discrete. Based on the different types of the response variable (e.g., continuous, binary or poisson), different models can be specified. Following Fitzmaurice et al. (2008), Fitzmaurice et al. (2012), let $\mu_{ij} = E(Y_{ij}|\boldsymbol{X}_{ij}, \boldsymbol{Z}_{ij})$ and $\text{Var}(Y_{ij}|\boldsymbol{X}_{ij}, \boldsymbol{Z}_{ij}) = \phi v(\mu_{ij})$, where ϕ is a scale parameter and $v(\mu_{ij})$ is a known variance function. We consider the following generalized heterogeneous regression model:

$$h^{-1}(\mu_{ij}) = \eta_i(\boldsymbol{X}_{ij};\ \boldsymbol{\beta}_i) + \boldsymbol{Z}_{ij}^T\boldsymbol{\gamma}, \tag{2}$$

where η_i is a function of $\boldsymbol{X}_{ij}$ identified by a set of parameters $\boldsymbol{\beta}_i$, and $\boldsymbol{\gamma}$ are coefficients of $\boldsymbol{Z}_{ij}$. Note that the heterogeneity of function η_i is decided by the parameters $\boldsymbol{\beta}_i$. Moreover, $h^{-1}(\cdot) \equiv g(\cdot)$ is a link function, which is a known real-valued monotonic and differentiable function used in GLMs (McCullagh & Nelder, 1989). Several commonly used link functions include identity, logit and log functions. For example, when the responses are continuous variables, we can let $h^{-1}(\mu_{ij}) = \mu_{ij}$, $v(\mu_{ij}) = 1$ and $\text{Var}(Y_{ij}|\boldsymbol{X}_{ij}, \boldsymbol{Z}_{ij}) = \phi$. For binary responses, $h^{-1}(\mu_{ij}) = \text{logit}(\mu_{ij})$, $\phi = 1$ and $\text{Var}(Y_{ij}|\boldsymbol{X}_{ij}, \boldsymbol{Z}_{ij}) = \mu_{ij}(1 - \mu_{ij})$. For poisson responses, $h^{-1}(\mu_{ij}) = \log(\mu_{ij})$, $\phi = 1$ and $\text{Var}(Y_{ij}|\boldsymbol{X}_{ij}, \boldsymbol{Z}_{ij}) = \mu_{ij}$. Moreover, $\eta_i(\boldsymbol{X}_{ij};\ \boldsymbol{\beta}_i)$, as a function of $\boldsymbol{X}_{ij}$, can be adapted into different forms. For instance, if there exists a linear relationship in $\boldsymbol{X}_{ij}$, it can be written as $\eta_i(\boldsymbol{X}_{ij};\ \boldsymbol{\beta}_i) = \boldsymbol{X}_{ij}^T\boldsymbol{\beta}_i$; η_i can also be an unknown function of $\boldsymbol{X}_{ij}$. In general, model (2) is quite flexible. It includes both parametric and semiparametric models as special cases.

Under model (2), the subject-specific coefficient $\boldsymbol{\beta}_i$ represents the heterogeneous causal effect of $\boldsymbol{X}_{ij}$ on the ith subject, and $\boldsymbol{\gamma}$ indicates the homogeneous effect of the covariates $\boldsymbol{Z}_{ij}$. Since the heterogeneity can be driven by unobserved latent factors, we assume that $\boldsymbol{\beta}_i$ in (2) are from $K (K \geq 1)$ different subgroups, in which the subjects from the same group will produce the same effect associated with $\boldsymbol{X}_{ij}$. More specifically, we have $\boldsymbol{\beta}_i = \boldsymbol{\alpha}_k$, if the ith subject belongs to the kth group $\mathcal{G}_k$, where $\boldsymbol{\alpha}_k$ is the common vector for all the $\boldsymbol{\beta}_i$'s in this group ($1 \leq k \leq K$), and $\mathcal{G} = (\mathcal{G}_1, \ldots, \mathcal{G}_K)$ is a mutually exclusive partition of $\{1, \ldots, n\}$. Hence, our goal is to cluster the heterogeneous population with respect to $\boldsymbol{\beta}_i$. Note that, the number of subgroups K is usually unknown in reality, and is much smaller than the number of subjects n.

3 Model Specifications

In this section, we introduce three specific heterogeneous regression models that can be derived from the proposed model (2), and they are of practical use in real applications..

3.1 Subject-Specific Coefficients Regression Model

When $\boldsymbol{X}_{ij}$ represents the treatment variables and $\boldsymbol{Z}_{ij}$ represents the observed baseline covariates, to detect the HTE, we can assume a linear structure in η_i, i.e. $\eta_i(\boldsymbol{X}_{ij}; \boldsymbol{\beta}_i) = \boldsymbol{X}_{ij}^T\boldsymbol{\beta}_i$, and consider the subject-specific coefficients regression model,

$$h^{-1}(\mu_{ij}) = \boldsymbol{X}_{ij}^T\boldsymbol{\beta}_i + \boldsymbol{Z}_{ij}^T\boldsymbol{\gamma}. \tag{3}$$

The first component in each $\boldsymbol{X}_{ij}$ is assumed to be 1, so that the intercept is always included in $\boldsymbol{\beta}_i$. The HTE is represented by the distribution of $\boldsymbol{\beta}_i$ across i. In reality, the collection of the observed covariates is often limited, and the heterogeneity can arise from unobserved latent factors. Therefore, the traditional regression models by incorporating HTE as interactions between treatment variables and observed baseline covariates may not be appropriate to measure the true HTE. For latent subgroup analysis, mixture modeling methods can also be applied. For example, Shen and He (2015) proposed a two-subgroup logistic-normal mixture model to assess the HTE, in which it requires a normal assumption in the response and also relates the group memberships to the observed covariates through a logistic regression model. However, this method needs to pre-specify the number of subgroups, and normal distributions to the responses. Moreover, how to apply this method to correlated data or discrete responses still needs investigation.

We consider the fusion penalized learning method that can be applied to more general settings. Similar to the assumptions imposed on the generalized heterogeneous regression model, we assume that $\boldsymbol{\beta}_i$'s are from K subgroups, where the subjects in each subgroup share the same treatment coefficients so that the heterogeneity can be driven by observed covariates, unobserved latent covariates or a combination of both types of covariates. That is, we have $\boldsymbol{\beta}_i = \boldsymbol{\alpha}_k$ for all $i \in \mathcal{G}_k$, $1 \leq k \leq K$, with $\boldsymbol{\alpha}_k$ being the common vector in this subgroup. Without assuming a specific form on $\boldsymbol{\beta}_i$, this model is considerably flexible. On the other hand, the numerical results based on this model can also provide insights for future studies in exploring the true HTE. For example, if the results indicate that only a small group of patients benefits from the new treatment, it can motivate the researchers to search for more possible covariates that can contribute to the HTE in future studies. It is worth noting that when we have cross-sectional data with continuous responses, model (3) can be modified into model (1) given in Ma et al. (2019).

3.2 Subject-Specific Intercepts Regression Model

Given model (3), when $\boldsymbol{X}_{ij}$ are unavailable, we can only include the intercept in $\boldsymbol{X}_{ij}^T\boldsymbol{\beta}_i$ such that $\boldsymbol{X}_{ij} = 1$ and $\boldsymbol{\beta}_i = a_i$. Then model (3) becomes the subject-specific intercepts regression model

$$h^{-1}(\mu_{ij}) = a_i + \boldsymbol{Z}_{ij}^T\boldsymbol{\gamma}, \tag{4}$$

where a_i's are unknown subject-specific intercepts, and the heterogeneity driven by unknown or unobserved latent factors is modeled through the intercept a_i.

Motivated by the Ocular Hypertension Treatment Study (OHTS) (Gordon et al., 1999; Kass et al., 2002, 2010), Liu et al. (2022) used model (4) with identity link to capture the heterogeneity in the progression rate of primary open-angle glaucoma (POAG) developed on left and right eyes of different individuals. Evidence shows that people with ocular hypertension (OH) have a high risk of developing POAG. One goal of the OHTS is to identify the risk factors that can predict which individuals with OH are most likely to develop visual field loss due to POAG. In Liu et al. (2022), among individuals who have developed the POAG endpoint, the slope of visual field measure (sVFM) of each eye is taken as the response variable. Therefore, some individuals may have only one eye reaching the POAG endpoint, while others have both. The sVFMs are also correlated with each other. When investigating the risk factors of developing visual field loss, apart from some baseline risk factors, such as race, age, gender, randomization assignment, etc., the heterogeneity among eyes across individuals can also be cause by some unobserved factors. Liu et al. (2022) used the subject-specific intercepts a_i's to characterize the heterogeneity.

Another popular approach to capture the heterogeneity that cannot readily be explained is to incorporate it into a random effects model, for example, the random intercept model, which has the same form with model (4). However, it assumes that the random intercepts follow a normal distribution. Different from this assumption, with the proposed method, Liu et al. (2021) assumed that a_i's belong to K subgroups, having K distinct values. That is, $a_i = \alpha_k$ for all $i \in \mathcal{G}_k$ with $1 \leq k \leq K$, where α_k is the common value for all the a_i's in this subgroup. Similarly, this subject-specific intercepts model is applicable to the cross-sectional data as well, which has been discussed in Ma (2017).

3.3 Subject-Specific Nonparametric Regression Model

If $\eta_i(\boldsymbol{X}_{ij}; \boldsymbol{\beta}_i)$ does not have a clear structure, i.e., η_i is an unknown function of $\boldsymbol{X}_{ij}$, $\mu_{ij} = \eta_i(\boldsymbol{X}_{ij};\ \boldsymbol{\beta}_i) + \boldsymbol{Z}_{ij}^T\boldsymbol{\gamma}$ can be considered as a partially linear model. For illustration convenience, we consider the case that $\boldsymbol{X}_{ij}$ only includes one covariate t_{ij}, so the model becomes

$$h^{-1}(\mu_{ij}) = \eta_i(t_{ij}; \boldsymbol{\beta}_i) + \boldsymbol{Z}_{ij}^T\boldsymbol{\gamma}, \tag{5}$$

where η_i is an unknown smooth function of t_{ij}. We can use B-splines (De Boo, 1978; Liu et al., 2010; Xue & Liang, 2010; Ma, 2014, 2012) to approximate the unknown function , i.e., $\eta_i(t_{ij}; \boldsymbol{\beta}_i) \approx \boldsymbol{B}(t_{ij})^T\boldsymbol{\beta}_i$, then (5) can be written as

$$h^{-1}(\mu_{ij}) \approx \boldsymbol{B}(t_{ij})^T\boldsymbol{\beta}_i + \boldsymbol{Z}_{ij}^T\boldsymbol{\gamma}, \tag{6}$$

where $\boldsymbol{B}(t) = (B_1(t), \ldots, B_p(t)))^T$ is the vector of normalized B-spline basis functions (see De Boo (1978)), and $\boldsymbol{\beta}_i$ is the vector of B-spline coefficients. It can be seen that, after the B-spline approximation, model (6) with identity link is similar to the subject-specific coefficients model given in (3). When the covariates Z_{ij} are unavailable, only the nonparametric part is included in the model, and then

$$h^{-1}(\mu_{ij}) = \eta_i(t_{ij}; \boldsymbol{\beta}_i) \approx \boldsymbol{B}(t_{ij})^T \boldsymbol{\beta}_i, \tag{7}$$

which can be viewed as a subject-specific nonparametric regression model.

Liu et al. (2022) utilized model (7) with identity link to study the disease progression of Alzheimer's. When there is no data information about the treatment variables and baseline covariates, understanding the heterogeneity in the disease progression is a key factor to the treatment development. Alzheimer's disease (AD), as a leading cause of dementia for adults, is a progressive disease that will get worse over time. The progression of AD varies from person to person, i.e., different people show different degrees of symptoms with memory loss, mental decline and delusion. As the progression of AD is often measured by cognitive scores at multiple time points, to identify the latent subgroups of individuals who share a common disorder progression, Liu et al. (2022) used the subject-specific nonparametric function η_i given in (7) to model each individual's cognitive scores over time, in which the heterogeneity can be driven by observed or unobserved covariates. Each functional curve can be approximated by B-splines. As a result, the heterogeneity represented by the subject-specific unknown functions η_i is reflected in the B-spline coefficients $\boldsymbol{\beta}_i$. Therefore, the goal of clustering the population based on the heterogeneous longitudinal trajectories can be transformed into clustering based on the heterogeneous B-spline coefficients $\boldsymbol{\beta}_i$'s.

Due to the trajectory heterogeneity, Liu et al. (2022) assumed that $\eta_i(t)$'s in (7) were from K different subgroups, with subjects in the same subgroup sharing the same function. Specifically, $\eta_i(t) = \alpha_k(t)$ for all the subjects belonging to the kth subgroup $\mathscr{G}_k$, where $\alpha_k(t)$ is the common function for all the $\eta_i(t)$'s from $\mathscr{G}_k$. Compared to the growth mixture modeling method (Fraley & Raftery, 2002; Song et al., 2007; Jung & Wickrama, 2008), which requires to specify the underlying distribution of the data, the subject-specific nonparametric model is more flexible.

4 A General Estimation Procedure

In this section, we introduce a general estimation procedure for estimating the proposed generalized heterogeneous model using the ADMM algorithm. For demonstration simplicity, we focus on studying the model with a linear structure in η_i, i.e.,

$$h^{-1}(\mu_{ij}) = \boldsymbol{X}_{ij}^T \boldsymbol{\beta}_i + \boldsymbol{Z}_{ij}^T \boldsymbol{\gamma}. \tag{8}$$

Let $\boldsymbol{Y}_i = (Y_{i1}, \ldots, Y_{im_i})^T$ be the $m_i \times 1$ vector of the response values, $\boldsymbol{X}_i = (\boldsymbol{X}_{i1}, \ldots, \boldsymbol{X}_{im_i})^T$ be the $m_i \times p$ matrix of the treatment variables and $\boldsymbol{Z}_i = (\boldsymbol{Z}_{i1}, \ldots, \boldsymbol{Z}_{im_i})^T$ be the $m_i \times q$ matrix of the covariates. We further denote $\boldsymbol{\mu}_i = (\mu_{i1}, \ldots, \mu_{im_i})^T$ and $h^{-1}(\boldsymbol{\mu}_i) = \left\{h^{-1}(\mu_{i1}), \ldots, h^{-1}(\mu_{im_i})\right\}^T$. Then, model (8) can be rewritten as

$$h^{-1}(\boldsymbol{\mu}_i) = \boldsymbol{X}_i\boldsymbol{\beta}_i + \boldsymbol{Z}_i\boldsymbol{\gamma}. \tag{9}$$

For longitudinal data, the measurements from different subjects are independent, but they are correlated within each subject. Following Ma (2013), Wang et al. (2005), Ma (2012), we allow $\boldsymbol{X}_i$ and $\boldsymbol{Z}_i$ to be dependent. Let $\boldsymbol{\Sigma}_i$ and $\boldsymbol{V}_i$ be the true and assumed working covariance of $\boldsymbol{Y}_i$, where $\boldsymbol{\Sigma}_i = \mathrm{Var}(\boldsymbol{Y}_i)$ and $\boldsymbol{V}_i = \boldsymbol{S}_i^{1/2}\boldsymbol{R}_i\boldsymbol{S}_i^{1/2}$, in which $\boldsymbol{S}_i$ is an $m_i \times m_i$ diagonal matrix with the marginal variances of Y_{ij} along the diagonal, and $\boldsymbol{R}_i$ is an invertible working correlation matrix depending on a correlation parameter ρ. Since the true covariance is often unknown in real applications, a working covariance matrix $\boldsymbol{V}_i$, specified by the working correlation matrix $\boldsymbol{R}_i$, is usually used to replace $\boldsymbol{\Sigma}_i$ in estimation. Some commonly used working correlation structures include exchangeable and AR(1) structures. It is noteworthy that our estimators are consistent even if the working covariance matrix is mis-specified, i.e. $\boldsymbol{V}_i \neq \boldsymbol{\Sigma}_i$.

To estimate model (9), following Ma (2017), Ma et al. (2019), we apply the fusion penalized method, and consider the objective function

$$Q_n(\boldsymbol{\beta}, \boldsymbol{\gamma}; \lambda) = \frac{1}{2}\sum_{i=1}^{n}(\boldsymbol{Y}_i - \boldsymbol{\mu}_i)^T\boldsymbol{V}_i^{-1}(\boldsymbol{Y}_i - \boldsymbol{\mu}_i) + \sum_{1\le i<j\le n} p\left(\left\|\boldsymbol{\beta}_i - \boldsymbol{\beta}_j\right\|_2, \lambda\right), \tag{10}$$

where $\boldsymbol{\mu}_i = \boldsymbol{\mu}_i(\boldsymbol{\beta}_i, \boldsymbol{\gamma}) = h(\boldsymbol{X}_i\boldsymbol{\beta}_i + \boldsymbol{Z}_i\boldsymbol{\gamma})$, $\boldsymbol{\beta} = (\boldsymbol{\beta}_1^T, \ldots, \boldsymbol{\beta}_n^T)$, $\|\cdot\|_2$ represents the L_2 norm with $\|\boldsymbol{a}\|_2 = (\sum a_i^2)^{1/2}$, and $p(\cdot, \lambda)$ is a concave penalty function with a tuning parameter $\lambda \ge 0$. For a given $\lambda > 0$, denote

$$(\hat{\boldsymbol{\beta}}(\lambda), \hat{\boldsymbol{\gamma}}(\lambda)) = \arg\min_{\boldsymbol{\beta},\boldsymbol{\gamma}} Q_n(\boldsymbol{\beta}, \boldsymbol{\gamma}; \lambda). \tag{11}$$

When λ is sufficiently large, the penalty term will shrink some pairs of $\boldsymbol{\beta}_i - \boldsymbol{\beta}_j$ to zero, so that we can partition the population into subgroups. For notation simplicity, we write $(\hat{\boldsymbol{\beta}}(\lambda), \hat{\boldsymbol{\gamma}}(\lambda)) \equiv (\hat{\boldsymbol{\beta}}, \hat{\boldsymbol{\gamma}})$. Let $\left\{\hat{\boldsymbol{\alpha}}_1, \ldots, \hat{\boldsymbol{\alpha}}_{\hat{K}}\right\}$ be the unique values of $\hat{\boldsymbol{\beta}}$, where $\hat{K}$ is the number of the unique values, i.e. the estimated number of subgroups. Then, in the kth subgroup $\hat{\mathscr{G}}_k$, we have $\hat{\mathscr{G}}_k = \left\{i : \hat{\boldsymbol{\beta}}_i = \hat{\boldsymbol{\alpha}}_k, 1 \le i \le n\right\}$ with $1 \le k \le \hat{K}$. The concave fusion penalized method can simultaneously estimate the parameters and achieve model selection (cluster the estimated parameters into subgroups). It is worth noting that although the dimension of the parameter vector $\boldsymbol{\beta}$ is high, we impose a sparsity assumption in the generalized heterogeneous model (2) such that

$\boldsymbol{\beta}_i = \boldsymbol{\alpha}_k$ for all $i \in \mathscr{G}_k$, $1 \leq k \leq K$, i.e. the actual number of parameters in the model is $Kp + q$, which is much smaller than the sample size n. As a result, the number of distinct values in $\hat{\boldsymbol{\beta}}$ obtained from the concave fusion penalized estimation procedure is only $\hat{K}p + q$. Moreover, under the sparsity assumption, the resulting concave penalized estimators of parameters enjoy estimation and model selection consistency (Fan & Li, 2001; Zhang, 2010). The rate of convergence of the fusion penalized estimators in longitudinal settings is provided in Liu et al. (2022).

As discussed in Ma (2017), Ma et al. (2019), instead of using the lasso penalty $p_\tau(t, \lambda) = \lambda\,|t|$ (Tibshirani, 1996), which over-shrinks the large coefficients, tends to generate more subgroups and leads to biased estimates, we apply the concave penalty functions, such as minimax concave penalty (MCP) (Zhang, 2010) and smoothly clipped absolute deviation penalty (SCAD) (Fan & Li, 2001). Both of them induce nearly unbiased estimates and enjoy the oracle property that provides a theoretical foundation for conducting statistical inference. Moreover, they yield similar results (Ma, 2017; Ma et al., 2019; Liu et al., 2021). The MCP penalty is defined as

$$p_\tau(t, \lambda) = \lambda \int_0^{|t|} (1 - x/(\tau\lambda))_+ dx, \quad \tau > 1,$$

and the SCAD penalty has the form

$$p_\tau(t, \lambda) = \lambda \int_0^{|t|} \min\left\{1, (\tau - x/\lambda)_+/(\tau - 1)\right\}, \quad \tau > 2,$$

where τ is a parameter that controls the concavity of the penalty function, and $(x)_+ = x$, if $x > 0$; $(x)_+ = 0$, otherwise.

We then compute $(\hat{\boldsymbol{\beta}}(\lambda), \hat{\boldsymbol{\gamma}}(\lambda))$ given in (11) in a given interval $[\lambda_{\min}, \lambda_{\max}]$, in which $\lambda_{\min}$ is a small positive number, and $\lambda_{\max}$ is the value that forces a constant $\hat{\boldsymbol{\beta}}$ solution, i.e. $\hat{\boldsymbol{\beta}}_i(\lambda_{\max}) = \hat{\boldsymbol{\beta}}_l(\lambda_{\max})$, $1 \leq i < l \leq n$. The details for computing the solution path of $\left\{\hat{\boldsymbol{\beta}}(\lambda) : \lambda \in [\lambda_{\min}, \lambda_{\max}]\right\}$ on a grid of λ values are provided later in this section. To select the optimal tuning parameter λ, a data-driven procedure such as BIC (Bayes Information Criterion) is considered.

4.1 ADMM Algorithm with Concave Penalties

To obtain the solution of (11), we apply an ADMM algorithm. Since the penalty function $p\left(\|\boldsymbol{\beta}_i - \boldsymbol{\beta}_j\|_2, \lambda\right)$ given in (10) is not separable with respect to $\boldsymbol{\beta}_i$'s, it is very difficult to compute the solutions in (11) directly. To solve this problem, we introduce a new set of parameters $\boldsymbol{\delta}_{ij} = \boldsymbol{\beta}_i - \boldsymbol{\beta}_j$. Then the minimization of function (10) can be reformulated as a constraint optimization problem by minimizing

$$L_0(\boldsymbol{\beta}, \boldsymbol{\gamma}, \boldsymbol{\delta}) = \tfrac{1}{2}\sum_{i=1}^{n}(\boldsymbol{Y}_i - \boldsymbol{\mu}_i)^T \boldsymbol{V}_i^{-1}(\boldsymbol{Y}_i - \boldsymbol{\mu}_i) + \sum_{1\le i<j\le n} p_\tau\left(\left\|\boldsymbol{\delta}_{ij}\right\|_2, \lambda\right),$$
$$\text{subject to } \boldsymbol{\beta}_i - \boldsymbol{\beta}_j - \boldsymbol{\delta}_{ij} = \mathbf{0},$$

where $\boldsymbol{\mu}_i = \boldsymbol{\mu}_i(\boldsymbol{\beta}_i, \boldsymbol{\gamma}) = h(\boldsymbol{X}_i\boldsymbol{\beta}_i + \boldsymbol{Z}_i\boldsymbol{\gamma})$ and $\boldsymbol{\delta} = \left\{\boldsymbol{\delta}_{ij}^T, i < j\right\}^T$. Let $\langle \boldsymbol{a}, \boldsymbol{b}\rangle = \boldsymbol{a}^T\boldsymbol{b}$ the inner product of two vectors. We have the augmented Lagrangian

$$L(\boldsymbol{\beta}, \boldsymbol{\gamma}, \boldsymbol{\delta}, \boldsymbol{v}) = L_0(\boldsymbol{\beta}, \boldsymbol{\gamma}, \boldsymbol{\delta}) + \sum_{i<j}\langle \boldsymbol{v}_{ij}, \boldsymbol{\beta}_i - \boldsymbol{\beta}_j - \boldsymbol{\delta}_{ij}\rangle + \frac{\vartheta}{2}\sum_{i<j}\left\|\boldsymbol{\beta}_i - \boldsymbol{\beta}_j - \boldsymbol{\delta}_{ij}\right\|_2^2, \tag{12}$$

where the dual variables $\boldsymbol{v} = \left\{\boldsymbol{v}_{ij}^T, i < j\right\}^T$ are the Lagrange multipliers and ϑ is a penalty parameter.

Let $\boldsymbol{Y} = (\boldsymbol{Y}_1^T, \ldots, \boldsymbol{Y}_n^T)^T$, $\boldsymbol{V} = \text{diag}(\boldsymbol{V}_1, \ldots, \boldsymbol{V}_n)$ and $\boldsymbol{\mu} = (\boldsymbol{\mu}_1^T, \ldots, \boldsymbol{\mu}_n^T)^T = h(\boldsymbol{X\beta} + \boldsymbol{Z\gamma})$, where $\boldsymbol{X} = \text{diag}(\boldsymbol{X}_1, \ldots, \boldsymbol{X}_n)$ and $\boldsymbol{Z} = (\boldsymbol{Z}_1^T, \ldots, \boldsymbol{Z}_n^T)^T$. Given $\boldsymbol{\delta}$ and $\boldsymbol{v}$, we can rewrite (12) as

$$\begin{aligned}
L(\boldsymbol{\beta}, \boldsymbol{\gamma}; \boldsymbol{\delta}, \boldsymbol{v}) &= \frac{1}{2}\sum_{i=1}^{n}(\boldsymbol{Y}_i - \boldsymbol{\mu}_i)^T \boldsymbol{V}_i^{-1}(\boldsymbol{Y}_i - \boldsymbol{\mu}_i) \\
&\quad + \frac{\vartheta}{2}\sum_{i<j}\left\|\boldsymbol{\beta}_i - \boldsymbol{\beta}_j - \boldsymbol{\delta}_{ij} + \vartheta^{-1}\boldsymbol{v}_{ij}\right\|_2^2 + C_0 \\
&= \frac{1}{2}(\boldsymbol{Y} - \boldsymbol{\mu})^T \boldsymbol{V}^{-1}(\boldsymbol{Y} - \boldsymbol{\mu}) + \frac{\vartheta}{2}\left\|\boldsymbol{A\beta} - \boldsymbol{\delta} + \vartheta^{-1}\boldsymbol{v}\right\|_2^2 + C_0,
\end{aligned}$$

where C_0 is a constant independent of $\boldsymbol{\beta}$ and $\boldsymbol{\gamma}$, $\boldsymbol{A} = \boldsymbol{D} \otimes \boldsymbol{I}_p$ (Kronecker product), $\boldsymbol{D} = \left\{\left(\boldsymbol{e}_i - \boldsymbol{e}_j\right), i < j\right\}^T$, $\boldsymbol{e}_i$ is a $n \times 1$ vector with the ith element equal to 1 and the rest ones being 0, and $\boldsymbol{I}_p$ is a $p \times p$ identity matrix. Then, we can obtain

$$\begin{aligned}
S_1(\boldsymbol{\beta}, \boldsymbol{\gamma}; \boldsymbol{\delta}, \boldsymbol{v}) &= \frac{\partial L(\boldsymbol{\beta}, \boldsymbol{\gamma}; \boldsymbol{\delta}, \boldsymbol{v})}{\partial \boldsymbol{\beta}} \\
&= -\boldsymbol{X}^T \boldsymbol{\Delta} \boldsymbol{V}^{-1}(\boldsymbol{Y} - \boldsymbol{\mu}) + \vartheta \boldsymbol{A}^T(\boldsymbol{A\beta} - \boldsymbol{\delta} + \vartheta^{-1}\boldsymbol{v}), \\
S_2(\boldsymbol{\beta}, \boldsymbol{\gamma}; \boldsymbol{\delta}, \boldsymbol{v}) &= \frac{\partial L(\boldsymbol{\beta}, \boldsymbol{\gamma}; \boldsymbol{\delta}, \boldsymbol{v})}{\partial \boldsymbol{\gamma}} = -\boldsymbol{Z}^T \boldsymbol{\Delta} \boldsymbol{V}^{-1}(\boldsymbol{Y} - \boldsymbol{\mu}),
\end{aligned}$$

where $\boldsymbol{\Delta} = \text{diag}\left\{\frac{\partial \mu_{ij}}{\partial h^{-1}(\mu_{ij})}\right\}$.

Let $S(\boldsymbol{\beta}, \boldsymbol{\gamma}; \boldsymbol{\delta}, \boldsymbol{\upsilon}) = (S_1(\boldsymbol{\beta}, \boldsymbol{\gamma}; \boldsymbol{\delta}, \boldsymbol{\upsilon})^T, S_2(\boldsymbol{\beta}, \boldsymbol{\gamma}; \boldsymbol{\delta}, \boldsymbol{\upsilon})^T)^T$. The estimators of $(\boldsymbol{\beta}, \boldsymbol{\gamma}, \boldsymbol{\delta}, \boldsymbol{\upsilon})$ can be computed iteratively using the ADMM algorithm. To be more specific, given the values of $\boldsymbol{\delta}^m$, $\boldsymbol{\upsilon}^m$ at step m, the estimates are updated as follows:

$$(\boldsymbol{\beta}^{m+1}, \boldsymbol{\gamma}^{m+1}) \Leftarrow S(\boldsymbol{\beta}, \boldsymbol{\gamma}; \boldsymbol{\delta}^m, \boldsymbol{\upsilon}^m) = \mathbf{0} \tag{13}$$

$$\boldsymbol{\delta}^{m+1} = \operatorname{argmin}_{\boldsymbol{\delta}} L\left(\boldsymbol{\beta}^{m+1}, \boldsymbol{\gamma}^{m+1}, \boldsymbol{\delta}, \boldsymbol{\upsilon}^m\right), \tag{14}$$

$$\boldsymbol{\upsilon}_{ij}^{m+1} = \boldsymbol{\upsilon}_{ij}^m + \vartheta\left(\boldsymbol{\beta}_i^{m+1} - \boldsymbol{\beta}_j^{m+1} - \boldsymbol{\delta}_{ij}^{m+1}\right). \tag{15}$$

In (13), the updates for $(\boldsymbol{\beta}, \boldsymbol{\gamma})$ are discussed later in Remark 2 and Remark 3 in detail.

In (14), given $\boldsymbol{\beta}^{m+1}$, $\boldsymbol{\gamma}^{m+1}$ and $\boldsymbol{\upsilon}^m$, the minimization problem is equivalent to minimizing

$$\frac{\vartheta}{2}\left\|\boldsymbol{\zeta}_{ij}^m - \boldsymbol{\delta}_{ij}\right\|_2^2 + p_\tau\left(\left\|\boldsymbol{\delta}_{ij}\right\|_2, \lambda\right)$$

over $\boldsymbol{\delta}_{ij}$, where $\boldsymbol{\zeta}_{ij}^m = \boldsymbol{\beta}_i^{m+1} - \boldsymbol{\beta}_j^{m+1} + \vartheta^{-1}\boldsymbol{\upsilon}_{ij}^m$. Following Ma et al. (2019), for the MCP penalty with $\tau > 1/\vartheta$, we can update $\boldsymbol{\delta}$ by

$$\boldsymbol{\delta}_{ij}^{m+1} = \begin{cases} \frac{\mathrm{ST}(\boldsymbol{\zeta}_{ij}^m, \lambda/\vartheta)}{1-1/(\tau\vartheta)} & \text{if } \|\boldsymbol{\zeta}_{ij}^m\|_2 \le \tau\lambda, \\ \boldsymbol{\zeta}_{ij}^m & \text{if } \|\boldsymbol{\zeta}_{ij}^m\|_2 > \tau\lambda, \end{cases} \tag{16}$$

and for the SCAD penalty with $\tau > 1/\vartheta + 1$,

$$\boldsymbol{\delta}_{ij}^{m+1} = \begin{cases} \mathrm{ST}(\boldsymbol{\zeta}_{ij}^m, \lambda/\vartheta) & \text{if } \|\boldsymbol{\zeta}_{ij}^m\|_2 \le \lambda + \lambda/\vartheta, \\ \frac{\mathrm{ST}(\boldsymbol{\zeta}_{ij}^m, \tau\lambda/((\tau-1)\vartheta))}{1-1/((\tau-1)\vartheta)} & \text{if } \lambda + \lambda/\vartheta < \|\boldsymbol{\zeta}_{ij}^m\|_2 \le \tau\lambda, \\ \boldsymbol{\zeta}_{ij}^m & \text{if } \|\boldsymbol{\zeta}_{ij}^m\|_2 > \tau\lambda, \end{cases} \tag{17}$$

where $\mathrm{ST}(\boldsymbol{x}, t) = (1 - t/\|\boldsymbol{x}\|_2)_+\boldsymbol{x}$ is the groupwise soft thresholding operator.

Based on the discussion above, we summarize the ADMM algorithm as follows:

Algorithm 1 ADMM algorithm

```
Initialize δ^0, υ^0.
for m = 0, 1, 2, ··· do
    Update (β^{m+1}, γ^{m+1}) using (13)
    Update δ^{m+1} using (16) or (17)
    Update υ^{m+1} using (15)
    if the convergence criterion is met, then
        Stop and denote the last iteration by (β̂(λ), γ̂(λ)),
    else
        m = m + 1.
    end if
end for
Ensure: Output
```

To start the ADMM algorithm, it is important to find a reasonable initial value. For this purpose, we fit a generalized linear regression model by using all the measurements within each subject. Then the initial estimates $\boldsymbol{\beta}_i^0$ and $\boldsymbol{\gamma}_i^0 (i = 1, \ldots, n)$ can be obtained. As a result, we have $\boldsymbol{\beta}^0 = ((\boldsymbol{\beta}_1^0)^T, \ldots, (\boldsymbol{\beta}_n^0)^T)^T$, and let $\boldsymbol{\gamma}^0 = \frac{1}{n}\sum_{i=1}^n \boldsymbol{\gamma}_i^0$. Moreover, set the initial values $\boldsymbol{\delta}_{ij}^0 = \boldsymbol{\beta}_i^0 - \boldsymbol{\beta}_j^0$ in $\boldsymbol{\delta}^0$ and $\boldsymbol{\upsilon}^0 = \mathbf{0}$.

The ADMM algorithm is stopped when the primal residual $\boldsymbol{r}^{m+1} = \boldsymbol{A}\boldsymbol{\beta}^{m+1} - \boldsymbol{\delta}^{m+1}$ is close to zero such that $\|\boldsymbol{r}^{m+1}\|_2 < \varepsilon$ for some small value $\varepsilon > 0$.

Remark 1 Let $\boldsymbol{R} = \text{diag}(\boldsymbol{R}_1, \ldots, \boldsymbol{R}_n)$ and $\boldsymbol{S} = \text{diag}(\boldsymbol{S}_1, \ldots, \boldsymbol{S}_n)$. Then we have $\boldsymbol{V} = \boldsymbol{S}^{1/2}\boldsymbol{R}\boldsymbol{S}^{1/2}$, where $\boldsymbol{S}$ and $\boldsymbol{V}$ depend on $(\boldsymbol{\beta}, \boldsymbol{\gamma})$, and $\boldsymbol{R}$ is specified by the correlation parameter ρ for a given correlation structure. Thus, $\boldsymbol{S}$ and $\boldsymbol{V}$ can also be updated in each step based on $\boldsymbol{\beta}^m$ and $\boldsymbol{\gamma}^m$ in the ADMM algorithm. To obtain the estimate of $\boldsymbol{V}$, we need to estimate the correlation parameter ρ required in $\boldsymbol{R}$. Given $(\boldsymbol{\beta}^m, \boldsymbol{\gamma}^m)$, we estimate ρ using the residual-based moment method given in Liang and Zeger (1986). Accordingly, the value of $\boldsymbol{V}$ is updated at step m as $\boldsymbol{V}^m = (\boldsymbol{S}^m)^{1/2}\boldsymbol{R}^m(\boldsymbol{S}^m)^{1/2}$. Note that, for the setting of continuous responses, since we assume equal marginal covariances of Y_{ij}, we use the same way given in Fitzmaurice et al. (2012) (see page 357) to estimate ϕ. This is one way to estimate $\boldsymbol{V}$ with its value updated at each step. We can also choose to not update the value of $\boldsymbol{V}$ in the estimation procedure, i.e., we only estimate ρ and ϕ once using the initial values $\boldsymbol{\beta}^0$ and $\boldsymbol{\gamma}^0$.

Remark 2 For continuous responses, since $h^{-1}(\mu_{ij}) = \mu_{ij}$, we have $\frac{\partial \mu_{ij}}{\partial h^{-1}(\mu_{ij})} = 1$, then $\boldsymbol{\Delta}$ is an identity matrix, and $\boldsymbol{\mu} = \boldsymbol{X\beta} + \boldsymbol{Z\gamma}$. By solving (13), the updates $(\boldsymbol{\beta}^{m+1}, \boldsymbol{\gamma}^{m+1})$ have the closed-form:

$$\begin{aligned}\boldsymbol{\beta}^{m+1} &= [\boldsymbol{X}^T(\boldsymbol{V}^m)^{-1}\boldsymbol{Q}_Z\boldsymbol{X} + \vartheta\boldsymbol{A}^T\boldsymbol{A}]^{-1} \\ &\quad\times\left[\boldsymbol{X}^T(\boldsymbol{V}^m)^{-1}\boldsymbol{Q}_Z\boldsymbol{Y} + \vartheta\boldsymbol{A}^T(\boldsymbol{\delta}^m - \vartheta^{-1}\boldsymbol{v}^m)\right], \\ \boldsymbol{\gamma}^{m+1} &= [\boldsymbol{Z}^T(\boldsymbol{V}^m)^{-1}\boldsymbol{Z}]^{-1}\boldsymbol{Z}^T(\boldsymbol{V}^m)^{-1}(\boldsymbol{Y} - \boldsymbol{X}\boldsymbol{\beta}^{m+1}),\end{aligned}$$

where $\boldsymbol{Q}_Z = \boldsymbol{I}_N - \mathbf{Z}[\mathbf{Z}^T(\boldsymbol{V}^m)^{-1}\mathbf{Z}]^{-1}\mathbf{Z}^T(\boldsymbol{V}^m)^{-1}$ and $N = \sum_{i=1}^n m_i$.

Remark 3 For binary responses, since $h^{-1}(\mu_{ij}) = \log(\frac{\mu_{ij}}{1-\mu_{ij}})$, then $\frac{\partial \mu_{ij}}{\partial h^{-1}(\mu_{ij})} = \mu_{ij}(1 - \mu_{ij}) = \text{Var}(Y_{ij}|\boldsymbol{X}_{ij}, \boldsymbol{Z}_{ij})$, and $\boldsymbol{\Delta} = \text{diag}\left\{\mu_{ij}(1-\mu_{ij})\right\} = \text{diag}(\boldsymbol{S}_1, \ldots, \boldsymbol{S}_n) = \boldsymbol{S}$. Similarly, for poisson responses, $h^{-1}(\mu_{ij}) = \log(\mu_{ij})$, $\frac{\partial \mu_{ij}}{\partial h^{-1}(\mu_{ij})} = \mu_{ij} = \text{Var}(Y_{ij}|\boldsymbol{X}_{ij}, \boldsymbol{Z}_{ij})$, and $\boldsymbol{\Delta} = \text{diag}\left\{\mu_{ij}\right\} = \text{diag}(\boldsymbol{S}_1, \ldots, \boldsymbol{S}_n) = \boldsymbol{S}$. Since there are no closed-form solutions for $(\boldsymbol{\beta}^{m+1}, \boldsymbol{\gamma}^{m+1})$ in these cases, we use

$$\begin{aligned}\boldsymbol{\beta}^{m+1} &= \boldsymbol{\beta}^m - [\frac{\partial S_1(\boldsymbol{\beta}, \boldsymbol{\gamma}; \boldsymbol{\delta}^m, \boldsymbol{v}^m)}{\partial \boldsymbol{\beta}}|_{(\boldsymbol{\beta}=\boldsymbol{\beta}^m, \boldsymbol{\gamma}=\boldsymbol{\gamma}^m)}]^{-1} S_1(\boldsymbol{\beta}^m, \boldsymbol{\gamma}^m; \boldsymbol{\delta}^m, \boldsymbol{v}^m) \\ &= \boldsymbol{\beta}^m - (\boldsymbol{X}^T\boldsymbol{M}_1^m\boldsymbol{X} + \vartheta\boldsymbol{A}^T\boldsymbol{A})^{-1} \\ &\quad\times[\boldsymbol{X}^T\boldsymbol{M}_2^m(\boldsymbol{\mu}^m - \boldsymbol{Y}) + \vartheta\boldsymbol{A}^T(\boldsymbol{A}\boldsymbol{\beta}^m - \boldsymbol{\delta}^m + \vartheta^{-1}\boldsymbol{v}^m)], \\ \boldsymbol{\gamma}^{m+1} &= \boldsymbol{\gamma}^m - [\frac{\partial S_2(\boldsymbol{\beta}, \boldsymbol{\gamma}; \boldsymbol{\delta}^m, \boldsymbol{v}^m)}{\partial \boldsymbol{\gamma}}|_{(\boldsymbol{\beta}=\boldsymbol{\beta}^m, \boldsymbol{\gamma}=\boldsymbol{\gamma}^m)}]^{-1} S_2(\boldsymbol{\beta}^m, \boldsymbol{\gamma}^m; \boldsymbol{\delta}^m, \boldsymbol{v}^m) \\ &= \boldsymbol{\gamma}^m - (\mathbf{Z}^T\boldsymbol{M}_1^m\mathbf{Z})^{-1}\mathbf{Z}^T\boldsymbol{M}_2^m(\boldsymbol{\mu}^m - \boldsymbol{Y}),\end{aligned}$$

where

$$\begin{aligned}\boldsymbol{M}_1^m &= \boldsymbol{\Delta}^m(\boldsymbol{V}^m)^{-1}\boldsymbol{\Delta}^m = (\boldsymbol{S}^m)^{1/2}\boldsymbol{R}^{-1}(\boldsymbol{S}^m)^{1/2}, \\ \boldsymbol{M}_2^m &= \boldsymbol{\Delta}^m(\boldsymbol{V}^m)^{-1} = (\boldsymbol{S}^m)^{1/2}\boldsymbol{R}^{-1}(\boldsymbol{S}^m)^{-1/2}.\end{aligned}$$

The whole computation procedure above is for one given λ value. To compute the solution path of $\boldsymbol{\beta}$ against a grid of λ values, referring to Ma et al. (2019), a warm start with a continuation strategy is used to update the solutions. Let $\lambda_{\min} = \lambda_0 < \lambda_1 < \cdots < \lambda_L = \lambda_{\max}$, where $0 \leq \lambda_{\min} < \lambda_{\max} < \infty$. We compute $\hat{\boldsymbol{\beta}}(\lambda_0)$ by using $\boldsymbol{\beta}^0$ as the initial value, and then compute $\hat{\boldsymbol{\beta}}(\lambda_l)$ by using $\hat{\boldsymbol{\beta}}(\lambda_{l-1})$ as the initial value ($l = 1, \ldots, L$). In order to select the optimal tuning parameter λ, we use the modified BIC (Wang et al., 2007) for high-dimensional data settings by minimizing

$$\mathrm{BIC}(\lambda) = \log\left[\sum_{i=1}^{n}(\boldsymbol{Y}_i - \hat{\boldsymbol{\mu}}_i(\lambda))^T \boldsymbol{V}_i^{-1}(\boldsymbol{Y}_i - \hat{\boldsymbol{\mu}}_i(\lambda))/N\right] + C_n\frac{\log N}{N}(\hat{K}(\lambda)p + q),$$

where $\hat{\boldsymbol{\mu}}_i(\lambda) = h(\boldsymbol{X}_i\hat{\boldsymbol{\beta}}_i(\lambda) + \boldsymbol{Z}_i\hat{\boldsymbol{\gamma}}(\lambda))$, $N = \sum_{i=1}^n m_i$ and C_n is a positive number which can depend on n. Following Ma (2017), Liu et al. (2021), Liu et al. (2022), Zhu (2021), we use $C_n = c\log(\log(np + q))$, and c is a positive constant.

In longitudinal data settings, the simulation results in Liu et al. (2022) show that the concave fusion penalized method performs well when c takes values around 1. In general, its performance is not sensitive to the choice of c values, and it yields similar estimation results when we use different values for c chosen from $[0.5, 2]$. In addition to the modified BIC method, we can also apply other data-driven approaches, such as the Calinski-Harabasz index (CH Index) method considered in Liu et al. (2022), to select the optimal value for λ. For the CH index method, we obtain the optimal λ value by maximizing $\mathrm{CH}(\lambda) = \frac{B_{\hat{K}(\lambda)}/(\hat{K}(\lambda)-1)}{W_{\hat{K}(\lambda)}/(n-\hat{K}(\lambda))}$, where $B_{\hat{K}(\lambda)}$ and $W_{\hat{K}(\lambda)}$ are the between and within group sum of square errors for a given λ value. This approach is a fully data-driven method for obtaining the optimal λ value without the need of pre-specifying any constants. Our numerical results in Liu et al. (2022) show that the two methods have comparable performance.

By applying the ADMM algorithm, we can obtain the parameter estimates $\hat{\boldsymbol{\beta}}(\lambda) = ((\hat{\boldsymbol{\beta}}_1(\lambda))^T, \ldots, (\hat{\boldsymbol{\beta}}_n(\lambda))^T)^T$ at each given λ. The concave penalty can automatically shrink the difference $\hat{\boldsymbol{\beta}}_i(\lambda) - \hat{\boldsymbol{\beta}}_j(\lambda)$ to be exactly zero when i and j belong to the same subgroup. As a result, the algorithm yields a small number of distinct values in $\hat{\boldsymbol{\beta}}(\lambda)$, denoted by $\hat{K}(\lambda)$, which is also the estimated number of subgroups. In practice, $\hat{K}(\lambda)$ is much smaller than n. When the pairwise difference is large, the concave penalty becomes inactive. The plot of the solution paths of parameter estimates given in Figure 1 of Liu et al. (2022) shows that the concave fusion penalized method is capable of identifying the correct number of subgroups when λ takes any value in a wide interval, and the optimal value of λ selected by BIC and CH index falls into this interval. Moreover, the resulting estimates of parameters remain nearly unbiased and close to the values of true parameters when λ's value lies in this interval. This is because that under the sparsity assumption, the concave penalized estimators enjoy the oracle property that they are asymptotically equivalent to the oracle estimators obtained by assuming that the true model (the true subgroups) is known (Fan & Li, 2001; Zhang, 2010; Ma, 2017; Liu et al., 2022), when λ satisfies an order condition. As a result, the concave penalized estimators are asymptotically normally distributed, based on which statistical inference can be conducted for the population parameters.

5 Simulation Studies

In this section, we investigate the performance of the proposed penalized method by presenting some simulation studies and partial numerical results from Liu et al. (2022) based on the subject-specific nonparametric regression model given in Sect. 3.3. More studies and results are shown in Liu et al. (2022). For the longitudinal data setting, we consider unbalanced data here. Fixed values are used for ϑ and τ in the ADMM algorithm: $\vartheta = 1$ and $\tau = 3$.

To evaluate the accuracy of the clustering results, Liu et al. (2022) provided three measures: Rand Index (RI) (Rand, 1971), Normalized Mutual Information (NMI) (Vinh et al., 2010) and accuracy percentage (%). The accuracy percentage (%) is defined as the proportion of subjects that are correctly identified. These three values are between 0 and 1, with higher values indicating better performance.

Notice that, the original model given in Liu et al. (2022) has the form

$$Y_{ij} = \eta_i(t_{ij}) + \varepsilon_{ij}, \quad i = 1, \cdots, n, \quad j = 1, \cdots, m_i, \tag{18}$$

where Y_{ij} is the jth measurement of the ith subject, η_i is an unknown function of t_{ij} and ε_{ij} is the random error. The random errors are independent across subjects, but they are correlated within each subject. Moreover, they have zero mean values. Clearly, this model is equivalent to the one given in Sect. 3.3.

5.1 Two Subgroups Example

The data are simulated from the heterogeneous nonparametric regression model (18) with two subgroups, where $\eta_i(t) = \alpha_1(t)$ if $i \in \mathscr{G}_1$ and $\eta_i(t) = \alpha_2(t)$ if $i \in \mathscr{G}_2$. We first generate a balanced dataset with $m_i = T$, and the time points t_{ij}'s are equally spaced on $[0, 1.2]$. We then generate an unbalanced dataset by randomly allowing 50% of the subjects to miss 30%, 40% or 50% of the time points. The random errors $\boldsymbol{\varepsilon}_i = (\varepsilon_{i1}, \cdots, \varepsilon_{iT})^T$ are simulated from $N(\mathbf{0}, \boldsymbol{\Sigma}_E)$, where $\boldsymbol{\Sigma}_E$ has an AR(1) covariance structure with $\rho = 0.3$ and $\sigma = 0.5$. Liu et al. (2022) considered four setups of (n, T), including $\{n = 100, T = 20\}$, $\{n = 100, T = 50\}$, $\{n = 150, T = 20\}$ and $\{n = 150, T = 50\}$. To generate the functional curves $\{\alpha_1(t), \alpha_2(t)\}$ for the two subgroups, we focus on the the Middle and Far cases given in Liu et al. (2022),

$$\text{Middle} \begin{cases} \alpha_1(t) = -0.5t^2 + 1.25t, \\ \alpha_2(t) = -1.3t^2 + 3.25t, \end{cases} \qquad \text{Far} \begin{cases} \alpha_1(t) = -0.5t^2 + 1.25t, \\ \alpha_2(t) = -2.5t^2 + 6.25t, \end{cases}$$

where the distance between the two functions is increased from the Middle case to the Far case. For more details about the simulation setting, please see Liu et al. (2022). In what follows, all the numerical results are for unbalanced data.

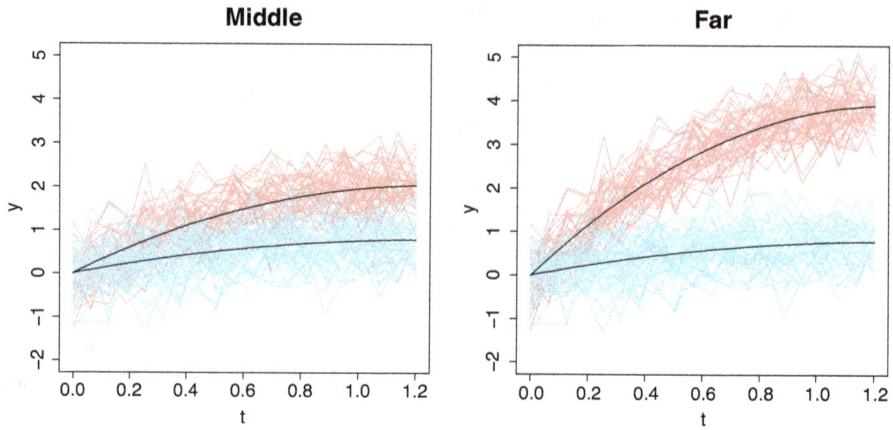

Fig. 1 The simulated trajectories (red, blue thin lines) and the true functions (black thick lines) of the two subgroups generated from one replication with $n = 100,\ T = 20$ for the unbalanced data

Figure 1 shows the simulated trajectories (blue and red thin lines) and true functions (black thick lines) of the two subgroups in the Middle and Far cases, respectively, based on one sample with $n = 100,\ T = 20$. We can observe that, compared to the Far case, there are more overlaps of the trajectories between the two subgroups in the Middle case, making it difficult to distinguish the two subgroups for the Middle case.

To implement the proposed fusion penalized method, we use quadratic splines to approximate the nonparametric components. Additionally, the optimal λ value is selected by minimizing the modified BIC given in Liu et al. (2022). Table 1 from Liu et al. (2022) reports the sample mean, median and per of the estimated number of subgroups $\hat{K}$, where per is the percentage of $\hat{K}$ equal to the true number of subgroups, as well as the average values of RI, NMI, % for the clustering accuracy for different setups of $\{n,\ T\}$ in the Middle and Far cases based on 100 replications. Note that when calculating RI, NMI and %, we only use the replications with $\hat{K}$ equal to the true number of subgroups K. We can observe that the mean and median of $\hat{K}$ are equal to 2, which is the true number of subgroups. Moreover, RI, NMI and % are close to 1 in the Middle case and even become to 1 in the Far case, indicating excellent clustering results by our method. Meanwhile, it can also be observed that as T increases in the Middle case, RI, NMI and % become larger. From the Middle case to the Far case, with the distance between the true functions becoming larger, the subgroups can be more accurately identified.

To further study the estimation accuracy, Table 2 from Liu et al. (2022) presents the mean value of the square root of the mean squared error (RMSE) for the estimated functions $\hat{\alpha}_1(t), \hat{\alpha}_2(t)$ by the proposed and Oracle methods. For the oracle (Oracle) method, the true group membership is utilized to calculate RMSE. From Table 2, it can be seen that the RMSE values obtained by the proposed method are comparable to those from the Oracle method for both Middle and Far cases for

Table 1 The sample mean and median of $\hat{K}$, the percentage (per) of $\hat{K}$ equal to the true number of subgroups, the Rand Index (RI), Normalized mutual information (NMI), and the clustering accuracy percentage (%) which is the proportion of subjects that are correctly identified based on 100 realizations in Two Subgroups Example for the unbalanced data for different $\{n,\ T\}$ setups and function distances

Functions	Setting	Mean	Median	Per	RI	NMI	%
Middle	n = 100, T = 20	2.00	2.00	1.00	0.9903	0.9664	0.9951
	n = 100, T = 50	2.00	2.00	1.00	0.9996	0.9985	0.9998
	n = 150, T = 20	2.00	2.00	1.00	0.9874	0.9535	0.9937
	n = 150, T = 50	2.00	2.00	1.00	0.9999	0.9995	0.9999
Far	n = 100, T = 20	2.00	2.00	1.00	1.0000	1.0000	1.0000
	n = 100, T = 50	2.00	2.00	1.00	1.0000	1.0000	1.0000
	n = 150, T = 20	2.00	2.00	1.00	1.0000	1.0000	1.0000
	n = 150, T = 50	2.00	2.00	1.00	1.0000	1.0000	1.0000

Table 2 The mean value of the square root of the MSE (RMSE) for the estimated functions $\hat{\alpha}_1(t), \hat{\alpha}_2(t)$ obtained by the proposed and Oracle methods in Two Subgroups Example for unbalanced data

		Middle		Far	
Setting	Method	$\hat{\alpha}_1(t)$	$\hat{\alpha}_2(t)$	$\hat{\alpha}_1(t)$	$\hat{\alpha}_2(t)$
n = 100, T = 20	Oracle	0.0385	0.0400	0.0385	0.0400
	Proposed	0.0394	0.0408	0.0385	0.0400
n = 100, T = 50	Oracle	0.0261	0.0243	0.0261	0.0243
	Proposed	0.0262	0.0243	0.0261	0.0243
n = 150, T = 20	Oracle	0.0337	0.0302	0.0337	0.0302
	Proposed	0.0344	0.0302	0.0337	0.0302
n = 150, T = 50	Oracle	0.0214	0.0221	0.0214	0.0221
	Proposed	0.0214	0.0221	0.0214	0.0221

different $\{n,\ T\}$ setups . All these results demonstrate that our proposed penalized method performs well for clustering the heterogeneous trajectories.

6 Real Data Application

In this section, we present partial numerical results obtained from Liu et al. (2022) to demonstrate our proposed penalized method using an Alzheimer's disease (AD) data application. The data were obtained from the Alzheimer's Disease Neuroimaging Initiative (ADNI) database (adni.loni.usc.edu). The ADNI was launched in 2003 as a public-private partnership, led by Principal Investigator Michael W. Weiner, MD. The primary goal of ADNI has been to test whether serial magnetic resonance imaging (MRI), positron emission tomography (PET), other biological markers, and clinical and neuropsychological assessment can be combined to measure the progression of mild cognitive impairment (MCI) and early Alzheimer's disease (AD).

As described in Sect. 3.3, the disease progression of Alzheimer's is heterogeneous. The goal is to cluster the AD patients based on their cognitive scores observed over time, and a subject-specific nonparametric regression model is considered. To conduct latent subgroup analysis, Liu et al. (2022) focused on the longitudinal data of ADASCOG13 (Alzheimer's Disease Assessment Scale-Cognitive Subscale) for patients from ADNI1, ADNIGO and ADNI2 measured at different time points. As some patients have missing measurements at some time points, the data is unbalanced. ADASCOG13 is widely used to assess the severity of the dementia in AD. Its values range from 0 to 85, with higher values indicating worst performance due to more cognitive errors. In total, there are 1253 patients used in the analysis. For more data information, please see the real data section in Liu et al. (2022). The goal here is to identify subgroups based on the heterogeneous longitudinal trajectories. To fit the heterogeneous model (7), ADASCOG13 is considered as the response. By using the fusion penalized method, Liu et al. (2022) identified two subgroups, one progression group with 361 patients and one non-progression group with 892 patients. Figure 2 from Liu et al. (2022) displays the trajectories of individual patients within each subgroup and the corresponding estimated mean curve for each subgroup. It can be seen that the subgroup in

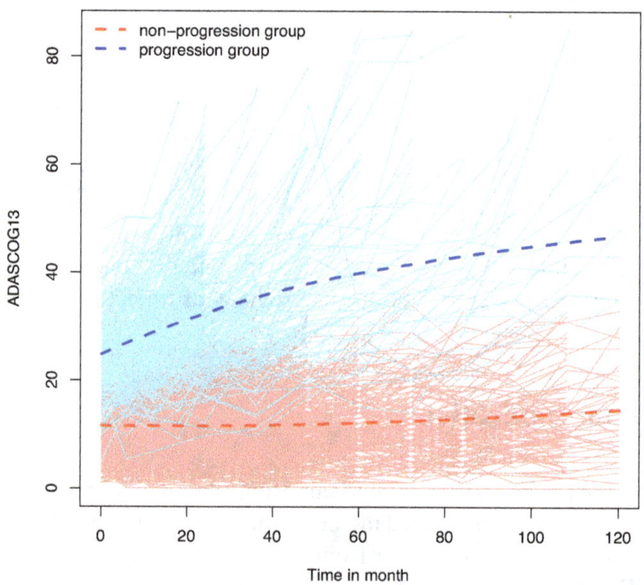

Fig. 2 The trajectories of individual patients within each identified subgroup (blue, red solid lines) and the estimated mean curve (dashed lines) for each subgroup based on ADASCOG13. The blue group is the progression group, with higher values of ADASCOG13, indicating faster cognition decline

blue represents the progression group, as the estimated mean curve of this group increases over time, indicating faster cognition decline. In contrast, the subgroup depicted in red is the non-progression group, with the values of the estimated mean curve almost remaining constant over time. Hence, in order to test whether the drug can slow down or halt the cognition decline, the progression group is potentially of more interest to be recruited in clinical trials. Apparently, two reasonable subgroups are successfully identified with their memberships covered by the fusion penalized method. We refer to Liu et al. (2022) for more detailed numerical results with the R codes available at https://doi.org/10.6084/m9.figshare.17275148.

7 Discussion

In this chapter, we introduce a pairwise fusion penalized learning method to conduct subgroup analysis based on a generalized heterogeneous model, in which the heterogeneity can be driven by observed or unobserved latent factors. Our method can be applied to both cross-sectional and longitudinal data. We present three specific models as special cases of the proposed generalized heterogeneous model for subgroup investigation in different application scenarios. The fusion penalized method can automatically identify the latent group memberships, and simultaneously estimate the parameters without a *prior* knowledge of the grouping information. It is quite flexible and can be applied for latent subgroup analysis for different types of responses, including continuous, binary and poisson data. For implementation, a general estimation procedure with the ADMM algorithm is presented. Through simulation studies and a real data application, we demonstrate that the proposed method has a promising performance. The fusion penalized method can also be applied to other settings such as the joint modeling of survival and longitudinal data. This interesting topic deserves a thorough investigation in future research studies.

Acknowledgments The authors are grateful to the editor and two anonymous reviewers for their insightful comments that have helped us improve the article substantially. The research is supported in part by the U.S. NSF grants DMS-17-12558 and DMS-20-14221 and the UCR Academic Senate CoR Grant.

Data collection and sharing for this project was funded by the Alzheimer's Disease Neuroimaging Initiative (ADNI) (National Institutes of Health Grant U01 AG024904) and DOD ADNI (Department of Defense award number W81XWH-12-2-0012). ADNI is funded by the National Institute on Aging, the National Institute of Biomedical Imaging and Bioengineering, and through generous contributions from the following: AbbVie, Alzheimer's Association; Alzheimer's Drug Discovery Foundation; Araclon Biotech; BioClinica, Inc.; Biogen; Bristol-Myers Squibb Company; CereSpir, Inc.; Cogstate; Eisai Inc.; Elan Pharmaceuticals, Inc.; Eli Lilly and Company; EuroImmun; F. Hoffmann-La Roche Ltd and its affiliated company Genentech, Inc.; Fujirebio; GE Healthcare; IXICO Ltd.; Janssen Alzheimer Immunotherapy Research & Development, LLC.; Johnson & Johnson Pharmaceutical Research & Development LLC.; Lumosity; Lundbeck; Merck & Co., Inc.; Meso Scale Diagnostics, LLC.; NeuroRx Research; Neurotrack Technologies; Novartis Pharmaceuticals Corporation; Pfizer Inc.; Piramal Imaging; Servier; Takeda Pharmaceutical

Company; and Transition Therapeutics. The Canadian Institutes of Health Research is providing funds to support ADNI clinical sites in Canada. Private sector contributions are facilitated by the Foundation for the National Institutes of Health (http://www.fnih.org). The grantee organization is the Northern California Institute for Research and Education, and the study is coordinated by the Alzheimer's Therapeutic Research Institute at the University of Southern California. ADNI data are disseminated by the Laboratory for Neuro Imaging at the University of Southern California.

References

Banfield, J. D., & Raftery, A. E. (1993). Model-based Gaussian and non-Gaussian clustering. *Biometrics, 49*, 803–821.

Boyd, S., Parikh, N., & Chu, E. (2011). Distributed optimization and statistical learning via the alternating direction method of multipliers. *Foundations and Trends in Machine Learning, 3*, 1–122.

De Boor, C. (1978). *A practical guide to splines* (vol. 27, p. 325). New York: Springer-verlag.

Everitt, B. S., & Hand, D. J. (1981). Mixtures of normal distributions. In *Finite Mixture Distributions* (pp. 25–57). Dordrecht: Springer.

Fan, J., & Li, R. (2001). Variable selection via nonconcave penalized likelihood and its oracle properties. *Journal of the American statistical Association, 96*, 1348–1360.

Fitzmaurice, G., Davidian, M., Verbeke, G., & Molenberghs, G. (Eds.). (2008). *Longitudinal data analysis*. CRC Press.

Fitzmaurice, G. M., Laird, N. M., & Ware, J. H. (2012). *Applied longitudinal analysis* (vol. 998). John Wiley & Sons.

Fraley, C., & Raftery, A. E. (2002). Model-based clustering, discriminant analysis, and density estimation. *Journal of the American statistical Association, 97*, 611–631.

Gordon, M. O., Kass, M. A., & Ocular Hypertension Treatment Study Group. (1999). The ocular hypertension treatment Study: design and baseline description of the participants. *Archives of Ophthalmology, 117*, 573–583.

Jung, T., & Wickrama, K. A. (2008). An introduction to latent class growth analysis and growth mixture modeling. *Social and Personality Psychology Compass, 2*, 302–317.

Kass, M. A., Gordon, M. O., Gao, F., Heuer, D. K., Higginbotham, E. J., Johnson, C. A., ... & Wilson, M. R. (2010). Delaying treatment of ocular hypertension: the ocular hypertension treatment study. *Archives of Ophthalmology, 128*, 276–287.

Kass, M. A., Heuer, D. K., Higginbotham, E. J., Johnson, C. A., Keltner, J. L., Miller, J. P., ... & Ocular Hypertension Treatment Study Group. (2002). The Ocular Hypertension Treatment Study: a randomized trial determines that topical ocular hypotensive medication delays or prevents the onset of primary open-angle glaucoma. *Archives of Ophthalmology, 120*, 701–713.

Kravitz, R. L., Duan, N., & Braslow, J. (2004). Evidence-based medicine, heterogeneity of treatment effects, and the trouble with averages. *The Milbank Quarterly, 82*, 661–687.

Liang, K. Y., & Zeger, S. L. (1986). Longitudinal data analysis using generalized linear models. *Biometrika, 73*, 13–22.

Liu, L., Gordon, M., Miller, J. P., Kass, M., Lin, L., Ma, S., & Liu, L. (2021). Capturing heterogeneity in repeated measures data by fusion penalty. *Statistics in Medicine, 40*, 1901–1916.

Liu, M., Yang, J., Liu, Y., Jia, B., Chen, Y. F., Sun, L., & Ma, S. (2022). A fusion learning method to subgroup analysis of Alzheimer's disease. *Journal of Applied Statistics, 50*(8), 1686–1708.

Liu, R., & Yang, L. (2010). Spline-backfitted kernel smoothing of additive coefficient model. *Econometric Theory, 26*, 29–59.

Ma, S. (2012). Two-step spline estimating equations for generalized additive partially linear models with large cluster sizes. *The Annals of Statistics, 40*, 2943–2972.

Ma, S. (2014). A plug-in the number of knots selector for polynomial spline regression. *Journal of Nonparametric Statistics, 26*, 489–507.

Ma, S., & Huang, J. (2017). A concave pairwise fusion approach to subgroup analysis. *Journal of the American Statistical Association, 112*, 410–423.

Ma, S., Huang, J., Zhang, Z., & Liu, M. (2019). Exploration of heterogeneous treatment effects via concave fusion. *The International Journal of Biostatistics, 16*, 1–26.

Ma, S., Song, Q., & Wang, L. (2013). Simultaneous variable selection and estimation in semiparametric modeling of longitudinal/clustered data. *Bernoulli, 19*, 252–274.

McCullagh, P., & Nelder, J. (1989). *Generalized linear models* (2nd ed.). Chapman and Hall/CRC Press.

Rand, W. M. (1971). Objective criteria for the evaluation of clustering methods. *Journal of the American Statistical Association, 66*, 846–850.

Rosenbaum, P. R., & Rubin, D. B. (1983). The central role of the propensity score in observational studies for causal effects. *Biometrika, 70*, 41–55.

Shen, J., & He, X. (2015). Inference for subgroup analysis with a structured logistic-normal mixture model. *Journal of the American Statistical Association, 110*, 303–312.

Song, J. J., Lee, H. J., Morris, J. S., & Kang, S. (2007). Clustering of time-course gene expression data using functional data analysis. *Computational Biology and Chemistry, 31*, 265–274.

Sørensen, T. I. (1996). Which patients may be harmed by good treatments? *Lancet (London, England), 348*, 351–352.

Tibshirani, R. (1996). Regression shrinkage and selection via the lasso. *Journal of the Royal Statistical Society: Series B (Methodological), 58*, 267–288.

Vinh, N. X., Epps, J., & Bailey, J. (2010). Information theoretic measures for clusterings comparison: variants, properties, normalization and correction for chance. *The Journal of Machine Learning Research, 11*, 2837–2854.

Wang, H., Li, R., & Tsai, C. L. (2007). Tuning parameter selectors for the smoothly clipped absolute deviation method. *Biometrika, 94*, 553–568.

Wang, N., Carroll, R. J., & Lin, X. (2005). Efficient semiparametric marginal estimation for longitudinal/clustered data. *Journal of the American Statistical Association, 100*, 147–157.

Wang, R., Lagakos, S., Ware, J., Hunter, D., & Drazen, J. (2007). Reporting of subgroup analyses in clinical trials. *The New England Journal of Medicine, 357*, 2189–2194.

Xue, L., & Liang, H. (2010). Polynomial spline estimation for a generalized additive coefficient model. *Scandinavian Journal of Statistics, 37*, 26–46.

Zeger, S. L., & Liang, K.Y. (1992). An overview of methods for the analysis of longitudinal data. *Statistics in Medicine, 11*, 1825–1839.

Zhang, C. H. (2010). Nearly unbiased variable selection under minimax concave penalty. *The Annals of Statistics, 38*, 894–942.

Zhu, X., Tang, X., & Qu, A. (2021). Longitudinal clustering for heterogeneous binary data. *Statistica Sinica, 31*, 603–624.

Part III
Precision Medicine in Clinic Trials and the Applications to EHR Data

Mining for Health: A Comparison of Word Embedding Methods for Analysis of EHRs Data

Emily Getzen, Yucheng Ruan, Lyle Ungar, and Qi Long

Abstract Electronic health records (EHRs), routinely collected as part of healthcare delivery, offer great promise for advancing precision health. At the same time, they present significant analytical challenges. In EHRs, data for individual patients are collected at irregular time intervals and with varying frequencies; they include both structured and unstructured data. Advanced statistical and machine learning methods have been developed to tackle these challenges, for example, for predicting diagnoses earlier and more accurately. One powerful tool for extracting useful information from EHRs data is word embedding algorithms, which represent words as vectors of real numbers that capture the words' semantic and syntactic similarities. Learning embeddings can be viewed as automated feature engineering, producing features that can be used for predictive modeling of medical events. Methods such as Word2Vec, BERT, FastText, ELMo, and GloVe have been developed for word embedding, but there has been little work on re-purposing these algorithms for the analysis of structured medical data. Our work seeks to fill this important gap. We extended word embedding methods to embed (structured) medical codes from a patient's entire medical history, and used the resultant embeddings to build prediction models for diseases. We assessed the performance of multiple embedding methods in terms of predictive accuracy and computation time using the Medical Information Mart for Intensive Care (MIMIC) database. We found that using Word2Vec, Fast- Text, and GloVe algorithms yield comparable models, while more recent contextual embeddings provide marginal further improvement. Our results provide insights and guidance to practitioners regarding the use of word embedding methods for the analysis of EHR data.

E. Getzen · Y. Ruan · L. Ungar · Q. Long (✉)
University of Pennsylvania, Philadelphia, PA, USA
e-mail: getzene@pennmedicine.upenn.edu; ycruan@alumni.upenn.edu; ungar@cis.upenn.edu; qlong@upenn.edu

Y. Zhao, D.-G. Chen (eds.), *Statistics in Precision Health*, ICSA Book Series in Statistics, https://doi.org/10.1007/978-3-031-50690-1_13

1 Introduction

In 2009, the Health Information Technology for Economic and Clinical Health (HITECH) Act was passed as a means to promote the adoption and meaningful use of health information technology in United States (Hoerbst & Ammenwerth, 2010). As a result, the use of EHR systems has significantly increased in hospitals and other points of care over the past decade. Similar trends have also been observed in other countries. Currently, there are many diseases that go undiagnosed by a physician until they have already caused considerable damage to the body. Other diseases share common symptoms, making it difficult for physicians to differentiate them (Farhan et al., 2016). Flaws in clinical reasoning have also been shown to lead to diagnostic errors—potentially leading to harm for the patient and unnecessary medical expenses (van den Berge & Mamede, 2013). Going beyond human capability and harnessing the power of these rich datasets could lead to significant improvements in patients' health and quality of life. EHRs that span years and detail the health history of patients are available for analysis, and the potential for their use in early stage disease detection, risk prediction, and treatment evaluation could prove to be an invaluable contribution to the advancement of precision health.

The original intention of recording medical data in EHR systems, however, was not for research purposes, but primarily for billing and data transfer, as indicated in Evans (2016). As such, complex EHRs data oftentimes come from heterogeneous sources with characteristics that make it difficult to analyze with classical statistical and machine learning methods (Farhan et al., 2016). Some of these challenges include high dimensionality, the presence of structured and unstructured data, and time-series data collected at irregular intervals and frequencies (see Fig. 1 for an example). Since data from individual patients are sequential in nature and have variable lengths, this makes it impossible to directly use classical dimension reduction and feature engineering methods such as principal component analysis (Abdi & Williams, 2010) and factor analysis (Mair, 2018). However, one way of extracting meaningful information from a complex sequential EHRs dataset is through the use of word embedding algorithms. In typical natural language processing settings, word embedding algorithms build a lower dimensional vector representation of a word from a corpus of text (Naili et al., 2017). Two words that have a higher probability of appearing in the same sentence have a smaller cosine distance between their vector representations. These methods can be extended to EHRs by treating medical events or medical codes as "words", and a collection of medical records as the "corpus". The "context" around an event is defined as the order of medical events happening before and after that event (Farhan et al., 2016). Structured medical codes are an important piece of EHRs data and could contribute significantly to early-stage disease detection and risk prediction. Since they are sequential in nature, the features are not well-defined and dependencies exist between the events, thus making word embedding algorithms particularly effective tools for analysis.

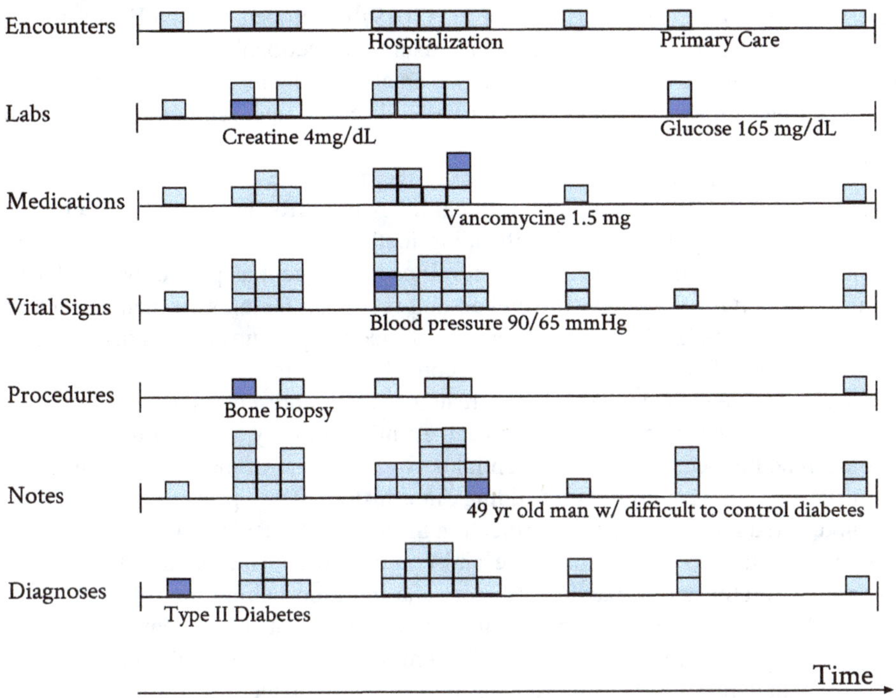

Fig. 1 Sample patient health record data illustrating the challenges for analysis of EHRs data

A growing body of literature explores the use of text mining and word embedding algorithms for analyzing EHR data. Miotto et al. (2016) sought to harness the rich information on patient data contained in clinical notes through deep learning, but raw text cannot be applied as meaningful features in such algorithms. After extracting words related to patient phenotype using standard ontologies for medical terminologies, the authors made use of word embedding algorithms like Word2Vec and Doc2Vec to convert the text into a form that the learning algorithms can understand. They also jointly embedded medical codes and words from clinical notes for EHR driven phenotyping (Miotto et al., 2016). Liu et al. (2015) utilized task-oriented resources to learn word embeddings for expanding ambiguous abbreviations in clinical notes, since the abbreviations are oftentimes difficult for both humans and machines to understand and utilize to improve patient care. In terms of accuracy, they achieved close to expert human performance (Liu et al., 2015). Pham et al. (2016) embedded hospital admissions (diagnoses, procedures, and medications) into vectors, created a patient representation, and showed that the patient timeline vectors resulted in better predictive power than classifiers trained on raw data (Pham et al., 2016). Word embedding can even be used to assist in de-identification of personal health information—Dernoncourt et al. (2016) created an automated de-identification system that generates embeddings for patient notes before feeding them into an artificial neural network (ANN) model

(Dernoncourt et al., 2016). Huang et al. (2019) developed ClinicalBERT which develops representations of clinical notes using bidirectional transformers (Huang et al., 2019), and (Li et al., 2020) developed an embedding model known as BEHRT which employs 4 key concepts from an EHR: diseases, age, segment, and position (Li et al., 2020).

The vast majority of work in this area, including the aforementioned papers, uses word-embedding methods for clinical notes in EHRs, but there has been very limited work on applying word embedding methods for the analysis of structured medical codes. Farhan et al. (2016) developed a sequential prediction model of clinical phenotypes by using word embedding vectors as features for their prediction model (Farhan et al., 2016). They used Word2Vec (skip-gram) to generate a vector representation of a structured medical code by feeding the algorithm with the medical code sequences. They then created vector representations of a patient's entire medical history by multiplying each event's vector by a temporal weight that was designed to account for the potential decay of impact of medical histories, and then summing across all the embeddings in a history. They proposed a prediction method called Patient-Diagnosis Projection Similarity (PDPS) in which they project patient vectors into the vector space and compute the cosine similarity between the patient vector and a diagnosis vector. The intuition behind the method is that the smaller the cosine distance, the more likely it is that the patient will get the diagnosis in the next visit. This allows for a multi-label classification problem. The authors found that PDPS achieved the highest performance in most cases when compared to logistic regression and other prediction methods. However, the logistic regression model did not incorporate the vector representations of medical events from Word2Vec (Farhan et al., 2016). Li et al. (2019) extended Word2Vec and PDPS to a distributed learning algorithm that can incorporate information from multiple data sources in one model without sharing individual-level data, and hence offer some level of data privacy protection (Li et al., 2019).

One important limitation of Farhan et al. (2016) and Li et al. (2019) is that they used only Word2Vec to embed structured EHR events, but there is now a wide variety of word embedding tools such as FastText, GLoVe, ELMo, and BERT. In this work, we conduct a systematic comparison of re-purposing word embedding methods for analysis of structured medical data. By generating patient sequences using the same process as described in Farhan et al. (2016) and using the MIMIC-III database, we build multiple prediction models for the same diseases using different word embedding algorithms, and assess prediction performance and computation time. We also compare our disease-specific prediction models to the PDPS method in Farhan et al. (2016) to see if we can attain improved results. Our work seeks to compare the current state-of-the-art embedding methods and provide valuable insights on which types of methods are better.

2 Methods

We first briefly review popular word embedding methods, namely, Word2Vec (Mikolov et al., 2013), FastText (Joulin et al., 2017), GloVe (Pennington et al., 2014), ELMO (Peters et al., 2018a), and BERT (Peters et al., 2018b). We then describe the approach for embedding EHRs data using the MIMIC III data as an example.

2.1 Word Embedding Methods

Word2Vec was proposed in 2013 by a research team led by Tomas Mikolov at Google. It is a two-layer neural network that is trained to convert words to vectors of real numbers such that the cosine distance between two vectors reflects the similarity of the two words. The input into the neural network is a corpus of text, and the output is a vector space with each unique word assigned a vector. It has two model formulations—continuous bag of words (CBOW) and skip-gram (see Fig. 2). CBOW utilizes the surrounding context (other words around the target word) to predict the target word, and skip-gram uses a single word to predict a target context. The hidden layer following training contains the vectors corresponding to each word. Word2Vec depends on two parameters—size and context window. Size refers to the dimension of each output vector and context window determines the number of words before and after the target word to be included as context when training the algorithm. The original authors of Word2Vec recommend a context window of 10 for skip-gram and 5 for CBOW, and a size between 100 and 1000.

FastText is an extension of Word2Vec which also supports CBOW or skip-gram models. It was developed in 2015 by Facebook's AI Research lab. The main difference is the fact that FastText represents each word as a collection of n-grams. The word "remember" for example, with n = 4, would have a FastText representation for the character n-grams as $< rem, reme, emem, memb, embe, mber, ber >$. The

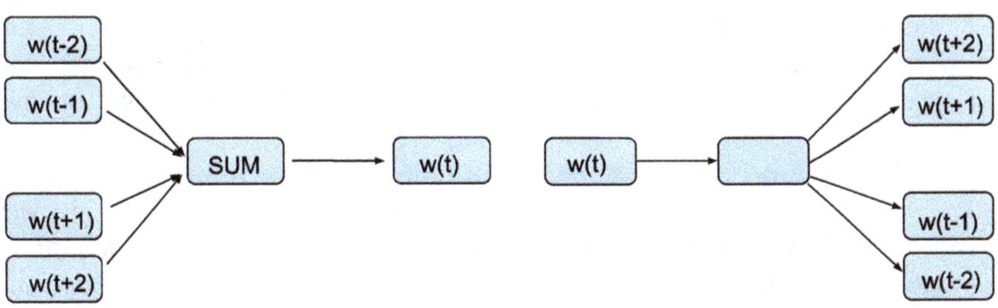

Fig. 2 Two training mechanisms for Word2Vec and FastText: Continuous Bag of Words (CBOW) and Skip-Gram (w(t-i) refers to the ith word before the target word, w(t+i) refers to the ith word after the target word)

Table 1 Sample co-occurrence matrix for sentence "The fox ate the mouse" to illustrate GloVe training

	The	Fox	Ate	Mouse
The	0	1	1	1
Fox	1	0	1	0
Ate	1	1	0	0
Mouse	1	0	0	0

Table 2 Log ratios of word co-occurrences from a sample corpus of text to illustrate how distances between word vectors in space are determined

Probability and ratio	k = solid	k = gas	k = water	k = fashion
$P(k \mid ice)$	1.9×10^{-4}	6.6×10^{-5}	3.0×10^{-3}	1.7×10^{-5}
$P(k \mid steam)$	2.2×10^{-5}	7.8×10^{-4}	2.2×10^{-3}	1.8×10^{-5}
$P(k \mid ice)/P(k \mid steam)$	8.9	8.5×10^{-2}	1.36	0.96

boundaries serve to protect the meanings of shorter words that might appear as character n-grams in a longer word, and are ignored during training. The algorithm generates a vector for each n-gram and sums across the vectors to get the resultant word embedding. The ability to more properly recognize rare words is a benefit of using character n-grams. With regard to medical data, it is unclear whether or not the same logic would hold since medical events are coded as individual entities without designated structures such as prefixes, suffixes, etc.

GloVe, or global vectors model, is an unsupervised learning algorithm developed in 2014 as an open-source project at Stanford. The main difference between GloVe and Word2Vec/FastText is that GloVe does not solely rely on local statistics (context windows used for training in CBOW and skip-gram) as it also incorporates global statistics (word co-occurrence) to generate the vectors. GloVe derives the semantic relationships between words from a co-occurrence matrix, which counts the number of times a given word has co-occured with another word. Table 1 provides an example on calculating the co-occurrence matrix.

The relationships between words is displayed by examining the ratio of co-occurrence probabilities with probe words. For example, let $P(k|w)$ be the conditional probability that word k appears in the context of word w. The word "ice" tends to co-occur more frequently with the word "solid" as opposed to "gas", whereas the word "steam" tends to co-occur more frequently with "gas" rather than "solid", but both words co-occur frequently with the word "water" and infrequently with the word "fashion". So for k = solid, $P(k|ice)$ will be high, $P(k|steam)$ will be low, and the ratio between the two will be very large. For k = water or k = fashion, the ratio will be close to 1; and for k = gas, the ratio will be very small (Pennington et al., 2014). The training objective is to learn word vectors such that the dot product equals the logarithm of words' probability of co-occurrence. The logarithm of the ratios is associated with the vector differences in the word vector space, thus capturing more information. Table 2 provides an example on calculating log ratios of word co-occurrences.

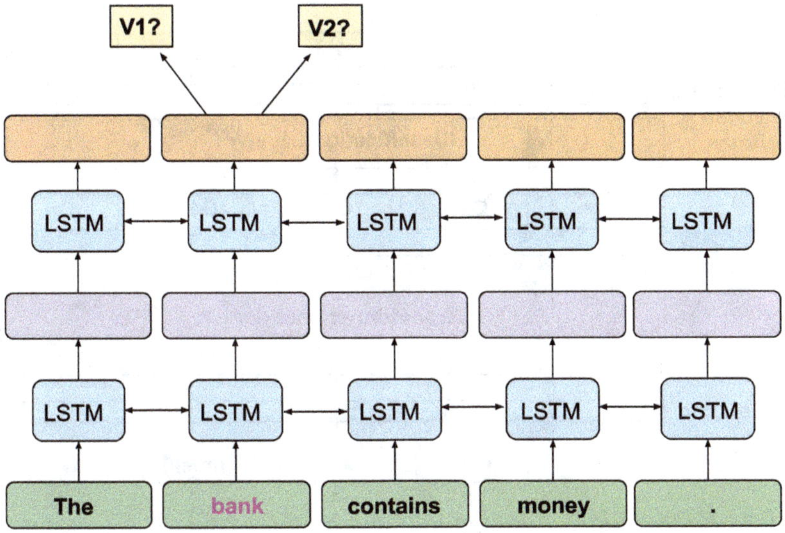

Fig. 3 ELMo bidirectional LSTM training mechanism can produce different embeddings for the same word depending on the context

ELMo word embeddings are learned from a deep bidirectional language model (biLM) that is pre-trained on a large text corpus. The mechanism seeks to address challenges in both modeling complex characteristics of words (syntax and semantics) as well as linguistic contexts (modeling polysemy). Unlike the other methods listed above, ELMo representations are deep– they are a function of the internal layers of the biLM (see Fig. 3). The higher level LSTM architecture captures more context-dependent parts of word meanings and the lower levels capture more of the aspects of syntax. It is a character-based model which further allows it to form representations of out-of-vocabulary words. ELMo can be customized for different NLP tasks by taking a linear combination of all hidden states of LSTM, with weights being task-specific (for example, one linear combination for a next word prediction task, and another linear combination for a text summarization task).

BERT, or Bidirectional Encoder Representations from Transformers, is a word embedding algorithm developed by Google in 2018. It incorporates a transformer neural network architecture– an encoder-decoder model which utilizes attention mechanisms. This architecture has been shown to have advantages over other sequential models such as LSTM, RNN, etc. BERT uses masked language modeling for training, meaning it masks 15% of the words in a sequence and uses the position information to infer what they should be (see Fig. 4). The BERT model uses 12 layers of transformer encoders and each output can be used as a word embedding (the authors identified that the best approach was to sum the last four layers). Out of the five embedding methods described, BERT is the only algorithm that can read the sentence bidirectionally since its self attention layer performs self attention on both directions.

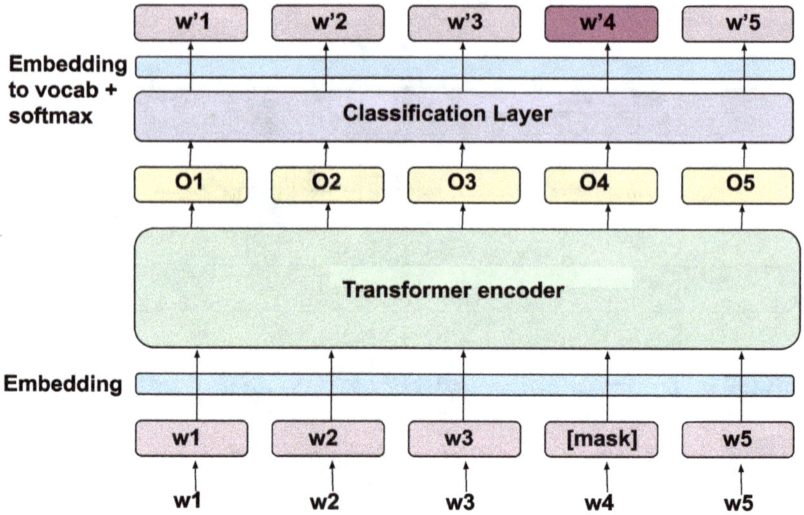

Fig. 4 To train, BERT masks a percentage of words and tries to predict them

There are two types of word embeddings that can be generated using a word embedding algorithm– static word embeddings and contextual word embeddings. Models that generate static embeddings create a single representation for a word. This is oftentimes problematic in languages since polysemous words must share a single vector. Word2Vec, FastText, and GloVe can only create static embeddings. Contextual embeddings, in contrast, are also a function of the "context" of the word–e.g., the other words in the sentence. This allows the vector representation for a word to differ depending on its meaning, as inferred from its context. For example, for the two sentences "let's swim by the river bank" and "I have to deposit money at the bank", the embedding for the word "bank" would differ for the two sentences since it has different meanings based on the context. Both ELMo and BERT have the ability to generate context-dependent word embeddings, on top of static embeddings. Contextual embeddings could prove to be useful with regard to medical data, since depending on how the physician input the codes, some ICD-9 codes could be generalized to levels which can lack some disease specificity (level 3 diagnoses could represent types of diseases and level 5 diagnoses could represent specific diseases). An example is hyperthyroidism and hypothyroidism falling under the same category when generalized to level 3. Contextual embeddings could help distinguish these diagnoses based on the medical context.

2.2 *Embedding of EHRs*

We used the Medical Information Mart for Intensive Care III (MIMIC III) database from (Johnson et al., 2016), which consists of de-identified health-related data of patients who stayed in critical care units. There were 3950 patients in the training set and 1692 patients in the test set. For each medical record we extracted codes for prescriptions, lab tests, symptoms, conditions, and diagnoses. The mean number of structured medical codes per patient was 315, with a minimum number of events of 7, and a maximum of 3463 (see Fig. 5 for distribution of events). The maximum number of medical visits in a patient's record is 42 (see Fig. 6 for distribution of medical visits). In using this data for word embedding, one can think of the "word" as a structured medical code in the patient record (a prescription, lab test, symptom, condition, or diagnosis), and a "sentence" as the chronological sequence of all structured events in a patient record (see Fig. 7 for data structure). Since we wish to predict future diagnoses, we split the sequence of events into the feature window and prediction window. We define target diagnosis as the ICD9 diagnosis code corresponding to the disease that we want to predict. If the target diagnosis is present in a patient's medical record, the feature window contains the sequence of structured medical events up to the visit that contains the first target diagnosis. If the target diagnosis is not present in a patient's medical record, the feature window contains all medical visits up to the most recent visit. The prediction window comprises either the most recent medical visit, or the visit containing the first occurrence of the target diagnosis. It is positive if it contains the target diagnosis, and negative if it does not

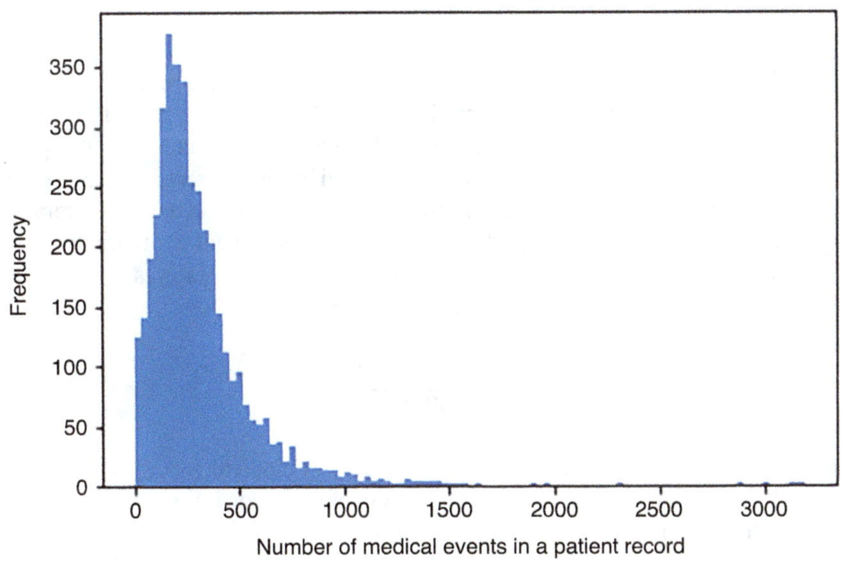

Fig. 5 Distribution of number of structured medical codes per patient medical record

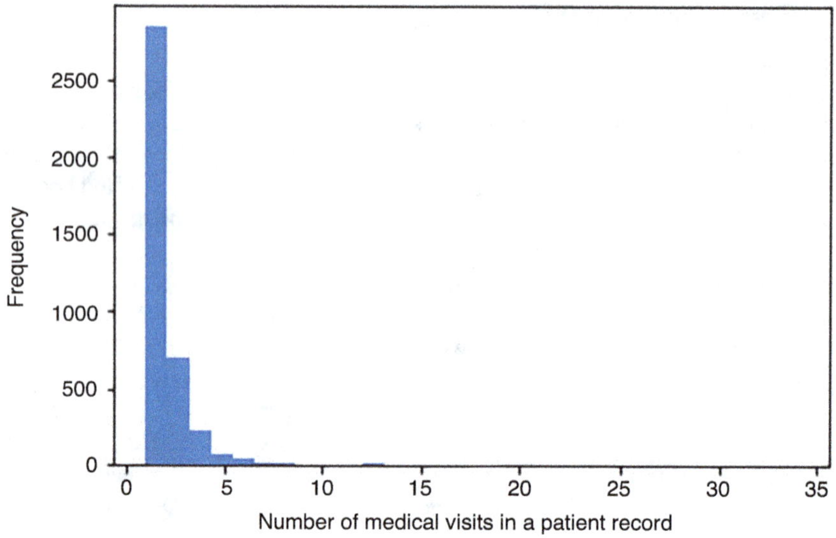

Fig. 6 Distribution of number of medical visits per patient medical record

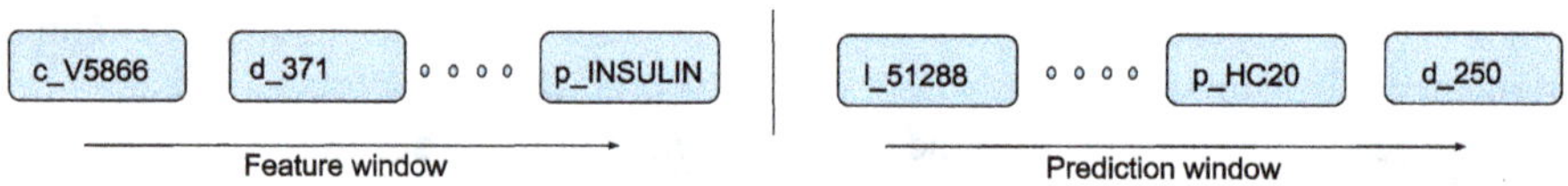

Fig. 7 MIMIC-III structured event data structure example. The structured events in an individual's medical record are split into the feature window and prediction window

contain the target diagnosis. The events in the feature window are used to generate the vector representation of the patient history (see Fig. 8 for the workflow).

There are a few ways to create an overall vector representation of the patient history. We follow Farhan et al. (2016) which gives more recent events greater weights in the overall representation by multiplying each event by a temporal weight. They then sum the weighted embeddings for each event to obtain the overall vector representation of the patient history to be used as the features for prediction (see Fig. 9). The equation that describes this process is as follows:

$$\frac{\sum V_i \times e^{\lambda \times (i-n+1)/n}}{\sum e^{\lambda \times (i-n+1)/n}} \tag{1}$$

where V_i is the embedding for event i in a patient's medical record, λ is the time decay factor, and n is the total number of events in the record. For our experiments, a decay factor of 5 is used.

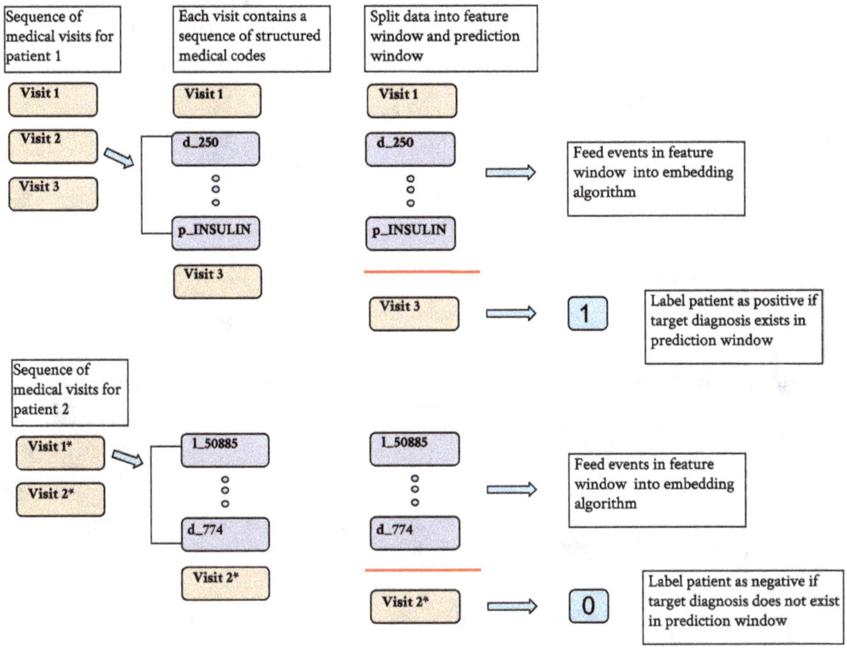

Fig. 8 Each patient may have a different number of medical visits that occur at different times (* indicates different visit time). Each medical visit has a variable sequence of medical codes indicating prescriptions, lab tests, diagnoses, conditions, or symptoms. Medical events prior to the visit containing the diagnosis of interest or the most recent visit comprise the feature window. These events are fed into the word embedding algorithm. The remaining visit is the prediction window and determines whether or not a patient is labeled positive or negative for having the diagnosis of interest present in this visit

Overall patient representations can also be generated without the temporal weights. This involves simply obtaining the embeddings for each medical event and summing across all of the embeddings in the patient record.

3 Numerical Experiments

To evaluate the performance of the embedding algorithms described in Sect. 2, we use the embeddings of patient medical history to build prediction models for disease diagnoses using EHRs data prior to the time of diagnosis and compare prediction performance.

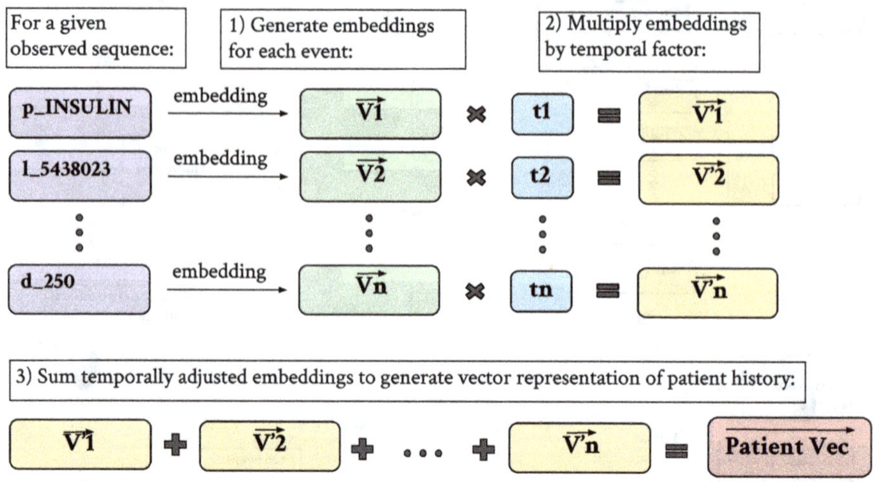

Fig. 9 Patient vector generation via temporal summation of individual event embeddings. $\overrightarrow{Vi}$ refers to the embedding for event i, ti refers to the temporal weight applied to its corresponding event embedding, $\overrightarrow{V'i}$ refers to the resultant embedding following temporal adjustment, $\overrightarrow{PatientVec}$ refers to the vector sum of the adjusted events in a patient history

3.1 Building Disease Prediction Models

To build disease prediction models, we first split patient medical records into the patient history (events prior to the most recent visit) and the events in the most recent visit (see Fig. 7). The outcome for the patient is labeled as positive if the target diagnosis is present in the most recent visit. If the target diagnosis is present in the previous medical history, we only use the chronological sequence of events leading up to the visit containing the first diagnosis, and exclude the subsequent visits to prevent the target itself being used as a feature in prediction for the next visit.

Farhan et al. (2016) developed a prediction model known as Patient Diagnosis Projection Similarity (PDPS) (see Fig. 10) (Farhan et al., 2016). This prediction model involves projecting patient sequences into the vector space while accounting for the temporal impact of events as described in the previous section. The method uses the cosine similarity between the projected patient vector and the diagnosis vector to predict whether or not the patient is at risk for developing that disease. The main benefit that this method provides is its ability to predict risks for multiple diseases simultaneously.

However, improvement in prediction accuracy could be achieved with building disease-specific prediction models instead. We develop such models based on each aforementioned word embedding algorithm using both penalized logistic regression and deep learning methods. Briefly, we use each embedding algorithm to generate

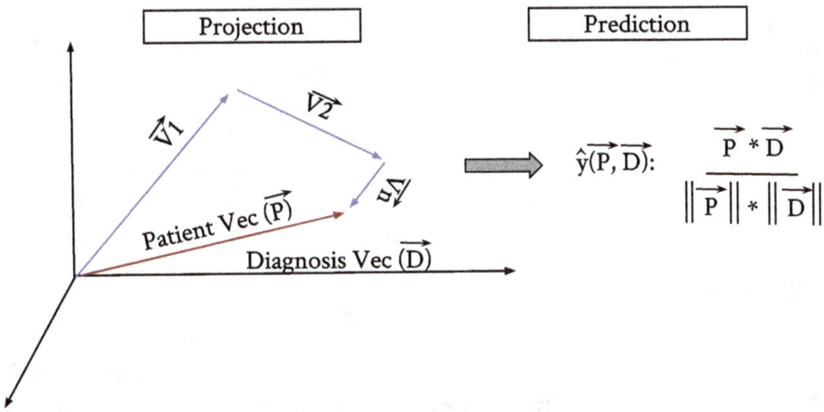

Fig. 10 Patient Diagnosis Projection Similarity (PDPS) involves predicting a future diagnosis based on the angle between the patient vector and the diagnosis vector. $\overrightarrow{Vi}$ refers to the embedding for event i, $\overrightarrow{P}$ refers to the patient vector following weighted event vector summation, $\overrightarrow{D}$ refers to the vector for the target diagnosis that we want to predict in the next visit, and $\hat{y}(\overrightarrow{P}, \overrightarrow{D})$ refers to the cosine similarity of the patient vector and target diagnosis vector

a patient history vector representation of dimension 100, where each element of a patient vector is used as a feature in prediction. We assess embedding algorithms by predicting a variety of diagnoses, and comparing the model performance.

We assess model performance using both F1 score and AUC. The equation for F1 score is as follows :

$$F = \frac{2 \times Precision \times Recall}{Precision + Recall} \tag{2}$$

where

$$Precision = \frac{TP}{TP + FP} \quad Recall = \frac{TP}{TP + FN} \tag{3}$$

Our experiments focus on predicting diseases that have a prevalence of > 0.08 in the MIMIC dataset. We include 6 of those diagnoses with varying prevalence in our results (see Table 3). These diagnoses are: Diabetes, Chronic Kidney Disorder, Heart Failure, Lipid Metabolism Disorder, Fluid/Electrolyte Disorder, and Cardiac Dysrhythmia.

Table 3 Diagnoses whose presence in the next visit is used as outcome variable in the numerical experiments, and their prevalence in the data

Diagnosis	Prevalence
Diabetes	0.3180
Chronic Kidney Disease	0.2027
Heart Failure	0.3717
Lipid Metabolism Disorder	0.2748
Fluid / Electrolyte Disorder	0.3587
Cardiac Dysrhythmia	0.3759

3.2 Experimental Set Up

All numerical experiments are carried out using Python version 3.6 and Tensorflow version > 1.10. To obtain the processed MIMIC-III sequences containing only diagnoses, prescriptions, lab tests, symptoms, and conditions, we first follow the instructions to install MIMIC-III in a local Postgres database using the tutorial on the MIMIC website. We then follow the data pre-processing steps from the SequentialPhenotypePredictor github page from (Farhan et al., 2016). We use the Gensim package to train Word2Vec and FastText, and the package glove_python from pypi to train our GloVe model. We train our medical ELMo model utilizing a Tensorflow implementation using the bilm-tf github page from (Peters et al., 2018a). As for the medical BERT pre-training process, we follow the instructions on the official BERT github (Peters et al., 2018b). We first transform the plain text corpus of structured medical data into a serialized format suitable for training in tensorflow. Using the provided scripts, we learn the WordPiece vocabulary and pre-train BERT models on our GPU and CPU servers separately. To extract the embeddings from the pre-trained BERT models, we can use the Python packages, bert-serving-server and bert-serving-client, to obtain the sentence (word) encodes from our pre-trained models. After completing embedding using one of the above-mentioned algorithms, we use the sklearn Python module to build our lasso prediction model (referred to as "Lasso" in our tables), and tensorflow.keras to build our artificial neural network prediction model with two hidden layers (referred to as "DL" in our tables). We do 50 runs for each method and average each AUC and F1 score. These averages are presented in the subsequent tables.

3.3 Hyperparameters

Our Word2Vec model was trained with the skip-gram method with a context window of 10, and a size of 100, and 70 epochs. Our FastText architecture uses the same parameters as the Word2Vec model. Our GloVe architecture trains with a window size of 10, 100 components, and 70 epochs. Our ELMo architecture trains using the bidirectional mechanism with 7 filters for the character CNN portion, 10% dropout, 2 LSTM layers with dimension 1024 and projection dimension 50, 10 epochs, and batch size of 128. Our BERT architecture has 12 hidden layers, 10 attention heads, 10% dropout and yields an output embedding dimension of 100. Our feed forward neural network ("DL") used for evaluation has two hidden layers. The first layer has 55 nodes and the second has 20. It uses mean squared error loss function and adam optimizer.

3.4 Results

We first compare model performance from the word embedding algorithms using static embeddings only. Recall, Word2Vec, FastText, and GloVe can only produce static embeddings, whereas ELMo and BERT can produce both static and contextual embeddings. For ELMo and BERT, we generate static embeddings for each unique medical event in the data by treating each event as its own sentence so as not to incorporate contextual information. In our first set of results, we incorporate the temporal decay into the construction of patient sequences (see Table 4 and Figs. 11, 12, 13, 14, 15, and 16).

Overall, it appears that the models that use Word2Vec, FastText, and GloVe embeddings tend to yield the best and most consistent results for predicting diagnoses at a patient's next visit. The models that use ELMo static embeddings work well for some diseases, but for others, the prediction performance worsens compared to the models that use other algorithms. The models that use BERT static embeddings have the worst performance compared to models that use other algorithms. We obtain an accuracy of 0.66 for our BERT Masked Language Model, which is significantly lower than Google's pre-trained models (ranging around 0.98). BERT is a more complex model with far more parameters than an algorithm like Word2Vec; therefore, it requires more data to build a better model. Of note, Google's BERT models are typically trained on billions of tokens with a vocabulary (unique tokens) in tens of thousands. Our model was trained on about 1.2 million tokens with a vocabulary size of 2730. The insufficient pre-training data could be the reason for the drop in model performance. This could also be the rationale for ELMo's worsened performance for certain diagnoses, since ELMo is a fairly complex model as well.

We also assess model performance without incorporating the temporal weights into embedding patient sequences. Table 5 shows that the time decay weights consistently improve prediction results for the diseases, but to varying degrees, noting larger improvements for predicting lipid metabolism disorder but smaller improvements for predicting cardiac dysrhythmia. Whether or not more recent events would have more of an impact on the diagnosis of a disease in the next visit would largely depend on the unique way the disease itself develops, so unsurprisingly, temporal weighting is likely more important for acute conditions than chronic conditions. Notably, many of the deep learning models have a significant drop in performance when time decay weighting is not incorporated as opposed to when it is.

Table 4 Prediction performance for various models that use patient vectors generated by each embedding algorithm based on static embeddings and temporal adjustment

Diabetes	Word2Vec	FastText	GloVe	ELMo	BERT
AUC (lasso)	0.7588	0.7451	0.7467	0.7243	0.6624
AUC (DL)	0.7444	0.7547	0.7329	0.6665	0.6028
F1 score (lasso)	0.0825	0.0843	0.1020	0.0810	0.0204
F1 (DL)	0.1244	0.1499	0.1651	0.1542	0.0088
Chronic Kidney Disease					
AUC (lasso)	0.7596	0.7479	0.7328	0.7234	0.6772
AUC (DL)	0.7951	0.7797	0.7859	0.7079	,0.6620
F1 score (lasso)	0.1067	0.0946	0.0403	0.0284	0.0007
F1 score (DL)	0.1231	0.1161	0.1726	0.1089	0.0000
Heart Failure					
AUC (lasso)	0.7201	0.7283	0.7282	0.6792	0.6739
AUC (DL)	0.7313	0.7329	0.7033	0.6594	0.6522
F1 score (lasso)	0.1757	0.1993	0.2215	0.1404	0.1478
F1 score (DL)	0.2770	0.2793	0.2951	0.2295	0.0116
Lipid Metabolism Disorder					
AUC (lasso)	0.7211	0.6950	0.7013	0.6600	0.6503
AUC (DL)	0.7072	0.7107	0.7038	0.6669	0.6447
F1 score (lasso)	0.1277	0.0885	0.0873	0.1086	0.0442
F1 score (DL)	0.1097	0.0761	0.1799	0.1647	0.0157
Fluid/Electrolyte Disorder					
AUC (lasso)	0.6510	0.6317	0.6507	0.6194	0.5909
AUC (DL)	0.6361	0.6474	0.6349	0.5976	0.6003
F1 score (lasso)	0.3701	0.3723	0.3813	0.3093	0.3083
F1 score (DL)	0.4201	0.4105	0.4149	0.4075	0.1182
Cardiac Dysrhythmia					
AUC (lasso)	0.6317	0.6353	0.6360	0.6175	0.6077
AUC (DL)	0.6400	0.6566	0.6152	0.5887	0.5758
F1 score (lasso)	0.0502	0.0641	0.0900	0.0568	0.0145
F1 score (DL)	0.1090	0.0818	0.1295	0.1950	0.0011

We assess whether or not models that use contextual embeddings generate better prediction results. Recall, some ICD9 diagnosis codes are generalized to level three specificity, which could be interpreted as having "multiple meanings". Contextual embeddings could prove to be useful in providing some specificity to such diagnosis codes. We develop models using both BERT and ELMo static and contextual embeddings. The results in Table 6 show that using contextual embeddings tends to improve prediction results for ELMo. However, for BERT, it tends to worsen results compared to using static embeddings.

We compare our disease-specific models to the Patient Diagnosis Projection Similarity (PDPS) from Farhan et al. (2016) by comparing prediction performance for the four main diagnoses used in their work. The authors in this paper do not

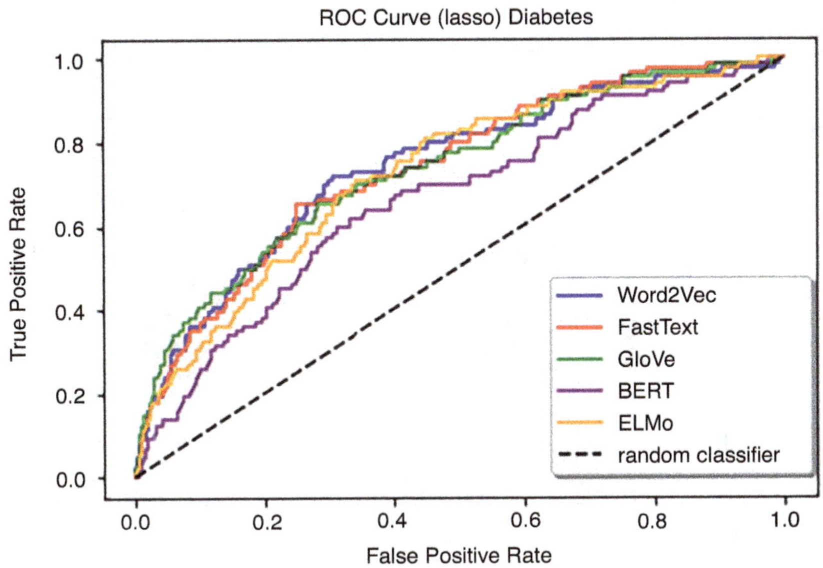

Fig. 11 ROC curves for predicting Diabetes (lasso)

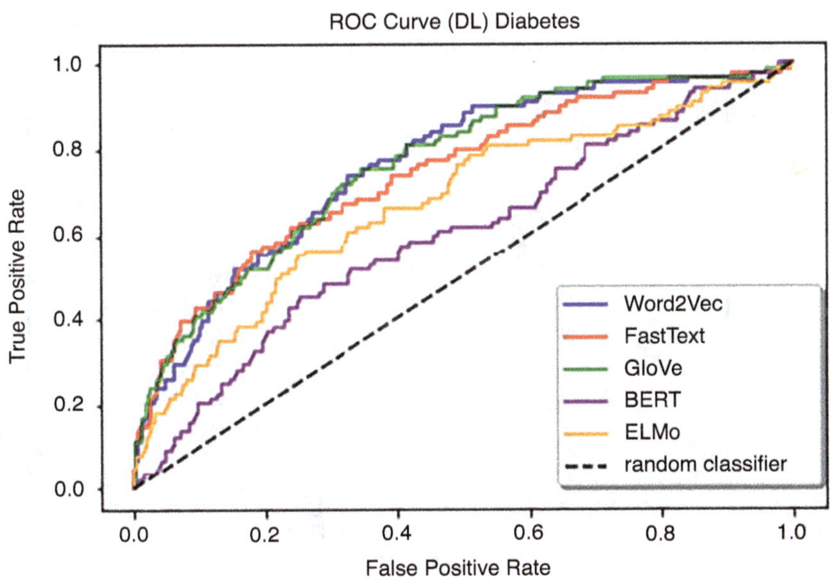

Fig. 12 ROC curves for predicting Diabetes (DL)

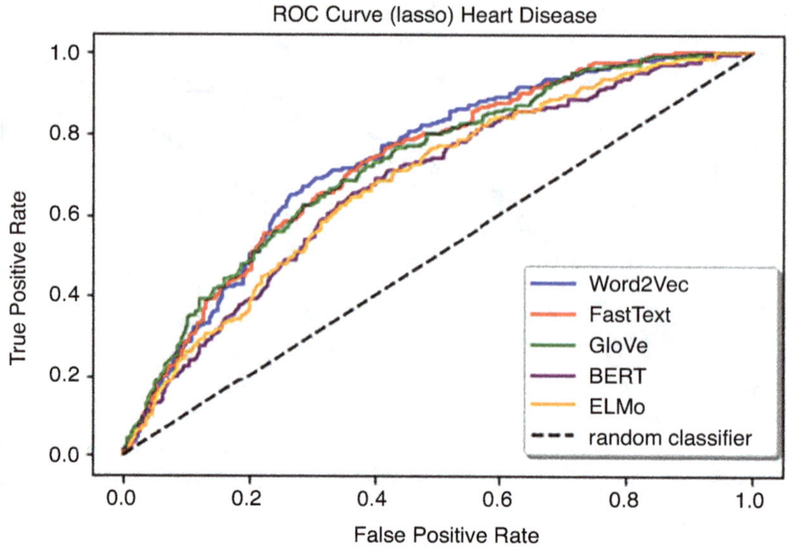

Fig. 13 ROC curves for predicting Heart Disease (lasso)

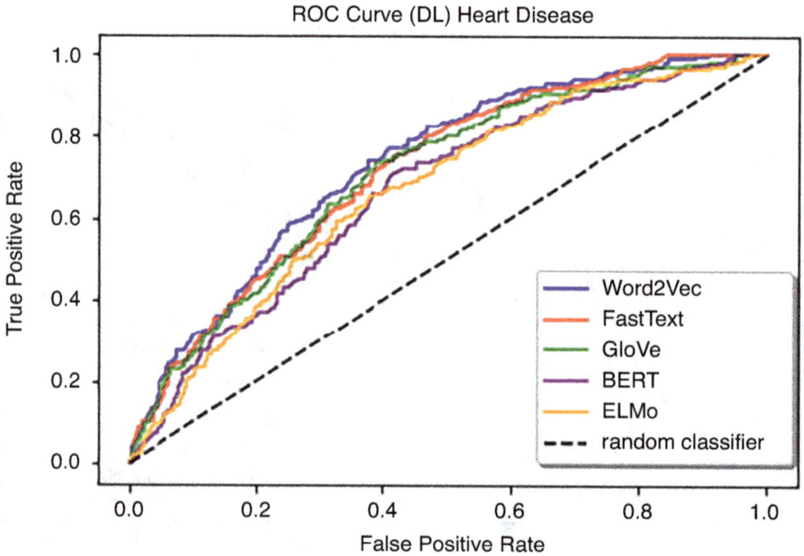

Fig. 14 ROC curves for predicting Heart Disease (DL)

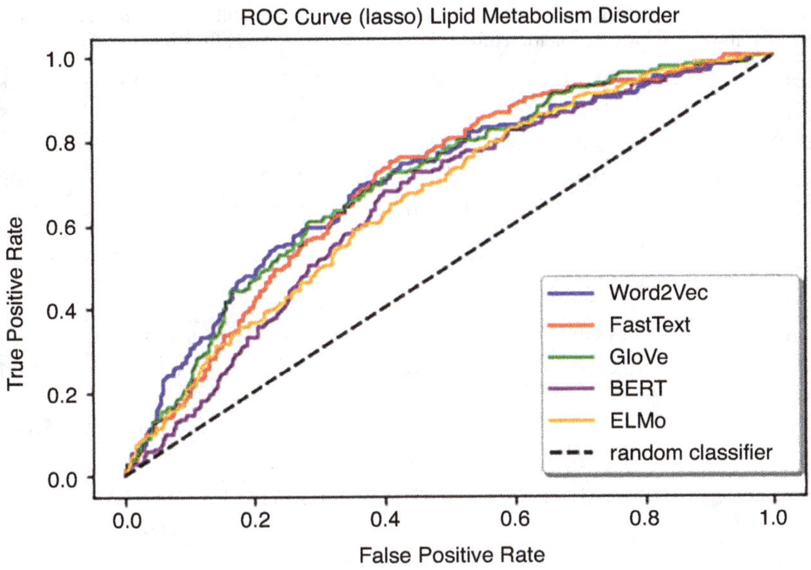

Fig. 15 ROC curves for predicting Lipid Metabolism Disorder (lasso)

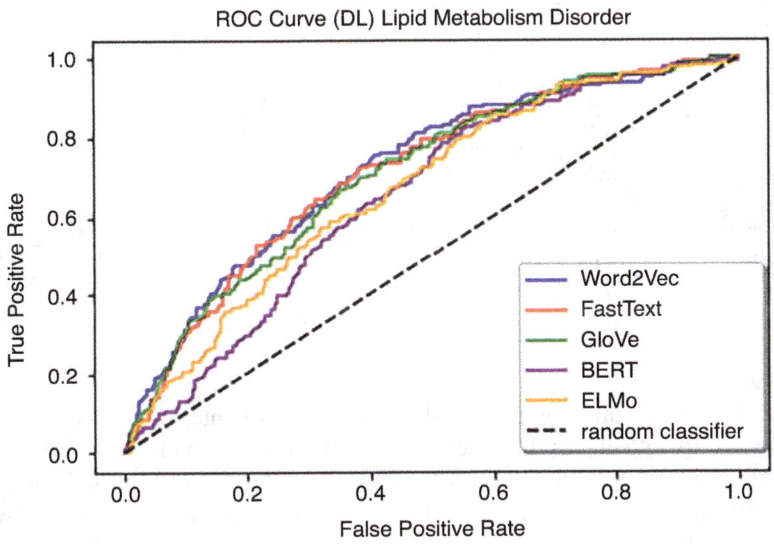

Fig. 16 ROC curves for predicting Lipid Metabolism Disorder (DL)

Table 5 Prediction performance for various models that use patient vectors generated by each embedding algorithm based on static embeddings without temporal adjustment

Diabetes	Word2Vec	FastText	GloVe	ELMo	BERT
AUC (lasso)	0.6486	0.6395	0.6334	0.6417	0.6516
AUC (DL)	0.4930	0.4760	0.5203	0.4992	0.4971
F1 score (lasso)	0.2056	0.1731	0.1372	0.1745	0.0824
F1 (DL)	0.0174	0.0010	0.0627	0.0021	0.0966
Chronic Kidney Disease					
AUC (lasso)	0.7431	0.7323	0.7197	0.7408	0.6772
AUC (DL)	0.4767	0.4798	0.4992	0.4935	0.4905
F1 score (lasso)	0.1843	0.1743	0.1973	0.1341	0.0007
F1 score (DL)	0.0000	0.0000	0.0125	0.0022	0.0000
Heart Failure					
AUC (lasso)	0.7121	,0.7044	0.7263	0.6790	0.6760
AUC (DL)	0.6530	0.6043	0.6880	0.4951	0.5018
F1 score (lasso)	0.1972	0.1987	0.2382	0.1980	0.1625
F1 score (DL)	0.2547	0.1063	0.2863	0.0050	0.0000
Lipid Metabolism Disorder					
AUC (lasso)	0.6292	0.6465	0.6539	0.6065	0.5993
AUC (DL)	0.6027	0.5771	0.6398	0.5042	0.5009
F1 score (lasso)	0.0556	0.0652	0.0555	0.0444	0.0449
F1 score (DL)	0.1083	0.0520	0.1602	0.00429	0.0117
Fluid/Electrolyte Disorder					
AUC (lasso)	0.6318	0.6071	0.6205	0.5741	0.6025
AUC (DL)	0.5918	0.5995	0.5894	0.5234	0.5002
F1 score (lasso)	0.3277	0.3277	0.3090	0.3221	0.3152
F1 score (DL)	0.3991	0.3391	0.41824	0.0692	0.1334
Cardiac Dysrhythmia					
AUC (lasso)	0.6198	0.6102	0.6044	0.6008	0.5884
AUC (DL)	0.6028	0.5843	0.5901	0.50321	0.5006
F1 score (lasso)	0.0697	0.0748	0.0654	0.0830	0.0741
F1 score (DL)	0.1411	0.0647	0.2188	0.00521	0.0194

remove prior visits that contain the diagnosis code of interest, so we adjust our experiments to reflect this in order to directly compare with their results. The prevalence for each diagnosis is listed in Table 7. The results in Table 8 demonstrate an overall significant improvement in model performance when using disease-specific models.

Table 6 Prediction performance for models that use patient vectors generated by contextual embeddings and by static embeddings

Diabetes	ELMo Cont.	ELMo Static	BERT Cont.	BERT Static
AUC (lasso)	0.7235	0.7243	0.6386	0.6624
AUC (DL)	0.7000	0.6665	0.5963	0.6028
F1 score (lasso)	0.1200	0.0810	0.07921	0.0204
F1 score (DL)	0.2254	0.1542	0.0216	0.0088
Chronic Kidney Disease				
AUC (lasso)	0.7590	0.7234	0.6671	0.6772
AUC (DL)	0.7136	0.7079	0.6365	0.6620
F1 score (lasso)	0.0000	0.0284	0.0425	0.0007
F1 score (DL)	0.2180	0.1089	0.0000	0.0000
Heart Failure				
AUC (lasso)	0.7133	0.6792	0.6276	0.6739
AUC (DL)	0.6731	0.6594	0.6182	0.6522
F1 score (lasso)	0.1940	0.1404	0.1025	0.1478
F1 score (DL)	0.2670	0.2295	0.0028	0.0116
Lipid Metabolism Disorder				
AUC (lasso)	0.6912	0.6600	0.6511	0.6503
AUC (DL)	0.6656	0.6669	0.6520	0.6447
F1 score (lasso)	0.1675	0.1086	0.0454	0.0442
F1 score (DL)	0.1923	0.1647	0.0122	0.0157
Fluid/Electrolyte Disorder				
AUC (lasso)	0.6424	0.6194	0.5776	0.5909
AUC (DL)	0.5920	0.5976	0.5780	0.6003
F1 score (lasso)	0.3537	0.3093	0.2606	0.3083
F1 score (DL)	0.4746	0.4075	0.0851	0.1182
Cardiac Dysrhythmia				
AUC (lasso)	0.6352	0.6175	0.5771	0.6077
AUC (DL)	0.5829	0.5887	0.5871	0.5758
F1 score (lasso)	0.0148	0.0568	0.0147	0.0145
F1 score (DL)	0.2773	0.1950	0.0000	0.0011

Table 7 Diseases presented in the main results of Farhan et al. (2016) and their prevalences in the data

Diagnosis	Prevalence
Systolic Heart Failure	0.3717
Acute Posthemorrhagic Anemia	0.2772
Hyperlipidemia	0.2748
Septicemia	0.1856

Table 8 Comparison of PDPS prediction results from Farhan et al. (2016) to our disease-specific models. Recall that in these experiments, visits that contain instances of the target diagnosis are included in the feature window

Systolic Heart Failure	PDPS	Lasso	DL
AUC	0.795	0.8998	0.9051
F1 score	0.213	0.7918	0.7994
Acute Posthemorrhagic Anemia			
AUC	0.618	0.8011	0.7982
F1 score	0.175	0.7197	0.7035
Hyperlipidemia			
AUC	0.723	0.8644	0.8755
F1 score	0.238	0.7166	0.7347
Septicemia			
AUC	0.652	0.7971	0.8141
F1 score	0.242	0.5626	0.6202

Table 9 Computing time in seconds for training each embedding algorithm using CPU and GPU

Embedding Algorithm	CPU (s)	GPU (s)
Word2Vec	89	24
FastText	262	48
GloVe	11	14
BERT	1830	60
ELMo	382	72

Table 10 Computing time in seconds for generating static and contextual embeddings for ELMo and BERT using CPU and GPU

Type of Embedding	CPU (s)	GPU (s)
ELMo (contextual)	8942	8688
ELMo (static)	67	70
BERT (contextual)	1301	343
BERT (static)	162	112

3.5 *Computational Efficiency*

We assess computational efficiency by measuring the amount of time that it takes to train each algorithm on the MIMIC data using CPU and GPU. Our CPU cluster consists of 2 x Intel Xeon E5-2660 v3 @ 2.60 GHz CPUs with 128 GB RAM installed. And the GPU cluster contains 2 x GenuineIntel Intel(R) Xeon(R) Silver 4216 CPUs @ 2.10GHz with 192 GB RAM, as well as 4 x NVIDIA 2080 TI.

On the CPU, GloVe is the fastest model to train followed by Word2Vec, FastText, ELMo, and BERT. BERT is faster to train than ELMo on the GPU (see Table 9). It is important to consider that these are strict training times as they do not include the time it takes to set up each training before submitting to either cluster. Word2Vec, FastText, and GloVe are far easier to set up than ELMo and BERT. Static embeddings are also much faster to generate than contextual embeddings (see Table 10). For static embeddings, one only needs to extract the unique events in the dataset. On the other hand, obtaining contextual embeddings requires use of all the data since each embedding is a function of the sentence that contains it.

4 Discussion

In this paper, we have re-purposed traditional word embedding technologies for sequences of structured medical codes rather than actual sentences or words, and compared advanced word embedding methods for analysis of EHRs data in extensive numerical experiments. Based on our findings, for this scale of data, we would recommend Word2Vec, FastText, or GloVe to practitioners, due to their relatively fast training and set-up time, and competitive prediction performance compared to the other algorithms. ELMo and especially BERT did not perform as well in our experiments when using static embeddings, likely due to the fact that the algorithms require substantially more training data. Both of these algorithms have far more parameters than Word2Vec, GloVe, or FastText. However, the embeddings that incorporated contextual information for ELMo led to improvements in model performance for many diagnoses.

Implementing a time decay weighting did improve results to varying degrees among the diagnoses. However, the extent to which there is an improvement depends on the unique progression of a diagnosis. It could be argued that model performance for predicting other medical conditions could decrease when using the time decay weighting. For example, the development of certain cancers such as lung cancer that can be attributed to cumulative impact of risk factors such as smoking over a long period of time including particularly, early in life.

A major finding of this work is the fact that the use of contextual embeddings as opposed to static embeddings can improve model performance across many disease prediction models. We see this in our comparisons of models that utilize ELMo contextual embeddings vs. ELMo static embeddings. This displays the potential advantages of using a more advanced model that can incorporate contextual information on structured medical events. We hypothesize that given more training data, one could achieve better results than Word2Vec, GloVe, or FastText when using contextual embedding. This is likely due to the fact that algorithms such as ELMo and BERT can generate embeddings that improve the specificity of some ambiguous medical codes, thus leading to better models. However, we do not see this improvement when using BERT contextual embeddings, likely due to a lack of enough training data. BERT is a larger model than ELMo.

We found that building disease-specific models significantly outperforms the PDPS prediction method. However, both of these approaches have their own strengths and weaknesses. For a practitioner interested in predicting one disease in particular, it would be beneficial to use a disease-specific model for improved prediction accuracy. However, if one is more interested in the ability to predict risk for multiple medical conditions, PDPS would be the better method to implement due to its ability to solve a multi-classification problem. With our methods, one would need to create a new model for each new diagnosis to predict, which is more inconvenient and time-consuming.

One limitation of this work is that EHRs tend to have irregular time intervals between visits and events, but only the ordering of the events was incorporated to capture the temporal information in our experiments. As such, this approach does not necessarily paint an accurate picture of the patient record. To mitigate this problems, we could instead feed the individual event embeddings into a recurrent neural network with the exact time stamps to generate the embeddings for patient sequences before building prediction models. This way, we could more effectively utilize the temporal information and create a better model for each medical event.

There is also the issue of the time decay weighting that yields varying results depending on the diagnosis. In this paper, we use the same equation as in Farhan et al. (2016) to generate the fixed time decay weighting. It is of interest to see how varying the time decay factor could optimize prediction results for various diagnoses. This could also elucidate the unique temporal progressions of various diseases as they are compared with one another.

In this work, we are focused on comparing (unsupervised) word embedding algorithms for structured medical codes that do not require the use of a clinical outcome to train. These embeddings can be subsequently used for a wide range of downstream learning tasks such as building prediction models for multiple disease outcomes or clustering for disease subtype identification etc. The prediction models in our numerical experiments are used as an extrinsic evaluation approach (Wang et al., 2019) to compare the word embedding algorithms in the settings of our interest, rather than the focus of the paper. We recognize that there is typically a trade-off between general applicability for multiple tasks and optimal performance for a specific task. For example, there exist sophisticated prediction models for medical events that incorporate medical code embedding, for example, RETAIN, a two-level neural network attention model for sequential data (Choi et al., 2017) and GRAM: a graph-based attention model for healthcare representation learning (Choi et al., 2016). The embeddings generated from these models are trained using a pre-specified clinical outcome through supervised learning, and may be sub-optimal for other downstream learning tasks such as predicting a different clinical outcome or multiple clinical outcomes. While it is beyond the scope of our current work, it is of future interest to compare these embeddings with those from the word-embedding methods investigated in this paper.

In addition, there are other embedding algorithms that can be used in analysis of EHRs data, e.g., Google's Universal Sentence Encoder (USE) and Generalized Auto-Regressive model for NLU (XLNet). It is of potential interest to compare these methods in future research with more data, when it becomes possible to train these models using the MIMIC-III data.

Acknowledgments This work is partly supported by NIH grant R01-GM124111. The content is the responsibility of the authors and does not necessarily represent the views of NIH. The MIMIC dataset used in this paper is publicly available on https://physionet.org/.

References

Abdi, H., & Williams, L. J. (2010). Principal component analysis. *Wiley Interdisciplinary Reviews: Computational Statistics, 2*(4), 433–459.

Choi, E., Bahadori, M. T., Kulas, J. A., Schuetz, A., Stewart, W. F., & Sun, J. (2017). RETAIN: An interpretable predictive model for healthcare using reverse time attention mechanism. arXiv:1608.05745.

Choi, E., Bahadori, M. T., Song, L., Stewart, W. F., & Sun, J. (2016). Graph-based attention model for healthcare representation learning. arXiv:1611.07012.

Dernoncourt, F., Lee, J. Y., Uzuner, O., & Szolovits, P. (2016). De-identification of patient notes with recurrent neural networks. *Journal of the American Medical Informatics Association, 24*, 596—606.

Evans, R. S. (2016). Electronic health records: then, now, and in the future. *Yearbook of Medical Informatics, Suppl 1*, S48–S61.

Farhan, W., Wang, Z., Huang, Y., Wang, S., Wang, F., & Jiang, X. (2016). A predictive model for medical events based on contextual embedding of temporal sequences. *JMIR Medical Informatics, 4*, e39.

Hoerbst, A., & Ammenwerth, E. (2010). Electronic health records. A systematic review on quality requirements. *Methods of Information in Medicine, 49*, 320–336.

Huang, K., Altosaar, J., & Ranganath, R. (2019). ClinicalBERT: modeling clinical notes and predicting hospital readmission. Pre-print. arxiv:1904.05342.

Johnson, A. E. W., Pollard, T. J., Shen, L., Lehman, L.-w. H., Feng, M., Ghassemi, M., Moody, B., Szolovits, P., Anthony Celi, L., & Mark, R. G. (2016). MIMIC-III, a freely accessible critical care database. *Scientific Data, 3*, 160035.

Joulin, A., Grave, E., Bojanowski, P., & Mikolov, T. (2017). Bag of tricks for efficient text classification. *Proceedings of the 15th Conference of the European Chapter of the Association for Computational Linguistics: Volume 2, Short Papers, 2*, 427–431.

Li, Z., Roberts, K., Jiang, X., & Long, Q. (2019). Distributed learning from multiple EHR databases: contextual embedding models for medical events. *Journal of Biomedical Informatics, 92*, 103138.

Li, Y., Solares, J. R. A., Hassaine, A., Ramakrishnan, R., Canoy, D., Zhu, Y., Rahimi, K., & Salimi-Khorshidi, G. (2020). BEHRT: transformer for electronic health records. *Scientific Reports, 10*, 7155.

Liu, Y., Tao, G., Matthews, K. S., Ji, H., & McGuinness, D. L. (2015). Exploiting task-oriented resources to learn word embeddings for clinical abbreviation expansion. In *Proceedings of BioNLP 15.*

Mair, P. (2018). Factor analysis. *Modern psychometrics with R* (pp. 17–61). Cham, Switzerland: Springer International.

Mikolov, T., Sutskever, I., Chen, K., Corrado, G. S., & Dean, J. (2013). Distributed representations of words and phrases and their compositionality. *Advances in Neural Information Processing Systems, 7*, 3111–3119.

Miotto, R., Li, L., Kidd, B. A., & Dudley, J. T. (2016). Deep patient: an unsupervised representation to predict the future of patients from the electronic health records. *Scientific Reports, 6*, 26094.

Naili, M., Habacha, A., Hajjami, H., & Ghezala, B. (2017). Comparative study of word embedding methods in topic segmentation. *Procedia Computer Science, 112*, 340–349.

Pennington, J., Socher, R., & Manning, C. D. (2014). GloVe: global vectors for word representation. *Empirical Methods in Natural Language Processing (EMNLP), 14*, 1532–1543.

Peters, M. E., Neumann, M., Iyyer, M., Gardner, M., Clark, C., Lee, K., & Zettlemoyer, L. (2018a). Deep contextualized word representations. In *Proceedings of the of NAACL.*

Peters, M. E., Neumann, M., Iyyer, M., Gardner, M., Clark, C., Lee, K., & Zettlemoyer, L. (2018b). Pre-training of deep bidirectional transformers for language understanding. Preprint. arXiv:1810.04805.

Pham, T., Tran, T., Phung, D., & Venkatesh, S. (2016). DeepCare: a deep dyanmic memory model for predictive medicine. *Advances in Knowledge Discovery and Data Mining, 9652*, 30–41.
van den Berge, K., & Mamede, S. (2013). Cognitive diagnosic error in internal medicine. *European Journal of Internal Medicine, 24*, 525–529.
Wang, B., Wang, A., Chen, F., Wang, Y., & Kuo, C. C. J. (2019). Evaluating word embedding models: Methods and experimental results. *APSIPA Transactions on Signal and Information Processing, 8*, e19.

Adaptive Designs for Precision Medicine in Clinical Trials: A Review and Some Innovative Designs

Yanping Qiu and Feifang Hu

Abstract Precision medicine proposes the customization of health-care tailored to a patient based on the individual characteristics. Adaptive randomization designs are effective ways to optimize patients' treatments by incorporating individual information. We illustrate the mathematical framework for adaptive randomization and provide a brief review of popular adaptive designs that incorporate covariates under this framework. We also provide a brief overview on covatiate-adaptive design and covariate-adjusted response-adaptive design. Currently no design has been proposed in literature to incorporate discrete and continuous prognostic covariates, and predictive covariates simultaneously. To fill the vacant, this chapter proposes a new class of covariate-adjusted response-adaptive procedures to optimize the treatments based on individual covariates. The proposed procedure can balance both discrete and continuous covariates to make valid and credible comparisons among different treatments. Theoretical properties for the new designs are investigated in some simplified scenarios. Extensive simulation studies have been conducted, which demonstrate the superiorities of new procedure over existing designs.

Keywords Adaptive design · Precision medicine · Innovative design · Covariate-adaptive randomization · Covariate-adjusted response-adaptive randomization

---

Y. Qiu
School of Statistics, Renmin University of China, Beijing, China

Statistics & Decision Sciences, Janssen Research & Development, Beijing, China
e-mail: qiuyp@ruc.edu.cn

F. Hu (✉)
Department of Statistics, George Washington University, Washington, DC, USA
e-mail: feifang@gwu.edu

Y. Zhao, D.-G. Chen (eds.), *Statistics in Precision Health*, ICSA Book Series in Statistics, https://doi.org/10.1007/978-3-031-50690-1_14

1 Introduction

Precision medicine, which is also called personalized medicine, is a desirable medical model that proposes the customization of health-care tailored to a patient based on the individual characteristics. This model allows researchers to assess and predict the treatment effects for certain groups of patients and thus makes it possible to provide the most appropriate treatments.

Lots of researchers provided their insights and came up with different procedures to achieve this ideal model. Hu et al. (2015a) stated that precision medicine is to use individual information to optimize that patient's preventative and therapeutic care and suggested that some subject characteristics or biomarkers can be helpful in the randomization procedure. To develop precision medicine in clinical trials, some biomarkers or covariates have been taken into consideration. As suggested by Buyse et al. (2011) and Krisam and Kieser (2015), biomarkers can be classified into two distinct types: prognostic and predictive covariates. These biomarkers have great impacts on the final responses of patients, but they could play completely different roles. Prognostic biomarkers provide information on the likely outcome of disease regardless of treatment and can be used to predict the most likely prognosis of an individual patient, while predictive biomarkers are linked to certain disease and likely to predict the response to a specific treatment. In the clinical practice, some biomarkers are identified as prognostic or predictive (Baker et al., 2004; Garde-Noguera et al., 2018; Gibney et al., 2016). Garde-Noguera et al. (2018) discussed the prognostic and predictive biomarkers for patients with non-small cell lung cancer after treated with nivolumab. Fukada et al. (2021) proposed that Presepsin can be a potential biomarker to identify the severity of the novel coronavirus pneumonia (COVID-19), which helps clinicians identify patients with high-risk neo-coronary pneumonia and determine treatment strategies at an early stage.

The precision medicine is much-needed in oncology therapeutic area considering the rapid progression of disease and high mortality rate of cancers and the heterogeneity of different cancer types and patient populations. The advances in technology including genotyping and biomarker testing have helped speed up the pace of the clinical research in this field. Scientists get to understand that cancer cells may have many different changes in genes, proteins, and other biomarkers that make the cancer cells grow and spread. Many biomarkers or covariates are used as predictive signals to predict the treatment effect of specific drug in certain cancer patients. For instance, the programmed death-ligand 1 (PD-L1) tumor proportion score and epidermal growth factor receptor (EGFR) mutation status are identified as predictive biomarkers of pembrolizumab in non-small cell lung cancer patients (Miyawaki et al., 2020). Microsatellite instability-high (MSI-H) or mismatch repair deficient (dMMR) are also identified and FDA granted accelerated approval of pembrolizumab for adults and pediatric patients with unresectable or metastatic solid tumors (U.S. Food and Drug Administration, 2017). The biomarker information are brought into the trial designs, like umbrella trial and basket trials, which will be further discussed in Sect. 3.

Since biomarkers are mathematically treated as covariates in statistical analyses, Zhao et al. (2022) suggested to classify covariates into two types accordingly: (i) prognostic covariate Z, and (ii) predictive covariate X. Predictive covariates could be used to predict the response to a specific treatment, while prognostic covariates predict the most likely prognosis of an individual patient and not linked to a specific treatment. Consequently, prognostic covariates should be balanced to provide a valid comparison, while predictive covariates may be used for the selection of suitable treatments. Some covariates could be both prognostic and predictive in practice, and in this chapter it's only classified to the type which is more clinical important or meaningful.

To evaluate treatment effect in clinical trials, randomization serves as the "gold standard" and provides a fair comparison among treatments regarding the selection bias. The use of complete randomization may result in highly variable numbers of patients in some strata for covariates. Various of designs have been developed to eliminate the negative effect of complete randomization and meet the needs for efficacy and ethics. Response-adaptive designs utilize the information of responses in previous patients to optimize the potential benefits for subsequent individuals. Covatiate-adaptive designs have been proposed to improve the balance of prognostic covariates across treatment groups and more details are illustrated in Sect. 4. With accumulated information and research in clinical trials, the roles of some covariates have been identified as prognostic or predictive covariates. Covariate-adjusted response-adaptive designs were proposed to include the information of predictive covariates for better treatment assignment and a new class of covariate-adjusted response-adaptive designs are proposed to satisfy the need to balance prognostic covariates and optimize treatment based on predictive covariates simultaneously in Sect. 5.

The remaining parts of this chapter are organized as follows. Section 2 presents a mathematical framework of adaptive randomization designs. Section 3 introduces some popular innovative designs in clinical trials under the mathematical framework. Section 4 briefly describes covatiate-adaptive designs. Section 5 introduces the developments of covariate-adjusted response-adaptive designs and proposes a new class of designs to fill the vacancy of applying both prognostic and predictive covariates in the clinical trials. We also provide the theories of the new procedure when the estimated allocation proportion is fixed and numerical results comparing with existing procedures for both continuous and binary responses. Finally Sect. 6 summarizes the chapter with a short discussion.

2 Mathematical Framework of Adaptive Randomization Designs

To have a comprehensive understanding of adaptive randomization procedures, we first introduce a mathematical framework and then illustrate some popular adaptive designs under this framework in Sect. 3. The framework was first suggested by

Hu and Rosenberger (2006), and it was generalized it to the scenario with both predicative and prognostic covariates in Zhao et al. (2022).

Consider a clinical trial of K treatments and n patients, each patient is assigned randomly to one of K treatments. Let $\boldsymbol{T} = (\boldsymbol{T}_1, \ldots, \boldsymbol{T}_n)^T$ be the random treatment assignments, where $\boldsymbol{T}_i = (T_{i,1} \ldots, T_{i,K})$ is a vector of zeros except for a one in the j-th position, indicating that the i-th patient receives the j-th treatment, $i = 1, \ldots, n$, $j = 1, \ldots, K$. Let $\boldsymbol{Y} = (\boldsymbol{Y}_1, \ldots, \boldsymbol{Y}_n)^T$ be a random matrix of response variables, where $\boldsymbol{Y}_i = (Y_{i,1}, \ldots, Y_{i,K})$ represents the sequence of responses that would be observed if each treatment were assigned to the i-th patient independently, and only one element of $\boldsymbol{Y}_i$ can be observed. $(\boldsymbol{Z}_i, \boldsymbol{X}_i)$ are the covariates of interest for i-th patient, where $\boldsymbol{Z}_i$ and $\boldsymbol{X}_i$ are the corresponding prognostic and predictive covariates.

A randomization procedure can be defined by certain allocation function $\boldsymbol{\phi}_n$, which is the conditional probability of assigning the n-th patient to treatments $1, \ldots, K$. Complete randomization (CR) assigns patients to each treatment with equal probability, that is,

$$\boldsymbol{\phi}_{n,CR} = \frac{1}{K}.$$

A covariate-adaptive randomization (CAR) design defined by $\boldsymbol{\phi}_{n,CAR}$ depends on the treatment assignments from the previous $n-1$ patients and the covariate vectors of these n patients, that is,

$$\boldsymbol{\phi}_{n,CAR} = E[\boldsymbol{T}_n | \boldsymbol{T}_1, \ldots, \boldsymbol{T}_{n-1}, (\boldsymbol{Z}_1, \boldsymbol{X}_1), \ldots, (\boldsymbol{Z}_n, \boldsymbol{X}_n)].$$

The covariate-adaptive designs proposed in the literature usually only balance prognostic covariates, without consideration of the predictive covariates. Therefore, the conditional allocation probabilities depend on prognostic covariates and

$$\boldsymbol{\phi}_{n,CAR} = E[\boldsymbol{T}_n | \boldsymbol{T}_1, \ldots, \boldsymbol{T}_{n-1}, \boldsymbol{Z}_1, \ldots, \boldsymbol{Z}_n].$$

A response-adaptive (RA) design incorporates responses in the randomization procedure and it's defined by $\boldsymbol{\phi}_{n,RA}$ depending on the treatment assignments and responses from the previous $n-1$ patients, that is,

$$\boldsymbol{\phi}_{n,RA} = E[\boldsymbol{T}_n | \boldsymbol{T}_1, \ldots, \boldsymbol{T}_{n-1}, \boldsymbol{Y}_1, \ldots, \boldsymbol{Y}_{n-1})].$$

By incorporating both responses and covariates in the randomization procedure, a covariate-adjusted response-adaptive (CARA) randomization design is defined by $\boldsymbol{\phi}_{n,CARA}$, which depends on the treatment assignments, responses, and covariate vectors from the previous $n-1$ patients, and the current patient's covariates, that is,

$$\boldsymbol{\phi}_{n,CARA} = E[\boldsymbol{T}_n | \boldsymbol{T}_1, \ldots, \boldsymbol{T}_{n-1}, \boldsymbol{Y}_1, \ldots, \boldsymbol{Y}_{n-1}, (\boldsymbol{Z}_1, \boldsymbol{X}_1), \ldots, (\boldsymbol{Z}_n, \boldsymbol{X}_n)].$$

However, the CARA designs proposed in literature are usually only able to deal with predictive covariates, that is,

$$\boldsymbol{\phi}_{n,CARA} = E[\boldsymbol{T}_n|\boldsymbol{T}_1, \ldots, \boldsymbol{T}_{n-1}, \boldsymbol{Y}_1, \ldots, \boldsymbol{Y}_{n-1}, \boldsymbol{X}_1, \ldots, \boldsymbol{X}_n].$$

Zhao et al. (2022) proposed a class of CARA designs which incorporate discrete prognostic covariates and predictive covariates. In Sect. 5 of this chapter, we provide a brief review of CARA designs and generalize the CARA designs that can simultaneously balance continuous and discrete prognostic covariates and optimize patients' treatment assignments based on predictive covariates and the responses from previous patients.

3 Some Popular Innovative Designs

In the practice of clinical trials, several popular innovative designs have been proposed to meet the needs of precision medicine during the last decades. This kind of practice is supported and recommended by agencies. FDA published a draft guidance on complex innovative trial designs in September 2019 and finalized it in December 2020 to support the goal of facilitating and advancing the use of these innovatice designs (U.S. Food and Drug Administration, 2020). This document provides guidance to sponsors and applicants on interacting with the FDA on complex innovative design proposals for drugs or biological products. One important type of innovative designs is the master protocol which investigates multiple hypotheses through concurrent sub-studies and FDA published a guidance on master protocols in March 2022 (Park et al., 2019; U.S. Food and Drug Administration, 2022). Basket trials, umbrella trials, and platform trials are novel designs under the framework of master protocols. Seamless designs and enrichment designs are also popular trial types of interest. This section introduces these designs under the mathematical framework in Sect. 2.

3.1 Basket Design

Traditionally clinical trials in oncology therapeutic area focus on the treatments of cancers with some specific genetic mutation or at certain locations such as lung, breast, etc. A more recent development for oncology clinical trials such as a basket study could include patients with a certain genetic mutation in common regardless of the location of cancer in the body. In general, basket trial investigates a targeted therapy on multiple diseases that have common molecular alternations to quickly validate the mechanism of action and identify the subtype of tumors on which the drug is most efficacious. If a substudy shows a strong response signal, it may allow for the expansion of this substudy to a larger confirmatory trial to support a marketing application. The graphical representation of basket trials can be seen in top-left part of Fig. 1.

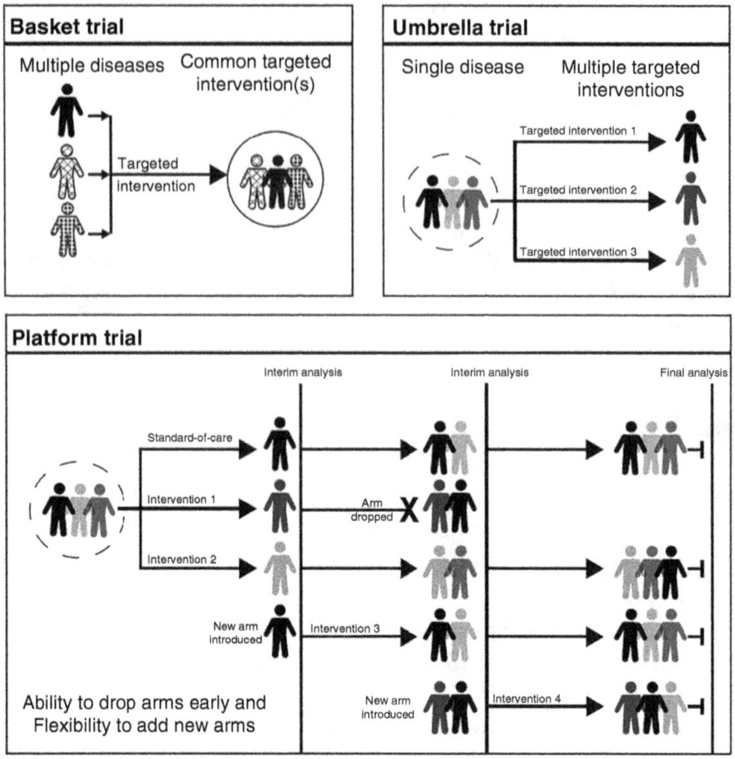

Fig. 1 Graphical representation of basket trials, umbrella trials, and platform trials from Park et al. (2019)

Patients in Basket trials usually have certain characteristics in common, which are supposed to share similar responses to the targeted therapy. However, patients with different subtypes of diseases could respond variously. Each substudy could include specific objectives and the data from each substudy can be analyzed independently. This independent approach is simple and avoids bias with clean data, and meanwhile it often lacks efficiency. Two general ways can be used to improve efficiency. One approach is to apply a more efficient design in each substudy, for instance, a more efficient and flexible Bayesian design instead of Simon's two-stage design. Another approach is to adaptively borrow information across substudies using Bayesian methods (Chu & Yuan, 2018; Jiang et al., 2021; Zhou et al., 2017).

Basket design has been applied to many clinical trials in recent years (Lopez-Chavez et al., 2015; NCI-Molecular, 2017; Oneal et al., 2018; Park et al., 2019). The majority of basket trials are exploratory in early phase of clinical trials, usually with a single intervention arm and have relatively small sample sizes (Park et al., 2019). The allocation probabilities for basket trials with single treatment are $\boldsymbol{\phi}_n = 1$ with $T = 1$ as the only treatment under the mathematical framework in Sect. 2. Basket studies have the advantage to investigate the drug targeting one mutation

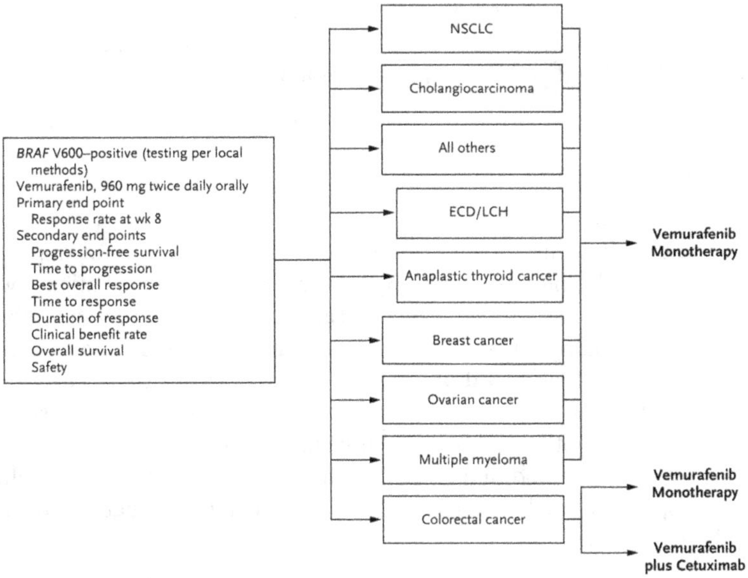

Fig. 2 Basket trial for Vemurafenib in Hyman et al. (2015)

across several tumor types simultaneously and could be helpful to determine if a drug targeting genetic mutation at a particular site may be effective in treating patients with same genetic mutation and cancer located in another site of the body. These designs also have some limitations and challenges in implementation. With small sample sizes and thus limited precision of efficacy estimations, the results from basket trials should be interpreted with cautious and may need additional confirmatory trials before final approval. Furthermore, basket trials only present the results for certain subpopulation with common biomarker values and we cannot distinguish predictive biomarkers from prognostic biomarkers without a comparative arm.

Basket trials have been widely accepted by health authorities. The US Food and Drug Administration (FDA) approved several new drugs based on basket studies (Oneal et al., 2018; U.S. Food and Drug Administration, 2018). For instance, Vemurafenib was approved by FDA in 2017 based on the data from Erdheim-Chester disease (ECD) cohort of 22 patients in a basket trial (Oneal et al., 2018). This trial is an open-label, non-randomized, multicohort phase II study conducted in BRAF V600 mutation-positive nonmelanoma cancers and included non-small-cell lung cancer, ovarian cancer, ECD and other cancers to explore the efficacy and safety of vemurafenib monotherapy (Hyman et al., 2015; Oneal et al., 2018). The study design can be found in Fig. 2. Also FDA granted accelerated approval to pembrolizumab in 2017 for adult and pediatric patients with unresectable or metastatic, microsatellite instability-high (MSI-H) or mismatch repair deficient (dMMR) solid tumors that have progressed following prior treatment and who have no satisfactory alternative treatment options or with MSI-H or dMMR colorectal cancer. The approval was based on data from 149 patients with MSI-H or dMMR cancers enrolled across five uncontrolled, multi-cohort, multi-center, single-arm clinical trials (U.S. Food and Drug Administration, 2017).

3.2 Umbrella Design

Umbrella trials evaluate multiple targeted investigational drugs for a single disease that may be stratified into subgroups by molecular alternation. Patients with the same disease type are enrolled to several parallel cohorts based on their predictive biomarkers. Each cohort can be considered as a biomarker enrichment design. The activities of the investigational drugs can be compared to a common control arm. FDA suggested that the drug chosen as the control arm should be the standard of care (SOC) for the target population, and this may change over time if newer drugs replace the SOC (U.S. Food and Drug Administration, 2022). For umbrella trials based on pre-specified predictive biomarkers, the allocation probability depends on these predictive biomarkers,

$$\boldsymbol{\phi}_n = E[\boldsymbol{T}_n|\boldsymbol{T}_1, \ldots, \boldsymbol{T}_{n-1}, \boldsymbol{X}_1, \ldots, \boldsymbol{X}_n].$$

If the predictive biomarkers are identified during the on-going trials, then the allocation probability also depends on responses and can be treated as CARA design, that is,

$$\boldsymbol{\phi}_n = E[\boldsymbol{T}_n|\boldsymbol{T}_1, \ldots, \boldsymbol{T}_{n-1}, \boldsymbol{Y}_1, \ldots, \boldsymbol{Y}_{n-1}, \boldsymbol{X}_1, \ldots, \boldsymbol{X}_n].$$

In general, comparative analyses in umbrella trials are between the investigational drug and the common control, and not among experimental treatment arms. Usually umbrella trials investigate experiential drugs and have larger sample size than basket trials (Park et al., 2019). The benefit of the investigational drug can be more directly attributed to the predictive biomarker, and the trial maximizes the possible benefits to participants. Also umbrella trials have the advantage to share the common control arm to improve efficiency, which is recommended by FDA (U.S. Food and Drug Administration, 2022). The disadvantages of such trials include large number of sample sizes, multiple intervention arms and longer duration.

An umbrella trial was conducted in Lung-MAP patients with previously-treated advanced squamous cell lung cancer (Ferrarotto et al., 2015). The study design can be found in Fig. 3 and it aimed to identify biomarker-drug pairs that would lead to successful therapeutic outcomes and registration of new agents. In this trial, eligible subjects were assigned to substudies based on their biomarkers or to a non-match therapy substudy for subjects not eligible for the biomarker-specific substudies. Subjects were randomized to a biomarker-driven target or to SOC therapy within each substudy. It facilitated the evaluation of multiple investigational therapies to independently conduct therapeutic studies under a single trial infrastructure in lung squamous cell cancer subpopulations.

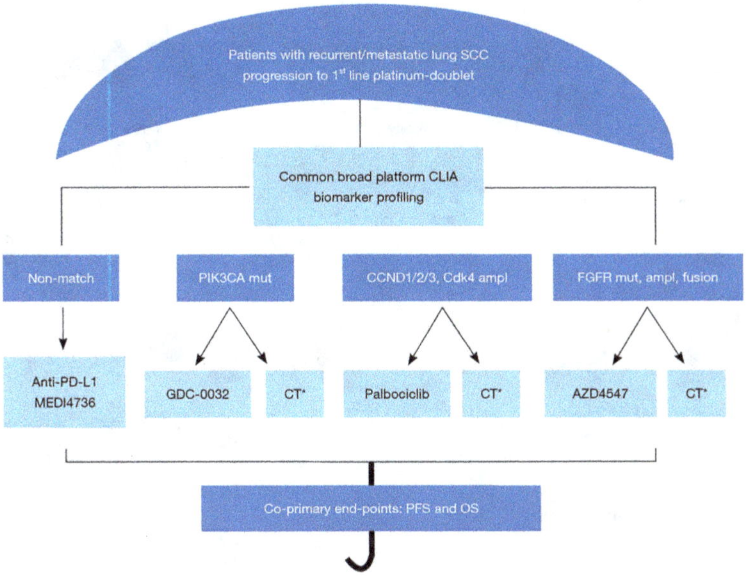

Fig. 3 Umbrella trial for Lung-MAP patients from Ferrarotto et al. (2015)

3.3 *Platform Design*

Platform trials refer to trials that evaluate several interventions against a common control group (Park et al., 2019). This design has pre-specified adaptation rules to allow dropping off ineffective intervention(s) and flexibility of adding new intervention(s) during the trial. During this procedure, several agents may graduate from the phase II evaluation with seamless transitions into phase III evaluations (Park et al., 2016; Sydes et al., 2018). The allocation probability $\boldsymbol{\phi}_n$ in platform trials usually depends on the treatment assignments, responses, and covariate vectors from the previous $n-1$ patients, and the current patient's covariates, that is,

$$\boldsymbol{\phi}_n = E[\boldsymbol{T}_n|\boldsymbol{T}_1, \ldots, \boldsymbol{T}_{n-1}, \boldsymbol{Y}_1, \ldots, \boldsymbol{Y}_{n-1}, (\boldsymbol{Z}_1, \boldsymbol{X}_1), \ldots, (\boldsymbol{Z}_n, \boldsymbol{X}_n)].$$

The inclusion of $(\boldsymbol{Z}_i, \boldsymbol{X}_i)$ depends on the actual procedures.

Platform trials permit relatively flexility in implementation regrading the addition or exclusion of treatment methods or patient populations, thereby enable an efficient transition to a confirmatory trial. It allows to share the control group and no need to add a new trial for every invention, which leads to less recruitment of patients. These features make platform designs very attractive because the biological knowledge advances in high speed and many new drugs and combination therapies are considered in trial evaluations during the ongoing trials. Since only a small percentage of drugs are tested in clinical trials and the overall success rate is very low, platform trials could drop off ineffective drugs in early stage of trials (Wong

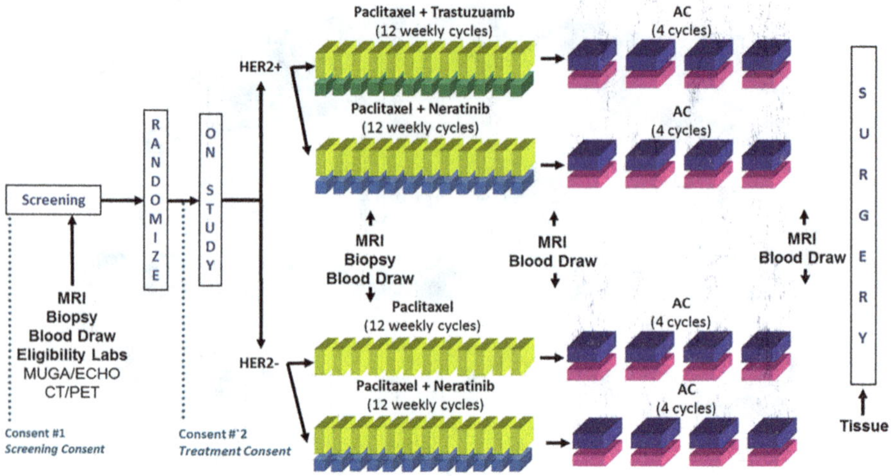

Fig. 4 Platform trial in Park et al. (2016)

et al., 2019). The drawbacks of platform trials include the challenges with large sample size, long study duration, high cost of managing and executing the trials. The rules for graduation and dropping off from the trials need to be clearly specified at the planning stage. Adaptive randomization method could be applied in the patient assignments to allocate patients to effective arms with high probabilities, which may increase the complexity of trial execution.

An adaptive phase II, platform trial with multiple experimental-therapy groups was designed to assess new agents combined with standard neoadjuvant therapy in patients with breast cancer who are at high risk for early recurrence and the study design can be seen in Fig. 4 (Park et al., 2016). Biomarker assessments according to human epidermal growth factor receptor 2 (HER2) status were performed at baseline and were used to stratify patients in the randomization and a common control group was used. It was found that Neratinib added to standard therapy was likely to result in higher rates of pathological complete response compared to standard chemotherapy with trastuzumab among patients with HER2-positive, hormone-receptor-negative breast cancer.

3.4 *Other Designs*

Seamless design and enrichment designs are also popular adaptive designs in the practice. They may be embedded or have overlap with other trial types like umbrella design or platform trial. For example, the umbrella trial in Lung-MAP patients was a Phase II/III seamless design.

Seamless designs consists of two types: operationally seamless and inferentially seamless. Operationally seamless designs combine two separate trials, eg, individual

Phase II and Phase III trials, into one trial. An inferentially seamless design makes use of the data from patients enrolled before and after adaptation, eg, pulls the data collected in both the Phase II and Phase III trials, into the final analysis. Therefore, seamless design expedites the overall study procedure and use patient data more efficiently. The allocation probability in seamless designs depends on the actual randomization procedures of individual trials. It also allows the flexibility to change or cut treatments during the study, which provides patients a greater chance to receive safe and efficacious treatment. Meanwhile, the data analysis of seamless trial must take account the biases introduced in the procedure, including the multiplicity due to multiple tests and potential bias of treatment effect due to combining data from different trials.

Enrichment designs use predictive biomarkers to define the enrollment criteria, which only include a subset of analysis population who are likely to benefit from the treatment when strong characterization of the target patient population is observed (Li et al., 2016). Thus the conditional allocation probability depends on predicative biomarkers and responses, that is,

$$\boldsymbol{\phi}_n = E[\boldsymbol{T}_n|\boldsymbol{T}_1, \ldots, \boldsymbol{T}_{n-1}, \boldsymbol{Y}_1, \ldots, \boldsymbol{Y}_{n-1}, \boldsymbol{X}_1, \ldots, \boldsymbol{X}_n].$$

If the trials also have balancing purpose for prognostic covariates, then $\boldsymbol{Z}_1, \ldots, \boldsymbol{Z}_n$ are also involved. This design may substantially increase power of the study, but the efficacy of the clinical trial relies heavily on the correct characterization of the target population. The analysis plan needs to be pre-specified to obtain statistically valid conclusions and the interpretation of the results need to be cautious when a small subset of patients drive treatment response (Wang & Hung, 2013). FDA have published guidance of enrichment strategies for clinical trials regarding the use of enrichment design (U.S. Food and Drug Administration, 2019).

4 Covariate-Adaptive Randomization Designs

Various of covariate-adaptive randomization procedures have been proposed in the existing literatures, and these randomization designs could greatly improve the balance of covariates across treatment groups, avoid confounding issue, make the comparison of different treatments persuasive and reliable, and also improve the efficiency of statistical inference.

In clinical trials containing only discrete covariates, stratified permuted block design is commonly used because it is easy to perform in clinical trials and performs well with a small number of strata. However, stratified permuted block design also has some shortcomings, for example, the design has potential selection bias, the predictability of random allocation when the block is fixed, and when the number of covariates is large or the number of strata is large, it leads to potential serious imbalance within the stratum and overall imbalance (Pocock, 1982).

To avoid these shortcomings of stratified permuted block design, minimization methods have been proposed in the literatures (Hu & Hu, 2012ba; Pocock &

Simon, 1975; Taves, 1974). The Pocock & Simon marginal method achieves the balance of covariates on a marginal basis, as well as on an overall basis, but not on a hierarchical basis (Kundt, 2009). Hu and Zhang (2020) has theoretically demonstrated that Pocock & Simon method can achieve the overall balance and marginal balance of covariates, but the degree of imbalance within the stratum is of the same order as $\sqrt{n}$ when the sample size n increases. Hu and Hu (2012ba) generalized the minimization method and proved that the design archives overall, marginal and within-stratum imbalance under some mild conditions. Lebowitsch et al. also presented a generalized multidimensional dynamic allocation method that simultaneously balances treatment assignments at three levels (Lebowitsch et al., 2012). Hu et al. provided literature review on minimization methods and the theoretical development for covariate-adaptive randomization (Hu et al., 2014). However, these designs only apply to the discrete covariates, and not applicable for continuous covariates.

Several methods have been published to balance continuous covariates. Frane (1998) proposed to minimize the differences in means by minimizing the p-values of t-tests. Nishi and Takaichi (2003) suggested to minimize the weighted sum of the differences in means and standard deviations of covariates and the differences in group size. Hoehler (1987) and Stigsby and Taves (2010) considered to use ranks instead of the actual covariate values. Su (2011) proposed to balance the distribution of continuous covariates through ensuring similar quantiles. Hu and Hu (2012a) proposed to measure the imbalance through the maximum group size difference over all possible divisions of covariate range. Lin and Su (2012) used the area between the empirical distributions as the imbalance measure. Ma and Hu (2013) proposed to balance continuous covariates based on kernel densities, and demonstrated that this method could achieve distribution balance of both continuous and discrete covariates and keep the group size balanced. Qin et al. (2024) proposed an optimal method for covariate balancing and obtained the theoretical properties. Liu et al. (2015) considered covariate-adaptive designs with missing covariates in clinical trials and Liu and Hu (2020) proposed to balance unobserved covariates with covariate-adaptive randomized experiments. Li et al. (2021a) considered the sample size re-estimation and Zhu and Hu (2019) evaluated the validity of sequential monitoring and provided theoretical properties for covariate-adaptive randomized clinical trials.

A number of clinical trials have employed covariate-adaptive randomization designs such as Weir and Lees (2003), Fakhry et al. (2015), Stott et al. (2017), and Sun et al. (2018). In the data analysis of clinical trial, some covariates used during adaptive randomization are not included in the final working model for certain reasons. For example, investigation center or country is widely used as a stratification factor in randomization to promote balanced distribution across sites or countries. However, we usually do not involve them in the final statistical model. Therefore, it's common in clinical trials that some covariates are not suitable for inclusion in the working model for analysis, or some relevant covariates are erroneously excluded or ignored. It is pointed out that if the linear working model only includes partial covariates used in the randomization, then in the process of statistical inference, such as applying t-test, we may encounter the phenomenon that the working model is conservative (Li et al., 2021b; Ma et al., 2015). Shao, Yu and Zhong stated that to obtain a valid test, the test statistic should take all covariates used in the randomization procedure into account. Excluding some covariates from the working model may distort the sampling distribution of the test statistics, so that statistical inference is no longer effective (Shao et al., 2010).

Although covariate-adaptive designs are widely used in clinical trials, the statistical inference available for efficacy in such designs is primarily directed at linear models or generalized linear models for non-survival data (Bugni et al., 2018; Li et al., 2019, 2021b; Ma et al., 2015, 2020; Shao & Yu, 2013). For survival data, Ye and Shao studied the statistical inference and theoretical properties in the context of covariate-adjusted randomization (Ye & Shao, 2020).

5 Covariate-Adjusted Response-Adaptive Randomization Designs

CARA designs were introduced to utilize the information of covariates and response variable for better treatment assignment. The literature review of existing CARA designs are included in Sect. 5.1 and we propose a new class of CARA designs in Sect. 5.2.

5.1 Some Existing CARA Designs

Bandyopadhyay and Biswas (2001) proposed CARA design for continuous response in the framework of linear models. Rosenberger, Vidyashankar and Agarwal considered CARA designs for binary responses using logistic regression models (Rosenberger et al., 2001). Furthermore, Biswas and Bhattacharya proposed a new CARA design for binary responses in Phase III clinical trials involving multiple treatment groups, whose target assignment is based on the combination of medical ethics and statistical efficiency in the presence of interaction between treatment and covariates (Biswas & Bhattacharya, 2018). Zhu et al. (2013) proposed a new adaptive design under the framework of a multiple linear regression model that investigates the interaction between covariates and treatment, and theoretically derived the optimal allocation that maximizes the power of the test.

Zhang et al. (2007) provided a general family of CARA designs with a broad class of models, under which different goals can be achieved by choosing different allocation functions. The main method is as follows: assume that a clinical trial has K treatment groups, the covariate of a patient is $\boldsymbol{Z}$, the response variable of the kth treatment group is Y_k, $k = 1, \ldots, K$, $\boldsymbol{T} = (T_1, \ldots, T_k)$ is the treatment allocation, in which only one component is 1 and the rest is 0, and the value 1 indicates that the patient is allocated to this treatment, $\boldsymbol{Y} = (Y_1, \ldots, Y_k)$ is the response of the treatment, and only observed in one component, we assume that

$$E(Y_k|\boldsymbol{Z}_k) = p_k(\boldsymbol{\theta}_k, \boldsymbol{Z}_k), \boldsymbol{\theta}_k \in \boldsymbol{\Theta}_k, k = 1, \ldots, K,$$

where $p_k(\cdot, \cdot)$ is a function with known format, $\boldsymbol{\theta}_k$ is unknown parameter, $\boldsymbol{\Theta}_k$ is the parameter space for $\boldsymbol{\theta}_k$, $k = 1, \ldots, K$. Assume that the first n patients have been assigned, when the $(n+1)$th patient is ready for randomization, the covariates $\boldsymbol{Z}_{n+1}$ are known, and the probability of assigning this patient to treatment k is

$$\psi_k = P(T_{n+1,k} = 1|\mathcal{F}_n, \boldsymbol{Z}_{n+1}) = \pi_k(\hat{\boldsymbol{\theta}}_m, \boldsymbol{Z}_{n+1}), k = 1, \ldots, K,$$

where $\mathcal{F}_n = \sigma(\boldsymbol{T}_1, \ldots, \boldsymbol{T}_n, \boldsymbol{Y}_1, \ldots, \boldsymbol{Y}_n, \boldsymbol{Z}_1, \ldots, \boldsymbol{Z}_n)$ is the $\sigma-$space constructed by the historical information. Under some general assumptions, Zhang et al. (2007) proved that the proposed design has good theoretical properties, such as that the probability of a patient being assigned to any treatment group converges to a certain constant, and the observed allocation ratio converges to the target allocation ratio at a rate of $O\left(\sqrt{\log\log n/n}\right)$ and has asymptotically normal distribution.

Zhang and Hu (2009) further introduced a new family of CARA designs and showed that the new designs have less variabilities and more efficient. Cheung et al. (2014) developed the theoretical properties of CARA designs to conduct hypothesis testing which are applicable to generalized linear models. Huang et al. (2013) proposed a general framework of longitudinal CARA procedures and studied their properties under widely satisfied conditions. Sverdlov et al. (2013) proposed CARA randomization procedures for survival intervention trials when the primary outcomes follow an exponential regression model. Hu et al. (2015b) considered two objectives of clinical trials including power and ethics, and derived a theoretical framework for a new class of CARA designs that obtain information on covariates and response variables, and combine power and ethics metrics into new allocation goal. Aletti et al. (2018) proposed a general class of CARA designs based on a new functional urn model, which allows the distribution of the responses conditioned on covariates to be estimated nonparametrically.

There are also some references for CARA designs from Bayesian perspectives (Ning & Huang, 2010; Yuan et al., 2011). They established a logistic regression model for the response variable based on patients' allocation and covariate information, and adjusted the response-adaptive designs through different measurements to evaluate covariate imbalance. Their regression models did not consider the interaction between covariates and allocation, and could not derive the theoretical properties of the allocation procedure.

These CARA designs haven't considered the different impacts of covariates on the response variable. We classify covariates into two types: prognostic or predictive covariates, and adjust the randomization procedure according to the roles of covariates. To achieve the goal of precision medicine and to maintain both efficiency and ethics, a good design should have two desirable properties: (a) balance over prognostic covariates to provide a valid and convincing comparison; (b) based on the responses and the predictive covariate, select the most suitable treatment for patients. Unfortunately, the CARA designs in published papers can only be applied for discrete prognostic covariates (Zhao et al., 2022). None of the existing CARA methods consider balancing and treatment selection simultaneously in general scenario. The fundamental difficulty lies in the complexity dealing with continuous prognostic covariates, predictive covariates and responses simultaneously. In this context, we propose a new class of CARA designs to fill this gap and extend the design to general types of covariates.

5.2 *New CARA Designs*

Consider a clinical trial with predictive covariates $\mathbf{X}$ and prognostic covariates $\mathbf{Z}$, our aim is to balance $\mathbf{Z}$ and meanwhile to allocate patients to better treatments. Without loss of generality, we assume two treatment groups ($T = 1$ for treatment 1, and 0 for treatment 2) are involved in the trials. Throughout this chapter, covariates are identified as predictive or prognostic before the conduction of clinical trials and their values are assumed to be known before randomization. This is consistent with practical experience in the industry. A new family of CARA procedure is proposed to achieve the goal of balancing over prognostic covariates and better treatment selection simultaneously. Intuitively, when a new patient meets the inclusion/exclusion criteria and is ready for randomization, we have 2 options, treatment 1 and 2, to choose. Then an imbalance measure could be applied to evaluate the degree of imbalance between two treatment groups in each scenario, then allocate new patient to the treatment group with smaller imbalance measure in a high probability. This is the basic idea of the new CARA procedure.

5.2.1 Introduction of New Designs

Let $(\mathbf{X}_i, \mathbf{Z}_i)$ be the covariates of ith patient. We first consider a simple case that $\mathbf{Z}$ only has discrete components, $\mathbf{Z} = (Z_1, \ldots, Z_I)^T$, and m_i levels for the i-th element. Denote $T = 1$ for treatment 1, and 0 for treatment 2, and T_i is the treatment assignment for ith patient. For convenience, we denote $(z_1, \ldots, z_I)$ as the stratum formed by patients who possess the same prognostic covariate profile $(z_1, \ldots, z_I)^T$, and $(j; z_j)$ as the margin formed by patients whose jth prognostic covariate is at level z_j.

Assume that $n-1$ ($n > n_0$) patients have been assigned to one of the treatments, and their responses are observed. Suppose nth patient is ready for randomization, and the prognostic covariates of this patient $\mathbf{Z}_n$ fall within the stratum $(z_1^*, \ldots, z_I^*)$.

Then by assuming that this patient were assigned to treatment 1, we define an imbalance measure $Imb_n^{(1)}(\mathbf{X}_n)$ by

$$Imb_n^{(1)}(\mathbf{X}_n) = w_o\left[D_n^{(1)}(\mathbf{X}_n)\right]^2 + \sum_{j=1}^{I} w_{m,j}\left[D_n^{(1)}(\mathbf{X}_n; j, z_j^*)\right]^2$$
$$+w_s\left[D_n^{(1)}(\mathbf{X}_n; z_1^*, \ldots, z_I^*)\right]^2,$$

where

$$D_n^{(1)}(\mathbf{X}_n) = N_{n1}(\mathbf{X}_n) - \hat{\rho}(\mathbf{X}_n)N_n(\mathbf{X}_n),$$
$$= \sum_{i=1}^{n} \mathbb{I}_{\{T_i=1, X_i=\mathbf{X}_n\}} - \hat{\rho}(\mathbf{X}_n)\sum_{i=1}^{n} \mathbb{I}_{\{X_i=\mathbf{X}_n\}},$$

$$D_n^{(1)}(\mathbf{X}_n; j, z_j^*) = N_{n1}(\mathbf{X}_n; j, z_j^*) - \hat{\rho}(\mathbf{X}_n)N_n(\mathbf{X}_n; j, z_j^*),$$

and

$$D_n^{(1)}(\mathbf{X}_n; z_1^*, \ldots, z_I^*) = N_{n1}(\mathbf{X}_n; z_1^*, \ldots, z_I^*) - \hat{\rho}(\mathbf{X}_n)N_n(\mathbf{X}_n; z_1^*, \ldots, z_I^*).$$

are differences between the observed and expected number of patients in treatment 1 group on the overall, the margin (j, z_j^*) and within the stratum $(z_1^*, \ldots, z_I^*)$, and w_o, $w_{m,j}$ $(j = 1, \ldots, I)$, w_s are nonnegative weights placed on overall, marginal and within-stratum differences, satisfying $w_o + \sum_{j=1}^{I} w_{m,j} + w_s = 1$. $N_n(\mathbf{X}_n)$ and $N_{n1}(\mathbf{X}_n)$ are the overall number of patients with the value of predictor covariate $\mathbf{X}_n$ in the first n patients, and the corresponding number of patients in the treatment 1 group, $\mathbb{I}_{\{\cdot\}}$ is the indicator function. $\rho(\cdot)$ is the allocation proportion for treatment 1 and $\hat{\rho}(\mathbf{X}_n)$ is its estimation for patients with predictive covariate $\mathbf{X}_n$, which is calculated based on $\mathbf{X}_n$ and the responses of first $(n-1)$ patients.

Similarly, we define $Imb_n^{(2)}(\mathbf{X}_n)$ by assuming the nth patient were assigned to treatment 2 group. Then conditional on the assignments of the first $(n-1)$ patients and the covariates of the first n patients, we assign the n-th patient to a treatment based on the comparison of $Imb_n^{(1)}(\mathbf{X}_n)$ and $Imb_n^{(2)}(\mathbf{X}_n)$. A general way to achieve this is to define the allocation probability with a function $g(\cdot)$,

$$P(T_n = 1|\mathcal{T}_{n-1}, \mathcal{X}_n, \mathcal{Z}_n) = g(Imb_n^{(1)} - Imb_n^{(2)}),$$

where $n > 1$, $g(\cdot)$ is a function which satisfies $0 < g(x) < 1$, $g(x) + g(-x) = 1$, and $g(x) \leq 0.5$ for $x \geq 0$. A widely used design is the Efron's biased coin design with $g(\cdot)$ defined as

$$g(x) = \begin{cases} q, & \text{if } x > 0, \\ \frac{1}{2}, & \text{if } x = 0, \\ p, & \text{if } x < 0, \end{cases}$$

where $p > 1/2$ and $p + q = 1$. In particular, $p \in [0.75, 0.95]$ was suggested in Hu and Hu (2012a).

The overall procedure is conducted as follows:

Step 1. Assign $n_0(n_0 = 2k)$ subjects to each treatment by using a restricted randomization procedure (e.g., permuted block design). In the sequel, we set $n_0 = 40$ for all scenarios in this article.;

Step 2. Assume that $n - 1(n > 2n_0)$ subjects have been assigned to one of the treatments, and their responses are observed. When the nth subject is ready for randomization, define the imbalance measure $Imb^{(1)}$ and $Imb^{(2)}$ if this subject will be assigned to Treatment k, $k = 1, 2$.

Step 3. Define allocation probability with a function $g(\cdot)$ and allocate nth patient to Treatment 1 with probability

$$P(T_n = 1|\mathcal{T}_{n-1}, \mathbf{X}_n, \mathbf{Z}_n) = g(Imb_n^{(1)} - Imb_n^{(2)});$$

Step 4. Repeat Step 2-3 until all subjects are randomized.

To assign more patients to better treatment based on the predictive covariates, the estimation of allocation proportion is the key to success. The estimated allocation proportion could be totally different for each $\mathbf{X}_n$ according to the responses of patients assigning to different treatments. Intuitively, if patients with $\mathbf{X}_n$ tend to have better responses in treatment group $T = 1$ than those in $T = 0$ group, the corresponding allocation proportion $\hat{\rho}(\mathbf{X}_n)$ will be greater than 0.5. Some recommended methods to calculate estimated allocation proportion in the simulation scenarios can be found in Sect. 5.2.3.

If only continuous prognostic covariates $\mathbf{Z} = (Z_1, \ldots, Z_S)^T$, we also proceed the steps similarly by including the overall and marginal differences to obtain the overall imbalance. Borrowing the idea of kernel method for continuous variables, the following randomization procedure is considered.

$$\begin{aligned}
&Imb_n^{(1)}(\mathbf{X}_n)\\
&= w_o\big[D_n^{(1)}(\mathbf{X}_n)\big]^2 + \sum_{j=1}^{S} w_{m,j}\big[D_n^{(1)}(\mathbf{X}_n; Z_{jn})\big]^2\\
&= w_o\big[D_n^{(1)}(\mathbf{X}_n)\big]^2 + \sum_{j=1}^{S} w_{m,j}\\
&\quad \times\Big\{\sum_{i=1}^{n} \mathbb{I}_{\{\mathbf{X}_i=\mathbf{X}_n\}}\big[\mathbb{I}_{\{T_i=1\}} - \hat{\rho}(\mathbf{X}_n)\big]\frac{1}{h_j(n)}K\Big(\frac{Z_{ji} - Z_{jn}}{h_j(n)}\Big)\Big\}^2,\\
&= w_o\big[D_n^{(1)}(\mathbf{X}_n)\big]^2 + \sum_{j=1}^{S} w_{m,j}\Big\{\sum_{i=1}^{n} \mathbb{I}_{\{\mathbf{X}_i=\mathbf{X}_n\}}\big[T_i - \hat{\rho}(\mathbf{X}_n)\big]\frac{1}{h_j(n)}K\Big(\frac{Z_{ji} - Z_{jn}}{h_j(n)}\Big)\Big\}^2,
\end{aligned}$$

where $K(\cdot)$ is a kernel function, $h_j(\cdot)$ is the bandwidth and it is a function of the sample size n. Similar to Ma and Hu (2013), we use the Scott's rule $h_j(n) = \hat{\sigma} n^{-0.2}$, where $\hat{\sigma}$ is the estimated standard deviation of the covariate values for Z. w_o and $w_{m,j}$ are the weights of the overall and marginal differences satisfying $w_o + \sum_{j=1}^{I} w_{m,j} = 1$. If the nth subject were assigned to treatment 2, $Imb_n^{(2)}(\mathbf{X}_n)$ can be defined in the same way as $Imb_n^{(1)}(\mathbf{X}_n)$.

The proposed design can be easily extended to the general cases with both continuous prognostic covariates $Z_1, \ldots, Z_S$, and discrete prognostic covariates $Z_{S+1}, \ldots, Z_I$, $I > S$, $S \in \mathbb{N}^+$, $I \in \mathbb{N}^+$, the overall imbalance can be defined as

$$
\begin{aligned}
&Imb_n^{(1)}(\mathbf{X}_n, Z_n)\\
&= w_o\big[D_n^{(1)}(\mathbf{X}_n)\big]^2 + \sum_{j=1}^{S} w_{m,j}\big[D_n^{(1)}(\mathbf{X}_n; Z_{jn})\big]^2\\
&\quad + \sum_{j=S+1}^{I} w_{m,j}\big[D_n^{(1)}(\mathbf{X}_n; Z_{jn})\big]^2 + w_s\big[D_n^{(1)}(\mathbf{X}_n; Z_{S+1,n}, \ldots, Z_{In})\big]^2\\
&= w_o\Big[\sum_{i=1}^{n} \mathbb{I}_{\{T_i=1, \mathbf{X}_i=\mathbf{X}_n\}} - \hat{p}(\mathbf{X}_n)\sum_{i=1}^{n} \mathbb{I}_{\{\mathbf{X}_i=\mathbf{X}_n\}}\Big]^2\\
&\quad + \sum_{j=1}^{S} w_{m,j}\Big\{\sum_{i=1}^{n} \mathbb{I}_{\{\mathbf{X}_i=\mathbf{X}_n\}}\big[\mathbb{I}_{\{T_i=1\}} - \hat{p}(\mathbf{X}_n)\big]\frac{1}{h_j(n)}K\Big(\frac{Z_{ji}-Z_{jn}}{h_j(n)}\Big)\Big\}^2\\
&\quad + \sum_{j=S+1}^{I} w_{m,j}\Big[\sum_{i=1}^{n} \mathbb{I}_{\{T_i=1, \mathbf{X}_i=\mathbf{X}_n, Z_{ji}=Z_{jn}\}} - \hat{p}(\mathbf{X}_n)\sum_{i=1}^{n} \mathbb{I}_{\{\mathbf{X}_i=\mathbf{X}_n, Z_{ji}=Z_{jn}\}}\Big]^2\\
&\quad + w_s\Big[\sum_{i=1}^{n} \mathbb{I}_{\{T_i=1, \mathbf{X}_i=\mathbf{X}_n, Z_{ji}=Z_{jn}, j=S+1,\ldots,I\}} - \hat{p}(\mathbf{X}_n)\\
&\quad \times \sum_{i=1}^{n} \mathbb{I}_{\{\mathbf{X}_i=\mathbf{X}_n, Z_{ji}=Z_{jn}, j=S+1,\ldots,I\}}\Big]^2.
\end{aligned}
$$

where $\mathbb{N}^+$ is the set of all positive integers, w_o, $w_{m,j}$ $(j = 1, \ldots, I)$, w_s are nonnegative weights placed on overall, marginal and within-stratum differences, satisfying $w_o + \sum_{j=1}^{I} w_{m,j} + w_s = 1$. and $h_j(\cdot)$ is the bandwidth for covariate Z_j.

In general case, if the randomization ratio for both treatments is $p : 1$ $(p > 0)$ based on prognostic covariates, the imbalance measure could be modified by incorporating p into the overall difference as follows, and similarly for marginal difference and within-stratum difference.

$$D_n^{(1)}(\mathbf{X}_n) = N_{n1}(\mathbf{X}_n) - p\hat{\rho}(\mathbf{X}_n)N_n(\mathbf{X}_n).$$

Moreover, the proposed design could be degenerated to response-adaptive design in the absence of prognostic covariates. For patients with certain value of predictive covariates, the allocation proportion only relies on the responses of previously assigned patients and the design becomes response-adaptive design.

5.2.2 Theoretical Properties with Fixed Estimated Allocation Proportion

Consider a simple scenario with fixed estimated allocation proportion $\hat{\rho}(\mathbf{X}) = \rho(\mathbf{X})$, discrete predictive covariate $\mathbf{X}$ and discrete prognostic covariate $\mathbf{Z} = (Z_1, \ldots, Z_I)^T$, where Z_j has m_j levels, $j = 1, \ldots, I$. Let $D_n(\mathbf{X}; k_1, \ldots, k_I)$ be the within stratum difference for stratum $(k_1, \ldots, k_I)$ and $\mathbf{X}$, and denote $\mathbf{D}_n = [D_n(\mathbf{X}; k_1, \ldots, k_I)]_{1\le k_1\le m_1,\ldots,1\le k_I\le m_I}$ as the vector of length $m = \Pi_{i=1}^I m_i$, which includes the within-stratum differences for all strata with predictive covariate $\mathbf{X}$. $\mathbf{Z}_n = (k_1, \ldots, k_I)$ can take m values and it follows an m-dimensional multinomial distribution with parameter $\mathbf{p} = \{p(k_1, \ldots, k_I)\}_{1\le k_1\le m_1,\ldots,1\le k_I\le m_I}$, where $p(k_1, \ldots, k_I)$ is the probability that the patient has prognostic covariate in stratum $(k_1, \ldots, k_I)$. For simplicity, we assume that $p(k_1, \ldots, k_I) > 0$ for all strata. In fact, if $p(k_1, \ldots, k_I) = 0$, we have

$$P\Big\{D_n(\mathbf{X}_n; k_1, \ldots, k_I) = D_{n-1}(\mathbf{X}_n; k_1, \ldots, k_I)\Big\} = 1.$$

In our randomization procedure, we have

$$D_n(\mathbf{X}_n; k_1, \ldots, k_I) = D_{n-1}(\mathbf{X}_n; k_1, \ldots, k_I) + \mathbb{I}_{\{T_n=1\}} - \rho.$$

For a general value $\mathbf{x}$,

$$D_n(\mathbf{x}; k_1, \ldots, k_I) = D_{n-1}(\mathbf{x}; k_1, \ldots, k_I) + \mathbb{I}_{\{\mathbf{X}_n=\mathbf{x}\}}(\mathbb{I}_{\{T_n=1\}} - \rho).$$

Define a state space

$$\mathcal{D} = \Big\{n_1 - n_2\rho | n_1, n_2 \in \mathbb{N}\Big\},$$

where $\mathbb{N}$ is the set of all nonnegative integers, and $\mathcal{D}$ is a countable set. We can find that $\mathbf{D}_n$ is a function of $(\mathbf{D}_{n-1}, \mathbf{Z_n}, \mathbf{X}_n, T_n)$, and conditionally independent of $(\mathbf{D}_1, \ldots, \mathbf{D}_{n-2})$ given $\mathbf{D}_{n-1}$. Therefore $\mathbf{D}_n$ is a Markov chain on the $\mathcal{D}$.

To investigate the properties of $\mathbf{D}_n$, firstly we calculate the following quantity which plays a critical role in the allocation,

$$\begin{aligned}
& Imb_n^{(1)}(\mathbf{X}_n) - Imb_n^{(2)}(\mathbf{X}_n) \\
&= w_o\,[D_{n-1}(\mathbf{X}_n) + 1 - \rho]^2 + \sum_{j=1}^{I} w_{m,j}\left[D_{n-1}(\mathbf{X}_n;\, j, k_j) + 1 - \rho\right]^2 \\
&\quad + w_s\,[D_{n-1}(\mathbf{X}_n; k_1, \ldots, k_I) + 1 - \rho]^2 - w_o\,[D_{n-1}(\mathbf{X}_n) - \rho]^2 \\
&\quad - \sum_{j=1}^{I} w_{m,j}\left[D_{n-1}(\mathbf{X}_n;\, j, k_j) - \rho\right]^2 + w_s\,[D_{n-1}(\mathbf{X}_n; k_1, \ldots, k_I) - \rho]^2 \\
&= 2\big[w_o D_{n-1}(\mathbf{X}_n) + \sum_{j=1}^{I} w_{m,j} D_{n-1}(\mathbf{X}_n;\, j, k_j) + w_s D_{n-1}(\mathbf{X}_n; k_1, \ldots, k_I)\big] \\
&\quad -2\rho + 1.
\end{aligned}$$

$$Imb_n^{(1)}(\mathbf{X}) - Imb_n^{(2)}(\mathbf{X}) = \mathbb{I}_{\{\mathbf{X}_n = \mathbf{X}\}}\big[Imb_n^{(1)}(\mathbf{X}_n) - Imb_n^{(2)}(\mathbf{X}_n)\big].$$

Denote $\Lambda_n(\mathbf{X}_n; k_1, \ldots, k_I) = w_o D_{n-1}(\mathbf{X}_n) + \sum_{j=1}^{I} w_{m,j} D_{n-1}(\mathbf{X}_n;\, j, k_j) + w_s D_{n-1}(\mathbf{X}_n; k_1, \ldots, k_I)$ and then we have $Imb_n^{(1)}(\mathbf{X}_n) - Imb_n^{(2)}(\mathbf{X}_n) = 2\Lambda_n(\mathbf{X}_n; k_1, \ldots, k_I) - 2\rho + 1$. In the proposed randomization procedure, the allocation probability is determined by $\Lambda_n(\mathbf{X}_n; k_1, \ldots, k_I)$. Denote

$$\mathbf{\Gamma}_n = \Big\{2\Lambda_n(\mathbf{X}_n; k_1, \ldots, k_I) + 1 - 2\rho\Big\}_{1 \le k_1 \le m_1, \ldots, 1 \le k_I \le m_I}$$

and it impacts the change raw of $\mathbf{D}_n$. Similar to the method in Liu et al. (2017), we can prove that $\mathbf{\Gamma}_n = L(\mathbf{D}_n)$ is a one-to-one linear map. The theoretical properties of $\mathbf{\Gamma}_n$ and $\mathbf{D}_n$ are provided by the following theorem and the proof is provided in the appendix.

Theorem 1 *Assume that the allocation proportion ρ is fixed. Consider the scenario with predictive biomarke $\mathbf{X}$ and discrete prognostic covariate $\mathbf{Z} = (Z_1, \ldots, Z_I)^T$, where Z_j has m_j levels, $I \ge 1$, $m_i \ge 1$ and $1 \le j \le I$. The weights w_o, $w_{m,j}$ and w_s are nonnegative with the sum $w_o + \sum_{j=1}^{I} w_{m,i} + w_s = 1$. $g(\cdot)$ is a function satisfies $0 < g(x) < 1$, $g(x) + g(-x) = 1$, and $g(x) \le 0.5$ when $x \ge 0$. Then $(\mathbf{\Gamma}_n)_{n\ge1}$ is a positive recurrent Markov Chain on the space $\mathcal{L}(\mathcal{D})$ and $E\|(\mathbf{\Gamma}_n)\| = O(1)$. For each value of $\mathbf{X}$, we have*

(a) if $w_s > 0$, then $[\mathbf{D}_n(\mathbf{X})]_{n\ge1}$ is a positive recurrent Markov chain on the space $\mathcal{D}$ and $E\|\mathbf{D}_n(\mathbf{X})\| = O(1)$.
(b) if $w_{m,j} > 0$, then $D_n(\mathbf{X}; i, k_i) = O_P(1)$ is a positive recurrent Markov chain on the space $\mathcal{D}$ and $E|\mathbf{D}_n(\mathbf{X};\, j, k_j)| = O(1)$.
(c) for all cases, $D_n(\mathbf{X}) = O_P(1)$ and $E|D_n(\mathbf{X})| = O(1)$.

Proof We apply the techniques in Meyn and Tweedie (2009) to prove this theorem. Let the increment of $\mathbf{D}_n$ be $\Delta\mathbf{D}_n$. Then $\Delta\mathbf{D}_n$ has only one non-zero element which is either $1-\rho$ or ρ. For each fixed stratum $\mathbf{k} = (k_1, \ldots, k_I)^T$, we have

$$\begin{aligned} P(\Delta D_n(\mathbf{k}; \mathbf{X}_n) = 1-\rho|\mathcal{F}_{n-1}) &= g(Imb^{(1)} - Imb^{(2)})p(\mathbf{k}) \\ &= g(2\Lambda_{n-1} - 2\rho + 1)p(\mathbf{k}), \\ P(\Delta D_n(\mathbf{k}; \mathbf{X}_n) = -\rho|\mathcal{F}_{n-1}) &= 1 - g(2\Lambda_{n-1} - 2\rho + 1)p(\mathbf{k}), \\ P(\Delta D_n(\mathbf{k}; \mathbf{X}_n) = 0|\mathcal{F}_{n-1}) &= 1 - p(\mathbf{k}). \end{aligned}$$

Similar to Liu et al. (2017), we have

$$\begin{aligned} P(\Delta\mathbf{D}_n(\mathbf{X}_n) = \Delta\mathbf{d})|\mathcal{F}_{n-1}) &= g(\frac{\Delta\mathbf{d}\cdot\mathbf{\Gamma}_{n-1}}{\|\Delta\mathbf{d}\|})\left|\frac{\Delta\mathbf{d}\cdot\mathbf{p}}{\|\Delta\mathbf{d}\|}\right| \\ &= g\Big(\frac{\Delta\mathbf{d}\cdot L(\mathbf{D}_{n-1})}{\|\Delta\mathbf{d}\|}\Big)\left|\frac{\Delta\mathbf{d}\cdot\mathbf{p}}{\|\Delta\mathbf{d}\|}\right|, \ \Delta\mathbf{d} \in \Delta\mathcal{D}, \end{aligned}$$

where $\Delta\mathcal{D}$ is the state space of $\Delta\mathbf{D}_n, n \geq 1$, $\|\cdot\|$ is the Euclid norm, and $\mathbf{a}\cdot\mathbf{b} = \sum_{\mathbf{k}} \mathbf{a}(\mathbf{k})\mathbf{b}(\mathbf{k}), \forall \mathbf{a}, \mathbf{b} \in \Delta\mathcal{D}$. Therefore $P(\Delta\mathbf{D}_n(\mathbf{X}_n) = \Delta\mathbf{d}|\mathcal{F}_{n-1})$ is positive and only depends on $\mathbf{\Gamma}_{n-1} = L(\mathbf{D}_{n-1})$. $\mathbf{D}_n$ is conditional independent of $(\mathbf{D}_1, \ldots, \mathbf{D}_{n-2})$ given $\mathbf{D}_{n-1}$, and thus $[\mathbf{D}_n(\mathbf{X})]_{n\geq 1}$ is an irreducible Markov chain on space $\mathcal{D}$.

If we define

$$\begin{aligned} V_n = w_o\big[D_n(\mathbf{X}_n)\big]^2 &+ \sum_{j=1}^{I}\sum_{k_j=1}^{m_I} w_{m,j}\big[D_n(\mathbf{X}_n; j, k_j)\big]^2 \\ &+ \sum_{k_1,\ldots,k_I} w_s\big[D_n(\mathbf{X}_n; k_1, \ldots, k_I)\big]^2, \end{aligned}$$

then V_n is a function of $\mathbf{D}_n$, and thus it's a function of $\mathbf{\Gamma}_n$.

Next we show that there exists a bounded set $\mathcal{C}$ and a constant b such that

$$P_\lambda V(\mathbf{\Gamma}) - V(\mathbf{\Gamma}) \leq -1 + b\mathbb{I}_{\mathbf{\Gamma}\in\mathcal{C}}, \quad (1)$$

where $P_\lambda V(\mathbf{\Gamma}) = \sum_{\mathbf{\Gamma}'\in L(\mathcal{D})} P_\lambda(\mathbf{\Gamma}, \mathbf{\Gamma}')V(\mathbf{\Gamma}')$ and $P_\lambda(\mathbf{\Gamma}, \mathbf{\Gamma}')$ is the transition probability from state $\mathbf{\Gamma}$ to $\mathbf{\Gamma}'$.

Furthermore, for any integer $r \geq 2$, there exists a bounded set $\mathcal{C}$ and a constant b such that

$$P_\lambda V^{r+1}(\mathbf{\Gamma}) - V^{r+1}(\mathbf{\Gamma}) \leq -[V(\mathbf{\Gamma}) + 1]^r + b\mathbb{I}_{\mathbf{\Gamma}\in\mathcal{C}}. \quad (2)$$

According to Theorem 11.3.4 in Meyn and Tweedie (2009), $\{\boldsymbol{\Gamma}_n\}_{n\geq 1}$ is a positive recurrent Markov chain based on condition (1), which implies that $\{\boldsymbol{\Gamma}_n\}_{n\geq 1}$ is bounded in probability and has an invariant probability measure π. From drift condition (2), we have the following conclusion by Theorem 14.3.7 in Meyn and Tweedie (2009),

$$\pi[V(\boldsymbol{\Gamma})+1]^r \leq b,$$

and it suggests that $\sup_n E[V(\boldsymbol{\Gamma})+1]^r < \infty$.

We have $\sup_n E\|\Gamma_n\|^2 < \infty$ by the following derivation,

$$\begin{aligned}|\Gamma_n(\mathbf{k})|^2 &\leq \Big[w_o|D_n(\mathbf{X}_n)| + \sum_{j=1}^{I} w_{m,j}|D_n(\mathbf{X}_n;\, j, k_j)| + w_s|D_n(\mathbf{X}_n;\, \mathbf{k})|\Big]^2 \\ &\leq \Big[w_o|D_n(\mathbf{X}_n)|^2 + \sum_{j=1}^{I} w_{m,j}|D_n(\mathbf{X}_n;\, j, k_j)|^2 + w_s|D_n(\mathbf{X}_n;\, \mathbf{k})|^2\Big],\end{aligned}$$

and thus $\|\boldsymbol{\Gamma}_n\|^2 \leq mV(\boldsymbol{\Gamma}_n)$. Therefore $\{\boldsymbol{\Gamma}_n\}_{n\geq 1}$ is a positive recurrent Markov chain and $E\|\boldsymbol{\Gamma}_n\| = O(1)$.

We elaborate the procedure to check drift conditions (1) and (2). When $\mathbf{Z_n} = \mathbf{k}$ and $\mathbf{X}_n = \mathbf{x}$, if $T_n = 1$, we get $D_n = D_{n-1} + 1 - \rho$, and

$$\begin{aligned}V_n - V_{n-1} &= w_o(D_{n-1}(\mathbf{x}) + 1 - \rho)^2 + \sum_{j=1}^{I} w_{m,j}\big[D_{n-1}(\mathbf{x};\, j, k_j) + 1 - \rho\big]^2 \\ &\quad + w_s\big[D_{n-1}(\mathbf{x};\, \mathbf{k}) + 1 - \rho\big]^2 \\ &\quad - w_o D_{n-1}^2(\mathbf{x}) - \sum_{j=1}^{I} w_{m,j} D_{n-1}^2(\mathbf{x};\, j, k_j) - w_s D_{n-1}^2(\mathbf{x};\, \mathbf{k}) \\ &= 2(1-\rho)\big[w_o D_{n-1}(\mathbf{x}) + \sum_{j=1}^{I} w_{m,j} D_{n-1}(\mathbf{x};\, j, k_j) + w_s D_{n-1}(\mathbf{x};\, \mathbf{k})\big] \\ &\quad + (1-\rho)^2 \\ &= 2(1-\rho)\boldsymbol{\Gamma}_{n-1}(\mathbf{x};\, \mathbf{k}) + (1-\rho)^2.\end{aligned}$$

Similarly, if $T_n = 0$, we have $D_n = D_{n-1} - \rho$, and $V_n - V_{n-1} = -2\rho\boldsymbol{\Gamma}_{n-1}(\mathbf{x};\, \mathbf{k}) + \rho^2$.

Then given $\mathbf{Z}_n = \mathbf{k}, \mathbf{X}_n = \mathbf{x}$ and $\mathcal{F}_{n-1}$, the expectation of $V_n - V_{n-1}$ is

$$\begin{aligned} E\big[V_n - V_{n-1}|Z_n = k, \mathbf{x}, \mathcal{F}_{n-1}\big] &= \Big[2(1-\rho)\mathbf{\Gamma}_{n-1}(\mathbf{x};\mathbf{k}) + (1-\rho)^2\Big]g(\mathbf{\Gamma}_{n-1})p(\mathbf{k}) \\ &\quad + \Big[-2\rho\mathbf{\Gamma}_{n-1}(\mathbf{x};\mathbf{k}) + \rho^2\Big](1 - g(\mathbf{\Gamma}_{n-1}))p(\mathbf{k}) \\ &= \Big[2(1-\rho)g(\mathbf{\Gamma}_{n-1}) - 2\rho(1 - g(\mathbf{\Gamma}_{n-1}))\Big]\mathbf{\Gamma}_{n-1} + \\ &\quad (1-\rho)^2 g(\mathbf{\Gamma}_{n-1}) + \rho^2(1 - g(\mathbf{\Gamma}_{n-1})), \end{aligned}$$

where $g(\mathbf{\Gamma}_{n-1}) = g(2\mathbf{\Gamma}_{n-1} - (2\rho - 1))$.

Let the above formula be $S(\mathbf{\Gamma}_{n-1})$ and then we have $S(\mathbf{\Gamma}_{n-1}) = E\big[V_n|\mathcal{F}_{n-1}\big] - V_{n-1}$, by the Markov property of $\mathbf{\Gamma}_n$, and thus $S(\mathbf{\Gamma}) = P_\lambda V(\mathbf{\Gamma}) - V(\mathbf{\Gamma})$. To ensure $S(\mathbf{\Gamma})$ is bounded in probability, we only need to show $\mathbf{\Gamma}$ is bounded in probability and this can be proved by the assumption $g(x) \le 0.5$. Thus drift condition (1) is satisfied.

5.2.3 Numerical Simulations

To further evaluate the performance of the proposed procedure, we conducted extensive simulations comparing it with existing methods in different scenarios and also provide the simulated type I errors for testing treatment effect under different designs. To be specific, the proposed method is is denoted as "Proposed" and compared with complete randomization (denoted as "CR") and the covariate-adaptive method using kernel in Ma and Hu (2013) (denoted as "CAR"). For the cases with only discrete covariates, the CARA design suggested by Zhang et al. (2007) is also included in the comparison (denoted as "CARA").

Scenario 1 (Linear Model with Two Discrete Prognostic Covariates) Suppose we have continuous responses $Y_i, i = 1, \ldots, n$, and they follow a linear regression model:

$$Y_i = \mu_1 T_i + \mu_2(1 - T_i) + \beta_1 X_i + \beta_2 Z_{1i} + \beta_3 Z_{2i} + \beta_4 X_i T_i + \varepsilon_i,$$

where treatment indicator $T_i = 0$ or 1, and predictive covariate $X_i = -1$ or 1, prognostic covariates Z_{1i} and Z_{2i} take value -1 or 1 with equal probability 1/2. Let the true values be $\mu_1 = \mu_2 = 0.5, \beta_1 = 1, \beta_2 = 0.5, \beta_3 = -0.5, \beta_4 = -1$, and random errors ε_i are independent and identically distributed with normal distribution $\mathcal{N}(0, 0.5^2)$. The first $n_0 = 20$ subjects are allocated using permuted block design, and the block size is set as 4. In each simulation, the random values of X, Z_1 and Z_2 for all subjects, and the responses of the first n_0 subjects are the same for the CAK method and the CARA method, and then the remaining subjects are randomized based on the mechanisms of different procedures. In the simulation, we set $p = 0.8$ for Efron's biased coin design and equal weights for the CARA method.

The following estimated target allocation proportions are used:

$$\hat{\rho}(X_n = 1) = \frac{\Phi(\hat{Y}_{n-1,1}|X_n = 1)}{\Phi(\hat{Y}_{n-1,1}|X_n = 1) + \Phi(\hat{Y}_{n-1,2}|X_n = 1)},$$

and

$$\hat{\rho}(X_n = -1) = \frac{\Phi(\hat{Y}_{n-1,1}|X_n = -1)}{\Phi(\hat{Y}_{n-1,1}|X_n = -1) + \Phi(\hat{Y}_{n-1,2}|X_n = -1)},$$

where Φ is the cumulative density function of a standard normal distribution and $\hat{Y}_{n-1,j}|X = x, j = 1, 2$, are the estimated response of treatment j for the nth patients with $X_n = x$ using estimated parameters based on the responses and treatment assignments of previous $n-1$ patients with $X_i = x, x = 1, -1, i = 1, \ldots, n-1$.

The following indexes are used to make comparisons of different procedures: (1) $N_{ij}, i = 1, -1, j = 0, 1$ are the numbers of subjects with $X = i$ that are randomized to $T = j$; (2) the probability of randomizing to a better treatment, P_1, the proportion of subjects with $X = 1$ which are randomized to $T = 0$, and P_{-1}, the proportion of subjects with $X = -1$ which are randomized to $T = 1$; (3) $\bar{Y}$ is the average success rate; (4) $Z1_{ij}$ and $Z2_{ij}, i = 1, -1, j = 0, 1$ are the numbers of subjects with $Z_1 = i$ or $Z_2 = i$ that are randomized to $T = j$. Mean and standard deviation of each criterion after m replicates of simulations are provided.

Table 1 summarizes the simulation results in Scenario 1 with sample size $n = 400$, and the number of simulations $m = 5000$. From this table, CR and CAR procedures allocate patients to $T = 0$ or $T = 1$ without consideration of the predictive covariate X, which is within our expectation. In contrast, CARA and the proposed method tends to assign more patients with $X = 1$ to treatment $T = 0$, and patients with $X = -1$ to treatment $T = 1$. From the values of P_1 and P_{-1}, we can see that both CR and CAR randomize patients to each treatment with almost the same probability, while CARA and the proposed method could apply the information of X and allocate them with a high probability to a more suitable treatment. From the success rate $\bar{Y}$, we could see that CR and CAR procedures have success rate around 0.5, while the success rates for the CARA and the proposed method are much higher for patients with either $X = 1$ or $X = -1$, i.e., the proposed method greatly increases the success rates for all patients. From the number of patients in each subgroup of Z_1 and Z_2, we could see that all the four methods could balance Z_1 and Z_2 pretty well. On the whole, the proposed method performs best.

We also consider the case with fixed ρ in Scenario 1 and assume that the estimated target allocation proportions are fixed with the true values of parameters:

$$\hat{\rho}(X_n = 1) = \frac{\Phi(\mu_1 + \beta_1 + \beta_4)}{\Phi(\mu_1 + \beta_1 + \beta_4) + \Phi(\mu_2 + \beta_1)} = 0.426,$$

Table 1 The average results for 4 methods with $n = 400$ and $m = 5000$ in Scenario 1

				Proposed	
Index	CR	CAR	CARA	Non-fixed $\hat{\rho}$	Fixed ρ
N_{11}	100.230(8.599)	99.995(6.976)	86.195(8.565)	85.378(4.683)	85.147(4.607)
N_{-11}	99.795(8.654)	100.008(6.977)	134.675(11.382)	137.338(8.004)	137.451(7.663)
N_{10}	99.996(8.719)	100.001(7.177)	113.827(9.643)	114.784(6.071)	114.615(5.982)
N_{-10}	99.978(8.633)	99.996(7.184)	65.303(9.702)	62.499(4.875)	62.787(4.646)
P_1	0.499(0.035)	0.500(0.025)	0.569(0.038)	0.573(0.009)	0.574(0.009)
P_{-1}	0.500(0.036)	0.500(0.025)	0.673(0.046)	0.687(0.019)	0.686(0.017)
$\bar{Y}$	0.500(0.056)	0.500(0.057)	0.621(0.054)	0.631(0.045)	0.629(0.045)
$Z1_{11}$	99.958(8.618)	100.041(4.975)	110.415(10.047)	111.588(6.185)	111.15(6.298)
$Z1_{-11}$	100.269(8.665)	99.962(4.984)	110.455(8.986)	111.128(6.392)	111.278(6.285)
$Z1_{10}$	99.809(8.501)	100.043(5.013)	89.523(9.641)	88.469(5.490)	88.59(5.375)
$Z1_{-10}$	99.964(8.647)	99.954(5.003)	89.607(8.261)	88.815(5.377)	88.672(5.391)
$Z2_{11}$	100.077(8.623)	99.892(5.086)	110.423(8.788)	111.271(6.596)	111.32(6.379)
$Z2_{-11}$	100.150(8.581)	100.110(5.075)	110.447(10.094)	111.445(6.377)	111.278(6.285)
$Z2_{10}$	99.648(8.476)	99.932(5.058)	89.481(8.212)	88.840(5.406)	88.73(5.325)
$Z2_{-10}$	100.125(8.643)	100.066(5.069)	89.649(9.521)	88.443(5.523)	88.672(5.391)

and

$$\hat{\rho}(X_n = -1) = \frac{\Phi(\mu_1 - \beta_1 - \beta_4)}{\Phi(\mu_1 - \beta_1 - \beta_4) + \Phi(\mu_2 - \beta_1)} = 0.691.$$

From the simulation results in Table 1, the allocation proportion for subjects with $X = 1$ is 0.4265, and 0.6874 for subjects with $X = -1$, which is quite close to the theoretical allocation proportions. We also calculate the marginal and within stratum imbalance, which are presented in Figs. 5, 6, 7. We could see that the within stratum differences for all strata are very stable as the allocation procedure goes on.

Scenario 2 (Linear Model with Two Continuous Prognostic Covariates) Suppose we have continuous responses $Y_i, i = 1, \ldots, n$, and they follow a linear regression model:

$$Y_i = \mu_1 T_i + \mu_2(1 - T_i) + \beta_1 X_i + \beta_2 Z_{1i} + \beta_3 Z_{2i} + \beta_4 X_i T_i + \varepsilon_i,$$

where treatment indicator $T_i = 0$ or 1, predictive covariate $X_i = -1$ or 1, prognostic covariate Z_{1i} follows a standard normal distribution $\mathcal{N}(0, 1)$ and Z_{2i} follow an exponential distribution with parameter $\lambda = 1$. Let the true values be $\mu_1 = \mu_2 = 0.5, \beta_1 = 1, \beta_2 = 0.5, \beta_3 = -0.5, \beta_4 = -1$, and random errors ε_i are i.i.d. with distribution $\mathcal{N}(0, 0.5^2)$.

To measure the imbalance of continuous covariate Z_1 between two treatment groups, the absolute differences between the mean, standard deviation and interquartile range (IQR) of Z_1 in two treatment groups can be applied, and we denote them

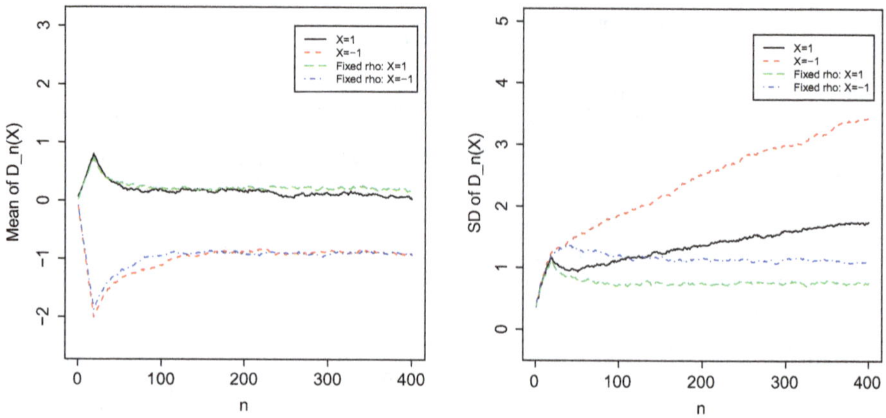

Fig. 5 Average marginal imbalances with $m = 5000$ simulations in Scenario 1

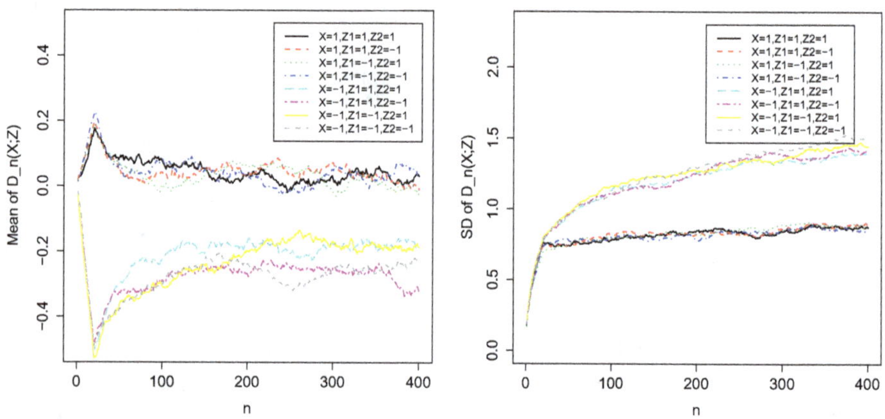

Fig. 6 Average within-stratum imbalances with $m = 5000$ simulations in Scenario 1

as DMZ, DSDZ and DIQRZ, respectively. Mean and standard deviation of each indexes after m replicates of simulations are provided.

In one simulation, the kernel density of Z_1 and Z_2 with $n = 400$ is presented in Fig. 8. We find that the proposed design and CAR balance the distribution of covariates, which perform much better than CR design. The average simulation results for Scenario 2 are presented in Table 2 based on $m = 5000$ simulations. Similar to Scenario 1, Table 2 also shows that CR and CAR procedures allocate patients to two treatments evenly without considering the values of predictive covariate X, while the New method randomizes more patients to the more suitable treatment. From the P_1, P_{-1} and $\bar{Y}$, the new procedure have much higher probabilities to allocate patients to a better treatment than CR and CAR based on individual characters. From the imbalance indexes DMZ, DSDZ and DIQRZ, one can see that CR and

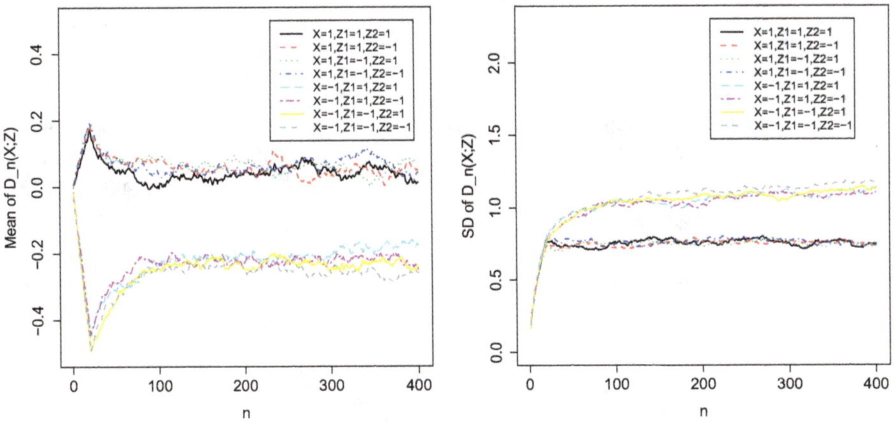

Fig. 7 Average within-stratum imbalances with fixed ρ and $m = 5000$ simulations in Scenario 1

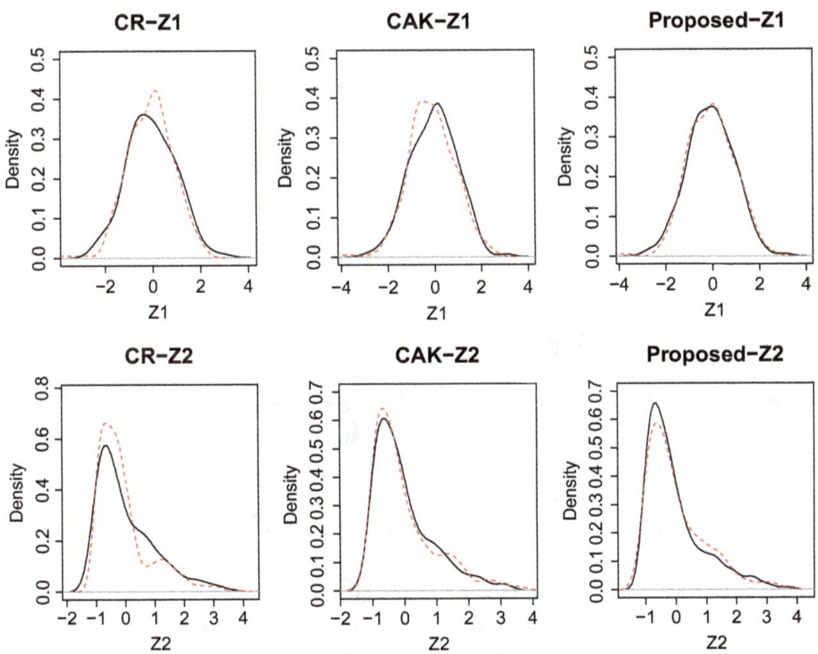

Fig. 8 The kernel density of Z_1 and Z_2 with $n = 400$ in Scenario 2

Table 2 The average results for 3 methods with $n = 400$ and $m = 5000$ in Scenario 2

			Proposed	
Index	CR	CAR	Non-fixed $\hat{\rho}$	Fixed ρ
N_{11}	99.777(8.429)	99.456(10.363)	74.890(4.562)	86.264(4.400)
N_{-11}	100.110(8.518)	99.836(10.184)	148.990(8.945)	135.564(6.893)
N_{10}	99.709(8.507)	100.030(10.240)	124.596(6.972)	114.386(5.773)
N_{-10}	100.404(8.330)	100.678(10.270)	51.524(5.484)	63.786(3.473)
P_1	0.500(0.034)	0.501(0.045)	0.625(0.014)	0.570(0.007)
P_{-1}	0.499(0.034)	0.498(0.045)	0.743(0.025)	0.680(0.008)
$\bar{Y}$	−0.001(0.054)	−0.003(0.055)	0.182(0.051)	0.627(0.049)
DMZ1	0.083(0.061)	0.084(0.061)	0.056(0.042)	0.052(0.039)
DSDZ1	0.057(0.043)	0.031(0.023)	0.046(0.035)	0.042(0.032)
DIQRZ1	0.126(0.097)	0.113(0.089)	0.089(0.068)	0.074(0.057)
DMZ2	0.079(0.059)	0.138(0.102)	0.059(0.046)	0.054(0.040)
DSDZ2	0.109(0.083)	0.068(0.061)	0.101(0.080)	0.096(0.076)
DIQRZ2	0.123(0.094)	0.203(0.145)	0.087(0.067)	0.079(0.064)

CAR methods balance the prognostic covariates pretty well, and the new method performs not worse than them.

Similar to Scenario 1, we also apply the fixed estimated target allocation proportions and the results can be seen in the last column of Table 2.

Scenario 3 (Linear Model with One Continuous and One Discrete Prognostic Covariates) Suppose we have continuous responses $Y_i, i = 1, \ldots, n$, and they follow a linear regression model:

$$Y_i = \mu_1 T_i + \mu_2(1 - T_i) + \beta_1 X_i + \beta_2 Z_{1i} + \beta_3 Z_{2i} + \beta_4 X_i T_i + \varepsilon_i,$$

where treatment indicator $T_i = 0$ or 1, predictive covariate $X_i = -1$ or 1 with equal probability, prognostic covariate Z_{1i} follows a standard normal distribution $\mathcal{N}(0, 1)$ and Z_{2i} takes value -1 or 1 with equal probability 0.5. Let the true values be $\mu_1 = \mu_2 = 0.5, \beta_1 = 1, \beta_2 = 0.5, \beta_3 = -0.5, \beta_4 = -1$, and random errors ε_i are i.i.d. with distribution $\mathcal{N}(0, 0.5^2)$. Table 3 presents the simulation results in Scenario 3 and we could obtain similar conclusions for the superiority of the proposed method over CR and CAR methods on the patient selection. Even though both continuous and discrete prognostic covariates are included in the model, the proposed procedure could balance both types very well, just as CR and CAR methods.

Scenario 4 (Logistic Model with One Continuous and One Discrete Prognostic Covariates) Suppose we have binary responses $Y_i, i = 1, \ldots, n$, and they follow a logistic regression model:

$$logit P(Y_i = 1) = \mu_1 T_i + \mu_2(1 - T_i) + \beta_1 X_i + \beta_2 Z_{1i} + \beta_3 Z_{2i} + \beta_4 X_i T_i, i = 1, \ldots, n,$$

Table 3 The average results for 3 methods with $n = 400$ and $m = 5000$ in Scenario 3

	CR	CAR	Proposed
N_{11}	100.001(8.733)	100.038(7.074)	85.334(4.646)
N_{-11}	100.046(8.592)	99.982(7.076)	137.171(7.973)
N_{10}	100.009(8.578)	99.972(7.109)	114.676(6.178)
N_{-10}	99.945(8.754)	100.009(7.111)	62.819(5.231)
P_1	0.500(0.035)	0.500(0.025)	0.573(0.010)
P_{-1}	0.500(0.035)	0.500(0.025)	0.686(0.020)
$\bar{Y}$	0.500(0.056)	0.500(0.055)	0.629(0.050)
DMZ1	0.079(0.060)	0.081(0.061)	0.047(0.036)
DSDZ1	0.056(0.043)	0.018(0.017)	0.038(0.029)
DIQRZ1	0.123(0.094)	0.111(0.085)	0.068(0.052)
$Z2_{11}$	100.177(8.694)	100.044(5.049)	111.312(6.425)
$Z2_{-11}$	99.869(8.552)	99.975(5.045)	111.193(6.301)
$Z2_{10}$	99.892(8.595)	100.025(5.046)	88.758(5.336)
$Z2_{-10}$	100.061(8.665)	99.955(5.049)	88.738(5.424)

where treatment indicator $T_i = 0$ or 1, predictive covariate $X_i = -1$ or 1 with equal probability, prognostic covariate Z_{1i} follows a standard normal distribution $\mathcal{N}(0, 1)$ and Z_{2i} takes value -1 or 1 with equal probability 0.5. Let the true values be $\mu_1 = \mu_2 = 0.5$, $\beta_1 = 1$, $\beta_2 = 0.5$, $\beta_3 = -0.5$, $\beta_4 = -2$. We use the optimal allocations proposed by Rosenberger et al. (2001), which minimize the expected number of treatment failures for a fixed asymptotic variance:

$$\hat{\rho}(X = 1) = \frac{\sqrt{P(Y = 1|T = 1, X = 1)}}{\sqrt{P(Y = 1|T = 1, X = 1)} + \sqrt{P(Y = 1|T = 0, X = 1)}},$$

and

$$\hat{\rho}(X = -1) = \frac{\sqrt{P(Y = 1|T = 1, X = -1)}}{\sqrt{P(Y = 1|T = 1, X = 1)} + \sqrt{P(Y = 1|T = 0, X = -1)}}.$$

The simulation results in Scenario 4 are summarized in Table 4. In the situations with binary responses, the proposed method could achieve a much higher success rate $\bar{Y}$ than the CR and CAR methods besides the good performance on the balance of Z_1 and Z_2. Thus it meets the two desirable design properties as we mentioned.

Scenario 5: Simulation for type I errors.

In this part, we consider the simulation for type I errors in the testing of treatment effect using different designs. To be more specific, we will test the null hypothesis $H_0 : \mu_1 = \mu_2$. The simulation data comes from four true models and different working models are included to show the performance in hypothesis testing at the 0.05 nominal level. Assume that the predictive covariate $X_i = -1$ or 1 with equal probability, prognostic covariate Z_{1i} follows a standard normal distribution $\mathcal{N}(0, 1)$

Table 4 The average results for 3 methods with $n = 400$ and $m = 5000$ in Scenario 4

	CR	CAR	Proposed
N_{11}	100.075(8.452)	99.829(7.090)	52.262(5.708)
N_{-11}	99.928(8.994)	100.214(7.086)	147.551(9.414)
N_{10}	99.986(8.879)	100.232(7.167)	147.799(9.131)
N_{-10}	100.011(8.885)	99.725(7.196)	52.388(5.374)
P_1	0.500(0.035)	0.501(0.025)	0.739(0.025)
P_{-1}	0.500(0.036)	0.501(0.025)	0.738(0.025)
$\bar{Y}$	0.500(0.067)	0.505(0.066)	0.976(0.063)
DMZ1	0.080(0.060)	0.078(0.058)	0.063(0.045)
DSDZ1	0.056(0.044)	0.018(0.016)	0.051(0.039)
DIQRZ1	0.124(0.095)	0.109(0.085)	0.098(0.076)
$Z2_{11}$	100.033(8.664)	99.818(5.085)	99.690(7.118)
$Z2_{-11}$	99.970(8.717)	100.225(5.020)	100.123(6.887)
$Z2_{10}$	99.560(8.532)	99.775(5.008)	99.903(6.933)
$Z2_{-10}$	100.437(8.676)	100.182(5.077)	100.284(7.111)

and Z_{2i} takes value -1 or 1 with equal probability 0.5, and random errors ε_i are i.i.d. with distribution $\mathcal{N}(0, 0.5^2)$.

As we can see from results with $m = 5000$ replicates in Table 5, when the working model is exactly the same as true model, the simulated type I errors are around 0.05 for these three designs. However, the type I errors seem to be conservative under CAR and the proposed design when some covariates are ignored in the working model, while CR method could retain the correct type I error. We obtain similar conclusion when the working models are mis-specified. The conservative results on type I errors for CAR and the proposed CARA design are consistent with the findings in some published papers and we could consider the adjustment for the variance of treatment effect similarly (Li et al., 2021b; Ma et al., 2015, 2020).

6 Discussion

Precision medicine is an ideal medical model and has become an important part of healthcare. Technical developments allow scientists to efficiently identify biomarkers that are associated with certain diseases and can be incorporated into clinical trials to select or optimize therapeutic care. Based on the different roles played in clinical trials, we classify the covariates of patients into prognostic covariates for balancing purposes and predictive covariates for treatment selection. Many innovative adaptive randomization designs have been proposed to handle covriates and thus promote precision medicine. In this chapter, we provide an

Table 5 The simulated type I errors (%) for 3 designs with $n = 200$ and $m = 5000$

Assumed working model	CR	CAR	Proposed
True model: $Y_i = 0.5T_i + 0.5(1 - T_i) + X_i + 0.5Z_{1i} - 0.5Z_{2i} - X_iT_i + \varepsilon_i$			
$Y_i = \mu_1 T_i + \mu_2(1 - T_i) + \beta_1 X_i + \beta_2 Z_{1i} + \beta_3 Z_{2i} + \beta_4 X_i T_i$	5.44	5.14	4.88
$Y_i = \mu_1 T_i + \mu_2(1 - T_i) + \beta_1 X_i + \beta_4 X_i T_i$	5.16	1.80	0.48
True model: $logit\,P(Y_i = 1) = 0.5T_i + 0.5(1 - T_i) + X_i + 0.5Z_{1i} - 0.5Z_{2i} - 2X_iT_i$			
$logit\,P(Y_i = 1) = \mu_1 T_i + \mu_2(1 - T_i) + \beta_1 X_i + \beta_2 Z_{1i} + \beta_3 Z_{2i} + \beta_4 X_i T_i$	5.06	5.08	5.22
$logit\,P(Y_i = 1) = \mu_1 T_i + \mu_2(1 - T_i) + \beta_1 X_i + \beta_4 X_i T_i$	5.02	4.74	4.66
True model: $Y_i = 0.5T_i + 0.5(1 - T_i) + 0.5Z_{1i} - 0.5Z_{2i} - X_iT_i + \varepsilon_i$			
$Y_i = \mu_1 T_i + \mu_2(1 - T_i) + \beta_1 X_i + \beta_2 Z_{1i} + \beta_3 Z_{2i} + \beta_4 X_i T_i$	4.90	4.84	5.04
$Y_i = \mu_1 T_i + \mu_2(1 - T_i) + \beta_1 X_i + \beta_4 X_i T_i$	5.18	1.40	0.48
True model: $Y_i = 0.5T_i + 0.5(1 - T_i) + 0.5Z_{1i}^2 - 0.5Z_{2i} - X_iT_i + \varepsilon_i$			
$Y_i = \mu_1 T_i + \mu_2(1 - T_i) + \beta_1 X_i + \beta_2 Z_{1i} + \beta_3 Z_{2i} + \beta_4 X_i T_i$	5.04	0.24	1.64
$Y_i = \mu_1 T_i + \mu_2(1 - T_i) + \beta_1 X_i + \beta_4 X_i T_i$	5.04	0.02	0.76

overview of the mathematical framework of adaptive randomizations and review some popular innovative designs under this framework.

We have reviewed the covariate-adaptive designs which only balance prognostic covariates, and covariate-adjusted response-adaptive designs which only deal with predictive covariates for reasons of efficiency and ethics. We also propose an new family of covariate-adjusted response-adaptive designs that enable us to balance both discrete and continuous prognostic covariates, and to allocate patients to the most suitable treatment simultaneously. When there is no prognostic covariate or no predictive covariate, the procedure can still proceed based the objectives of clinical trials, and it's simplified as covariate-adaptive randomization or response-adaptive randomization, thus it can be considered as a generalization of classical covariate-adaptive design and response-adaptive design.

One limitation of the proposed design is that it requires patients to be randomized sequentially and obtain the treatment outcome before the randomization of next patient, which could be relaxed to allow patients to be enrolled in batches and use the statistical methods in Hu and Rosenberger (2006) for delayed response. Another one is that covariates and their impact on treatment effect should be identified before the conduction of clinical trials in order to implement the procedure. If certain influential covariates are unknown in advance, directly applying the proposed method might not be the most efficient way and we may need additional steps to identify them.

The theoretical properties of the proposed design in general situation are still under development due to the complexity of sequentially updating the allocation proportion and the non-parametric kernel method. This will be an interesting research topic to be investigated. Also statistical inference part has been studied for adaptive designs and we showed that the type I errors for the proposed covariate-adjusted response-adaptive designs are conservative, which is similar to the case for covariate-adaptive designs. The adjustment for conservativeness is still an open question and worth for further investigation.

References

Aletti, G., Ghiglietti, A., & Rosenberger, W. F. (2018). Nonparametric covariate-adjusted response-adaptive design based on a functional urn model. *The Annals of Statistics, 46*(6B), 3838–3866.

Baker, S., Ten. Tije, A, Carducci, M., Gelderblom, H., Dawkins, F., McGuire, W., et al. (2004). Evaluation of cyp3a activity as a predictive covariate for docetaxel clearance. *Journal of Clinical Oncology, 22*(14_{suppl}), 2006–2006.

Bandyopadhyay, A., & Biswas, A. (2001). Adaptive designs for normal responses with prognostic factors. *Biometrika, 88*(2), 409–419.

Biswas, A., & Bhattacharya, R. (2018). A class of covariate—adjusted response—adaptive allocation designs for multitreatment binary response trials. *Journal of Biopharmaceutical Statistics, 28*(5), 809–823.

Bugni, F. A., Canay, I. A., & Shaikh, A. M. (2018). Inference under covariate-adaptive randomization. *Journal of the American Statistical Association, 113*, 1784–1796.

Buyse, G. M., Goemans, N., Van den Hauwe, M., Thijs, D., de Groot, I. J., Schara, U., Ceulemans, B., Meier, T., & Mertens, L. (2011). Idebenone as a novel, therapeutic approach for duchenne muscular dystrophy: Results from a 12 month, double-blind, randomized placebo-controlled trial. *Neuromuscular Disorders, 21*, 396–405.

Cheung, S. H., Zhang, L. X., Hu, F., & Chan, W. S. (2014). Covariate-adjusted response- adaptive designs for generalized linear models. *Journal of Statistical Planning and Inference, 149*, 152–161.

Chu, Y., & Yuan, Y. (2018). A Bayesian basket trial design using a calibrated Bayesian hierarchical model. *Clinical Trials, 15*(2), 149–158.

Fakhry, F., Spronk, S., van der Laan, L., Wever, J. J., Teijink, J. A., Hoffmann, W. H., Smits, T. M., van Brussel, J. P., Stultiens, G. N., Derom, A., den Hoed, P. T., Ho, G. H., van Dijk, L. C., Verhofstad, N., Orsini, M., van Petersen, A., Woltman, K., Hulst, I., van Sambeek, M. R., Rizopoulos, D., Rouwet, E. V., & Hunink, M. G. (2015). Endovascular revascularization and supervised exercise for peripheral artery disease and intermittent claudication: a randomized clinical trial. *Journal of the American Statistical Association, 314*, 1936–1944.

Ferrarotto, R., Redman, M. W., Gandara, D. R., Herbst, R. S., Papadimitrakopoulou, V. A. (2015). Lung–MAP–framework, overview, and design principles. *Chinese Clinical Oncology, 4*(3), 36.

Frane, J. W. (1998). A method of biased coin randomization, its implementation, and its validation. *Drug Information Journal, 32*, 423–432.

Fukada, A., Kitagawa, Y., Matsuoka, M., Sakai, J., et al. (2021). Presepsin as a predictive biomarker of severity in COVID-19: A case series. *Journal of Medical Virology, 93*(1), 99–101.

Garde-Noguera, J., Martin-Martorell, P., De, Julin M., et al. (2018). Predictive and prognostic clinical and pathological factors of nivolumab efficacy in non-small–cell lung cancer patients. *Clinical & Translational Oncology, 20*(8), 1072–1079.

Gibney, G. T., Weiner, L. M., & Atkins, M. B., (2016). Predictive biomarkers for checkpoint inhibitor-based immunotherapy. *The Lancet Oncology, 17*(12), e542–e551.

Hoehler, F. K. (1987). Balancing allocation of subjects in biomedical research: a minimization strategy based on ranks. *Computers and Biomedical Research, 20*, 209–213.

Hu, F., & Rosenberger, W. (2006). *The theory of response-adaptive randomization in clinical trials*. New York: Wiley and Sons.

Hu, F., Hu, Y., Ma, W., Zhang, L., & Zhu, H. (2015a). Statistical inference of adaptive randomized clinical trials for personalized medicine. *Clinical Investigation, 5*, 415–425.

Hu, F., & Zhang, L. (2020). On the theory of covariate-adaptive designs. arXiv:2004.02994.

Hu, J., Zhu, H., & Hu, F. (2015b). A unified family of covariate-adjusted response-adaptive designs based on efficiency and ethics. *Journal of the American Statistical Association, 110*, 357–367.

Huang, T., Liu, Z., & Hu, F. (2013). Longitudinal covariate-adjusted response-adaptive randomization design. *Journal of Statistical Planning and Inference, 143*, 1816–1827.

Hu, Y., & Hu, F. (2012a). Balancing treatment allocation over continuous covariates: a new imbalance measure for minimization. *Journal of Probability and Statistics* (2012). https://doi.org/10.1155/2012/842369.

Hu, Y., & Hu, F. (2012b). Asymptotic properties of covariate-adaptive randomization. *The Annals of Statistics, 40*, 1794–1815.

Hu, F., Hu, Y., Ma, Z., & Rosenberger, W. F. (2014). Adaptive Randomization for balancing covariates. *Wires Computational Statistics, 6*(4), 288–303.

Hyman, D. M., Puzanov, I., Subbiah, V., Faris, J. E., et al. (2015). Vemurafenib in multiple nonmelanoma cancers with BRAF V600 mutations. *The New England Journal of Medicine, 373*, 726–736.

Jiang, L., Nie, L., Yan, F., & Yuan, Y. (2021). Optimal Bayesian hierarchical model to accelerate the development of tissue-agnostic drugs and basket trials. *Contemporary Clinical Trials, 107*, 106460.

Krisam, J., & Kieser, M. (2015). Optimal decision rules for biomarker-based subgroup selection for a targeted therapy in oncology. *International Journal of Molecular Sciences, 16*, 10354–10375.

Kundt, G. (2009). Comparative evaluation of balancing properties of stratified randomization procedures. *Methods of Information in Medicine, 48*, 129–134.

Lebowitsch, J., Ge, Y., Young, B., & Hu, F. (2012). Generalized multidimensional dynamic allocation method. *Statistics in Medicine, 31*, 3537–3544.

Li, J., Zhao, L., Tian, L., Cai, T., Claggett, B., Callegaro, A., Dizier, B., Spiessens, B., Ulloa-Montoya, F., & Wei, L. J. (2016). A predictive enrichment procedure to identify potential responders to a new therapy for randomized, comparative controlled clinical studies. *Biometrics, 72*, 877–887.

Li, X., Zhou, J., & Hu, F. (2019). Testing hypotheses under adaptive randomization with continuous covariates in clinical trials. *Statistical Methods in Medical Research, 28*, 1609–1621.

Li, X., Ma, W., & Hu, F. (2021a). Sample size re-estimation for covariate-adaptive randomized clinical trials. *Statistics in Medicine, 40*(12), 2839–2858.

Li, Y., Ma, W., Qin, Y., & Hu, F. (2021b). Testing for treatment effect in covariate-adaptive randomized trials with generalized linear models and omitted covariates. *Statistical Methods in Medical Research, 30*(9), 2148–2164.

Lin, Y., & Su, Z. (2012) Balancing continuous and categorical baseline covariates in sequential clinical trials using the area between empirical cumulative distribution functions. *Statistics in Medicine, 31*, 1961–1971.

Liu, Y., & Hu, F. (2020). Balancing unobserved covariates with covariate-adaptive randomized experiments. *Journal of the American Statistical Association*. https://doi.org/10.1080/01621459.2020.1825450.

Liu, Z., Yin, J., & Hu, F. (2015). Covariate-adaptive designs with missing covariates in clinical trials. *Science in China, Series A, 58*, 1191–1202.

Liu, Z., Ban, T., & Huang, T. (2017). General covariate-adaptive randomization targeting unequal allocation ratio. *Journal of Statistical Planning and Inference, 191*, 68–80.

Lopez-Chavez, A., Thomas, A., Rajan, A., Raffeld, M, Morrow, B., Kelly, R., Carter, C.A., Guha, U., Killian, K., Lau, C.C., Abdullaev, Z., Xi, L., Pack, S., Meltzer, P.S., Corless, C.L.,Sandler, A., Beadling, C., Warrick, A.,Liewher, D.J., Steinberg, S.M., Berman, A., Doyle, A., Szabo, E., Wang, Y., & Giaccone, G. (2015). Molecular profiling and targeted therapy for advanced thoracic malignancies: a biomarker–derived, multiarm, multihistology phase II basket trial. *Journal of Clinical Oncology, 33*, 1000–1007.

Ma, W., Hu, F., & Zhang, L. (2015). Testing hypotheses of covariate-adaptive randomized clinical trials. *Journal of the American Statistical Association, 110*, 669–680.

Ma, W., Qin, Y., Li, Y., & Hu, F. (2020). Statistical inference for covariate-adaptive randomization procedures. *Journal of the American Statistical Association, 115*(531), 1488–1497.

Ma, Z., & Hu, F. (2013). Balancing continuous covariates based on Kernel densities. *Contemporary Clinical Trials, 34*, 262–269.

Meyn, S., & Tweedie, R. (2009). *Markov chains and stochastic stability* (2nd edn.). Cambridge University Press.

Miyawaki, E., Murakami, H., Mori, K., Mamesaya, N., Kawamura, T., Kobayashi, H., Omori, S., Wakuda, K., Ono, A., Kenmotsu, H., Naito, T., & Takahashi, T. (2020) PD-L1 expression and response to pembrolizumab in patients with EGFR-mutant non-small cell lung cancer. *Japanese Journal of Clinical Oncology, 50*(5), 617–622.

NCI-Molecular Analysis for Therapy Choice (NCI-MATCH) Trial (2017). http://www.cancer.gov/about-cancer/treatment/clinical-trials/nci-supported/nci-match.

Ning, J., & Huang, X. (2010). Response-adaptive randomization for clinical trials with adjustment for covariate imbalance. *Statistics in Medicine, 29*, 1761–1768.

Nishi, T., & Takaichi, A. (2003). An extended minimization method to assure similar means of continuous prognostic variables between treatment groups. *Japanese Journal of Biometrics, 24*, 43–55.

Oneal, P. A., Kwitkowski, V., Luo, L., Shen, Y. L., Subramaniam, S., Shord, S., Goldberg, K. B., McKee, A. E., Kaminskas, E., Farrell, A., & Pazdur, R. (2018). FDA approval summary: Vemurafenib for the treatment of patients with erdheim-chester disease with the BRAFV600 mutation. *The Oncologist, 23*(12), 1520–1524.

Park, J. W., Liu, M. C., Yee, D., Yau, C., vant Veer, L. J., Symmans, W. F., et al. (2016). Adaptive randomization of Neratinib in early breast cancer. *New England Journal of Medicine, 375*, 11–22.

Park, J. J. H., Siden, E., Zoratti, M. J., Dron, L., Harari, O., Singer, J., Lester, R. T., Thorlund, K., Mills, E. J. (2019). Systematic review of basket trials, umbrella trials, and platform trials: a landscape analysis of master protocols. *Trials, 20*(1), 572.

Pocock, S., & Simon, R. (1975). Sequential treatment assignment with balancing for prognostic factors in the controlled clinical trial. *Biometrics, 3*, 103–115.

Pocock, S. J. (1982). Statistical aspects of clinical trial design. *Statistician, 31*, 1–18.

Qin, Y., Li, Y., Ma, W., & Hu, F. (2024). An optimal method for covariate balancing and its properties. *Statistica Sinica*, To appear.

Rosenberger, W., Vidyashankar, A., & Agarwal, D. (2001). Covariate-adjusted response-adaptive designs for binary response. *Journal of Biopharmaceutical Statistics, 11*, 227–236.

Shao, J., Yu, X., & Zhong, B. (2010). A theory for testing hypotheses under covariate-adaptive randomization. *Biometrika, 97*, 347–360.

Shao, J., & Yu, X. (2013). Validity of tests under covariate-adaptive biased coin randomization and generalized linear models. *Biometrics, 69*, 960–969.

Stigsby, B., & Taves, D. R. (2010). Rank-minimization for balanced assignment of subject in clinical trials. *Contemporary Clinical Trials, 31*, 147–150.

Stott, D. J., Rodondi, N., Kearney, P. M., Ford, I., Westendorp, R. G., et al. (2017). Thyroid hormone therapy for older adults with subclinical hypothyroidism. *New England Journal of Medicine, 376*, 2534–2544.

Su, Z. (2011) Balancing multiple baseline characteristics in randomized clinical trials. *Contemporary Clinical Trials, 32*, 547–550.

Sun, J. M., Lee, K. H., Kim, B. S., Kim, H.G., Min, Y. J., Yi, S. Y., Yun., H. J., Jung, S. H., Lee, S. H., Ahn, J. S., Park, K., & Ahn, M. J. (2018). Pazopanib maintenance after first-line etoposide and platinum chemotherapy in patients with extensive disease small-cell lung cancer: a multicentre, randomised, placebo-controlled phase II study (kcsg-lu12-07). *British Journal of Cancer, 118*, 648–653.

Sverdlov, O., Rosenberger, W., & Ryeznik, Y. (2013). Utility of covariate-adjusted response-adaptive randomization in survival trials. *Statistics in Biopharmaceutical Research, 5*(1), 38–53.

Sydes, M. R., Spears, M. R., Mason, M. D., Clarke, N. W., Dearnaley, D. P., de Bono, J. S., et al. (2018). Adding abiraterone or docetaxel to long-term hormone therapy for prostate cancer: directly randomised data from the STAMPEDE multi-arm, multi-stage platform protocol. *Annals of Oncology, 29*, 1235–48.

Taves, D. R. (1974). Minimization: a new method of assigning patients to treatment and control groups. *Clinical Pharmacology & Therapeutics, 15*, 443–453.

U.S. Food and Drug Administration. (2017). FDA grants accelerated approval to pembrolizumab for first tissue/site agnostic indication. Available at https://www.fda.gov/drugs/resources-information-approved-drugs/fda-grants-accelerated-approval-pembrolizumab-first-tissuesite-agnostic-indication.

U.S. Food and Drug Administration. (2018). FDA approves larotrectinib for solid tumors with NTRK gene fusions. Available at https://www.fda.gov/drugs/fda-approves-larotrectinib-solid-tumors-ntrk-gene-fusions.

U.S. Food and Drug Administration. (2019). Guidance for industry: Enrichment strategies for clinical trials to support determination of effectiveness of human drugs and biological products. https://www.fda.gov/media/121320/download.

U.S. Food and Drug Administration. (2022). Guidance for industry: Interacting with the FDA on Complex Innovative Trial Designs for Drugs and Biological Products. Available at https://www.fda.gov/media/130897/download.

U.S. Food and Drug Administration. (2020). Guidance for industry: Master protocols: efficient clinical trial design strategies to expedite development of oncology drugs and biologics. Available at https://www.fda.gov/media/78495/download.

Wang, S. J., & Hung, H. M. (2013). Adaptive enrichment with subpopulation selection at interim: Methodologies, applications and design considerations. *Contemporary Clinical Trials, 36*, 673–681.

Weir, C. J., & Lees, K. R. (2003). Comparison of stratification and adaptive methods for treatment allocation in an acute stroke clinical trial. *Statistics in Medicine, 22*, 705–726.

Wong, C. H., Siah, K. W., Lo, A. W. (2019). Estimation of clinical trial success rates and related parameters. *Biostatistics, 20*(2), 273–286.

Ye, T., & Shao, J. (2020). Robust tests for treatment effect in survival analysis under covariate-adaptive randomization. *Journal of the Royal Statistical Society: Series B, 82*(5), 1301–1323.

Yuan, Y., Huang, X., & Liu, S. (2011). A bayesian response-adaptive covariate-balanced randomization design with application to a leukemia clinical trial. *Statistics in Medicine, 30*, 1218–1229.

Zhang, L., Hu, F., Cheung, S. H., & Chan, W. S. (2007). Asymptotic properties of covariate-adjusted response-adaptive designs. *The Annals of Statistics, 35*(3), 1166–1182.

Zhang, L., & Hu, F. (2009). A new family of covariate-adjusted response adaptive designs and their asymptotic properties. *Applied Mathematics–A Journal of Chinese Universities, 24*, 1–13.

Zhao, W., Ma, W., Wang, F., & Hu, F. (2022). Incorporating covariates information in adaptive clinical trials for precision medicine. *Pharmaceutical Statistics, 21*, 176–195.

Zhou, H., Lee, J. J., & Yuan, Y. (2017). BOP2: Bayesian optimal design for phase II clinical trials with simple and complex endpoints. *Statistics in Medicine, 36*(21), 3302–3314.

Zhu, H., Hu, F., & Zhao, H. (2013). Adaptive clinical trial designs to detect interaction between treatment and a dichotomous biomarker. *Canadian Journal of Statistics, 41*(3), 525–539.

Zhu, H., & Hu, F. (2019). Sequential monitoring of covariate-adaptive randomized clinical trials. *Statistica Sinica, 29*, 265–282.

Maximum Likelihood Estimation and Design and Inference Considerations for Sequential Multiple Assignment Randomized Trials

Robert A. Tumasian III and Bin Cheng

Abstract This chapter provides a discussion of the sequential multiple assignment randomized trial (SMART) design, its operational characteristics and advantages compared to alternative design options, and details on the statistical analyses associated with SMARTs, with motivating examples. Inference and sample size calculation in the SMART setting are addressed, focusing on the maximum likelihood estimation (MLE) approach, which helps to reveal some unique properties in statistical inference in SMARTs and is generalizable to other semi-parametric estimation methods. The chapter ends with final remarks about the benefits of our proposed methods and closing comments on the utility of SMARTs.

Keywords Adaptive intervention · Sequential multiple assignment randomized trial · Gate-keeping approach · Multiple comparison · Sample size calculation

1 Introduction and Rationale

Traditional clinical practice consists of an initial diagnosis and choosing a matching intervention based on clinical judgment (Bierman et al., 2006). However, this approach has become antiquated due to several biases (Arkes, 1981; Bierman et al., 2006; Garb, 1998) and the growing real-world evidence and precision medicine paradigms. Clinical decision-making has become largely data-driven, which has had numerous positive implications, including the ability to account for the complexities of patient-level differences. In clinical research, this translates to investigating individualized interventions based on patient demographics and other characteristics beyond disease status. In standard randomized controlled trials (RCTs), patients are randomly assigned to a particular treatment and followed for a pre-specified amount of time to determine if the active treatment outperforms the control based

R. A. Tumasian III · B. Cheng (✉)
Department of Biostatistics, Columbia University, New York, NY, USA
e-mail: bc2159@cumc.columbia.edu

Y. Zhao, D.-G. Chen (eds.), *Statistics in Precision Health*, ICSA Book Series in Statistics, https://doi.org/10.1007/978-3-031-50690-1_15

on a primary endpoint measure. However, in many settings, treatment responses can be highly heterogeneous among patients and single interventions may become ineffective over time (e.g., when treating chronic diseases), which has led to the incorporation of secondary subgroup analyses into RCTs to gain insights into individualized treatment decision rules and to the development (Almirall et al., 2012; Collins et al., 2007) and broad application (Kasari et al., 2014; Tamura et al., 2016) of sequential multiple assignment randomized trials (SMARTs) across numerous therapeutic areas.

SMARTs provide an innovative design comprised of a collection of decision rules, or adaptive interventions (AIs). An AI is a series of stage-wise treatment decisions that are informed by a patient's historical and ongoing clinical information, including history of treatment and interim treatment responses. In a SMART, participants' prior medical information and accrued data are used to inform treatment selection for future patients in a manner that maximizes the outcome for every patient (Zhong et al., 2019). A SMART is often utilized to determine if an AI should be investigated further in a phase III trial (Murphy, 2005).

AIs have commonly been used for treating chronic diseases, such as depression. An example of this is the CODIACS depression management trial, which assigned patients to receive either a medication or problem-solving therapy at baseline, with the possibility of patients changing to a different treatment depending on an intermediate outcome measure in order to identify the optimal, or highest performing, AI (Cheung et al., 2015). The primary objective of the study was to maximize depression reduction over 6 months using the Beck Depression Inventory. We will revisit the CODIACS trial in Sect. 2.4.

There are various estimation methods available for identifying the optimal AI embedded in a SMART. One example is G-computation (Robins, 1986), which uses a counterfactual framework to obtain unbiased treatment effect estimates for each AI. Specifically, a model is first selected to regress the outcome of interest on the AIs, adjusting for important covariates, to predict the outcomes for each subject under every possible AI. Then, these predictions, or counterfactuals, are regressed on the AIs to extract the marginal treatment effects for each AI (Snowden et al., 2011).

Another technique is to use inverse probability weighting (IPW) (Murphy et al., 2001), which employs logistic regression to estimate the probability of a subject being assigned to the AIs based on measured covariates. Then, the inverse of the estimate is used as a weight in downstream analyses to account for confounding, loss to follow-up, and restricted randomization (e.g., non-responders are underrepresented in some AI comparisons if they have been re-randomized, while responders are not). In SMARTs, this probability is usually known, but estimating it can provide efficiency gains. More estimation options can be found in Wallace and Moodie (2014).

In contrast to G-computation and IPW, the MLE approach outlined in this chapter is not a model-based procedure (avoids model misspecification that could lead to substantial bias) and relies on less strong statistical assumptions. The MLE method is also very beneficial since it produces estimates whose asymptotic distribution can be easily derived and understood, and yields a closed form asymptotic variance so that the contribution of each parameter is easier to deduce, enabling better understanding of how each parameter affects the asymptotic variance and hence the sample size determination.

By comparing all of the estimated AI values, adjusting for multiple pairwise comparisons, the optimal AI can be selected. When protecting against a high false positive (type I error) rate in RCTs, the most popular adjustment procedure for multiple comparisons is Bonferroni's correction, which is known to be conservative as the number of embedded AIs increases (Hochberg & Ajit, 1987; Zhong et al., 2019). Most adjustment techniques used to compare non-adaptive interventions, including those developed by Hsu (1981) and Tukey (1949), require complete specification of the correlation structure between the estimated treatment effects. However, in the SMART setting, these traditional methods cannot be implemented, since the treatment effect estimates are often unknown a priori (Zhong et al., 2019). In Sect. 2.4, we describe a gate-keeping approach that uses an estimate of the correlation structure that incorporates the SMART design.

The SMART design has many advantages over the general RCT and group sequential (GS) designs. Most notably, SMARTs promote accelerated evaluation, which can expedite treatment delivery. The use of AIs in SMARTs allows investigators to more quickly help patients who are not responding or adhering well to a particular treatment. Unlike in a standard RCT or GS design, patients are able to move to a different treatment in a SMART, given that they experience a pre-specified intermediate outcome suggesting treatment inadequacy. This is beneficial for both researchers and patients, as it conveys important insights into treatment efficacy over time and provides patients access to a potentially more effective intervention. SMARTs have the unique ability to identify an optimal decision rule, or AI, based on simultaneously testing sequences of stage-wise factors (e.g., administration method, dosage, and/or length of treatment) that may impact treatment efficacy, rather than running separate RCTs or conducting separate tests at each stage in a GS design (Lei et al., 2012).

Finally, although SMARTs were initially proposed for estimating and identifying AIs, small sample or "small n" SMARTs (snSMARTs) have also been developed to investigate treatments for rare diseases (Tamura et al., 2016), which are geared toward strengthening inference around the first stage treatments (Fang et al., 2020; Wei et al., 2018). Furthermore, the idea of a SMART with adaptive randomization (SMART-AR) has been explored for circumstances when recruitment is staggered or spread across time. In a SMART-AR, accrued data is used to determine randomization probabilities for future participants and more informative inferences can be made at later stages when the most participants are enrolled (Cheung et al., 2015). These extensions of the general SMART are important but beyond the scope of this chapter, which will focus on the traditional usage of SMARTs for AI inference using maximum likelihood estimation. Future work could investigate likelihood-based methods for situations in which a snSMART or SMART-AR may be suitable.

2 Inference for SMARTs

This section begins by presenting the general SMART framework and specifying notation, and then provides an overview of maximum likelihood estimation (MLE) in the SMART setting. This is followed by the construction of a Wald-type statistic for inference, using its asymptotic properties. The section closes with a discussion of the gate-keeping and simultaneous confidence interval approaches for selecting the optimal AI in a SMART.

2.1 Design Setup

For simplicity, we will consider the standard two-stage SMART design provided in Fig. 1. However, all methodology reported in this chapter, which is adapted from Zhong et al. (2021, 2019), can be extended to SMARTs with three or more stages, as appropriate.

Suppose there are I treatment options $a_1, \ldots, a_I$ at stage 1 and under each treatment a_i, there are J_i possible intermediate outcome categories, denoted by $o_{i1}, \ldots, o_{iJ_i}$ with $i = 1, \ldots, I$ and $j = 1, \ldots, J_i$. Further, suppose a patient receiving treatment a_i at stage 1 with intermediate outcome o_{ij} has K_{ij} treatment options at stage two, represented by $b_{ij1}, \ldots, b_{ijK_{ij}}$ with $k = 1, \ldots, K_{ij}$.

Next, let V_l, X_l, W_l, and Y_l represent the stage one treatment, intermediate outcome, stage two treatment, and primary outcome, respectively, for patient l, where $l = 1, \ldots, n$ and n denotes the total sample size of the SMART. Now, suppose $\pi_i = \Pr(V_l = a_i)$ is the randomization probability of assigning patient l to treatment a_i at stage one, and $\pi_{ijk} = \Pr(W_l = b_{ijk} | V_l = a_i, X_l = o_{ij})$ is the randomization probability of assigning patient l to treatment b_{ijk} at stage two given the clinical history of patient l $(V_l = a_i, X_l = o_{ij})$. We can now specify the randomization scheme for a general two-stage SMART as $\{\pi_i, \pi_{ijk} : i = 1, \ldots, I; j = 1, \ldots, J_i; k = 1, \ldots, K_{ij}\}$, where π_i and π_{ijk} denote vectors of randomization probabilities for stages one and two, respectively. The data for completer patient l can be summarized as (V_l, X_l, W_l, Y_l), which are assumed to be independent and identically distributed for all l with $\Pr(V_l = a_i) = \pi_i$, $\Pr(X_l = o_{ij} | V_l = a_i) = p_{ij}$, $\Pr(W_l = b_{ijk} | V_l = a_i, X_l = o_{ij}) = \pi_{ijk}$, and $Y_l | (V_l = a_i, X_l = o_{ij}, W_l = b_{ijk}) \sim f(y_l | \phi_{ijk}, \tau_{ijk})$, where ϕ_{ijk} for all i, j, k are the parameters of interest and τ_{ijk} denotes a single nuisance parameter or a vector of nuisance parameters. A typical example of ϕ_{ijk} in a SMART is described in Sect. 2.2 below and τ_{ijk} can denote the parameters of the within-subject variance-covariance structure over the study period (it can also include other parameters that are necessary but not of direct interest, such as external factors that may influence the trial).

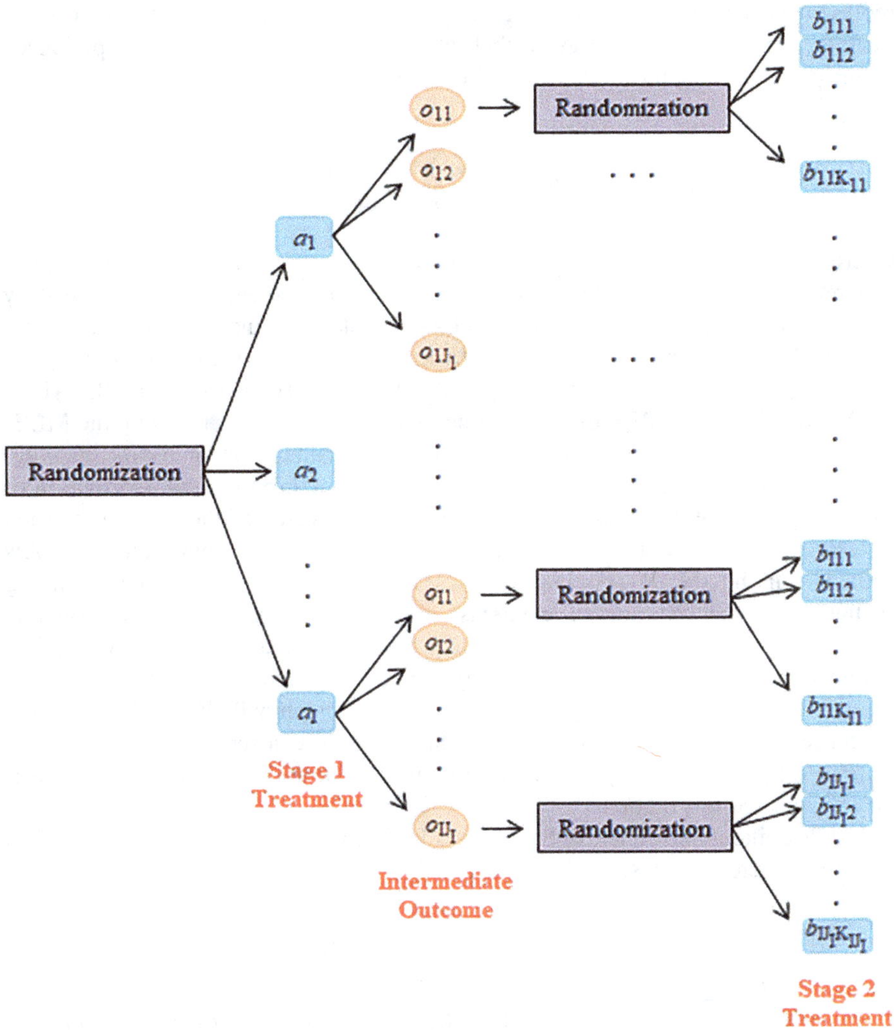

Fig. 1 General layout of a two-stage SMART design, where a_i, o_{ij}, and b_{ijk} denote the stage one treatment options ($i = 1, \ldots, I$), the intermediate outcomes ($j = 1, \ldots, J_i$), and the stage two treatment options ($k = 1, \ldots, K_{ij}$), respectively. Adapted from Zhong et al. (2019)

2.2 *Maximum Likelihood Estimation*

To obtain an asymptotically efficient maximum likelihood estimator (MLE) for (ϕ_{ijk},τ_{ijk}), $f(y_l|\phi_{ijk}, \tau_{ijk})$ must satisfy all of the regularity conditions outlined in Appendix A of Van der Vaart (2000). We will assume these conditions hold true for ease of exposition. Let

$$A_{i;k_{i1},\ldots,k_{iJ_i}} = (a_i; b_{i1k_{i1}}, \ldots, b_{iJ_ik_{iJ_i}})$$

represent an AI, where a patient receives treatment a_i at stage one and receives treatment $b_{ijk_{ij}}$ at stage two given that intermediate outcome o_{ij} was experienced by the patient. Then, the value of an AI can be denoted as

$$\theta_{i;k_{i1},\ldots,k_{iJ_i}} = \sum_{j=1}^{J_i} p_{ij}\phi_{ijk_{ij}},$$

where p_{ij} are the intermediate outcome rates. Here, and in most SMART settings, ϕ_{ijk} represents the expected primary outcome Y given the patient's clinical history throughout the trial (we refer to this clinical history as the patient's "treatment sequence"). Therefore, in this case, an AI value can be interpreted as the marginal expected primary outcome across all possible treatment sequences under the AI.

We can obtain the MLE of an AI value, denoted $\hat{\theta}_{i;k_{i1},\ldots,k_{iJ_i}}$, by using the MLEs of each p_{ij} and the sequence-specific mean of the primary outcome $\phi_{ijk_{ij}}$ as plug-in estimators, which can be approximated by maximizing the joint distribution of (V_l, X_l, W_l, Y_l). Suppose $\Theta = (\theta_1, \ldots, \theta_D)^{\mathrm{T}}$ is a vector of the AI values listed in lexicographical order of $\{i; k_{i1}, \ldots, k_{iJ_i}\}$, where D is the total number of AIs embedded in the SMART. For our purposes, we will define lexicographical ordering as the numerical ordering of elements with multiple indices. As an example, suppose a two-stage SMART has two treatment options at stage one (a_1,a_2), four intermediate outcomes $(o_{11}, o_{12}, o_{21}, o_{22})$, and six stage two treatment options $(b_{111}, b_{112}, b_{121}, b_{211}, b_{212}, b_{221})$. Note that this design will have six treatment sequences and four AIs. In lexicographical order, we can represent the AI values as $(\theta_{1;1,1}, \theta_{1;2,1}, \theta_{2;1,1}, \theta_{2;2,1})$, which is similar to how they would be ordered if they were listed in a dictionary.

The log-likelihood function based on $\{(V_l, X_l, W_l, Y_l); l = 1, \ldots, n\}$, the data for all completer patients, can be represented as

$$\begin{aligned}
&\log(L(p_{ij}, \phi_{ijk}, \tau_{ijk}))\\
&\quad= \sum_{l=1}^{n}\sum_{i=1}^{I}\sum_{j=1}^{J_i}\sum_{k=1}^{K_{ij}} \log f(y_l|\phi_{ijk}, \tau_{ijk})\mathrm{I}(V_l = a_i, X_l = o_{ij}, W_l = b_{ijk})\\
&\quad+ \sum_{l=1}^{n}\sum_{i=1}^{I}\sum_{j=1}^{J_i}\sum_{k=1}^{K_{ij}} \log(p_{ij})\mathrm{I}(V_l = a_i, X_l = o_{ij}, W_l = b_{ijk}) + constant,
\end{aligned}$$

where $\mathrm{I}(\cdot)$ is the indicator function. The MLEs for $\{p_{ij}, \phi_{ijk}, \tau_{ijk}\}$, denoted $\{\hat{p}_{ij}, \hat{\phi}_{ijk}, \hat{\tau}_{ijk}\}$, can be obtained by setting the respective score functions (the first derivatives of the log-likelihood function above) equal to zero and solving for the parameter of interest. All of the derivatives of the score function must be less than zero to ensure maximization. Generally, there are no closed form solutions for $\hat{\phi}_{ijk}$ and $\hat{\tau}_{ijk}$, but the MLE of p_{ij} can be expressed as

$$\hat{p}_{ij} = \frac{\sum_{l=1}^{n} \sum_{k=1}^{K_{ij}} \mathrm{I}(V_l = a_i, X_l = o_{ij}, W_l = b_{ijk})}{\sum_{l=1}^{n} \sum_{j'=1}^{J_i} \sum_{k=1}^{K_{ij'}} \mathrm{I}(V_l = a_i, X_l = o_{ij'}, W_l = b_{ij'k})}.$$

Each AI value $\theta_{i;k_{i1},\ldots,k_{iJ_i}}$ can be determined by

$$\mathbf{p}_i = \begin{bmatrix} p_{i1} \\ \vdots \\ p_{iJ_i} \end{bmatrix}, \quad \boldsymbol{\phi}_i = \begin{bmatrix} \phi_{i1} \\ \vdots \\ \phi_{iJ_i} \end{bmatrix}, \quad \boldsymbol{\phi}_{ij} = \begin{bmatrix} \phi_{ij1} \\ \vdots \\ \phi_{ijK_{ij}} \end{bmatrix},$$

where the MLEs of $\mathbf{p}_i$ and $\boldsymbol{\phi}_i$ can be represented as $\hat{\mathbf{p}}_i$ and $\hat{\boldsymbol{\phi}}_i$, respectively. Define $\boldsymbol{\theta}_i$ as the vector of AI values arranged in lexicographical order, which can be expressed in two identical ways as $\boldsymbol{\theta}_i = \mathbf{A}_i \Lambda_i(\mathbf{p}_i)\boldsymbol{\phi}_i = \mathbf{A}_i \Gamma_i(\boldsymbol{\phi}_i)\mathbf{p}_i$, where

$$\mathbf{A}_i = \Big[\mathbf{I}_{K_{i1}} \otimes \mathbf{1}_{K_{i2}} \otimes \cdots \otimes \mathbf{1}_{K_{iJ_i}} | \mathbf{1}_{K_{i1}} \otimes \mathbf{I}_{K_{i2}} \otimes \cdots \otimes \mathbf{1}_{K_{iJ_i}} | \\ \cdots | \mathbf{1}_{K_{i1}} \otimes \cdots \otimes \mathbf{1}_{K_{i(J_i-1)}} \otimes \mathbf{I}_{K_{iJ_i}}\Big]$$

is a G_i x m_i matrix with $\otimes$ representing the Kronecker product, $G_i = \prod_{j=1}^{J_i} K_{ij}$, and $m_i = \sum_{j=1}^{J_i} K_{ij}$. Further, $\mathbf{1}_k$ denotes a $k \times 1$ vector of ones, $\mathbf{I}_k$ denotes the $k \times k$ identity matrix, $\Lambda_i(\mathbf{p}_i) = bldiag\{p_{ij}\mathbf{I}_{ij}\}$ is an $m_i \times m_i$ block diagonal matrix, and $\Gamma_i(\boldsymbol{\phi}) = bldiag\{\boldsymbol{\phi}_{ij}\}$ is a G_i x J_i block diagonal matrix. Therefore, the MLE of $\boldsymbol{\theta}_i$ can be represented as $\hat{\boldsymbol{\theta}}_i = \mathbf{A}_i \Lambda_i(\hat{\mathbf{p}}_i)\hat{\boldsymbol{\phi}}_i = \mathbf{A}_i \Gamma_i(\hat{\boldsymbol{\phi}}_i)\hat{\mathbf{p}}_i$. Let $\Sigma_{\mathbf{p}_i} = \pi_i^{-1}(diag\{\mathbf{p}_i\} - \mathbf{p}_i\mathbf{p}_i^{\mathrm{T}})$, $\Sigma_{\boldsymbol{\phi}_i} = bldiag\{\Sigma_{\phi_{i1}}, \ldots, \Sigma_{\phi_{iJ_i}}\}$, $\Sigma_{\boldsymbol{\phi}_{ij}} = (\pi_i p_{ij} \pi_{ijk})^{-1} bldiag\{\sigma^2(\boldsymbol{\phi}_{ijk}, \boldsymbol{\tau}_{ijk})\}$, and $\sigma^2(\boldsymbol{\phi}_{ijk}, \boldsymbol{\tau}_{ijk}) = (i_{\phi_{ijk}\phi_{ijk}} - i^{\mathrm{T}}_{\phi_{ijk}\tau_{ijk}} \cdot i^{-1}_{\tau_{ijk}\tau_{ijk}} \cdot i_{\phi_{ijk}\tau_{ijk}})^{-1}$, where "$diag\{\cdot\}$" denotes a diagonal matrix containing the diagonal entries of "$\cdot$" and the block Fisher information matrix of $f(y|\phi_{ijk}, \tau_{ijk})$ is

$$\begin{bmatrix} i_{\phi_{ijk}\phi_{ijk}} & i^{\mathrm{T}}_{\phi_{ijk}\tau_{ijk}} \\ i_{\phi_{ijk}\tau_{ijk}} & i_{\tau_{ijk}\tau_{ijk}} \end{bmatrix}.$$

Suppose $\Sigma = bldiag\{\Sigma_1, \ldots, \Sigma_I\}$, where

$$\Sigma_i = \mathbf{A}_i[\Gamma_i(\boldsymbol{\phi}_i)\Sigma_{\mathbf{p}_i}\Gamma_i(\boldsymbol{\phi}_i)^{\mathrm{T}} + \Lambda_i(\mathbf{p}_i)\Sigma_{\boldsymbol{\phi}_i}\Lambda_i(\mathbf{p}_i)]\mathbf{A}_i^{\mathrm{T}}.$$

Under these regularity conditions,

$$\sqrt{n}\begin{bmatrix}\hat{\mathbf{p}}_i - \mathbf{p}_i \\ \hat{\boldsymbol{\phi}}_i - \boldsymbol{\phi}_i\end{bmatrix} \xrightarrow{d} N\left(\begin{bmatrix}\mathbf{0}\\ \mathbf{0}\end{bmatrix}, \begin{bmatrix}\Sigma_{\mathbf{p}_i} & \mathbf{0} \\ \mathbf{0} & \Sigma_{\boldsymbol{\phi}_i}\end{bmatrix}\right).$$

Then, using the identical forms of $\boldsymbol{\theta}_i$ given above and the delta method,

$$\sqrt{n}(\hat{\boldsymbol{\theta}}_i - \boldsymbol{\theta}_i) = (\mathbf{A}_i\Gamma_i(\boldsymbol{\phi}_i)|\mathbf{A}_i\Lambda_i(\mathbf{p}_i))\sqrt{n}\begin{bmatrix}\hat{\mathbf{p}}_i - \mathbf{p}_i \\ \hat{\boldsymbol{\phi}}_i - \boldsymbol{\phi}_i\end{bmatrix} + \mathbf{o}_p(1)$$

$$\xrightarrow{d} N(\mathbf{0}, \mathbf{A}_i[\Gamma_i(\boldsymbol{\phi}_i)\Sigma_{\mathbf{p}_i}\Gamma_i(\boldsymbol{\phi}_i)^{\mathrm{T}} + \Lambda_i(\mathbf{p}_i)\Sigma_{\boldsymbol{\phi}_i}\Lambda_i(\mathbf{p}_i)]\mathbf{A}_i^{\mathrm{T}}) = N(\mathbf{0}, \Sigma_i)$$

as $n \to \infty$. Therefore,

$$\sqrt{n}(\hat{\Theta} - \Theta) \xrightarrow{d} N(\mathbf{0}, \Sigma),$$

where Σ is a block diagonal matrix with i-th block Σ_i, defined above as the asymptotic covariance matrix of the MLEs of the AI values sharing stage one treatment a_i.

Two remarks are in order. First, the conditional distribution $Y_l|(V_l = a_i, X_l = o_{ij}, W_l = b_{ijk})$ does not need to be parametric. In fact, it can be assumed to be semi-parametric, in which case the ϕ_{ijk} parameters could be estimated via some estimating equations and the asymptotic distribution could be established similarly. This means that the MLE method is not essential to our approach.

Second, it can be shown that $rank(\Sigma_i) = \sum_{j=1}^{J_i} K_{ij} - J_i + 1$. When $J_i \geq 2$, Σ_i will not be full rank. In fact, when J_i is much larger than 2, the asymptotic variance Σ of the MLE $\hat{\Theta}$ is degenerate. That is, its rank is much smaller than its dimension. This is a phenomenon that seems unique to SMARTs and does not appear in traditional designs. As will be seen in future sections, this degeneracy of the asymptotic variance has a crucial impact on statistical inference in SMARTs, including sample size determination. One heuristic explanation is that every AI value is a linear combination of the sequence-specific endpoints. Since SMARTs generally contain more AIs than treatment sequences, the AI values will be linearly dependent and thus their estimates will also be asymptotically linearly dependent, resulting in a degenerate asymptotic covariance matrix (Zhong et al., 2019).

Expanding on this further, consider a SMART design that randomizes participants to two treatment options at the first stage, assesses response versus non-response at interim, and then re-randomizes responders to two options and non-responders to two different options, yielding eight embedded regimes. There are 15 parameters to estimate in this design (eight treatment sequence means and seven probabilities, some of which are known by design) and 28 pairwise comparisons of embedded regimes. However, only 20 of those pairwise comparisons utilize completely independent samples, yielding $20 - 15 = 5$ degrees of freedom. Therefore, the degenerate asymptotic covariance is related to the fact that some individuals are involved in multiple embedded regimes (the covariance matrix would be diagonal if every regime was comprised of a unique set of individuals).

2.3 Wald-type Test

Suppose we are interested in testing the following global hypotheses:

$$H_0: \theta_1 = \theta_2 = \cdots = \theta_D \text{ versus } H_1: \theta_d\text{'s are not all equal,}$$

where $D = \sum_{i=1}^{I} \prod_{j=1}^{J_i} K_{ij}$. Let $C = [\mathbf{1}_{D-1} | - \mathbf{I}_{D-1}]$ denote a $(D-1) \times D$ contrast matrix, where the first column is a $(D-1)$ vector of 1's and the remaining $(D-1) \times (D-1)$ matrix is diagonal with -1 as the diagonal entries. For instance, a SMART with four AIs would have the following contrast matrix:

$$\mathbf{C} = \begin{bmatrix} 1 & -1 & 0 & 0 \\ 1 & 0 & -1 & 0 \\ 1 & 0 & 0 & -1 \end{bmatrix}.$$

Suppose Σ is the covariance matrix of $\hat{\Theta}$ and let $\hat{\Sigma}$ be the plug-in estimator for Σ, obtained using the MLEs of p_{ij}, ϕ_{ijk}, and τ_{ijk} described in the previous section. We can express a Wald-type test statistic as

$$Q = n(\mathbf{C}\hat{\Theta})^{\mathrm{T}}(\mathbf{C}\Sigma\mathbf{C}^{\mathrm{T}})^{-}(\mathbf{C}\hat{\Theta}),$$

where $(\mathbf{C}\Sigma\mathbf{C}^{\mathrm{T}})^{-}$ denotes the generalized inverse of the square matrix $\mathbf{C}\Sigma\mathbf{C}^{\mathrm{T}}$. Under H_0, as $n \to \infty$, $Q \xrightarrow{d} \chi^2_v$, a central chi-squared random variable with v degrees of freedom, where

$$v = \sum_{i=1}^{I} \sum_{j=1}^{J_i} K_{ij} - \sum_{i=1}^{I} J_i + I - 1.$$

Furthermore, given a sequence of local alternative hypotheses $\{\Theta_n\}$ such that

$$\lim_{n\to\infty} n(\mathbf{C}\hat{\Theta}_n)^{\mathrm{T}}(\mathbf{C}\Sigma\mathbf{C}^{\mathrm{T}})^{-}(\mathbf{C}\hat{\Theta}_n) = \lambda^* > 0,$$

$Q \xrightarrow{d} \chi^2_v(\lambda^*)$, where λ^* denotes a non-centrality parameter. Thus, we have an asymptotic level α test for testing H_0 versus H_1, where H_0 is rejected if $Q > \chi^2_{(v,\alpha)}$, the $(1-\alpha)$th percentile of χ^2_v. This test reduces to the general Wald test when comparing non-adaptive interventions, since $rank(\Sigma_i) = 1$ (Zhong et al., 2019).

In a traditional design, to test the above global hypotheses, the asymptotic chi-squared test should have $D - 1$ degrees of freedom. Therefore, had we disregarded the degeneracy of the asymptotic variance, we would have claimed under the SMART design that our test has $D - 1$ degrees of freedom, while it actually has v degrees of freedom, where v could be much smaller.

In applications where τ_{ijk} represents the variance parameters, the restricted maximum likelihood (REML) as opposed to full maximum likelihood is often used to improve the estimation of τ_{ijk}. In this case, the asymptotic results for REML can be obtained for SMARTs similarly with a slight modification (i.e., treating the nuisance parameters as known and establishing the asymptotic distributions for the parameters of interest, and then plugging in the REML estimates when conducting the Wald test).

2.4 AI Selection via the Gate-Keeping Approach

In SMART research, although the current practice typically focuses on comparing a pre-specified pair of AIs, gate-keeping strategies are used to expeditiously identify the highest performing AIs in early phase developmental trials to advance into phase III studies, enabling accelerated assessment and treatment availability. The Wald-type test proposed in the previous section is an example of a gate-keeping method. When H_0 is rejected, the AI with the highest estimated value is selected and recommended for further investigation. However, if we fail to reject H_0, comparison of the AIs is stopped due to insufficient evidence supporting any superiority among the AIs being evaluated. In summary, the primary goal of the gate-keeping approach is to assess several candidate AIs simultaneously, controlling for a strict, pre-specified type I error rate, to efficiently select the outperforming AI that can be studied in a future phase III trial (Zhong et al., 2019). However, it does not control the error of recommending a non-best AI.

As an example, the proposed Wald-type test has been applied to data from 108 patients in the CODIACS trial, with the goal of identifying which AI should be selected in a depression treatment program based on a 6-month post-baseline reduction in depression, according to the Beck Depression Inventory (Cheung et al., 2015). Interventions resulting in higher depression reduction were considered more effective, so the value of an AI was defined as the expected reduction in depression at 6 months. The two-stage CODIACS design is provided in Fig. 2, which has eight AIs and follows the general two-stage SMART structure shown in Fig. 1 with $i = \{1, 2\}$, $j = \{1, 2\}$, and $k = \{1, 2\}$. The MLEs for all eight AIs, along with their standard errors, are provided in Table 1.

Applying the Wald-type test under the hypotheses in Sect. 2.3 with eight total AIs ($D = 8$) yielded a test statistic of $Q = 36.0$. Under H_0, $Q \sim \chi_5^2$, which resulted in $p < 0.001$. Thus, the AI values were significantly different at the $\alpha = 0.05$ level, and the fifth AI ($d = 5$) in Table 1 was selected for further investigation, since it was the best (i.e., had the highest MLE). A traditional pairwise testing technique could have been used to compare the AIs to the observed best AI, such as Bonferroni's correction, but this would have failed to identify an overall difference between the

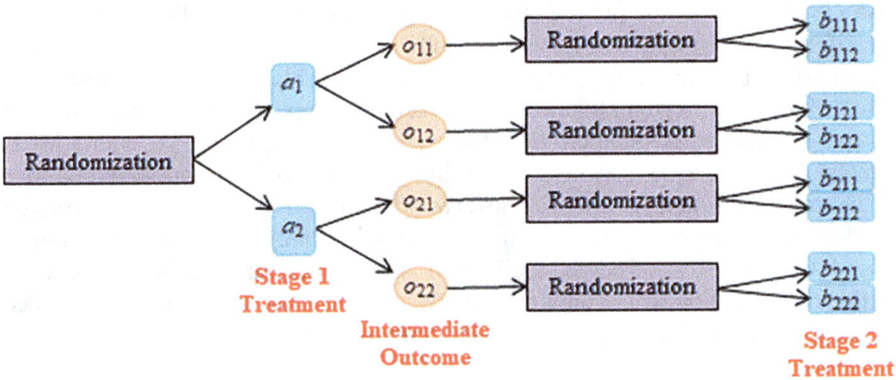

Fig. 2 The two-stage CODIACS trial design, where a_i, o_{ij}, and b_{ijk} denote the two stage one treatment options ($i = 1, 2$), the two intermediate outcome categories ($j = 1, 2$), and the two stage two treatment options ($k = 1, 2$), respectively

Table 1 MLEs and p-values of pairwise comparisons between each AI and the observed best AI ($d = 5$) in the CODIACS trial. The estimated asymptotic standard error (SE) of the MLE using the inverse Fisher information matrix is reported. PST denotes problem-solving therapy. Adapted from Zhong et al. (2019)

AI (d)	Stage 1 Treatment	Stage 2 Treatment for Responders	Stage 2 Treatment for Non-responders	MLE (SE)	P-value
1	Medication	Medication	Medication	6.3(1.1)	0.135
2	Medication	Medication	PST	3.3(1.2)	0.049
3	Medication	PST	Medication	10.7(0.6)	0.434
4	Medication	PST	PST	7.8(1.1)	0.210
5	PST	Medication	Medication	15.5(6.0)	–
6	PST	Medication	PST	9.5(1.0)	0.320
7	PST	PST	Medication	14.2(6.1)	0.201
8	PST	PST	PST	8.2(1.1)	0.236

AIs and thus no AI would have been recommended for further examination. For instance, the second AI in Table 1 had a p-value less than 0.05, but adjusting for multiple comparisons with Bonferroni's correction would require a p-value less than $0.05/\binom{8}{2} = 0.0018$) (Zhong et al., 2019).

2.5 AI Selection via Simultaneous Confidence Intervals

Another strategy that has been presented in the literature to select the optimal AI embedded in a SMART is simultaneous confidence intervals, which compare AI values to that of the unknown best AI using methods developed in Edwards and

Hsu (1983). This technique is particularly appealing for SMARTs geared toward identifying promising or nearly best AIs and removing inferior AIs. These intervals are an extension of multiple comparison with the best (MCB) intervals (Edwards & Hsu, 1983), which account for the design structure as opposed to traditional multiple comparison procedures that are only for single stage designs. An AI with a MCB interval excluding zero (or with an upper limit less than zero) is considered inferior to the true best AI at a pre-specified level α, and thus this AI would not be explored any further (Zhong et al., 2021). If an AI has a MCB interval with an upper limit of zero, it is potentially considered the best AI but not necessarily the unique best AI. In the ideal situation where only one AI has an upper bound of zero, the inference is that all other AIs have been statistically demonstrated to be inferior to that single AI.

We will now describe how to construct the proposed simultaneous confidence intervals for multiple comparisons of AIs in SMARTs by expanding on the MCB interval approach. Similarly to Sect. 2.3, let C_d be a $(D-1)\times D$ contrast matrix such that its d-th column is a $(D-1)$ vector of ones, its j-th column is a $(D-1)$ vector whose j-th entry is -1 and all other entries are zero if $j \leq d-1$, and its $(j\text{-}1)$ entry is -1 and all other entries are zero if $j \geq d+1$. Then, from Sect. 2.2, as $n \to \infty$,

$$C_d\sqrt{n}(\hat{\theta}-\theta) \xrightarrow{d} N(\mathbf{0}, C_d\Sigma C_d^{\mathrm{T}}).$$

Letting σ_{id}^2 denote the i-th diagonal entry of $C_d\Sigma C_d^{\mathrm{T}}$, we can express the asymptotic variance with respect to the correlation coefficient matrix R_d as

$$C_d\Sigma C_d^{\mathrm{T}} = diag\{\sigma_{id}\}\cdot R_d \cdot diag\{\sigma_{id}\},$$

where $diag\{\cdot\}$ denotes a diagonal matrix with diagonal entries "$\cdot$" and all off-diagonal entries equal to zero. For a given level α, let $\zeta_d(\alpha) > 0$ be the unique solution to the equation

$$\Pr\left(|Z_{id}| \leq \zeta_d(\alpha); i=1,\ldots,D-1\right) = \Pr\left(\max_{i\in\{1,\ldots,D-1\}}|Z_{id}| \leq \zeta_d(\alpha)\right) = 1-\alpha,$$

where $(Z_{1d},\ldots,Z_{D-1,d})^{\mathrm{T}} \sim N(0,\hat{R}_d)$ and $\hat{R}_d$ is a consistent estimator for R_d (Zhong et al., 2021).

To find $\hat{R}_d$, we know that R_d is a continuous function of p_i, ϕ_i, Σ_{p_i}, and Σ_{ϕ_i} for $i=1,\ldots,I$. We also know that the MLEs of $\hat{p}_i$, $\hat{\phi}_i$, $\hat{\Sigma}_{p_i}$, and $\hat{\Sigma}_{\phi_i}$ are consistent estimators. In addition, the plug-in estimator $\hat{R}_d$ is consistent by the continuous mapping theorem. Further, Σ and thus R_d will not be full rank, and the rank of Σ implies that all $D-\sum_{i=1}^I m_i + \sum_{i=1}^I J_i - I$ smallest eigenvalues of R_d will be zero. To improve efficiency on $\hat{R}_d$, the $D-\sum_{i=1}^I m_i + \sum_{i=1}^I J_i - I$ smallest eigenvalues of $\hat{R}_d$ are set equal to zero to match those of R_d. For example, consider the eigendecomposition $\hat{R}_d = \hat{U}\hat{E}\hat{U}^{\mathrm{T}}$, where $\hat{U}$ is an orthogonal matrix of the eigenvalues of $\hat{R}_d$ and $\hat{E}$ is a diagonal matrix of the eigenvalues of $\hat{R}_d$. Let E^* be a diagonal matrix after replacing the $D - \sum_{i=1}^{I} m_i + \sum_{i=1}^{I} J_i - I$ smallest eigenvalues of $\hat{E}$ with zeros. Then, $R_d^* = \hat{U} E^* \hat{U}^{\mathrm{T}}$ is a more efficient estimator of R_d than $\hat{R}_d$ (Zhong et al., 2021).

Now, let $\mathscr{D} = \{d : \hat{\theta}_d - \hat{\theta}_i + \zeta_d(\alpha)\hat{\sigma}_{id}/\sqrt{n} > 0, \forall i \neq d\}$, where n denotes the total sample size of the study and $\hat{\sigma}_{id}$ is the plug-in estimator for σ_{id}, the standard deviation of the i-th diagonal entry of $C_d \Sigma C_d^{\mathrm{T}}$. Then, the lower and upper bounds of the MCB interval $[L_i, U_i]$ for the i-th AI are calculated as $L_i = \min_{d \in \mathscr{D}} L_{id}$ and $U_i = \min_{d \in \mathscr{D}} U_{id}$, where

$$L_{id} = \begin{cases} 0, & d = i \\ \hat{\theta}_i - \hat{\theta}_d - \frac{\zeta_d(\alpha)\hat{\sigma}_{id}}{\sqrt{n}}, & d \neq i \end{cases}$$

and

$$U_{id} = \begin{cases} 0, & d = i \\ \min\{0, \hat{\theta}_i - \hat{\theta}_d + \frac{\zeta_d(\alpha)\hat{\sigma}_{id}}{\sqrt{n}}\}, & d \neq i. \end{cases}$$

It has been shown that the MCB intervals defined above are for comparison to the true best $\max_j \theta_j$ without assuming which AI is truly the best, and have asymptotic coverage of at least $1 - \alpha$ (Zhong et al., 2021). By construction, no upper limit can exceed zero (i.e., $U_i \leq 0$ for all i), and an AI with a negative U_i is considered inferior to the best AI with $(1 - \alpha) \cdot 100\%$ confidence. Additionally, if the set $\mathscr{D} = \{d^*\}$ is a singleton, then the MCB interval corresponding to that AI will be [0,0]. The converse is also true (Zhong et al., 2021).

Simulation studies have shown that this MCB interval method outperforms the standard Bonferroni-adjusted multiple comparison technique based on the width of the confidence intervals used for estimation (Zhong et al., 2021). The improved utility of this method has also been shown using patient data from the CODIACS trial, with the goal of identifying effective depression management regimens (AIs) and discontinuing those that are inferior (Cheung et al., 2015). Recall that a reduction in the Beck Depression Inventory at 6 months was the primary endpoint used in the trial, and the value of an AI was measured by the expected reduction in depression score. Table 2 provides the 80% MCB intervals for the eight possible two-stage AIs embedded in the study. Since the upper bound of the MCB interval for the second AI is negative, it is considered inferior to the others. Note that the purpose of this technique is not to determine the best AI, but to remove inferior AIs from further investigation (Zhong et al., 2021).

Pairwise Bonferroni-adjusted confidence intervals were also calculated for comparison to the MCB interval approach and are included in Table 2. Since all eight of the Bonferroni-adjusted confidence intervals contain zero, this method failed to differentiate the AIs. Notice that the MCB intervals yield more precise estimates, since the average widths of the eight MCB intervals were shorter than their corresponding Bonferroni-adjusted confidence intervals. No Bonferroni-adjusted

Table 2 MLEs of pairwise comparisons between each AI and the observed best AI ($d = 5$) in the CODIACS trial. The estimated asymptotic standard error (SE) of the MLE using the inverse Fisher information matrix is reported. The 80% CIs obtained using the MCB method compares each AI to the true, unknown best AI. The 80% CIs via Bonferroni-adjustment compare each AI to the observed best AI. PST denotes problem-solving therapy. Adapted from Zhong et al. (2021)

AI (d)	Stage 1 Treatment	Stage 2 Treatment for Responders	Stage 2 Treatment for Non-responders	MLE (SE)	80% CI via MCB	80% CI via Bonferroni
1	Medication	Medication	Medication	6.3(1.1)	[−19.7, 0.0]	[−25.7, 7.3]
2	Medication	Medication	PST	3.3(1.2)	[−22.7, −0.3]	[−28.7, 4.5]
3	Medication	PST	Medication	10.7(0.6)	[−15.2, 0.0]	[−21.1, 11.8]
4	Medication	PST	PST	7.8(1.1)	[−18.2, 0.0]	[−24.2, 8.8]
5	PST	Medication	Medication	15.5(6.0)	[−7.6, 0.0]	–
6	PST	Medication	PST	9.5(1.0)	[−16.3, 0.0]	[−22.2, 10.2]
7	PST	PST	Medication	14.2(6.1)	[−8.9, 0.0]	[−3.8, 1.4]
8	PST	PST	PST	8.2(1.1)	[−17.6, 0.0]	[−23.6, 9.1]

interval was calculated for the fifth AI ($d = 5$), because this observed best AI was treated as the optimal AI (i.e., the MLE was the largest). The remaining seven MCB intervals were built by comparing them to the fifth AI. Thus, the fifth AI, considered as the best, had no associated confidence interval (Zhong et al., 2021).

We make three remarks concerning the proposed MCB intervals. First, the confidence intervals constructed are for comparison with the true best, not the estimated best. Second, the proposed method generalizes the traditional MCB method to the case where the estimated AIs are not only correlated but also heavily degenerated. Third, a careful investigation of the theoretical justification of the proposed MCB intervals in Zhong et al. (2021) suggests that the proposed interval could be further improved by estimating the number of AIs that tie as the true best. This issue is currently under investigation.

3 Sample Size Determination

When designing a SMART to select an AI for further clinical assessment, it is essential to derive the smallest sample size required to detect a minimal clinically meaningful effect size within a certain pre-specified margin of error; however, powering SMARTs is an ongoing challenge. For instance, sample size determination may rely on knowledge of the response rate at stage one in order to sufficiently power stage two, and some subjects may contribute to the treatment effects of more than one AI. These complications underscore the importance of design simplification (e.g., minimizing the number of treatment options evaluated at each stage and the number of possible intermediate outcomes) (Murphy, 2005). Unfortunately, power procedures for comparing complex AIs are not widely available and simulation studies to inform the development of such methods are limited (Wallace et al., 2016), but this remains an active area of research.

Prior work has been conducted for the purpose of comparing simple main effects (e.g., comparing the main effect of the first stage treatment while controlling for the second stage treatment) (Nahum-Shani et al., 2012) or comparing AIs via test statistics from marginal mean model variances (Seewald et al., 2020) or other calculations (Dawson & Lavori, 2010). Methods have also been derived for ascertaining the optimal AI (Chakraborty & Moodie, 2013; Kosorok & Moodie, 2015); however, the power implications of some of these approaches have not been evaluated in depth (Wallace et al., 2016). Recent studies, including some that have utilized closely related methods to those presented in this chapter, have helped to address this problem (Artman et al., 2020; Ogbagaber et al., 2016). The impact of the desired power on the required sample size when using the Wald-type test introduced in Sect. 2.3 can be summarized by the following general procedure.

First, based on the structure of the intended SMART, determine the degrees of freedom v using the equation in Sect. 2.3. Then, with a pre-specified type I error rate α and targeted statistical power β, calculate the non-centrality parameter λ^* needed under the alternative by solving

$$\chi^2_{(v,\beta)}(\lambda^*) = \chi^2_{(v,\alpha)}(0),$$

where $\chi^2_{(v,\beta)}(\lambda^*)$ is the $(1-\beta)$th percentile of the χ^2_v distribution with non-centrality parameter λ^* and $\chi^2_{(v,\alpha)}(0)$ is the $(1-\alpha)$th percentile of the central χ^2_v distribution. Next, using the design parameters (π_i,π_{ijk}), assumed intermediate outcome probabilities (p_{ij}), and primary outcome parameter values (ϕ_{ijk},τ_{ijk}) from the conditional outcome distribution f, determine the targeted AI values $(\hat{\Theta}^*)$ and their covariance structure (Σ^*). Then, the standardized effect size (Δ) can be calculated as

$$\Delta = (\mathbf{C}\Theta^*)^{\mathrm{T}}(\mathbf{C}\Sigma^*\mathbf{C}^{\mathrm{T}})^{-}(\mathbf{C}\Theta^*).$$

Finally, the total required sample size (n) for the anticipated SMART will be

$$n = \lceil \lambda^*/\Delta \rceil,$$

where $\lceil x \rceil$ denotes the ceiling of x, or the smallest integer larger than x. Some commonly used values for α, β, λ^*, and v are provided in Table 3. Larger power, smaller type I error, and larger degrees of freedom (which are all associated with the number of treatment options and intermediate outcomes in the SMART design) require larger non-centrality parameters and hence larger sample sizes (Zhong et al., 2019).

This sample size determination procedure is based on testing the global hypotheses in Sect. 2.3 via the proposed Wald-type test. In principle, it is similar to the sample size calculation based on the ANOVA F-test under a parallel group design.

Table 3 Non-centrality parameters (λ^*) under different degrees of freedom (DF), family-wise error (α), and targeted power conditions that can be used to calculate the total required sample size (n) of a SMART. Adapted from Zhong et al. (2019)

	($\boldsymbol{\alpha}$, power)					
DF ($\boldsymbol{v}$)	(0.01, 0.90)	(0.01, 0.80)	(0.05, 0.90)	(0.05, 0.80)	(0.10, 0.90)	(0.10, 0.80)
2	17.42	13.88	12.65	9.63	10.45	7.71
3	19.24	15.45	14.17	10.90	11.79	8.80
4	20.73	16.75	15.41	11.94	12.88	9.68
5	22.02	17.87	16.47	12.83	13.81	10.44
6	23.18	18.87	17.42	13.62	14.65	11.13
7	24.23	19.78	18.28	14.35	15.41	11.75
8	25.20	20.63	19.08	15.02	16.11	12.32
9	26.12	21.42	19.81	15.65	16.76	12.86
10	26.98	22.17	20.53	16.24	17.38	13.36
11	27.79	22.88	21.20	16.80	17.96	13.84
12	28.57	23.56	21.83	17.34	18.52	14.30
13	29.31	24.21	22.44	17.85	19.05	14.74
14	30.03	24.83	23.02	18.34	19.56	15.16
15	30.71	25.43	23.58	18.81	20.06	15.56
16	31.38	26.01	24.13	19.27	20.53	15.95
17	32.02	26.57	24.65	19.71	20.99	16.33
18	32.65	27.11	25.16	20.14	21.43	16.69
19	33.25	27.64	25.65	20.56	21.87	17.05
20	33.84	28.16	26.13	20.96	22.29	17.39

Some may argue that the sample size for a SMART can be obtained to detect a pre-specified difference between the best and the worst (or the control) AI. However, this approach may unduly inflate the total sample size required because many AIs other than the best are involved and hence must be accounted for. In this sense, our approach usually yields smaller sample sizes for the same targeted power.

It must be emphasized that due to the sequential nature of AIs, the sample size for a SMART is typically larger than what would be required for a single stage design. By design, the second stage sample sizes are smaller than the first stage sample sizes; therefore, if the second stage treatment does not result in further separation in the endpoint of interest, there is a loss in power in delineating differences between the AIs. This especially becomes an issue when more than two stages are considered. Therefore, we suggest that the number of stages be limited to no more than three, unless strong evidence indicates otherwise.

4 Discussion

Statistical methods for SMART data inference have been developed by many authors, including those mentioned in the Introduction, and software is available to implement SMART methodology (Zhong et al., 2021). Our approach has several advantages. First, our method reveals a unique feature of SMARTs (i.e., the degeneracy of the asymptotic variance of the AI estimates). Accounting for this degeneracy allows us to produce methods that are more powerful than competing strategies. Second, our framework allows the primary endpoint Y to be of a different type. This aligns more closely with real practice, where the evaluation method varies with treatment; some treatments may produce continuous measures while others may yield categorical or count data.

Compared to the alternative methods discussed in the Introduction, a major advantage of the proposed Wald test is that it facilitates the determination of the degrees of freedom of the test, which could be much smaller than the number of embedded AIs. Note that although the asymptotic distribution of the AI estimates under other methods may not be the same as ours, the degrees of freedom of the global tests should be the same irrespective of the method used. A complete proof of the degrees of freedom under the MLE method can be found in Zhong et al. (2019); this proof may become more difficult, or even impossible, using other techniques.

In summary, the SMART design has many advantages over alternative design options. From the clinicians' perspective, it can help to identify better treatments for further exploration in phase III studies by simultaneously considering and pruning multiple treatment options. The design is equally appealing to patients, because it opens up the opportunity to personalize treatment to maximize efficacy. Additionally, it seems plausible that a likelihood-based method could be developed for "small n" SMARTs (snSMARTs), which warrants further investigation.

The SMART design also has some limitations. For example, analyzing data from SMARTs can pose some challenges. As a SMART design increases in structural complexity (e.g., high number of stages and/or treatments), the number of possible AIs increases exponentially, which can lead to computational intensity and inferential challenges. Accounting for a very large number of pairwise comparisons could result in a considerable loss in power and inflated sample sizes, yielding a highly conservative design. This conundrum is known as "the curse of dimensionality". However, the gate-keeping and simultaneous confidence interval approaches introduced in this chapter help to alleviate this issue (Zhong et al., 2019).

Finally, SMARTs are not always appropriate or feasible due to time constraints. In a SMART, measurement of the intermediate outcome must be done within a relatively small window to ensure that findings are not outdated. Since all trial stages and follow-up must be completed in a reasonable amount of time, conducting a SMART may be more difficult than utilizing a general single stage trial in some cases. Moreover, prudently selecting treatment sequences, and an appropriate number of them, can also complicate SMARTs, since business, clinical, and statistical insights may clash. Although SMARTs can test more realistic interventions than standard RCTs, it is still challenging to practically test several different decision points when there are copious treatment options. Dropout at the intermediate outcome can also complicate SMARTs. Developing pre- and post-randomization re-engagement strategies to address subject attrition is often difficult. Lastly, SMARTs can be more costly than traditional trial designs and require very rigorous planning (Kidwell et al., 2016).

References

Almirall, D., et al. (2012). Designing a pilot sequential multiple assignment randomized trial for developing an adaptive treatment strategy. *Statistics in Medicine, 31*(17), 1887–1902.

Arkes, H. R. (1981). Impediments to accurate clinical judgment and possible ways to minimize their impact. *Journal of Consulting and Clinical Psychology, 49*, 323–330.

Artman, W. J., et al. (2020). Power analysis in a SMART design: sample size estimation for determining the best embedded dynamic treatment regime. *Biostatistics, 21*(3), 432–448.

Bierman, K. L., et al. (2006). Examining clinical judgment in an adaptive intervention design: The fast track program. *Journal of Consulting and Clinical Psychology, 74*(3), 468.

Chakraborty, B., & Moodie, E. E. M. (2013). *Statistical methods for dynamic treatment regimes*. New York: Springer.

Cheung, Y. K., Chakraborty, B., & Davidson, K. W. (2015). Sequential multiple assignment randomized trial (SMART) with adaptive randomization for quality improvement in depression treatment program. *Biometrics, 71*(2), 450–459.

Collins, L. M., Murphy, S. A., & Strecher, V. (2007). The multiphase optimization strategy (MOST) and the sequential multiple assignment randomized trial (SMART): new methods for more potent eHealth interventions. *American Journal of Preventive Medicine, 32*(5), S112–S118.

Dawson, R., & Lavori, P. W. (2010). Sample size calculations for evaluating treatment policies in multi-stage designs. *Clinical Trials, 7*, 643–652.

Edwards, D. G., & Hsu, J. C. (1983). Multiple comparisons with the best treatment. *Journal of the American Statistical Association, 78*(384), 965–971.

Fang, F., et al. (2020). Bayesian methods to compare dose levels with placebo in a small n, sequential, multiple assignment, randomized trial. *Statistics in Medicine, 40*, 963–977.

Garb, H. N. (1998). *Studying the clinician: Judgment research and psychological assessment*. Washington: American Psychological Association.

Hochberg, Y., & Ajit, T. C. (1987). *Multiple comparison procedures*. New York: Wiley.

Hsu, J. C. (1981). Simultaneous confidence intervals for all distances from the "best". *The Annals of Statistics, 9*, 1026–1034.

Kasari, C., et al. (2014). Communication interventions for minimally verbal children with autism: A sequential multiple assignment randomized trial. *Journal of the American Academy of Child & Adolescent Psychiatry, 53*(6), 635–646.

Kidwell, K. M., et al. (2016). Adaptive interventions and SMART designs: application to child behavior research in a community setting. *American Journal of Evaluation, 37*(3), 344–363.

Kosorok, M. R., & Moodie, E. E. M. (2015). Adaptive treatment strategies in practice: Planning trials and analyzing data for personalized medicine. In *Society for Industrial and Applied Mathematics*.

Lei, H., et al. (2012). A "SMART" design for building individualized treatment sequences. *Annual Review of Clinical Psychology, 8*, 21–48.

Murphy, S. A. (2005). An experimental design for the development of adaptive treatment strategies. *Statistics in Medicine, 24*(10), 1455–1481.

Murphy, S. A., et al. (2001). Marginal mean models for dynamic regimes. *Journal of the American Statistical Association, 96*(456), 1410–1423.

Nahum-Shani, I., et al. (2012). Experimental design and primary data analysis methods for comparing adaptive interventions. *Psychological Methods, 17*, 457–477.
Ogbagaber, S. B., et al. (2016). Design of sequentially randomized trials for testing adaptive treatment strategies. *Statistics in Medicine, 35*(6), 840–858.
Robins, J. (1986). A new approach to causal inference in mortality studies with a sustained exposure period—application to control of the healthy worker survivor effect. *Mathematical Modelling, 7*(9–12), 1393–1512.
Seewald, N. J., et al. (2020). Sample size considerations for comparing dynamic treatment regimens in a sequential multiple-assignment randomized trial with a continuous longitudinal outcome. *Statistical Methods in Medical Research, 29*(7), 1891–1912.
Snowden, J., et al. (2011). Implementation of G-computation on a simulated data set: Demonstration of a causal inference technique. *American Journal of Epidemiology, 173*(7), 731–738.
Tamura, R. N., et al. (2016). A small n sequential multiple assignment randomized trial design for use in rare disease research. *Contemporary Clinical Trials, 46*, 48–51.
Tukey, J. W. (1949). Comparing individual means in the analysis of variance. *Biometrics, 5*, 99–114.
Van der Vaart, A. W. (2000). *Asymptotic statistics* (Vol. 3). Cambridge: Cambridge University.
Wallace, M. P., et al. (2016). SMART thinking: A review of recent developments in sequential multiple assignment randomized trials. *Current Epidemiology Reports, 3*(3), 225–232.
Wallace, M. P., & Moodie, E. E. M. (2014). Personalizing medicine: a review of adaptive treatment strategies. *Pharmacoepidemiology and Drug Safety, 23*(6), 580–585.
Wei, B., et al. (2018). A Bayesian analysis of small n sequential multiple assignment randomized trials (snSMARTs). *Statistics in Medicine, 37*, 3723–3732.
Zhong, X., et al. (2021). SMARTAR: An R package for designing and analyzing Sequential Multiple Assignment Randomized Trials. *PeerJ, 9*, e10559.
Zhong, X., et al. (2021). Comparing adaptive interventions under a general sequential multiple assignment randomized trial design via multiple comparisons with the best. *Journal of Statistical Planning and Inference, 211*, 143–153.
Zhong, X., et al. (2019). A gate-keeping test for selecting adaptive interventions under general designs of sequential multiple assignment randomized trials. *Contemporary Clinical Trials, 85*, 105830.

Precision Medicine Designs for Cancer Clinical Trials

Beibei Guo and Ying Yuan

Abstract This chapter reviews Bayesian adaptive designs for precision oncology clinical trials. We focus on two aspects of precision medicine: how to more accurately capture the multi-dimensional nature of clinical outcomes, and how to more accurately account for patient's heterogeneity. We use the Bayesian optimal interval phase I-II (BOIN12) design as an example to illustrate the first aspect. The BOIN12 design uses the utility function to explicitly account for the risk-benefit tradeoff of the treatment and identify the optimal biological dose (OBD), and thus more accurately reflects the multiple clinical effects of the treatment underlying clinical decisions in practice. We then use a novel Bayesian adaptive design to illustrate how to determine subgroup OBD by accounting for patient's genomic profile and heterogeneity. The design first uses a dimension-reduction technique to reduce the dimension of genomic biomarkers, and then incorporates this condensed biomarker information into the clinical outcome model and adaptive decision making. In the presence of patient heterogeneity, another effective approach to achieving precision medicine is to use the enrichment strategy. We describe a novel Bayesian adaptive enrichment design that simultaneously achieves two primary goals of precision medicine: to identify the sensitive subgroup and to test if the treatment is effective in that subgroup. The design uses the interim data to continuously and adaptively identify the subgroup of patients who are expected to benefit from the treatment, and actively enriches such subgroup by adaptively modifying patient enrollment criteria during the trial. The design has substantially higher power than the conventional designs to detect the treatment effect and identify the target subpopulation.

B. Guo
Department of Experimental Statistics, Louisiana State University, Baton Rouge, LA, USA
e-mail: beibeiguo@lsu.edu

Y. Yuan (✉)
Department of Biostatistics, The University of Texas MD Anderson Cancer Center, Houston, TX, USA
e-mail: yyuan@mdanderson.org

Y. Zhao, D.-G. Chen (eds.), *Statistics in Precision Health*, ICSA Book Series in Statistics, https://doi.org/10.1007/978-3-031-50690-1_16

Keywords Precision medicine · Clinical trial · Risk-benefit tradeoff · Bayesian adaptive design · Optimal biological dose

1 Precision Medicine

Precision medicine provides emerging approaches for disease treatment and prevention that take into account each person's individual variability in genes, environment, and lifestyle (Collins & Varmus, 2015; Hodson, 2016). It is revolutionizing cancer research, with the premise that patients are heterogeneous and may have different responses to treatments. Precision medicine is a broad medical paradigm and optimizing the treatment benefit under this paradigm requires the clinical trial design to account for the patient's individual characteristics when searching for the optimal treatment or dose that is personalized for each patient. In this chapter, we will focus on two aspects of precision medicine: how to more accurately capture the multi-dimensional nature of clinical outcomes, and how to more accurately account for patient's heterogeneity. In what follows, we will describe

- (Section 2) Designs to capture the multi-dimensional nature of clinical outcomes. In this section, we will review a Bayesian optimal interval phase I-II design to consider the risk-benefit tradeoff of the treatment based on multiple endpoints.
- (Section 3) Designs to account for patient's heterogeneity. In this section, we will review a Bayesian adaptive design to determine subgroup optimal dose by accounting for patient's genomic profile and a Bayesian adaptive enrichment design to identify the sensitive subgroup and to test if the treatment is effective in that subgroup.

2 Design to Capture Multi-Dimensional Clinical Outcomes

Conventional clinical trials aim to identify the maximum tolerated dose (MTD), defined as the dose with a toxicity probability that is closest to the target toxicity rate specified by physicians. Under this paradigm, phase I and phase II trials are typically performed separately, with phase I designed to determine the MTD and subsequent phase II evaluating the potential efficacy of the drug at the MTD. This is based on the assumption that both toxicity and efficacy increase with dose, which is often true for conventional cytotoxic agents.

The characteristics of precision medicine make this traditional paradigm increasingly inefficient and dysfunctional. For example, novel molecularly targeted agents and immunotherapy aim to match patients with targeted therapies to achieve maximum clinical benefit, so they naturally fit the rationale of precision medicine. For these agents, efficacy may plateau, or even decrease at high doses. In addition, some of these therapies demonstrate minimal toxicity at high doses, making the MTD unlikely to be reached. Thus, these agents are not necessarily administered at the MTD to achieve optimal treatment effects. As a result, to optimize the treatment effect of targeted therapy and immunotherapy, precision medicine requires the trial to take into account both toxicity and efficacy outcomes. The objective of dose-finding trials in precision medicine should be to determine the optimal biological dose (OBD) that optimizes patient's risk-benefit tradeoff. In this section, we present a design that accounts for such tradeoff based on multiple endpoints, serving as the basis for Sect. 3, which covers designs that consider patient's heterogeneity based on multiple endpoints.

For example, consider five doses of a targeted or immunotherapy agent, $d = 1, 2, 3, 4, 5$, with true toxicity probabilities $p_T = (0.05, 0.12, 0.27, 0.35, 0.50)$. If the true efficacy probabilities are $p_E = (0.20, 0.35, 0.36, 0.37, 0.38)$, then efficacy reaches a plateau of about 0.35 at $d = 2$. The conventional dose-finding methods (e.g., 3 + 3 algorithm or continuous reassessment method (CRM) with a target toxicity probability of 0.3) are most likely to select $d = 3$ as the MTD and use it for cohort expansion or a subsequent phase II trial. However, $d = 2$ is obviously more desirable than the MTD (i.e., $d = 3$) with much lower toxicity and virtually identical efficacy. This example shows that for targeted therapy and immunotherapy, it is important to consider both toxicity and efficacy concurrently to identify the OBD, rather than the MTD.

Most of the phase I/II dose-finding designs for identifying OBD assume statistical models for the dose-toxicity and dose-efficacy curves. As toxicity and efficacy data accrue during the trial, they continuously update the estimates of the models after each cohort, and then use them to determine the dose for the next cohort, see the book by Yuan et al. (2017) for comprehensive reviews of phase I/II dose-finding designs. Unfortunately, these novel model-based designs are rarely used in practice, because they are often difficult to understand and require complicated statistical modeling and computation, as well as expensive infrastructure for implementation.

2.1 Bayesian Optimal Interval Phase I/II (BOIN12) Design for Immunotherapy and Targeted Therapies

To remove these barriers, Lin et al. (2020) propose the BOIN12 design to find the OBD that optimizes the risk-benefit tradeoff. The BOIN12 design makes the decision of dose escalation and de-escalation by simultaneously taking account of efficacy and toxicity, and adaptively allocates patients to the dose that optimizes the toxicity-efficacy tradeoff. BOIN12 is transparent, easy to implement, and its adaptation rule can be pre-tabulated.

2.1.1 Toxicity-Efficacy Tradeoff

BOIN12 uses utility to measure the toxicity-efficacy tradeoff, whose value should be elicited from clinicians to reflect the clinical desirability of each possible toxicity and efficacy outcome. For ease of exposition, we assume toxicity and efficacy endpoints are binary, but the BOIN12 design can be generalized to multi-level or multiple outcomes, see Lin et al. (2020) for an example of accounting for severe immune-related adverse events using a three-level toxicity endpoint. Let Y_T and Y_E be the toxicity and efficacy outcomes, respectively. Given any patient in the trial, there are four possible outcomes: 1 = $(Y_T = 0, Y_E = 1)$, 2 = $(Y_T = 0, Y_E = 0)$, 3 = $(Y_T = 1, Y_E = 1)$, and 4 = $(Y_T = 1, Y_E = 0)$, as shown in Table 1. Clearly, $(Y_T = 0, Y_E = 1)$ is the most desirable, $(Y_T = 1, Y_E = 0)$ is the least desirable, and the other two are in between. Let $u_1, \cdots, u_4$ denote utilities ascribed to the four possible toxicity-efficacy outcomes, respectively. To elicit the utility for the four outcomes, we assign the most desirable outcome $(Y_T = 0, Y_E = 1)$ a score of u_1 =100, and the least desirable outcome $(Y_T = 1, Y_E = 0)$ a score of u_4 =0, and then ask clinicians to use them as a reference to specify scores to the other two outcomes. The scores ascribed to the other two outcomes (i.e., u_2 and u_3) should be elicited from clinicians to reflect their clinical desirability. For example, if patients are willing to tolerate toxicity in exchange for efficacy, then $u_3 \geq u_2$ can be used; otherwise, $u_3 < u_2$ should be used. From the dose-finding perspective, setting $u_3 \geq u_2$ usually leads to a more thorough exploration of the dose space than setting $u_3 < u_2$. Table 1 provides an example of elicited utility. The proposed design does not depend on the scale of the utility, and other scoring system can also be used, e.g., use 1 and 10 to indicate the least and most desirable outcomes, respectively.

Given a specific dose d, let $p_1, \cdots, p_4$ denote the respective probabilities of observing the four possible outcomes at d, which typically vary across doses. The desirability (or mean utility) of d is defined as

$$u(d) = p_1 u_1 + p_2 u_2 + p_3 u_3 + p_4 u_4. \tag{1}$$

A higher value of $u(d)$ indicates a higher desirability of dose d in the risk-benefit tradeoff, and the dose with the highest desirability is the OBD. Revisit the scenario where true toxicity probabilities $p_T = (0.05, 0.12, 0.27, 0.35, 0.50)$ and efficacy probabilities $p_E = (0.20, 0.35, 0.36, 0.37, 0.38)$, respectively, for five investigational doses. Assuming the utility in Table 1 and the correlation between p_T and p_E is 0.25, the desirability for $d = 3$ is $u(d = 3) = 50.8$, and the desirability for $d = 2$ is $u(d = 2) = 56.2$. By using the utility approach, we naturally account for the risk-benefit tradeoff and correctly identify $d = 2$ as the OBD.

Table 1 Utility table for binary efficacy and toxicity endpoints

	$Y_E = 1$	$Y_E = 0$
$Y_T = 0$	$u_1 = 100$	$u_2 = 40$
$Y_T = 1$	$u_3 = 60$	$u_4 = 0$

To safeguard patients from toxic or futile doses, dose acceptability criteria are used to determine admissible doses that can be used to treat patients. Let ϕ_T and ϕ_E be the clinician-specified toxicity upper limit and efficacy lower limit, respectively. Generally, ϕ_E can take the value of the target response rate specified for a standard phase II trial. Because BOIN12 considers the toxicity-efficacy tradeoff, the value of ϕ_T should be set slightly higher than the target toxicity rate used in conventional toxicity-based phase I designs. For example, if 30% is an appropriate target toxicity rate that the conventional phase I design uses, then $\phi_T = 35\%$ or 40% is a reasonable choice for BOIN12. Let $p_T(d)$ and $p_E(d)$ denote the toxicity and efficacy probabilities of dose d, respectively. Dose d is defined as admissible if the observed data indicates that d is reasonably safe and efficacious (e.g., $\Pr(p_T(d) > \phi_T|data) < 0.95$ and $\Pr(p_E(d) < \phi_E|data) < 0.9$). During the trial, doses that are not admissible should be eliminated and incoming patients are assigned only to admissible doses. If all doses are eliminated, the trial is terminated without any dose selected.

2.1.2 Dose Escalation/de-escalation Rule

The dose-finding rule of the BOIN12 design is given below and described in Fig. 1.

1. Treat the first cohort of patients at the lowest or a prespecified starting dose.
2. Suppose at the current dose d, n patients have been treated, among which n_T patients had toxicity and n_E patients had efficacy. Define the observed toxicity rate as $\hat{p}_T(d) = n_T/n$, and let λ_e and λ_d denote the optimal dose escalation and de-escalation boundaries adopted from the BOIN design (Liu and Yuan, 2015, Yuan et al. (2016)) (see Fig. 1 for the values of λ_e and λ_d).
   (a) If $\hat{p}_T(d) > \lambda_d$, de-escalate the dose to $d - 1$ to treat the next cohort of patients.
   (b) If $\lambda_e < \hat{p}_T(d) \leq \lambda_d$ and $n \geq 6$, determine the desirability of d and $d - 1$ using the prespecified rank-based desirability score (RDS) table (see the details in Lin et al. (2020)) and choose the one that has the higher desirability to treat the next cohort.
   (c) Otherwise, i.e., $\hat{p}_T(d) \leq \lambda_e$, or $\lambda_e < \hat{p}_T(d) \leq \lambda_d$ and $n < 6$, determine the desirability of $d - 1$, d, and $d + 1$, using the prespecified RDS table and choose the one that has the highest desirability to treat the next cohort.
3. Repeat step 2 until the maximum sample size is reached. The final OBD is selected based on the following two-step procedure:
   (a) Identify the MTD as the dose level that has the isotonically estimated toxicity probability closest to the toxicity upper limit ϕ_T.
   (b) The final OBD is chosen as the dose level that has the highest estimated utility among the doses that are not higher than the MTD.

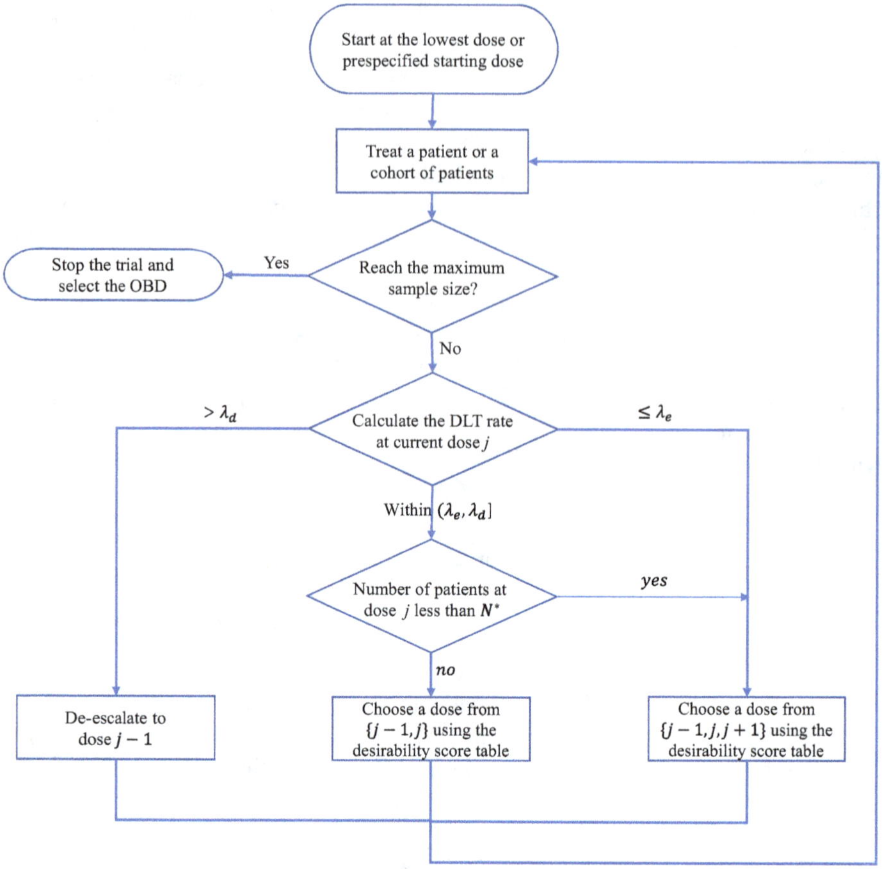

Target toxicity rate ϕ_T	0.20	0.25	0.30	0.35	0.40
Escalation boundary λ_e	0.157	0.197	0.236	0.276	0.316
De-escalation boundary λ_d	0.238	0.298	0.359	0.419	0.480

Fig. 1 The schema of the BOIN12 design, where (λ_e, λ_d) are a pair of optimized dose escalation and de-escalation boundaries adopted from the BOIN design, and N^* is a prespecified sample size cutoff (e.g., $N^* = 6$)

A key innovation of BOIN12 is that its calculation of dose desirability (in Step 2b and 2c above) can be pre-tabulated and included in the trial protocol before the trial starts, which greatly simplifies design implementation. This is achieved by using the quasi-beta-binomial model (Yuan et al., 2007), which converts complicated desirability calculations (that involve bivariate toxicity and efficacy outcomes) into a simple beta-binomial modeling. The RDS table can be easily generated using the BOIN12 companion software. During the trial, the desirability of a dose can be simply determined as follows: count the number of patients treated at that dose, the number of patients who experienced toxicity, the number of patients who experienced efficacy, and the number of patients who experienced efficacy without toxicity (not needed when $u_2 + u_3 = 100$), then look up the RDS table to determine its RDS. Technical details of the quasi-beta-binomial model, RDS tables under two common clinical settings, and an example showing how to use the RDS table to determine dose assignment can be found in Lin et al. (2020) and its Supplementary Data. Of note, the RDS is not the raw value of desirability, defined in Eq. (1), but the desirability rank among all possible outcomes. A larger value of RDS means a higher utility. This allows for a simpler presentation of the desirability of a dose using integers.

For ethical consideration, BOIN12 always assigns the next cohort of patients to the dose with the highest estimate of desirability. To avoid the algorithm to get stuck at a locally optimal dose, we impose a dose exploration rule: if (i) the number of patients treated at the current dose d is greater than 8, (ii) the observed toxicity rate $\hat{p}_T(d)$ is less than the de-escalation boundary λ_d, and (iii) the next higher dose has never been used for treating patients, then we escalate the dose for treating the next cohort of patients.

Simulation results provided in Lin et al. (2020) show that compared with existing phase I/II dose-finding designs, the BOIN12 design is simpler to implement, has higher accuracy to identify the OBD, and allocates more patients to the OBD. For more details of the simulation studies, please refer to Lin et al. (2020).

One limitation of the BOIN12 design is that it does not consider patient heterogeneity and so it aims to identify the OBD for the whole patient population. However, in reality, patients form subgroups stratified by biomarkers. The optimal dose for treating patients with novel targeted agents may differ according to the patient's biomarker status, which requires the dose-finding designs to account for patient's biomarker profiles.

3 Designs that Account for Patient's Heterogeneity

Like BOIN12 design, other conventional clinical trials also aim to find an optimal dose/treatment ignoring patient variability, that is, the optimal dose/treatment is the same for all patients of a specific disease (Storer, 1989; O'Quigley, Pepe, and Fisher, 1990; Whitehead & Brunier, 1995; Thall & Russell, 1998; Stylianou & Flournoy, 2002; Thall & Cook, 2004). However, due to inter-patient heterogeneity, a treatment can be very successful for some patients but not for others. For example, some drugs may benefit only 2% of the patients and there are even drugs that are harmful to certain ethnic groups (Schork, 2015). To overcome this "one-size-fits-all" approach and to acknowledge patient heterogeneity, in January 2015, a new \$215-million national Precision Medicine Initiative was launched. Optimizing the treatment benefit under the paradigm of precision medicine requires clinical trials to account for the patient's individual characteristics when searching for the optimal dose/treatment. This approach can yield treatment options for patients whose cancer is not responding favorably to standard therapy.

In this section, we will discuss two clinical trial designs that take into account patient's heterogeneity. The first one is a Bayesian adaptive design to determine subgroup OBD by accounting for patient's biomarker profile, and the second one is a Bayesian adaptive enrichment design that simultaneously identifies the sensitive subgroup and tests if the treatment is effective in that subgroup.

3.1 Bayesian Biomarker-Based Phase I/II Design

To account for biomarker effect, the most natural approach is to include the patient's biomarkers as well as the interactions between those biomarkers and the dose as covariates in the dose-finding model (Babb and Rogatko, 2001; O'Quigley and Paoletti, 2003; Ivanova and Wang, 2006; Thall et al., 2008). A common limitation of these methods is that they can accommodate only a very small number (e.g., 1 or 2) of patient characteristics into the dose-finding process in order to avoid the curse of dimensionality when working with the small sample sizes that are typical of early-phase trials. For example, the method of Thall, Nguyen and Estey (2008) considered only two prognostic factors and their interactions with the dose. In practice, however, the number of patient characteristics that may affect the dose-response relationship can be easily more than two, and moreover these characteristics (e.g., some biomarkers) may be highly correlated. The direct inclusion of a large number of potentially highly correlated patient characteristics as covariates in regression models leads to extremely unreliable estimates or identification problems.

To overcome these difficulties, Guo and Yuan (2017) describe a Bayesian phase I/II personalized dose-finding method that can accommodate a more realistic number of biomarkers. To circumvent the curse of dimensionality and potential high correlations among biomarkers, the canonical partial least squares (CPLS) method is used to extract a small number of components from the covariate matrix consisting of dose, biomarker profile, and dose-by-biomarker interactions. These CPLS components capture most of the covariate information that is predictive of the response variable (Garthwaite, 1994). Conditional on the CPLS components, a latent-variable approach is used to model the ordinal toxicity and efficacy outcomes.

3.1.1 Canonical Partial Least Squares

Partial least squares (PLS) is a dimension-reduction technique for modeling the relationship between response variable(s) and explanatory variables (Stone & Brooks, 1990; Frank & Friedman, 1993; Garthwaite, 1994). The basic idea is to construct a few PLS components, which are linear combinations of the explanatory variables, to capture most of the information contained in the explanatory variables for predicting the response variable(s). These PLS components can be used to replace the original high-dimensional explanatory variables in a subsequent regression analysis. PLS determines components using both the explanatory variables and the response variable(s), and so is more relevant in the regression context and provides better prediction, compared with the well-known principle component analysis. PLS was originally developed for continuous response variables (Garthwaite, 1994). To handle categorical response variables, Indahl et al. (2008) proposed the CPLS method, which integrates the PLS methodology with canonical correlation analysis. The CPLS algorithm is a sequential procedure that iteratively extracts components of $\boldsymbol{X}$ in such a way that the first component contains the most information for predicting Y, the second component contains less information than the first, and so on. Indahl et al. (2008) show that CPLS is able to extract more information from the first few components than the ordinary PLS approach. This property is particularly attractive for early-phase clinical trials because of the small sample sizes. A parsimonious model with fewer CPLS components is favored. Details on how to determine CPLS components can be found in Guo and Yuan (2017) and Indahl et al. (2008).

3.1.2 Probability Models

Consider a phase I/II clinical trial with R prespecified doses $(d_1, \cdots, d_R)$ of a molecularly targeted agent. We define $Y_T = 1, 2, \cdots, J$ corresponding to increasing levels of severity of toxicity, and $Y_E = 1, 2, \cdots, K$ corresponding to increasingly desirable efficacy. There are several motivations to model ordinal toxicity and efficacy endpoints. First, the ordinal variables reflect the actual scale used for measuring toxicity and efficacy in clinical studies. For example, in many trials, toxicity is scored as grades 1 to 5 and coded as a three-level ordinal variable (i.e., grades 1-2, grade 3, and grade $\geq$ 4), and efficacy is scored as PD, SD, or PR/CR. These refined scales contain more information than the dichotomized toxicity and efficacy outcomes for decision making (Yuan et al., 2007; Houede et al., 2010). Second, and more importantly, using binary toxicity and efficacy outcomes is often problematic for the novel molecularly targeted agents. For example, the binary efficacy outcome typically defines CR/PR as a response, and PD/SD as a nonresponse. However, many novel targeted agents function by stabilizing the tumor rather than shrinking it. For these agents, SD is often regarded as a favorable outcome, although not as desirable as CR/PR, and thus should be distinguished from PD.

Let n denote the number of patients enrolled in the trial at an interim decision time, $\boldsymbol{X} = (X_1, \cdots, X_p)'$ denote the vector of p binary biomarkers, and Z denote the dose with $Z \in (d_1, \cdots, d_R)$. Define an $n \times (1 + 2p)$ covariate matrix $\boldsymbol{V} = (Z, X_1, \cdots, X_p, ZX_1, \cdots, ZX_p)$, where $ZX_1, \cdots, ZX_p$ represent the interactions between dose Z and biomarkers X_1 to X_p. To overcome issues of small sample size and potential multicollinearity, CPLS is used to reduce the dimension of the covariate matrix $\boldsymbol{V}$. Specifically, CPLS is applied to extract the first q_1 CPLS components $\boldsymbol{W}_T(Z, \boldsymbol{X}) = (W_{T,1}(Z, \boldsymbol{X}), \cdots, W_{T,q_1}(Z, \boldsymbol{X}))$ from $\boldsymbol{V}$ for predicting Y_T, and the first q_2 CPLS components $\boldsymbol{W}_E(Z, \boldsymbol{X}) = (W_{E,1}(Z, \boldsymbol{X}), \cdots, W_{E,q_2}(Z, \boldsymbol{X}))$ for predicting Y_E. Each of these components is a linear combination of biomarkers $\boldsymbol{X}$, dose Z and their interactions. As biomarkers that are predictive of toxicity may differ from those that are predictive of efficacy, the CPLS components are drawn independently for toxicity and efficacy. In practice, some biomarkers could affect the treatment efficacy but are less likely to affect the toxicity. The CPLS method will automatically detect such differences and assign different weights to different biomarkers for predicting efficacy and toxicity.

The latent variable approach is use to model the relationship between Y_k and $\boldsymbol{W}_k$, where $k = T$ or E. Let subscript i index patients. Assume there are two latent normal variables $Y^*_{T,i}$ and $Y^*_{E,i}$ for patient i, which follow linear models

$$Y^*_{T,i}|\boldsymbol{W}_{T,i}, \boldsymbol{\beta}_T, \theta_i, \tau^2_{T,i} \sim N(\theta_i + \boldsymbol{W}_{T,i}\boldsymbol{\beta}_T, \tau^2_{T,i})$$

$$Y^*_{E,i}|\boldsymbol{W}_{E,i}, \boldsymbol{\beta}_E, \theta_i, \tau^2_{E,i} \sim N(\theta_i + \boldsymbol{W}_{E,i}\boldsymbol{\beta}_E, \tau^2_{E,i}),$$

where $\boldsymbol{\beta}_T$ and $\boldsymbol{\beta}_E$ are vectors of the regression coefficients, and $\theta_i \sim N(0, \sigma^2)$ is a patient-specific random effect that is used to induce correlation between $Y^*_{T,i}$ and $Y^*_{E,i}$.

The observed toxicity and efficacy outcomes $Y_{T,i}$ and $Y_{E,i}$ for the ith patient are related to $Y^*_{T,i}$ and $Y^*_{E,i}$ as

$$Y_{T,i} = j \quad \text{if} \quad \gamma_{T,j-1} < Y^*_{T,i} \le \gamma_{T,j}$$

$$Y_{E,i} = k \quad \text{if} \quad \gamma_{E,k-1} < Y^*_{E,i} \le \gamma_{E,k},$$

where $-\infty \equiv \gamma_{T,0} < \gamma_{T,1} < \cdots < \gamma_{T,J} \equiv \infty$ and $-\infty \equiv \gamma_{E,0} < \gamma_{E,1} < \cdots < \gamma_{E,K} \equiv \infty$ are unknown cutoffs. To identify the toxicity model, we set $\tau^2_{T,i} = 1$ (Albert & Chib, 1993). This results in the familiar cumulative probit model for toxicity:

$$\Phi^{-1}[\Pr(Y_{T,i} \le j|\boldsymbol{W}_{T,i}, \theta_i, \gamma_{T,j}, \boldsymbol{\beta}_T)] = \gamma_{T,j} - \theta_i - \boldsymbol{W}_{T,i}\boldsymbol{\beta}_T \tag{2}$$

for $j = 1, \cdots, J-1$. Although equivalent, taking the latent variable approach greatly simplifies the posterior sampling using the Gibbs sampler (Albert & Chib, 1993).

It is tempting to take the same approach to using a cumulative probit model for Y_E by setting $\tau^2_{E,i} = 1$. However, the cumulative probit model such as (2) implies a stochastic ordering of the effects of the covariates on the outcome; that is, for any two doses d_r and $d_{r'}$, either $\Pr(Y_{T,i} \le j|\boldsymbol{W}_T(Z = d_r, \boldsymbol{X})) \le \Pr(Y_{T,i} \le j|\boldsymbol{W}_T(Z = d_{r'}, \boldsymbol{X}))$ or $\Pr(Y_{T,i} \le j|\boldsymbol{W}_T(Z = d_r, \boldsymbol{X})) \ge \Pr(Y_{T,i} \le j|\boldsymbol{W}_T(Z = d_{r'}, \boldsymbol{X}))$ for all $j = 1, \cdots, J-1$. Such a monotonic assumption could be problematic for efficacy because the efficacy behavior of molecularly targeted agents is much more complex. The dose-efficacy curve may plateau or sometimes be lower at higher doses (Postel-

Vinay et al., 2009). To accommodate this feature of molecularly targeted agents, the following dispersed cumulative probit model is used by setting $\tau_{E,i} = \exp(\tau Z_i)$ (McCullagh, 1980, Cox, 1995, and Agresti, 2002),

$$\Phi^{-1}[\Pr(Y_{E,i} \le k | \boldsymbol{W}_{E,i}, \theta_i, \gamma_{E,k}, \boldsymbol{\beta}_E)] = \frac{\gamma_{E,k} - \theta_i - \boldsymbol{W}_{E,i}\boldsymbol{\beta}_E}{\exp(\tau Z_i)}, \quad k = 1, \cdots, K-1. \tag{3}$$

By introducing the dispersion parameter, τ, Y_E^* will have different dispersions across doses Z, and thus the model does not impose the stochastic ordering of the dose effect on efficacy. When $\tau = 0$, the model reduces to the standard cumulative probit model.

Based on the above formulation, the likelihood function for patient i is

$$\begin{aligned} p(y_{T,i}, y_{E,i}) &= \Pr(Y_{T,i} = y_{T,i}, Y_{E,i} = y_{E,i}) \\ &= \Pr\Big(\gamma_{T,y_{T,i}-1} < Y_{T,i}^* \le \gamma_{T,y_{T,i}}, \gamma_{E,y_{E,i}-1} < Y_{E,i}^* \le \gamma_{E,y_{E,i}}\Big) \\ &= \int_{\gamma_{T,y_{T,i}-1}}^{\gamma_{T,y_{T,i}}} \int_{\gamma_{E,y_{E,i}-1}}^{\gamma_{E,y_{E,i}}} \phi_2(Y_{T,i}^*, Y_{E,i}^* | \boldsymbol{\mu}_Y, \boldsymbol{\Sigma}_Y) dY_T^* dY_E^* \end{aligned}$$

where $\phi_2(\cdot|\boldsymbol{\mu}, \boldsymbol{\Sigma})$ is the density of a bivariate normal distribution with mean vector $\boldsymbol{\mu}$ and covariance matrix $\boldsymbol{\Sigma}$,

$$\boldsymbol{\mu}_Y = \begin{pmatrix} \boldsymbol{W}_{T,i}\boldsymbol{\beta}_T \\ \boldsymbol{W}_{E,i}\boldsymbol{\beta}_E \end{pmatrix}, \text{ and } \boldsymbol{\Sigma}_Y = \begin{pmatrix} \tau_{T,i}^2 + \sigma^2 & \sigma^2 \\ \sigma^2 & \tau_{E,i}^2 + \sigma^2 \end{pmatrix}.$$

Letting $\Theta = (\gamma_{T,1}, \cdots, \gamma_{T,J}, \beta_T, \gamma_{E,1}, \cdots, \gamma_{E,K}, \beta_E, \tau, \sigma^2)$ denote the parameters, and $\mathscr{D}_n = \{(Y_{T,i}, Y_{E,i}, Z_i, \boldsymbol{W}_i), i = 1, \cdots, n\}$ denote the data from the n patients when an interim decision is to be made, the likelihood based on n patients is $\mathscr{L}(\mathscr{D}_n|\Theta) \propto \prod_{i=1}^{n} p(y_{T,i}, y_{E,i})$. Letting $f(\Theta)$ be the joint prior distribution for Θ, the joint posterior distribution of Θ is $f(\Theta|\mathscr{D}_n) \propto f(\Theta)\mathscr{L}(\mathscr{D}_n|\Theta)$.

A practical approach to determining the number of CPLS components (i.e., q_1 and q_2) is to use the scree plot as a guidance. In CPLS analysis, the scree plot displays the root mean squared error of prediction (RMSEP) for the response variable (Y_T or Y_E) versus the number of CPLS components. Given a specific number of CPLS components, the RMSEP can be obtained using cross-validation. The ideal pattern in a scree plot is a steep curve, followed by a flat line. We retain those CPLS components in the steep curve before the first point that starts the flat line trend. When the scree plot suggests a large number of CPLS components, we recommend restricting $q_1, q_2 \le 4$ because of the small sample size of phase I/II trials. In numerical studies described in Guo and Yuan (2017), 2 CPLS components generally work well under a variety of scenarios.

3.1.3 Personalized Optimal Dose

Consider the situation where both toxicity and efficacy outcomes are categorized as ordinal variables with 3 categories, with $Y_T = 1, 2, 3$ representing grade 1 or 2, grade 3, and grade ≥ 4 adverse events, respectively, and $Y_E = 1, 2, 3$ representing PD, SD, and PR/CR, respectively. To define the personalized optimal dose for each patient, the set of acceptable doses is first defined to safeguard against using doses that are overly toxic or futile. Let $\pi_T(3) \equiv \Pr(Y_T = 3|Z, \boldsymbol{X})$, $\pi_E(3) \equiv \Pr(Y_E = 3|Z, \boldsymbol{X})$, $\pi_E(2, 3) \equiv \Pr(Y_E = 2 \text{ or } 3|Z, \boldsymbol{X})$, ϕ_T be the upper bound on the probabilities of toxicity, and $\phi_{E,3}$ and $\phi_{E,2}$ be the lower bounds on the probabilities of efficacy. The values of ϕ_T, $\phi_{E,3}$ and $\phi_{E,2}$ should be elicited from physicians. At an interim time with observed data $\mathscr{D}_n$, dose Z is defined as having acceptable toxicity for a patient with biomarker vector $\boldsymbol{X}$, if

$$\Pr\big(\pi_T(3) \geq \phi_T | W(Z, \boldsymbol{X}), \mathscr{D}_n\big) \leq \delta_T, \tag{4}$$

and acceptable efficacy if

$$\Pr\big(\pi_E(3) \leq \phi_{E,3} | W(Z, \boldsymbol{X}), \mathscr{D}_n\big) \leq \delta_{E,3} \tag{5}$$

or

$$\Pr\big(\pi_E(2, 3) \leq \phi_{E,2} | W(Z, \boldsymbol{X}), \mathscr{D}_n\big) \leq \delta_{E,2}, \tag{6}$$

where $\delta_{E,3}$, $\delta_{E,2}$ and δ_T are non-negative cutoffs that should be calibrated to obtain good design operating characteristics. The definition of acceptable efficacy (5) and (6) accounts for the practical consideration that for a molecularly targeted agent, a dose is considered promising if it can achieve a certain rate of CR/PR or a certain rate of SD (if the rate of CR/PR is low).

Given a dose that has acceptable toxicity and efficacy, its desirability is measured using a utility function that is similar to that in the BOIN12 design: the desirability of the most desirable outcome pair (i.e., $y_T = 1$, $y_E = 3$) is fixed as $\zeta(1, 3) = 100$ and the desirability of the least desirable outcome pair (i.e., $y_T = 3$, $y_E = 1$) is fixed as $\zeta(3, 1) = 0$, and then use these two pairs as references to score the desirability of the other elementary outcome pairs $\{\zeta(y_T, y_E), y_T = 1, 2, 3, y_E = 1, 2, 3\}$ using the scale of (0, 100). One example is the following utility table provided by Guo and Yuan (2017).

	y_E=1	y_E=2	y_E=3
y_T=1	10	50	100
y_T=2	5	20	60
y_T=3	0	10	20

For patients with biomarker profile $\boldsymbol{X}$, the *personalized optimal dose* is defined as the dose that maximizes the true mean utility

$$U_{true}(Z, \boldsymbol{X}) = \sum_{y_T=1}^{3} \sum_{y_E=1}^{3} \zeta(y_T, y_E)\Pr(y_T, y_E|Z, \boldsymbol{X}, \Theta),$$

while also having both acceptable toxicity and efficacy defined by (4)–(6). Note that $U_{true}(Z, \boldsymbol{X})$ depends on biomarker profile $\boldsymbol{X}$, and thus the target doses may be different for patients who have different biomarker profiles. During the trial, given the interim data $\mathscr{D}_n$ collected from n patients, we evaluate the desirability of dose Z for a patient with biomarker profile $\boldsymbol{X}$ based on the posterior mean utility,

$$\begin{aligned} &U_n(W_T(Z, \boldsymbol{X}), W_E(Z, \boldsymbol{X})) \\ &= \sum_{y_T=1}^{3} \sum_{y_E=1}^{3} \zeta(y_T, y_E) E_\Theta[\Pr\{y_T, y_E \mid W_T(Z, \boldsymbol{X}), W_E(Z, \boldsymbol{X}), \Theta\}|\mathscr{D}_n]. \end{aligned}$$

3.1.4 Personalized Dose-Finding Algorithm

Due to the small amount of available data at the beginning of the trial, a two-stage dose-finding scheme is adopted. In stage I, conventional dose escalation is conducted with a maximum sample size of N_1 patients, based only on the toxicity outcome without considering the patients' biomarker profiles. The objective of this stage is to preliminarily evaluate the safety of the doses and collect some toxicity and efficacy data to facilitate the model fitting in stage II. To be consistent with the conventional approach, at this stage, we dichotomize toxicity into dose-limiting toxicity or not, and employ the Bayesian optimal interval design (BOIN; Liu and Yuan, 2015) to conduct dose escalation.

Once stage I is completed, stage II starts and it performs personalized dose finding with a maximum sample size of N_2 patients. Since the optimal dose depends on the patient's individual biomarker profile, patients are treated one by one in stage II, and the dose-response curves are continuously updated to determine the optimal dose for each patient. Specifically, assuming that $n-1$ patients have been enrolled in the trial, a dose is assigned to the nth patient with biomarker profile $\boldsymbol{X}_n$ as follows.

1. Based on the observed data $\mathscr{D}_{n-1}$, the CPLS components are extracted to fit the model, and the acceptable dose set $\mathscr{A}(\boldsymbol{X}_n)$ is determined for the nth patient. Note that as $\mathscr{A}(\boldsymbol{X}_n)$ depends on the patient's individual biomarker profile, it varies from patient to patient.
2. If $\mathscr{A}(\boldsymbol{X}_n)$ is an empty set, that is, no dose is acceptable for that patient, then the patient is treated off protocol. If a certain number, say 10, contiguous patients are treated off protocol, then the trial is terminated with the conclusion that no dose is acceptable.

3. If $\mathscr{A}(\boldsymbol{X}_n)$ is not empty, adaptive randomization is used to choose a dose in $\mathscr{A}(\boldsymbol{X}_n)$. The patient is randomized to dose j with probability λ proportional to its posterior expected utility, i.e.,

$$\lambda = \frac{U_{n-1}(W_T(Z = d_j, \boldsymbol{X}), W_E(Z = d_j, \boldsymbol{X}))}{\sum_{d_j \in \mathscr{A}(\boldsymbol{X}_n)} U_{n-1}(W_T(Z = d_j, \boldsymbol{X}), W_E(Z = d_j, \boldsymbol{X}))}.$$

4. Steps 1–3 are repeated until reaching the maximum sample size of N_2. Dose skipping is not allowed when a dose escalation occurs.

After stage II is completed, the personalized optimal dose, for each possible biomarker profile $\boldsymbol{X}$, is defined as the dose in $\mathscr{A}(\boldsymbol{X})$ with the largest posterior mean utility $U_N(W_T(Z, \boldsymbol{X}), W_E(Z, \boldsymbol{X}))$, where $N = N_1 + N_2$.

3.1.5 Simulation Studies

Extensive simulation studies were performed in Guo and Yuan (2017) to evaluate the performance of the design. Graphical presentations of the main effects and interaction effects of the biomarkers on efficacy and toxicity of the simulation scenarios are provided in Figs. 2 and 3. As shown by Figs. 2 and 3, the scenarios were chosen to cover various situations of biomarker effects. For example, in scenario 1, biomarkers 1 and 5 have main effects on toxicity, and biomarkers 1, 2, 4, and 5 interact with the dose to affect toxicity; biomarkers 1 and 3 have main effects on efficacy, and biomarkers 1 and 4 interact with the dose to affect efficacy.

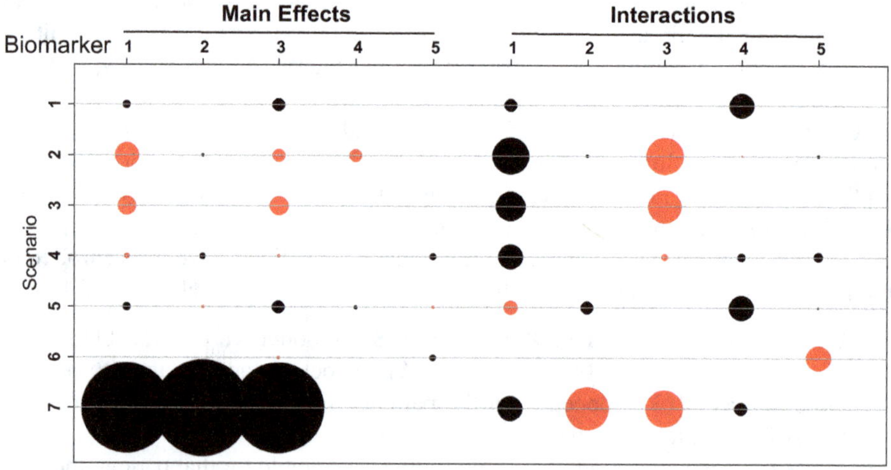

Fig. 2 Main effects and interaction effects on efficacy probabilities for the scenarios in Guo and Yuan (2017). The radius of a circle is proportional to the absolute value of the coefficient. Red and black circles respectively represent positive and negative coefficients

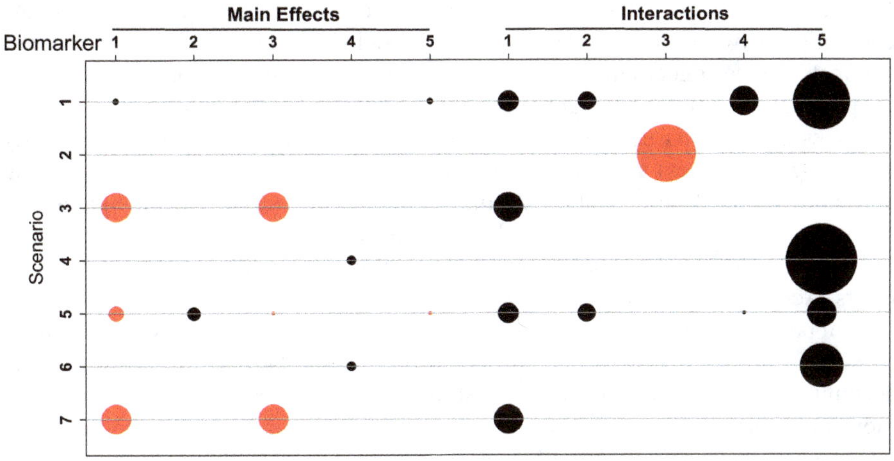

Fig. 3 Main effects and interaction effects on toxicity probabilities for the scenarios in Guo and Yuan (2017). The radius of a circle is proportional to the absolute value of the coefficient. Red and black circles respectively represent positive and negative coefficients

In scenario 2, none of the biomarkers has a main effect on toxicity, only biomarker 3 interacts with the dose to affect toxicity, and all 5 biomarkers are involved in the efficacy probability. Scenario 7 shows an interesting and common case with targeted agents: no treatment effect is observed except in the presence of some combination of biomarkers. The drug is effective only for patients for whom the first 3 biomarkers are positive, regardless of the status of the other two biomarkers. Simulation results show that the proposed design has desirable operating characteristics. Please see Guo and Yuan (2017) for more details.

We evaluated the sensitivity of the proposed design to the specification of the utility function. We considered the following 2 alternative utility functions: The first new utility gives a desirability score of 0 to (y_T, y_E) as long as $y_E = 1$ (i.e., PD) regardless of the value of y_T. That is, if patients experience disease progression, toxicity does not affect the desirability of the drug. The second utility assigns higher desirability scores to (y_T, y_E) when $y_E = 2$ or 3 (i.e., SD or PR/CR). As shown in Table 4 in Guo and Yuan (2017), the results under the two alternative utilities are generally similar to the original results, suggesting that the proposed design is not sensitive to the elicited utility.

	Utility 1			Utility 2		
	y_E=1	y_E=2	y_E=3	y_E=1	y_E=2	y_E=3
y_T=1	0	50	100	0	65	100
y_T=2	0	20	60	0	35	75
y_T=3	0	10	20	0	25	35

In order to capture the important features of toxicity and efficacy, and the correlation between the two endpoints, our model has a relatively large number of parameters. One concern may be that at the inception of the trial, data are sparse and the parameter estimates are highly variable and mainly driven by the prior. However, this does not cause issues because the number of investigational doses is typically small (e.g., <6 doses) and our dose-finding algorithm does not allow for skipping untried doses for dose escalation. At the beginning of the trial when the parameter estimates are highly variable, the dose-finding algorithm acts somewhat "semi-randomly" by trying the doses sequentially from low to high, guided largely by the priors. Actually, in some anti-intuitive sense, such uncertainty and "randomness" are helpful because it provides the design freedom to move around and explore the dose space, and to avoid being stuck at a local dose. When the trial proceeds and data accumulate, we obtain more reliable estimates and the dose assignment becomes more stable and converges to the target dose. Therefore, as long as at the middle or late stage of the trial, we have adequate data to make reasonable estimates, we are likely to make the correct dose assignment and select the target dose at the end of the trial. In addition, the primary objective of the phase I/II trial is to identify the optimal dose among a set of prespecified doses, not to obtain accurate estimates. This also renders the design higher tolerance to the variability of parameter estimates. As long as the method obtains the rank of estimated desirability correctly, it will correctly select the target dose. As the simulation studies suggest, our design is reliable in identifying the optimal doses.

3.2 Adaptive Enrichment Design

In the presence of patient heterogeneity, the enrichment design is another effective approach to achieving precision medicine. The enrichment design is motivated by the fact that heterogeneity of patient response to an experimental treatment, E, may be due to biological covariates that modify the effects of E at the cellular or molecular level. In that case, only a subset of "E-sensitive" patients defined by those biological covariates may respond favorably to E. Precision medicine uses biological covariates to restrict use of drug to an identified subset of E-sensitive patients, avoiding administration of E in non-sensitive patients who are unlikely to benefit from E.

An efficient approach in such settings is the enrichment design that focuses on E-sensitive patients. Most existing enrichment designs assume that an E-sensitive subgroup is known based on pre-clinical studies or limited phase II trial data, and moreover that E is substantively efficacious in that subpopulation. However, due to differences between pre-clinical settings and human biology, data often show that one or both of these assumptions are false. Therefore, the key problems are how to use biological covariates to (1) determine whether an E-sensitive subgroup exists, and if so identify it, and (2) determine whether E provides an improvement over a standard control therapy, C, in the subgroup.

Park et al. (2022) propose a Bayesian group sequential (GS) enrichment design (AED) that compares E to C based on survival time Y, and uses an early response Z as an ancillary outcome to assist with variable selection and enrichment. Initially, the design enrolls patients under broad eligibility criteria. At each interim decision, submodels for regression of Y and Z on baseline covariates and treatment are fit, variable selection is used to identify covariates that characterize treatment-sensitive patients and determines a personalized benefit index (PBI), and superiority and futility decisions are made. Enrollment of each cohort is restricted to the most recent adaptively identified treatment-sensitive patients. Group sequential decision cutoffs are calibrated to control overall type I error and account for the adaptive enrollment restriction.

3.2.1 Design Structure

Consider a comparative clinical trial with patients randomized to E $(G = 1)$ or C $(G = 0)$ in a fixed ratio. For each patient, we observe a covariate vector $\boldsymbol{x} \in \mathbb{R}^p$ at enrollment, a short-term response indicator Z, and a time-to-event endpoint Y. In cancer trials, Y is typically the progression-free survival or the overall survival time, and Z may be the indicator of $\geq 50\%$ tumor shrinkage after treatment. For right-censoring of Y at follow up time U when an interim decision is to be made, we define the observed event time $Y^o = \min(Y, U)$ and event indicator $\delta = I(Y \leq U)$.

In treatment arm $G = 0$ or 1, let $\pi(\boldsymbol{x}, G, \boldsymbol{\theta}_Z) = \Pr(Z = 1|G, \boldsymbol{x}, \boldsymbol{\theta}_Z)$ and $h_G(y|\boldsymbol{x}, Z, \boldsymbol{\theta}_Y)$ denote the response rate of Z for a patient with covariates $\boldsymbol{x}$ and the hazard function of Y at time y for a patient with covariates $\boldsymbol{x}$ and response indicator Z, respectively, where $\boldsymbol{\theta}_Z$ and $\boldsymbol{\theta}_Y$ are the model parameter vectors. At each interim decision, the AED design adaptively selects two subvectors of $\boldsymbol{x}$ to identify patients expected to benefit more from E than C in terms of Z or Y. The first subvector, $\boldsymbol{x}_Z$, is determined based on the difference in response probabilities $\Delta_Z(\boldsymbol{x}, \boldsymbol{\theta}_Z) = \pi(\boldsymbol{x}, 1, \boldsymbol{\theta}_Z) - \pi(\boldsymbol{x}, 0, \boldsymbol{\theta}_Z)$ in the regression model for $(Z \mid G, \boldsymbol{x})$. The second subvector, $\boldsymbol{x}_Y$, is identified based on the hazard ratio $\Delta_Y(\boldsymbol{x}, \boldsymbol{\theta}_Y) = h_1(y|\boldsymbol{x}, Z, \boldsymbol{\theta}_Y)/h_0(y|\boldsymbol{x}, Z, \boldsymbol{\theta}_Y)$ in the regression model for $(Y \mid Z, G, \boldsymbol{x})$. Since a covariate predictive of a higher tumor response probability often is predictive of a longer survival, $\boldsymbol{x}_Z$ and $\boldsymbol{x}_Y$ may share common terms. To account for association, selection of these two subvectors is based on correlated vectors of latent variable selection indicators. This will be described in Sect. 3.2.3.

The schema of the AED design is shown in Fig. 4. The AED design enrolls a maximum of N patients sequentially in cohorts of sizes $c_1, \ldots, c_K$, with $\sum_{k=1}^{K} c_k = N$. The design uses a Bayesian group sequential procedure with both superiority and futility stopping rules for comparing the two treatment arms in the most recently identified E-sensitive subpopulation. The trial begins by enrolling patients under broad eligibility criteria for the first cohort. When the first cohort of patients' outcomes have been evaluated, the subvectors $\boldsymbol{x}_Z$ and $\boldsymbol{x}_Y$ are determined and used to compute a PBI, given formally below. The PBI is used to define the E-sensitive patients, and the comparative tests are defined in terms of the E-sensitive

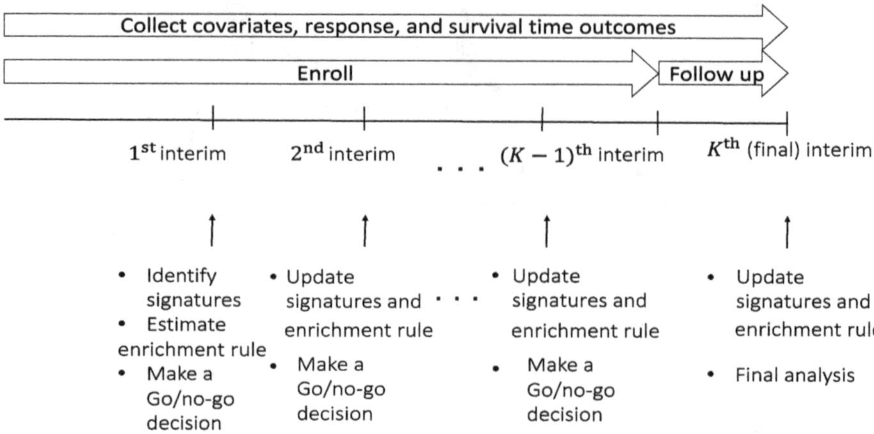

Fig. 4 Schema of the proposed design

patients. If the trial is not terminated early due to superiority or futility, then only E-sensitive patients are enrolled in the second cohort. This process is repeated group sequentially until the end of the trial. If the maximum sample size N is reached, a final analysis is performed when all patients' follow-ups have completed.

3.2.2 Probability Model

The joint distribution $(Y, Z \mid G, \boldsymbol{x})$ is modeled as a mixture of the conditional distribution of $(Y \mid Z, G, \boldsymbol{x})$ weighted by the marginal distribution of $(Z \mid G, \boldsymbol{x})$. For simplicity, here we assume that Z always is observed before Y.

AED uses the probit regression for the marginal distribution of $(Z \mid \boldsymbol{x}, G)$,

$$\pi(\boldsymbol{x}_i, G_i, \boldsymbol{\theta}_Z) = \Phi\big(\tilde{\boldsymbol{x}}_i^\top \boldsymbol{\beta}_Z + G_i \tilde{\boldsymbol{x}}_i^\top \boldsymbol{\gamma}_Z\big),$$

where i indexes patients, $\tilde{\boldsymbol{x}} = (1, \boldsymbol{x}^\top)^\top$, $\boldsymbol{\beta}_Z$ is the vector of covariate main effects, and $\boldsymbol{\gamma}_Z = (\gamma_{Z,0}, \gamma_{Z,1}, \ldots, \gamma_{Z,p})^\top$ is the vector of E-versus-C treatment-covariate interactions, with the main experimental versus control effect $\gamma_{Z,0}$. Denoting $\mathbf{Z}_n = (Z_1, \ldots, Z_n)$, $\mathbf{G}_n = (G_1, \ldots, G_n)$, $\mathbb{X}_n = (\boldsymbol{x}_1, \ldots, \boldsymbol{x}_n)^\top$, and $\boldsymbol{\theta}_Z = \big(\boldsymbol{\beta}_Z^\top, \boldsymbol{\gamma}_Z^\top\big)^\top$, the marginal likelihood of Z for the first n patients is

$$\mathscr{L}_n(\mathbf{Z}_n, \mathbf{G}_n, \mathbb{X}_n, \boldsymbol{\theta}_Z) = \prod_{i=1}^{n} \Phi\big(\tilde{\boldsymbol{x}}_i^\top \boldsymbol{\beta}_Z + G_i \tilde{\boldsymbol{x}}_i^\top \boldsymbol{\gamma}_Z\big)^{Z_i} \big\{1 - \Phi\big(\tilde{\boldsymbol{x}}_i^\top \boldsymbol{\beta}_Z + G_i \tilde{\boldsymbol{x}}_i^\top \boldsymbol{\gamma}_Z\big)\big\}^{1-Z_i}.$$

AED uses a proportional piecewise exponential (PE) hazard model (Ibrahim et al., 2005; Kim et al., 2007; McKeague & Tighiouart, 2000; Sinha et al., 1999) for the conditional distribution of $(Y \mid Z, G, \boldsymbol{x})$. With a partition of the time axis into M

intervals $I_m = (\tau_{m-1}, \tau_m]$ for $m = 1, \ldots, M$, with $0 = \tau_0 < \tau_1 < \cdots < \tau_M < \infty$, the proportional PE hazard function is

$$h(y|Z, G, \boldsymbol{x}, \boldsymbol{\theta}_Y) = \phi_m \exp\big(\boldsymbol{x}^\top \boldsymbol{\beta}_Y + G\tilde{\boldsymbol{x}}^\top \boldsymbol{\gamma}_Y + \alpha_Y Z\big) I(y \in I_m), \quad m = 1, \ldots, M, \tag{7}$$

where $\phi_m > 0$ is the hazard in the mth subinterval, α_Y is the main effect of response $(Z = 1)$ on the hazard of Y, and $\boldsymbol{\theta}_Y = \big(\phi_1, \ldots, \phi_M, \boldsymbol{\beta}_Y^\top, \boldsymbol{\gamma}_Y^\top, \alpha_Y\big)^\top$ is the parameter vector.

Denote $\mathbf{Y}_n^o = (Y_1^o, \ldots, Y_n^o)$, $\boldsymbol{\delta}_n = (\delta_1, \ldots, \delta_n)$, and $\mathscr{O}_n = (\mathbf{Z}_n, \mathbf{Y}_n^o, \boldsymbol{\delta}_n)$, the observed data from the first n patients in the trial. The joint likelihood function of the two endpoints Z and Y is

$$\begin{aligned} &\mathscr{L}_n(\mathscr{O}_n, \mathbf{G}_n, \mathbb{X}_n, \boldsymbol{\theta}_Z, \boldsymbol{\theta}_Y) \\ &\quad = \mathscr{L}_n(\mathbf{Z}_n, \mathbf{G}_n, \mathbb{X}_n, \boldsymbol{\theta}_Z) \prod_{i=1}^{n} \{f(Y_i^o | Z_i, G_i, \boldsymbol{x}_i, \boldsymbol{\theta}_Y)\}^{\delta_i} \\ &\qquad \times \left\{1 - F(Y_i^o | Z_i, G_i, \boldsymbol{x}_i, \boldsymbol{\theta}_Y)\right\}^{1-\delta_i}, \end{aligned} \tag{8}$$

where $f(Y|Z, G, \boldsymbol{x}, \boldsymbol{\theta}_Y)$ and $F(Y|Z, G, \boldsymbol{x}, \boldsymbol{\theta}_Y)$ are the conditional density function and cdf of Y, respectively. The marginal likelihood function of Y is obtained by averaging the joint likelihood function of (Y, Z) over Z.

3.2.3 Sequentially Adaptive Variable Selection

To identify $\boldsymbol{x}_Z$ and $\boldsymbol{x}_Y$ that identify patients more likely to benefit from E, joint variable selection is performed on $\boldsymbol{x}$ in the two submodels $(Y \mid Z, G, \boldsymbol{x})$ and $(Z \mid G, \boldsymbol{x})$, to accommodate the possibility that a covariate predictive of one outcome may also be predictive of the other. For each $t = Z$ or Y, let $\boldsymbol{\psi}_t$ denote the regression coefficient vector, excluding the intercept, so $\boldsymbol{\psi}_Z = (\beta_{Z,1}, \cdots, \beta_{Z,p}, \gamma_{Z,0}, \gamma_{Z,1}, \cdots, \gamma_{Z,p})$ and $\boldsymbol{\psi}_Y = (\beta_{Y,1}, \cdots, \beta_{Y,p}, \gamma_{Y,0}, \gamma_{Y,1}, \cdots, \gamma_{Y,p})$.

For each submodel $t = Z$ or Y, AED performed Bayesian spike-and-slab variable selection (Mitchell & Beauchamp, 1988; George & McCulloch, 1993; Ishwaran et al., 2005) to determine variables to be included in the submodel by using sparse posterior coefficient estimates. For each t, let $\boldsymbol{\lambda}_t = (\lambda_{t,1}, \ldots, \lambda_{t,2p+1})^\top$ be a vector of latent variable selection indicators corresponding to $(\boldsymbol{x}, G, G\boldsymbol{x})$. If $\lambda_{t,j} = 1$, then the jth variable in $(\boldsymbol{x}, G, G\boldsymbol{x})$ is included in the submodel for outcome t and if $\lambda_{t,j} = 0$, it is excluded. In each submodel, we restrict the variable selection algorithm to satisfy the *strong hierarchy interaction* constraint (Liu et al., 2015) so that, if the interaction term Gx_j is included, then the main effect terms x_j and G corresponding to Gx_j also must be included.

AED adopted a "joint variable selection" strategy by endowing $\boldsymbol{\lambda}_Z$ and $\boldsymbol{\lambda}_Y$ with a joint distribution to borrow information, which accommodates the fact that some covariates may be predictive of both Z and Y. To account for correlation, AED assumes a bivariate Bernoulli distribution $(\lambda_{Z,j}, \lambda_{Y,j})$, for $j = 1, \ldots, 2p+1$. Let $p_{Z,j} = \Pr(\lambda_{Z,j} = 1)$ and $p_{Y,j} = \Pr(\lambda_{Y,j} = 1)$ denote the marginal probabilities that the jth variable is included in the submodels $(Z \mid G, \boldsymbol{x})$ and $(Y \mid Z, G, \boldsymbol{x})$, respectively, and let

$$\rho_j = \frac{\Pr(\lambda_{Y,j} = 1, \lambda_{Z,j} = 1)/\Pr(\lambda_{Y,j} = 0, \lambda_{Z,j} = 1)}{\Pr(\lambda_{Y,j} = 1, \lambda_{Z,j} = 0)/\Pr(\lambda_{Y,j} = 0, \lambda_{Z,j} = 0)}$$

denote the odds ratio for the jth pair of latent variables. We denote by $\mathscr{B}(p_{Z,j}, p_{Y,j}, \rho_j)$ the joint Bernoulli distribution of $(\lambda_{Z,j}, \lambda_{Y,j})$.

AED uses the spike-and-slab prior model for the joint variable selection,

$$\psi_{Z,j}|\lambda_{Z,j} \sim (1-\lambda_{Z,j})\mathcal{N}(0, \tau_{Z,j}^2) + \lambda_{Z,j}\mathcal{N}(0, u_{Z,j}^2\tau_{Z,j}^2), \quad j = 1, \ldots, 2p+1 \tag{9}$$

$$\psi_{Y,j}|\lambda_{Y,j} \sim (1-\lambda_{Y,j})\mathcal{N}(0, \tau_{Y,j}^2) + \lambda_{Y,j}\mathcal{N}(0, u_{Y,j}^2\tau_{Y,j}^2), \quad j = 1, \ldots, 2p+1, \tag{10}$$

where $u_{Z,j}, \tau_{Z,j}^2, u_{Y,j}, \tau_{Y,j}^2$, $j = 1, \ldots, 2p+1$ are prespecified hyperparameters. AED chooses large $u_{Z,j}$ and small $\tau_{Z,j}$ in (9) so that $\lambda_{Z,j} = 1$ implies a nonzero estimate of $\psi_{Z,j}$, whereas $\lambda_{Z,j} = 0$ implies that the covariate corresponding to $\psi_{Z,j}$ has a negligible effect on Z. Similar choices are applied to (10) to obtain sparse vectors of coefficient estimates for Y. The latent indicator variables are assumed to follow the prior distributions $(\lambda_{Z,j}, \lambda_{Y,j})|p_{Z,j}, p_{Y,j}, \rho_j \sim \mathscr{B}(p_{Z,j}, p_{Y,j}, \rho_j)$ for $j = 1, \ldots, p+1$ and $\lambda_{Z,j} \sim \text{Bernoulli}(p_{Z,j})$ and $\lambda_{Y,j} \sim \text{Bernoulli}(p_{Y,j})$, for $j = p+2, \ldots, 2p+1$. To ensure the strong hierarchical property, the following constraints are imposed

$$p_{Z,j} = p_{Z,j-p-1}p_{Z,p+1}\min\{p_{Z,j-p-1}, p_{Z,p+1}\}, \quad j = p+2, \ldots, 2p+1 \tag{11}$$

$$p_{Y,j} = p_{Y,j-p-1}p_{Y,p+1}\min\{p_{Y,j-p-1}, p_{Y,p+1}\}, \quad j = p+2, \ldots, 2p+1. \tag{12}$$

AED performes joint variable selection at each interim decision during the trial to obtain the subvectors $\boldsymbol{x}_Z$ and $\boldsymbol{x}_Y$. While this procedure may miss important covariates in $\boldsymbol{x}_Y$ early in the trial, as the trial progresses the probabilities of identifying truly important covariates increase.

3.2.4 Adaptive Enrichment

For each cohort $k = 1, \dots, K$, let d_k be the accumulated number of events (i.e., $Y_i = Y_i^o$) at the time when the kth adaptive enrichment is performed, and let $n_k = \sum_{j=1}^{k} c_j$ be the total number of patients enrolled in the first k cohorts. Let $\mathscr{D}_k = \{(Y_i^o, \delta_i, Z_i, G_i, \boldsymbol{x}_i),\ i = 1, \dots, n_k\}$ be the accumulated data, and $\boldsymbol{x}_Z^{(k)}$ and $\boldsymbol{x}_Y^{(k)}$ the selected subvectors at the kth interim decision. AED defines the *personalized benefit index* (PBI) for a patient with covariate vector $\boldsymbol{x}$ as

$$\Omega(\boldsymbol{x} \mid \mathscr{D}_k) = (1 - \omega_k) \Pr\{\Delta_Z(\boldsymbol{x}_Z^{(k)}, \boldsymbol{\theta}_Z) > \epsilon_1 \mid \mathscr{D}_k\} + \omega_k \Pr\{\Delta_Y(\boldsymbol{x}_Y^{(k)}, \boldsymbol{\theta}_Y) < \epsilon_2 \mid \mathscr{D}_k\}, \tag{13}$$

where the weight is $\omega_k = d_k/n_k$. The cutoffs ϵ_1 and ϵ_2 be design parameters specified to quantify minimal clinically significant improvements in response probability and survival, respectively. The PBI depends on $\boldsymbol{x}$ only through the selected subvectors $\boldsymbol{x}_Z^{(k)}$ and $\boldsymbol{x}_Y^{(k)}$, that is, $\Omega(\boldsymbol{x} \mid \mathscr{D}_k) = \Omega(\boldsymbol{x}_Z^{(k)}, \boldsymbol{x}_Y^{(k)} \mid \mathscr{D}_k)$. It is a weighted average of the posterior probabilities that a patient with covariates $\boldsymbol{x}$ will benefit from E more than C, defined in terms of the comparative treatment effect parameters $\Delta_Z(\boldsymbol{x}_Z^{(k)}, \boldsymbol{\theta}_Z)$ and $\Delta_Y(\boldsymbol{x}_Y^{(k)}, \boldsymbol{\theta}_Y)$. Early in the trial, when there are few observed event times, the PBI will depend on Z more than on Y. Later into the trial when more events occur, the weight $(1 - \omega_k)$ for the response probability difference becomes smaller and the weight ω_k for the survival hazard ratio component becomes larger, so the PBI depends more on the survival time data. To use the PBI for decision making, we consider a patient with biomarker profile $\boldsymbol{x}$ to be eligible for enrollment into the $k + 1^{st}$ cohort of the trial if their PBI is sufficiently large, formalized by the rule

$$\Omega(\boldsymbol{x} \mid \mathscr{D}_k) = \Omega\left(\boldsymbol{x}_Z^{(k)}, \boldsymbol{x}_Y^{(k)} \mid \mathscr{D}_k\right) > \upsilon \left(\frac{n_k}{N}\right)^g \tag{14}$$

for $k = 1, \dots, K - 1$.

3.2.5 Bayesian Sequential Monitoring Rules

The kth interim decisions are based on $\mathscr{D}_k$, the accumulated data from k successive cohorts. Patients within each cohort are homogeneous since they satisfy the same eligibility criteria, but patients may be heterogeneous between cohorts since different cohorts may have different eligibility criteria because the variable selection is repeated and the PBI is refined during the trial. Let

$$\mathscr{X}_k = \{\boldsymbol{x} : \Omega(\boldsymbol{x}_Z^{(k)}, \boldsymbol{x}_Y^{(k)} \mid \mathscr{D}_k) > \upsilon\,(n_k/N)^g\}$$

denote the set of the covariates satisfying the eligibility criteria used for the $k+1^{st}$ cohort. At this point in the trial, the covariate-averaged long-term outcome treatment effect is defined as

$$T_{Y,k}(\boldsymbol{\theta}) = \int_{\mathscr{X}_k} \Delta_Y(\boldsymbol{x}_Y^{(k)}, \boldsymbol{\theta}_Y)\hat{p}_k(\boldsymbol{x}_Y^{(k)})d\boldsymbol{x}_Y^{(k)},$$

where $\hat{p}_k(\boldsymbol{x}_Y^{(k)})$ denotes the empirical distribution of $\boldsymbol{x}_Y^{(k)}$ on the set $\mathscr{X}_k$. Since these expectations are computed over the selected enrichment set $\mathscr{X}_k$, i.e., the patients who are expected to benefit more from E than C in the kth cohort, $T_{Y,k}(\boldsymbol{\theta})$ is a treatment effect in the sense of precision medicine. Note that E is more effective than C for patients with $\boldsymbol{x} \in \mathscr{X}_k$ if $T_{Y,k}(\boldsymbol{\theta})$ is sufficiently small.

Denoting the number of events in the jth cohort by e_j, AED defines the test statistic at the kth analysis as the weighted average of the treatment effects, $\overline{T}_{Y,k}(\boldsymbol{\theta}) = \sum_{j=1}^{k} w_{Y,j} T_{Y,j}(\boldsymbol{\theta})$, where the jth weight is $w_{Y,j} = e_j / \sum_{l=1}^{k} e_l$ (Lehmacher & Wassmer, 1999). As Y is a time-to-event endpoint, $T_{Y,j}(\boldsymbol{\theta})$ must be updated at each interim decision time. Let b_1 denote clinician-specified hazard ratio (e.g., ≤ 1) under which E is deemed superior to C in the long-term endpoint Y, and let b_2 denote the clinician-specified hazard ratio (e.g., ≥ 1) under which E is deemed inferior to C. Let (B_1, B_2) be prespecified probability cutoffs that should be tuned through simulation studies. For the interim analysis at each $k = 1, \cdots, K-1$, the decision rules are as follows:

1. **Superiority**: Stop the trial for superiority of E over C in $\mathscr{X}_k$ if $Pr\{\overline{T}_{Y,k}(\boldsymbol{\theta}) < b_1 | \mathscr{D}_k\} > B_1$.
2. **Futility**: Stop the trial for futility of E over C in $\mathscr{X}_k$ if $Pr\{\overline{T}_{Y,k}(\boldsymbol{\theta}) > b_2 | \mathscr{D}_k\} > B_2$.
3. **Final Decision**: If the trial is not stopped early, at the last analysis ($k = K$), conclude that E is superior to C in the final E-sensitive subset $\mathscr{X}_K$ if $Pr\{\overline{T}_{Y,K}(\boldsymbol{\theta}) < b_1 | \mathscr{D}_K\} > B_1$, and otherwise conclude that E is not superior to C in $\mathscr{X}_K$.

To account for the possibility that only a very small percentage of patients may benefit from E, the following additional futility stopping rule may be included at each interim. For a prespecified threshold $0 < q < 1$, the futility rule stops the trial if the estimated proportion of E-sensitive patients in the trial is $< q$. Park et al. (2022) recommend to use a value in the range 0.01–0.10 in practice.

Since the Bayesian spike-and-slab variable selection method does not necessarily drop covariates with little or no contribution to identify $\mathscr{X}_K$, identification of the final E-sensitive subset $\mathscr{X}_K$ based on PBI involves all covariates at the end of the trial. To facilitate practical use, one can simplify the E-sensitive subset identification rule by dropping covariates that have low posterior probability (i.e., <0.10) of being selected in the prediction models.

An important practical issue is to decide when to begin the adaptive enrichment. This depends on several factors, including the number of covariates, their information-to-noise ratio, the percentage of sensitive patients, the treatment difference between sensitive and insensitive patients, and the variances of the outcomes Z and Y. In practice, logistical limitations will often limit the number of interim decisions to 1, 2, or 3. Based on these considerations, as a rule of thumb, a reasonable time to initiate the adaptive enrichment is after 1/3 to 1/2 of the maximum number of patients have been enrolled.

We refer the reader to the original article by Park et al. (2022) for more details of the design (e.g., prior distributions on the parameters, estimation procedure, a model elaboration for situations where Y is observed before Z can be evaluated, the generalization of the joint variable selection algorithm to the case where the predictors may be correlated, details of determining the design parameters $\upsilon > 0$ and $g > 0$, procedure to calibrate the values of (B_1, B_2), and simulation studies).

4 Summary

In this chapter, we have discussed several adaptive design examples under the paradigm of precision medicine. Considering multiple endpoints and risk-benefit tradeoff better reflects the medical decisions in practice. In the presence of patient's genomic profile, e.g., biomarker profile, Bayesian adaptive designs that utilize dimension-reduction technique can be used to determine subgroup OBD. The enrichment design is another effective approach to achieving precision medicine by using biomarkers to adaptively identify the sensitive subgroup and determine whether the treatment is effective in that subgroup. The design strategies and methodologies described in this chapter can be modified to meet the need of specific clinical trials. Software for the designs discussed in this chapter can be found on Github website https://github.com/beibeiguo/precision-medicine-book-chapter.git. No datasets were analyzed in this chapter as all data are computer simulated.

References

Agresti, A. (2002). *Categorical data analysis*. New York: Wiley.

Albert, J., & Chib, S. (1993). Bayesian analysis of binary and polychotomous reponse data. *Journal of the American Statistical Association, 88*, 669–679.

Babb, J. S., & Rogatko, A. (2001). Patient specific dosing in a phase I cancer trial. *Statistics in Medicine, 20*, 2079–2090.

Collins, F., & Varmus, H. (2015). A new initiative on Precision medicine. *The New England Journal of Medicine, 372*, 793–795.

Cox, C. (1995). Location-scale cumulative odds models for ordinal data: a generalized non-linear model approach. *Statistics in Medicine, 14*, 1191–1203.

Frank, I., & Friedman, J. (1993) A statistical view of some chemometrics regression tools. *Technometrics, 35*, 109–135.

Garthwaite, P. (1994). An interpretation of partial least squares. *Journal of the American Statistical Association, 89*, 122–127.

George, E. I., & McCulloch, R. E. (1993). Variable selection via gibbs sampling. *Journal of the American Statistical Association, 88*, 881–889.

Guo, B., & Yuan, Y. (2017). Bayesian phase I/II biomarker-based dose finding for precision medicine with molecularly targeted agents. *Journal of the American Statistical Association, 112*, 508–520.

Hodson, R. (2016). Precision medicine. *Nature, 537*, S49.

Houede, N., Thall, P., Nguyen, H., Paoletti, X., & Kramar, A. (2010). Utility-based optimization of combination therapy using ordinal toxicity and efficacy in phase I/II trials. *Biometrics, 66*, 532–540.

Ibrahim, J. G., Chen, M. H., & Sinha, D. (2005). Bayesian survival analysis. In *Wiley online library*.

Indahl, U., Liland, K., & Nas, T. (2008). Canonical partial least squares—a unified PLS approach to classification and regression problems. *Journal of Chemometrics, 23*, 495–504.

Ishwaran, H., & Rao, J. S. (2005). Spike and slab variable selection: frequentist and Bayesian strategies. *The Annals of Statistics, 33*, 730–773.

Ivanova, A., & Wang, K. (2006). Bivariate isotonic design for dose-finding with ordered groups. *Biometrics, 25*, 2018–2026.

Kim, S., Chen, M., Dey, D. K., & Gamerman, D. (2007). Bayesian dynamic models for survival data with a cure fraction. *Lifetime Data Analysis, 13*, 17–35.

Lehmacher, W., & Wassmer, G. (1999). Adaptive sample size calculations in group sequential trials. *Biometrics, 55*, 1286–1290.

Lin, R., Zhou, Y., Yan, F., Li, D., &Yuan, Y. (2020). BOIN12: Bayesian optimal interval phase I/II Trial design for utility-based dose finding in immunotherapy and targeted therapies. In *JCO Precision Oncology*. In Press.

Liu, S., & Yuan, Y. (2015). Bayesian optimal interval designs for phase I clinical trials. *Journal of the Royal Statistical Society Series C-Applied Statistics, 64*, 507–523.

Liu, C., Ma, J., & Amos, C. I. (2015). Bayesian variable selection for hierarchical gene-environment and gene-gene interactions. *Human genetics, 134*, 23–36.

McCullagh, P. (1980). Regression models for ordinal data. *Journal of the Royal Statistical Society, Series B, 42*, 109–142.

McKeague, I. W., & Tighiouart, M. (2000). Bayesian estimators for conditional hazard functions. *Biometrics, 56*, 1007–1015.

Mitchell, T. J., & Beauchamp, J. J. (1988). Bayesian variable selection in linear regression. *Journal of the American Statistical Association, 83*, 1023–1032.

O'Quigley, J., & Paoletti, X. (2003). Continual reassessment method for ordered groups. *Biometrics, 59*, 430–440.

O'Quigley, J., Pepe, M., & Fisher, L. (1990). Continual reassessment method: a practical design for phase I clinical trials in cancer. *Biometrics, 46*, 33–48.

Park, Y., Liu, S., Thall, P., & Yuan, Y. (2022). Bayesian group sequential enrichment designs based on adaptive regression of response and survival time on baseline biomarkers. *Biometrics, 78*, 60–71.

Postel-Vinay S., Arkenau, H. T., Olmos D., et al. (2009). Clinical benefit in phase I trials of novel molecularly targeted agents: Does dose matter?. *British Journal of Cancer, 100*, 1373–1378.

Schork, N. J. (2015). Personalized medicine: Time for one-person trials. *Nature, 520*, 609–611.

Sinha, D., Chen, M. H., & Ghosh, S. K. (1999). Bayesian analysis and model selection for interval-censored survival data. *Biometrics, 55*, 585–590.

Stone, M., & Brooks, R. J. (1990). Continuum regression: cross-validated sequentially constructed prediction embracing ordinary least squares, partial least squares and principal component regression. *Journal of the Royal Statistical Society. Series B, 52*, 237–269.

Storer, B. E. (1989). Design and analysis of phase I clinical trials. *Biometrics, 45*, 925–937.

Stylianou, M., & Flournoy, N. (2002). Dose finding using the biased coin up-and-down design and isotonic regression. *Biometrics, 58*, 171–177.

Thall, P., & Russell, K. (1998). A strategy for dose-finding and safety monitoring based on efficacy and adverse outcomes in phase I/II clinical trials. *Statistics in Medicine, 27*, 4895–4913.

Thall, P., & Cook, J. (2004). Dose-finding based on efficacy-toxicity trade-offs. *Biometrics, 60*, 684–693.

Thall, P., Nguyen, H., & Estey, E. (2008). Patient-specific dose finding based on bivariate outcomes and covariates. *Biometrics, 64*, 1126–1136.

Whitehead, J., & Brunier, H. (1995). Bayesian decision procedures for dose determining experiments. *Statistics in Medicine, 14*, 885–893.

Yuan, Z., Chappell, R., & Bailey, H. (2007). The continual reassessment method for multiple toxicity grades: a Bayesian quasi-likelihood approach. *Biometrics, 63*, 173–179.

Yuan, Y., Nguyen, H. Q., & Thall, P. F. (2017). Bayesian designs for phase I-II clinical trials. New York: CRC Press.

Yuan, Y., Hess, K. R., Hilsenbeck, S. G., et al. (2016). Bayesian optimal interval design: A simple and well-performing design for phase I oncology trials. *Clinical Cancer Research, 22*, 4291–4301.

Part IV
Precision Medicine in Survival Analysis and Genomic Studies

Variant Selection and Aggregation of Genetic Association Studies in Precision Medicine

Jianchang Hu, Shiying Wang, and Heping Zhang

Abstract Precision medicine provides tailored treatments to patients based on their demographic and genomic profiles, and is a promising approach to improving treatment efficiency. Genetic biomarkers identified from association studies are useful in disease risk prediction, subgroup identification and heterogenous drug responses, all which are important steps in precision medicine. Marginal tests of association are powerful for common and relatively major variant identification, but variants with low frequency or modest association can explain additional disease risk and may be important for pharmaceutical intervention. In this chapter, we review set-based association methods with a focus on variant selection and aggregation involving low frequency and signal variants. We present a unifying framework as well as a detailed classification of methods. Also discussed in this article are the control of type I error and replication of discoveries. Simulation studies are conducted to demonstrate the effectiveness of data-driven methods under different assumptions of causal variants. We conclude with some discussions on the challenges and future research directions.

Keywords Genetic association · Precision medicine · Low-frequency variant · Super-variant · Permutation inference · Sample re-use

1 Introduction

Precision medicine is a promising approach to improving treatment efficiency as it provides tailored treatments to patients based on their demographic and genomic profiles (Jameson & Longo, 2015; Ashley, 2016; Matchett et al., 2017). Genetic

J. Hu
Institute of Medical Biometry and Statistics, University of Lübeck, Lübeck, Germany
e-mail: jianchang.hu@yale.edu

S. Wang · H. Zhang (✉)
Department of Biostatistics, Yale University School of Public Health, New Haven, CT, USA
e-mail: shiying.wang@yale.edu; heping.zhang@yale.edu

Y. Zhao, D.-G. Chen (eds.), *Statistics in Precision Health*, ICSA Book Series in Statistics, https://doi.org/10.1007/978-3-031-50690-1_17

information helps in at least two areas. Firstly, using genetic biomarkers to form a predictive polygenic score can not only predict patients' risk of disease, but also help identify subgroups of patients having certain sub-types of a disease (Udler et al., 2019; Harrison et al., 2020). Once a subgroup is identified, a tailored treatment could be designed and provided. Secondly, genetic information is also vital for pharmacogenomics (Ashley, 2016; Ritchie, 2012; Crews et al., 2012; Mooney, 2015). The goal of pharmacogenomics is to search for genetic variants associated with drug response phenotypes such as drug efficacy and adverse side effects. With such information, effective treatment strategies can be designed, evaluated, and implemented.

Genetic association studies are critical to both aforementioned areas in precision medicine. The genetic biomarkers used in the construction of polygenic scores are often those identified in association studies (Udler et al., 2019; Harrison et al., 2020; Warren et al., 2017). Pharmacogenomics uses association studies in which the phenotypes are related to drug response (Ritchie, 2012).

Genetic association studies search for heritable genetic variants underlying traits of interest, including quantitative phenotypes such as height and body mass index (BMI), and dichotomous disease status (Tam et al., 2019). Genotypes of single nucleotide polymorphisms (SNPs) are commonly used genetic variants. The most straightforward association analysis is to marginally test the association of each SNP with the phenotype, and then with a proper multiple testing adjustment. SNPs showing significant effects are identified and can be further analyzed in terms of gene mapping and functionality. These genome-wide association (GWA) studies have led to successfully pinpointing genetic factors for many complex diseases, including age-related macular degeneration (Klein et al., 2005; Consortium, A.G., 2013), Type-2 diabetes (Morris et al., 2012; Xue et al., 2018) and Crohn's disease (Franke et al., 2010; De Lange et al., 2017).

The marginal approach works well when searching for common variants whose minor allele frequency (MAF) are at least 5% when hundreds of thousands of samples are available. When the association is modest, or when rare variants are concerned, this marginal approach may not be powerful enough to identify associated genetic factors (Asimit & Zeggini, 2010).

However, from the theory of evolution point of view, alleles are likely to be deleterious if they are highly inclined toward disease, and thus are often kept at low frequency (Kryukov et al., 2007; Pritchard, 2001; Gibson, 2012; MacArthur et al., 2012). In addition, from the perspective of precision medicine, the study of low-frequency variants may reveal protective alleles that could be the target of pharmaceutical intervention (Ritchie, 2012). Variants with low frequency and modest association are also believed to contribute to the "missing heritability" (Eichler et al., 2010; Maher, 2008). Finding them may help build a better polygenic risk prediction model. Therefore, methods other than the marginal testing approach are needed to search for variants with low frequency and modest association.

With the advancement of high-throughput next-generation sequencing technology (Metzker, 2010; Meyerson et al., 2010), exome sequencing and even whole-genome sequencing data can be generated (more discussions on study design and data collection are detailed in the review (Lee et al., 2014)). This enables us to assess the role of low-frequency variants in complex traits. A common approach to dealing with the weak association or rare variants is to combine information at multiple loci (Lee et al., 2014; Zuk et al., 2014; Li & Leal, 2008). Set-based association studies, where a set could be a gene or a target region on a chromosome, are especially useful (Lee et al., 2014; Van Hout et al., 2020; Cirulli et al., 2020).

Broadly, there are two categories of methods. For a given gene or target region, the first category of methods combines information from all SNPs within the set. Examples include principal component analysis (PCA)-based gene analysis (Gauderman et al., 2007), burden tests and the sequence kernel association test (SKAT) (Wu et al., 2011). The other category of methods explicitly selects only important loci within the set to form a new biomarker for the test of association with the trait of interest. Methods in this category include annotation-based selection, adaptive burden tests with MAF thresholds such as the cohort allelic sums test (CAST) (Morgenthaler & Thilly, 2007), the step-up method (Hoffmann et al., 2010), and the super-variant based method (Song & Zhang, 2014; Hu et al., 2021; Hu et al., 2020).

The first strategy focuses on detecting if there is any association with the trait of interest within the set of interests. It does not explicitly indicate which SNPs contribute more to the association. In contrast, the second one provides more interpretable results in terms of biomarkers as these methods select a subset of SNPs that account for the association within the set. In the context of precision medicine, the second strategy is more informative as it provides candidate SNPs for later polygenic score construction as well as potential targets for pharmaceutical analysis. Therefore, in this chapter, we emphasize the second strategy.

We first provide a unifying modeling framework for those methods before discussing the specifics. We will address practical issues of type I error control and replication probability, and present some simulation studies to compare the effectiveness of those methods under different assumptions of causal variants. We conclude with some remarks on challenges and future research directions.

2 A Unifying Model

The genetic association study with variant selection and aggregation can be unified by the following model. Suppose that the study includes N subjects from a population, the demographic or clinical information is collected, the trait of interest or phenotype Y is measured, and in total K bi-allelic SNPs are genotyped. Let X_{ik} be the additive coding for variant k and Z_{im} be the m-th covariate of subject i. Then a disease model can be given by

$$g\left(\mu_i\right) = \alpha_0 + \sum_m \alpha_m Z_{im} + \sum_k \beta_k X_{ik}, \tag{1}$$

where $\mu_i = E(Y_i)$, Y_i is the phenotype for subject i, and can be either dichotomous or continuous, and g is the link function where it is logit for logistic regression and identity for linear regression. When the association signal is modest or a variant has a low MAF, the estimation of individual variant's effect on Y may be unstable and inconsistent (Yan & Bien, 2020). This motivates us to select important loci within the set to form a new biomarker for the test of association. To this end, the selection and aggregation model is as follows:

$$g(\mu_i) = \alpha_0 + \sum_m \alpha_m Z_{im} + \beta \tilde{X}_i, \tag{2}$$

where $\tilde{X}_i$ either takes additive coding as $\sum_k w_k X_{ik}$ or indicator coding as $I\left(\sum_k v_k X_{ik} > 0\right)$, and $w_k = a_k s_k v_k$ with a_k being a continuous weight to adjust for MAF of variant k, $s_k \in \{-1, 1\}$ determining the direction of the variant effect, and $v_k \in \{0, 1\}$ being a binary indicator for the inclusion of variant k into the model.

The additive coding for $\tilde{X}_i$ is the model used in (Hoffmann et al., 2010). The only difference between models (1) and (2) for the additive coding is that $\sum_k \beta_k X_{ik}$ in (1) is replaced with $\beta \sum_k w_k X_{ik}$ in (2). Specifically, model (2) assumes that the effect sizes of all causal variants are known subject to an unknown common constant.

Model (2) creates a single aggregated biomarker through weighing the variants. However, not all variants within the pre-specified set are included in forming this aggregated biomarker. As discussed in (Zuk et al., 2014), the inclusion of many neutral variants will only dilute the overall strength of association, and this is especially vital for rare variant analysis. From the perspective of precision medicine, it is essential to narrow the number of SNPs for further experimental analyses so that the target of pharmaceutical intervention may be identified. The inclusion of a variant is controlled by the binary indicator v_k. We will discuss different methods for determining v_k.

For additive coding, the weight of a variant is related to the MAF of the variant. As mentioned in (Tam et al., 2019), there is a common belief that for a deleterious variant, the lower the MAF, the higher the effect size. To incorporate such allele information, the continuous weight a_k is used, and in the literature, two commonly used weights are (1) $1/\sqrt{p_k(1-p_k)}$ (Madsen & Browning, 2009; Price et al., 2010) and (2) Beta(p_k, a, b) (Lee et al., 2014; Wu et al., 2011), where p_k is the estimated MAF and Beta(p, a, b) is the p-th quantile of beta distribution with parameters a and b. Common choices of parameters in the second weighing scheme (beta distribution) are $a = 1$, $b = 25$, which gives more weight to rare variants, and $a = b = 1$, which considers equal weights for all variants. When $a = 0.5$, $b = 0.5$, it reduces to the first weighting scheme.

Moreover, for additive coding, we need to estimate the effect direction of each variant. If we believe that all variants are deleterious, then we can set $s_k = 1$. All methods in the literature rely on the marginal estimation. Han and Pan (Han & Pan, 2010) fit a marginal regression for the association between the variant and disease to determine the direction. Hoffmann et al. (2010) consider allele frequency in case and control groups for dichotomous phenotype to estimate if the variant is more likely to be deleterious. Song and Zhang (2014) use the magnitudes of marginal t-statistics to rank the importance of variants and the signs of the t-statistics to distinguish the deleterious variants from the protective variants.

The indicator coding focuses on the presence or absence of any minor allele among selected SNPs. This form of aggregation can indirectly and conveniently incorporate interaction between different SNPs. Take two variants as an example. We have

$$I(X_1 + X_2 > 0) = I(X_1 > 0) + I(X_2 > 0) - I(X_1 > 0) * I(X_2 > 0).$$

Thus, the statistical interaction (Cordell, 2009) is reflected by the indicator coding. However, as more SNPs are included, the binary indicator becomes closer to constant 1 which is uninformative, so the indicator coding may need to restrict the number of SNPs that can be selected. Moreover, this way of aggregation assumes that the effects of all variants are in the same direction, and if common and rare variants are considered at the same time, the variation of aggregated biomarkers is largely dominated by common variants.

The additive coding, on the other hand, can adjust the effect sizes due to different MAFs of variants. With a reasonable estimation on the direction of variants' effects, SNPs having different directional effects can be combined and identified at the same time. Also, the number of SNPs to include solely depends on the selection procedure. However, the additive coding assumes that the disease risk is proportional to the number of minor alleles (Lewis, 2002), so it may not account for the interactive effects between variants, which may contribute to the "missing heritability" (Cordell, 2009) and may affect the approach for variant detection (Zuk et al., 2012).

Lastly, for extremely rare variants such as those in exome sequencing studies, aggregation with indicator coding is very close to, if not the same as, additive coding.

3 Specific Variant Selection Methods

Model (2) uses a subset of SNPs to form an aggregated biomarker, and not all v_k are assigned to 1. We describe two categories of set-based association methods for variant selection. One is based on deterministic information such as functional annotations and physical position. The other is data-driven.

3.1 Deterministic Methods

For deterministic methods, v_k is determined by biological characteristics such as functional annotation and physical location on a chromosome. Rare variant analysis often takes this approach. Variants in the protein-coding region are classified as synonymous, missense and nonsense variants, and often only nonsense variants are aggregated to form biomarkers for association tests (Van Hout et al., 2020; Cirulli et al., 2020). Once the variants are selected and aggregated into a new biomarker, then we test the null hypothesis $\beta = 0$, where β is given in model (2). Because the selection is determined *a priori*, the test can be performed on the aggregated biomarkers as if they are given. If multiple regions are considered at the same time, then the Bonferroni adjustment is needed for proper family-wise error rate (FWER) control based on the number of the aggregated biomarkers, not at the original SNP level. These methods are easy to use and interpret because the selection makes use of known biological knowledge. However, how to make proper use of known biological knowledge is generally challenging and can be difficult to assess. This decision affects the power of the deterministic approaches (Byrnes et al., 2013).

3.2 Data-Driven Methods

In contrast, for data-driven methods, v_k is determined by a learning procedure involving variable selection on all SNPs in a pre-specified set. Methods in this category include the cohort allelic sums test (CAST) (Morgenthaler & Thilly, 2007), the variable minor allele frequency threshold (VT) method (Price et al., 2010), step-up method (Hoffmann et al., 2010), and super-variants (Song & Zhang, 2014; Hu et al., 2020).

CAST and VT consider various thresholds on MAF to determine which SNPs to include for aggregation. Specifically, for a predetermined threshold u on MAF, $v_k = I(p_k < u)$ where p_k is the estimated minor allele frequency of variant k. CAST considers the indicator coding for variant aggregation. VT considers several different MAF thresholds and combines the test statistics from all thresholds into one single statistic to test the association. With some asymptotic-based approaches for test statistics combination (Liu & Xie, 2020), these methods are easy to apply and computationally less expensive. In addition, because the indicator coding is used for aggregation, interactions within variants are considered. However, the focus of VT is to enhance the test of association instead of explicit variant selection. Moreover, using only thresholds on MAF to select variants could include many neutral variants, which may dilute the strength of association between an aggregated biomarker and the trait of interest. In addition, it poses some challenges on a replication study as the estimated MAF of a variant may change from one dataset to another.

The step-up method considers a greedy forward selection procedure to determine the SNPs for aggregation. In particular, at each step, it selects a variant into the model such that the test statistic of the aggregated biomarker improves the most compared with that of the previous step. For instance, in the first step, the univariate test statistic T_k for each variant k is calculated, and the variant with the "best" statistic (i.e., $\max T_k$) is selected. In the second step, each variant is aggregated with the selected one in the first step, and the univariate test statistic of the aggregated biomarker is calculated. The variant which leads to the most improved test statistic of the aggregated biomarker is then selected. The procedure continues until no further improvements can be made. Compared with CAST, the selection procedure takes into account the variants' association with the phenotype. Therefore, less neutral variants may be included. However, this method can be computationally intensive as in each step all variants need to be investigated.

The methods of identifying super-variants use an importance measure that indicates how strongly a variant is associated with the phenotype. Commonly used importance measures include the marginal t-statistic (Song & Zhang, 2014) and depth importance measure (Hu et al., 2020), where the former emphasizes the main effect of each variant, and the latter also takes interactive effects among variants into consideration. SNPs are then ranked according to their importance measures. If all variants are considered deleterious, then the SNPs are ordered based on their importance measures from high to low. If the direction of effect can be estimated, for instance, the use of the sign of the marginal t-statistic, then SNPs with different effect directions are separately ordered, both in the order of the absolute magnitude, from high to low. Once the SNPs are ordered, a forward inclusion procedure is run to decide how many top SNPs are used to form an aggregated biomarker. We defer further details to (Song & Zhang, 2014; Hu et al., 2020).

Both the step-up and super-variant methods utilize forward inclusion processes. It is possible to consider additional backward deletion procedures. If a forward inclusion procedure provides a set of variants $\{X_1, \ldots, X_K\}$, the backward deletion removes one SNP at a time from a set of selected SNPs, and the weakest SNP based on an association test or any other criterion is subject to removal. The procedure ends if no candidate SNP can be removed.

4 Control of Type I Error

One major statistical issue for the data-driven methods is that the test of association is a post-selection inference problem. As a result, the significance of the association tends to be over-optimistic and inflate type I error.

The following two approaches are commonly used to control type I error. Sample re-use is a simple and straightforward approach. We use one part of a dataset to select variants and form possible biomarkers, and use the rest to test the association of the biomarkers with the phenotype of interest. The variant selection procedure is independent of the part of the data used for the test. So, for the samples used in the test of association, the selection is determined *a priori*, and the test can be performed on the aggregated biomarkers as if they are given. Thus, the overall type

I error of the association tests can be controlled. Hu et al. (2020) and Li et al. (Hu et al., 2021) adopt this approach in their analyses. The disadvantage of this approach is that it decreases the effective sample size, potentially leading to a loss of power in the detection of signals.

The other approach is to consider the permutation-based inference. Morgenthaler and Thilly (2007), Hoffmann et al. (2010) and Price et al. (2010) take this approach to testing association. The idea is to obtain an empirical distribution for the p-values of the association test under the null hypothesis that all genetic variants are not associated with the phenotype, and then use this empirical distribution to assess the significance of the association for the aggregated biomarker. For instance, for a dichotomous phenotype, the data-driven method can be first applied on the original dataset, and it may select a set of SNPs to form an aggregated biomarker. A p-value can be obtained by testing the association between this aggregated biomarker and the phenotype. Then the dichotomous phenotypes in the dataset can be randomly permuted among all subjects, say, 500 or 1000 times. Each time, the data-driven method is applied on the permuted dataset in the same way as it is applied to the original dataset. We obtain a p-value from every permutation. Then, the p-values from all permutations provide an empirical distribution of the p-values to assess the significance of the observed p-value obtained from the original dataset. This approach is conceptually simple but computationally time-consuming. The accuracy of the empirical null distribution for the p-values depends on the number of permutations performed; the more permutations, the better estimation. In addition, the larger the dataset, often the more computational time needed in each permutation. Therefore, this approach may not be recommended for analyzing large datasets.

5 Replication Study

Replication is essential to genetic association study. The idea is to reproduce the associations found in the discovery study on an independent dataset to reduce the likelihood of spurious findings (Kraft et al., 2009; Marigorta et al., 2018; König, 2011). For precision medicine, replicated SNPs may be more promising targets for pharmaceutical intervention.

Two types of replication studies exist (Kraft et al., 2009; Marigorta et al., 2018). One is "intra-population replication", where the samples in the discovery and replication studies are coming from the same population or very similar ones. It provides more evidence of the association results within the same population. The other type is "inter-population replication" where the samples in the discovery and replication studies are coming from different populations. One example is using European participants in a discovery study and using Chinese participants in a replication study. The "inter-population replication" tries to extend the findings to another population. However, population structure may affect the replication results in this case. There is no guarantee that the association pattern would be the same across different populations. In fact, Greene et al. (2009) demonstrate that a small change in MAF or marginal penetration of a SNP can greatly reduce the power to replicate its association. Therefore, we emphasize the importance of establishing solid association via "intra-population replication" before moving onto the stage of "inter-population replication".

From the perspective of precision medicine, the "intra-population replication" would be more informative. The goal of precision medicine is to design efficient treatment strategies for a subgroup of patients based on their demographic and genetic profiles. Therefore, validating the association within the same or similar population is essential because ethnicity is generally a criterion to form subgroups. What the best treatment strategy is for a population warrants a detailed study within that population.

The exact replication requires genotyping the same set of SNPs identified in the discovery study. Then the aggregated biomarkers are formed in the same way as in the discovery study. The tests of association with aggregated biomarkers are performed, and assessment of the statistical significance of test results is subject to multiple testing adjustment if several biomarkers are tested at the same time.

If the data of identified SNPs are not available, the replication studies may go through the same analysis procedure adopted in the discovery study on the replication dataset to see if the same gene or region can be identified. But even if the same gene or region is verified to have significant associations, SNPs contribute to the association may differ from those identified in the discovery study and further analysis needs to be conducted to understand the differences. Many other factors can influence the replication study and detailed discussions can be found in (Kraft et al., 2009; Marigorta et al., 2018; Huffman, 2018).

6 Simulation

In this section, we present simulation studies to demonstrate the performance of data-driven methods under various assumptions of causal variants.

We use the imputed genotype data from the UK Biobank database (Field ID: 22801-22822). Specifically, we extract subjects without self-reported cancer or non-cancer diseases and use their SNP data on chromosome 17, which is the same dataset used in the simulation of Hu et al. (2020). We remove SNPs with low call rate (missing probability $\geq$0.05), extremely rare variants (minor allele frequency $<$0.005), and disrupted Hardy-Weinberg equilibrium (p-value $<1 \times 10^{-7}$). After genotype quality control, we divide the SNPs on Chromosome 17 into 80 non-overlapping local sets by the physical position so that each set consists of SNPs within a segment of physical length 1 Mb (see Hu et al. (2020) for a detailed description on this segmentation). In each repetition, we randomly sample 10,000 subjects. We then randomly sample one SNP set out of the 80 sets, and within the sampled SNP set, we randomly sample 400 rare variants ($0.005 < \text{MAF} < 0.05$) and 100 common variants ($0.05 < \text{MAF} < 0.5$). The selected SNPs form the pre-specified set for the set-based association analysis. The physical order of selected SNPs is kept to maintain the linkage disequilibrium pattern. We assume that the minor alleles of causal SNPs within the pre-specified SNP set have deleterious effects and increase the penetrance of a disease. A disease status is generated based on the model Y_i~ Bernoulli(m_i), where m_i is the disease penetrance depending on the randomly sampled sets of the rare and common SNPs as specified below. We randomly split the generated dataset into a discovery set and a replication set of equal size. All methods are performed on the discovery set to detect the association, and the replication set is used to check if their results can be replicated by an independent dataset from the same population. The whole process is repeated 500 times.

We consider the adaptive burden test with MAF thresholds, the step-up method, and the super-variant based method. The importance measure used in the super-variant based method is the p-value of the marginal score test between a SNP and the phenotype. For each method, we consider both the indicator coding and additive coding to aggregate SNPs, and we use weight $1/\sqrt{p_k\,(1-p_k)}$ in additive coding, where p_k is the sample MAF. We compare the performance of the methods above in control of type I error, power of detection, and replication probability.

To control type I error for the step-up and super-variant based methods, we consider both sample re-use and permutation approaches which are commonly used for data-driven methods as discussed in the previous section. For the sample re-use approach, we equally split the discovery set into two sets. The first set is used for variant selection, and the second set is used for testing the association of aggregated SNPs with phenotype. For the permutation approach, we randomly permute the disease status 500 times to construct the empirical distribution of p-values. As suggested by Hoffmann et al. (2010) for the step-up method, 500 permutations are run for speed and should be accurate for the simulation. We also consider adding a backward deletion procedure to the step-up and super-variant based methods. For the adaptive burden tests, we consider MAF thresholds at 0.01 and 0.05 which are commonly used thresholds for grouping rare and low-frequency variants, respectively, and conduct 500 permutations to construct the empirical distribution of p-values. R codes to implement simulation are included in supplementary materials.

6.1 When There Is No Causal Variant

In this subsection, we consider the situation where no variants are associated with phenotype, and the focus is to see if sample re-use and permutation strategy can help control the type I error rate at the pre-determined significance level. We generate the disease status from the following null model.

$$\log\left(\frac{\mathrm{m}_i}{1-\mathrm{m}_i}\right) = -1, \mathrm{i} = 1, \ldots, 10,000.$$

Table 1 Type I error rates for different methods at the significance level = 0.05

	Indicator coding	Additive coding
Sample re-use		
Super-variant	0.048	0.046
Super-variant with deletion	0.043	0.052
Step-up	0.050	0.050
Step-up with deletion	0.050	0.042
Permutation		
Super-variant	0.060	0.050
Super-variant with deletion	0.058	0.054
Step-up	0.056	0.050
Step-up with deletion	0.054	0.058
Adaptive burden with MAF threshold 0.05	0.028	0.046
Adaptive burden with MAF threshold 0.01	0.046	0.056

The type I error rates of each method, presented in Table 1, are evaluated at the significance level $\alpha = 0.05$.

Table 1 shows that both the sample re-use and permutation strategies work well in controlling type I error for the step-up and super-variant based methods. For the adaptive burden test with MAF thresholds, the permutation approach works well for most scenarios. However, when using MAF threshold 0.05 along with the indicator coding, the method is too conservative. The main reason is that using this MAF threshold includes all rare variants for aggregation. This makes the aggregated biomarker close to constant 1 in the indicator coding. Therefore, the estimated distribution of p-values under the null hypothesis becomes inaccurate.

6.2 When There Exist Causal Variants

Next, we investigate the situation where causal variants exist. We consider two scenarios of causal variants. The first scenario is for a dense signal, where a relatively large fraction of SNPs is causal. The second scenario is for a sparse signal, where only a small fraction of SNPs is causal. Both scenarios are widely considered in the literature of rare variant analysis (Lee et al., 2012, 2014; Chen et al., 2012; Derkach et al., 2013).

Moreover, under each scenario, we vary the relative effect sizes of common variants to rare variants, from low to high. However, we assign larger absolute effect sizes to rare variants than common variants in all scenarios, which reflects the belief that for a deleterious variant, the lower the MAF, the higher the effect size (Kryukov et al., 2007; Park et al., 2011). In addition, we vary the effect directions of causal variants to represent the possible situations in which they have only deleterious effects or could be both deleterious and protective.

We compare the power of all methods as above. We also record the average number of SNPs selected by the step-up and super-variant based methods in both the dense and sparse signal scenarios. We further record the accuracy of causal variant selection of these two methods under the sparse signal scenario.

6.2.1 Dense Signal Scenario

The dense signal scenario assumes that a large fraction of SNPs within the pre-specified set is causal, and their marginal effect sizes are small (Lee et al., 2014; Zuk et al., 2014). We consider the disease model given by model (1) with only genetic factors; that is,

$$\log\left(\frac{\mathrm{m}_i}{1-\mathrm{m}_i}\right) = \alpha_0 + \sum_k \beta_k X_{ik}.$$

We randomly extract 30% SNPs within the pre-specified SNP set as causal variants. Following the simulation setting of Hoffmann et al. (2010), we define the effect size of each causal variant as $\beta_k = \log\left(1 + (\varphi - 1)\left(\frac{\min(\boldsymbol{p})}{p_k}\right)^{1/F}\right)$, where p_k is the sample MAF of the k-th variant, $\boldsymbol{p}$ is a vector of the sample MAF's of all causal variants, φ is the common odds ratio. The effect sizes of common variants relative to rare variants are controlled by the parameter F. When $F = 1$, common variants have almost no effect. For larger values of F, common variants have an increasing effect on phenotype. In the simulation, we consider two sets of coefficients $\varphi = 1.1$, $F = 1$ and $\varphi = 1.1$, $F = 3$. Under the setting where all the causal variants have deleterious effects, we assign positive effect size β_k to the k-th variant and use D (deleterious) to denote this setting. While, under the setting where causal variants have both deleterious and protective effects, we use the same effect size β_k for the k-th variant, but randomly sample 10% causal variants and change their effect directions to negative. This setting is denoted as DP (deleterious and protective).

6.2.2 Sparse Signal Scenario

The sparse signal scenario assumes that within the pre-specified set, only a small number of SNPs are causal variants (Lee et al., 2014; Wu et al., 2011). We randomly extract two rare variants and one common variant within the SNP set as causal variants. We generate the disease status from the following model:

$$\begin{aligned}\log\left(\frac{\mathrm{m}_i}{1-\mathrm{m}_i}\right) &= -1 + \beta_1 * I\left(Rare\_SNP1_i + Rare\_SNP2_i > 0\right) \\ &\quad + \beta_2 * I\left(Common\_SNP_i > 0\right).\end{aligned}$$

We consider two sets of coefficients: $\beta_1 = 0.7$, $\beta_2 = 0.3$ and $\beta_1 = 0.6$, $\beta_2 = 0.4$. This setting is similar to the simulation setting of Pritchard (2001) and Hu et al. (2020). To account for the setting where causal variants have both deleterious and protective effects, we also consider coefficients: $\beta_1 = 0.7$, $\beta_2 = -0.3$ and $\beta_1 = 0.6$, $\beta_2 = -0.4$.

6.2.3 Power of Association Detection

Figures 1 and 2 present the power of each method at the significance level $\alpha = 0.05$. Although this nominal significance level is too large for GWA studies, for the purpose of assessing the relative power of different methods, this choice is reasonable and reduces the computation burden.

We first discuss the results under the setting where all the causal variants have only deleterious effects. Both Figs. 1a and 2a show that when common variants have increasing effects, all methods have increased power. This is expected because the common variants with relatively larger effects are easier to find and improve the overall strength of association between the aggregated biomarker and phenotype, so the association between the pre-specified SNP set and the phenotype is easier to detect.

We also see from Fig. 1a that in the dense signal scenario, the aggregation with additive coding enjoys higher power than the indicator coding. The disease model under the dense signal scenario is additive, so higher power from the additive coding is expected. In addition, the indicator coding limits the number of variants to include for aggregation (see Fig. 3 for a comparison of the number of variants selected). Under the dense scenario, this may leave out many causal variants and could affect the strength of association between the aggregated biomarkers with phenotype.

The advantage of the additive coding is not clear under the sparse signal scenario for the step-up and super-variant based methods. As can be seen in Fig. 2a, when the sample re-use strategy is used, the aggregation with additive coding has a slightly higher power. But when the permutation strategy is used to control type I error, the additive coding yields a lower power.

Figure 2a also shows that under the sparse signal scenario, a backward deletion procedure may help the sample re-use strategy, but could be ineffective for the permutation strategy. This is especially true if the additive coding is used for aggregation.

In the dense signal scenario, when causal variants have both deleterious and protective effects, we observe reduced power when rare variant have stronger effects shown in the left panel of Fig. 1b compared to the power when all the causal variants have deleterious effects (left panel of Fig. 1a). The power loss is expected when the assumption of the methods used in this simulation that all the variants have the same effect direction is mis-specified. However, when common variants have increasing effects, we find comparable power in the right panel of Figs. 1b and 1a. Given the ratio of the rare and common variants is 4:1 in this simulation study, the causal variants randomly sampled to have protective effects are more likely be rare

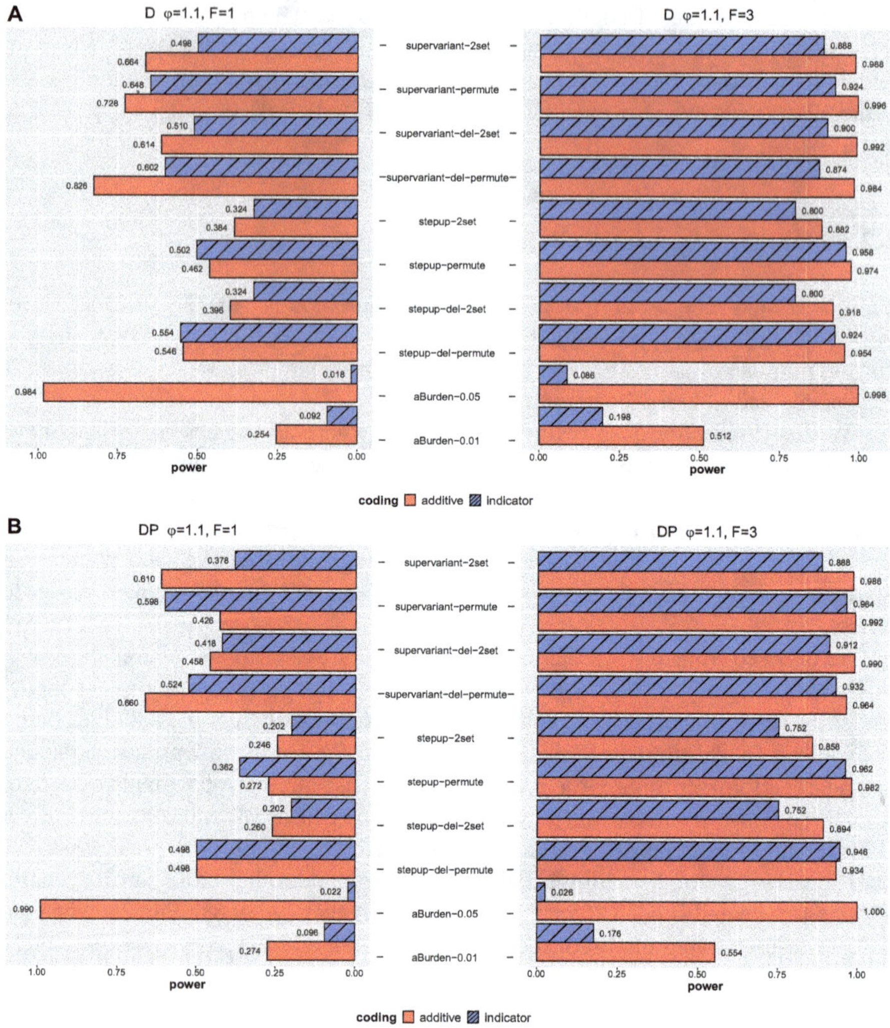

Fig. 1 Power of different methods under the dense signal scenario. (**a**) All the causal variants have deleterious effects. (**b**) Causal variants have both deleterious and protective effects. "aBurden-*u*" stands for adaptive burden test with MAF threshold *u*. "-2set" and "-permute" indicate the use of the sample re-use strategy and the permutation strategy, respectively. D represents deleterious and DP stands for deleterious and protective effects

variants (it is not necessarily true in the real data setting), which may explain no obvious power loss when common variants have stronger effects.

Under the sparse signal scenario, Fig. 2b reveals consistently lower power when rare and common variants have different effect directions than Fig. 2a. We also notice that using additive coding suffers greater power loss than indicator coding.

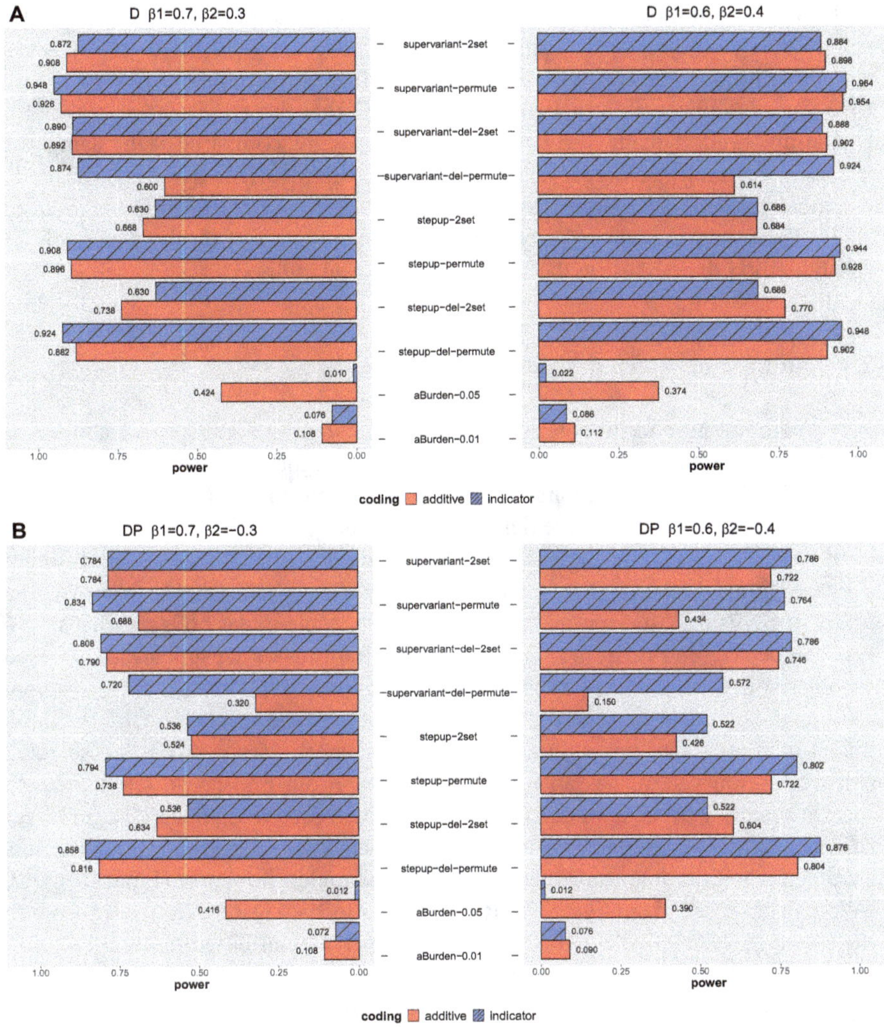

Fig. 2 Power of different methods under the sparse signal scenario. (**a**) All the causal variants have deleterious effects. (**b**) Causal variants have both deleterious and protective effects

The possible reason is that the aggregated variants may suffer more from the cancellation of different directional effects in additive coding.

In general, we observe that the step-up and super-variant based methods are more powerful than the adaptive burden test with MAF thresholds. This is reasonable because variant selection in the adaptive burden test with MAF thresholds is not phenotype-dependent. Therefore, the selected SNPs for aggregation may include many neutral variants that dilute the strength of association. The only exception is in the dense signal scenario when MAF threshold 0.05 is chosen along with the additive coding. In this case, all rare variants are included for aggregation. Because this is the dense signal setting and rare variants are assigned larger effect sizes, aggregating all rare variants with the additive coding enjoys a higher power.

We can also observe from both Figs. 1 and 2 that the reduction of power for the step-up method using the sample re-use strategy compared with using the permutation strategy is bigger than that for the super-variant based selection. In the sample re-use strategy, the effective sample size for variant selection is reduced to half. Therefore, the sample size can have a bigger impact on the power of the step-up method than it has on the super-variant based method.

6.2.4 Number of Selected Variants

We compare the number of selected SNPs for the step-up and super-variant based methods under two scenarios of causal variants. Figure 3 presents the average number of selected SNPs among all repetitions where the association is detected when all the causal variants have deleterious effects.

Figure 3 reveals that the super-variant based method selects more variants under the dense signal scenario than under the sparse signal scenario. In contrast, the number of SNPs that the step-up method selects is stable under both scenarios.

As can be seen in Fig. 3, the variant selection patterns of the super-variant based method are different under two scenarios of causal variants. Under the dense signal scenario, the number of selected SNPs increases as the effect size of common variants increases. But under the sparse signal scenario, fewer SNPs are selected when a common variant has an increasing effect. From Fig. 6, we know that the selection accuracy of the common causal variant increases when it has a larger effect size. This contributes to the reduction in the number of selected SNPs; fewer variants are needed when the true causal variant is selected under the sparse signal scenario.

We can also see from Fig. 3 that adopting the backward deletion procedure reduces the number of selected SNPs for the super-variant based method by a lot. This can be beneficial for precision medicine, especially for pharmacogenomics. The potential targets for drug intervention are narrowed down to a few candidate variants, which could save the costs for further experimental analyses.

Figure 4 presents the average number of selected SNPs when causal variants have both deleterious and protective effects. It shows a similar pattern to that in Fig. 3, but the number of selected SNPs decreases. This is expected because the effects of variants with opposite effect directions may cancel out, then a smaller number of variants are selected.

6.2.5 Accuracy of Causal Variant Selection

In the sparse signal scenario, we also compare the accuracy of variant selection between the step-up and super-variant based methods. Specifically, when the aggregated SNPs achieve significance at the level $\alpha = 0.05$, we record if causal

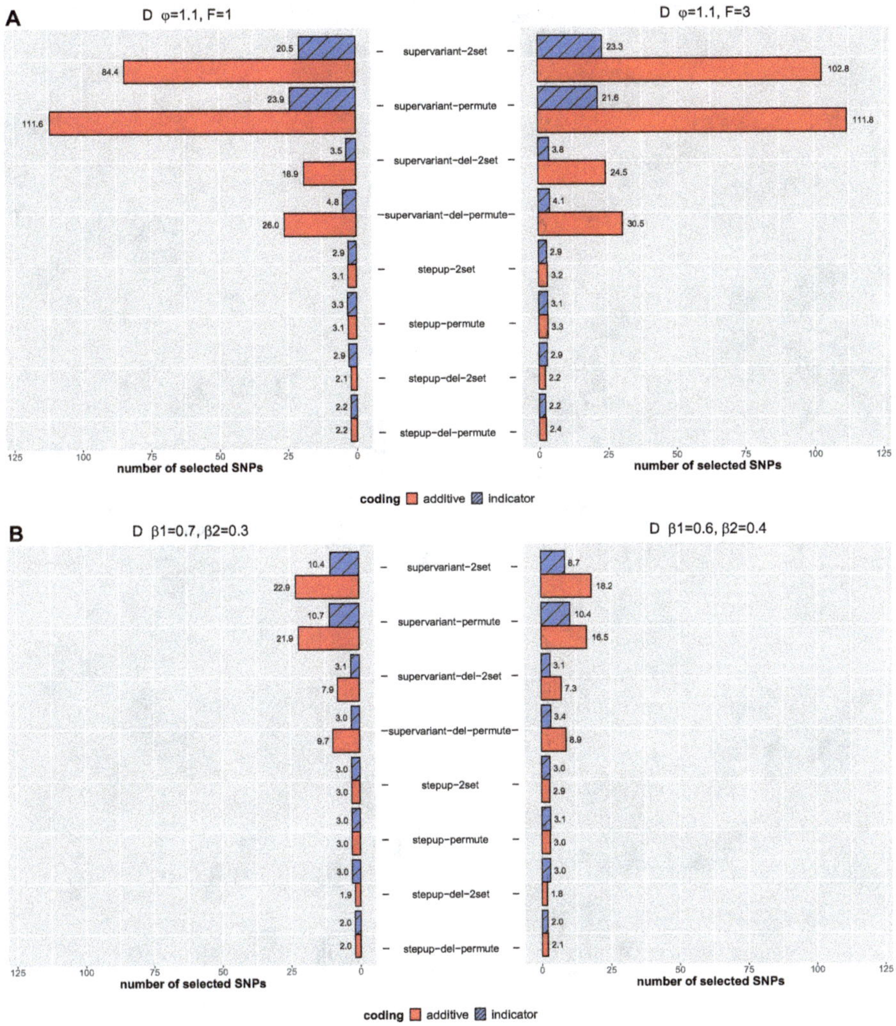

Fig. 3 Average number of selected SNPs for the step-up and super-variant based methods when all the causal variants have deleterious effects. (**a**) Under the dense signal scenario. (**b**) Under the sparse signal scenario

SNPs are selected. Figures 5 and 6 display the rate of causal variant selection when all the causal variants have deleterious effects.

Figures 5 and 6 reveal that both methods have higher selection accuracy with the additive coding and with permutation strategy compared to the indicator coding and the sample re-use strategy. In addition, when a common variant has an increasing effect, both methods can identify it with a higher frequency. But at the same time, the selection accuracy for rare variants decreases.

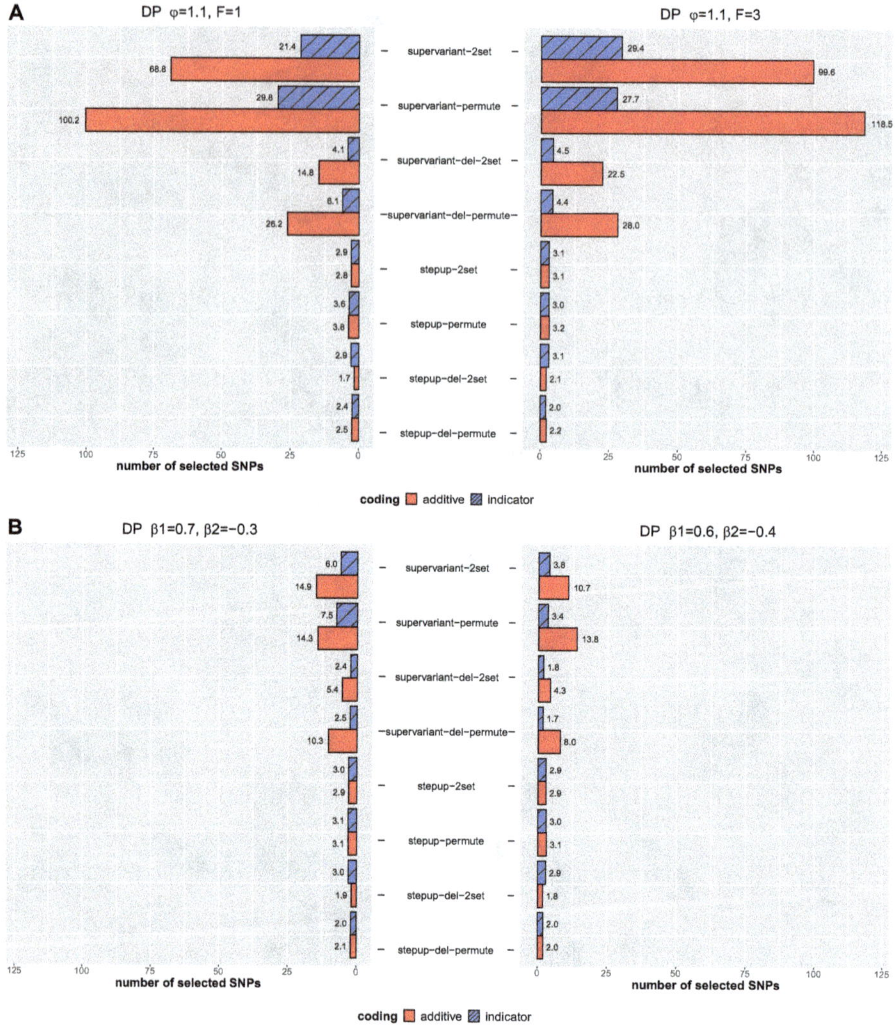

Fig. 4 Average number of selected SNPs for the step-up and super-variant based methods when causal variants have both deleterious and protective effects. (**a**) Under the dense signal scenario. (**b**) Under the sparse signal scenario

We also find that the super-variant based method can identify causal variants with higher frequency than the step-up method, especially when the permutation strategy is used.

Furthermore, the backward deletion procedure barely affects the selection accuracy for the step-up method. Figure 3 confirms this result as the number of selected SNPs with and without the backward deletion differs at most by 1. In contrast, the

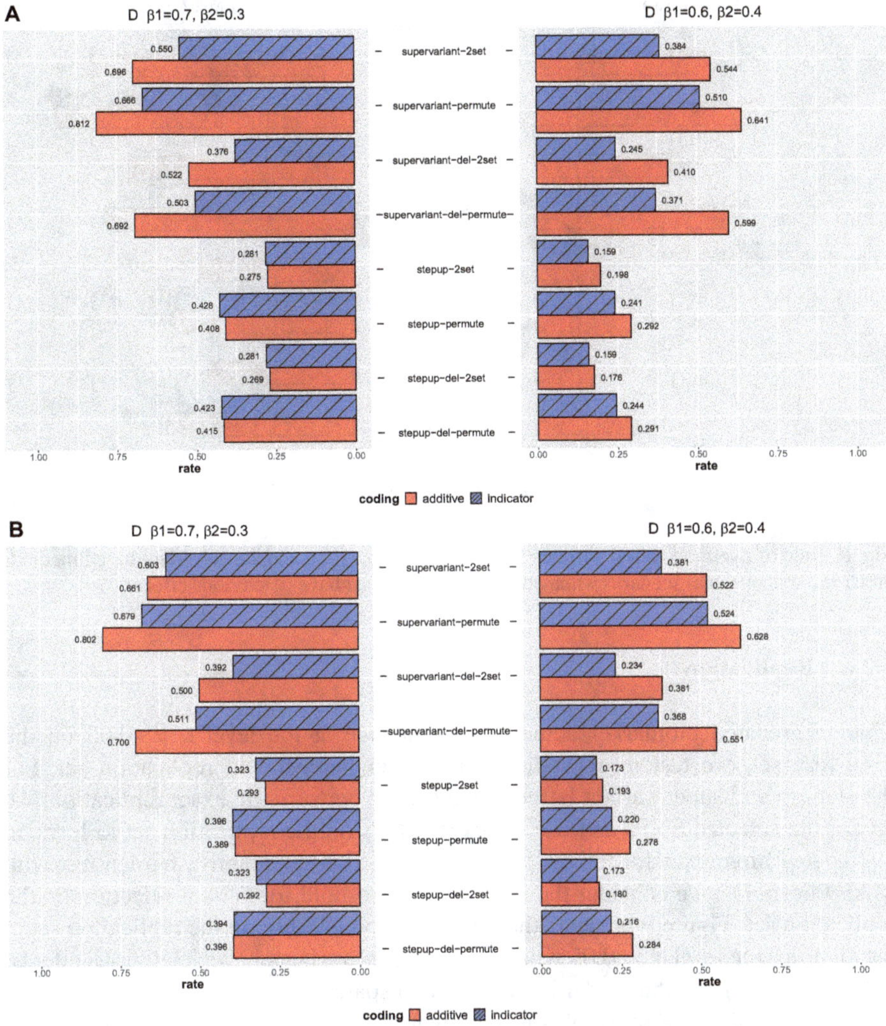

Fig. 5 Rate of rare causal variants selection for the step-up and super-variant based methods under the sparse signal scenario when all the causal variants have deleterious effects. (**a**) Rate of selection for variant *Rare* _ *SNP*1. (**b**) Rate of selection for variant *Rare* _ *SNP*2

backward deletion reduces the selection accuracy of the super-variant based method. Overall, it can still maintain a higher accuracy than the step-up method.

Figures 7 and 8 display the rate of causal variant selection when causal variants have both deleterious and protective effects. We observe a lower or comparable detect rate compared to Figs. 5 and 6. The patterns of detect rate between additive coding and indicator coding, as well as permutation and sample re-use approach are similar to that when all the causal variants have deleterious effects.

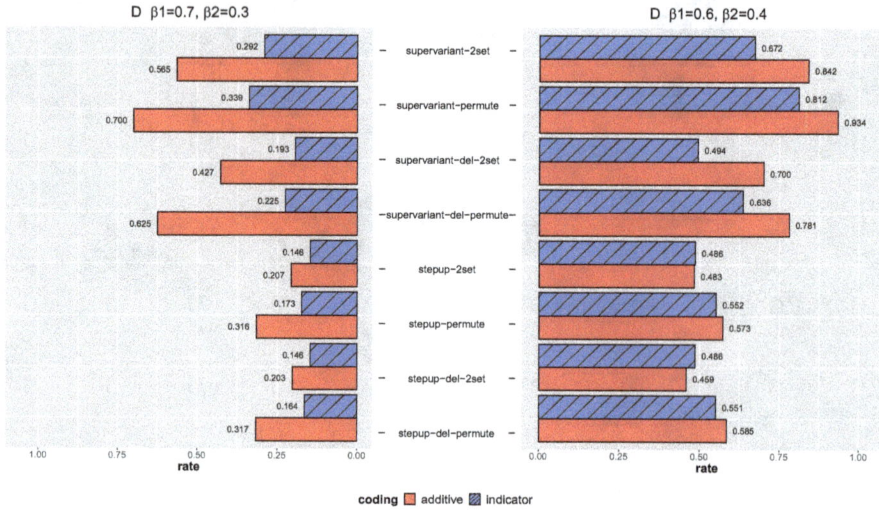

Fig. 6 Rate of common causal variants selection for the step-up and super-variant based methods under the sparse signal scenario when all the causal variants have deleterious effects

6.2.6 Replication

If an aggregated biomarker achieves significance at the level $\alpha = 0.05$ on the discovery set, we test its reproducibility on an independent replication set. For the step-up and super-variant based methods, we perform the exact replication; we record the selected SNPs and use the exact SNPs on the replication set to form an aggregated biomarker for the test of association. For the adaptive burden test with MAF thresholds, we consider the same MAF threshold for variant selection on the replication set. Figure 9 presents the rate of reproducibility on the replication set at the significance level $\alpha = 0.05$ when all the causal variants have deleterious effects.

Figure 9 suggests that under both dense and sparse signal scenarios, the adaptive burden test with MAF thresholds has the worst reproducibility. This is understandable as the variant selection of the adaptive burden test with MAF thresholds is not phenotype-dependent. Again, the only exception occurs under the dense signal scenario and when the method uses the MAF threshold 0.05 along with the additive coding.

We can see from Fig. 9 that the reproducibility of the selected SNPs from the super-variant based method is higher than that from the step-up method. This is expected from the selection accuracy results in Figs. 5 and 6. The super-variant based method selects more true causal variants, which boosts its reproducibility within the same population.

It is also interesting to see that the backward deletion procedure improves the reproducibility for both step-up and super-variant based methods. This is consistent

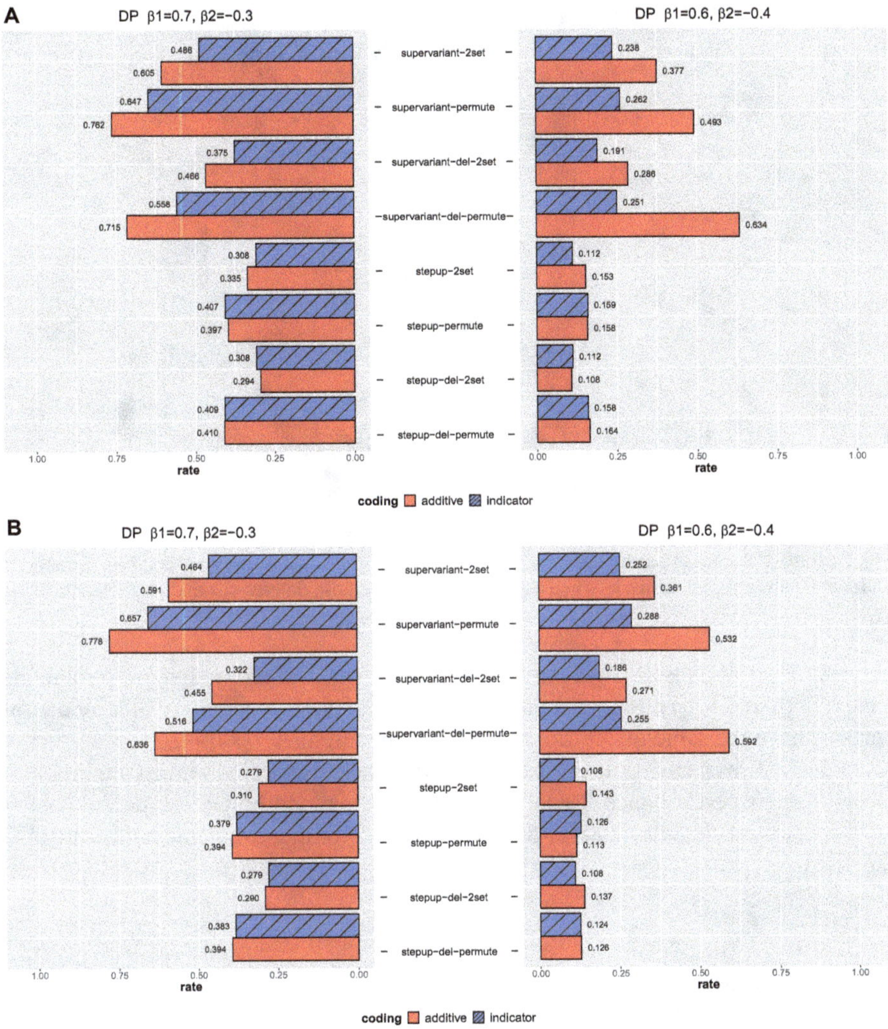

Fig. 7 Rate of rare causal variants selection for the step-up and super-variant based methods under the sparse signal scenario when causal variants have both deleterious and protective effects. (**a**) Rate of selection for variant $Rare\_SNP1$. (**b**) Rate of selection for variant $Rare\_SNP2$

with the possibility that aggregating many neutral variants may dilute the strength of association.

It is noteworthy from Fig. 9 that under the dense signal scenario, the permutation strategy may not be as desirable as the sample re-use strategy in terms of reproducibility, especially when the common variants have almost no effects.

Figure 10 presents the rate of reproducibility on the replication set when causal variants have both deleterious and protective effects. There is no obvious difference

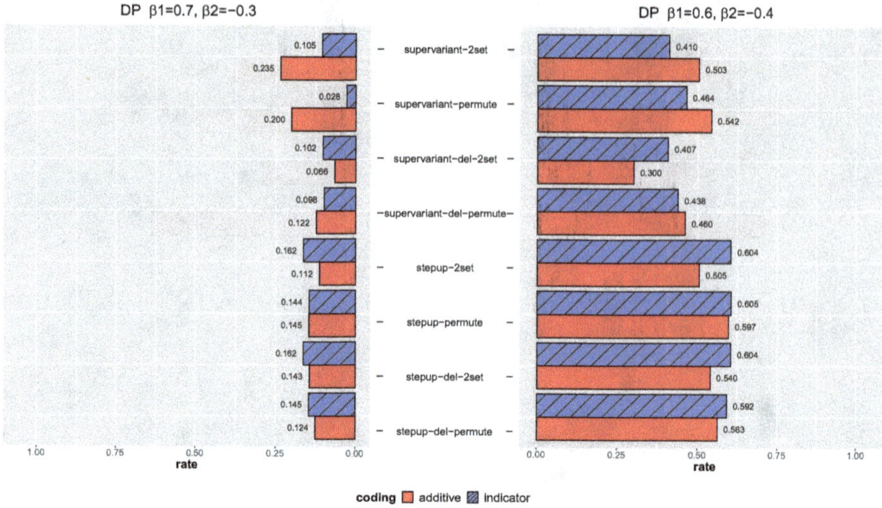

Fig. 8 Rate of common causal variants selection for the step-up and super-variant based methods under the sparse signal scenario when causal variants have both deleterious and protective effects

between Figs. 9 and 10, suggesting that super-variant based and step-up methods have a high and robust rate of reproducibility even if the assumption that all causal variants have the same effect direction is mis-specified.

In Table 2, we provide a summary of all methods considered in the simulation in terms of their performance. Recommendations when and how these methods may be used along with the available software packages for their implementation are also included.

7 Conclusions

We review the set-based genetic association methods with a focus on variant selection and aggregation involving variants with low frequency and modest association. Combining information from multiple loci could boost the overall strength of association. At the same time, by not using all variants avoids including many neutral variants, which may dilute the strength of association with the trait of interest. Precision medicine can use the identified SNPs from these association studies in constructing risk prediction scores and in searching for targets of drug intervention.

The variant selection could be based on external information such as functional annotation. Using known biological knowledge leads to interpretable results, but how to properly use such knowledge is generally challenging and can affect the power of this approach. In addition, functional annotations for variants within

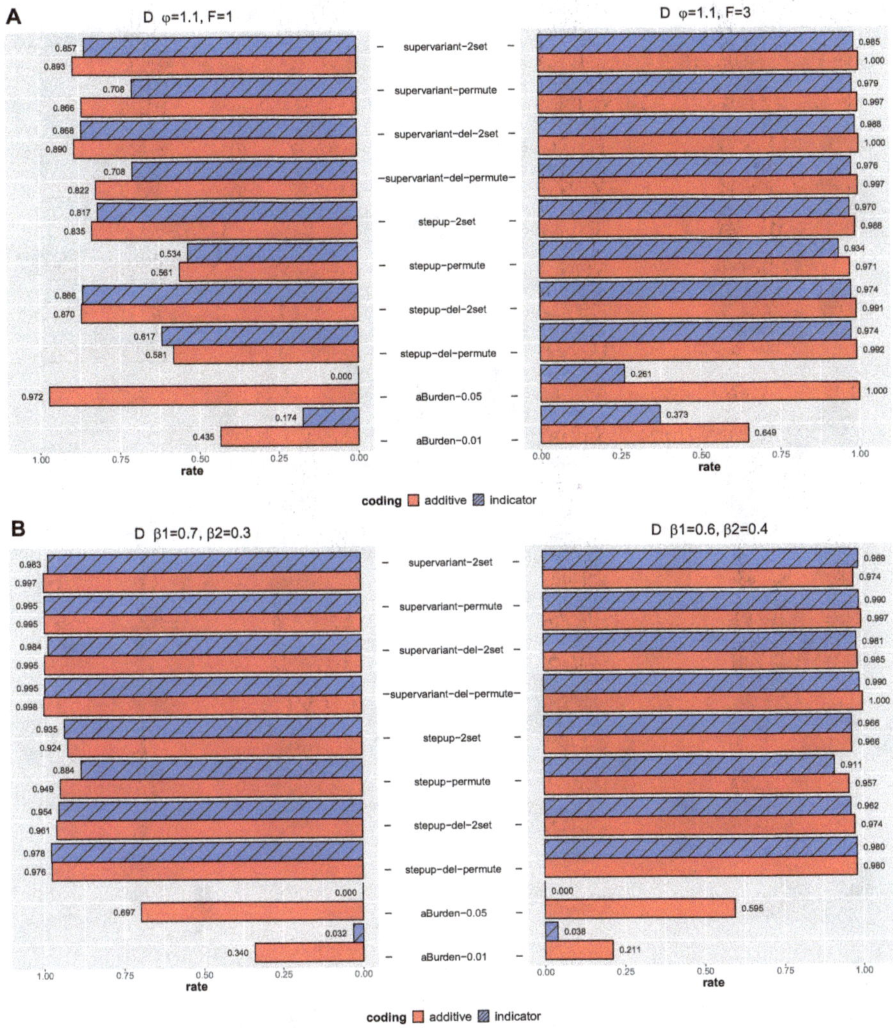

Fig. 9 Rate of reproducibility for all methods when all the causal variants have deleterious effects. (**a**) Under the dense signal scenario. (**b**) Under the sparse signal scenario

non-exome regions is limited. This would be a challenge for the whole-genome association study. We expect that with the cost of whole-genome sequencing reduces, more effort will be spent on annotating those non-exome variants (Boyle et al., 2012; Consortium, E.P., 2012).

Data-driven methods can select SNPs in high linkage disequilibrium with causal variants. The super-variant based method may also differentiate underlying settings of causal variants. However, most of the existing methods only consider the main effects during variant selection. Depth importance measure (Chen et al.,

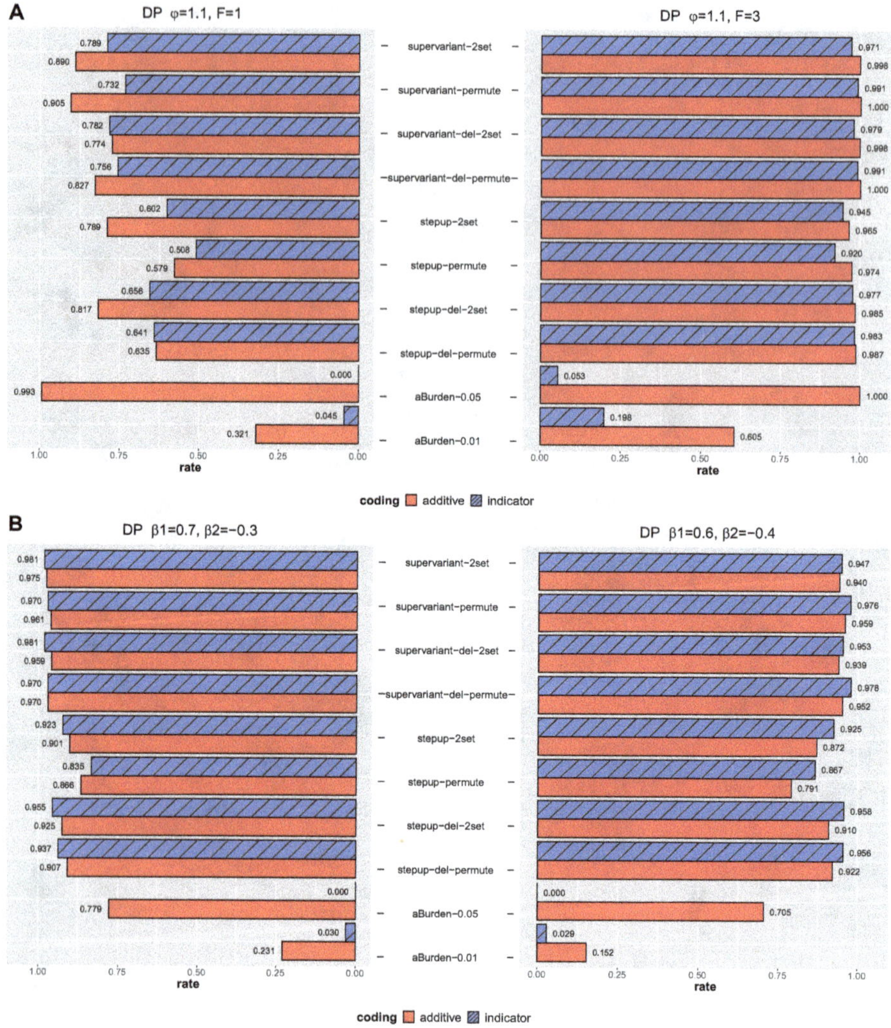

Fig. 10 Rate of reproducibility for all methods when causal variants have both deleterious and protective effects. (**a**) Under the dense signal scenario. (**b**) Under the sparse signal scenario

2007) is an example to incorporate interactive effects into importance measure construction. Future developments in this direction are useful to properly account for interactive effects during variant selection. In addition, reflecting interactive effects in aggregated biomarkers may boost the strength of association. Aggregation with the indicator coding is one way to achieve this goal, but when low-frequency variants are involved, aggregating with the indicator coding is close to the additive coding. In the latter case, it is more challenging to detect the existence of interactive effects before they are incorporated into modeling and/or test of association.

Table 2 Performance summary of all methods considered and their implementation

Methods	Super-variant	Step-up	Adaptive burden with MAF threshold
Power	High power with additive coding in both dense and sparse signal scenarios when sample-reuse strategy is adopted	High power with indicator coding in dense signal scenario when common variants have relatively large effects and in sparse signal scenario when permutation strategy is considered	High power only in dense scenario when MAF threshold 0.05 is used
Variant selection	Have the ability to distinguish dense and sparse signal scenarios; higher selection accuracy with additive coding and permutation strategy; may suffer loss of accuracy when causal variants can be both deleterious and protective	Cannot distinguish between dense and sparse signal scenarios; low selection accuracy overall; common variant selection accuracy is robust against the effect direction of causal variants	No variant selection when there are only a small amount of causal variants
Reproducibility	Reproducibility is overall high with additive coding	Reproducibility is high with additive coding when sample-reuse strategy is adopted	Reproducibility is only high in the dense signal scenario when MAF threshold 0.05 is used
Recommendation	Can be used to distinguish between dense and sparse signal scenarios; use with additive coding and sample-reuse strategy for detection and variant selection	Only use in the sparse signal scenario with permutation strategy for common variant identification	Only use to detect whether there is any association between the set of variants and phenotype of interest under dense signal scenario, and use MAF threshold 0.05
Software	diTARV	AssotesteR	AssotesteR

With the advancement of sequencing technology, we are able to obtain extreme rare variants in the exome sequencing datasets where many variants could be singletons or doubletons (Taliun et al., 2021; Backman et al., 2021). Such variants pose great challenges for association study. One current approach is to first aggregate these extreme rare variants according to their functional annotations, and then perform marginal test of association with phenotype of interest (Backman et al., 2021). This approach may also be considered by methods reviewed in this chapter where aggregated biomarkers are put together with common and low-frequency variants for set-based analysis. This could be an interesting direction for future research on exome sequencing association study.

In this review, we focus on population-based association analysis. For the study of low-frequency variants, family-based studies may enable multiple copies of rare mutations to be observed (Veltman & Brunner, 2012) to facilitate the detection. When family-based data are available, analysis of such data may provide complementary information. Several methods of performing rare-variant association tests in families have been developed (Chen et al., 2013; Ionita-Laza et al., 2013; Schifano et al., 2012; Wang et al., 2013), and further developments are warranted. Integrating results from both population- and family-based studies is also an interesting direction for variant detection.

Acknowledgments The work of Heping Zhang is supported in part by NIH grants HG010171 and MH116527. This study has been conducted using the UK Biobank Resource under Application Number 42009. The authors would like to thank the Yale Center for Research Computing for the guidance and use of the research computing infrastructure.

Conflict of interests The authors declare no conflict of interest.

References

Ashley, E. A. (2016). Towards precision medicine. *Nature Reviews Genetics, 17*(9), 507.

Asimit, J., & Zeggini, E. (2010). Rare variant association analysis methods for complex traits. *Annual Review of Genetics, 44*, 293–308.

Backman, J. D., Li, A. H., Marcketta, A., Sun, D., Mbatchou, J., Kessler, M. D., Benner, C., Liu, D., Locke, A. E., & Balasubramanian, S. (2021). Exome sequencing and analysis of 454,787 UK Biobank participants. *Nature, 599*(7886), 628–634.

Boyle, A. P., Hong, E. L., Hariharan, M., Cheng, Y., Schaub, M. A., Kasowski, M., Karczewski, K. J., Park, J., Hitz, B. C., & Weng, S. (2012). Annotation of functional variation in personal genomes using RegulomeDB. *Genome Research, 22*(9), 1790–1797.

Byrnes, A. E., Wu, M. C., Wright, F. A., Li, M., & Li, Y. (2013). The value of statistical or bioinformatics annotation for rare variant association with quantitative trait. *Genetic Epidemiology, 37*(7), 666–674.

Chen, L. S., Hsu, L., Gamazon, E. R., Cox, N. J., & Nicolae, D. L. (2012). An exponential combination procedure for set-based association tests in sequencing studies. *American Journal of Human Genetics, 91*(6), 977–986.

Chen, X., Liu, C.-T., Zhang, M., & Zhang, H. (2007). A forest-based approach to identifying gene and gene–gene interactions. *Proceedings of the National Academy of Sciences, 104*(49), 19199–19203.

Chen, H., Meigs, J. B., & Dupuis, J. (2013). Sequence kernel association test for quantitative traits in family samples. *Genetic Epidemiology, 37*(2), 196–204.

Cirulli, E. T., White, S., Read, R. W., Elhanan, G., Metcalf, W. J., Tanudjaja, F., Fath, D. M., Sandoval, E., Isaksson, M., & Schlauch, K. A. (2020). Genome-wide rare variant analysis for thousands of phenotypes in over 70,000 exomes from two cohorts. *Nature Communications, 11*(1), 1–10.

Consortium, A.G. (2013). Seven new loci associated with age-related macular degeneration. *Nature Genetics, 45*(4), 433.

Consortium, E.P. (2012). An integrated encyclopedia of DNA elements in the human genome. *Nature, 489*(7414), 57.

Cordell, H. J. (2009). Detecting gene–gene interactions that underlie human diseases. *Nature Reviews Genetics, 10*(6), 392–404.

Crews, K. R., Hicks, J. K., Pui, C. H., Relling, M. V., & Evans, W. E. (2012). Pharmacogenomics and individualized medicine: Translating science into practice. *Clinical Pharmacology and Therapeutics, 92*(4), 467–475.

De Lange, K. M., Moutsianas, L., Lee, J. C., Lamb, C. A., Luo, Y., Kennedy, N. A., Jostins, L., Rice, D. L., Gutierrez-Achury, J., & Ji, S.-G. (2017). Genome-wide association study implicates immune activation of multiple integrin genes in inflammatory bowel disease. *Nature Genetics, 49*(2), 256–261.

Derkach, A., Lawless, J. F., & Sun, L. (2013). Robust and powerful tests for rare variants using Fisher's method to combine evidence of association from two or more complementary tests. *Genetic Epidemiology, 37*(1), 110–121.

Eichler, E. E., Flint, J., Gibson, G., Kong, A., Leal, S. M., Moore, J. H., & Nadeau, J. H. (2010). Missing heritability and strategies for finding the underlying causes of complex disease. *Nature Reviews Genetics, 11*(6), 446–450.

Franke, A., McGovern, D. P., Barrett, J. C., Wang, K., Radford-Smith, G. L., Ahmad, T., Lees, C. W., Balschun, T., Lee, J., & Roberts, R. (2010). Genome-wide meta-analysis increases to 71 the number of confirmed Crohn's disease susceptibility loci. *Nature Genetics, 42*(12), 1118–1125.

Gauderman, W. J., Murcray, C., Gilliland, F., & Conti, D. V. (2007). Testing association between disease and multiple SNPs in a candidate gene. *Genetic Epidemiology, 31*(5), 383–395.

Gibson, G. (2012). Rare and common variants: Twenty arguments. *Nature Reviews Genetics, 13*(2), 135–145.

Greene, C. S., Penrod, N. M., Williams, S. M., & Moore, J. H. (2009). Failure to replicate a genetic association may provide important clues about genetic architecture. *PLoS One, 4*(6), e5639.

Han, F., & Pan, W. (2010). A data-adaptive sum test for disease association with multiple common or rare variants. *Human Heredity, 70*(1), 42–54.

Harrison, J. R., Mistry, S., Muskett, N., & Escott-Price, V. (2020). From polygenic scores to precision medicine in Alzheimer's disease: A systematic review. *Journal of Alzheimer's Disease* (Preprint), 1–13.

Hoffmann, T. J., Marini, N. J., & Witte, J. S. (2010). Comprehensive approach to analyzing rare genetic variants. *PLoS One, 5*(11), e13584.

Hu, J., Li, C., Wang, S., Li, T., & Zhang, H. (2021). Genetic variants are identified to increase risk of COVID-19 related mortality from UK Biobank data. *Human Genomics, 15*(1), 1–10.

Hu, J., Li, T., Wang, S., & Zhang, H. (2020). Supervariants identification for breast cancer. *Genetic Epidemiology, 44*(8), 934–947.

Huffman, J. (2018). Examining the current standards for genetic discovery and replication in the era of mega-biobanks. *Nature Communications, 9*(1), 1–4.

Ionita-Laza, I., Lee, S., Makarov, V., Buxbaum, J. D., & Lin, X. (2013). Family-based association tests for sequence data, and comparisons with population-based association tests. *European Journal of Human Genetics, 21*(10), 1158–1162.

Jameson, J. L., & Longo, D. L. (2015). Precision medicine—personalized, problematic, and promising. *Obstetrical & Gynecological Survey, 70*(10), 612–614.

Klein, R. J., Zeiss, C., Chew, E. Y., Tsai, J.-Y., Sackler, R. S., Haynes, C., Henning, A. K., SanGiovanni, J. P., Mane, S. M., & Mayne, S. T. (2005). Complement factor H polymorphism in age-related macular degeneration. *Science, 308*(5720), 385–389.

König, I. R. (2011). Validation in genetic association studies. *Briefings in Bioinformatics, 12*(3), 253–258.

Kraft, P., Zeggini, E., & Ioannidis, J. P. (2009). Replication in genome-wide association studies. *Statistical Science*. A review journal of The Institute of Mathematical Statistics, *24*(4), 561.

Kryukov, G. V., Pennacchio, L. A., & Sunyaev, S. R. (2007). Most rare missense alleles are deleterious in humans: Implications for complex disease and association studies. *American Journal of Human Genetics, 80*(4), 727–739.

Lee, S., Abecasis, G. R., Boehnke, M., & Lin, X. (2014). Rare-variant association analysis: Study designs and statistical tests. *American Journal of Human Genetics, 95*(1), 5–23.

Lee, S., Wu, M. C., & Lin, X. (2012). Optimal tests for rare variant effects in sequencing association studies. *Biostatistics, 13*(4), 762–775.

Lewis, C. M. (2002). Genetic association studies: Design, analysis and interpretation. *Briefings in Bioinformatics, 3*(2), 146–153.

Li, B., & Leal, S. M. (2008). Methods for detecting associations with rare variants for common diseases: Application to analysis of sequence data. *American Journal of Human Genetics, 83*(3), 311–321.

Liu, Y., & Xie, J. (2020). Cauchy combination test: A powerful test with analytic p-value calculation under arbitrary dependency structures. *Journal of the American Statistical Association, 115*(529), 393–402.

MacArthur, D. G., Balasubramanian, S., Frankish, A., Huang, N., Morris, J., Walter, K., Jostins, L., Habegger, L., Pickrell, J. K., & Montgomery, S. B. (2012). A systematic survey of loss-of-function variants in human protein-coding genes. *Science, 335*(6070), 823–828.

Madsen, B. E., & Browning, S. R. (2009). A groupwise association test for rare mutations using a weighted sum statistic. *PLoS Genetics, 5*(2), e1000384.

Maher, B. (2008). Personal genomes: The case of the missing heritability. *Nature, 456*(7218), 18–21.

Marigorta, U. M., Rodríguez, J. A., Gibson, G., & Navarro, A. (2018). Replicability and prediction: Lessons and challenges from GWAS. *Trends in Genetics, 34*(7), 504–517.

Matchett, K. B., Lynam-Lennon, N., Watson, R. W., & Brown, J. A. (2017). Advances in precision medicine: Tailoring individualized therapies. *Cancers, 9*(11), 146.

Metzker, M. L. (2010). Sequencing technologies—the next generation. *Nature Reviews Genetics, 11*(1), 31–46.

Meyerson, M., Gabriel, S., & Getz, G. (2010). Advances in understanding cancer genomes through second-generation sequencing. *Nature Reviews Genetics, 11*(10), 685–696.

Mooney, S. D. (2015). Progress towards the integration of pharmacogenomics in practice. *Human Genetics, 134*(5), 459–465.

Morgenthaler, S., & Thilly, W. G. (2007). A strategy to discover genes that carry multi-allelic or mono-allelic risk for common diseases: A cohort allelic sums test (CAST). *Mutation Research/Fundamental and Molecular Mechanisms of Mutagenesis, 615*(1–2), 28–56.

Morris, A. P., Voight, B. F., Teslovich, T. M., Ferreira, T., Segre, A. V., Steinthorsdottir, V., Strawbridge, R. J., Khan, H., Grallert, H., & Mahajan, A. (2012). Large-scale association analysis provides insights into the genetic architecture and pathophysiology of type 2 diabetes. *Nature Genetics, 44*(9), 981.

Park, J.-H., Gail, M. H., Weinberg, C. R., Carroll, R. J., Chung, C. C., Wang, Z., Chanock, S. J., Fraumeni, J. F., & Chatterjee, N. (2011). Distribution of allele frequencies and effect sizes and their interrelationships for common genetic susceptibility variants. *Proceedings of the National Academy of Sciences, 108*(44), 18026–18031.

Price, A. L., Kryukov, G. V., de Bakker, P. I., Purcell, S. M., Staples, J., Wei, L.-J., & Sunyaev, S. R. (2010). Pooled association tests for rare variants in exon-resequencing studies. *American Journal of Human Genetics, 86*(6), 832–838.

Pritchard, J. K. (2001). Are rare variants responsible for susceptibility to complex diseases? *American Journal of Human Genetics, 69*(1), 124–137.

Ritchie, M. D. (2012). The success of pharmacogenomics in moving genetic association studies from bench to bedside: Study design and implementation of precision medicine in the post-GWAS era. *Human Genetics, 131*(10), 1615–1626.

Schifano, E. D., Epstein, M. P., Bielak, L. F., Jhun, M. A., Kardia, S. L., Peyser, P. A., & Lin, X. (2012). SNP set association analysis for familial data. *Genetic Epidemiology, 36*(8), 797–810.

Song, C., & Zhang, H. (2014). TARV: Tree-based analysis of rare variants identifying risk modifying variants in CTNNA2 and CNTNAP2 for alcohol addiction. *Genetic Epidemiology, 38*(6), 552–559.

Taliun, D., Harris, D. N., Kessler, M. D., Carlson, J., Szpiech, Z. A., Torres, R., Taliun, S. A. G., Corvelo, A., Gogarten, S. M., & Kang, H. M. (2021). Sequencing of 53,831 diverse genomes from the NHLBI TOPMed Program. *Nature, 590*(7845), 290–299.

Tam, V., Patel, N., Turcotte, M., Bossé, Y., Paré, G., & Meyre, D. (2019). Benefits and limitations of genome-wide association studies. *Nature Reviews Genetics, 20*(8), 467–484.

Udler, M. S., McCarthy, M. I., Florez, J. C., & Mahajan, A. (2019). Genetic risk scores for diabetes diagnosis and precision medicine. *Endocrine Reviews, 40*(6), 1500–1520.

Van Hout, C. V., Tachmazidou, I., Backman, J. D., Hoffman, J. D., Liu, D., Pandey, A. K., Gonzaga-Jauregui, C., Khalid, S., Ye, B., & Banerjee, N. (2020). Exome sequencing and characterization of 49,960 individuals in the UK Biobank. *Nature, 586*(7831), 749–756.

Veltman, J. A., & Brunner, H. G. (2012). De novo mutations in human genetic disease. *Nature Reviews Genetics, 13*(8), 565–575.

Wang, X., Lee, S., Zhu, X., Redline, S., & Lin, X. (2013). GEE-based SNP set association test for continuous and discrete traits in family-based association studies. *Genetic Epidemiology, 37*(8), 778–786.

Warren, H. R., Evangelou, E., Cabrera, C. P., Gao, H., Ren, M., Mifsud, B., Ntalla, I., Surendran, P., Liu, C., & Cook, J. P. (2017). Genome-wide association analysis identifies novel blood pressure loci and offers biological insights into cardiovascular risk. *Nature Genetics, 49*(3), 403.

Wu, M. C., Lee, S., Cai, T., Li, Y., Boehnke, M., & Lin, X. (2011). Rare-variant association testing for sequencing data with the sequence kernel association test. *American Journal of Human Genetics, 89*(1), 82–93.

Xue, A., Wu, Y., Zhu, Z., Zhang, F., Kemper, K. E., Zheng, Z., Yengo, L., Lloyd-Jones, L. R., Sidorenko, J., & Wu, Y. (2018). Genome-wide association analyses identify 143 risk variants and putative regulatory mechanisms for type 2 diabetes. *Nature Communications, 9*(1), 1–14.

Yan, X., & Bien, J. (2020). Rare feature selection in high dimensions. *Journal of the American Statistical Association*, 1–14.

Zuk, O., Hechter, E., Sunyaev, S. R., & Lander, E. S. (2012). The mystery of missing heritability: Genetic interactions create phantom heritability. *Proceedings of the National Academy of Sciences, 109*(4), 1193–1198.

Zuk, O., Schaffner, S. F., Samocha, K., Do, R., Hechter, E., Kathiresan, S., Daly, M. J., Neale, B. M., Sunyaev, S. R., & Lander, E. S. (2014). Searching for missing heritability: Designing rare variant association studies. *Proceedings of the National Academy of Sciences, 111*(4), E455–E464.

Leveraging Functional Annotations Improves Cross-Population Genetic Risk Prediction

Yixuan Ye, Leqi Xu, and Hongyu Zhao

Abstract The clinical use of polygenic risk scores (PRS) in non-European populations is hindered by the Eurocentric biases in genetic studies and the poor transferability of genetic results across populations. As functional annotations are commonly used to prioritize signals in post-GWAS analysis and have shown potential in improving the portability of the standard PRS (P+T PRS), here we further examine the benefits of leveraging functional annotations in more sophisticated and accurate PRS methods for cross-population genetic risk prediction. We collected five tiers of annotations and employed AnnoPred, which is a Bayesian framework leveraging functional annotations on the basis of LDPred. Through the analysis of 339,584 unrelated individuals in the UK Biobank, we showed that both LDPred and AnnoPred were superior than P+T PRS; and after leveraging functional annotations, AnnoPred consistently outperformed LDPred across eight traits/diseases and four populations that were investigated. These results illustrate the superiority of more sophisticated PRS models over the standard PRS and further demonstrate the benefit of incorporating functional annotations in improving cross-population prediction.

Keywords Polygenic risk score · AnnoPred · Cross-population · Functional annotations · UK Biobank

---

These authors "Yixuan Ye and Leqi Xu " contribute equally to the work

---

Y. Ye
Program of Computational Biology and Bioinformatics, Yale University, New Haven, CT, USA
e-mail: yixuan.ye@yale.edu

L. Xu · H. Zhao (✉)
Department of Biostatistics, Yale School of Public Health, New Haven, CT, USA
e-mail: leqi.xu@yale.edu; hongyu.zhao@yale.edu

Y. Zhao, D.-G. Chen (eds.), *Statistics in Precision Health*, ICSA Book Series in Statistics, https://doi.org/10.1007/978-3-031-50690-1_18

1 Introduction

Polygenic risk score (PRS), the weighted sum of risk allele counts, has been widely adopted as a useful tool to predict human complex traits based on genetic information (Dudbridge, 2013). This is largely due to the remarkable success of genome-wide association studies (GWAS) and improved PRS algorithms that have been shown to have substantial stratification capacity and potential for broad-scale clinical use Mavaddat et al. (2019); Natarajan et al. (2017); Sharp et al. (2019). However, due to the Eurocentric biases in GWAS, the well-established PRSs are mostly derived from and applied to European populations (EUR), leaving other populations underrepresented and understudied (Martin et al., 2019; Petrovski & Goldstein, 2016).

To bridge the equity gap, several studies have attempted to build PRS for non-EUR populations by either (1) developing the PRS weights based on non-EUR training data (Bentley et al., 2020; Blechter et al., 2021; Shieh et al., 2020) or (2) applying PRS weights derived from EUR training data to non-EUR populations; yet both lead to considerably less accurate PRS (Duncan et al., 2019; Lam et al., 2019), where the first strategy lacks power given the limited sample size of non-EUR GWASs, and the second strategy is hindered by the poor transferability of EUR-derived PRS due to the differences in linkage disequilibrium (LD), allele frequencies, and genetic architectures across populations (Amariuta et al., 2020; Wang et al., 2020).

While the most straightforward solution to increase PRS performance in the understudied populations is to build well-powered GWAS resources, it is however very expensive and time-consuming (Need & Goldstein, 2009). An alternative way is to increase the transferability of EUR-derived PRS by leveraging more known information. A recent study has shown that functional annotations have the potential to improve the cross-population portability of genetic data (Amariuta et al., 2020). Specifically, the authors of this study introduced 707 cell-type specific regulatory annotations to prioritize variants in building PRS. Through the analysis of 21 phenotypes, they observed that annotation-informed PRS increased trans-ancestry portability from EUR to East Asians (EAS) over standard PRS (clumping and thresholding PRS, P+T) by 49.9% with regard to R^2. While their results illustrated the important role of functional annotations in trans-population PRS, their work was however limited to the simple P+T PRS algorithm and single version of functional annotation. Besides, during their application, they only considered one single non-EUR population (EAS), which further hindered a broader conclusion they could make for the role of functional annotations in trans-population prediction.

Here we extend previous work mainly in three ways to better examine the benefits of leveraging functional annotations in improving cross-population genetic risk prediction. First, instead of using P+T PRS (Chang et al., 2015), we employ AnnoPred (Hu et al., 2017) to leverage annotations on the basis of LDPred (Vilhjálmsson et al., 2015), which is a Bayesian method that explicitly models linkage disequilibrium (LD) among SNPs to improve predictive power. More specifically, AnnoPred shares a similar Bayesian framework with LDPred but uses a different prior based on variants' memberships in different functional categories. Second, in addition to investigating the transferability of PRS from EUR to EAS, we also consider other non-EUR populations, including South-Asian (SAS) and African (AFR). Third, we collect and examine five tiers of functional annotations where the 707 cell-type specific regulatory annotations are organized as one tier.

2 Methods

2.1 UK Biobank

The UK Biobank (UKBB) is a large prospective study designed to be a resource for research into the causes of complex diseases among middle-old aged adults (Sudlow et al., 2015). The study protocol, details of the study design, and data access are available online (Bycroft et al., 2018). In brief, UKBB recruited 502,618 participants aged 40–69 years from 22 assessment centers across the United Kingdom between 2006 and 2010; participants were genotyped using two genotyping arrays, the Affymetrix UK BiLEVE AxiomTM Array and UKBB AxiomTM Array. We excluded variants with missing rate >0.01, minor allele frequency <0.05, Hardy-Weinberg equilibrium P value <1e-5, and imputation quality score <0.3. Sample exclusion criteria included having related individuals within UKBB, lack of genetic data (sufficient DNA could not be extracted from the blood samples of 3% participants), discordance between reported and genotype inferred sex, poor heterozygosity or missingness, sex chromosome aneuploidy, and withdrawal of informed consent.

We used the "online augmentation decomposition and Procrustes transformation" (OADP) algorithm implemented in the Python package, FRAPOSA, to infer the ancestry of each individual in the UKBB (Zhang et al., 2020). The Phase 3 release of the 1000 Genomes Project (1KG) data served as our reference panel (Auton et al., 2015). After excluding related individuals and intersecting the high-quality genotyped SNPs with the UKBB, the 1KG reference panel included 145,282 variants and 2492 individuals from five super-populations: 498 EUR, 657 Africans, 347 admixed Americans, 503 East Asians, and 487 South Asians. Each UKBB individual was then projected onto to the first four genotypic principal components (PCs) calculated in 1KG and was accordingly assigned to the closest ancestry, resulting in 316,736 EUR, 8654 SAS, 2364 EAS, and 6992 AFR in our analysis. And we randomly selected 306,736 EUR as the training set while the remaining as the testing set.

In this study, we considered four binary diseases, including atrial fibrillation (AF), breast cancer (BC), coronary artery disease (CAD), type 2 diabetes (T2D); and four quantitative traits, including high-density lipoprotein (HDL), low-density lipoprotein cholesterol (LDL-C), total cholesterol (TC), and triglyceride (TG).

Identification of cases for binary diseases was based on both self-reports in an interview with a trained nurse, and electronic health record information including both inpatient International Classification of Diseases (ICD-10, ICD-9) diagnosis codes and the Office of Population Censuses and Surveys (OPCS-4) procedure codes as reported before. Lipid levels were measured in blood samples collected at recruitment (Sudlow et al., 2015).

2.2 Overview of Polygenic Risk Score

We derived PRS using publicly available large-scale GWAS summary statistics (Consortium TC, the CARDIoGRAMplusC4D Consortium, 2015; Christophersen et al., 2017; Michailidou et al., 2017; Scott et al., 2017; Willer et al., 2013) and the LD reference panel of 498 EUR samples from the 1KG (Auton et al., 2015) based on three polygenic prediction methods: P+T (Chang et al., 2015), LDPred (Vilhjálmsson et al., 2015), and AnnoPred (Hu et al., 2017).

P+T denotes the LD clumping and p-value thresholding method, which was conducted using PLINK version 1.90b (using the—clump flag). Specifically, P+T removes the SNP with the less significant effect-size for each pair of SNPs that have a physical distance smaller than 250 kb and pairwise correlation greater than the specified R_T^2 threshold. PRS is then built as the marginal effect-weighted sum of the remaining variants whose statistical significance level is below a certain p-value threshold, P_T. In our study, we varied the P_T value from 1.0, 0.5, 0.05, 5E-4, 5E-6, to 5E-8, and the R_T^2 value from 0.2, 0.4, 0.6, 0.8 to 1.0, resulting in a total of 30 combinations.

LDPred is a Bayesian approach that infers the posterior mean effect size of each variant from GWAS summary statistics while accounting for LD. It places an independent point-normal prior on the effect-size of each variant and shrinks it based on LD information from a reference panel. The prior has two parameters: the heritability parameter, which is estimated from GWAS summary statistics while considering LD, and the tuning parameter p, which is the proportion of causal variants. We considered p within the set of {1.0, 0.3, 0.1, 0.03, 0.01, 3E-4, 1E-4, 3E-5, 1E-5, 3E-6, 1E-6}, which is the default setting suggested by the LDPred.

AnnoPred is also a Bayesian framework but leverages diverse types of genomic and epigenomic functional annotations to improve risk prediction accuracy on the basis of LDPred. Specifically, it uses an empirical prior of SNP effect size based on functional annotations of the SNPs and signal enrichment in different annotation classes estimated from GWAS summary statistics. To increase flexibility against different genetic architectures, two different priors relating the annotations are proposed, with one assuming the same proportion of causal SNPs but different effect sizes across annotation categories, and another assuming different proportions of causal SNPs but the same effect size across annotation categories. The empirical prior is then jointly modeled with GWAS summary statistics and LD estimated from an external reference panel to infer the posterior effect size of each variant. Here the fraction of causal variants is also set as a tuning parameter, p. For fair comparison with LDPred, we also considered p within the set of {1.0, 0.3, 0.1, 0.03, 0.01, 3E-4, 1E-4, 3E-5, 1E-5, 3E-6, 1E-6}. Each combination of p, prior and annotation tier would generate a candidate PRS. Considering there were 11 values of p, two priors and five tiers of annotations, 110 candidate PRS were generated from AnnoPred for each trait.

2.3 Annotations

We incorporated a number of functional annotations in this study, including 53 baseline annotations for diverse genomic features (Baseline Annotation) (Finucane et al., 2015), GenoCanyon general functionality scores (GenoCanyon Annotation) (Lu et al., 2015), GenoSkyline tissue-specific functionality scores of 7 tissue types (brain, gastrointestinal tract, lung, heart, blood, muscle, epithelial) (GenoSkyline Annotation) (Lu et al., 2016), GenoSkyline-Plus tissue-specific functionality scores of 7 tissue types (immune, brain, cardiovascular, muscle, gastrointestinal tract, epithelial, other) (GenoSkyline-Plus Tissue Annotation), GenoSkyline-Plus cell-specific functionality scores of 66 cell types (GenoSkyline-Plus Cell Annotation) (Lu et al., 2017), and IMPACT cell-specific functionality scores of 707 cell types (IMPACT Cell Annotation) (Amariuta et al., 2020). Specifically, as the original IMPACT Cell Annotation scores were continuous, we dichotomized the scores based on the median score across the genome and removed annotation categories with a proportion of SNPs less than $5 \times 10 - 4$ in the genome, which resulted in 651 annotation categories for IMPACT Cell Annotation.

Integrating functional annotations with higher tissue and cell type specificity may further increase risk prediction accuracy if the tissue/cell type is biologically relevant to the disease; but on the other hand, it may also bring more noise and reduce the prediction accuracy. Hence here we consider five different tiers of annotations with different granularity. And since different traits might have different annotation enrichment patterns, we trained AnnoPred PRS for each trait by comparing results based on these five different tiers of annotations:

1. tier0: 53 Baseline Annotation + 1 GenoCanyon Annotation.
2. tier1: 53 Baseline Annotation + 1 GenoCanyon Annotation + 7 GenoSkyline Annotation.
3. tier2: 53 Baseline Annotation + 1 GenoCanyon Annotation + 7 GenoSkyline-Plus Tissue Annotation.
4. tier3: 53 Baseline Annotation + 1 GenoCanyon Annotation + 66 GenoSkyline-Plus Cell Annotation.
5. tier4: 53 Baseline Annotation + 651 IMPACT Cell Annotation.

2.4 Statistical Analysis

As parameter tuning is required for all three PRS methods, for each method and its associated candidate tuning parameter(s), we estimated the effect sizes of all candidate SNPs using external European GWAS summary statistics and computed PRS for all individuals in the UKBB training dataset (306,736 EUR). For each binary trait, the optimal tuning parameter(s) was (were) selected based on the maximal area under the receiver operator curve (AUC) in a Cox regression model with the disease status as outcome, the onset age as follow-up time, and the disease-specific candidate PRS as the predictor in the training dataset. The reason we use Cox regression here instead of Logistic regression is that: as suggested by Mars's study (Mars et al., 2020), PRS may have different impact on people at different ages, we believe incorporating age in a Cox-model can better represent the prediction capacity of PRS. For each quantitative trait, the optimal tuning parameter(s) was (were) selected based on the maximal predictive R^2 in a linear regression model with the scaled traits' measurements as the outcome and the corresponding candidate PRS as the covariate in the training dataset.

The PRS model built using the "optimal" tuning parameter(s) was then applied to the independent testing sets (10,000 EUR, 8654 SAS, 2364 EAS, and 6992 AFR), and AUC or R^2 were calculated to compare the prediction performance of the three models across different populations and traits. For all regression models, we adjusted for sex, age at recruitment, and the first four genetic PCs. All statistical analyses were conducted using R software v.3.4.3.

2.5 Enrichment Analysis

To further investigate the role of different annotations played in cross-population genetic risk prediction, we also applied LD score regression (LDSC) (Bulik-Sullivan et al., 2015; Finucane et al., 2015) to perform tissue and cell-type enrichment analysis for each trait in each population. We firstly used each tier of annotations and 1KG reference panel to estimate annotation-stratified LD scores. Enrichment was then defined as the ratio of the percentage of heritability explained by variants in each annotated category versus the percentage of variants covered by that category.

In LDSC analysis for EUR, we used publicly available UKBB-external GWAS summary statistics; while for non-EUR, given the limited availability of external GWAS summary statistics, we conducted GWAS based on UKBB individuals of SAS, EAS, and AFR ancestry, respectively. Sex, age at recruitment, and the first four genetic PCs were adjusted when performing each of the GWASs using PLINK. Bonferroni multiple-testing correction was applied for each tier/population/trait of enrichment analysis, separately.

3 Results

3.1 Participants Characteristics

A total of 306,736 participants of EUR ancestry from UKBB were used in the training stage (mean age 57 years [SD 8], 164,530 (54%) were females). The testing set included 10,000 EUR (mean age 57 years [SD 8], 5403 (54%) were females), 8654 SAS (mean age 53 years [SD 9], 4026 (47%) were females), 2364 EAS (mean age 52 years [SD 8], 1585 (67%) were females), and 6992 AFR (mean age 52 years [SD 8], 3989 (57%) were females) (Fig. 1). The prevalence of each binary disease (AF, BC, CAD, T2D) and the average level of quantitative trait (HDL, LDL, TC, TG) across different populations are summarized in Table 1.

3.2 Genetic Risk Prediction Accuracy

Consistent with previous studies (Hu et al., 2017; Ye et al., 2021), when predicting the genetic risk for EUR, AnnoPred provided the best prediction performance for both disease phenotypes and quantitative traits, with the AUC being 0.629 (95% CI: 0.616–0.642) for AF, 0.642 (95% CI: 0.626–0.658) for BC, 0.652 (95% CI: 0.638–0.662) for CAD, and 0.664 (95% CI: 0.653–0.675) for T2D, respectively and the R^2 being 0.114 (95% CI: 0.108–0.200) for HDL, 0.085 (95% CI: 0.080–0.090) for LDL-C, 0.080 (95% CI: 0.075–0.085) for TC, and 0.071 (95%CI: 0.066–0.076) for TG, respectively. Compared with LDPred, the improved percentage of AUC ranged from 1.1% (T2D) to 3.4% (BC), and R^2 ranged from 6.1% (LDL-C) to 28.3% (HDL) (Figs. 2 and 3, Table 2).

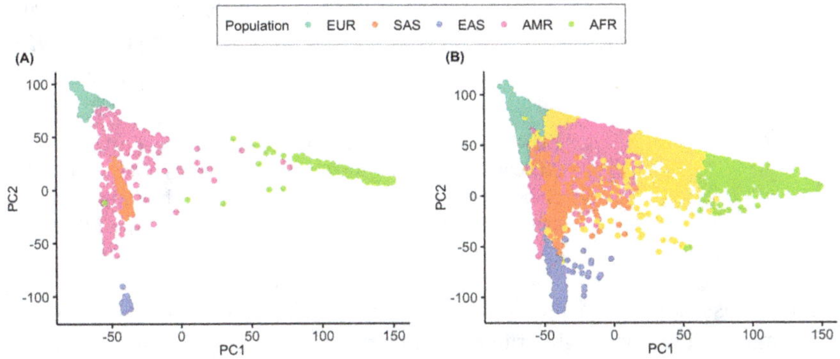

Fig. 1 Participants' ancestries. X-axes and Y-axes are the first and second genetic principal components for samples from (**a**) the 1000G reference panel and (**b**) the UK Biobank (UKBB). The results are consistent with the reference analysis by Zhang et al. (2020). (**a**) 1KG samples. (**b**) UKBB samples

Table 1 Number of individuals from different populations for different traits

Traits/Diseases	Training dataset	Testing dataset
HDL-C	EUR: 267701	EUR: 8728 EAS: 2052 SAS: 7474 AFR: 6068
LDL-C	EUR: 291831	EUR: 9525 EAS: 2252 SAS: 8197 AFR: 6543
TC	EUR: 292388	EUR: 9538 EAS: 2254 SAS: 8197 AFR: 6543
TG	EUR: 292143	EUR: 9532 EAS: 2252 SAS: 8209 AFR: 6553
AF (case/control)	EUR: 13888/292848	EUR: 476/9524 EAS: 36/2328 SAS: 210/8444 AFR: 131/6861
BC (case/control)	EUR: 10905/295831	EUR: 347/9653 EAS: 79/2285 SAS: 180/8474 AFR: 128/6864
CAD (case/control)	EUR: 16470/290266	EUR: 61/9439 EAS: 53/2311 SAS: 815/7839 AFR: 180/6812
T2D (case/control)	EUR: 16672/290064	EUR: 581/9419 EAS: 139/2225 SAS: 1631/7023 AFR: 822/6170

EUR: European; SAS: South Asian; EAS: East Asian; AFR: African; HDL-C: high-density lipoprotein cholesterol; TC: total cholesterol; TG: triglyceride; AF: atrial fibrillation; BC: breast cancer; CAD: coronary artery disease; T2D: type 2 diabetes

As for cross population predictions, AnnoPred also consistently outperformed LDpred and P+T across eight traits and three non-EUR populations, with only a few exceptions. Specifically, for SAS, the AUC increased by 0.1%–3.4% for disease phenotypes and R^2 increased by 0.2%–28.5% for quantitative traits; for EAS, the AUC increased by 0.3%–8.3% for disease phenotypes and R^2 increased by 4.0%–16.7% for quantitative traits; and for AFR, the AUC increased by −0.5%–7.8%

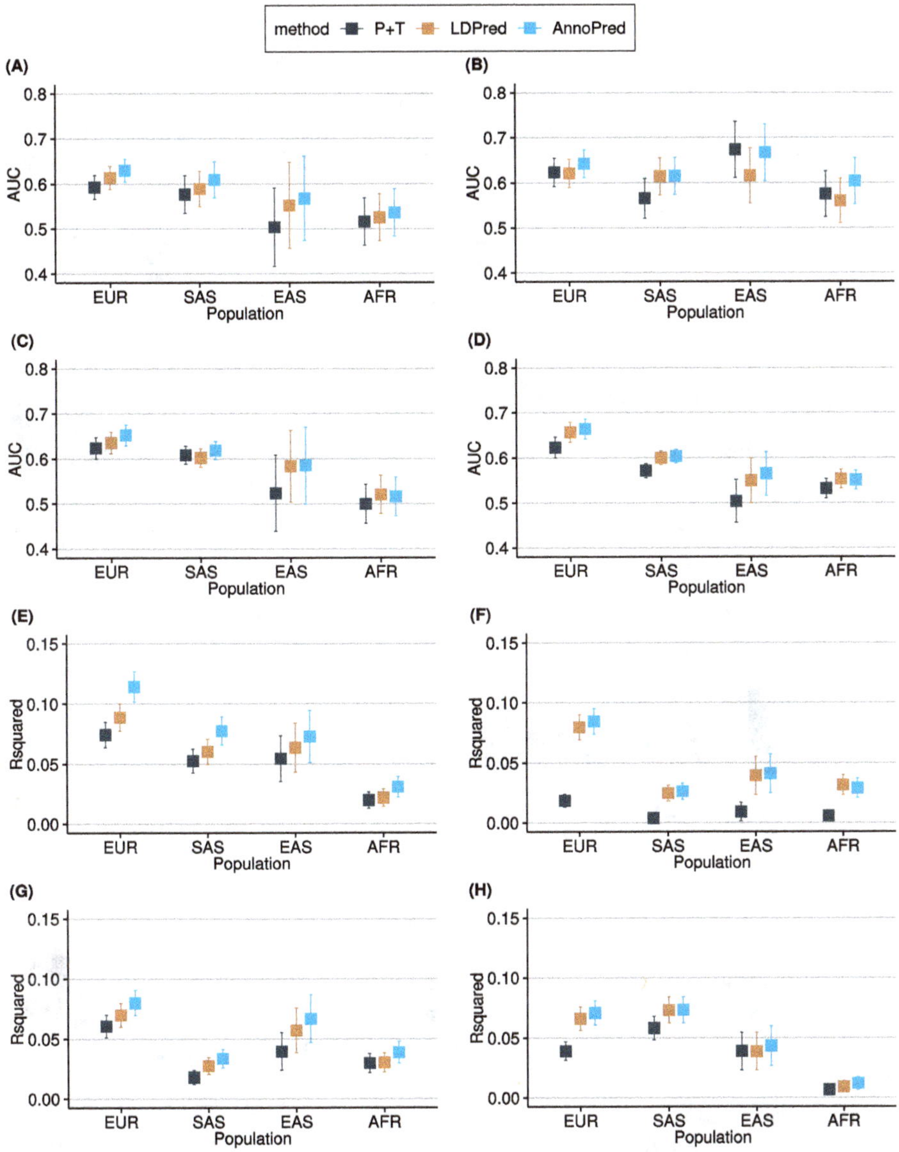

Fig. 2 Performance of polygenic risk scores (PRS). Candidate PRS were generated using three PRS methods (P+T, LDpred, AnnoPred). Tuning parameters of each method were selected in the European training set, and the predictive performance using the optimal tuning parameters was then assessed in the testing set across different populations. The prediction accuracy was measured by area under the receiver operator curve (AUC) for (**a**) atrial fibrillation (AF), (**b**) breast cancer (BC), (**c**) coronary artery disease (CAD), and (**d**) type 2 diabetes (T2D); by R squared for (**e**) High-density lipoprotein (HDL), (**f**) Low-density lipoprotein cholesterol (LDL), (**g**) total cholesterol (TC), and (**h**) triglyceride (TG). Both AUC and R squared are shown with 95% CI. (**a**) AF. (**b**) BC. (**c**) CAD. (**d**) DIA. (**e**) HDL. (**f**) LDL. (**g**) TC. (**h**) TG

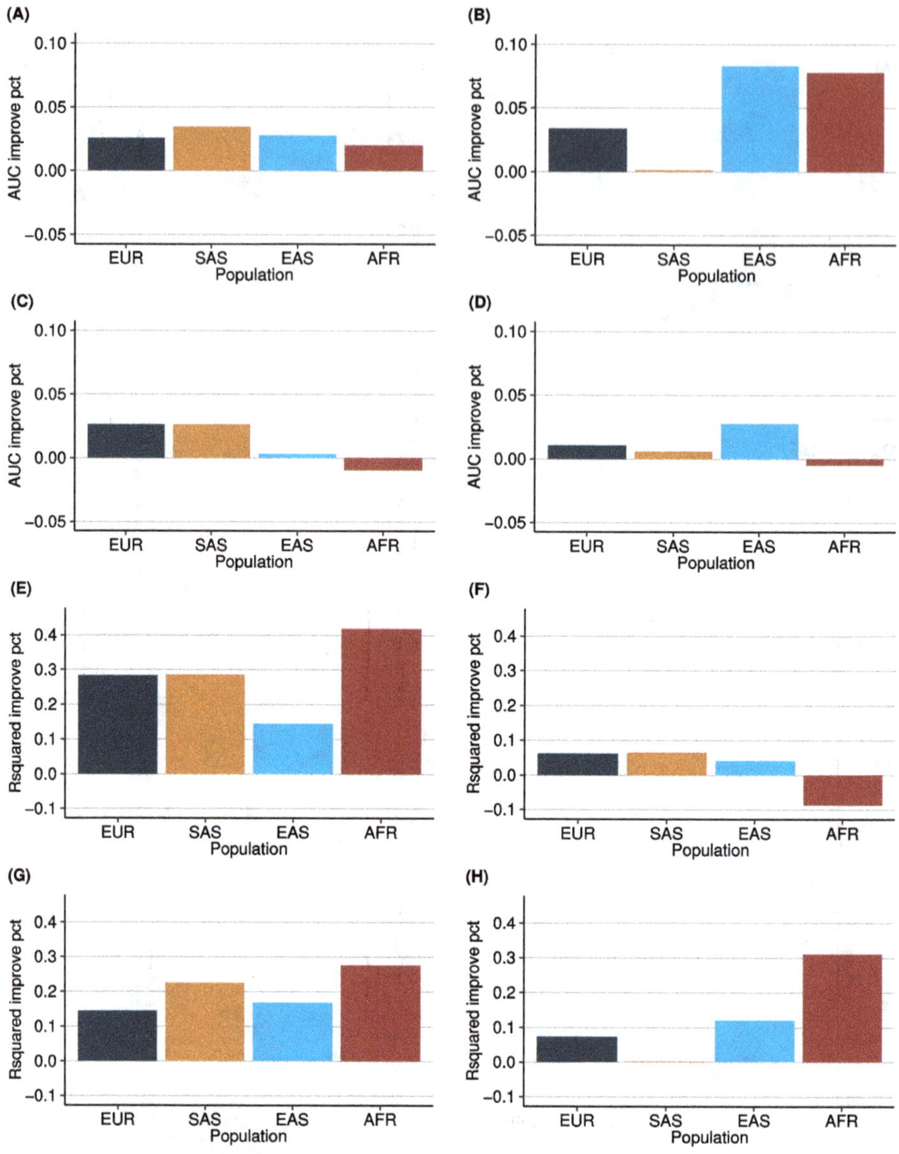

Fig. 3 Prediction improvement by integrating annotations. The prediction improvement of AnnoPred over LDpred for (**a**) atrial fibrillation (AF), (**b**) breast cancer (BC), (**c**) coronary artery disease (CAD), (**d**) type 2 diabetes (T2D), (**e**) High-density lipoprotein (HDL), (**f**) Low-density lipoprotein cholesterol (LDL), (**g**) total cholesterol (TC), and (**h**) triglyceride (TG). (**a**) AF. (**b**) BC. (**c**) CAD. (**d**) DIA. (**e**) HDL. (**f**) LDL. (**g**) TC. (**h**) TG

Table 2 Prediction accuracy of PRS for different traits in different methods

Traits/Diseases	Pop	P+T (R^2/AUC)	Ldpred (R^2/AUC)	AnnoPred (R^2/AUC)	P+T (SD)	Ldpred (SD)	AnnoPred (SD)
HDL (R^2)	EUR	0.074	0.089	**0.114**	0.005	0.006	0.006
	SAS	0.053	0.060	**0.077**	0.005	0.005	0.006
	EAS	0.054	0.064	**0.073**	0.010	0.010	0.011
	AFR	0.020	0.022	**0.031**	0.004	0.004	0.004
TC (R^2)	EUR	0.061	0.070	**0.080**	0.005	0.005	0.005
	SAS	0.018	0.028	**0.034**	0.003	0.004	0.004
	EAS	0.040	0.057	**0.067**	0.008	0.009	0.010
	AFR	0.030	0.030	**0.039**	0.004	0.004	0.005
LDL (R^2)	EUR	0.018	0.080	**0.085**	0.003	0.005	0.005
	SAS	0.004	0.025	**0.026**	0.001	0.003	0.003
	EAS	0.009	0.040	**0.041**	0.004	0.008	0.008
	AFR	0.006	**0.032**	0.029	0.002	0.004	0.004
TG (R^2)	EUR	0.039	0.066	**0.071**	0.004	0.005	0.005
	SAS	0.058	0.073	**0.073**	0.005	0.006	0.006
	EAS	0.039	0.039	**0.043**	0.008	0.008	0.008
	AFR	0.007	0.009	**0.012**	0.002	0.002	0.003
AF (AUC)	EUR	0.593	0.614	**0.629**	0.013	0.013	0.013
	SAS	0.577	0.589	**0.609**	0.021	0.020	0.020
	EAS	0.503	0.552	**0.567**	0.044	0.049	0.048
	AFR	0.515	0.525	**0.536**	0.027	0.027	0.027
CAD(AUC)	EUR	0.624	0.636	**0.652**	0.012	0.012	0.012
	SAS	0.608	0.602	**0.618**	0.010	0.010	0.010
	EAS	0.524	0.584	**0.585**	0.043	0.040	0.043
	AFR	0.500	**0.521**	0.516	0.022	0.022	0.022
BC(AUC)	EUR	0.623	0.621	**0.642**	0.016	0.016	0.016
	SAS	0.565	0.614	**0.615**	0.022	0.021	0.021
	EAS	**0.673**	0.616	0.667	0.032	0.031	0.032
	AFR	0.575	0.560	**0.603**	0.026	0.025	0.026
T2D(AUC)	EUR	0.623	0.657	**0.664**	0.012	0.011	0.011
	SAS	0.571	0.600	**0.603**	0.008	0.008	0.008
	EAS	0.504	0.550	**0.565**	0.024	0.025	0.025
	AFR	0.532	**0.553**	0.550	0.011	0.011	0.011

EUR: European; SAS: South Asian; EAS: East Asian; AFR: African; HDL-C: high-density lipoprotein cholesterol; TC: total cholesterol; TG: triglyceride; AF: atrial fibrillation; BC: breast cancer; CAD: coronary artery disease; T2D: type 2 diabetes. The best performed metrics are marked by the bolder fonts for disease phenotypes and R^2 increased by -8.7%–41.8% for quantitative traits when compared with LDPred. The exceptions where LDPred was better include predicting for LDL-C, CAD, and T2D out of the eight diseases/traits in the AFR population, and for these diseases, both LDPred and AnnoPred provided moderate to low prediction accuracy.

Even after incorporating annotations, we still observed substantial loss of prediction accuracy for most traits when applying the EUR-derived PRS to non-EUR populations. For instance, the AUC of CAD PRS in EUR was 0.652 (95% CI: 0.638–0.662), but reduced to 0.618 (95% CI: 0.608–0.628) in SAS, 0.585 (95% CI: 0.542–0.628) in EAS, and 0.521 (95% CI: 0.494–0.538) in AFR; the AUC of AF PRS in EUR was 0.629 (95% CI: 0.616–0.642), but reduced to 0.609 (95% CI: 0.589–0.629) in SAS, 0.567 (95% CI: 0.519–0.615) in EAS, and 0.535 (95% CI: 0.509–0.563) in AFR; and the AUC of T2D PRS in EUR was 0.664 (95% CI: 0.653–0.675), but reduced to 0.603 (95% CI: 0.595–0.611) in SAS, 0.565 (95% CI: 0.540–0.590) in EAS, and 0.550 (95% CI: 0.539–0.561) in AFR. Similar reduction pattern was observed for HDL, LDL-C, and TC where the loss of prediction accuracy increased monotonically with the increased genetic distance (FST) (Scutari et al., 2016) between each non-EUR population and EUR population.

We also compared the prediction performance of AnnoPred when using different tiers of annotations. While the best-performed tier varied across traits, we noticed that tier4 outperformed other tiers in 19 out of the 32 scenarios with eight diseases/traits and four populations. On the other hand, tier1 never outperformed the other tiers in the 32 scenarios. Considering that tier4 and tier1 contain the largest and smallest number of annotation categories, respectively, these results together suggest that integrating more functional annotations with higher tissue and cell type specificity can increase the cross-population genetic risk prediction accuracy (Fig. 4, Table 3).

3.3 Enrichment Across Annotations

To better understand these traits' genetic basis as related to annotated PRS, we further investigated their enrichment patterns across different annotation categories. As illustrated in Fig. 5, only a few enrichment signals could be identified in non-EUR population due to the limited power of GWAS with small sample size. But for the EUR population, we could observe substantial enrichment signals consistent with previous biological reports (Amariuta et al., 2020; Finucane et al., 2015; Lu et al., 2015, 2017, 2016). Specifically, BC was the only trait that was significantly enriched in breast tissue; AF and CAD were both significantly enriched in heart, GI, and muscle; T2D, which was highly related with obesity, was mainly enriched in brain tissue; and lipids traits were all significantly enriched in immune and liver tissues.

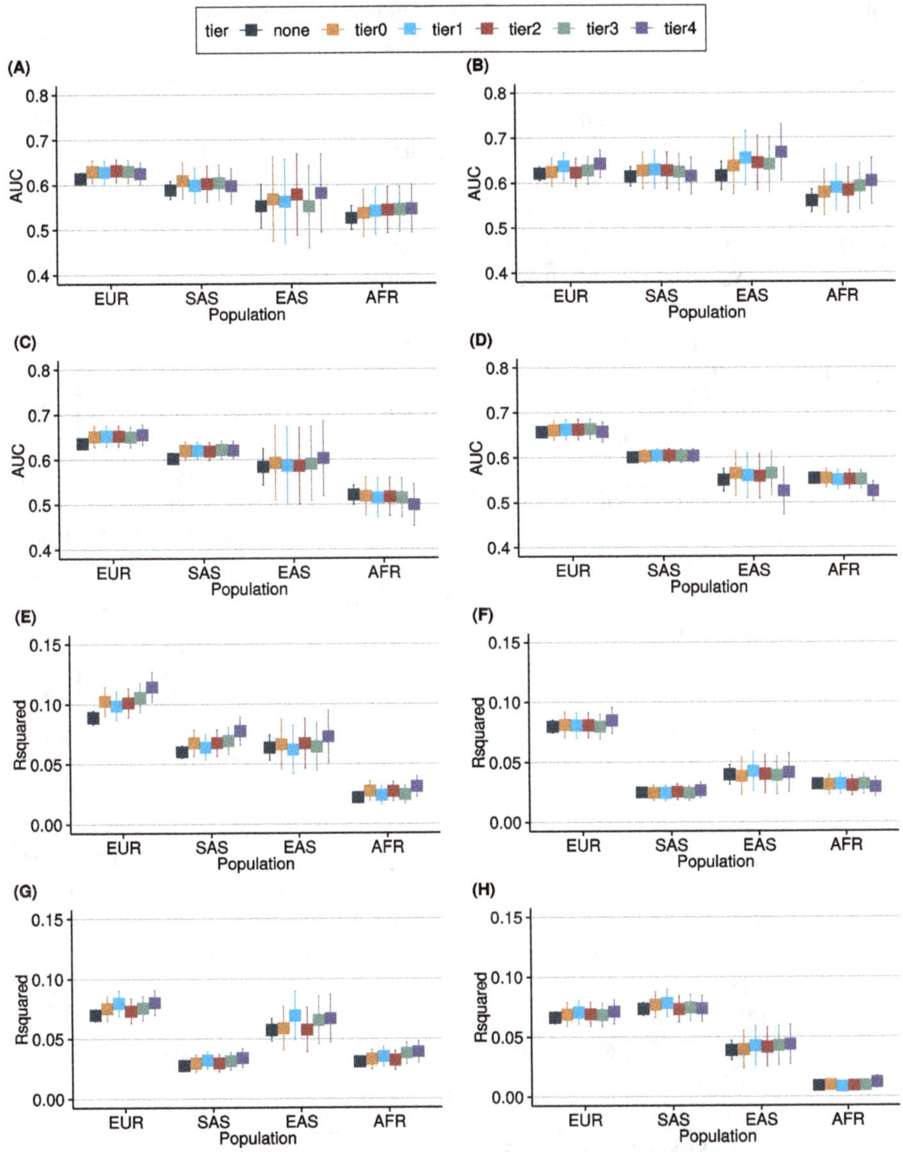

Fig. 4 Prediction accuracy of AnnoPred PRS based on different tiers. The prediction performance of AnnoPred based on different function annotations for (**a**) atrial fibrillation (AF), (**b**) breast cancer (BC), (**c**) coronary artery disease (CAD), and (**d**) type 2 diabetes (T2D); by R squared for (**e**) High-density lipoprotein (HDL), (**f**) Low-density lipoprotein cholesterol (LDL), (**g**) total cholesterol (TC), and (**h**) triglyceride (TG). Both AUC and R squared are shown with 95% CI. Specifically, "none" tier indicates results from LDPred where no annotations involved. (**a**) AF. (**b**) BC. (**c**) CAD. (**d**) DIA. (**e**) HDL. (**f**) LDL. (**g**) TC. (**h**) TG

Table 3 Prediction accuracy of PRS for different traits in different methods

Traits/ Diseases	Pop	tier0	tier1	tier2	tier3	tier4	tier0 (SD)	tier1 (SD)	tier2 (SD)	tier3 (SD)	tier4 (SD)
HDL (R^2)	EUR	0.103	0.098	0.101	0.105	**0.114**	0.006	0.006	0.006	0.006	0.006
	SAS	0.068	0.064	0.068	0.069	**0.077**	0.006	0.005	0.006	0.006	0.006
	EAS	0.066	0.062	0.067	0.064	**0.073**	0.011	0.010	0.011	0.010	0.011
	AFR	0.027	0.024	0.027	0.024	**0.031**	0.004	0.004	0.004	0.004	0.004
TC (R^2)	EUR	0.075	0.080	0.073	0.076	**0.080**	0.005	0.005	0.005	0.005	0.005
	SAS	0.029	0.032	0.029	0.031	**0.034**	0.004	0.004	0.004	0.004	0.004
	EAS	0.059	**0.069**	0.057	0.066	0.067	0.010	0.010	0.009	0.010	0.010
	AFR	0.033	0.035	0.032	0.038	**0.039**	0.004	0.004	0.004	0.005	0.005
LDL (R^2)	EUR	0.081	0.080	0.081	0.080	**0.085**	0.005	0.005	0.005	0.005	0.005
	SAS	0.024	0.024	0.025	0.024	**0.026**	0.003	0.003	0.003	0.003	0.003
	EAS	0.038	**0.042**	0.040	0.039	0.041	0.008	0.008	0.008	0.008	0.008
	AFR	0.031	0.032	0.030	**0.032**	0.029	0.004	0.004	0.004	0.004	0.004
TG (R^2)	EUR	0.069	0.070	0.069	0.068	**0.071**	0.005	0.005	0.005	0.005	0.005
	SAS	0.077	**0.078**	0.073	0.074	0.073	0.006	0.006	0.006	0.006	0.006
	EAS	0.039	0.043	0.041	0.043	**0.043**	0.008	0.008	0.008	0.008	0.008
	AFR	0.010	0.008	0.009	0.009	**0.012**	0.002	0.002	0.002	0.002	0.003
AF (AUC)	EUR	0.629	0.628	**0.632**	0.630	0.625	0.013	0.013	0.013	0.013	0.013
	SAS	**0.609**	0.598	0.602	0.604	0.597	0.020	0.020	0.020	0.020	0.020
	EAS	0.567	0.562	0.578	0.551	**0.580**	0.048	0.048	0.046	0.047	0.044
	AFR	0.536	0.540	0.543	0.544	**0.545**	0.027	0.027	0.026	0.026	0.027
CAD(AUC)	EUR	0.651	0.652	0.652	0.650	**0.655**	0.012	0.012	0.012	0.012	0.012
	SAS	0.620	0.620	0.618	**0.621**	0.620	0.012	0.010	0.010	0.010	0.010
	EAS	0.592	0.586	0.585	0.591	**0.602**	0.043	0.044	0.043	0.042	0.043
	AFR	**0.518**	0.513	0.516	0.514	0.498	0.022	0.022	0.022	0.022	0.024
BC(AUC)	EUR	0.624	0.637	0.623	0.628	**0.642**	0.016	0.016	0.016	0.016	0.016
	SAS	0.627	**0.629**	0.627	0.624	0.615	0.021	0.021	0.021	0.021	0.021
	EAS	0.637	0.654	0.644	0.641	**0.667**	0.031	0.031	0.031	0.031	0.032
	AFR	0.578	0.588	0.582	0.592	**0.603**	0.026	0.026	0.025	0.026	0.026
T2D(AUC)	EUR	0.660	0.662	0.662	**0.664**	0.657	0.011	0.011	0.011	0.011	0.011
	SAS	0.602	**0.604**	0.604	0.603	0.603	0.008	0.008	0.008	0.008	0.008
	EAS	**0.564**	0.559	0.557	0.565	0.525	0.025	0.025	0.025	0.025	0.027
	AFR	**0.553**	0.549	0.550	0.550	0.524	0.011	0.011	0.011	0.011	0.011

EUR: European; SAS: South Asian; EAS: East Asian; AFR: African; HDL-C: high-density lipoprotein cholesterol; TC: total cholesterol; TG: triglyceride; AF: atrial fibrillation; BC: breast cancer; CAD: coronary artery disease; T2D: type 2 diabetes. The best performed metrics are marked by the bolder fonts

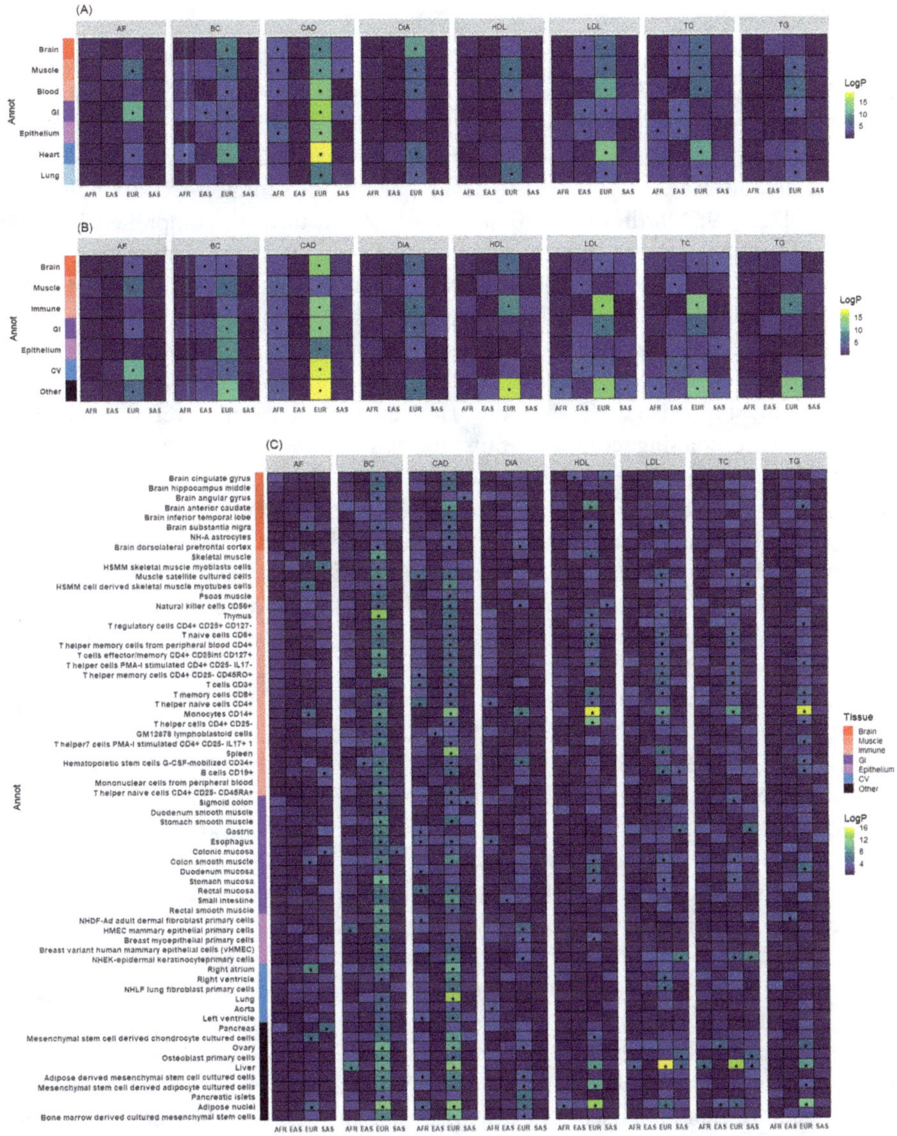

Fig. 5 Tissue and cell-type enrichment across different tiers. (**a**) Tissue GenoSkyline Enrichment result. (**b**) Tissue GenoSkyline-Plus Enrichment result. (**c**) Cell GenoSkyline-Plus Enrichment result

4 Discussions

PRS has great potential in guiding precision medicine, yet its usage in non-EUR populations is largely hindered by the small sample size of non-EUR GWAS and poor transferability of EUR-derived PRS, which leads to concern of health disparity and motivates researchers to develop better PRS for non-EUR populations (Lam et al., 2019; Martin et al., 2019). In this study, through comprehensive real data analyses across eight traits and four populations, we have demonstrated that by leveraging annotations in LDPred, AnnoPred may substantially improve the prediction accuracy of PRS in samples from diverse populations.

The importance of annotation in cross-population PRS we observed here is in line with a recent study by Amariuta etc. (2020), where the authors introduced a functionally informed P+T PRS and showed its superiority over the standard P+T PRS in predicting complex traits for EAS. However, as their study was intended to investigate the ability of the newly-developed IMPACT annotations in improving the transferability of PRS instead of developing optimal PRS for different populations, they only considered P+T method which was simple conceptually and computationally but with limited predictive capacity. And to incorporate functional annotations in P+T, their PRS was restricted to a subset of top-ranked IMPACT SNPs that could explain a specific proportion of heritability, where the proportion was determined ad hoc. In our study, in addition to show the usefulness of annotations in cross-population genetic risk prediction, we also developed PRS for eight diseases/traits (HDL, LDL, TC, TG, AF, BC, CAD, and T2D) in different populations (EUR, EAS, SAS, and AFR) with significantly better performance than previous reports (Hu et al., 2017). With better risk prediction performance than P+T, we showed that AnnoPred offers a principled approach to incorporating annotations and increasing the cross-population prediction capacity of PRS.

Our results suggest that including more functional annotations with higher tissue and cell type specificity can further improve PRS prediction accuracy. As our results showed that different traits had different annotation enrichment patterns, we hypothesized that annotations with higher specificity allowed us to better capture the trait-specific shared proportions of heritability across different populations, and thus lead to better cross-population PRS prediction performance; which was consistent with previous results between EUR and EAS (Amariuta et al., 2020). However, including more annotations didn't always result in higher PRS accuracy, which might be due to the similarity between some of the functional regions which caused multicollinearity when applying the annotation-stratified LD score regression and resulted in numerically unstable heritability estimates (Hu et al., 2017).

We note that although AnnoPred can improve the cross-population risk prediction performance, there was still substantial loss of PRS accuracy. Similar to previous theoretical estimation (Wang et al., 2020), there was more reduction in prediction accuracy as the genetic distance between EUR and the target population increased, with the lowest performance for AFR. These results suggest that although sophisticated model with annotations may help, the improvement of the prediction accuracy is still limited by adding annotations to single-population PRS method. One potential future direction would be adding the functional annotations to multiple-population PRS methods, where several recent studies (Cai et al., 2021; Jiashun et al., 2022; Ruan et al., 2022) have shown that aggregating GWAS summary statistics from multiple populations can boost the power of PRS prediction, especially in the minority population. Meanwhile, developing large-scale GWAS for diverse populations is still necessary to fully bridge the equity gap.

We also note some other limitations of this work. First, by tuning parameters (causal fraction of SNPs) of AnnoPred in EUR training set and applying the best parameters to non-EUR, we implicitly assumed that the proportion of causal SNPs was the same across different populations, which might be violated for certain diseases or traits (Durvasula & Lohmueller, 2021; Wang et al., 2020). Second, the functional annotations we considered in this study were mostly generated based on the EUR datasets, which might neglect some population-specific biological functions from other populations. Lastly, in this study we only used EUR samples for training, as more diverse GWAS samples become available, future frameworks integrating annotations and multi-populations data could be more powerful to build better-performed PRS across different populations.

5 Conclusion

By leveraging functional annotations and LD information, AnnoPred could significantly improve the cross-population genetic risk prediction.

Data and Code Availability

The full PRS weights produced by AnnoPred in this study are freely available on figshare: https://figshare.com/s/1008ac61ac273519ed6e. Our analyses codes are provided on Github: https://github.com/yeyixuan/Springer_PRS_analyses. Other resources can be found in the following webs:
AnnoPred, https://github.com/yiminghu/AnnoPred
LDPred, https://github.com/bvilhjal/ldpred
PRScs, https://github.com/getian107/PRScs
PLINK, https://www.cog-genomics.org/plink
GNOVA, https://github.com/xtonyjiang/GNOVA

Acknowledgments We sincerely thank many GWAS consortia for making their GWAS summary data publicly accessible. The individual-level data that support the findings of this study are available with the permission of UK Biobank (https://www.ukbiobank.ac.uk). We conducted the research using the UK Biobank resource under an approved data request (ref: 29900). This study was supported in part by the NIH grant R01 GM134005, and National Science Foundation (NSF) grants DMS 1713120 and DMS 1902903.

References

1000 Genomes Project Consortium, Auton, A., Brooks, L. D., et al. (2015). A global reference for human genetic variation. *Nature, 526*, 68–74.

Amariuta, T., Ishigaki, K., Sugishita, H., et al. (2020). Improving the trans-ancestry portability of polygenic risk scores by prioritizing variants in predicted cell-type-specific regulatory elements. *Nature Genetics, 52*, 1346–1354.

Bentley, A. R., Callier, S. L., & Rotimi, C. N. (2020). Evaluating the promise of inclusion of African ancestry populations in genomics. *NPJ Genomic Medicine, 5*(1), 5.

Blechter, B., Wong, J. Y. Y., Agnes Hsiung, C., et al. (2021). Sub-multiplicative interaction between polygenic risk score and household coal use in relation to lung adenocarcinoma among never-smoking women in Asia. *Environment International, 147*, 105975.

Bulik-Sullivan, B. K., Loh, P.-R., Finucane, H. K., et al. (2015). LD Score regression distinguishes confounding from polygenicity in genome-wide association studies. *Nature Genetics, 47*, 291–295.

Bycroft, C., Freeman, C., Petkova, D., et al. (2018). The UK Biobank resource with deep phenotyping and genomic data. *Nature, 562*, 203–209.

Cai, M., Xiao, J., Zhang, S., et al. (2021). A unified framework for cross-population trait prediction by leveraging the genetic correlation of polygenic traits. *The American Journal of Human Genetics, 108*(4), 632–655.

Chang, C. C., Chow, C. C., Tellier, L. C., et al. (2015). Second-generation PLINK: Rising to the challenge of larger and richer datasets. *Gigascience, 4*, 7.

Christophersen, I. E., Rienstra, M., Roselli, C., et al. (2017). Large-scale analyses of common and rare variants identify 12 new loci associated with atrial fibrillation. *Nature Genetics, 49*, 946–952.

Consortium TC, the CARDIoGRAMplusC4D Consortium (2015) A comprehensive 1000 Genomes–based genome-wide association meta-analysis of coronary artery disease. *Nature Genetics, 47*, 1121–1130.

Dudbridge, F. (2013). Power and predictive accuracy of polygenic risk scores. *PLoS Genetics, 9*, e1003348.

Duncan, L., Shen, H., Gelaye, B., et al. (2019). Analysis of polygenic risk score usage and performance in diverse human populations. *Nature Communications, 10*, 3328.

Durvasula, A., & Lohmueller, K. E. (2021) Negative selection on complex traits limits phenotype prediction accuracy between populations. *The American Journal of Human Genetics, 108*, 620–631.

Finucane, H. K., Bulik-Sullivan, B., Gusev, A., et al. (2015). Partitioning heritability by functional annotation using genome-wide association summary statistics. *Nature Genetics, 47*, 1228–1235.

Hu, Y., Lu, Q., Powles, R., et al. (2017). Leveraging functional annotations in genetic risk prediction for human complex diseases. *PLoS Computational Biology, 13*, e1005589.

Lam, M., Chen, C.-Y., Li, Z., et al. (2019). Comparative genetic architectures of schizophrenia in East Asian and European populations. *Nature Genetics, 51*, 1670–1678.

Lu, Q., Hu, Y., Sun, J., et al. (2015). A statistical framework to predict functional non-coding regions in the human genome through integrated analysis of annotation data. *Scientific Reports, 5*, 10576.

Lu, Q., Powles, R. L., Abdallah, S., et al. (2017). Systematic tissue-specific functional annotation of the human genome highlights immune-related DNA elements for late-onset Alzheimer's disease. *PLoS Genetics, 13*, e1006933.

Lu, Q., Powles, R. L., Wang, Q., et al. (2016). Integrative tissue-specific functional annotations in the human genome provide novel insights on many complex traits and improve signal prioritization in genome wide association studies. *PLoS Genetics, 12*, e1005947.

Martin, A. R., Kanai, M., Kamatani, Y., et al. (2019). Clinical use of current polygenic risk scores may exacerbate health disparities. *Nature Genetics, 51*, 584–591.

Mars, N., Koskela, J. T., Ripatti, P., et al. (2020). Polygenic and clinical risk scores and their impact on age at onset and prediction of cardiometabolic diseases and common cancers. *Nature Medicine, 26*, 549–557.

Mavaddat, N., Michailidou, K., Dennis, J., et al. (2019). Polygenic risk scores for prediction of breast cancer and breast cancer subtypes. *American Journal of Human Genetics, 104*, 21–34.

Michailidou, K., Lindström, S., Dennis, J., et al. (2017). Association analysis identifies 65 new breast cancer risk loci. *Nature, 551*, 92–94.

Natarajan, P., Young, R., Stitziel, N. O., et al. (2017). Polygenic risk score identifies subgroup with higher burden of atherosclerosis and greater relative benefit from statin therapy in the primary prevention setting. *Circulation, 135*, 2091–2101.

Need, A. C., & Goldstein, D. B. (2009). Next generation disparities in human genomics: concerns and remedies. *Trends in Genetics, 25*, 489–494.

Petrovski, S., & Goldstein, D. B. (2016). Unequal representation of genetic variation across ancestry groups creates healthcare inequality in the application of precision medicine. *Genome Biology, 17*, 157.

Ruan, Y., Lin, Y. F., Feng, Y. C. A., et al. (2022) Improving polygenic prediction in ancestrally diverse populations. *Nature Genetics, 54*, 573–580.

Scott, R. A., Scott, L. J., Mägi, R., et al. (2017). An expanded genome-wide association study of type 2 diabetes in Europeans. *Diabetes, 66*, 2888–2902.

Scutari, M., Mackay, I., & Balding, D. (2016). Using genetic distance to infer the accuracy of genomic prediction. *PLoS Genetics, 12*, e1006288.

Sharp, S. A., Rich, S. S., Woodm A. R., et al. (2019). Development and standardization of an improved type 1 diabetes genetic risk score for use in newborn screening and incident diagnosis. *Diabetes Care, 42*, 200–207.

Shieh, Y., Fejerman, L., Lott, P. C., et al. (2020). A polygenic risk score for breast cancer in US Latinas and Latin American women. *JNCI: Journal of the National Cancer Institute, 112*, 590–598.

Sudlow, C., Gallacher, J., Allen, N., et al. (2015). UK biobank: An open access resource for identifying the causes of a wide range of complex diseases of middle and old age. *PLoS Medicine, 12*, 1001779.

Vilhjálmsson, B. J., Yang, J., Finucane, H. K., et al. (2015). Modeling linkage disequilibrium increases accuracy of polygenic risk scores. *American Journal of Human Genetics, 97*, 576–592.

Wang, Y., Guo, J., Ni, G., et al. (2020), Theoretical and empirical quantification of the accuracy of polygenic scores in ancestry divergent populations. *Nature Communications, 11*, 3865.

Willer, C. J., Schmidt, E. M., Sengupta, S., et al. (2013). Discovery and refinement of loci associated with lipid levels. *Nature Genetics, 45*, 1274–1283.

Xiao, J., Cai, M., Hu, X., et al. (2022). XPXP: Improving polygenic prediction by cross-population and cross-phenotype analysis. *Bioinformatics, 38*(7), 1947–1955

Ye, Y., Chen, X., Han, J., et al. (2021) Interactions between enhanced polygenic risk scores and lifestyle for cardiovascular disease, diabetes mellitus and lipid levels. *Circulation: Genomic and Precision Medicine, 14*(1), e003128. https://doi.org/10.1161/CIRCGEN.120.003128.

Zhang, D., Dey, R., Lee, S. (2020). Fast and robust ancestry prediction using principal component analysis. *Bioinformatics, 36*, 3439–3446.

A Soft-Thresholding Operator for Sparse Time-Varying Effects in Survival Models

Yuan Yang, Jian Kang, and Yi Li

Abstract We consider a class of Cox models with time-dependent effects that may be zero over certain unknown time regions or, in short, sparse time-varying effects. The model is particularly useful for biomedical studies as it conveniently depicts the gradual evolution of effects of risk factors on survival. Statistically, estimating and drawing inference on infinite dimensional functional parameters with sparsity (e.g., time-varying effects with zero-effect time intervals) present enormous challenges. To address them, we propose a new soft-thresholding operator for modeling sparse, piecewise smooth and continuous time-varying coefficients in a Cox time-varying effects model. Unlike the common regularized methods, our approach enables one to estimate non-zero time-varying effects and detect zero regions simultaneously, and construct a new type of sparse confidence intervals that accommodate zero regions. This leads to a more interpretable model with a straightforward inference procedure. We develop an efficient algorithm for inference in the target functional space, show that the proposed method enjoys desired theoretical properties, and present its finite sample performance by way of simulations. We apply the proposed method to analyze the data of the Boston Lung Cancer Survivor Cohort, an epidemiological cohort study investigating the impacts of risk factors on lung cancer survival, and obtain clinically useful results.

Keywords Time-varying effects · Soft-thresholding operator · Sparse confidence intervals · Survival models

Y. Yang
Parexel, Waltham, MA, USA
e-mail: Anna.Yang@parexel.com

J. Kang · Y. Li (✉)
Department of Biostatistics, University of Michigan, Ann Arbor, MI, USA
e-mail: jiankang@umich.edu; yili@umich.edu

Y. Zhao, D.-G. Chen (eds.), *Statistics in Precision Health*, ICSA Book Series in Statistics, https://doi.org/10.1007/978-3-031-50690-1_19

1 Introduction

The Cox proportional hazards model, proposed by the late Sir D.R. Cox (1972), has dominated the analysis of survival studies for decades and has also become an indispensable analytical tool in the era of precision medicine, because of its elegant estimation and inference framework and ease of interpretation (Hong et al., 2019). One key assumption of the proportional hazards model is that the effect of a given covariate remains constant over time, which, however, may not always hold. In fact, non-proportionality has been commonly observed and sparked much interest (Schoenfeld, 1982; Huang, 1999; Martinussen et al., 2002; Marzec, 1997; Murphy, 1993; Winnett & Sasieni, 2003), which led to the development of time-dependent coefficients Cox models (Hastie & Tibshirani, 1993).

An often overlooked feature in time-dependent effects Cox models is the sparsity associated with time-varying effects, meaning that the covariate effects can be zero on some specific time intervals and non-zero but time-varying on the others. For example, Anderson and Senthilselvan (1982) noticed the effects of some covariates disappeared in the later follow-up in a vulvar cancer study; Gore et al. (1984) found that the influence of signs recorded at diagnosis waned with time in a breast cancer study; Tian et al. (2005) noted sparsity in the edema effect during the early stage and also showed the effect of prothrombin on survival diminished over time in a biliary cirrhosis study. In the motivating Boston Lung Cancer Survivor Cohort (Christiani, 2017), an epidemiological study investigating the impacts of clinical and molecular risk factors on lung cancer survival, chemotherapy and radiotherapy did not seem to increase or decrease patients' overall survival, leading to the notion of detecting zero-effect regions for these treatment options.

It is rather challenging to detect no-effects periods and estimate non-zero effects simultaneously, as the existing methods for fitting the time-dependent Cox models cannot achieve these goals. For example, the commonly used penalized spline models (Zucker & Karr, 1990; Yan & Huang, 2012; Lian et al., 2013; He et al., 2017) and kernel weighted likelihood approaches (Cai & Sun, 2003; Tian et al., 2005) can detect or label covariates as time-varying or time-constant, but cannot detect no-effects periods within each covariate effect's trajectory.

We propose a new statistical method that can efficiently model sparse time-varying effects in a survival setting, by using a soft-thresholding operator to represent the time-varying effects in the Cox model. Both soft-thresholding and hard-thresholding approaches can be applied in this setting and have their own merits. However, we opt for soft-thresholding because it respects the continuity of the effect with respect to time and may conveniently depict the gradual evolution of effects of risk factors on survival. Indeed, the concept of soft thresholding was introduced by Donoho and Johnstone (1994); Donoho (1995), who applied this estimator to the coefficients of a wavelet transform of a function measured with noise. Since then, the use of soft-thresholding for effect shrinkaging has been flourishing: Chang et al. (2000) proposed an adaptive, data-driven thresholding method for image denoising in a Bayesian framework; Tibshirani (1996) pointed out that the Lasso estimator is a soft-thresholding estimator when the covariate matrix has an orthonormal design. Kang et al. (2018) used a soft-thresholding operator for modeling sparse, continuous, and piecewise smooth functions in image data analysis; however, as their method was not developed for survival analysis, it is unclear whether its extension to a survival setting is feasible.

We propose a soft-thresholding operator to model time-varying effects of covariates in a survival regression setting, and use the B-splines to approximate the nonparametric parts. Estimation is carried out by maximizing a penalized and smoothed partial likelihood. We prove the asymptotic properties of our proposed estimator, and introduce a new class of sparse confidence intervals for quantifying the uncertainty of the sparse functional estimates.

This chapter is organized as follows. In Sect. 2, we introduce the proposed soft-thresholding operator for a Cox model with sparse time-varying effects and derive an algorithm for fitting the model. Section 3 lists the theoretical properties of the method and proposes the sparse confidence intervals for inferring from the estimated sparse time-varying effects. We present simulation results in Sect. 4 to assess the finite sample performance of our methods, and analyze the aforementioned Boston Lung Cancer study in Sect. 5. Section 6 concludes this chapter with a brief summary. We defer all the technical proofs to the Appendix.

2 Methods

2.1 Model

Let T_i^u and T_i^c represent the survival and censoring times, respectively, for the ith patient. Observed are $T_i = T_i^u \wedge T_i^c$ with $a \wedge b = \min\{a, b\}$, and the death indicators $\Delta_i = \mathscr{I}(T_i^u \le T_i^c)$ with $\mathscr{I}(A)$ indicating whether condition A holds ($= 1$) or not ($= 0$). Let $\mathbf{Z}_i = (Z_{i1}, \ldots, Z_{ip})^\top$ be a p-dimensional covariate vector for sample i. The observed data consist of n independent vectors, $(T_i, \Delta_i, \mathbf{Z}_i)$, which are identical and independently distributed (i.i.d.) copies of $(T, \Delta, \mathbf{Z})$. Further, $(T_i^u, T_i^c), i = 1, \ldots, n$, are i.i.d. copies of (T^u, T^c).

Denote by $\lambda(t|\mathbf{Z}_i)$ the hazard function at t given $\mathbf{Z}_i$. A time-varying effects Cox model stipulates that

$$\lambda(t \mid \mathbf{Z}_i) = \lambda_0(t) \exp\{\mathbf{Z}_i^\top \boldsymbol{\beta}(t)\},$$

where $\lambda_0(t)$ is the baseline hazard, and $\boldsymbol{\beta}(t) = \{\beta_1(t), \ldots, \beta_p(t)\}^\top$ are the p time-dependent coefficients corresponding to $\mathbf{Z}_i$.

The log partial likelihood with time-varying coefficients is

$$\mathrm{PL}(\boldsymbol{\beta}) = \sum_{i=1}^{n} \Delta_i \left\{ \sum_{j=1}^{p} Z_{ij}\beta_j(T_i) - \log\left[\sum_{l \in R_i} \exp\left\{ \sum_{j=1}^{p} Z_{lj}\beta_j(T_i) \right\} \right] \right\}, \tag{1}$$

where $R_i = \{l : T_l > T_i\}$ is the risk set at T_i.

We assume that $\beta_j(\cdot)$, $j = 1, \ldots, p$, is continuous everywhere, with zero-effect regions ($\mathscr{R}_0$) consisting of at least one interval, and is smooth over regions (positive $\mathscr{R}_+$ and negative $\mathscr{R}_-$) where its effect is non-zero. On each interval of the non-zero regions, the dth derivative of $\beta_j(t)$ exists and satisfies the Lipschitz condition. That is, for any s, t in the interval, there exists a constant $C > 0$ such that

$$|\beta_j^{(d)}(s) - \beta_j^{(d)}(t)| \leq C|s - t|^w, \tag{2}$$

where d is a non-negative integer, and $w \in (0, 1]$ such that $m \equiv d + w > 0 \cdot 5$. Let $\mathbb{H}$ be the set of all such functions. Often, we use $d = 3$ (as in our later simulations and data analysis) corresponding to piecewise cubic functions. Let $\boldsymbol{\beta}_0(\cdot) = \{\beta_{01}(\cdot), \ldots, \beta_{0p}(\cdot)\}^\top$ be the true coefficient vector in the model that generates the observed data and $\beta_{0j} \in \mathbb{H}$.

We use the soft-thresholding operator ζ to represent a varying coefficient with zero regions:

$$\zeta\{\theta(t), \alpha\} = \{\theta(t) - \alpha\}\mathscr{I}\{\theta(t) > \alpha\} + \{\theta(t) + \alpha\}\mathscr{I}\{\theta(t) < -\alpha\},$$

where $\alpha > 0$ is the thresholding parameter and $\theta(t)$ is a real-valued function.

To avoid technicalities at the tail of the distribution of T^u, we estimate $\beta_j(\cdot)$ over a finite interval $(0, \tau)$ and base the estimation on the partial likelihood over the same interval, where τ is within the support of T. By doing so, we need to effectively replace T_i and Δ_i in likelihood (1) (and also the modified partial likelihoods thereafter) by $T_i = \min(T_i^u, T_i^c, \tau)$ and $\Delta_i = I(T_i^u \leq T_i^c \wedge \tau)$. In practice, if τ is chosen to be the maximum observed survival time in the data, no such replacements are needed.

It can be shown that for any function $\beta(t) \in \mathbb{H}$ and any $\alpha > 0$, there exists at least one $\theta(t) \in \mathbb{F}_0$ such that $\beta(t) = \zeta\{\theta, \alpha\}(t)$, where $\mathbb{F}_0$ is the class of functions θ defined on $[0, \tau]$, with the dth derivative $\theta^{(d)}$ satisfying the Lipschitz condition (2). As such, we introduce a new penalized likelihood for estimation

$$\begin{aligned} &\mathrm{PL}(\boldsymbol{\theta}) \\ &= \sum_{i=1}^{n} \Delta_i \left\{ \sum_{j=1}^{p} Z_{ij}\zeta\{\theta_j(T_i), \alpha_j\} - \log\left[\sum_{l \in R_i} \exp\left\{ \sum_{j=1}^{p} Z_{lj}\zeta\{\theta_j(T_l), \alpha_j\} \right\} \right] \right\} \\ &-\rho||\boldsymbol{\theta}||_2^2, \end{aligned} \tag{3}$$

where $\boldsymbol{\theta}(t) = \{\theta_1(t), \ldots, \theta_p(t)\}^\top$ and $\rho > 0$ is the predetermined penalization coefficient.

With the soft-thresholding representation, we can convert the problem from estimating non-smooth functions to estimating smooth functions. Among many approaches for modeling smooth functions, we will utilize the B-spline basis approach because of its convenience and numerical stability (Wahba, 1980); other alternatives may include P-splines and smoothing splines. Let $\mathbb{F}$ be the B-spline function sieve space, $K = O(n^\nu)$ be an integer with $0 < \nu < 0 \cdot 5$, and $B_k(t)(1 \le k \le q$, and $q = K + d)$ be the B-spline basis functions of degree $d + 1$ associated with the knots $0 = t_0 < t_1 < \cdots < t_{K-1} < t_K = 1$, satisfying $\max_{1 \le k \le K}(t_k - t_{k-1}) = O(n^{-\nu})$. Let $\boldsymbol{B}(t) = \{B_1(t), \ldots, B_q(t)\}^\top$ be a functional vector of the B-spline bases; with $d = 3$, this corresponds to a vector of cubic B-spline bases. Then, we have

$$\mathbb{F} = \left\{\theta(t) : \theta(t) = \sum_{k=1}^{q} \gamma_k B_k(t), t \in [0, \tau], \gamma_k \in \mathbb{R}, k = 1, \ldots, q\right\}.$$

For given α and q, we define the thresholding sieve space

$$\mathbb{S}_{q,\alpha} = \left\{\beta(t) = \zeta\{\theta(t), \alpha\} : \theta(t) = \sum_{k=1}^{q} \gamma_k B_k(t), t \in [0, \tau], \gamma_k \in \mathbb{R}, k = 1, \ldots, q\right\}.$$

Let $\boldsymbol{\gamma}_j = (\gamma_{j1}, \ldots, \gamma_{jq})^\top$ be the basis coefficients for $\theta_j(t)$. Then we represent $\theta_j(t) = \boldsymbol{B}(t)^\top \boldsymbol{\gamma}_j$. The penalized log partial likelihood can be written as

$$\begin{aligned} \mathrm{PL}(\boldsymbol{\gamma}) = \sum_{i=1}^{n} \Delta_i & \left\{\sum_{j=1}^{p} Z_{ij} \zeta\{\boldsymbol{B}(T_i)^\top \boldsymbol{\gamma}_j, \alpha_j\} \right. \\ & \left. - \log\left[\sum_{l \in R_i} \exp\left\{\sum_{j=1}^{p} Z_{lj} \zeta\{\boldsymbol{B}(T_i)^\top \boldsymbol{\gamma}_j, \alpha_j\}\right\}\right]\right\} \\ & - \rho \sum_{j=1}^{p} \sum_{i=1}^{n} \{\boldsymbol{B}(T_i)^\top \boldsymbol{\gamma}_j\}^2, \end{aligned} \tag{4}$$

where $\boldsymbol{\gamma} = (\boldsymbol{\gamma}_1, \ldots, \boldsymbol{\gamma}_p)$.

2.2 Estimation

It is challenging to directly maximize the likelihood function (4) as the thresholding operator $\zeta(\theta, \alpha)$ is non-smooth. We therefore consider a smooth approximation of $\zeta(\theta, \alpha)$:

$$h_\eta\{\theta(t), \alpha\} = \frac{1}{2}\Bigg(\left[1 + \frac{2}{\pi}\arctan\{\theta_-(t)/\eta\}\right]\theta_-(t) + \\ \left[1 - \frac{2}{\pi}\arctan\{\theta_+(t)/\eta\}\right]\theta_+(t)\Bigg),$$

where $\alpha > 0$, $\eta > 0$ and $\theta_\pm(t) = \theta(t) \pm \alpha$. Noting $\lim_{\eta\to 0} h_\eta\{\theta(t), \alpha\} = \xi(\theta, \alpha)$, we define $h_0\{\theta(t), \alpha\} = \xi(\theta, \alpha)$. As such, $h_\eta\{\theta(t), \alpha\}$ is a sufficiently smooth function in η and, in particular, in a small neighborhood, e.g., $\eta \in [0, \epsilon)$ where $\epsilon > 0$ is small. Taking a Taylor expansion of $h_\eta\{\theta(t), \alpha\}$ at $\eta = 0$ within this neighborhood, we can show that the approximation error between $h_\eta\{\theta(t), \alpha\}$ and $\zeta(\theta, \alpha)$ is bounded by $\eta + O(\eta^3)$. We drop η hereafter for simplicity of notation. Then, we obtain a smoothed log partial likelihood function:

$$\begin{aligned} \mathrm{PL}(\boldsymbol{\gamma}) = \sum_{i=1}^{n} \Delta_i \Bigg\{ & \sum_{j=1}^{p} Z_{ij} h\{\boldsymbol{B}(T_i)^\top \boldsymbol{\gamma}_j, \alpha_j\} \\ & - \log\left[\sum_{l\in R_i} \exp\left\{\sum_{j=1}^{p} Z_{lj} h\{\boldsymbol{B}(T_i)^\top \boldsymbol{\gamma}_j, \alpha_j\}\right\}\right]\Bigg\} \\ & - \rho \sum_{j=1}^{p}\sum_{i=1}^{n} \{\boldsymbol{B}(T_i)^\top \boldsymbol{\gamma}_j\}^2, \end{aligned} \tag{5}$$

forming the basis for estimation and inference.

Let $\tilde{\boldsymbol{\gamma}} = \arg\max_{\boldsymbol{\gamma}} \mathrm{E}_{T,\Delta,\boldsymbol{Z}}\mathrm{PL}(\boldsymbol{\gamma})$, where the expectation is taken with respect to the joint distribution of T, Δ and $\boldsymbol{Z}$ under the true parameter $\boldsymbol{\beta}_0(t)$. An estimate of $\tilde{\boldsymbol{\gamma}}$ is obtained by maximizing the likelihood (5) so that $\hat{\boldsymbol{\gamma}} = \arg\max_{\boldsymbol{\gamma}} \mathrm{PL}(\boldsymbol{\gamma})$. Then an estimate of $\boldsymbol{\beta}(t)$ is given by $\hat{\boldsymbol{\beta}}(t) = \{\hat{\beta}_1(t), \ldots, \hat{\beta}_p(t)\}^\top$ with $\hat{\beta}_j(t) = \zeta(\boldsymbol{B}(t)^\top \hat{\boldsymbol{\gamma}}_j, \alpha_j)$.

Optimizing $\mathrm{PL}(\boldsymbol{\gamma})$ can be implemented by gradient-based methods (Chau et al., 2014) and coordinate descent algorithms (Wright, 2015). With appropriate initial values, global optimizers can be reached. Specifically, for each $j = 1, \ldots, p$, we obtain the non-varying coefficients $(a_1, \ldots, a_p)^\top$ from the Cox model, then we set the initial $\boldsymbol{\gamma}_j^{(0)}$ to be a vector of a_j with length q. In practice, we recommend to vary the initial values and check the robustness of the final results. We choose the pre-specified parameters as follows. In theory, our method works for any α; however, in practice, a value of α comparable to the scale of true coefficients works best. Thus, we set α_j to be $0.5 \times |a_j|$. The choices of η and ρ can be specified in accordance with Condition C6. The knots of B-spline are equally spaced over $[0, \tau]$. The number of basis functions, q, can be determined through R-fold cross-validation. That is, partition the full data D into R equal-sized groups, denoted by D_r, for $r = 1 \ldots, R$, and let $\hat{\boldsymbol{\beta}}_{-r}^{(q)}(t)$ be the estimate obtained with q bases using all the data except for D_r. We obtain the optimal q by minimizing the cross-validation error, which is the average of the negative objective function (1) evaluated at $\hat{\boldsymbol{\beta}}_{-r}^{(q)}(t)$ on D_r with r running from 1 to R.

3 Inference

We begin with some needed notation. First, for a $p_1 \times q_1$ matrix $A = (a_{ij})$ and a $p_2 \times q_2$ matrix $B = (b_{ij})$, their Kronecker product is defined to be

$$A \otimes B = \begin{pmatrix} a_{11}B & \ldots & a_{1q_1}B \\ \ldots & \ldots & \ldots \\ a_{p_1 1}B & \ldots & a_{p_1 q_1}B \end{pmatrix}.$$

With that, we define the following:

$$\begin{aligned}
g(\boldsymbol{\beta}, \boldsymbol{Z}, t) &= \sum_{j=1}^{p} Z_j \beta_j(t), \\
g_n(\boldsymbol{\gamma}, \boldsymbol{Z}, t) &= \sum_{j=1}^{p} Z_j h_j(\boldsymbol{B}(t)\boldsymbol{\gamma}_j), \\
S_{0n}(\tilde{\boldsymbol{\gamma}}, t) &= \frac{1}{n}\sum_{i=1}^{n} Y_i(t)\exp(g_n(\tilde{\boldsymbol{\gamma}}, \boldsymbol{Z}_i, t)), \\
S_0(t) &= \mathrm{E}Y(t)\exp(g(\boldsymbol{\beta}, \boldsymbol{Z}, t)), \\
S_{1n}(\tilde{\boldsymbol{\gamma}}, t) &= \frac{1}{n}\sum_{i=1}^{n} Y_i(t)\exp(g_n(\tilde{\boldsymbol{\gamma}}, \boldsymbol{Z}_i, t))\boldsymbol{Z}_i \otimes \boldsymbol{B}_i, \\
S_1(t) &= \mathrm{E}Y(t)\exp(g(\boldsymbol{\beta}, \boldsymbol{Z}, t))\boldsymbol{Z} \otimes \boldsymbol{B}, \\
S_{2n}(\tilde{\boldsymbol{\gamma}}, t) &= \frac{1}{n}\sum_{i=1}^{n} Y_i(t)\exp(g_n(\tilde{\boldsymbol{\gamma}}, \boldsymbol{Z}_i, t))(\boldsymbol{Z}_i\boldsymbol{Z}_i^\top) \otimes (\boldsymbol{B}_i\boldsymbol{B}_i^\top), \\
\text{and} \quad S_2(t) &= \mathrm{E}Y(t)\exp(g(\boldsymbol{\beta}, \boldsymbol{Z}, t))(\boldsymbol{Z}\boldsymbol{Z}^\top) \otimes (\boldsymbol{B}\boldsymbol{B}^\top),
\end{aligned}$$

followed by some key sufficient conditions that guarantee the properties of our estimator.

C1 The failure time T^u and the censoring time T^c are conditionally independent given the covariate $\mathbf{Z}$.

C2 τ is chosen so that $\Pr(T^u > \tau|\mathbf{Z}) > 0$ almost surely and $\tau < \infty$; at τ, the baseline cumulative hazard function $\Lambda_0(\tau) \equiv \int_0^\tau \lambda_0(s)ds < \infty$.

C3 The covariates $\mathbf{Z}$ takes value in a bounded subset of $\mathbb{R}^p$ and $\Pr(Z_j = 0) < 1$. Also, $\sum_{j=1}^p |Z_j| = O_p(1)$.

C4 There exists a small positive constant ϵ such that $\Pr(\Delta = 1|\mathbf{Z}) > \epsilon$ and $\Pr(T^c > \tau|\mathbf{Z}) > \epsilon$ almost surely.

C5 Let $0 < c_1 < c_2 < \infty$ be two constants. The joint density $f(t, z, \Delta = 1)$ of $(T, \mathbf{Z}, \Delta = 1)$ satisfies $c_1 \leq f(t, z, \Delta = 1) < c_2$ for all $(t, z) \in [0, \tau] \times \mathbb{R}^p$.

C6 $\eta = o(q^{-m})$, $\rho = O(n^a)$ with $a \leq -1$, and $q = o(n)$.

C7 There exist a neighborhood Θ of $\tilde{\boldsymbol{\gamma}}$ and scalar, vector and matrix functions s_0, s_1 and s_2 defined on $\boldsymbol{\gamma} \times [0, \tau]$ such that for $j = 0, 1, 2$,

$$\sup_{0 \leq t \leq \tau, \boldsymbol{\gamma} \in \Theta} ||S_j(\boldsymbol{\gamma}, t) - s_{(j)}(\boldsymbol{\gamma}, t)|| \to_p 0.$$

C8 Let Θ, s_0, s_1 and s_2 be as in Condition C7 and define $e = s_1/s_0$ and $v = s_2/s_0 - e^{\otimes 2}$. For all $\boldsymbol{\gamma} \in \Theta$, $t \in [0, \tau]$:

$$s_1(\boldsymbol{\gamma}, t) = \frac{\partial}{\partial \boldsymbol{\gamma}} s_0(\boldsymbol{\gamma}, t), \quad s_2(\boldsymbol{\gamma}, t) = \frac{\partial^2}{\partial \boldsymbol{\gamma} \partial \boldsymbol{\gamma}^\top} s_0(\boldsymbol{\gamma}, t),$$

$s_0(\cdot, t)$, $s_1(\cdot, t)$, $s_2(\cdot, t)$ are continuous functions of $\boldsymbol{\gamma} \in \Theta$, uniformly in $t \in [0, \tau]$, s_0, s_1, and s_2 are bounded on $\Theta \times [0, \tau]$, and the matrix

$$\Sigma(\tilde{\boldsymbol{\gamma}}, \tau) = \int_0^\tau v(\tilde{\boldsymbol{\gamma}}, t) s_0(\tilde{\boldsymbol{\gamma}}, t) \tilde{\boldsymbol{\gamma}}(t) dt$$

is positive definite.

C9 There exists a $\delta > 0$ such that

$$n^{-1/2} \sup_{i,t} ||\mathbf{Z}_i||_\infty |Y_i(t) \mathscr{I}\{\mathbf{Z}_i^\top \boldsymbol{\beta} > -\delta ||\mathbf{Z}_i||_\infty\}| \to_p 0.$$

Condition C1 is commonly assumed in survival analysis for non-informative censoring. The finite τ condition of C2 is assumed in many studies, including Anderson and Senthilselvan (1982). Condition C3 is often assumed in nonparametric regression and is reasonable in practical situations as we do not observe infinite covariates. Condition C4 controls the censoring rate so that the data have adequate information (Sasieni, 1992). Condition C5 is needed for model identifiability and used in Huang (1999). Condition C6 controls estimation biases and ensures the convergence. Conditions C7, C8, and C9 are regularity conditions, which can be found in Anderson and Senthilselvan (1982).

3.1 Asymptotic Theory

Theorem 1 *Suppose Conditions C1–C6 hold. If* $\beta_{0j}(t) \in \mathbb{S}_{q,\alpha_j}$ *for* $j = 1, \ldots, p$ *with* q *and* α_j *being the same as in* $PL(\boldsymbol{\theta})$*, then*

$$||\hat{\boldsymbol{\beta}} - \boldsymbol{\beta}_0||_2 = O_p\left((q/n)^{1/2}\right);$$

if $\beta_{0j}(t) \notin \mathbb{S}_{q,\alpha_j}$ *for* $j = 1, \ldots, p$,

$$||\hat{\boldsymbol{\beta}} - \boldsymbol{\beta}_0||_2 = O_p\left(r_n^{1/2}\right),$$

where $r_n = q/n + q^{-2m}$.

Theorem 1 implies convergence of $\hat{\boldsymbol{\beta}}$ by Condition C6 and $m > 0 \cdot 5$. If the true curves are in the thresholding sieve space, there is no approximation error; and if q is $O(1)$, Theorem 1 suggests root-n consistency.

Let $\boldsymbol{e}_j$ be a directional vector of length p with jth entry as 1 and others 0. For any $t \in [0, \tau]$, let $\boldsymbol{a}(t) = \boldsymbol{e}_j \otimes \boldsymbol{B}(t)$, then $\hat{\theta}_j(t) = \boldsymbol{a}(t)^\top \hat{\boldsymbol{\gamma}}$.

Theorem 2 *Under Conditions C1–C9, we have for any* $t \in [0, \tau]$ *and* $j = 1, \ldots, p$,

$$\frac{\hat{\theta}_j(t) - \theta_j(t)}{\sigma_{nj}(t)} \rightarrow_d N(0, 1), \quad as\ n \rightarrow \infty,$$

where $\sigma^2_{nj}(t) = n\boldsymbol{a}(t)^\top \left[\{PL\}''(\tilde{\boldsymbol{\gamma}})\right]^{-1} \Sigma(\tilde{\boldsymbol{\gamma}}, 1) \left[\{PL\}''(\tilde{\boldsymbol{\gamma}})\right]^{-1} \boldsymbol{a}(t)$.

With Theorem 2, we can then obtain the asymptotic distribution of $\hat{\beta}_j(t)$ based on $\hat{\beta}_j(t) = \zeta\{\hat{\theta}_j(t), \alpha_j\}$.

Theorem 3 *Under Conditions C1–C9, for any* $t \in [0, \tau]$*, the limiting distribution of* $\hat{\beta}_j(t)$ $(j = 1, \ldots, p)$ *satisfies*

$$\lim_{n\rightarrow\infty} \left| Pr(\hat{\beta}_j(t) \leq x) - G_{nj}(x) \right| = 0,$$

where $G_{nj}(x) = \left[\Phi\left\{\frac{x+\alpha_j-\tilde{\theta}_j(t)}{\sigma_{nj}(t)}\right\}\mathscr{I}(x \geq 0) + \Phi\left\{\frac{x-\alpha_j-\tilde{\theta}_j(t)}{\sigma_{nj}(t)}\right\}\mathscr{I}(x < 0)\right]$ *and* $\Phi(\cdot)$ *is the cumulative distribution function of* $N(0, 1)$.

The limiting distribution in Theorem 3 guarantees that our proposed estimator can detect the zero-effect regions, because the probability of $\hat{\beta}_j(t) = 0$ can be greater than 0 even with a finite sample size.

3.2 Sparse Confidence Intervals

We introduce the sparse confidence intervals to gauge the uncertainty of the point estimates and make valid statistical inferences on the selection and the zero-effect region detection.

Given a $\xi \in (0, 1)$, for any $t \in [0, \tau]$ we construct a pointwise $(1-\xi)$ level asymptotic sparse confidence interval for $\beta_j(t)$, denoted by $[u_{nj}(t), v_{nj}(t)]$. Let $z_{\xi/2}$ and Φ be the $(1-\xi/2)$ quantile and the cumulative distribution function of $N(0,1)$, respectively. Let $P_+ = \Pr\{\hat{\beta}_j(t) > 0\}$ and $P_- = \Pr\{\hat{\beta}_j(t) < 0\}$, which can be estimated by $\hat{P}_+ = 1 - \Phi\{(\alpha_j - \hat{\theta}_j)/\hat{\sigma}_{nj}(t)\}$ and $\hat{P}_- = \Phi\{(-\alpha_j - \hat{\theta}_j)/\hat{\sigma}_{nj}(t)\}$ using Theorem 3. Here, $\hat{\sigma}_{nj}(t)$ is as defined in Theorem 2. We construct $[u_{nj}(t), v_{nj}(t)]$ as follows:

- if $\hat{P}_+ + \hat{P}_- \le \xi$, $u_{nj}(t) = v_{nj}(t) = 0$;
- else if $\hat{P}_+ < \xi/2$ and $\hat{P}_- < 1-\xi/2$, $[u_{nj}(t), v_{nj}(t)] = \left[\hat{\beta}_j(t) - \hat{\sigma}_{nj}(t)\hat{B}, 0\right]$ with $\hat{B} = \Phi^{-1}\left\{1 - \xi + \Phi(-\hat{\sigma}_{nj}^{-1}(t)\alpha_j + \hat{\sigma}_{nj}^{-1}(t)\hat{\theta}_j)\right\}$;
- else if $\hat{P}_- < \xi/2$ and $\hat{P}_+ < 1-\xi/2$, $[u_{nj}(t), v_{nj}(t)] = \left[0, \hat{\beta}_j(t) + \hat{\sigma}_{nj}(t)\hat{A}\right]$ with $\hat{A} = -\Phi^{-1}\left\{\xi - 1 + \Phi(\hat{\sigma}_{nj}^{-1}(t)\alpha_j + \hat{\sigma}_{nj}^{-1}(t)\hat{\theta}_j)\right\}$;
- else $[u_{nj}(t), v_{nj}(t)] = \left[\hat{\beta}_j(t) - \hat{\sigma}_{nj}(t)z_{\xi/2}, \hat{\beta}_j(t) + \hat{\sigma}_{nj}(t)z_{\xi/2}\right]$.

Theorem 4 *Under Conditions C1–C9, $[u_{nj}(t), v_{nj}(t)]$ is a $(1-\xi)$ level sparse confidence interval of $\beta_j(t)$ for $j = 1, \ldots, p$ and any $t \in [0, \tau]$.*

We omit its proof as it is a straightforward application of Theorem 3.

4 Simulations

We compare the proposed model with the regular time-varying effects Cox model. With $p = 3$, we design some special varying coefficient functions containing zero-effect regions as follows:

$$\begin{aligned} \beta_1(t) &= (-t^2 + 3)\mathscr{I}(t \le \sqrt{3}), \\ \beta_2(t) &= 2\log(t + 0.01)\mathscr{I}(t \ge 1), \\ \text{and} \quad \beta_3(t) &= \left(\frac{-6}{t+1} + 2\right)\mathscr{I}(t \le 2). \end{aligned} \tag{6}$$

We first simulate $\mathbf{Z}_i = (Z_{i1}, Z_{i2}, Z_{i3})^\top \sim \mathrm{N}(\mathbf{0}, \Sigma)$, where Σ has the following three structures: independent (Ind) with $\mathrm{cov}(Z_{ij}, Z_{ij^*}) = \mathscr{I}(j = j^*)$, autoregressive [AR(1)] with $\mathrm{cov}(Z_{ij}, Z_{ij^*}) = 0 \cdot 5^{|j-j^*|}$, and compound symmetry (CS) with $\mathrm{cov}(Z_{ij}, Z_{ij^*}) = \mathscr{I}(j = j^*) + 0 \cdot 5\mathscr{I}(j \ne j^*)$. We simulate $U_i \sim U(0,1)$, and solve T_i^u using $U_i = 1 - \exp\left\{-\int_0^{T_i^u} \lambda_0(u)\exp(\sum_{j=1}^3 Z_j\beta_j(u))du\right\}$, where $\lambda_0(u)$ is set to be some constant in (0, 1). The censoring times C_i are generated from $U(0, 10)$, and $T_i^c = C_i \wedge 3$.

We choose sample sizes $n = 500$, 2000 and 5000, and generate 200 independent datasets for each setting. For implementing our proposed model, we set $d = 3$, $\eta = 0 \cdot 001$, and α_j to be half of the absolute values of the least-squares estimates. The number of knots, K, is selected via cross-validation. Specifically, we tune K by conducting 10-fold cross-validation over a reasonable range such as {3, 5, 9, 13, 17, 21}. We note that ρ can also be selected via cross-validation; however, that would increase the computational burden, given that K needs to be tuned. Based on our numerical experience, we have found that specifying $\rho = 1/n^2$ that satisfies Condition C6 would give a good performance. Therefore, we set $\rho = 1/n^2$ in simulations and our later data analysis.

For comparison, we fit time-varying effects Cox models by following Zucker and Karr (1990); Yan and Huang (2012). For evaluation criteria, we use the integrated squared errors (ISE) and the averaged integrated squared errors (AISE), defined as $\text{ISE}(\beta_j) = n_g^{-1} \sum_{g=1}^{n_g} \{\hat{\beta}_j(t_g) - \beta_j(t_g)\}^2$ and $\text{AISE} = p^{-1} \sum_{j=1}^{p} \text{ISE}(\beta_j)$, respectively, where t_g ($g = 1, \ldots, n_g$) are the grid points on (0, 3). Table 1 shows that the soft-thresholding time-varying effects Cox model has a better accuracy than the regular time-varying effects Cox model by presenting smaller integrated squared errors and averaged integrated squared errors.

Table 1 Comparisons of estimation accuracy for the soft-thresholding time-varying effects Cox model and the regular time-varying effects Cox model

Covariance	n	Model	ISE(β_1)	ISE(β_2)	ISE(β_3)	AISE
Ind	500	STTV	62.6(77.1)	53.1(43.5)	58.7(59.5)	58.1(39.7)
		RegTV	75.5(94.6)	56.6(44.3)	61.9(60.6)	65.4(46.2)
	2000	STTV	12.4(9.7)	12.0(8.5)	13.1(10.4)	12.5(5.7)
		RegTV	13.9(8.2)	11.8(8.6)	12.4(8.8)	12.7(5.1)
	5000	STTV	4.2(3.2)	4.1(2.8)	4.0(2.7)	4.1(1.7)
		RegTV	5.6(3.0)	4.2(2.8)	4.5(2.7)	4.7(1.6)
AR(1)	500	STTV	16.2(16.0)	18.2(47.1)	15.2(11.1)	16.5(18.7)
		RegTV	16.3(14.1)	20.9(50.3)	13.4(8.2)	16.9(19.4)
	2000	STTV	3.6(2.2)	2.6(2.0)	3.9(2.3)	3.3(1.5)
		RegTV	3.7(2.2)	2.8(2.5)	3.1(1.6)	3.2(1.4)
	5000	STTV	1.3(1.0)	1.1(0.9)	1.2(0.8)	1.2(0.6)
		RegTV	1.9(0.9)	1.3(0.9)	1.3(0.8)	1.5(0.6)
CS	500	STTV	18.9(24.6)	19.1(30.2)	16.5(14.6)	18.2(16.2)
		RegTV	19.1(15.5)	20.4(30.3)	17.0(12.2)	18.8(13.2)
	2000	STTV	3.6(2.6)	2.7(2.5)	3.8(2.7)	3.4(1.8)
		RegTV	4.0(2.3)	2.8(2.4)	3.2(1.6)	3.4(1.4)
	5000	STTV	1.2(0.8)	1.1(0.9)	1.0(0.6)	1.1(0.5)
		RegTV	1.8(0.7)	1.1(0.9)	1.2(0.7)	1.4(0.5)

STTV: the soft-thresholding time-varying effects Cox model; RegTV: the regular time-varying effects Cox model; ISE: the integrated squared errors; AISE: the averaged integrated squared errors. All numbers are after being multiplied by 100

Figure 1, which plots the estimation curves and their median for the soft-thresholding time-varying effects Cox model and the regular time-varying effects Cox model, shows the medium estimation curves from the soft-thresholding time-varying effects Cox model coincide with the truth and the soft-thresholding approach has the zero-effect detection ability. In contrast, the regular time-varying effects Cox model fails to estimate zero effects.

Figure 2 compares the estimated coverage probabilities from the soft-thresholding time-varying effects Cox model and the regular time-varying effects Cox model, and shows that the soft-thresholding time-varying Cox model has a reasonable coverage probability in both zero-effect regions and non-zero-effect regions. In the region around the transition point, the soft thresholding time-varying effects Cox model has a higher coverage probability estimation than the regular time-varying effects Cox model. All of the results confirm that the soft-thresholding time-varying effects Cox model draws better inference than the regular time-varying effects Cox model.

With $|A|$ being the cardinality of set A, we next compare zero-effect region detection using the following criteria:

$$\text{Estimation-based true positive ratio: ETPR}(\beta)$$
$$= \frac{|\{t : \hat{\beta}(t) \neq 0 \text{ and } \beta(t) \neq 0\}|}{|\{t : \beta(t) \neq 0\}|},$$

$$\text{Estimation-based true negative ratio: ETNR}(\beta)$$
$$= \frac{|\{t : \hat{\beta}(t) = 0 \text{ and } \beta(t) = 0\}|}{|\{t : \beta(t) = 0\}|},$$

$$\text{Inference-based true positive ratio: ITPR}(\beta)$$
$$= \frac{|\{t : 0 \notin \text{CI}\{\hat{\beta}(t)\} \text{ and } \beta(t) \neq 0\}|}{|\{t : \beta(t) \neq 0\}|},$$

and

$$\text{Inference-based true negative ratio: ITNR}(\beta)$$
$$= \frac{|\{t : 0 \in \text{CI}\{\hat{\beta}(t)\} \text{ and } \beta(t) = 0\}|}{|\{t : \beta(t) = 0\}|},$$

where $\text{CI}\{\hat{\beta}(t)\}$ is the 95% confidence interval of $\beta(t)$.

We set a total of 100 grid points on $[0, 3]$, counting the number of t_g in each set as its cardinality. Table 2 shows that the soft-thresholding time-varying effects Cox model has a higher inference-based true negative ratio than the regular time-varying effects Cox model. Although the inference-based true positive and negative ratios are more reliable with controlled false discovery rates, their computational burden increases when the sample size increases. Therefore, the estimation-based true positive and negative ratios are favorable for large datasets as their calculation

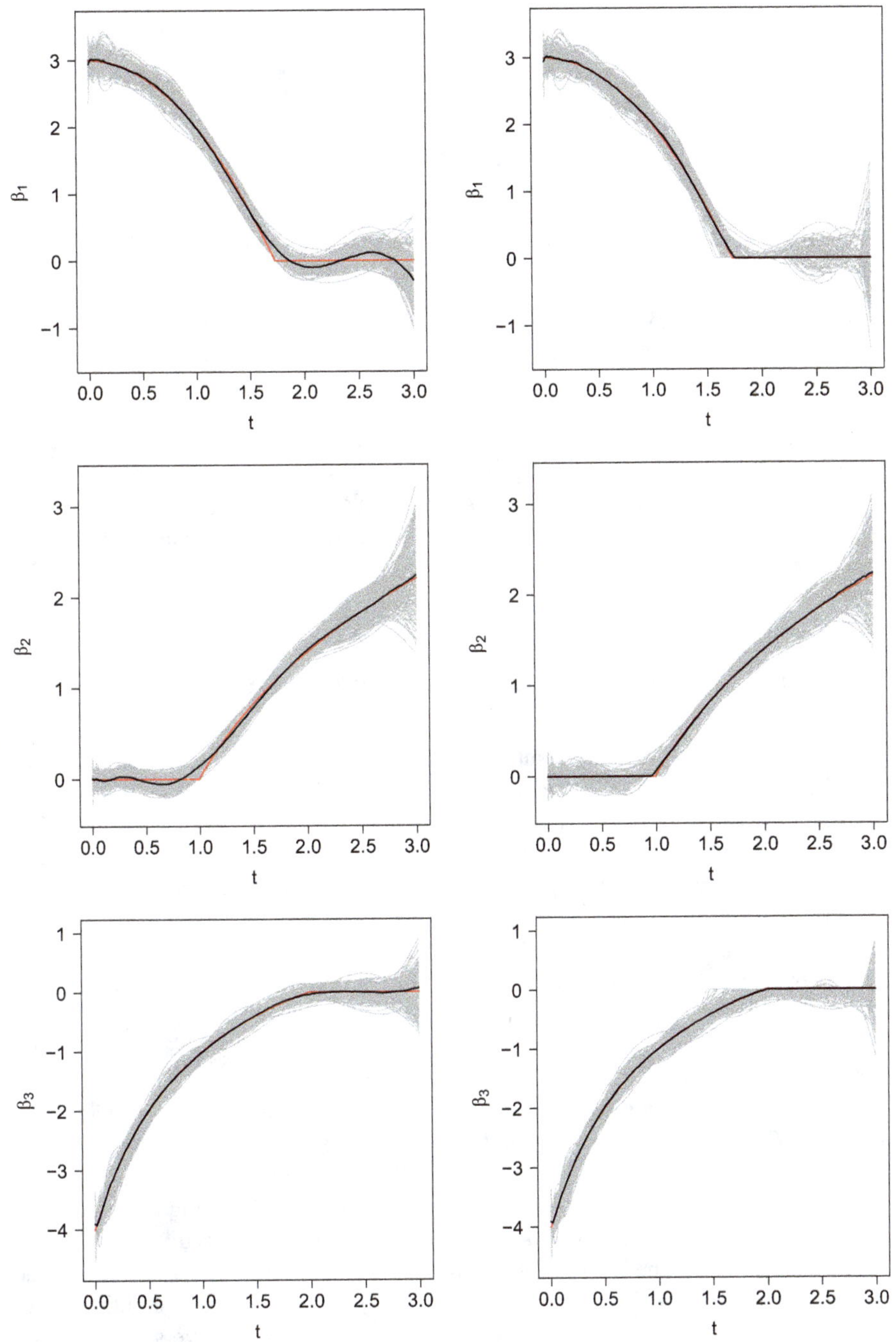

Fig. 1 Comparisons of the results obtained from the soft-thresholding time-varying effects Cox model (right panel) and the regular time-varying effects Cox model (left panel); the gray curves are 200 estimated curves based on 200 simulations, the black curves are the median estimates, and the red curves are the truth; the sample size is 5000 and the average censoring rate is 0.12

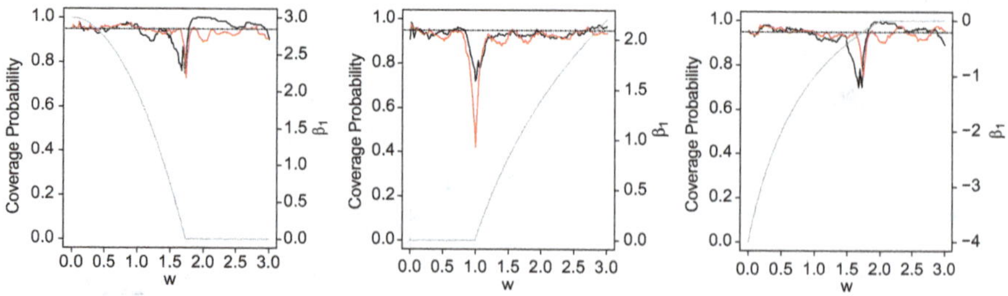

Fig. 2 Comparisons of coverage probability from the regular time-varying effects Cox model (RegTV) (the black curve) and the soft-thresholding time-varying effects Cox model (STTV) (the red curve). The sample size is 5000 and the average censoring rate is 0.12

merely depends on the estimations. First, our method presents a better estimation-based true negative ratio, indicating our method can detect zero-effect regions well. Second, as documented in Table 2, our method also presents a higher estimation-based true positive ratio than the inference-based true positive ratio, indicating a better performance of our method in inferring non-zero effects.

5 Analysis of the Boston Lung Cancer Survivor Cohort

We apply our method to study a subset of the Boston Lung Cancer Survivor Cohort (BLCSC) (Christiani, 2017). The data consist of $n = 599$ individuals, among whom 148 (24.7%) were alive and 451 (75.3%) were dead by the end of the follow up. The primary endpoint was overall survival measuring the time lag from the diagnosis of lung cancer to death or the end of the study, which ever came first. The range of the observed survival time was from 6 days to 8584 days, and the restricted mean survival and censoring times at $\tau = 8584$ days were 2124 (SE: 105) and 4397 (SE: 187) days, respectively. The observed survival time was skewed to the right. Patients who were alive were younger than those of those who died (average age in years: 55.4 vs. 61.2), and were slightly less likely to be Caucasian (89.9% vs. 95.8%). With early-stage lung cancer including stages lower than II, e.g., 1A, 1B, IIA, and IIB, 64.2% of the alive patients had early-stage lung cancer, slightly higher than those who died (62.3%). The percentage of the alive patients who had surgery was 83.8%, higher than that of the dead patients (63.0%). See Table 3 for more details.

Included in our analysis are age, race, education, sex, smoking status, cancer stage, and treatments received (surgery, chemotherapy, and radiotherapy). For comparisons, we fit the data by using the Cox proportional hazards model, the regular time-varying effects Cox model (RegTV) and the soft-thresholding time-varying effects Cox model (STTV). When implementing STTV, we set the needed parameters such as ρ, α_j and K as in done in the simulation section. In particular, with respect to the choice of K and ρ, we have determined that $K = 5$ by

Table 2 Comparisons of true positive ratios and true negative ratios for zero-effect region detection

			STTV				RegTV	
	n	β	ETPR	ETNR	ITPR	ITNR	ITPR	ITNR
Ind	500	β_1	0.96 (0.08)	0.44 (0.25)	0.81 (0.10)	0.94 (0.11)	0.81 (0.09)	0.95 (0.11)
		β_2	0.96 (0.05)	0.22 (0.13)	0.57 (0.17)	0.94 (0.08)	0.57 (0.18)	0.94 (0.12)
		β_3	0.95 (0.12)	0.37 (0.28)	0.55 (0.12)	0.94 (0.12)	0.55 (0.12)	0.94 (0.12)
	2000	β_1	0.95 (0.06)	0.61 (0.23)	0.89 (0.07)	0.95 (0.10)	0.91 (0.06)	0.94 (0.10)
		β_2	0.97 (0.04)	0.34 (0.16)	0.85 (0.08)	0.94 (0.08)	0.86 (0.08)	0.95 (0.08)
		β_3	0.96 (0.12)	0.50 (0.24)	0.70 (0.10)	0.94 (0.11)	0.72 (0.10)	0.95 (0.13)
	5000	β_1	0.98 (0.03)	0.64 (0.27)	0.95 (0.04)	0.94 (0.10)	0.96 (0.04)	0.93 (0.11)
		β_2	0.98 (0.03)	0.46 (0.18)	0.92 (0.05)	0.94 (0.09)	0.93 (0.04)	0.94 (0.10)
		β_3	0.97 (0.09)	0.50 (0.31)	0.81 (0.09)	0.96 (0.10)	0.80 (0.08)	0.96 (0.10)
AR(1)	500	β_1	0.96 (0.05)	0.60 (0.22)	0.90 (0.07)	0.95 (0.10)	0.93 (0.06)	0.93 (0.12)
		β_2	0.98 (0.04)	0.32 (0.18)	0.85 (0.08)	0.92 (0.13)	0.86 (0.08)	0.93 (0.12)
		β_3	0.97 (0.14)	0.51 (0.27)	0.69 (0.14)	0.95 (0.13)	0.73 (0.12)	0.95 (0.12)
	2000	β_1	0.97 (0.04)	0.71 (0.19)	0.94 (0.04)	0.95 (0.08)	0.99 (0.02)	0.92 (0.11)
		β_2	0.99 (0.02)	0.49 (0.19)	0.95 (0.04)	0.94 (0.10)	0.97 (0.03)	0.93 (0.11)
		β_3	0.96 (0.09)	0.62 (0.25)	0.77 (0.10)	0.92 (0.13)	0.86 (0.07)	0.94 (0.13)
	5000	β_1	1.00 (0.01)	0.79 (0.17)	0.98 (0.02)	0.96 (0.08)	1.00 (0.00)	0.85 (0.11)
		β_2	1.00 (0.01)	0.56 (0.17)	0.98 (0.02)	0.94 (0.09)	1.00 (0.01)	0.87 (0.11)
		β_3	0.97 (0.05)	0.63 (0.30)	0.90 (0.05)	0.97 (0.09)	0.91 (0.05)	0.96 (0.10)
CS	500	β_1	0.96 (0.06)	0.58 (0.23)	0.90 (0.07)	0.96 (0.10)	0.92 (0.07)	0.94 (0.12)
		β_2	0.98 (0.03)	0.32 (0.19)	0.85 (0.07)	0.93 (0.12)	0.86 (0.07)	0.94 (0.13)
		β_3	0.98 (0.13)	0.51 (0.29)	0.70 (0.13)	0.96 (0.11)	0.71 (0.12)	0.95 (0.11)
	2000	β_1	0.97 (0.04)	0.68 (0.21)	0.94 (0.04)	0.96 (0.08)	0.98 (0.02)	0.92 (0.12)
		β_2	0.99 (0.02)	0.48 (0.18)	0.96 (0.04)	0.94 (0.09)	0.97 (0.03)	0.94 (0.09)
		β_3	0.97 (0.11)	0.65 (0.25)	0.78 (0.11)	0.95 (0.09)	0.86 (0.07)	0.94 (0.13)
	5000	β_1	0.99 (0.01)	0.73 (0.18)	0.98 (0.02)	0.96 (0.08)	1.00 (0.01)	0.87 (0.11)
		β_2	1.00 (0.01)	0.55 (0.16)	0.98 (0.02)	0.92 (0.10)	1.00 (0.01)	0.89 (0.10)
		β_3	0.96 (0.06)	0.66 (0.30)	0.89 (0.06)	0.97 (0.08)	0.90 (0.05)	0.96 (0.11)

STTV: the soft-thresholding time-varying effects Cox model; RegTV: the regular time-varying effects Cox model

minimizing a 10-fold cross-validation error over a candidate set of {3, 5, 9, 13, 17, 21}, while setting the penalty parameter ρ to be $1/n^2$. We fit RegTV by using the penalized B-spline approach of Zucker and Karr (1990); Yan and Huang (2012). See the results as summarized in Figs. 3, 4, and 5.

Compared with the regular time-varying effects Cox model, the soft-thresholding time-varying effects Cox model agrees more to the Cox proportional hazards model. For some non-significant coefficients in the constant effect Cox model, STTV estimates those to be all zero over the time, such as for chemotherapy and radiotherapy. The results seem to be reasonable: insofar as surgery had a strong protective effect for this group of lung cancer patients, adding chemotherapy or

Table 3 Summary of the patient characteristics

	Alive ($n = 148$)	Dead ($n = 451$)
Age	55.4 (10.1)	61.2 (10.8)
Pack years	34.4 (29.7)	51.6 (38.5)
Race		
White (ref)	133 (89.9%)	432 (95.8%)
Others	15 (10.1%)	19 (4.2%)
Education		
Under high school (ref)	10 (6.8%)	72 (16%)
High school graduate	30 (20.3%)	113 (25.1%)
Above high school	108 (73.0%)	266 (59.0%)
Sex		
Female (ref)	113 (76.4%)	256 (56.8%)
Male	35 (23.6%)	195 (43.2%)
Smoking status		
Ever or never (ref)	96 (64.9%)	281 (62.3%)
Current	52 (35.1%)	170 (37.7%)
Cancer stage		
Early (ref)	95 (64.2%)	190 (42.1%)
Late	53 (35.8%)	261 (57.9%)
Surgery	124 (83.8%)	284 (63.0%)
Chemotherapy	48 (32.4%)	206 (45.7%)
Radiotherapy	35 (23.6%)	184 (40.8%)

Continuous variables are presented in mean (standard deviation), and categorical variables are presented in count (percentage). Due to rounding, some summations of percentages for one variable are not one. Reference groups are marked

radiotherapy did not seem to be associated with additional protective or harmful impacts on lung cancer patients' survival. This is consistent with the clinical practice that surgery is often the first line therapy for operable lung cancer patients (Zheng et al., 2018). Interestingly, smoking at diagnosis was associated with higher short-term (in the first 3 years post-diagnosis) and long-term (after 7 years post-diagnosis) mortality, but was not significantly associated with mortality between 3 and 7 years post-diagnosis, possibly a stabilization period for patients. The result highlights the importance of early cessation of smoking (Barbeau et al., 2006).

The other results are equally interesting. Adjusting for all the other factors, the expected hazard was significantly higher among male patients than female patients; older patients had a significantly higher hazard than younger patients; non-white patients had a lower hazard than white patients; a later cancer stage was strongly associated with worse lung cancer mortality. However, there were no significant associations between education levels and lung cancer mortality. In conclusion, the results of STTV are consistent with those obtained by using the Cox model, but STTV can more accurately capture the time-varying effect of each factor.

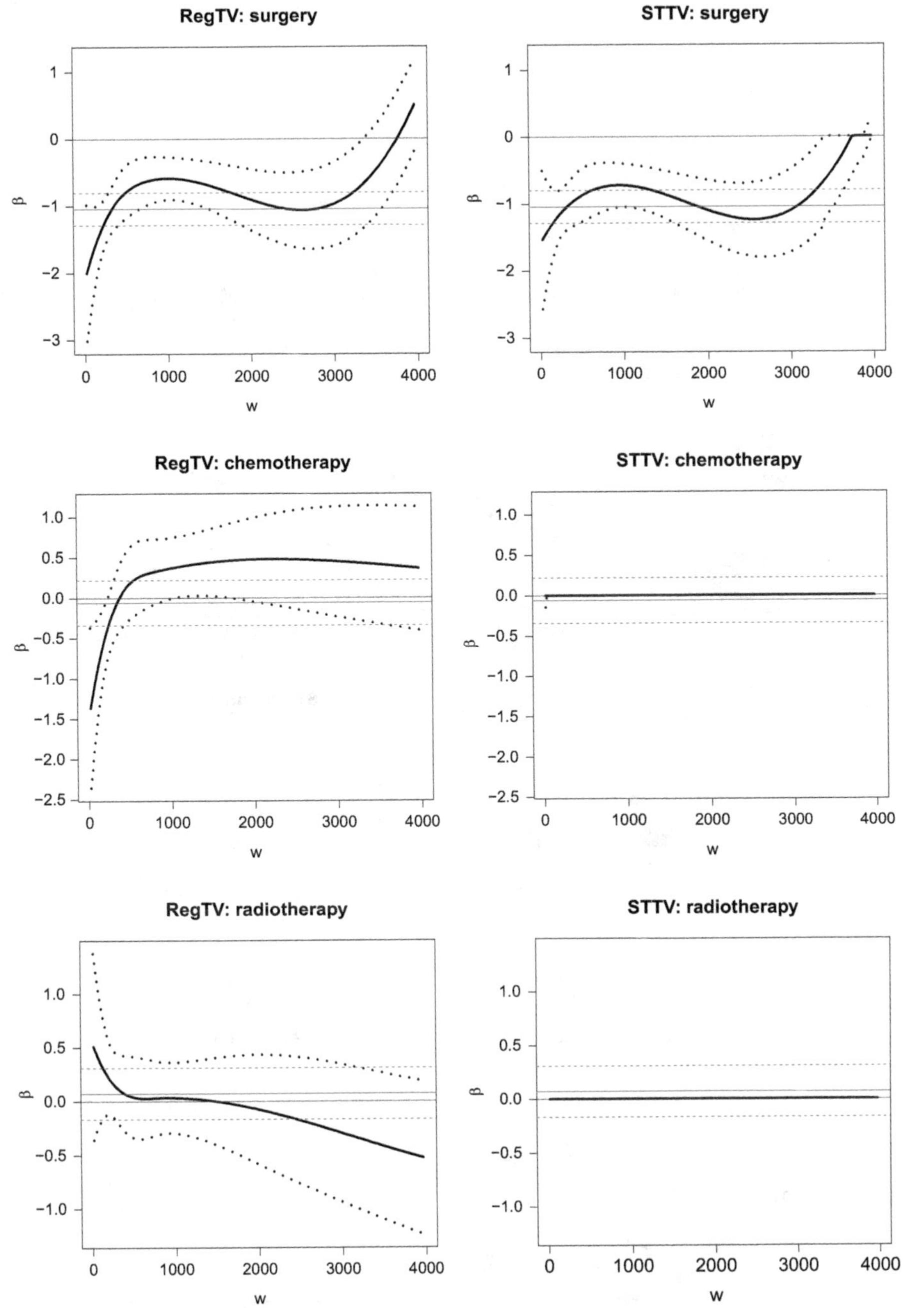

Fig. 3 Estimation results (part I) for the BLCSC data using the regular time-varying effects Cox model (RegTV) and the soft-thresholding time-varying effects Cox model (STTV): the solid lines are the estimated coefficient function curves; the dotted lines are the pointwise (sparse) confidence intervals; black lines are from varying coefficient models; red lines are from the constant effect Cox model

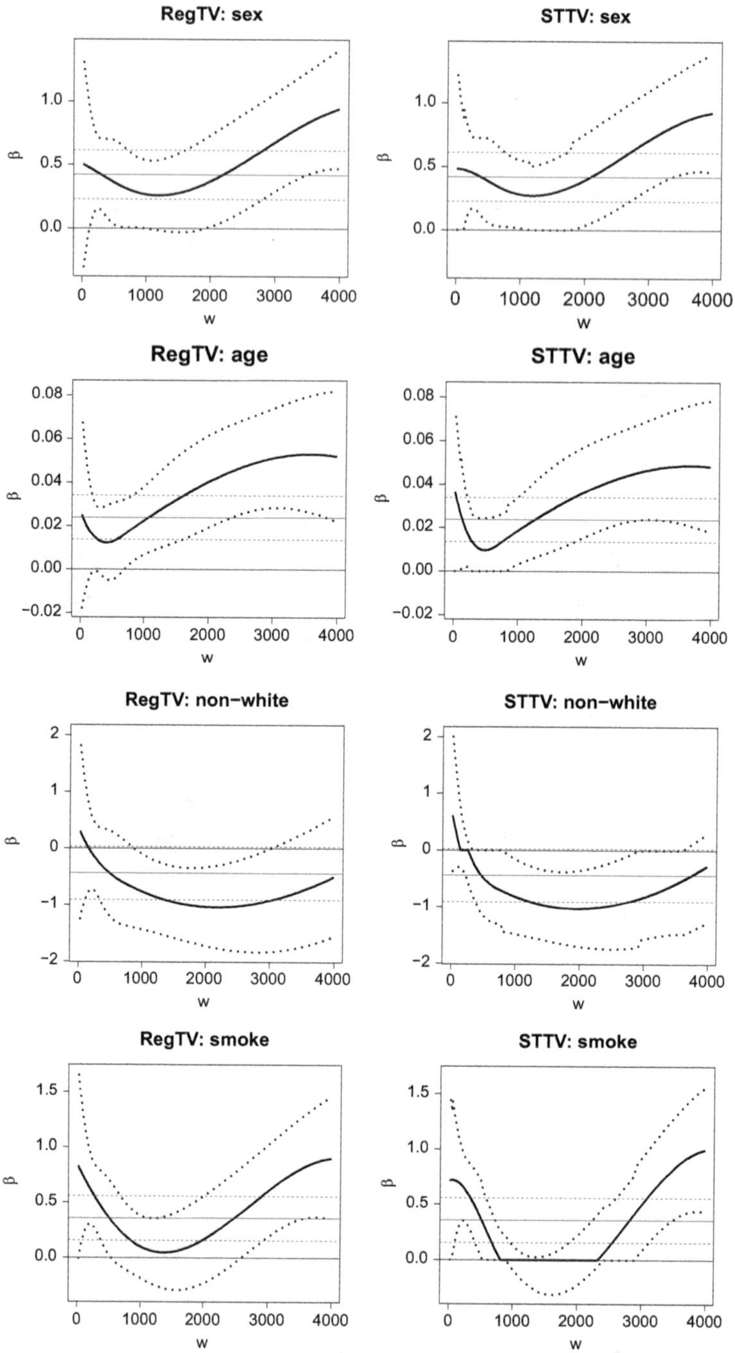

Fig. 4 Estimation results (part II) for the BLCSC data using the regular time-varying effects Cox model (RegTV) and the soft-thresholding time-varying effects Cox model (STTV): the solid lines are the estimated coefficient function curves; the dotted lines are the pointwise (sparse) confidence intervals; the black lines are from varying coefficient models and the red lines are from the Cox proportional hazards model

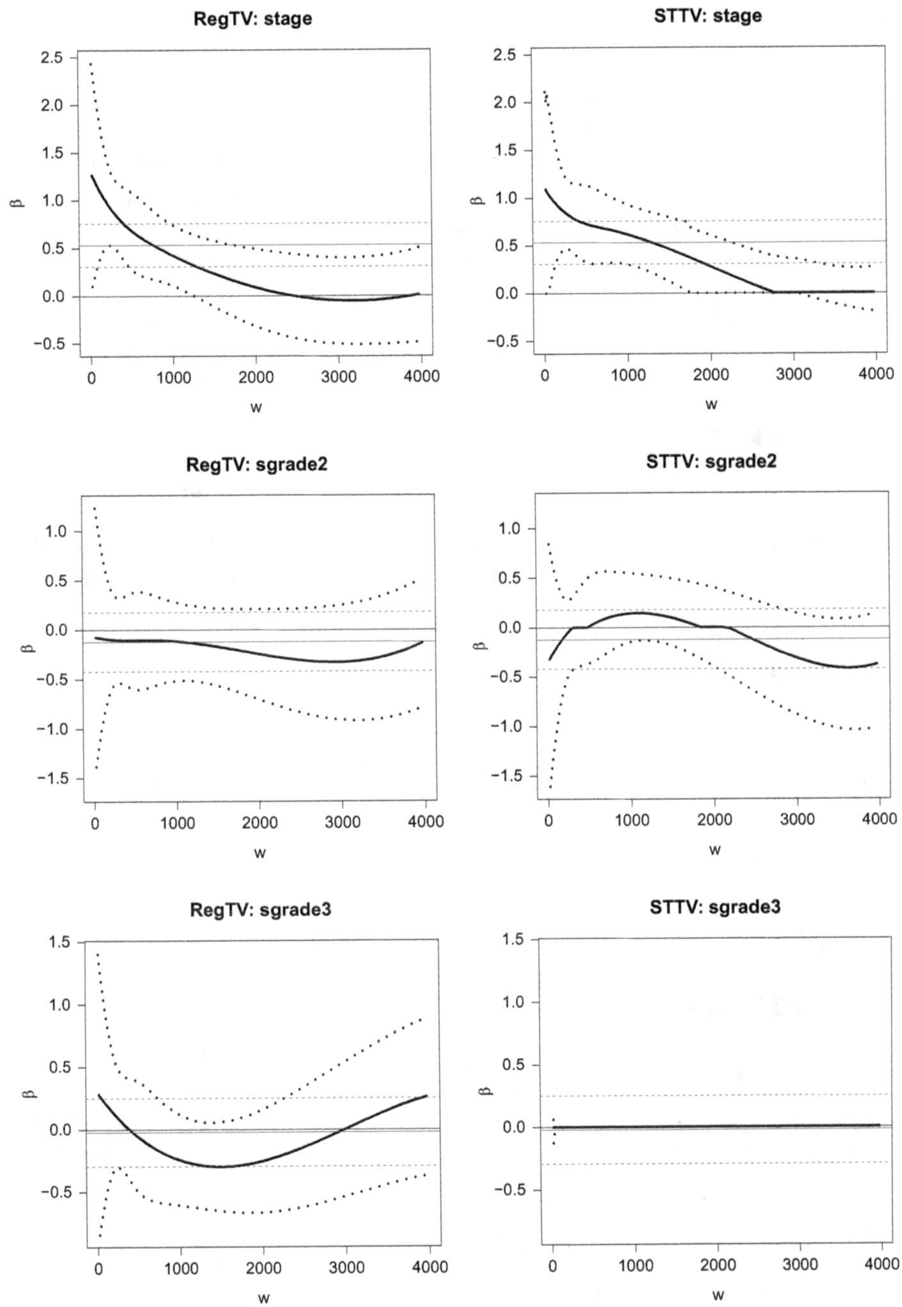

Fig. 5 Estimation results (part III) for the BLCSC data using the regular time-varying effects Cox model (RegTV) and the soft-thresholding time-varying effects Cox model (STTV): the solid lines are the estimated coefficient function curves; the dotted lines are the pointwise (sparse) confidence intervals; the black lines are from varying coefficient models and the red lines are from the Cox proportional hazards model

6 Discussion

To address the challenge of modeling time-varying coefficients with zero-effect regions in survival analysis, we proposed a new soft-thresholding time-varying coefficient model, where the varying coefficients are piecewise smooth with zero-effect regions. To quantify uncertainty of the estimates, we have designed a new type of sparse confidence intervals, which extend classical confidence intervals by accommodating exact zero estimates. Our framework enables us to estimate non-zero time-varying effects and detect zero-effect regions simultaneously, extending the already-widely-used Cox models to a new territory. The work pays tribute to Sir D.R. Cox, whose work has fundamentally influenced modern biomedical research.

Code and Data

The code for the simulation study is available at https://github.com/kangjian2016/STVCox. The access to the BLCSC data is privileged and needs a data use agreement (DUA) with Harvard School of Public Health. Interested readers can obtain the information about the DUA process via https://hcsra.sph.harvard.edu/data-use-agreement-dua.

Acknowledgments We thank Dr. David Christiani for providing the BLCSC data. We thank two referees for their insightful comments that have improved the presentation and the quality of the submission. The work is partially supported by grants from NIH (R01CA249096 and U01CA209414).

Appendix

Proof of Theorem 1

For every $f \in \mathbb{F}_0$, by Corollary 6.21 of Schumaker (2007), there exists an $f_n \in \mathbb{F}$, $||f_n - f||_\infty = O(q^{-m})$. For any $\delta_1 > 0$ and $\delta_2 > 0$, there exists an η (when constructing h) such that $||h(f_n) - \zeta(f_n)|| < \delta_1$ and $||h(f) - \zeta(f)|| < \delta_2$. Then we have,

$$\begin{aligned} ||h(f_n) - h(f)|| &< ||h(f_n) - \zeta(f_n)|| + ||\zeta(f_n) - \zeta(f)|| + ||\zeta(f) - h(f)|| \\ &= A_1 + A_2 + A_3. \end{aligned}$$

Let $\delta_1 = O(q^{-m})$ and $\delta_2 = O(q^{-m})$, then $A_1 < \delta_1 = O(q^{-m})$ and $A_2 < \delta_2 = O(q^{-m})$. We have $A_3 \le ||f_n - f|| = O(q^{-m})$ because the Lipschitz continuous property in Lemma 1 of Kang et al. (2018). Therefore, $||h(f_n) - h(f)||_\infty = O(q^{-m})$. For simplicity of notation, let h_{nj} denote $h(f_{nj})$ and h_{0j} denote $h(f_{0j})$.

Let $g_n = \sum_{j=1}^{p} Z_j h_{nj}$. Then, given $\mathbf{Z}$, $|g_n - g_0| = |\sum_{j=1}^{p} Z_j(h_{nj} - h_{0j})| \leq \sum_{j=1}^{p} |Z_j||(h_{nj} - h_{0j})| = O_p(q^{-m})$. Thus, we have $||g_n - g_0|| = O_p(q^{-m})$.

By Lemma 5.1 of Huang (1999), $||\hat{g}_n - g_n||_2^2 = o_p(1)$. We then only need to prove

$$\begin{aligned} &\mathrm{E} \sup_{\delta/2<||g-g_n||\leq\delta} |M_n(g) - M_n(g_n) - (M_0(g) - M_0(g_n))| \\ &= O_p(n^{-\frac{1}{2}}\delta(q^{\frac{1}{2}} + log^{\frac{1}{2}}(1/\delta))). \end{aligned} \tag{7}$$

It follows that

$$\begin{aligned} &M_n(g) - M_n(g_n) - \{M_0(g) - M_0(g_n)\} \\ =&P_{\Delta_n} m_n(\cdot, g) - P_{\Delta_n} m_n(\cdot, g_n) - P_\Delta m_0(\cdot, g) + P_\Delta m_0(\cdot, g_n) \\ =&P_{\Delta_n} m_n(\cdot, g) - P_\Delta m_0(\cdot, g) - P_{\Delta_n} m_n(\cdot, g_n) + P_\Delta m_0(\cdot, g_n) \\ =&P_{\Delta_n}\{\log S_{0n}(\cdot, g) - \log S_0(\cdot, g)\} - P_{\Delta_n}\{\log S_0(\cdot, g_n) - \log S_0(\cdot, g_n)\} \\ =&J_{1n} + J_{2n}. \end{aligned}$$

For any $\beta \in \mathbb{H}_n$ and any $\alpha > 0$, we can find at least one $f \in \mathbb{F}_n$ such that $\beta = \zeta(f, \alpha)$, then $\log N_{[]}(\epsilon, \mathbb{H}_n, \delta) \leq \log N_{[]}(\epsilon, \mathbb{F}_n, \delta) \lesssim c_1 q \log(\delta/\epsilon)$ by calculation in Shen and Wong (1994). Therefore, we can also obtain $\log N_{[]}(\epsilon, \mathbb{H}_n, \delta) \lesssim c_2 q \log(\delta/\epsilon)$ according to its construction. Because both exp and log are monotone functions, we have $\log N_{[]}(\epsilon, E_{n,\delta}, \delta) \lesssim c_2 q \log(\delta/\epsilon) + c_3 q \log(\delta/\epsilon) \lesssim c_4 q \log(\delta/\epsilon)$, where $c_4 = \max(c_2, c_3)$.

Therefore, $J_{[]}(\delta, \epsilon_{n,\delta}, \rho) = \int_0^\delta \sqrt{1 + \log N_{[]}(\epsilon, E_{n,\delta}, \rho)} d\epsilon \lesssim \delta q^{\frac{1}{2}}$. By Lemma 3.4.2 of Van Der Vaart and Wellner (1996), we have

$$\mathrm{E}||J_{1n}|| \lesssim n^{-\frac{1}{2}} q^{\frac{1}{2}} \delta (1 + \frac{q^{\frac{1}{2}}\delta}{\delta^2 \sqrt{n}} c_5) = O(n^{-\frac{1}{2}} q^{\frac{1}{2}} \delta). \tag{8}$$

On the other hand, we have

$$\begin{aligned} \sup_{||g-g_n||\leq\delta} |J_{2n}| &\leq 2 \sup_{0\leq t\leq\tau, ||g-g_n||\leq\delta} \left| log \frac{S_{0n}(\cdot, g)}{S_{0n}(\cdot, g_n)} - log \frac{S_0(\cdot, g)}{S_0(\cdot, g_n)} \right| \\ &\lesssim \sup_{0\leq t\leq\tau, ||g-g_n||\leq\delta} \left| \frac{S_{0n}(\cdot, g)}{S_{0n}(\cdot, g_n)} - \frac{S_0(\cdot, g)}{S_0(\cdot, g_n)} \right| \\ &\lesssim \sup_{0\leq t\leq\tau, ||g-g_n||\leq\delta} \left| \frac{S_{0n}(\cdot, g) S_0(\cdot, g_n) - S_{0n}(\cdot, g_n) S_0(\cdot, g)}{S_{0n}(\cdot, g_n) S_0(\cdot, g_n)} \right|. \end{aligned}$$

Since the denominator is bounded away from 0 with probability approaching to 1, we only need to consider the numerator. It follows that

$$
\begin{aligned}
&S_{0n}(\cdot, g)S_0(\cdot, g_n) - S_{0n}(\cdot, g_n)S_0(\cdot, g)\\
=&S_0(t, g_n)\{S_{0n}(t, g) - S_{0n}(t, g_n) - S_0(t, g) + S_0(t, g_n)\}-\\
&\{S_{0n}(t, g_n) - S_0(t, g_n)\}\{S_0(t, g) - S_{0n}(t, g)\}\\
=&I_{1n} - I_{2n}.
\end{aligned}
$$

Since $I_{1n} = S_0(t, g_n)Y(t)[\exp(g(z)) - \exp(g_n(z))]$, we consider the class of function $Y(t)\exp(g(z))$. Since exp is monotone and the entropy of the class of indicator function $Y(t) = I[0 \leq t \leq \tau]$ is $\delta \log^{\frac{1}{2}}(1/\delta)$, we have that the entropy of the class of function $Y(t)exp(g(z))$ is $\delta(q^{\frac{1}{2}} + log^{\frac{1}{2}}(1/\delta))$. By Lemma 3.4.2 of Van Der Vaart and Wellner (1996), $I_{1n} \lesssim n^{-\frac{1}{2}}\delta(q^{\frac{1}{2}} + \log^{\frac{1}{2}}(1/\delta))$.

By Taylor's expansion and Jensen's inequality, we have

$$
\begin{aligned}
|S_0(t, g) - S_0(t, g_n)| &\leq E(Y(t)[\exp(g) - \exp(g_n)])\\
&\leq \mathrm{E}(\exp(g_n)|g - g_n|)\\
&\lesssim (E(g - g_n)^2)^{\frac{1}{2}} = O_p(\delta).
\end{aligned}
$$

Since $S_n(t, g_n) - S_0(t, g_n) = O_p(n^{-\frac{1}{2}}q^{\frac{1}{2}})$, we obtain $I_{2n} = O_p(n^{-\frac{1}{2}}q^{\frac{1}{2}}\delta)$.

Therefore, $\sup_{||g-g_n||\leq\delta} |J_{2n}| \lesssim n^{-\frac{1}{2}}\delta(q^{\frac{1}{2}} + \log^{\frac{1}{2}}(1/\delta))$. Thus, we have $M_n(g) - M_n(g_n) - \{M_0(g) - M_0(g_n)\} = O_p(n^{-\frac{1}{2}}\delta(q^{\frac{1}{2}} + \log^{\frac{1}{2}}(1/\delta)))$.

By Theorem 3.4.1 of Van Der Vaart and Wellner (1996), the key function $\phi(\delta)$ takes the form of $\phi_n(\delta) = \delta(q^{\frac{1}{2}} + log^{\frac{1}{2}}(1/\delta))$. Therefore, $||(\hat{g}_n - g_n)||_2 = O_p((q/n)^{\frac{1}{2}})$. Therefore, we have

$$
\begin{aligned}
||\hat{g}_n - g_0||_2^2 &\leq ||\hat{g}_n - g_n||_2^2 + ||g_n - g_0||_2^2\\
&\leq O_p(q/n) + O_P(q^{-2m})\\
&\leq O_p(r_n),
\end{aligned}
\tag{9}
$$

where $r_n = q/n + q^{-2m}$.

Then by Lemma 1 of Stone (1985), we have

$$
E(Z_j\hat{h}_j(t) - Z_jh_j(t))^2 = O_p(r_n), \quad 1 \leq j \leq p. \tag{10}
$$

By Condition C3, there exists $\delta, \epsilon > 0$, $\Pr(|Z_j| > \delta) > \epsilon$. Then

$$
\begin{aligned}
E(Z_j\hat{h}_j(t) - Z_jh_j(t))^2 &> \Pr(|Z_j| > \delta)\delta^2(\hat{h}_j(t) - h_j(t))^2\\
&> \epsilon\delta^2(\hat{h}_j(t) - h_j(t))^2.
\end{aligned}
\tag{11}
$$

Therefore for any t, we have $(\hat{\beta}_j(t) - \beta_j(t))^2 = O_p(r_n)$, i.e. $|\hat{\beta}_j(t) - \beta_j(t)| = O_p(r_n^{1/2})$. Then we have $||\hat{\beta}_j - \beta_j||_\infty = O_p(r_n^{1/2})$ for $j = 1, \ldots, p$. ■

Proof of Theorem 2

We show Theorem 2 is true when $\tau = 1$. The extension to any $\tau < \infty$ satisfying condition C2 is straightforward and is omitted.

Following the counting process notation in Anderson and Senthilselvan (1982), we let

$$
\begin{aligned}
C(\boldsymbol{\gamma}, t) = &\sum_{i=1}^{n} \int_0^{\top} \sum_{j=1}^{p} Z_{ij} h_j(\boldsymbol{\gamma}_j, s) \mathrm{d}N_i(s) \\
&- \int_0^{\top} \log \left\{ \sum_{i=1}^{n} Y_i(s) \exp\{\sum_{j=1}^{p} Z_{ij}(s) h_j(\boldsymbol{\gamma}_j, s)\} \right\} \mathrm{d}\bar{N}(s),
\end{aligned}
$$

then we have,

$$
\mathrm{PL}(\boldsymbol{\gamma}) = C(\boldsymbol{\gamma}, 1) - \rho ||\boldsymbol{\theta}||_2^2.
$$

Then for any $\boldsymbol{\gamma}$,

$$
\mathrm{PL}^{'}(\boldsymbol{\gamma}) = C^{'}(\boldsymbol{\gamma}, 1) - \rho \sum_{i=1}^{n} \boldsymbol{\theta} \otimes \boldsymbol{B}(T_i).
$$

By Taylor's expansion, we have that

$$
\{\mathrm{PL}\}^{'}(\hat{\boldsymbol{\gamma}}) - \mathrm{PL}^{'}(\tilde{\boldsymbol{\gamma}}) = \{\mathrm{PL}\}^{''}(\boldsymbol{\gamma}^*)(\boldsymbol{\gamma} - \tilde{\boldsymbol{\gamma}}),
$$

where $\boldsymbol{\gamma}^*$ is on the line segment between $\hat{\boldsymbol{\gamma}}$ and $\tilde{\boldsymbol{\gamma}}$. Since $\{\mathrm{PL}\}^{'}(\hat{\boldsymbol{\gamma}}) = 0$, we have

$$
\begin{aligned}
\boldsymbol{\gamma} - \tilde{\boldsymbol{\gamma}} &= -\left[\{\mathrm{PL}\}^{''}(\boldsymbol{\gamma}^*)\right]^{-1} \mathrm{PL}^{'}(\tilde{\boldsymbol{\gamma}}) \\
&= -\left[\{\mathrm{PL}\}^{''}(\boldsymbol{\gamma}^*)\right]^{-1} \left\{ C^{'}(\tilde{\boldsymbol{\gamma}}, 1) - \rho \sum_{i=1}^{n} \boldsymbol{\theta}_0 \otimes \boldsymbol{B}(T_i) \right\} \\
&= -\left[\{\mathrm{PL}\}^{''}(\boldsymbol{\gamma}^*)\right]^{-1} C^{'}(\tilde{\boldsymbol{\gamma}}, 1) + \rho \left[\{\mathrm{PL}\}^{''}(\boldsymbol{\gamma}^*)\right]^{-1} \sum_{i=1}^{n} \boldsymbol{\theta} \otimes \boldsymbol{B}(T_i).
\end{aligned}
$$

The goal is to prove that for any non-zero $\boldsymbol{a}$,

$$\frac{\boldsymbol{a}^\top(\hat{\boldsymbol{\gamma}} - \tilde{\boldsymbol{\gamma}})}{\hat{\sigma}(\boldsymbol{a})} \to_d N(0, 1),$$

where $\hat{\sigma}(\boldsymbol{a}) = n\boldsymbol{a}^\top \left[\{\text{PL}\}''(\tilde{\boldsymbol{\gamma}})\right]^{-1} \Sigma(\tilde{\boldsymbol{\gamma}}, 1) \left[\{\text{PL}\}''(\tilde{\boldsymbol{\gamma}})\right]^{-1} \boldsymbol{a}$.

We claim that

$$\frac{\boldsymbol{a}^\top \left[-\{\text{PL}\}''(\boldsymbol{\gamma}^*)\right]^{-1} C'(\tilde{\boldsymbol{\gamma}}, 1)}{\hat{\sigma}(\boldsymbol{a})} \to_d N(0, 1) \tag{12}$$

and

$$\rho\boldsymbol{a}^\top \left[\{\text{PL}\}''(\boldsymbol{\gamma}^*)\right]^{-1} \sum_{i=1}^{n} \boldsymbol{\theta} \otimes \boldsymbol{B}(T_i)/\hat{\sigma}(\boldsymbol{a}) \to_p 0. \tag{13}$$

To show (12), we will utilize the martingale theories in Anderson and Senthilselvan (1982) to prove that $\boldsymbol{a}^\top \left[-\{\text{PL}\}''(\boldsymbol{\gamma}^*)\right]^{-1} C'(\tilde{\boldsymbol{\gamma}}, t)/\hat{\sigma}(\boldsymbol{a})$ is converging to a Gaussian process. Indeed,

$$C'(\tilde{\boldsymbol{\gamma}}, t) = \sum_{i=1}^{n} \int_0^\top \{A_i(\tilde{\boldsymbol{\gamma}}, s) - E(\tilde{\boldsymbol{\gamma}}, s)\}\, dM_i(s),$$

where $A_i(\tilde{\boldsymbol{\gamma}}, s) = \boldsymbol{U}_i \otimes \boldsymbol{B}_i$ and $E(\tilde{\boldsymbol{\gamma}}, s) = S_1(\tilde{\boldsymbol{\gamma}}, s)/S_0(\tilde{\boldsymbol{\gamma}}, s)$. Then we have

$$\begin{aligned}&\frac{\boldsymbol{a}^\top \left[-\{\text{PL}\}''(\boldsymbol{\gamma}^*)\right]^{-1}}{\hat{\sigma}(\boldsymbol{a})} C'(\tilde{\boldsymbol{\gamma}}, t)\\ &= \sum_{i=1}^{n} \int_0^\top \frac{\boldsymbol{a}^\top \left[-\{\text{PL}\}''(\boldsymbol{\gamma}^*)\right]^{-1}}{\hat{\sigma}(\boldsymbol{a})} \{A_i(\tilde{\boldsymbol{\gamma}}, s) - E(\tilde{\boldsymbol{\gamma}}, s)\}\, dM_i(s).\end{aligned}$$

Let

$$H_i(s) = \frac{\boldsymbol{a}^\top \left[-\{\text{PL}\}''(\boldsymbol{\gamma}^*)\right]^{-1}}{\hat{\sigma}(\boldsymbol{a})} \{A_i(\tilde{\boldsymbol{\gamma}}, s) - E(\tilde{\boldsymbol{\gamma}}, s)\},$$

we then can show claim 12 is true by applying Theorem I.2 in Anderson and Senthilselvan (1982). Condition (I.3) of Theorem I.2 is valid, because by Conditions C2, C7, and C8, we have

$$\begin{aligned}\int_0^{\top} \sum_{i=1}^{n} H_i^2(s)\lambda_i(s)\mathrm{d}s =&\boldsymbol{a}^{\top}\left[-\{\mathrm{PL}\}''(\boldsymbol{\gamma}^*)\right]^{-1}\cdot\\ &\int_0^{\top} \sum_{i=1}^{n} \{A_i(\tilde{\boldsymbol{\gamma}}, s) - E(\tilde{\boldsymbol{\gamma}}, s)\}\\ &\{A_i(\tilde{\boldsymbol{\gamma}}, s) - E(\tilde{\boldsymbol{\gamma}}, s)\}^{\top}\lambda_i(s)\mathrm{d}s\cdot\\ &\left[-\{\mathrm{PL}\}''(\boldsymbol{\gamma}^*)\right]^{-1}\boldsymbol{a}/\hat{\sigma}^2(\boldsymbol{a})\\ \rightarrow_p&\, r(t),\end{aligned}$$

where $r(t)$ is some positive function of t and $r(1) = 1$.

By similar arguments in Anderson and Senthilselvan (1982), condition (I.4) of Theorem I.2 is true by Conditions C2, C7, and C9. Then claim (12) is valid.

Claim (13) is valid because

$$\rho\left|\boldsymbol{a}^{\top}\left[\{\mathrm{PL}\}''(\boldsymbol{\gamma}^*)\right]^{-1}\sum_{i=1}^{n}\boldsymbol{\theta}\otimes\boldsymbol{B}(T_i)/\hat{\sigma}(\boldsymbol{a})\right| \le O_p(n\rho) \rightarrow_p 0 \tag{14}$$

by Condition C6. Therefore, for any non-zero $\boldsymbol{a}$,

$$\frac{\boldsymbol{a}^{\top}(\hat{\boldsymbol{\gamma}} - \tilde{\boldsymbol{\gamma}})}{\hat{\sigma}(\boldsymbol{a})} \rightarrow_d N(0, 1),$$

where $\hat{\sigma}(\boldsymbol{a}) = n\boldsymbol{a}^{\top}\left[\{\mathrm{PL}\}''(\tilde{\boldsymbol{\gamma}})\right]^{-1}\Sigma(\tilde{\boldsymbol{\gamma}}, 1)\left[\{\mathrm{PL}\}''(\tilde{\boldsymbol{\gamma}})\right]^{-1}\boldsymbol{a}$.

Since for any $t \in [0, \tau]$, $\hat{\theta}_j(t) = (\boldsymbol{e}_j \otimes \boldsymbol{B}(t))^{\top}\hat{\boldsymbol{\gamma}}$, then let $\boldsymbol{a} = \boldsymbol{e}_j \otimes \boldsymbol{B}(t)$, we have for any $t \in [0, \tau]$,

$$\frac{\hat{\theta}_j(t) - \theta_j(t)}{\sigma_{nj}(t)} \rightarrow_d N(0, 1),$$

where $\sigma_{nj}^2(t) = n\{\boldsymbol{e}_j \otimes \boldsymbol{B}(t)\}^{\top}\left[-\{\mathrm{PL}\}''(\boldsymbol{\gamma}^*)\right]^{-1}\Sigma(\tilde{\boldsymbol{\gamma}}, 1)\left[-\{\mathrm{PL}\}''(\boldsymbol{\gamma}^*)\right]^{-1}\{\boldsymbol{e}_j \otimes \boldsymbol{B}(t)\}$.

Finally, denote by $\hat{\sigma}_{nj}^2(t) = n\{\boldsymbol{e}_j \otimes \boldsymbol{B}(t)\}^\top \left[-\{\text{PL}\}''(\hat{\boldsymbol{\gamma}})\right]^{-1} \Sigma(\hat{\boldsymbol{\gamma}}, 1) \left[-\{\text{PL}\}''(\hat{\boldsymbol{\gamma}})\right]^{-1} \{\boldsymbol{e}_j \otimes \boldsymbol{B}(t)\}$. As $||\hat{\boldsymbol{\gamma}} - \tilde{\boldsymbol{\gamma}}||_2 \to_p 0$, it follows that $\hat{\sigma}_{nj}^2(t)/\sigma_{nj}^2(t) \to_p 1$ for $t > 0$. Hence, by the Slutsky theorem,

$$\frac{\hat{\theta}_j(t) - \theta_j(t)}{\hat{\sigma}_{nj}(t)} \to_d N(0, 1),$$

which justifies the use of $\hat{\sigma}_{nj}(t)$ as a consistent estimate of the variance. ■

References

Anderson, J. A., & Senthilselvan, A. (1982). A two-step regression model for hazard functions. *Journal of the Royal Statistical Society Series C (Applied Statistics), 31*(1), 44–51. https://doi.org/10.2307/2347073. http://www.jstor.org/stable/2347073

Barbeau, E. M., Li, Y., Calderon, P., Hartman, C., Quinn, M., Markkanen, P., Roelofs, C., Frazier, L., & Levenstein, C. (2006). Results of a union-based smoking cessation intervention for apprentice iron workers (United States). *Cancer Causes and Control, 17*(1), 53–61.

Cai, Z., & Sun, Y. (2003) Local linear estimation for time-dependent coefficients in Cox's regression models. *Scandinavian Journal of Statistics, 30*(1), 93–111. https://doi.org/10.1111/1467-9469.00320. papers2://publication/uuid/14A8538C-13F5-4307-8491-4505CAFCB8F9.

Chang, S. G., Yu, B., & Vetterli, M. (2000). Adaptive wavelet thresholding for image denoising and compression. *IEEE Transactions on Image Processing, 9*(9), 1532–1546.

Chau, M., Fu, M. C., Qu, H., Ryzhov, I. O. (2014). Simulation optimization: A tutorial overview and recent developments in gradient-based methods. In: *Proceedings of the winter simulation conference 2014* (pp. 21–35). New York: IEEE.

Christiani, D. C. (2017). The Boston lung cancer survival cohort. Tech. rep., Washington: NIH. http://grantome.com/grant/NIH/U01-CA209414-01A1.

Cox, D. R. (1972). Regression models and life-tables. *Journal of the Royal Statistical Society: Series B (Methodological), 34*(2), 187–202.

Donoho, D. L. (1995). De-Noising by Soft-Thresholding. *IEEE Transactions on Information Theory, 41*(3), 613–627. https://doi.org/10.1109/18.382009, 0611061v2

Donoho, D. L., & Johnstone, I. M. (1994). Ideal spatial adapatation by wavelet shrinkage. *Biometrika, 81*(3), 425–455.

Gore, S. M., Pocock, S. J., & Kerr, G. R. (1984). Regression models and non-proportional hazards in the analysis of breast cancer survival. *Journal of the Royal Statistical Society Series C (Applied Statistics), 33*(2), 176–195.

Hastie, T. J., & Tibshirani, R. (1993). Varying-coefficient Models. *Journal of the Royal Statistical Society, 55*(4), 757–796. https://doi.org/10.2307/2345993.

He, K., Yang, Y., yan, L., Zhu, J., & Li, Y. (2017). Modeling time-varying effects with large-scale survival data: An efficient Quasi-Newton approach. *Journal of Computational and Graphical Statistics, 26*(3), 635–645.

Hong, H. G., Christiani, D. C., & Li, Y. (2019). Quantile regression for survival data in modern cancer research: Expanding statistical tools for precision medicine. *Precision Clinical Medicine, 2*(2), 90–99.

Huang, J. (1999). Efficient estimation of the partly linear additive Cox model. *Annals of Statistics, 27*(5), 1536–1563.

Kang, J., Reich, B. J., & Staicu, A. M. (2018). Scalar-on-image regression via the soft-thresholded gaussian process. *Biometrika, 105*(1), 165–184.

Lian, H., Lai, P., & Liang, H. (2013). Partially linear structure selection in cox models with varying coefficients. *Biometrics, 69*(2), 348–357. https://doi.org/10.1111/biom.12024

Martinussen, T., Scheike, T. H., & Skovgaard, I. M. (2002). Efficient estimation of fixed and time-varying covariate effects in multiplicative intensity models. *Scandinavian Journal of Statistics, 29*(1), 57–74. https://doi.org/10.1111/1467-9469.00060

Marzec, L. (1997). On fitting Cox's regression model with time-dependent coefficients. *Biometrika, 84*(4), 901–908. https://doi.org/10.1093/biomet/84.4.901. http://biomet.oupjournals.org/cgi/doi/10.1093/biomet/84.4.901

Murphy, S. (1993). Testing for a time dependent coefficient in Cox's regression model. *Scandinavian Journal of Statistics, 20*(1), 35–50. http://www.jstor.org/stable/10.2307/4616258.

Sasieni, P. (1992). Non-orthogonal projections and their application to calculating the information in a partly linear Cox model. *Scandinavian Journal of Statistics, 19*, 215–233.

Schoenfeld, D. (1982). Partial residuals for the proportionnal hazards regression model. *Biometrika, 69*(1), 239–241. http://www.jstor.org/stable/2335876.

Schumaker, L. (2007), *Spline functions: Basic theory*. Cambridge: Cambridge University Press.

Shen, X., & Wong, W. H. (1994). Convergence rate of sieve estimates. *The Annals of Statistics, 22*, 580–615.

Stone, C. J. (1985). Additive regression and other nonparametric models. *The Annals of Statistics, 13*(2), 689–705.

Tian, L., Zucker, D., & Wei, L. J. (2005). On the cox model with time-varying regression coefficients. *Journal of the American Statistical Association, 100*(469), 172–183. https://doi.org/10.1198/016214504000000845.

Tibshirani, R. (1996). Regression shrinkage and selection via the lasso. *Journal of the Royal Statistical Society Series B: Statistical Methodology, 58*(1), 267–288. https://doi.org/10.1111/j.1467-9868.2011.00771.x, 11/73273.

Van Der Vaart, A. W., & Wellner, J. A. (1996). Weak convergence. In: *Weak convergence and empirical processes*. Berlin: Springer, pp. 16–28.

Wahba, G. (1980). Ill posed problems: Numerical and statistical methods for mildly, moderately and severely ill posed problems with noisy data. Tech. rep., Madison: Wisconsin Univ-Madison Department of Statistics.

Winnett, A., & Sasieni, P. (2003) Iterated residuals and time-varying covariate effects in Cox regression. *Journal of the Royal Statistical Society Series B: Statistical Methodology, 65*(2), 473–488. https://doi.org/10.1111/1467-9868.00397

Wright, S. J. (2015). Coordinate descent algorithms. *Mathematical Programming, 151*(1), 3–34.

Yan, J., & Huang, J. (2012). Model selection for cox models with time-varying coefficients. *Biometrics, 68*(2), 419–428. https://doi.org/10.1111/j.1541-0420.2011.01692.x, NIHMS150003.

Zheng, D., Ye, T., Hu, H., Zhang, Y., Sun, Y., Xiang, J., & Chen, H. (2018). Upfront surgery as first-line therapy in selected patients with stage IIIA non–small cell lung cancer. *The Journal of Thoracic and Cardiovascular Surgery, 155*(4), 1814–1822.

Zucker, D. M., & Karr, A. F. (1990). Nonparametric survival analysis with time-dependent covariate effects: A penalized partial likelihood approach. *The Annals of Statistics, 18*(1), 329–353. https://doi.org/10.1214/aos/1176347503.

Discovery of Gene-Specific Time Effects on Survival

Hyokyoung G. Hong

Abstract The success of variable screening of time-to-event data can be enhanced by identifying variables which influence the timing of mortality. However, most existing screening approaches based on a uniform screening criterion may fail to capture such subtle survival differences especially at the early or late event time due to the imposed assumptions on the survival function. By artificially inflating the difference in survival functions across strata defined by each candidate covariate, we propose a general class of screening criteria that can detect biomarkers with early or delayed impact on survival distributions, providing new perspectives for high-dimensional survival analysis.

Keywords Hybrid framework · Multiple myeloma · Non-parametric screening · Powered density · Time-to-event data

1 Introduction

The rapidly evolving collection of large-scale genetic/genomic and clinical data sets present opportunities for discovering important knowledge on patients' survival. In the era of precision medicine, identifying a set of biomarkers that can perform, as valuable prognostic tools for patients' survival, have been an important task for researchers (Colburn et al., 2001). Efficient and reliable variable screening methods that can preserve the necessary information of a high-dimensional feature space are key to predicting patient survival outcomes and classifying the population at risk with high accuracy. However, high-dimensional biomarkers challenge the existing statistical tools when analyzing time-to-event data with censoring.

When the outcome data are censored, it is unclear how the screening approaches proposed for noncensored outcome would work in terms of implementation, inter-

H. G. Hong (✉)
Biostatistics Branch, Division of Cancer Epidemiology and Genetics, National Cancer Institute, Bethesda, MD, USA
e-mail: grace.hong@nih.gov

Y. Zhao, D.-G. Chen (eds.), *Statistics in Precision Health*, ICSA Book Series in Statistics, https://doi.org/10.1007/978-3-031-50690-1_20

pretation and theoretical justifications. Compared to the great success of variable selection and feature screening approaches (Tibshirani, 1996; Fan & Lv, 2008), progress has been made on developing predictive models for time-to-event data in high-dimensional settings. Some related works include Fan et al. (2010), Tibshirani (2009), Zhao and Li (2012), and Gorst-Rasmussen and Scheike (2013). However, the reliability, validity, and usability of these methods often hinge upon rather restrictive assumptions. Model-free screening methods have recently emerged as a more robust and useful tool to avoid stringent conditions. Some works include a censored rank independence screening (Song et al., 2014), a quantile adaptive model-free screening (He et al., 2013), and a survival impact index screening (Li et al., 2016). These methods are typically robust against outliers in predictors and are applicable to a wide range of survival models.

A biomarker might have a heterogeneous effect on the time-to-event, and ignoring this characteristic would lead to larger numbers of false positives and negatives in discovering the important biomarkers. This information is crucial, and it would advance the current research on precision medicine a step further. Inspired by the complementary features of Cramér–von Mises and Kolmogorov criteria, Hong et al. (2020) proposed a model-free L_q-norm learning procedure, which includes the well-known Cramér–von Mises and Kolmogorov criteria as two special cases. Kolmogorov statistic measures the maximal distributional difference across the subdomain of each covariate, and thus it would be ideal to detect the short-term impacts of the covariate. On the other hand, the Cramér–von Mises statistic may be more powerful than the Kolmogorov statistic in detecting long term impacts of biomarkes on the survival outcome rather than short term impacts (Conover & Conover, 1980; Razali & Wah, 2011; Woodruff & Moore, 1988; Arnold & Emerson, 2011; Chiu & Liu, 2009). The work provided an integrative framework for detecting predictors with heterogeneous impact on survival outcome according to the varying time intervals, such as short term or long term, controlled by a single time-range parameter.

In this paper, we detail another class of nonparametric screening based on Hong et al. (2018a), termed power screening, to discover biomarkers with potential heterogeneous impact on early or delayed survival time, which could potentially greatly impact the precision medicine. In addition, we discuss how to construct the hybrid screening statistics when we are more concerned about controlling false positives or false negatives. This paper is organized as follows. Section 2 introduces notation for this paper and explains motivation. Section 3 presents the variable selection criteria and estimation methods. Section 4 investigates the statistical properties of the proposed screening through simulation. Section 5 offers concluding remarks.

2 Notation and Motivation

Let $(\Omega, \mathscr{F}, P)$ be the probability space that underlies all the random variables in this paper, where Ω is the sample space, $\mathscr{F}$ is the σ-algebra and P is the probability measure and assume that there are n independent subjects with p covariates with $p \gg n$. Let i and j indicate subjects and covariates, respectively. We denote $\mathbf{X}_i = (X_{i1}, \ldots, X_{ip})^{\mathrm{T}}$ as the covariate vector for subject i and X_{ij} as a covariate j for subject i. Let T_i be the survival time and C_i be the censoring time. We observe $Y_i = \min\{T_i, C_i\}$ and $\delta_i = I(T_i \leq C_i)$, where $I(\cdot)$ is an indicator function. We generally assume that T_i and C_i are independent given $\mathbf{X}_i$. We further assume that $(T_i, C_i, \mathbf{X}_i)$ are independently and identically distributed. Let $(T_i, X_{ij}, \mathbf{X}_i)$, $i = 1, \ldots, n$, be independent and identically distributed copies of $(T, X_j, \mathbf{X})$, where $\mathbf{X} = (X_1, \ldots, X_p)$. Denote by $S(\cdot)$ the marginal survival function of T and by $S(t \mid \mathbf{X})$ the conditional survival function of T given $\mathbf{X}$. We define the set of active covariates as

$$\mathscr{M} = \{j : S(t|\mathbf{X}) \text{ functionally depends on } X_j \text{ for some } t \in (0, \infty)\}.$$

In genetic and genomic studies, it is commonly assumed that only a small number of biomarkers are relevant to the disease-specific survival. That is, the cardinality of $\mathscr{M}$ is relatively small compared to p.

Our goal is to identify $\mathscr{M}$ given a large number of candidate variables, which can be continuous or categorical. We propose our method by first considering a categorical variable, X_j, with K_j categories such that $X_j \in \{1, \ldots, K_j\}$.

Existing non-parametric variable screening approaches in a high-dimensional survival setting have considered how each candidate biomarker influences overall survival functions. However, they have limitations in evaluating the impact of each gene on survival that occurs in the early or delayed stage of the time course. In addition, as the survival function converges to 1 and 0 in the beginning and end of the timeline, it is very challenging to gauge any difference in survival near time boundaries. By modifying the density function, $f(\cdot)$, we define the integrated powered density as

$$\int_0^t f^{\gamma}(s)ds, \tag{1}$$

where $\gamma > 0$ and $t \in (0, \infty)$. The proposed integrated powered density resembles the cumulative density function except that it does not necessarily approach 1 when $t \to \infty$. The added power index γ in (1) may inflate early (with $\gamma > 1$) or late differences (with $\gamma < 1$) and thus gives more flexibility in detecting the distributional differences. When $\gamma \neq 1$, Model (1) is closely related to the Renyi entropy (Cover & Thomas, 2012).

3 Proposed Method

We assume that each candidate variable is a categorical variable, i.e., $X_j \in \{1, \ldots, R_j\}$. When X_j is continuous we assume, without loss of generality, that the support of X_j is the real line $\mathbb{R}$ and slice X_j into R_j intervals by using the percentiles of the empirical distribution of X_j.

To measure the association between a covariate X_j and the survival time T, the following screening criteria is proposed to characterize the variability of the integrated powered density given γ across different categories of X_j. First, we compute the difference of integrated powered density between each pair of categories, $X_j = r_1$ and $X_j = r_2$, for $r_1, r_2 \in \{1, \ldots, R_j\}$. Secondly, we take the maximum of difference over all pairs of r_1, r_2:

$$\Phi_j^{(\gamma)} = \max_{r_1, r_2 \in \{1, \ldots, R_j\}} \sup_{t \in [0,\tau]} \left| \int_0^t f_{T|X_j}^{\gamma}(s|X_j = r_1) ds - \int_0^t f_{T|X_j}^{\gamma}(s|X_j = r_2) ds \right|, \tag{2}$$

where $f_{T|X_j}(s|X_j = r)$ denotes the conditional density function of T given $X_j = r$. Hereafter, we refer to the proposed power screening in (2) as POS. The proposed framework is general. When $\gamma = 1$, (2) is simply $\max_{k_1,k_2} \sup_t |F_{T|X_j}(t|X_j = r_1) - F_{T|X_j}(t|X_j = r_2)|$, which is the well-known Kolmogorov difference. However, the proposed POS is distinct from the Kolmogorov filter introduced by Mai and Zou (2015) in that the Kolmogorov filter slices the outcome variable, while POS slices the features.

We denote $0 \equiv t_0 < t_1 < \cdots < t_M$ as the ordered observed failure times, the marginal survival function of T at time t as $S_T(t)$, and the Kaplan-Meier estimation of $S_T(t)$ as $\hat{S}_T(t)$. We estimate the density of survival time by the following kernel density estimation:

$$\hat{f}_T(t) = -\int K\left(\frac{t-s}{h_n}\right) d\hat{S}_T(s) = \sum_{i=1}^{M} K\left(\frac{t-t_i}{h_n}\right) \left\{\hat{S}_T(t_{i-1}) - \hat{S}_T(t_i)\right\},$$

where $h_n > 0$ is the bandwidth and $K(\cdot)$ is a kernel function. The bandwidth h_n can be chosen based on plug-in or bootstrap methods. Finally, (2) can be estimated by

$$\widehat{\Phi}_j^{(\gamma)} = \max_{r_1, r_2} \sup_{t \in [0,\tau]} \left| \int_0^t \hat{f}_{T|X_j}^{\gamma}(s|X_j = r_1) ds - \int_0^t \hat{f}_{T|X_j}^{\gamma}(s|X_j = r_2) ds \right|. \tag{3}$$

The following criterion is used to select active variables

$$\widehat{\mathscr{M}_1} = \left\{ j : \widehat{\Phi}_j^{(\gamma)} > cn^{-\nu},\ j = 1, \ldots, p \right\},$$

where $c > 0$ and ν are constants for pre-determined thresholds.

When X_j is continuous, we use the following slicing scheme which allocates sufficient samples within each slice. Our preliminary analysis indicates that the choice of slices does little affect variable screening results, but an ensemble approach that combines the results from different slicing schemes can achieve significant improvement. Suppose that there are N different ways of slicing X_j. Each slicing scheme, $\Lambda_{ju}, u = 1, \ldots, N$, contains R_{ju} intervals. Specifically,

$$\Lambda_{ju} = \left\{ [\hat{q}_{ju(r-1)}, \hat{q}_{ju(r)}) : r = 1, \ldots, R_{ju} \text{ and } \bigcup_{r=1}^{R_{ju}} [\hat{q}_{ju(r-1)}, \hat{q}_{ju(r)}) = \mathbb{R} \right\},$$

where the slicing point $\hat{q}_{ju(r)}$ is the r/R_{ju}th percentile of the empirical distribution of X_j under partition Λ_{ju}. We then replace X_j by its discretized version $\tilde{X}_{ju}$ under Λ_{ju}, that is, $\tilde{X}_{ju} = r$ if $X_j \in [\hat{q}_{ju(r-1)}, \hat{q}_{ju(r)})$.

Let $\widehat{\Phi}^{(\gamma)}_{j,\Lambda_{ju}}$ be the screening statistic corresponding to the slicing scheme of Λ_{ju} for covariate X_j, where

$$\widehat{\Phi}^{(\gamma)}_{j,\Lambda_{ju}} = \max_{r_1, r_2} \sup_{t \in [0,\tau]} \left| \int_0^t \hat{f}^{\gamma}_{T|X_j}(s|\tilde{X}_{ju} = r_1) ds - \int_0^t \hat{f}^{\gamma}_{T|X_j}(s|\tilde{X}_{ju} = r_2) ds \right|.$$

The ensemble screening statistic $\widetilde{\Phi}^{(\gamma)}_j$ for continuous covariates is:

$$\widetilde{\Phi}^{(\gamma)}_j = \sum_{u=1}^{N} \widehat{\Phi}^{(\gamma)}_{j,\Lambda_{ju}},$$

leading to the following screening criterion

$$\widehat{\mathcal{M}_2} = \left\{ j : \widetilde{\Phi}^{(\gamma)}_j > cn^{-\nu},\ j = 1, \ldots, p \right\}, \tag{4}$$

where $c > 0$ and ν are constants for pre-determined thresholds.

4 Simulation Study Design

We investigate the finite sample performance of the following various screening methods for survival outcome: a principled Cox sure independence screening (PSIS) by Zhao and Li (2012); correlation screening, an improvised extension of the sure independence screening (Fan & Lv, 2008) to the censored outcome data with inverse probability weighting (CORS); censored rank independence screening (CRIS) proposed by Song et al. (2014); conditional screening for survival outcome (CS) by Hong et al. (2018b); the proposed power screening (POS) with power index γ, ranging from 0.7 to 1.3 with an increment of 0.1. Note that $\gamma = 1.0$ corresponds to the Kolmogorov statistic.

We consider the following two examples for our numerical experiments.

Example 2 The survival time was generated from the following Cox model,

$$\lambda(t \mid \mathbf{X}) = 2t(|X_1| + |X_2|),$$

where all the covariates $X_j, j = 1, \ldots, p$, were generated from an independent standard normal distribution. The censoring times C_i were generated from the following Cox model

$$\lambda_C(t \mid \mathbf{X}) = c \times \exp(\boldsymbol{\beta}^{\mathrm{T}}\mathbf{X}),$$

where $\boldsymbol{\beta} = (\mathbf{0.3}_2^{\mathrm{T}}, \mathbf{0}_{p-5}^{\mathrm{T}})^{\mathrm{T}}$, and c was chosen to give approximately 20% and 50% of censoring proportions. As the covariate X_j is continuous, we applied different number of slices $(3, \ldots, \lceil \log(n) \rceil)$ and reported the ensemble results (Table 1).

Example 1 The underlying random vector x^* was generated from a multivariate normal distribution with a mean vector of 0 and an exchangeable correlation structure with an equal correlation of 0.5. The observed covariate is defined as $X_j = I(x_j^* > 1)$, where I is an indicator function. The survival time was generated from an accelerated failure time model with a baseline hazard function, $\lambda_0(t) = 0.1(t-2)^2$, and $\boldsymbol{\beta} = (-\mathbf{0.5}_5^{\mathrm{T}}, \mathbf{0}_{p-5}^{\mathrm{T}})^{\mathrm{T}}$ (Table 2).

Table 1 Numerical results for Example 2 with $n = 500$, $p = 1000$ and censoring proportions (20% and 50%)

	CR = 20%			CR = 50%		
Methods	MMS (IQR)	TPR	PIT	MMS (IQR)	TPR	PIT
POS						
$\gamma = 0.7$	2.00 (0.00)	1.00	1.00	4.00 (13.00)	0.96	0.92
$\gamma = 0.8$	2.00 (0.00)	1.00	1.00	4.00 (12.00)	0.96	0.92
$\gamma = 0.9$	2.00 (0.00)	1.00	1.00	4.00 (13.00)	0.96	0.92
$\gamma = 1.0$	2.00 (0.00)	1.00	1.00	5.00 (13.00)	0.96	0.92
$\gamma = 1.1$	2.00 (0.00)	1.00	1.00	5.00 (15.00)	0.96	0.92
$\gamma = 1.2$	2.00 (0.00)	1.00	1.00	6.00 (16.25)	0.96	0.92
$\gamma = 1.3$	2.00 (0.00)	1.00	1.00	7.00 (16.00)	0.95	0.91
PSIS	697.50 (372.70)	0.11	0.01	463.50 (502.20)	0.28	0.07
CORS	830.50 (240.70)	0.01	0.00	833.50 (231.00)	0.01	0.00
CRIS	719.00 (356.40)	0.11	0.00	527.50 (471.70)	0.23	0.06
CS	356.00 (551.50)	0.59	0.18	169.00 (413.25)	0.68	0.36

MMS the minimum model size, *IQR* the interquartile range *TPR* the true positive rate, *PIT* the probability of including all active variables, *PSIS* a principled Cox sure independence screening by Zhao and Li (2012), *CORS* an improvised extension of the sure independence screening by Fan and Lv (2008) to the censored outcome data with inverse probability weighting, *CRIS* censored rank independence screening proposed by Song et al. (2014), *CS* conditional screening for survival outcome by Hong et al. (2018b), *POS* the proposed power screening with power index γ

Table 2 Numerical results for Example 1 with $n = 500,\ p = 1000$ and censoring proportions (20% and 50%)

	CR = 20%			CR = 50%		
Methods	MMS (IQR)	TPR	PIT	MMS (IQR)	TPR	PIT
POS						
$\gamma = 0.7$	8.00 (12.00)	0.99	0.96	14.50 (27.00)	0.98	0.89
$\gamma = 0.8$	8.00 (11.00)	0.99	0.96	13.50 (25.25)	0.98	0.91
$\gamma = 0.9$	7.00 (10.00)	0.99	0.96	13.00 (24.00)	0.98	0.91
$\gamma = 1.0$	7.00 (9.00)	0.99	0.96	13.00 (22.00)	0.98	0.91
$\gamma = 1.1$	7.00 (8.00)	0.99	0.96	11.00 (21.00)	0.98	0.92
$\gamma = 1.2$	7.00 (7.00)	0.99	0.96	12.00 (21.25)	0.98	0.92
$\gamma = 1.3$	7.00 (7.00)	0.99	0.96	12.00 (22.00)	0.98	0.92
PSIS	5.00 (1.00)	1.00	1.00	8.00 (1.00)	0.99	0.97
CORS	22.00 (44.00)	0.96	0.84	710.00 (334.40)	0.41	0.00
CRIS	1000.00 (0.00)	0.00	0.00	1000.00 (0.00)	0.00	0.00
CS	5.00 (0.00)	1.00	1.00	5.00 (2.00)	1.00	1.00

MMS the minimum model size, *IQR* the interquartile range, *TPR* the true positive rate, *PIT* the probability of including all active variables, *PSIS* a principled Cox sure independence screening by Zhao and Li (2012), *CORS* an improvised extension of the sure independence screening by Fan and Lv (2008) to the censored outcome data with inverse probability weighting, *CRIS* censored rank independence screening proposed by Song et al. (2014), *CS* conditional screening for survival outcome by Hong et al. (2018b), *POS* the proposed power screening with power index γ

Note that CS screening (Hong et al., 2018b) requires prior information on the conditioning variable. Thus, X_1 was arbitrarily chosen as the known informative feature for both examples. A total of 500 simulated datasets were analyzed per each scenario. For the comparison we consider the following metrics: the minimum model size (MMS) which measures the effectiveness of the screening tool by counting the number of covariates to ensure inclusion of all true covariates; the true positive rate (TPR) for the percentage of selected true covariates in the top $[n/\log n]$ variables according to the screening criteria; and the probability of including all active variables (PIT) among the top $[n/\log n]$ variables.

Example 2 is designed to have no marginal correlation between each of the active variables and the survival time. Although the proposed method identified all active variables, other competing methods, including CS which assumed one of the active variables was known, did not perform well.

In Example 1, PSIS, which is a parametric approach, worked best. While other non-parametric screening methods performed poorly with an extremely large MMS, the proposed POS still worked well. Moreover, we observed that the selected variables with different γ varied, implying that the proposed method is adaptive to assess and identify a covariate with early or delayed time-to-event.

The program code is available at: https://github.com/younghhk/software.

5 An Application of POS in Identifying Genetic Effects on Varying Survival Time Window

A better understanding of the biomarkers associated with time-to-event is essential to developing interventions that can seek to promote quality of life and increase life expectancy. In this section, we perform an exploratory analysis of multiple myeloma (MM) by applying the proposed POS to identify important genes associated with time-to-event. The expression profiles were obtained from a Gene Expression Omnibus (accession GSE24080) comprising of 554 newly diagnosed MM patients with 54,675 probes.

The outcome is progression-free survival. Each gene expression is standardized to have mean 0 and variance 1. We split the data randomly into training ($n_1 = 332$) and validation ($n_2 = 222$) sets. The median follow-time was 41.1 and 43.1 for training and validation sets, respectively. We performed POS on the training set with γ varying from 0.7 to 1.3 and selected top $[n_1/\log(n_1)] = 57$ genes. The number of overlapping genes between the differing timing parameter γ are reported in Table 3. Although there are common genes selected for different γ's, our results suggest that distinct genes might be associated with early or late survival.

Especially at $\gamma = 0.7$ and 1.3, based on the selected genes, we computed a risk score for each subject in the testing set based on Cox proportional hazards regression coefficients which were computed from the training set. We divided the patients into low- and high-risk groups. Figure 1a demonstrates the Kaplan-Meier curve illustrating more precise inference and a narrower survival difference at the late follow-up time periods when the risk prediction model was developed based on the genes selected via POS with $\gamma = 0.7$. On the other hand, Fig. 1b shows that the survival curves of the low- and high-risk groups obtained from the POS with $\gamma = 1.3$ were closer compared to POS with $\gamma = 0.7$ at the early time periods.

Overall, Fig. 1 indicates the potential advantage of POS in better identifying genes which impact mortality in different time windows by artificially inflating the survival difference.

Table 3 The number of overlapping genes in the MM data, selected by POS with varying γ

	$\gamma = 0.7$	$\gamma = 0.8$	$\gamma = 0.9$	$\gamma = 1.0$	$\gamma = 1.1$	$\gamma = 1.2$	$\gamma = 1.3$
$\gamma = 0.7$	57						
$\gamma = 0.8$	53	57					
$\gamma = 0.9$	49	53	57				
$\gamma = 1.0$	43	47	51	57			
$\gamma = 1.1$	39	43	47	53	57		
$\gamma = 1.2$	35	39	43	49	53	57	
$\gamma = 1.3$	32	36	40	46	49	53	57

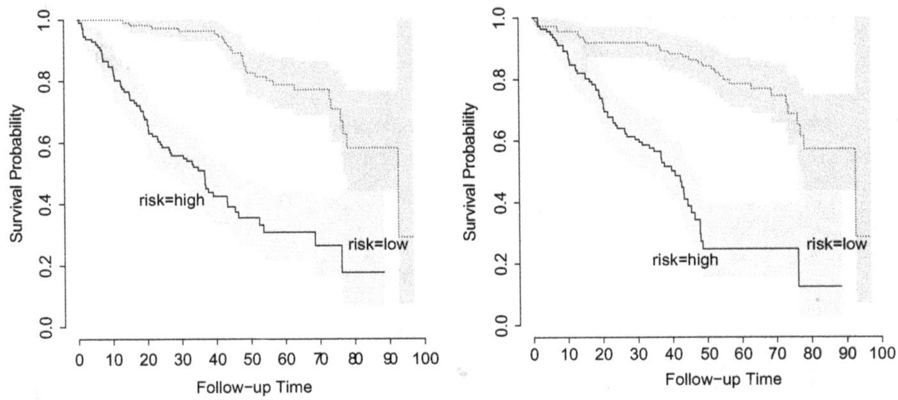

(a) POS with $\gamma = 0.7$ (b) POS with $\gamma = 1.3$

Fig. 1 Kaplan-Meier curves for low- and high-risk groups. (**a**) POS with $\gamma = 0.7$. (**b**) POS with $\gamma = 1.3$

6 Discussion

The introduced power screening intends to provide flexibility in existing one-size-fits-all screening approaches. This new class of model-free screening approaches with the embedded parameter amplifies the difference in survival along the time course. This feature is distinct from other existing non-parametric variable screening procedures in high-dimensional survival data as our procedure produces a series of different variable selection results depending on the embedded parameter.

Simulation studies indicate that a choice of the optimal γ depends on a simulation configuration. For a specific dataset, it is unclear which γ should be used. Instead of making an effort to choose the optimal parameter which is unverifiable, we direct our interest on how to utilize the different feature selection results. One of the important yet unresolved problems in feature selection in high-dimensional data is how to combine results from different screening procedures (Liu et al., 2015). Our issue here lies along the same line.

In particular, by retaining all covariates or biomarkers chosen by any of the considered screening procedures, we can reduce the false negative, which would be ideal for knowledge discovery in the exploratory phase. On the other hand, keeping only the common covariates chosen by different procedures would control false positives.

We provide a brief sketch of three ways to construct hybrid screening statistics that can further summarize the proposed power screening whose results depend on γ, denoted by $\widetilde{\mathcal{M}}_{H_1}$, $\widetilde{\mathcal{M}}_{H_2}$, and $\widetilde{\mathcal{M}}_{H_3}$. First, $\widetilde{\mathcal{M}}_{H_1} = \bigcup_{l=1}^{L} \widetilde{\mathcal{M}}^{(\gamma_l)}$, where $\widetilde{\mathcal{M}}^{(\gamma_l)} = \left\{ j : \widetilde{\Psi}_j^{(\gamma_l)} > cn^{-\nu_l},\ j = 1, \ldots, p \right\}$ and ν_l is a positive constant that depends on γ_l. We recommend this model when the research priority is on controlling the false negatives. Second, $\widetilde{\mathcal{M}}_{H_2} = \bigcap_{l=1}^{L} \widetilde{\mathcal{M}}^{(\gamma_l)}$. Here, $\widetilde{\mathcal{M}}_{H_2}$ includes common covariates that are selected by all γ_l $(l = 1, \ldots, L)$. This method can guarantee exclusion of unimportant covariates to the greatest extent, but this rather restrictive criterion may lead to many false negatives. Third, via normalization by rescaling between 0 and 1, we can compromise these two extreme forms of the hybrid approaches, i.e., $\widetilde{\mathcal{M}}_{H_1}$ and $\widetilde{\mathcal{M}}_{H_2}$, as following: $\widetilde{\mathcal{M}}_{H_3} = \left\{ j : \widetilde{\Psi}_j^{(\gamma_l)*} > cn^{-v},\ j = 1, \ldots, p \right\}$, where

$$\widetilde{\Psi}_j^{(\gamma_l)*} = \frac{\widetilde{\Psi}_j^{(\gamma_l)} - \min_{1 \le l \le L} \widetilde{\Psi}_j^{(\gamma_l)}}{\max_{1 \le l \le L} \widetilde{\Psi}_j^{(\gamma_l)} - \min_{1 \le l \le L} \widetilde{\Psi}_j^{(\gamma_l)}}.$$

This hybrid framework may help address different needs and enhance the benefits of the integrated framework.

References

Arnold, T. B., & Emerson, J. W. (2011). Nonparametric goodness-of-fit tests for discrete null distributions. *R Journal, 3*(2), 34–39.

Chiu, S. N., & Liu, K. I. (2009). Generalized Cramér–von Mises goodness-of-fit tests for multivariate distributions. *Computational Statistics & Data Analysis, 53*(11), 3817–3834.

Colburn, W., DeGruttola, V. G., DeMets, D. L., Downing, G. J., Hoth, D. F., Oates, J. A., Peck, C. C., Schooley, R. T., Spilker, B. A., & Woodcock, J. (2001). Biomarkers and surrogate endpoints: Preferred definitions and conceptual framework. Biomarkers definitions working group. *Clinical Pharmacol & Therapeutics, 69*, 89–95.

Conover, W. J., & Conover, W. J. (1980). *Practical Nonparametric Statistics*. Wiley: New York.

Cover, T. M., & Thomas, J. A. (2012). *Elements of Information Theory*. Wiley.

Fan, J., & Lv, J. (2008). Sure independence screening for ultrahigh dimensional feature space (with discussion). *Journal of Royal Statistical Society B, 70*(5), 849–911.

Fan, J., Feng, F., & Wu, Y. (2010). High-dimensional variable selection for Cox's proportional hazards model. In *IMS Collections Borrowing Strength: Theory Powering Applications- A Festschrift for Lawrence D. Brown*, (pp. 70–86). Institute of Mathematical Statistics.

Gorst-Rasmussen, A., & Scheike, T. (2013). Independent screening for single-index hazard rate models with ultrahigh dimensional features. *Journal of the Royal Statistical Society B, 75*(2), 217–245.

He, X., Wang, L., & Hong, H. G. (2013). Quantile-adaptive model-free variable screening for high-dimensional heterogeneous data. *The Annals of Statistics, 41*(1), 342–369.

Hong, H. G., Chen, X., Christiani, D. C., & Li, Y. (2018a). Integrated powered density: Screening ultrahigh dimensional covariates with survival outcomes. *Biometrics, 74*(2), 421–429.

Hong, H. G., Kang, J., & Li, Y. (2018b). Conditional screening for ultra-high dimensional covariates with survival outcomes. *Lifetime Data Analysis, 24*(1), 45–71.

Hong, H., Chen, X., Kang, J., & Li, Y. (2020). The l_q-norm learning for ultrahigh-dimensional survival data: an integrative framework. *Statistica Sinica, 30*(3), 1213–1233.

Li, J., Zheng, Q., Peng, L., & Huang, Z. (2016). Survival impact index and ultrahigh-dimensional model-free screening with survival outcomes. *Biometrics, 72*(4), 1145–1154.

Liu, J., Zhong, W., & Li, R. (2015). A selective overview of feature screening for ultrahigh-dimensional data. *Science China Mathematics, 58*(10), 2033–2054.

Mai, Q., & Zou, H. (2015). The fused Kolmogorov filter: A nonparametric model-free screening method. *The Annals of Statistics, 43*(4), 1471–1497.

Razali, N. M., & Wah, Y. B. (2011). Power comparisons of Shapiro–Wilk, Kolmogorov–Smirnov, Lilliefors and Anderson–Darling tests. *Journal of Statistical Modeling and Analytics, 2*(1), 21–33.

Song, R., Lu, W., Ma, S., & Jeng, X. J. (2014). Censored rank independence screening for high-dimensional survival data. *Biometrika, 101*(4), 799–814.

Tibshirani, R. (1996). Regression shrinkage and selection via the lasso. *Journal of the Royal Statistical Society. Series B (Methodological), 58*, 267–288.

Tibshirani, R. J. (2009). Univariate shrinkage in the Cox model for high dimensional data. *Statistical Applications in Genetics and Molecular Biology, 8*(1), 3498–3528.

Woodruff, B. W., & Moore, A. H. (1988). 7 application of goodness-of-fit tests in reliability. In *Handbook of Statistics* (Vol. 7, pp. 113–120). Elsevier

Zhao, S. D., & Li, Y. (2012). Principled sure independence screening for Cox models with ultra-high-dimensional covariates. *Journal of Multivariate Analysis, 105*(1), 397–411.

Modeling and Optimizing Dynamic Treatment Regimens in Continuous Time

Yanxun Xu and Zhiyue Zhang

Abstract Modeling and optimizing sequential clinical actions (dynamic treatment regimes, DTRs) in continuous time are crucial in many biomedical applications, especially in patient care and treatment for chronic diseases. Traditional statistical and machine learning methods usually focus on estimating a sequence of optimal treatments at given stages of intervention. However, optimizing clinical actions in continuous time has rarely been studied. In this chapter, we develop a model-based decision framework to optimize the personalized clinical decisions in continuous time including scheduling follow-up visits and assigning an optimal treatment at each visit. Specifically, the proposed framework has two steps. The first step is a Bayesian method that jointly models follow-up visits and treatments using a marked temporal point process (MTPP), which is then linked to clinical measurements through parameter sharing. The second step is to optimize the sequential clinical decisions while accounting for uncertainties in the clinical measurements using policy optimization methods. We demonstrate the usefulness of the proposed framework by applying it to a dataset from electronic medical records of patients after kidney transplantation, yielding interpretable and clinically useful results.

Keywords Dynamic treatment regimen · Kidney transplantation · Marked temporal point process · Reinforcement learning · Survival

1 Introduction

In a wide variety of biomedical applications, patient care and treatment often consist of a sequence of decision making in continuous time. For example, patients with chronic health conditions such as diabetes, HIV infection, cancer, and organ transplantation, are required to visit the clinic frequently for monitoring

Y. Xu (✉) · Z. Zhang
Department of Applied Mathematics and Statistics, Johns Hopkins University, Baltimore, MD, USA
e-mail: yanxun.xu@jhu.edu; zzhan179@jhu.edu

Y. Zhao, D.-G. Chen (eds.), *Statistics in Precision Health*, ICSA Book Series in Statistics, https://doi.org/10.1007/978-3-031-50690-1_21

their diseases. At each clinic visit, physicians collect the patient's health-related measurements, and then make decisions about what to do next (e.g., assigning the treatment and scheduling the next visit time) to optimize patient outcomes given a patient's individual characteristics and treatment history.

One typical application is caring for patients after kidney transplantation, the most common type of organ transplantation and the primary therapy for patients with end-stage kidney disease (Arshad et al., 2019). To prevent graft rejection, patients are usually hospitalized for a few days initially to monitor signs of complications, then required to have frequent checkups at an outpatient center after being released. At each visit, they are administered immunosuppressive drugs, such as tacrolimus, to keep their immune systems from attacking and rejecting the new kidney (Kasiske et al., 2010). Serious adverse events have been reported to be associated with tacrolimus dosage that is either very high or too low. Higher tacrolimus levels are associated with adverse effects such as neurotoxicity, nephrotoxicity, and cancers (Naesens et al., 2009). Lower tacrolimus levels are associated with an increased likelihood of graft rejection (Staatz et al., 2001). Besides assigning the right dosage of tacrolimus, another crucial medical decision is to determine the follow-up visits timing for patients after transplantation. While follow-up visit frequency varies from 0–12 months (Israni et al., 2014), patients with stable kidney function usually have less frequent follow-ups compared to non-stable patients. Another example is people living with HIV, who are recommended to follow up with physicians regularly by the current HIV treatment guidelines. At each visit, their sociodemographics, medication use, and laboratory test results are collected. Then physicians prescribe antiretroviral drugs for them to take daily until the next visit and schedule the next visit time. For some chronic conditions, there exist treatment guidelines to recommend follow-up frequency (e.g., semiannually for HIV infection). However, for certain diseases, there is no consensus or any clear treatment guideline. Therefore, the follow-up schedule can vary from person to person, and may need to be adaptive with the uniquely evolving disease progression of each patient. Consequently, optimizing personalized follow-up schedule and prescribing the right treatment tailored to each patient at each visit (i.e., precision medicine) are critical and can have a significant impact on patient outcomes such as survival and life of quality.

The gold standard of optimizing sequential treatments for discrete stages is to conduct randomized clinical trials and estimate the effect of different treatment strategies, with e.g. sequential multiple assignment randomized trial (SMART) designs (Lavori & Dawson, 2004; Murphy, 2005). However, when clinical decisions are made in continuous time, designing such a randomized trial becomes unfeasible. In contrast, modeling and optimizing sequential treatments from observational data such as electronic health/medical records, have drawn growing interest in both statistics and machine learning. For the kidney transplantation example mentioned above, kidney transplantation databases such as French computerized and validated data in transplantation (DIVAT), provide us opportunities and challenges to determine personalized optimal follow-up schedules and tacrolimus dosages. DIVAT data are collected from the date of transplantation until graft failure, defined as

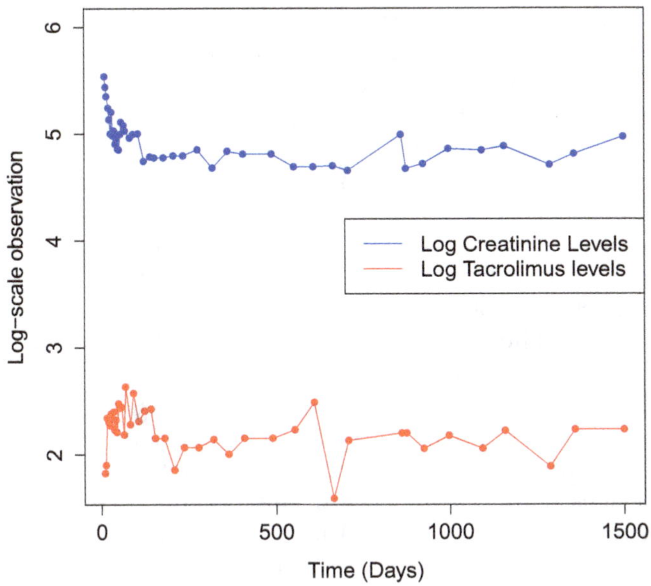

Fig. 1 Example data for a single patient's creatinine and tacrolimus blood levels across time. The points represent the actual visit times

either returning to dialysis or death with a functioning graft. At each scheduled follow-up visit, patients' creatinine levels, an important biomarker for measuring kidney function, and tacrolimus levels are collected longitudinally to investigate therapeutic strategies and determine the next follow-up time by physicians. For example, Fig. 1 presents one randomly selected patient's longitudinal creatinine levels, tacrolimus levels versus his/her follow-up visits from DIVAT. In the first several visits after kidney transplantation, this patient's creatinine levels were high, indicating the kidney was not working well, therefore, the clinician scheduled a high frequency of follow-up visits and prescribed high levels of tacrolimus. As time went by, this patient's kidney function became stable, as represented by slowly decreasing creatinine levels, then the prescribed tacrolimus levels were also decreasing gradually, accompanied by a lower frequency of visits.

Optimizing a sequence of treatments from observational data has been widely studied in the statistical literature on dynamic treatment regimens (DTR) (Robins, 1986; Murphy, 2003; Chakraborty, 2013; Zhao et al., 2015; Xu et al., 2016; Luckett et al., 2019). Most DTR methods assume that treatments are given either at specific stages (e.g., a salvage treatment is given when one patient becomes resistant to the frontline chemotherapy), or the treatment schedule is known a priori (e.g., scheduled follow-up visits every three months). However, optimizing sequential treatments in continuous time (i.e., determining the optimal timing for clinic visits) remains challenging. The time interval between two visits may carry significant information about the disease progression. For instance, patients with more stable conditions should have less frequent clinic visits compared to those who are less stable for reduction of costs and burden when intensive visits are unnecessary. To date, few studies have investigated optimizing visit times. For instance, Guan et al. (2019) developed a Bayesian nonparametric method to recommend a recall time for each patient with periodontal disease at each visit depending on their clinical characteristics. However, they only considered few fixed choices for recall times (e.g., 3 months, 6 months, or 9 months), making the proposed method inapplicable to a more general setup. For example, in the kidney transplantation case, the next follow-up time can be anytime between 0–12 months in general (Israni et al., 2014).

Attempts on sequential decision optimization in continuous time have also been made in the field of reinforcement learning (RL). Upadhyay et al. (2018) considered marked temporal point processes (MTPPs) (Aalen et al., 2008), which have become increasingly popular for modeling event data in continuous time (Tabibian et al., 2019), and characterized both the decisions made by an agent and the environment feedback with deep MTPPs that do not assume explicit functional forms for intensity functions or mark distributions. Following the development of neural ordinary differential equations (neural ODEs) (Chen et al., 2018) which serves as a powerful method to model continuous-time events, Du et al. (2020) and Yildiz et al. (2021) proposed to use neural ODEs to model continuous-time dynamics and learn optimal policies with model-based RL approaches. However, these recent deep RL methods all reside in the online RL domain where interactions with the environment are required and therefore cannot learn optimal policies solely from previously collected data (i.e., observational data), which is typical in the medical setting since interacting with patients is either infeasible or unethical (Levine et al., 2020; Tang & Wiens, 2021). Further, while these deep RL methods enjoy the model flexibility of deep networks, they lack model interpretability that is crucial to medical agents (e.g., physicians).

In this chapter, we develop a two-step model-based decision framework for modeling and optimizing dynamic treatment regimens in continuous time. In the first step, we propose a Bayesian method that jointly models the clinical measurements and clinical actions, including the follow-up visit schedule and treatment at each visit. Specifically, an MTPP is employed to jointly model follow-up schedules with the time of each visit as the occurrence of an event and the treatment at each follow-up visit as the associated mark. Then, this MTPP is linked to a model of clinical measurements through parameter sharing. We choose to use a Bayesian approach as it can naturally quantify parameter uncertainty through posterior inference. Such a joint model can also be used to learn how doctors treat patients in the observational data. In the second step, a policy optimization approach is proposed to optimize clinical decisions with the goal of maximizing patients' health outcomes while taking into account parameter uncertainties from the Bayesian joint model in the first step. In summary, the proposed framework determines personalized optimal clinical decisions in continuous time using a joint modeling framework of clinical measurements and clinical decisions with a policy optimization procedure.

The rest of the chapter proceeds as follows. In Sect. 2, we present the proposed two-step decision framework. In Sect. 3, our approach is demonstrated using the motivating kidney transplantation example by introducing a specific model under the proposed framework with the goal of maximizing patients' survival. We conduct simulation studies in Sect. 4 and apply the proposed method to the kidney transplantation dataset in Sect. 5. In Sect. 6, we conclude the chapter with a discussion.

2 Model-Based Decision Framework

In this section, we first introduce the notation, and then outline the proposed two-step model-based decision framework.

For each patient i, let $\boldsymbol{x}_i$ denote the patient's baseline covariates for $i = 1, \ldots, I$. At each visit time $t_{i,j}$ for patient i at the j-th follow-up visit, denote $y_{i,j}$ to be the longitudinal clinical measurement of interest: in our kidney transplantation application, it is the logarithm of the creatinine level (μmol/l), $j = 0, \ldots, J_i$. Then the physician would prescribe the dosage $d_{i,j}$, and schedule the next visit time $t_{i,j+1}$ based on the patient's baseline covariates $\boldsymbol{x}_i$ and the longitudinal creatinine measurements $\overline{y_{i,j}} = \{y_{i,j'} : j' \leq j\}$. In the kidney transplantation application, $d_{i,j}$ is the tacrolimus level (ng/ml). We represent the i-th patient's sequence of visits and assigned dosages up to time T_i by $\boldsymbol{e}_{i,T_i} = \{(t_{i,0}, d_{i,0}), \ldots, (t_{i,J_i}, d_{i,J_i})\}$. Denote $\boldsymbol{y}_i = (y_{i,0}, \ldots, y_{i,J_i})$. Our data can be summarized as $\mathscr{D} = \{\boldsymbol{x}_i, \boldsymbol{y}_i, \boldsymbol{e}_{i,T_i}\}_{i=1}^{I}$. Throughout this chapter, we use $p(\cdot)$ to denote the probability density. In order to identify causal effects, we make the standard assumptions of sequential ignorability and consistency (Murphy et al., 2003). The positivity assumption (i.e., the probability of receiving each treatment regimen of interest for each patient is greater than zero) is satisfied in our model due to the use of the marked temporal point process (MTPP) for modelling stochastic clinical decisions in continuous time.

2.1 *First Step*

In the first step, we propose a Bayesian joint model for observed clinical measurements and decisions: $p(\boldsymbol{y}_i, \boldsymbol{e}_{i,T_i} \mid \boldsymbol{x}_i), i = 1, \ldots, I$. We first introduce a marked temporal point process (MTPP) to model clinical decisions $\boldsymbol{e}_{i,T_i}$ including the follow-up schedule and dosages.

Modeling event data with marker information is important to learn the latent mechanisms that govern the observed stochastic event patterns over time in many domains, such as social science (Butts & Marcum, 2017), medical analytics (Liu et al., 2018), and finance (Hawkes, 2018). For example, when analyzing electronic health records of a patient in intensive care units, an event can be a new lab test or a procedure started, and the measurements associated with the event are marks

(Islam et al., 2017). Marked temporal point processes (Aalen et al., 2008) are a general framework for modeling such event data. Formally, a marked temporal point process is a random process, the realization of which consists of a sequence of events localized in time, i.e., $\mathscr{H} = \{(t_0, d_0), (t_1, d_1), \ldots, (t_n, d_n)\}$ with the occurrence time of event i being $t_i \in \mathbb{R}^+$ and $d_i \in \mathbb{R}^+$ being the associated mark. In the motivating application, t_i represents the time when a patient visits an outpatient center and d_i represents the tacrolimus dosage assigned by the physician. The first event is defined as the day of transplantation at $t_0 = 0$ with an initial dose d_0.

Denoting the event history up to time t to be $\mathscr{H}_t = \{(t_i, d_i) \in \mathscr{H} \mid t_i < t\}$, under MTPP the instantaneous rate of the events is characterized by the conditional intensity function $\lambda(t)$, namely

$$\lambda(t) = \lim_{dt \to 0} \frac{Pr\{\text{event happens in } [t, t+dt) \mid \mathscr{H}_t\}}{dt}.$$

Common forms of the conditional intensity function $\lambda(t)$ include Poisson process (Zhu & Li, 2018), Gamma process (Shibue & Komaki, 2020), Hawkes process (Hawkes, 1971) and its neural extension (Mei & Eisner, 2017). How to model the conditional intensity function $\lambda(t)$ depends on specific applications. For example, we can model the conditional intensity function for the motivating kidney example by incorporating known clinical knowledge that the patient's baseline covariates and longitudinal clinical measurements can affect the follow-up schedule: $\lambda_i(t_{i,j} \mid \overline{y_{i,j-1}}, \boldsymbol{x}_i)$. When a patient has higher creatinine levels, indicating unstable kidney function, more frequent visits need to be scheduled compared to patients with stable kidney function. Next, we model the dosage $d_{i,j}$ at the j-th visit of patient i as the "mark" of this visitation event by $p(d_{i,j} \mid \boldsymbol{x}_i, \overline{y_{i,j}}, \overline{t_{i,j}}, \overline{d_{i,j-1}})$, where $\overline{t_{i,j}} = \{t_{i,j'} : j' \le j\}$ and $\overline{d_{i,j}} = \{d_{i,j'} : j' \le j\}$.

Thus, the probability density of the i-th patient's sequence of clinic visits and assigned dosages $\boldsymbol{e}_{i,T_i} = \{(t_{i,0}, d_{i,0}), \ldots, (t_{i,J_i}, d_{i,J_i})\}$ up to time T_i can be written as

$$p(\boldsymbol{e}_{i,T_i} \mid \boldsymbol{y}_i, \boldsymbol{x}_i) = \underbrace{\exp\Big(-\int_0^{T_i} \lambda_i(t \mid \boldsymbol{y}_i, \boldsymbol{x}_i)dt\Big)}_{\text{Prob. of no visits at } t\in[0,T_i]\setminus\{t_{i,j}\}_{j=1}^{J_i}} \prod_{j=0}^{J_i} \underbrace{p(d_{i,j} \mid \boldsymbol{x}_i, \overline{y_{i,j}}, \overline{t_{i,j}}, \overline{d_{i,j-1}})}_{\text{Prob. of dosage}}$$
$$\times \prod_{j=1}^{J_i} \underbrace{\lambda_i(t_{i,j} \mid \overline{y_{i,j-1}}, \boldsymbol{x}_i)}_{\text{Prob. of a visit at } t_{i,j}}, \tag{1}$$

where $\boldsymbol{y}_i = (y_{i,0}, \ldots, y_{i,J_i})$.

To complete the joint model for observed clinical measurements and decisions $p(\boldsymbol{y}_i, \boldsymbol{e}_{i,T_i} \mid \boldsymbol{x}_i)$, we then link the MTPP model of clinical decisions to the model of the longitudinal measurements by carefully designing parameter sharing in order to capture the mutual influence between clinical observations and decisions. For example, we can use a linear dynamic model for longitudinal clinical measurements: $p(y_{i,j} \mid \boldsymbol{x}_i, \overline{y_{i,j-1}}, \overline{t_{i,j}}, \overline{d_{i,j-1}})$. Therefore, the joint model of the observed clinical measurements and decisions can be written as

$$\prod_{i=1}^{I} p(\boldsymbol{y}_i, \boldsymbol{e}_{i,T_i} \mid \boldsymbol{x}_i) = \quad p(\boldsymbol{e}_{i,T_i} \mid \boldsymbol{y}_i, \boldsymbol{x}_i) \prod_{i=1}^{I} \prod_{j} \left[p(y_{i,j} \mid \boldsymbol{x}_i, \overline{y_{i,j-1}}, \overline{t_{i,j}}, \overline{d_{i,j-1}}) \right]. \tag{2}$$

The detailed parametrization for each part of (2) can be application-specific. We will describe the details of the chosen parametrization for the motivating kidney transplantation application in Sect. 3.

Assume that the joint model (2) is parameterized with $\boldsymbol{\theta}$ and $\boldsymbol{\phi}$, where $\boldsymbol{\theta}$ denotes the set of parameters related to clinical decisions that control patients' follow-up schedules and dosages, and $\boldsymbol{\phi}$ denotes the remaining parameters. We complete the Bayesian joint model by assigning priors to $\boldsymbol{\theta}$ and $\boldsymbol{\phi}$. Then posterior inference can be obtained through Markov chain Monte Carlo (MCMC) simulations. Posterior inference of $\boldsymbol{\theta}$ will help us learn how physicians treat patients in the observational data, and posterior inference of $\boldsymbol{\phi}$ will facilitate our understanding of how the patient's clinical measurement evolves over time.

2.2 Second Step

Our goal is to optimize personalized clinical decisions including scheduling the patient's follow-up visits and prescribing dosages to maximize the patient's health outcome, e.g., the graft survival time in the kidney transplantation application. That means, we aim to learn the parameters $\boldsymbol{\theta}$ of the clinical decision model such that, for any (future/hypothetical) patient i, a certain reward R_i (e.g., graft survival time) can be maximized by prescribing the right amount of dosages and scheduling the visits at the right times. Formally, we aim to maximize the expected reward for any patient i with baseline covariates $\boldsymbol{x}_i$:

$$G_i(\boldsymbol{\theta}) = \int E_{(\boldsymbol{y}_i, \boldsymbol{e}_{i,T_i}) \sim p(\boldsymbol{y}_i, \boldsymbol{e}_{i,T_i} \mid \boldsymbol{\theta}, \boldsymbol{\phi}, \boldsymbol{x}_i)}[R_i] p(\boldsymbol{\phi} \mid \mathscr{D}) d\boldsymbol{\phi}. \tag{3}$$

Here the expectation is taken over all possible stochastic realizations of the clinical measurements $\boldsymbol{y}_i$ and the clinical decisions $\boldsymbol{e}_{i,T_i}$ from their joint distribution $p(\boldsymbol{y}_i, \boldsymbol{e}_{i,T_i} \mid \boldsymbol{\theta}, \boldsymbol{\phi}, \boldsymbol{x}_i)$; $\mathscr{D}$ denotes the observed data; and $p(\boldsymbol{\phi} \mid \mathscr{D})$ is the posterior distribution of $\boldsymbol{\phi}$ obtained from posterior inference of the Bayesian joint model (2). Note that the reward is stochastic, because the clinical measurements and decisions (i.e., dosages and visitation times) and their influence on the reward are all stochastic.

To find the optimal clinical decision parameter $\widetilde{\boldsymbol{\theta}}_i$ that maximizes the expected reward $G_i(\boldsymbol{\theta})$ after integrating out the uncertainty in the clinical measurements, we can employ policy optimization methods such as policy gradient method (Ruder, 2016) to optimize personalized clinical decisions. The uncertainties in the clinical measurements will be incorporated by integrating over the posterior distribution of $p(\boldsymbol{\phi} \mid \mathscr{D})$ when calculating the reward (3). The estimated posterior mean of $\boldsymbol{\theta}$ from the first step can be used as an initial value in the optimization procedure for efficient learning. Such a strategy is also related to imitation learning in RL (Hussein et al., 2017).

3 A Demonstration Application to Kidney Transplantation

For patients with kidney transplantation, graft survival time, defined as the time between the transplantation and the first graft failure, is a major clinical event of interest. Our goal is to optimize personalized clinical decisions including scheduling a patient's follow-up visitations and prescribing dosages to maximize the patient's graft survival time. In this section, we demonstrate the application of the proposed model-based decision framework to the motivating kidney transplantation data.

3.1 First Step: A Bayesian Joint Model

In this subsection, we describe the proposed Bayesian joint model for both clinical decisions and measurements. We first propose a marked temporal point process (MTPP) to model follow-up schedules, which can be seen as stochastic discrete events in continuous time, and tacrolimus dosages at follow-up visits as the corresponding marks. Then we develop a Bayesian joint model for survival event and longitudinal creatinine data, which will be linked to the MTPP model through sharing certain parameters.

Modeling event data with marker information using MTPP can be characterized by a conditional intensity function. To design a flexible conditional intensity function that takes into account patients' kidney condition measured by creatinine levels and captures patients' heterogeneity in visit times, we first plot the empirical intensity of the amount of time between visits for different ranges of creatinine levels across all visits from all patients in the motivating DIVAT dataset in Fig. 2a, indicating that patients with higher creatinine levels have shorter duration between follow-up visits. Furthermore, the intensity functions are highly skewed with higher peak values for patients with higher creatinine levels. However, the time at which the peak occurs is similar across different creatinine levels. Since the empirical intensity function is not monotonic, commonly-used Poisson processes that assume a constant intensity or Gamma processes that assume a monotone intensity will

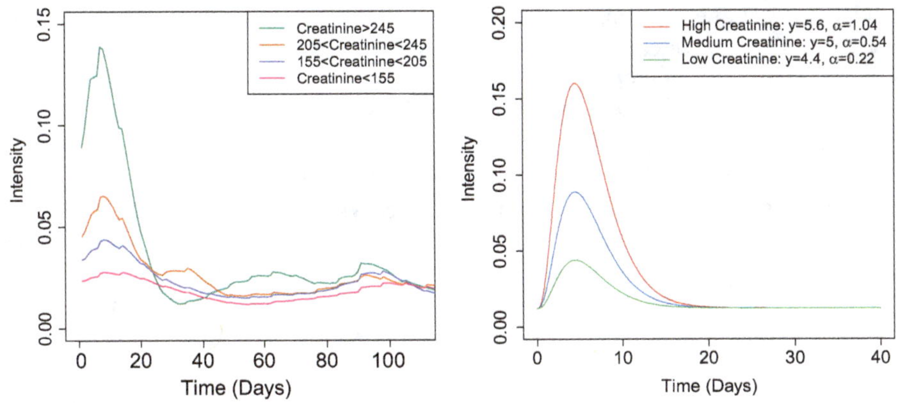

(a) (b)

Fig. 2 Panel (**a**) shows the empirical intensity plot for the amount of time (in days) between follow-up visitations. Panel (**b**) plots an example of how creatinine levels and model parameters affect the visitation intensity, where $\xi = 2$, $\boldsymbol{\beta}_\alpha = (10, -1.8)^T$, $\mu = -4.4$, $\nu_1 = 1.5$, and $\nu_2 = 1$

not apply. The Hawkes process assumes that the past events always elevate the intensities of future events and this "self-exciting" effect is additive—it is apparently not the dynamics that the visits in our data actually follow. Its neural extensions (Du et al., 2016; Mei & Eisner, 2017) are flexible enough to fit complex data but unable to explicitly incorporate human expert knowledge. This inspires us to model the conditional intensity function with a Gamma density function as follows:

$$\lambda_i(t) = \underbrace{\exp(\mu)}_{\text{Baseline Intensity}} + \underbrace{\alpha_{i,j}\,(t - t_{i,j})^{\kappa-1} e^{-\gamma(t-t_{i,j})} \frac{\gamma^\kappa}{\Gamma(\kappa)}}_{\text{Gamma density}} \text{ for } t \in (t_{i,j}, t_{i,j+1}], \tag{4}$$

where $\alpha_{i,j} > 0, \gamma > 0, \kappa \geq 1$. The parameter $\alpha_{i,j}$ is patient-specific. To ensure the existence of a peak, we set $\kappa = \exp(\beta_{v2}) + 1 \geq 1$: otherwise, the Gamma density part of (4) would be monotonically decreasing. The intensity function would eventually decrease to the "baseline level" $\exp(\mu)$ unless the next visitation happens and sets up a new intensity curve. For easy interpretation, we parameterize γ as $\gamma = \exp(\nu_2 - \nu_1)$ such that the "peak time" (i.e., when the peak of the intensity function occurs) can be easily computed as $\frac{\kappa-1}{\gamma} = \exp(\nu_1)$. Moreover, since the intensity often depends on the clinical measurement (e.g., a higher creatinine level implies a higher intensity), we model $\alpha_{i,j}$, which controls the peak intensity for patient i between time $t_{i,j}$ and $t_{i,j+1}$, by

$$\alpha_{i,j} = \frac{\xi}{1 + \exp((1, y_{i,j})\boldsymbol{\beta}_\alpha)}.$$

Under this parametrization, $\alpha_{i,j}$ ranges over $(0, \xi)$ and depends on the creatinine level $y_{i,j}$. Our design can be easily extended to incorporate other measurements by simply augmenting them to the vector $(1, y_{i,j})$. Figure 2b shows how the visitation intensity under our model is affected by the most recent creatinine level $y_{i,j}$ (and thus the magnitude parameter $\alpha_{i,j}$) given a specific set of parameter values.

We model the tacrolimus dosage (in log scale) $d_{i,j}$ at the j-th visit of patient i as the marks associated with each event (scheduled visit). As the dosage assigned by the physician usually depends on the patient's baseline covariates and the creatinine level at the current visit, we assume that

$$d_{i,j} = (1, y_{i,j}, \boldsymbol{x}_i)\boldsymbol{\beta}_d + \zeta_{i,j}, \tag{5}$$

where $\zeta_{i,j} \overset{\text{i.i.d.}}{\sim} \text{Normal}(0, \sigma_d^2)$. This Gaussian error assumption is suitable for our dosage data, which has continuous values, but can be easily modified for other dosage/treatment types (e.g., generalized linear regression models for discrete treatment choices). Thus, we can write the probability density of the i-th patient's sequence of visitations and assigned dosages $\boldsymbol{e}_{i,T_i}$ as $p(\boldsymbol{e}_{i,T_i} \mid \boldsymbol{y}_i, \boldsymbol{x}_i, \boldsymbol{\beta}_v, \boldsymbol{\beta}_d, \sigma_d^2)$ following (1), where $\boldsymbol{\beta}_v = \{\mu, \nu_1, \nu_2, \xi, \boldsymbol{\beta}_\alpha\}$.

Next, we link the submodel of clinical measurements including the longitudinal measurements (i.e., creatinine levels) and time-to-event data (i.e., graft survival time after kidney transplantation) to the MTPP of the clinical decisions through parameter sharing. Specifically, we use a linear-mixed effects model for longitudinal creatinine levels by assuming

$$y_i(t) = y_i^*(t) + \varepsilon_{i,j} = \boldsymbol{z}_i(t)\boldsymbol{\beta}_l + \boldsymbol{r}_i(t)\boldsymbol{b}_i + \varepsilon_i(t), \tag{6}$$

where $y_i^*(t)$ is the underlying true but unobserved longitudinal process at time $t \geq 0$, $\boldsymbol{\beta}_l$ is the fixed-effect coefficient, $\varepsilon_i(t) \overset{\text{i.i.d.}}{\sim} \text{Normal}(0, \sigma_l^2)$ and $\boldsymbol{b}_i \sim \text{Normal}(\boldsymbol{0}, \boldsymbol{\Sigma}_b)$. The covariate vectors $\boldsymbol{z}_i(t)$ and $\boldsymbol{r}_i(t)$ are associated with fixed and random effects, respectively:

$$\boldsymbol{z}_i(t) = (1, d_i(t), \boldsymbol{x}_i, t, t^2) \text{ and } \boldsymbol{r}_i(t) = (1, d_i(t), t),$$

where $d_i(t)$ at time t is the dosage assigned by the physician at the most recent visitation, i.e., $d_i(t) = d_{i,j}$ for $t \in (t_{i,j}, t_{i,j+1}]$. The reason why $d_i(t)$ is the same value $d_{i,j}$ over the interval $(t_{i,j}, t_{i,j+1}]$ is that the tacrolimus is taken orally on a daily basis. Denoting $\boldsymbol{d}_i = (d_{i,0}, \ldots, d_{i,J_i})$, the density of the observed sequence of creatinine measurements $\boldsymbol{y}_i$ is

$$p(\boldsymbol{y}_i \mid \boldsymbol{d}_i, \boldsymbol{x}_i, \boldsymbol{\beta}_l, \sigma_l^2, \boldsymbol{b}_i) = \prod_{j=1}^{J_i} p(y_{i,j} \mid t_{i,j}, d_{i,j-1}, \boldsymbol{x}_i, \boldsymbol{\beta}_l, \sigma_l^2, \boldsymbol{b}_i). \tag{7}$$

Lastly, we construct the survival submodel depending on the underlying true longitudinal trajectory $y_i^*(t)$ and the MTPP that models clinical decisions. We consider a proportional hazards model with a Weibull baseline hazard as follows:

$$h_i(t) = \exp\Big(-(\underbrace{\beta_{s1} y_i^*(t)}_{\text{longitudinal effect}} + \underbrace{\beta_{s2} d_i(t) + \beta_{s3}\text{Tox}_i(t)}_{\text{dosage effect}} + \underbrace{\beta_{s4}\alpha_i(t)}_{\text{visitation effect}} + h_0)\Big)\omega t^{\omega-1}, \tag{8}$$

where ω is the shape parameter. If desired, more complex survival models such as Bayesian nonparametric survival models (Xu et al., 2019) can be explored. The dosage effect term in (8) measures the overall drug effect on the patient: $\beta_{s2} d_i(t)$ is the "instantaneous" effect while $\beta_{s3}\text{Tox}_i(t)$ is the "accumulated" effect:

$$\text{Tox}_i(t) = \int_0^t d_i(\tau)\eta_{tox} \exp(-(t-\tau)/\eta_{tox}) d\tau,$$

where the parameter η_{tox} controls the rate of the exponential weighting for the past dosages. In practice, the instantaneous effect is usually beneficial (e.g., tacrolimus reduces the likelihood of graft rejection or death) while the accumulated effect is often toxic (and that is why we name it Tox): e.g., a prolonged high dosage of tacrolimus might have adverse effects on kidneys, central nervous system, and gastrointestinal tract, thereby worsening a patient's survival (Randhawa et al., 1997). The visitation and survival submodels are linked by sharing $\alpha_i(t) = \alpha_{i,j}$ for $t \in (t_{i,j}, t_{i,j+1}]$, since a high visitation intensity (i.e., larger $\alpha_{i,j}$) typically implies a higher risk, e.g., graft failure and thus shorter expected survival time.

For each patient i, let T_i and C_i denote the graft survival and administration censoring times, respectively, $i = 1, \ldots, I$. We observe only $\widetilde{T}_i = \min(T_i, C_i)$ and the censoring indicator $\delta_i = \mathbf{1}(T_i \leq C_i)$. Denote $f_i(t)$ and $S_i(t)$ to be the corresponding density and survival functions of the hazard function (8):

$$S_i(t) = \exp(-\int_0^t h_i(u)du),\ f_i(t) = h_i(t)S_i(t).$$

We can write the survival likelihood for patient i as

$$p(\widetilde{T}_i, \delta_i \mid \boldsymbol{y}_i, \boldsymbol{x}_i, \boldsymbol{e}_{i,\widetilde{T}_i}, \boldsymbol{\beta}_l, \boldsymbol{b}_i, \boldsymbol{\beta}_s) = f_i(\widetilde{T}_i \mid \boldsymbol{y}_i, \boldsymbol{x}_i, \boldsymbol{e}_{i,\widetilde{T}_i}, \boldsymbol{\beta}_l, \boldsymbol{b}_i, \boldsymbol{\beta}_s)^{\delta_i}$$
$$\times S_i(\widetilde{T}_i \mid \boldsymbol{y}_i, \boldsymbol{x}_i, \boldsymbol{e}_{i,\widetilde{T}_i}, \boldsymbol{\beta}_l, \boldsymbol{b}_i, \boldsymbol{\beta}_s)^{1-\delta_i}, \tag{9}$$

where $\boldsymbol{\beta}_s = \{\omega, \beta_{s1}, \beta_{s2}, \beta_{s3}, \beta_{s4}, h_0, \eta_{tox}, \boldsymbol{\beta}_\alpha, \xi\}$.

In summary, we propose a joint model consisting of an MTPP for clinical decisions including follow-up visitation schedules and dosages, a linear mixed effects model for longitudinal clinical measurements, and a time-to-event model for the patient survival; they are inter-connected by sharing structures and parameters. The joint density of the clinical observations and decisions can then factor as

$$\prod_{i=1}^{I} p(\boldsymbol{y}_i, \boldsymbol{e}_{i,\widetilde{T}_i}, \widetilde{T}_i, \delta_i \mid \boldsymbol{x}_i, \boldsymbol{\beta}_l, \boldsymbol{\beta}_d, \boldsymbol{\beta}_v, \boldsymbol{\beta}_s, \boldsymbol{b}_i, \sigma_l^2, \sigma_d^2)$$

$$\propto \prod_{i=1}^{I} \Bigg(\underbrace{p(\boldsymbol{e}_{i,\widetilde{T}_i} \mid \boldsymbol{y}_i, \boldsymbol{x}_i, \boldsymbol{\beta}_v, \boldsymbol{\beta}_d, \sigma_d^2)}_{(1)} \underbrace{p(\boldsymbol{y}_i \mid \boldsymbol{d}_i, \boldsymbol{x}_i, \boldsymbol{\beta}_l, \sigma_l^2, \boldsymbol{b}_i)}_{(7)}$$

$$\times \underbrace{p(\widetilde{T}_i, \delta_i \mid \boldsymbol{y}_i, \boldsymbol{x}_i, \boldsymbol{e}_{i,\widetilde{T}_i}, \boldsymbol{\beta}_l, \boldsymbol{b}_i, \boldsymbol{\beta}_s)}_{(9)} \Bigg). \tag{10}$$

Note that the density factorization in (10) may appear to give circular definitions of $\boldsymbol{d}_i \mid \boldsymbol{y}_i$ in (1) and $\boldsymbol{y}_i \mid \boldsymbol{d}_i$ in (7). However, it is a valid factorization because the dependencies are temporal: each dosage $d_{i,j}$ is conditional on the *current* measurement $y_{i,j}$ in $\boldsymbol{d}_i \mid \boldsymbol{y}_i$, while the *next* measurement $y_{i,j+1}$ is conditional on the current dosage $d_{i,j}$ in $\boldsymbol{y}_i \mid \boldsymbol{d}_i$.

The model is completed by imposing the following priors: $\boldsymbol{\beta}_d \sim$ Normal $(\boldsymbol{\beta}_{d0}, \boldsymbol{\Sigma}_{\boldsymbol{\beta}_d})$, $\sigma_d^2 \sim$ InverseGamma(π_{d1}, π_{d2}), $\boldsymbol{\beta}_l \sim$ Normal$(\boldsymbol{\beta}_{l0}, \boldsymbol{\Sigma}_{\boldsymbol{\beta}_l})$, $\sigma_l^2 \sim$ InverseGamma(π_{l1}, π_{l2}) for conjugacy. We assume a flat prior for $\boldsymbol{\Sigma}_b$. When conjugacy is unattainable for the visitation and survival parameters, we assume $\beta_{s1}, \beta_{s2}, \beta_{s3}, \beta_{s4}, h_0 \sim$ Normal$(\beta_{s0}, \sigma_{s0}^2)$, $\eta_{tox} \sim$ Gamma(π_{s1}, π_{s2}), $\omega \sim$ Gamma(π_{s3}, π_{s4}), $\mu, \nu_1, \nu_2 \sim$ Normal$(\beta_{v0}, \sigma_{v0}^2)$, $\boldsymbol{\beta}_\alpha \sim$ Normal$(\boldsymbol{\beta}_{\alpha 0}, \boldsymbol{\Sigma}_{\boldsymbol{\beta}_\alpha})$, and $\xi \sim$ Gamma(π_{v1}, π_{v2}). The specific prior numerical values are chosen so that the priors are generic and noninformative. We carry out posterior inference using the Markov chain Monte Carlo (MCMC) sampler. The computational details can be found in Hua et al. (2021).

3.2 Second Step: Optimize Clinical Decisions

In this subsection, we propose a policy gradient method using stochastic gradient descent (SGD) (Ruder, 2016) to optimize personalized clinical decisions including scheduling a patient's follow-up visitations and prescribing dosages to maximize the median graft survival time.

Let $\boldsymbol{\theta} = (\nu_1, \nu_2, \mu, \boldsymbol{\beta}_d, \sigma_d^2)$ denote the set of "policy" parameters related to clinical decisions, i.e., the parameters that only appear in the conditional intensity function (4) and the mark distribution (5), which control patients' follow-up schedules and dosages at follow-up visitations. We borrow the term "policy" from RL since the setting is similar: in RL, the "policy" refers to the distribution from which an intelligent agent samples its actions and that distribution is optimized to achieve the highest expected reward (Kaelbling et al., 1996; Sutton & Barto, 2018).

Let $\boldsymbol{\phi} = (\boldsymbol{\beta}_s, \boldsymbol{b}_i, \boldsymbol{\beta}_l, \sigma_l^2)$ denote the set of "observation" parameters, i.e., all other parameters in the joint model (10). Recall that in Sect. 2, we define the expected reward for any future/hypothetical patient i with baseline covariates $\boldsymbol{x}_i$ to be $G_i(\boldsymbol{\theta})$. In the kidney transplantation application, we define a personalized reward function R_i as the log-scaled median survival time: $R_i = \log(\widehat{T}_i)$, where $S_i(\widehat{T}_i) = 0.5$. If desired, other reward functions can be considered. Our goal is to find $\widetilde{\boldsymbol{\theta}}_i$ that maximizes the expected reward $G_i(\boldsymbol{\theta})$ after integrating out the uncertainty in the clinical measurements: $\text{maximize}_{p(\boldsymbol{e}_{i,T_i}|\boldsymbol{\theta})} G_i(\boldsymbol{\theta})$.

We use stochastic gradient descent (SGD) (Robbins & Monro, 1951), i.e.,

$$\boldsymbol{\theta}_{i,m+1} = \boldsymbol{\theta}_{i,m} + s_{i,m} \nabla_{\boldsymbol{\theta}} G_i(\boldsymbol{\theta}) \mid_{\boldsymbol{\theta}=\boldsymbol{\theta}_{i,m}},$$

which requires computing the gradient of the expected reward: $\nabla_{\boldsymbol{\theta}} G_i(\boldsymbol{\theta})$, where the operator $\nabla_{\boldsymbol{\theta}}$ is a vector differential operator with respect to each one of the policy parameters in $\boldsymbol{\theta}$. As the expectation is taken over realizations of the joint distribution $p(\boldsymbol{y}_i, T_i, \boldsymbol{e}_{i,T_i} \mid \boldsymbol{\theta}, \boldsymbol{\phi}, \boldsymbol{x}_i)$, it is intractable to directly compute $\nabla_{\boldsymbol{\theta}} G_i(\boldsymbol{\theta})$. Fortunately, we can indirectly compute this gradient by taking the expectation of the reward-weighted gradient of log-policy. Specifically, we first sample $\boldsymbol{y}_i, T_i, \boldsymbol{e}_{i,T_i}$ from $p(\boldsymbol{y}_i, T_i, \boldsymbol{e}_{i,T_i} \mid \boldsymbol{\theta}, \boldsymbol{\phi})$ and calculate R_i from the generated samples. We sample the j-th follow-up visitation time $t_{i,j}$ and the survival time T_i using an inverse transform sampling method: we first compute the cumulative distribution function (CDF) of the distribution, sample a random number U from Uniform(0, 1), and then invert the CDF function at U to yield the visitation/survival time (Giesecke et al., 2011). If the j-th visitation time occurs before the survival time, i.e., $t_{i,j} < T_i$, we sample $y_{i,j}$ and $d_{i,j}$ from their respective distributions and continue to sample the $(j+1)$-th visitation time and the survival time. We iteratively sample follow-up visitation times, survival times, longitudinal measurements, and dosages until the sampled survival event occurs before the next visit time. After obtaining samples of $\boldsymbol{y}_i, T_i, \boldsymbol{e}_{i,T_i}$, we can easily compute R_i. We then compute $\nabla_{\boldsymbol{\theta}} G_i(\boldsymbol{\theta})$ using the parametrization defined in (4) and (5) and the Monte Carlo method in order to integrate out $\boldsymbol{\phi}$. We use an adaptive step size algorithm and choose the step size to be:

$$s_{i,m} = \frac{0.01}{\sqrt{\sum_{l=m-50}^{m-1} \nabla_{\boldsymbol{\theta}} G_i(\boldsymbol{\theta}_{i,l})^2}}.$$

The details of sampling $\boldsymbol{y}_i, T_i, \boldsymbol{e}_{i,T_i}, R_i$ for the reward being the log median survival time, and the entire SGD algorithm for finding the optimal parameter $\widetilde{\boldsymbol{\theta}}_i$ can be found in Hua et al. (2021).

4 Simulation Study

We evaluate the proposed decision framework through simulation studies and demonstrate that the optimized clinical decision obtained from the proposed approach yields longer predictive median survival compared to alternative strategies. The simulated dataset and code can be found at https://github.com/YanxunXu/doct.

4.1 Simulation Setup

We simulate a synthetic DIVAT dataset composed of longitudinal creatinine measurements, tacrolimus dosages, follow-up schedules, and survival events/censoring for a total of $I = 600$ patients. We consider three baseline covariates in $\boldsymbol{x}_i$: donor age (AgeD), delayed graft function (DGF), and body mass index (BMI). DGF is a Bernoulli(0.4) random variable with 1 indicating that the patient uses dialysis within the first week of the transplant, 0 otherwise. For each patient, AgeD and BMI are sampled from Normal$(52.5, 15.8^2)$ and Normal$(24.3, 4.5^2)$, respectively, and then standardized. In the MTPP model for follow-up schedules, the simulated true parameters are set to be $\nu_1 = 2.5$, $\nu_2 = 1.5$, $\mu = -4.8$, $\xi = 2$, and $\boldsymbol{\beta}_\alpha = (9.5, -1.5)^T$ so that a higher creatinine level results in a higher visitation intensity; for assigning dosages, the simulated true $\boldsymbol{\beta}_d$ is set to be $(1, 0.2, 0.15, 0.2, 0.15)^T$ and $\sigma_d = 0.3$. In modeling log-transformed longitudinal creatinine levels, the simulated true parameters are set to be $\boldsymbol{\beta}_l = (5.3, 0.1, 0.3, 0.4, 0.25, -1 \times 10^{-4}, 3 \times 10^{-8})^T$, $\sigma_l = 0.1$, and $\boldsymbol{\Sigma}_b$ a diagonal matrix with diagonal elements $(0.04, 0.0049, 10^{-8})$. Patients' initial log-transformed creatinine levels right after transplantation $y_{i,0}$'s are independently generated from Normal$(5, 0.1^2)$. In the survival submodel (9), we set the true parameters $h_0 = 5$, $\omega = 1.05$, $\beta_{s1} = 1$, $\beta_{s2} = 0.9$, $\beta_{s3} = -0.75$, $\beta_{s4} = -5$, and $\eta_{tox} = 50$. The censoring times C_i's are independently generated from Weibull(3, 8000). Based on the proposed Bayesian joint model in Sect. 3.1, we generate the data $\boldsymbol{y}_i, \boldsymbol{e}_{i,\widetilde{T}_i}, \widetilde{T}_i, \delta_i$ for each patient i, $i = 1, \ldots, I$.

The simulated dataset has a total of 16,913 follow-up visits for 600 patients with a 9.5% censoring rate. The median survival time is 1557 days with the shortest being 11 days and the longest being 9977 days.

4.2 Simulation Results: Model Fitting

We apply the proposed Bayesian joint model to the simulated dataset. The hyperparameters are set to be $\boldsymbol{\beta}_{d0} = \boldsymbol{\beta}_{l0} = \boldsymbol{\beta}_{\alpha 0} = \boldsymbol{0}$, $\boldsymbol{\Sigma}_{\boldsymbol{\beta}_d} = \boldsymbol{\Sigma}_{\boldsymbol{\beta}_l} = \boldsymbol{\Sigma}_{\boldsymbol{\beta}_\alpha} = 100^2 I$, $\pi_{d1} = \pi_{d2} = \pi_{l1} = \pi_{l2} = \pi_{s3} = \pi_{s4} = 0.01$, $\pi_{s1} = \pi_{s2} = 0.01$, $\beta_{s0} = \beta_{v0} = 0$, $\sigma_{s0}^2 = \sigma_{v0}^2 = 100^2$, $\pi_{v1} = 400$, $\pi_{v2} = 200$. We run 20,000 MCMC iterations

Table 1 Parameter estimation with the Bayesian joint model

	Truth	Posterior mean (95% CI)	Squared errors mean (std)
β_{s1}	1	1.1(0.92, 1.26)	0.022 (0.022)
β_{s2}	0.9	0.75(0.38, 1.19)	0.073 (0.086)
β_{s3}	−0.75	−0.59(−0.99, −0.15)	0.077 (0.10)
β_{s4}	−5	−4.33(−4.91, −3.73)	0.54 (0.42)
h_0	5	5.82(4.52, 7.12)	1.20 (1.29)
k	2	2.1(1.94, 2.26)	0.017 (0.022)
ω	1.05	1.16(1.08, 1.23)	0.014 (8.7×10^{-3})
σ_l^2	0.01	0.01(0.01, 0.01)	2.5×10^{-8} (3.1×10^{-8})
σ_d^2	0.09	0.09(0.09, 0.09)	8.3×10^{-6} (5.8×10^{-6})
μ	−4.8	−4.75(−4.77, −4.73)	2.9×10^{-3} (1.1×10^{-3})
ν_1	2.5	2.51(2.49, 2.53)	1.8×10^{-4} (2.0×10^{-4})
ν_2	1.5	1.49(1.42, 1.56)	1.7×10^{-3} (2.1×10^{-3})

with an initial burn-in of 5000 iterations and a thinning factor of 50. We evaluate the parameter estimation of the proposed Bayesian joint model by considering two metrics: the posterior mean point estimate with 95% credible intervals (CIs), and the mean squared error (MSE) obtained by the averaged squared errors between the post-burn-in MCMC posterior samples and the simulated true values. We report results with both metrics on 12 critical model parameters in Table 1. Clearly, the proposed joint model can accurately estimate parameters in the sense that all posterior means are close to the ground truth.

4.3 Simulation Results: Policy Optimization

We now apply the proposed decision framework to the simulated dataset with the goal of maximizing patients' median survival time. We use the MCMC posterior mean of $\boldsymbol{\theta}$ as the starting value in the optimization procedure, which can be seen as how a physician normally would have treated patients in the simulated data, and apply the optimization algorithm with $M = 1000$ steps.

We randomly select two patients S1 and S2 from the simulated dataset and show their expected survival increases as the optimization algorithm continuously improves their personalized treatment policies in Fig. 3. Figure 3a shows patient S1, whose median survival time was estimated to be 2587 days before any policy optimization, and who now has an expected survival time of 3100 days with our improved treatment policy, attaining an increase in expected survival time of 20%. Patient S2 shown in Fig. 3b has an expected survival time trajectory that exhibits a similar increasing behavior – it has increased from 979 days to 1139 days, about a 16% improvement.

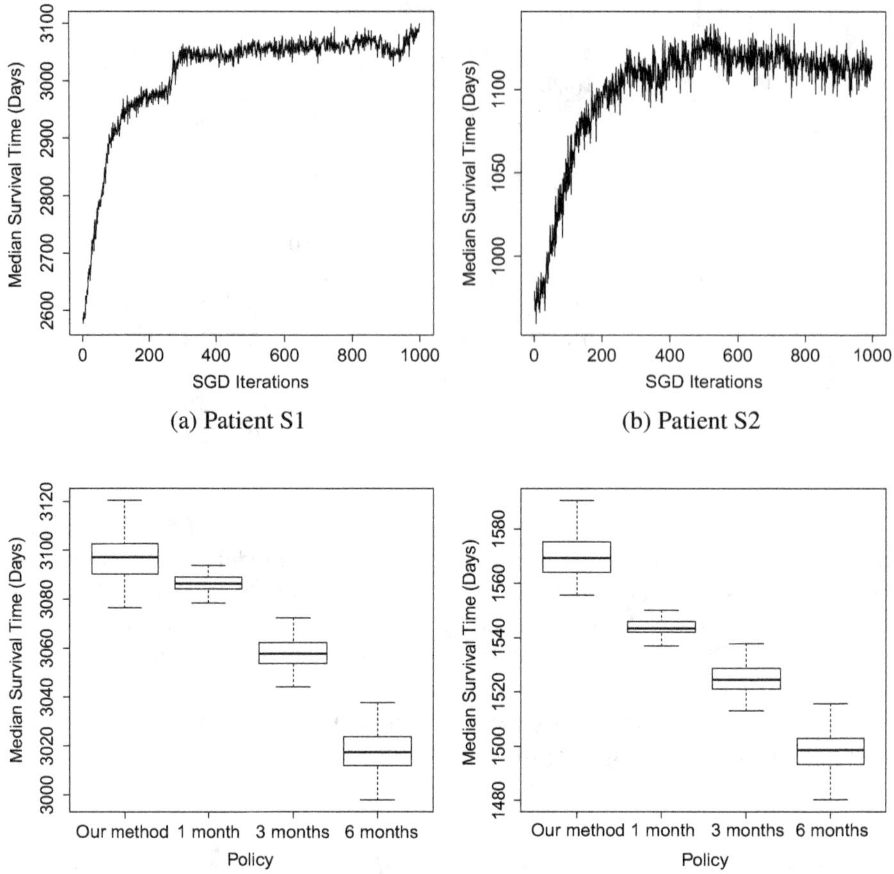

(a) Patient S1 (b) Patient S2

(a) Patient S1 (b) Patient S2

Fig. 3 (**a**), (**b**): predictive median survival time versus SGD iterations for two randomly selected patients S1 and S2. (**c**), (**d**): boxplots for median survival times under our method and the three alternatives based on regular visitation schedules for patients S1 and S2

To interpret how the optimized policies are beneficial to patients S1 and S2, we compare the initial values of policy parameters (i.e., posterior means from the first-step Bayesian joint model) to the SGD optimized values and summarize in Table 2. We define the personalized dosage intercepts to be $\tilde{\beta}_d = (1, \boldsymbol{x}_i)(\beta_{d1}, \beta_{d3}, \ldots, \beta_{dL})^T$, so that $\boldsymbol{\beta}_d = (\beta_{d2}, \tilde{\beta}_d)$. As shown in Table 2, the optimized dosage parameters $\tilde{\beta}_d$ and β_{d2} for patient S1 are both higher than the initial posterior means, suggesting that for the same observed creatinine level a higher dosage would benefit patient S1. On the other hand, the optimal $\tilde{\beta}_d$ and β_{d2} for patient S2 are both lower than the posterior means, indicating S2 would respond more favorably to lower dosages. The optimal dosage errors σ_d^2 for both patients

Table 2 Stochastic gradient descent optimal parameter results for the simulation data

	$\widetilde{\theta}_0$	$\widetilde{\theta}_{S1}$	$\widetilde{\theta}_{S2}$
$\tilde{\beta}_d$: personalized dosage intercept	S1: 1.196, S2:1.191	1.455	0.863
β_{d2}: dosage effect of creatinine	0.192	0.323	0.055
σ_d^2: dosage error	0.0927	0.0336	0.0293
μ: baseline visitation intensity	−4.747	−4.371	−4.515
ν_1: visitation intensity peak	2.510	2.478	2.391
ν_2: visitation intensity shape	1.487	2.119	2.101

have been significantly lowered, indicating that lower dosage variability can benefit their survival.

To demonstrate the importance of jointly optimizing follow-up schedules together with dosages, we further compare our algorithm to three alternative strategies with regular visitation schedules of every 1 month, 3 months, and 6 months, respectively, and dosage at each visitation is optimized in the same way as our joint optimization method. We plot the predictive median survival times of the two patients S1 and S2 under the four policies in Fig. 3c, d. While three alternative policies show that the more frequently a patient visits the physician the more beneficial it is to the patient, our method which flexibly schedules the next clinical visit achieves the highest predictive median survival time, thus highlighting the necessity of visit schedule optimization based on patients' longitudinal measurements and prescribed dosages.

5 DIVAT Data Application

We extract clinical records of $I = 947$ patients who received a first or second renal graft transplanted from a living or heart-beating deceased donor between 2000 and 2014 from Nantes University Hospital Centers in the DIVAT cohort (`www.divat.fr`). Immediately after transplantation, several baseline covariates as risk factors for graft failure were collected: donor age (AgeD), recipient age (AgeR), delayed graft function (DGF) defined as the indicator of the use of dialysis within the first week of transplant (1 = used dialysis, 0 = didn't use dialysis), diabetes history (Diab) with 1 indicating the patient has a history of diabetes and 0 otherwise, type of donor (Type), and body mass index (BMI). Baseline covariates further included the type of donors, where Type = 1 marked donation after brain death but with heart beating and Type = 0 corresponded to donation by a living donor. The longitudinal creatinine measurements and tacrolimus dosages for each patient were recorded at each follow-up visit since the date of transplantation until graft failure or censoring time.

5.1 Model Fitting

We set the baseline covariates $\boldsymbol{x}_i = (\text{AgeD}_i, \text{AgeR}_i, \text{DGF}_i, \text{BMI}_i, \text{Diab}_i, \text{Type}_i)$. Using the same hyperparameters in the prior distributions as those in the simulation study, we run the MCMC for a total of 20,000 iterations, with a burn-in period of 5000 iterations and a thinning factor of 50. We plot the estimated posterior means with 95% CIs for the critical model parameters in the dosage, longitudinal, and survival submodels in the top, middle, and bottom of Fig. 4, respectively.

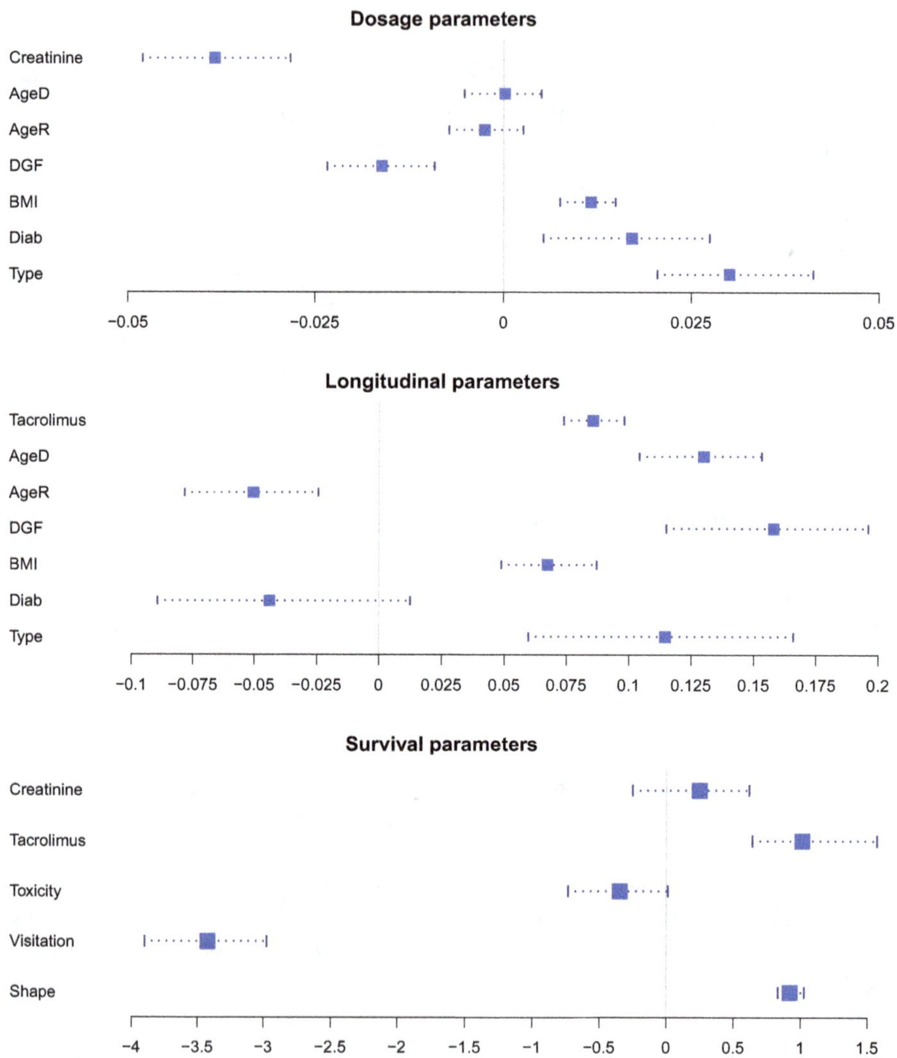

Fig. 4 Estimated posterior means (solid square) and 95% credible intervals for parameters in the dosage, longitudinal, and survival submodels

Our posterior inference has meaningful interpretations and agrees with previous studies: in the dosage submodel, the linear coefficient corresponding to DGF is negative, which suggests that patients treated with dialysis within the first week of transplant would tend to be assigned lower dosages; BMI however, is positively associated with the dosage, which is reasonable since it is standard practice to assign tacrolimus dosing based on body weight (Andrews et al., 2017). In addition, donor type is positively correlated with the dosage, suggesting that patients who received a kidney from a non-living donor were more likely to be assigned higher dosages. In the longitudinal creatinine submodel, the dosage, donor age, DGF, BMI, and donor type are positively associated with the creatinine level, agreeing with Katari et al. (1997), Gerchman et al. (2009), and Foucher et al. (2016), whereas the recipient age is negatively associated with the creatinine level, indicating that younger patients are likely to have lower creatinine levels (Maraghi et al., 2016). Diabetes history also decreases the creatinine level, which is consistent with the finding in Hjelmeseth et al. (2010). Finally, in the survival submodel, the dosage level is positively associated which is in contrast with the accumulated toxicity, suggesting that while a higher tacrolimus dosage reduces the hazard instantaneously, the built-up toxicity increases the hazard, agreeing with findings in Randhawa et al. (1997) and Böttiger et al. (1999).

5.2 *Personalized Treatment Optimization*

We now apply the policy gradient method to optimize tailored treatment plans in order to maximize each individual patient's median survival time. Similar to the simulation study, we initialize the optimization starting point $\boldsymbol{\theta}_0$ to be the posterior means, and perform stochastic gradient descent for $M = 1000$ for two arbitrarily selected patients R1 and R2. Patient R1 at transplantation was 60 years old with a BMI of 17, no history of diabetes, no DGF, and received donation from a 61-year-old non-living donor. Patient R2 at transplantation was 28 years old with a BMI of 25.5, no history of diabetes, no DGF, and received a kidney from a living 29-year-old donor. Patient R1 had an observed survival time of 1527 days, while patient R2 had a censored survival time of 4487 days. With the proposed policy optimization method, patient R1's predictive median survival time increased from 1793 to 1895 days at the maximum, a 5.7% improvement; patient R2's predictive median survival time increased from 5191 to 5628, an 8.4% improvement.

Moreover, to demonstrate the benefit of optimizing the follow-up visitation schedules and dosages together, we further compare the predictive median survival times under the non-optimized initial policy with three versions of optimized policies: (1) only visitation schedules are optimized; (2) only dosages are optimized; and (3) both visitation schedules and dosages are optimized. Figure 5 shows boxplots for 100 realizations of the predictive median survival times under each of the four policies. Comparing the four policies, we can see that while visitation schedule optimization accounts for more survival improvement for patient R1 than

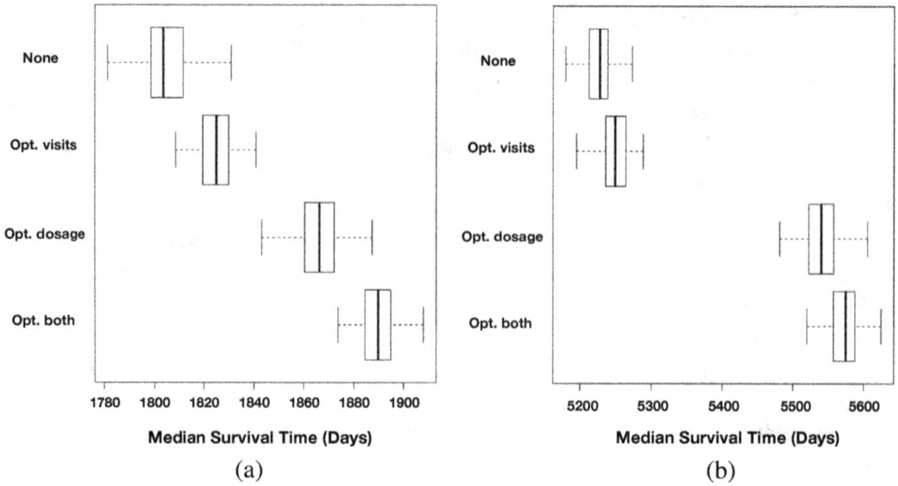

Fig. 5 Predictive median survival times boxplots for patients R1 and R2 under policy with both dosages and visitation schedules optimized (Opt. both), policy with only dosages optimized (Opt. dosage), policy with only visitation schedules optimized (Opt. visits), and non-optimized initial policy (none). (**a**) Patient R1. (**b**) Patient R2

for patient R2, optimizing treatment schedules is clearly beneficial to both patients, revealing that optimizing both visitation schedules and dosages is necessary to maximize patients' survival.

6 Conclusion

In this chapter, we developed a model-based decision framework to model and optimize sequential clinical decisions in continuous time. The analysis of the kidney transplantation data yields meaningful and interpretable results, showing that the proposed approach has the potential to assist physicians' decisions on personalized treatment.

There are a few potential extensions. First, the patient population can be heterogeneous, meaning that patients in different subgroups may respond differently to treatments or have different disease progression patterns depending on their clinical characteristics. The proposed Bayesian joint model can be extended to account for patients' heterogeneity by considering more flexible priors, such as Bayesian nonparametrics (Ferguson, 1973). Second, the policy gradient method is used in the second step to optimize sequential clinical decisions. Other optimization methods such as actor-critic algorithms (Konda & Tsitsiklis, 1999) can be explored. Third, standard MCMC techniques may be replaced with more sophisticated sampling algorithms such as Hamiltonian Monte Carlo (Duane et al., 1987) to improve sampling efficiency. Finally, the policy optimization relies on the learned dynamics of the clinical measurements from the Bayesian joint model in the first step. When the dynamics under the optimized policy deviates from the one under the "behavior policy" that generates the observational data (i.e., *distribution shift* happens), the policy learning can become unstable (Fujimoto et al., 2019). Since the proposed Bayesian approach naturally quantifies uncertainties in the dynamic model, certain model-based uncertainty-penalized policy optimization methods such as MOPO (Yu et al., 2020) can be considered.

References

Aalen, O., Borgan, O., & Gjessing, H. (2008). *Survival and event history analysis: A process point of view*. Springer Science & Business Media.

Andrews, L. M., de Winter, B. C., Tang, J.-T., Shuker, N., Bouamar, R., van Schaik, R. H., Koch, B. C., van Gelder, T., & Hesselink, D. A. (2017). Overweight kidney transplant recipients are at risk of being overdosed following standard bodyweight-based tacrolimus starting dose. *Transplantation Direct, 3*(2), e129.

Arshad, A., Anderson, B., & Sharif, A. (2019). Comparison of organ donation and transplantation rates between opt-out and opt-in systems. *Kidney International, 95*(6), 1453–1460.

Böttiger, Y., Brattström, C., Tydén, G., Säwe, J., & Groth, C. G. (1999). Tacrolimus whole blood concentrations correlate closely to side-effects in renal transplant recipients. *British Journal of Clinical Pharmacology, 48*(3), 445–448.

Butts, C. T., & Marcum, C. S. (2017). A relational event approach to modeling behavioral dynamics. In *Group Processes* (pp. 51–92). Cham: Springer.

Chakraborty, B. (2013). *Statistical methods for dynamic treatment regimes*. Springer.

Chen, R. T., Rubanova, Y., Bettencourt, J., & Duvenaud, D. K. (2018). Neural ordinary differential equations. *Advances in Neural Information Processing Systems, 31*.

Du, N., Dai, H., Trivedi, R., Upadhyay, U., Gomez-Rodriguez, M., & Song, L. (2016). Recurrent marked temporal point processes: Embedding event history to vector. In *Proceedings of the 22nd ACM SIGKDD International Conference on Knowledge Discovery and Data Mining*.

Du, J., Futoma, J., & Doshi-Velez, F. (2020). Model-based reinforcement learning for semi-Markov decision processes with neural odes. *Advances in Neural Information Processing Systems, 33*, 19805–19816.

Duane, S., Kennedy, A. D., Pendleton, B. J., & Roweth, D. (1987). Hybrid Monte Carlo. *Physics Letters B, 195*(2), 216–222.

Ferguson, T. S. (1973). A Bayesian analysis of some nonparametric problems. *The Annals of Statistics, 1*(2), 209–230.

Foucher, Y., Blanche, P., & Buron, F. (2016). A joint model for longitudinal and time-to-event data to better assess the specific role of donor and recipient factors on long-term kidney transplantation outcomes. *European Journal of Epidemiology, 31*(5), 469–479.

Fujimoto, S., Meger, D., & Precup, D. (2019). Off-policy deep reinforcement learning without exploration. In K. Chaudhuri & R. Salakhutdinov (Eds.), *Proceedings of the 36th International Conference on Machine Learning* (Proceedings of Machine Learning Research, Vol. 97, pp. 2052–2062). PMLR.

Gerchman, F., Tong, J., Utzschneider, K. M., Zraika, S., Udayasankar, J., McNeely, M. J., Carr, D. B., Leonetti, D. L., Young, B. A., de Boer, I. H., et al. (2009). Body mass index is associated with increased creatinine clearance by a mechanism independent of body fat distribution. *The Journal of Clinical Endocrinology & Metabolism, 94*(10), 3781–3788.

Giesecke, K., Kakavand, H., & Mousavi, M. (2011). Exact simulation of point processes with stochastic intensities. *Operations Research, 59*(5), 1233–1245.

Guan, Q., Reich, B. J., Laber, E. B., & Bandyopadhyay, D. (2019). Bayesian nonparametric policy search with application to periodontal recall intervals. *Journal of the American Statistical Association 115*(531), 1–13.

Hawkes, A. G. (1971). Spectra of some self-exciting and mutually exciting point processes. *Biometrika, 58*(1), 83–90.

Hawkes, A. G. (2018). Hawkes processes and their applications to finance: A review. *Quantitative Finance, 18*(2), 193–198.

Hjelmeseth, J., Røislien, J., Nordstrand, N., Hofsø, D., Hager, H., & Hartmann, A. (2010). Low serum creatinine is associated with type 2 diabetes in morbidly obese women and men: A cross-sectional study. *BMC Endocrine Disorders, 10*(1), 6.

Hua, W., Mei, H., Zohar, S., Giral, M., & Xu, Y. (2021). Personalized dynamic treatment regimes in continuous time: A Bayesian approach for optimizing clinical decisions with timing. *Bayesian Analysis, 1*(1), 1–30.

Hussein, A., Gaber, M. M., Elyan, E., & Jayne, C. (2017). Imitation learning: A survey of learning methods. *ACM Computing Surveys (CSUR), 50*(2), 1–35.

Islam, K. T., Shelton, C. R., Casse, J. I., & Wetzel, R. (2017). Marked point process for severity of illness assessment. In *Machine Learning for Healthcare Conference* (pp. 255–270).

Israni, A., Dean, C. E., Salkowski, N., Li, S., Ratner, L. E., Rabb, H., Powe, N. R., & Kim, S. J. (2014). Variation in structure and delivery of care between kidney transplant centers in the United States. *Transplantation, 98*(5), 520.

Kaelbling, L. P., Littman, M. L., & Moore, A. W. (1996). Reinforcement learning: A survey. *Journal of Artificial Intelligence Research, 4*, 237–285.

Kasiske, B. L., Zeier, M. G., Chapman, J. R., Craig, J. C., Ekberg, H., Garvey, C. A., Green, M. D., Jha, V., Josephson, M. A., Kiberd, B. A., Kreis, H. A., McDonald, R. A., Newmann, J. M., Obrador, G. T., Vincenti, F. G., Cheung, M., Earley, A., Raman, G., Abariga, S., Wagner, M., & Balk, E. M. (2010). KDIGO clinical practice guideline for the care of kidney transplant recipients: A summary. *Kidney International, 77*(4), 299–311.

Katari, S., Magnone, M., Shapiro, R., Jordan, M., Scantlebury, V., Vivas, C., Gritsch, A., McCauley, J., Starzl, T., Demetris, A., et al. (1997). Clinical features of acute reversible tacrolimus (fk 506) nephrotoxicity in kidney transplant recipients. *Clinical Transplantation, 11*(3), 237.

Konda, V., & Tsitsiklis, J. (1999). Actor-critic algorithms. *Advances in Neural Information Processing Systems, 12*, 1008–1014.

Lavori, P. W., & Dawson, R. (2004). Dynamic treatment regimes: Practical design considerations. *Clinical Trials, 1*(1), 9–20.

Levine, S., Kumar, A., Tucker, G., & Fu, J. (2020). Offline reinforcement learning: Tutorial, review, and perspectives on open problems. arXiv preprint arXiv:2005.01643.

Liu, H., Carlson, N. E., Grunwald, G. K., & Polotsky, A. J. (2018). Modeling associations between latent event processes governing time series of pulsing hormones. *Biometrics, 74*(2), 714–724.

Luckett, D. J., Laber, E. B., Kahkoska, A. R., Maahs, D. M., Mayer-Davis, E., & Kosorok, M. R. (2019). Estimating dynamic treatment regimes in mobile health using v-learning. *Journal of the American Statistical Association 115*(530), 1–34.

Maraghi, E., Foroushani, A. R., Younespour, S., Rostami, Z., Einollahi, B., Eshraghian, M. R., Akhoond, M. R., & Mohammad, K. (2016). Longitudinal assessment of serum creatinine levels on graft survival after transplantation: Joint modeling approach. *Nephro-Urology Monthly, 8*(4), 1–5.

Mei, H., & Eisner, J. (2017). The neural Hawkes process: A neurally self-modulating multivariate point process. *Advances in Neural Information Processing Systems, 2017*, 6755–6765.

Murphy, S. A. (2003). Optimal dynamic treatment regimes. *Journal of the Royal Statistical Society: Series B (Statistical Methodology), 65*(2), 331–355.

Murphy, S. A. (2005). An experimental design for the development of adaptive treatment strategies. *Statistics in Medicine, 24*(10), 1455–1481.

Murphy, S. A., Arjas, E., Jennison, C., Dawid, A. P., Cox, D. R., Senn, S., Cowell, R. G., Didelez, V., Gill, R. D., Kadane, J. B., & Robins, J. M. (2003). Optimal dynamic treatment regimes. *Journal of the Royal Statistical Society. Series B: Statistical Methodology, 65*(2), 331–366.

Naesens, M., Kuypers, D. R., & Sarwal, M. (2009). Calcineurin inhibitor nephrotoxicity. *Clinical Journal of the American Society of Nephrology, 4*(2), 481–508.

Randhawa, P. S., Starzl, T. E., & Demetris, A. J. (1997). Tacrolimus (FK506)-Associated Renal Pathology. *Advances in Anatomic Pathology, 4*(4), 265.

Robbins, H., & Monro, S. (1951). A Stochastic Approximation Method. *The Annals of Mathematical Statistics, 22*(3), 400–407.

Robins, J. (1986). A new approach to causal inference in mortality studies with a sustained exposure period–application to control of the healthy worker survivor effect. *Mathematical Modelling, 7*(9–12), 1393–1512.

Ruder, S. (2016). An overview of gradient descent optimization algorithms (pp. 1–14). arXiv preprint arXiv:1609.04747

Shibue, R., & Komaki, F. (2020). Deconvolution of calcium imaging data using marked point processes. *PLoS Computational Biology, 16*(3), e1007650.

Staatz, C., Taylor, P., & Tett, S. (2001). Low tacrolimus concentrations and increased risk of early acute rejection in adult renal transplantation. *Nephrology Dialysis Transplantation, 16*(9), 1905–1909.

Sutton, R. S., & Barto, A. G. (2018). *Reinforcement learning: An introduction*. MIT Press.

Tabibian, B., Upadhyay, U., De, A., Zarezade, A., Schölkopf, B., & Gomez-Rodriguez, M. (2019). Enhancing human learning via spaced repetition optimization. *Proceedings of the National Academy of Sciences, 116*(10), 3988–3993.

Tang, S., & Wiens, J. (2021). Model selection for offline reinforcement learning: Practical considerations for healthcare settings. In *Machine Learning for Healthcare Conference* (pp. 2–35). PMLR.

Upadhyay, U., De, A., & Gomez Rodriguez, M. (2018). Deep reinforcement learning of marked temporal point processes. *Advances in Neural Information Processing Systems, 31*.

Xu, Y., Müller, P., Wahed, A. S., & Thall, P. F. (2016). Bayesian nonparametric estimation for dynamic treatment regimes with sequential transition times. *Journal of the American Statistical Association, 111*(515), 921–950.

Xu, Y., Thall, P. F., Hua, W., & Andersson, B. S. (2019). Bayesian non-parametric survival regression for optimizing precision dosing of intravenous busulfan in allogeneic stem cell transplantation. *Journal of the Royal Statistical Society: Series C (Applied Statistics), 68*(3), 809–828.

Yildiz, C., Heinonen, M., & Lähdesmäki, H. (2021). Continuous-time model-based reinforcement learning. In *International Conference on Machine Learning* (pp. 12009–12018). PMLR.

Yu, T., Thomas, G., Yu, L., Ermon, S., Zou, J. Y., Levine, S., Finn, C., & Ma, T. (2020). Mopo: Model-based offline policy optimization. *Advances in Neural Information Processing Systems, 33*, 14129–14142.

Zhao, Y.-Q., Zeng, D., Laber, E. B., Song, R., Yuan, M., & Kosorok, M. R. (2015). Doubly robust learning for estimating individualized treatment with censored data. *Biometrika, 102*(1), 151–168.

Zhu, Y., & Li, S. (2018). One-dimensional Poisson marked point process model and Its Random Characteristic Analysis in Haze Weather. *IOP Conference Series: Earth and Environmental Science, 199*(5), p. 052041.

Index

Y. Zhao, D.-G. Chen (eds.), *Statistics in Precision Health*, ICSA Book Series in Statistics, https://doi.org/10.1007/978-3-031-50690-1

T

U

Printed in the USA
CPSIA information can be obtained
at www.ICGtesting.com
CBHW072151260924
14991CB00003B/33

9 783031 506895